AF517116

MICROBIAL DIVERSITY
Exploration & Bioprospecting

Editors

S Ram Reddy

M A Singara Charya

S Girisham

Published by:

Scientific Publishers (India)
5 A, New Pali Road, P.O. Box 91
Jodhpur 342 001 (India)

Branch Office
Scientific Publishers (India)
4806/24, Ansari Road, Daryaganj
New Delhi - 110 002 (India)

E-mail: info@scientificpub.com
Website: www.scientificpub.com

Print : 2021

ISBN: 978-81-7233-782-7

Typeset by Rajesh Ojha
Printed in India

Contents

Prof S M Reddy FBS, FIPS, FSMPP
Emeritus Professor,
Formerly Head, Dept. of Botany,
Kakatiya University, Warangal

Susheela Nivas
12-12-3/1
Ravindra Nagar
Opp. Sithafalmandi Rly. Station
Secunderabad

Foreword

Microbial diversity refers to the number of different microorganisms (richness) and their relative abundance (evenness) in an environment. It is one of the greatest knowledge gaps in the biological sciences, despite the fact that the working of the biosphere depends absolutely on the activity of microorganisms. Studies on microbial diversity are hampered not only by the technical inability to assess the species numbers, but also by the high heterogeneity of the environment, with its changing temporal and spatial microhabitats. The gold standard, plate count method with its inherent limitations does not reveal even a smallest fraction of actual species diversity of an environment. Fortunately, a large number of methods are now available that provide deeper insights of diversity and researchers can choose an appropriate method for their studies. The sequence based methods, however, now provide a way to survey biodiversity rapidly and comprehensively. Ribosomal RNA genes gathered from the environment are snapshots of organisms, representatives of different types of genomes. In the pursuit of unfolding of microbial diversity, discovery of new organisms and the development of resources based on microbial diversity are greater than ever before.

Among the methods designed to gain access to genetic and functional diversity of uncultured organisms, metagenomics, the genomic analysis of a population of microorganisms, has emerged as a centerpiece. This robust technology, apart from opening the 'black box' of microbial diversity, ushered the new vistas in microbial technology in terms of discovery of new drugs.

Microbial hot spots are of particular interest for the future study. At micro level guts of soil microfauna and rhizosphere are some examples of these hot spots. At macro level there are many ecosystems around the world whose microbial composition and role are yet to be studied. India with its diverse habitats and ecosystems is a treasure trove of biodiversity. Microbial diversity of Indian ocean, glaciers of Himalayas, desert sands, hot water springs of Western Himalayas, sediments of lakes, estuarines, soils of tropical rain forests and many other habitats remain largely unexplored. Metagenomic technology which is yet to be employed in the Indian context has boundless potential for exploring and exploiting microbial diversity. Needless to emphasize, such type of studies spanning over length and breadth of India are the urgent need of the hour, which hopefully unfold rich microbial diversity and also unravel many of the microorganisms of biotechnological importance.

At right time, the Department of Microbiology, Kakatiya University has taken an initiative and organized a National Seminar on such important theme. I compliment the Head, Staff of the department for successful organisation the seminar and also bringing the deliberations of the seminar in the form of proceedings. A good number of reviews and research articles contributed by eminent researchers make this book useful to all those who are interested in microbial diversity in particular and microbiology in general.

I wish many more successful academic ventures by the Microbiology Department.

Prof. S M Reddy

Preface

United Nations Conference on Environment and Development (UNCED, 1992) defines the biodiversity as "the variability among living organisms from all sources including inter alia, terrestrial, marine and other aquatic ecosystems and the ecological complexes of which they are part; this includes diversity within species, between species and of ecosystems".

Microbial diversity encompasses a spectrum of microscopic organisms including bacteria, fungi, algae etc. An estimated 50% of all living protoplasm on earth is microbial. There may be as many as 1 million species of bacteria, but only about 5,000 have been described. Microbial diversity is a vast frontier and potential gold mine for the biotechnology industry as it offers countless new genes and biochemical pathways for enzymes, antibiotics and other useful molecules. Unfortunately, microbial diversity is one of the greatest knowledge gaps in the biological sciences and remains largely unexplored and unexploited. Despite the improvements in cultivation methods there has not been significant change in our perception of microbial diversity. Molecular phylogenetic methods have recently provided a means for identifying diversity of organisms that occur in microbial communities without the need for cultivation.

Recent advances in exploration of microbial diversity, molecular biology, genomics, bioinformatics and genetic engineering have brought into focus several unforeseen applications of the microorganisms. With the availability of new molecular marker techniques, amazing microbial diversity of extreme habitats and unexplored habitats is unfolding. Consequently, there is a renewed interest shown in the applications of new microbes and their products. Recently a tremendous amount of research input is made on exploration and exploitation of microbes for biotechnological purposes. The microbial genetic resources are being harnessed for its potential to meet various human needs. With the advent of recombinant DNA technology and molecular tools, it is possible to exploit microbes for processes of economic development and environment management.

Keeping in view the importance of microbial diversity and its immense applications, the Department of Microbiology Kakatiya University has organized a National Seminar on *"Microbial Diversity - Exploration and Bioprospecting"* during 26th, 27th March, 2011 with an aim to bring scientists, active researchers, students and industrial entrepreneurs on to the common platform so that they can discuss and exchange the latest developments in the field and apply to suit their needs. The response to seminar by scientists, teachers and research scholars was overwhelming and fruitful discussions were made. The present proceedings is an outcome of the seminar deliberations. We thank all the invited speakers and researchers for contributing their full length papers with a short notice. It is our earnest hope that this proceedings will be highly useful to researchers and students to explore the latest developments in microbial diversity.

We express our sincere thanks to our former and present Vice-Chancellors, Prof N. Linga Murthy and Prof B. Venkatarathnam for their encouragement in organizing the seminar and helping us in bringing out these proceedings. The seminar was sponsored by UGC (SAP) DRS, DBT, DST and CSIR. With a sense of gratitude we thank these organizations for supporting us in this endeavour. Our grateful thanks are also due to UGC, New Delhi for providing the partial Financial Assistance under publication grant (2011-2012) for the publication of this proceedings. Finally, we are thankful to Shri Pawan Kumar Sharma, M/s Scientific Publishers (India) Jodhpur for coming forward to publish this book timely and in an excellent format and design.

S Ram Reddy
M A Singara Charya
S Girisham

Microbial World and Human Welfare

C. Manoharachary

Department of Botany, Osmania University, Hyderabad- 500 007.
E-mail: cmchary@rediffmail.com

ABSTRACT

Nature is bountiful of living biota which ranges from 3-50 million. Around 5% only are known to the world and 95% remain unnamed and hidden. One third of global biodiversity exists in India. Biodiversity is the resource material for biotechnology. India is confronted with population growth, poverty, hunger, energy crisis, food security, health security, environmental security and climate change. All these national issues have an impact on the development, growth and economy of the country. The consequences of loosing biodiversity are many. If the *in-situ* and *ex-situ* conservations including legal restriction are not followed, world's biota may become extinct. 21st century is the era of proteomics, genomics and metabilomics. Approximately 1.5 lakh microbes belonging to protista, algae, bacteria, viruses etc. are known. 2 million fungi are estimated and out of which only 1 lakh fungi are known and only 5-7% are cultured. In other words 95% of microbial world remains to be explored and identified properly. Data base, inventorization and gene banks need to be established both at national and regional level. Microbes include cyanobacteria, mycoplasma, archaebacteria, viruses, actinobacteria, some protozoans. chlamydiae, rickettisiae, prions, virioids. virusoides, phytoplasma, spiroplasma and others. Microbial technology includes understanding of microbial genetics, genetic recombination, waste recycling, mineral transformation pollution abatement, biofertilizer production, elucidation of antibiotics, enzyme production, metabolites of pharmaceuticals, industrial, agricultural and nutritional importance. Microbes also play an important role in food industry, water quality, dairy quality, sewage and waste disposal. There are some microbes which are pathogenic to human beings and cannot be neglected. Viruses are of phylogenetic and evolutionary importance. However, some viruses like bacteriophages are used as biocontrol agents. The transfer of genetic material has been amply demonstrated through bacteriophages. Halophilic bacteria are of immense importance and occur in habitats having high salt concentration. Indian oceanic waters harbor a great variety of salt loving bacteria, sulfate reducing hacteria. carotenoid bearing *Dunaniella. Haloferax. Haiomonas* etc. Halophilic bacteria contribute for proteins. energy conversion, production of food additive, enzymes useful in cosmetic industry and also help to understand the role of sea microbes. Rhizosphere is the hidden ecological niche and harbors a variety of microbes. *Rhizobium* and *Bradyrhizobium* form root nodules which convert atmospheric nitrogen into ammonia which is then used as nitrogen source by the plant. Fungi are important in biotechnology market and play significant role in human welfare.

Algae are found abundantly in diversified habitats. Algal biotechnology includes biofertilizers, feed and food, single cell protein, ceramic industry, useful in defense, space, cosmetics, antibiotic production, bioremediation and other processes.

Cyanobacteria are a group of extraordinarily diverse gram negative prokaryotes that originated 3.5 billion years ago. Cyanobacteria are potential organisms which can be useful to mankind in various ways. Similarly actinomycetes, *Frankia*, viruses, Mycoplasma, Chlamydae, Rickettsia, earliest prokaryotes and other microbes show varied morphology, biochemical pathways, metabolism, primary and secondary metabolites, diversification in habitat colonization and ecological adaptation. The microbial metabolism also differs from one organism to another. All the microbes have potential of biotechnological importance and market. Therefore, aspects and prospects of microbial world in relation to human welfare are discussed in this paper.

Key words: Microbes, Diversity, Human welfare

Microorganisms are minute organisms and cannot be seen with naked eye. Such organisms have to be observed with a microscope ('Micro' means small, 'scope' means to see). Microbes can be observed with clarity as the present compound microscopes have magnifications from ten to thousands of times greater than those of Robert Hooke built microscope with multiple lenses.

This group includes bacteria, viruses, fungi (yeasts and molds), protozoan, microscopic algae, actinomycetes etc. Viruses are regarded as being at the border line of life and non-life. Microbes are ubiquitous and occur in varied habitats such as soil, litter, water, air, animal excreta, extreme environments and others. The microbes colonize survive and multiply in/on the above substrates in the presence of favourable environmental conditions. Marine and freshwater microorganisms form the basis for the food chain in oceans, lakes and rivers. Microbes have ability to breakdown complex chemical substances including hydrocarbons, pesticides and wastes and convert them into simpler substance. Microorganisms are known to help in the recycling of chemical elements and incorporate them into the soil, water and air. Autotrophic bacteria and algae play an important role in photosynthesis which is a food and oxygen generating process. This process is critical to the existence of life on earth. Human intestine also contains bacteria that help in digestion and synthesis of some vitamins required by our bodies (B vitamin & K vitamin).

Microorganisms have applied value and play very important role in humam welfare. Microorganisms are used in the synthesis of products such as acetone, glycerine, organic acids, enzymes, alcohols, drugs and others. Industrial products also include vinegar, krait, pickles, alcoholic beverages, antibiotics, biopesticides, soysauce, cheese, yogurt and bread.. However some microorganisms are pathogenic; hence cause diseases in humans, animals and plants. The absence of chlorophyll in some has enforced them to become disease causing agents while some are saprophytic.

Louis Pasteur (1861), the French scientist has proved that microorganisms are indeed present in the air and they can contaminate seemingly, sterile solutions and argued that air itself does not give rise to microbial life. Pasteur has conducted series of experiments to prove this. He began by filling short necked flasks with beef broth and boiling them and few were left open. Within few days, the open flasks were found contaminated with microbes. Flasks having boiled, cooled and sealed, remained free of microorganisms. Pasteur proved that life can be destroyed by heat. Since then aseptic techniques were developed. Pasteur's work provided sufficient proof that microorganisms can not originate spontaneously. After this there were numerous discoveries in microbiology. Actually the period from 1857 – 1914 has been termed as Golden Age.

Pasteur later discovered that microorganisms called yeasts which convert the sugars to alcohol in the absence of air. The same is now called fermentation, now employed in making wine, beer etc. The process of killing potentially harmful bacteria of microorganism in milk or in alcohol etc., is called pasteurization. It is a process to heat the milk etc. just enough to kill most of the bacteria without affecting nutrient levels and quality.

Yeasts play an important role in fermentation. It is a link between microorganisms activity and physiological and chemical changes in organic materials. This discvovery made the scientists to think that the microorganisms might have similar relationships with plants and animals, in particular that the microbes might cause disease. This was called the germ theory of disease.

The main proof that bacteria cause disease came from Robert Koch in 1876. Koch, a young German physician was able to discover the bacterium (*Bacillus anthracis*), responsible for anthrax disease in cattle and sheep in Europe. He cultured the bacterium artificially in nutrients and then injected into healthy animals. Later he found that these animals became sick and died. Koch isolated the same anthrax bacterium from the blood and compared it with original bacterial isolates. Thus Koch has proved that a pathogenic bacterium was responsible for anthrax and it could be isolated artificially. The artificially isolated bacterium on introduction into healthy animals produced anthrax disease. The blood of the same, when examined showed the anthrax bacteria which was similar to that was observed in blood samples of natural animals died because of anthrax. These steps are called as Koch's postulates.

When numerous diseases appeared both on plants and animals including in humans, scientists started working on prevention and control of diseases. Edward Jenner a young British physician conducted few experiments to find protection from smallpox. Edward Jenner collected scrapings from small pox blisters. Then he made inoculations with cow pox material by scratching arm of a healthy person with small pox contaminated needle. Within few days, though the patient because mildly sick but recovered and never had small pox. This process was called vaccination.

Alexander Fleming, a Scottish physician and bacteriologist observed that a fungus was able to control the growth of a bacterium. The mold was identified as *Penicillium notatum*. In 1928 Fleming named the active inhibitor from this mould as penicillin. This wonder drug has paved the way to the discovery of many new antibiotics. The synthetic drugs and antibiotics though kill the pathogenic microbes; they also have damaging side effects on infected host. Few successful antiviral drugs were also discovered.

In recent times recombinant DNA technology has revolutionized research and could find practical applications in all areas of microbiology. Vaccines are now available for small pox, measles, mumps, polio and Hepatitis B. The small pox vaccine was so effective that not a single case was reported since 1977. Interferon, a substance generated by the body's own immune system was discovered. This inhibits the replication of virus and cancerous growth also.

Microbial diversity encompasses a spectrum of microscopic organisms including bacteria, fungi, algae etc. An estimated 50% of all living protoplasm on earth is microbial. There may be as many as one million species of bacteria, but only about 5,000 have been described. Microbial diversity is a vast frontier and potential gold mine for the biotechnology industry as it offers countless new genes and biochemical pathways for enzymes, antibiotics and other useful molecules. Microbial genetic resources constitute important components of the ecosystems and contribute vitally in maintaining ecological balance (Manoharachary *et al.* 2009).

The microbial genetic resources are being harnessed for its potential to meet various human needs. Microorganisms have much higher and a wide range of physiological and biochemical potentials as compared with all other organisms concerned. In all fields of human endeavour, including industry, agriculture and food, the microbes lay a predominant role. Also remarkable success has been achieved in exploiting the useful microorganisms in combating the harmful ones. With the advent of recombinant DNA technology and molecular tools, it is possible to exploit microbes for processes of economic development and environment management.

As microorganisms are integral to the entire biosphere, they influence the processes of all living organisms. Soil is a heterogeneous mass of all kinds of microbes and they develop a strong relationship with plants. This relationship has been exploited for improving the nutrient status and uptake, growth of plants and control of certain pests and diseases. Plant growth promoting microorganisms (PSM) are extensively studied for their potential applications in agriculture. They are commonly considered to promote plant growth by competing with plant pathogens or by producing plant growth regulators or by fixing nitrogen and improving the availability of soil phosphorus and other elements etc. (Tilak *et al.,* 2006; Manoharachary *et al.,* 2009). Their interactions with various diazotrophs, phosphate solubilizing microorganisms and mycorrhizae have been extensively reviewed (Dey *et al.*, 2005).

Use of such microbes in agriculture is not only economical but also eco-friendly. Certain aspects of diversity of soil microorganisms and their relationship with plants and their contribution on plant growth and soil productivity

Bacteria like *Bacillus thuringiensis* has been extensively used to control caterpillars, boll worms, tobacco bud worms and mosquitoes. Microbes are also used in the degradation of chemical compounds, which cause damage to environment. Two important microbial products have attracted world attention namely single cell protein and genetic engineered microbes (GEMS) Bacteria and fungi are used to produce single cell protein. Genetic engineering could find potential use in agriculture, medicine and others. We all live in a microbial world from birth until death. We have a variety of microbes on and inside our bodies. However normal flora usually benefits us but at times caused illness. Important natural resistance is provided by animals, plants and humans to infections.

Division of Microbial World

E.H. Haeckel, a German zoologist has suggested that the third kingdom called as 'Protista', be formed to include those unicellular microorganisms that are typically neither plants nor animals. These organisms called, the protists, include bacteria, algae, fungi and protozoa. Of these, bacteria were referred to as lower protists and others as higher protists. Viruses are not included in protists, because, they are not cellular organisms and therefore made a separate kingdom.

The concept of bacteria and blue-green algae as prokaryotic organisms was established in 1962 by Stanier and Van Niel. With this the concept of 'Prokaryotes' and 'Eukaryotes' has come into use for all living organisms.

Whittaker's Five Kingdom Concept

The characterization and taxonomical grouping of living organisms till early 1960s was based on the cellular organization as revealed by microscopic (light microscope and/or electron microscope) studies. A comprehensive system of classification based on the nutrition and absorption of nutrients by the living organisms was proposed by R.H. Whittaker in 1969. This is called as Whittaker's Five Kingdom concept (Fig 1). He proposed the classification of all living organisms into five kingdoms.

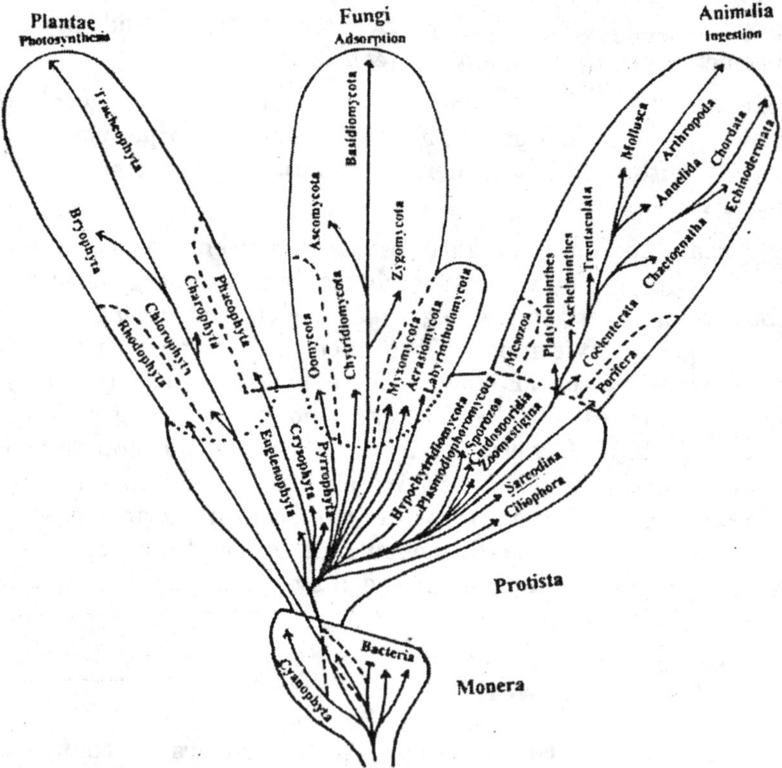

Fig. 1. Simplified Whittaker's five kingdom system

Carl Woese Concept

Developments in molecular biology techniques lead to more deeper understanding of living organisms at molecular level. These studies were done mainly in terms of type of genetic materials, the DNA and RNA and their sequence, structure and function of biomacromolecules in different living organisms and their molecular structure, stability of genetic structures and genes has come into understanding. Scientists have studied the genomic structures and nucleotide sequence of genetic materials in different organisms. The DNA base composition, G + C percentage ratio, DNA homology and gene sequence of more stable genes of prokaryotic and eukaryotic organisms were analyzed. The 16S rRNA and 18S rRNA gene structures and sequence of prokaryotes and eukaryotes respectively, were proposed to be taken into the major criteria for classification and identification of living organisms. This concept was proposed by Carl Woese (Fig 2). Carl Woese and his group in 1990 noted that bacteria are distant from plants and animals and, by contrast, plants and animals are not so far from each other. All this is based on the 16S rRNA and 18S rRNA gene sequencing and homology studies, DNA-DNA hybridization studies, and the DNA base analysis or DNA finger printing by restriction fragment length polymorphism (RFLP) analysis.

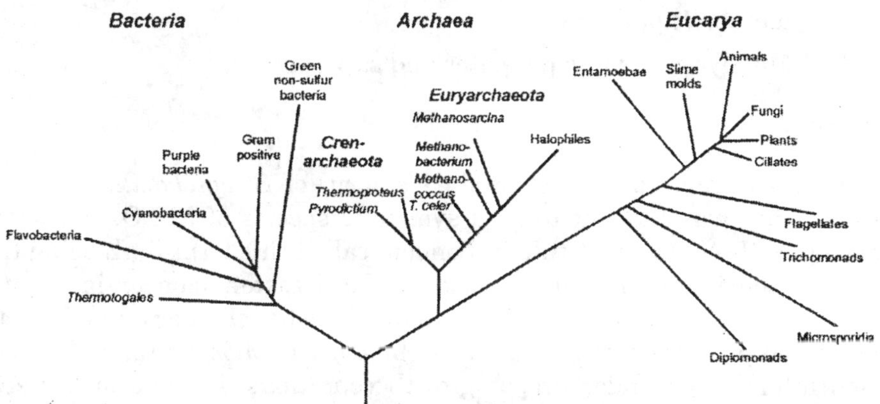

Fig 2. Carl Woese concept for grouping of all living organisms in to three domains (kingdom) based on 16S or 18S rRNA gene sequence

Diversity of Microorganisms

Mycoplasmas

In 1898 the French scientist, E. Nocard and E.R. Roux, studying pleural fluids of cattle suffering from bovine pleuropeneumonia, discovered the organisms that were entirely different from any other microorganisms then known. When cultivated on rich organic media containing about 20% of animal serum, the organisms were found in different forms as spheroid, thin, branching filaments, stellate or asteroid structures and other irregular forms. Similar pleomorphic (Gr. *Pleo*=many; *morphe*= forms) organisms were later isolated from other animals such as sheep, goat, dogs, rats, mice, human beings.

The characteristics are:

1. Unicellular, prokaryotic, usually non-motile and form fried-egg shaped colonies
2. Highly pleomorphic, form varying with culture conditions. Under different microscopy, they appear small coccoid bodies, ring forms and fine filaments, some of which are branched.
3. Filterable through bacterial filters.
4. Cell wall absent. Cells delimited by a triple layered lipoproteinaceous unit membrane, the plasma membrane.
5. Both DNA and RNA present. DNA base composition ranges from 23 to 36 (39 to 41 in *Mycoplasma pneumoniae*) mole per cent GC (cf L-forms).
6. Resistant to antibiotics as penicillins that act on cell walls.
7. Inhibited by tetracyclines and similar antibiotics that act on metabolic pathways.
8. Mostly free-living; parasites and saprophytes.

L-Forms

Klieneberger (1935) first studied a form of *Streptobacillus moniliformis*, what was thought to be a symbiotic species of PPLO. He, while working at the Lister Institute in London, called this form L_1, L being the initial of the Lister Institute named for Lord Lister, pioneer in aseptic surgery. The L_1 organism was later shown by Klieneberger and others to be a stable protoplast of L-form of *Streptobacillus moniliformis*. The two forms interchange under proper growth conditions. L-forms of bacteria are entirely different from L cells, a clone of animal tissue culture cells.

Protoplasts

Some microorganisms, both eukaryotic and prokaryotic (eg. yeasts, molds, bacteria, some higher plants) that have a distinct cell wall can at times exist without a cell wall in what is designated as wall-less state. The cell membrane and its intact contents are then called a protoplast. Unless the solute concentration of the suspending fluid is osmotically protective i.e. high enough (e.g. 3-20% glucose; 2-5% Nacl; 10-20% serum) to balance the intracellular osmotic pressure, protoplasts no longer retained by their thick walls, usually burst. Protoplast are therefore, said to be osmotically fragile.

An important property of bacteria cell wall is sensitivity to antibiotic penicillin. Penicillin inhibits synthesis of peptidoglycans. Young actively growing Gram-positive bacteria are, therefore, sensitive to penicillin. Thus these bacteria can be made protoplasts by treating them with penicillin. Animal cells including human cells, are unsusceptible to penicillin, because there are no peptidoglycans there.

Sphaeroplasts

Gram-negative bacteria, where the bulk of cell wall is made by lipopolysaccharides and lipoproteins are usually partly or wholly resistant to penicillin (as it inhibits only peptidoglycan).

These Gram-negative bacteria are also protected by their lipopolysaccharides and lipoproteins of cell wall (in outer layers) from the action of a group of enzymes, lysozyme, an enzyme discovered by Fleming, Lysozyme is found in egg white, secretions of skin and mucous membranes, tears and elsewhere. It attacks superficially the glycosidic bonds in the polysaccharide backbone of peptidoglycan of bacterial cell walls. When Gram-positive bacteria are treated with lysozyme, they are rapidly denuded of their cell walls and become naked protoplasts. In contrast, to make Gram-negative bacteria (though peptidoglycan present in their cell wall, but it is protected by outer layers of lipcomplexes) vulnerable to lysozyme it is necessary first to remove the lipocomplex of cell wall. These are removed with lipd solvents as NaOH or ethylenediaminetetraacetate (EDTA). Even then the wall is not completely removed and the result is osmotically fragile cells, still retaining some remnants of cell wall. Such only partially denuded cells are called spheroplast.

Archaebacteria

The archaebacteria are predominantly terrestrial and aquatic microbes, occurring in anaerobic or hypher or hydro-thermally and geothermally heated environments; also, some occur as symbionts in animals digestive tracts. They consist of aerobes, anaerobes and facultative anaerobes that grow chemolithoautotrophically, organotrophically, or facultatively organotrophically. Archaebacteria may be mesophiles or thermophiles, with some species growing even above 100ºC.

A unique biochemical feature of archaeobacteria is the presence of glycerol isopranyl ether lipids. The lack of murein (peptidoglycan-containing muramic acid) in cell walls makes archaebacteria insensitive to ß-lactam antibiotics. The "common arm" of the tRNA contains pseudouridine or 1-methylpseudoridine instead of ribothymidine. The sequences of 5S, 16S and 23S rRNAs are very different from the corresponding ones in eubacteria and eucaryotes.

Archaebacteria share some molecular features with eukaryotes: (a) the elongation factor 2(EF-2) contains the amino acid diphthamide and is therefore, ADP-ribosylable by diphtheria toxin, (b) amino acid sequences of the ribosomal "A" protein exhibit sequence homologies with the corresponding eukaryotic protein, (c) the methionyl initiator tRNA is not formylated, (d) some Trna genes contains introns, (e) the aminoacyl stem of the initiator tRNA terminates with the base pair "AU", (f) the DNA dependent RNA polymerases are multicomponent enzymes and are insensitive to the antiobiotic rifampicin and streptolydigin, (g) like the α-DNA polymerases of eucaryotes, the replicating, archaeobacteria DNA polymerases are not inhibited by aphidicolin or butylphenyl-dGTP, and (h) protein synthesis is inhibited by anisomycin but not by chloramphenicol.

Autotrophic archaebacteria do not assimilate carbon dioxide via the Calvin cycle. In *Methanobacterium*, CO_2 is fixed via an acety-CoA pathway, whereas in *Acidianus* and *Thermoproteus*, utotrophic CO_2 is fixed via a reductive tricarboxylic acid pathway. Fixation of N_2 has been demonstrated by some methanogens.

Gram stain results may be positive or negative within the same order because of very different types of cells envelopes. Gram-positive species possess pseudomurein, methanochondroitin and heteropolysaccharide cell walls; Gram-negative cells have (glyco-) protein surface layers. The cells may have a diversity of shapes, including spherical, spiral, plate- or rod-shaped; unicellular and multicellular forms in filaments or aggregates also occur. The diameter of an individiual cell may be 0.1 - >15 μm and the length of the filaments can be upto 200 μm. Multiplication is by binary fission, budding, constriction, fragmentation or unknown mechanisms. Colour of cell masses may be red, purple, pink, orange-brown, yellow, green, greenish black, gray or white.

The major groups of archaebacteria include (a) the methanogenic archaebacteria, (b) the archaebacterial sulfate, reducers, (c) the extremely halophilic archaebacteria, (d) the cell- wall -less archaebacteria and (e) the extremely thermophilic S-metabolizers.

Actinomycetes

The members of this large and important group are mainly soil inhabitants. They were regarded as intermediate forms between bacteria and fungi on account of their mycelial structure, but are now grouped with eubacteria because of their prokaryotic nature and cell wall composition. When growing on agar media, they produce mycelium which is partly aerial and partly submerged in the medium, formed from much branched but thin hyphae. Most members also produce single-celled spores. Thus they resemble fungal colonies in culture. In liquid media also they form individual colonies and do not make the medium turbid like other bacteria. The spores are formed either externally as conidia or inside sporangia as sporangiospores. The hyphae usually divide into long cells with more than one nucleoid in them. The spores are easily detached and disbursed. Many actinomycetes produce pigments of various shades. The cell walls are chemically distinguishable in to four types. While most actinomycetes are saprophytes, a few of them are pathogens also.

Actinomycetes, particularly the genus *Streptomyces* are economically important. Majority of the clinically useful antibiotics are produced by actinomycetes. Taxonomically the following six groups are recognized:

Nocardioforms (e.g. *Nocardia, Faenia*). They develop substrate mycelium, which easily breaks up into individual cells. They are mostly saprophytes in soil and d water but *Nocardia asteroids* causes a lung disease in humans and animals of low resistance.

Actinomycetes with multilocular sporangia. This section includes *Frankia* and *Geodermatophilas*. The hypha divides in two planes to produce clusters of spores. *Frankia* forms root nodules in non – leguminous trees, in symbiotic association. The spores of *Geodermatophilus* are motile.

Actinoplanates. In this group, extensive substrate mycelia are formed but aerial mycelium is absent or rudimentary, with sporangiospores rising above the agar surface. The spores are arranged in coils are parallel rows in the sporangium (e.g. *Actinoplanes*). *Micromonospora* produces single sporangiospores on branched sporophores.

Streptomyces and related forms: Streptomyces comprises a large genus with nearly 380 species. They are aerobic, produce aerial and subtrate mycelia and chains of nonmotile spores (conidia) enclosed in a thin fibrous sheeth. Different species are distinguished on the basis of morphologicaland biochemical features such as colour of mycelium, arrangement of conidiophores and conidia, spore surface characters, utilization of various sugars, nitrate reduction melanin production, antibiotic production etc. Soil is their natural habitat and they occur in large numbers, especially in tropic soils as they can decompose a large variety of organic substrates and also withstand desiccation. The characteristic earthy smell that soil gives out after a shower is due to volatile substances (geosmin etc.) that they produce.

Rickettsiae

These are spherical of rod shaped cells occurring as intracellular parasites. They have a cell and plasma membrane and are gram negative reaction. The cells are small and range from 0.3 to 5 mm in diameter and

0.8 to 2.0 mm in length. They have both DNA and RNA cells divide by binary-fusion like in bacteria. Rickettsias are intracellular prarasites that replicate in the host cytoplasm. Rock mountain spotted fever and Q-fever are the examples of the diseases caused in humans and are transmitted by arthropod vectors.

Prions

There are simplest living entities called prions and are made up of only a small protein but are capable of infection and replication. Prions modify a host protein to replicate the protein molecule. The prions cause few diseases in animals and man. Scrapie disease in sheep and kuru disease and mad cow disease in human are examples.

Viroids

Viroids are infectious agents consists of single stranded RNA molecule and are without a protein coat. They do not have a definite shape. Cadang-Cadang disease is caused by them. The RNA molecule is in the range of 240-375 nucleotides and shows a circular configuration. The viroid molecule is known to replicate in the host nucleus utilizing the host RNA polymerase.

Chlamydia

The psittacosis group, which comprises the genus *Chlamydia*, probably represents a further stage in degenerate evolution. The chlamydias are obligate parasites in which there has been an even greater loss of metabolic functions. Infact, for many years the chlamydias were considered to be large viruses rather than bacteria and only since the nature of virus replication has been well understood thus the bacterial nature of the chlamydias been firmly established. Many chlamydias are smaller than some of the true viruses, such as the smallpox virus, but the chalmydias divide by binary fission and do not replicate in the manner of viruses. Of the diseases caused by chalmydias, pisttacosis, an epidemic disease of birds that is transmitted occasionally to humans, is the most important, although a number of other diseases of birds and humans and other mammals are known.

Chlamydia cells have a chemical composition similar to that of other bacteria. Both RNA and DNA are present, in approximately equal proportions. The DNA content of a chlamydia corresponds to a molecular weight of approximately 4 X 10^8, about twice that of vaccinia virus and one-tenth that of *Escherichia coli*. The base composition is about 29 percent G–C. At least some of the RNA is in the form of ribosomes and like those of other prokaryotes, the ribosomes are 70-S particles composed of one 50-S and one 30-S unit.

The small, dense clamydial form has a typical rigid bacterial cell wall and muramic acid has been shown to be present. With the small form reorganizes to the large form it loses its cell-wall rigidity and there is some evidence that the cross-linking of the peptidoglycan is reduced in extent.

Viruses

The designation 'virus' (poison) was used originally for pathogenic agents about which very little was known. The term eventually came to mean the group of causative agents that could pass through bacterial filters (as discovered by Iwanovski in 1982) known as 'filterable viruses' or simply 'viruses'.

The viruses differ from microorganisms in the following properties. (1) They contain only one kind of nuclei acid, either RNA or DNA . They are unable to reproduce outside living cells. Outside the host cell the virus is called 'virus particle' or 'virion' are used. The virus particle consists of nucleic acid and a protein coat, known as the 'capsid'. Sometimes there is also an envelope.

Viruses become demonstrable by the consequences of their development in the host. They destroy whole cell complexes and produce tissue damage, necrotic areas, and lytic halos. Plants, animals, and microorganisms are the natural hosts for viruses.

Plant viruses

Plant viruses gain entrance to their host through lesions; they do no actively penetrate into plant tissues. The quantitative determination of plant viruses is based on the development of necrotic spots around artificially produced primary lesions. In nature the viruses are spread by vectors or by direct contact. Frequently viruses enter into leaf tissue through injuries produced by rubbing. Many viruses are spread by insects. In some cases, the virus multiplies in the alimentary tract of the insect (persistent viruses); new infections of plants can only occur after a certain incubation period in the insect. Non-persistent viruses are transmitted directly and passively by the insect sting or bite. Many plant diseases are caused by viruses. Potato viruses are of great economic importance. The most intensively investigated and best-known plant virus is the tobacco mosaic virus (TMV). Most plant viruses have RNA as their genetic material.

Animal viruses

In humans and other animals, viruses cause many diseases: smallpox, chicken-pox, measles, rabies, poliomyelitis, influenza, the common cold, foot and mouth disease of cattle, etc. Animal viruses are also transmitted by direct contract or via insects, and they apparently gain entry into the host cell by phagocytosis or pinocytosis. To investigate viruses in the laboratory it is necessary to use experimental animals or fertilized chicken eggs. Some viruses, however, can be cultivated in tissue cultures and their presence can be determined quantitatively. The genetic material of animal viruses can be either DNA or RNA. Whilst DNA is usually present as a double-stranded helix, RNA is found as single-strand polynucleotide chain.

Bacterial viruses

Viruses that use bacterial cells as hosts are called bacteriophages. There is hardly a single species of bacteria where sufficient evidence has not been found to be a phage. The presence of bacteriophages is recognized by the appearance of 'plaques' or lytic holes in a continuous bacterial growth. In a bacterial suspension, phage multiplication can lead to complete lysis in a short time. Phage nucleic acid occurs either as double- or single-stranded DNA, or as single stranded RNA. The phages of *E.coli* have served as models for bacteriophages generally. Use of phages plays animportant role on the transmission of genetic material from cell to cell.

Fungi

The fungi are highly specialized organisms and vary from soil to soil and in the same soil with an increase in depth. The fungal composition and distribution also depends on the extent of availability and utilization of the organic matter present in soil. A substantial number of fungal species are reported from Indian soils.

India with its different geographic regions, varied vegetation and wide variety of soils harbour a good number of fungal species which differ both quantitatively and qualitatively. Cultivated soils which receive high N, P, K amounts and organic matter are found harbouring penicillia, aspergilli, hyphomycetes etc., in large number. The forest/wild soils are found associated with qualitatively rich fungal flora including the occurrence of new and interesting fungi. Vegetation / crop cover has got greater impact on qualitative and qualitative composition of fungi.

Mankind is dependent on both harmful and beneficial microbes including fungi have around world wide interest because of international agreement on IPR and patenting. Biodiversity of soil fungi forms an important component as it forms resource material for biotechnology. 1.5 million is the estimated total fungal species count and of which only 79000 species are reported from world. Around 29000 fungal species are reported from India indicating that 1/3 of global fungal diversity exists in India. Further 50% of fungal wealth is hidden under different ecological niches and habitats. It is also known that 5-7% fungi are cultivable condition. Out of 5-7% of cultivable fungi 70% are soil fungi. However, it is reported that many soil fungi remain in sterile state. In view of the biotechnological importance attached to soil fungi, it becomes essential to identify and conserve fungal diversity (Manoharachary *et al*. 2005, 2009).

More number of genera and species and fungi are reported from soil habitat than in any other environment. Further soil is exposed to various weather conditions, hence harbour highly adaptable fungal species. Soil fungi along with other microbes are the primary decomposers, agents of elemental cycling, recycling of organic wastes, help in transformation, nutrient cycling, degradation of pesticides, hydrocarbons, xenobiotic compounds and other materials besides enhancing plant growth through the secretion of growth promoters like gibberellic acid, IAA, cytokinins and others. Soil fungi also play an important role not only as scavengers but also in complex food chain through their interactions with other soil biotic communities.

Numerous pharmaceutical compounds such as penicillin, cyclosporins, griseofulvin and others are reported from soil fungi. Fungal species belonging to *Trichoderma, Verticillium* and others serve as bio-control agents because of their role as mycoparasites and autogonistic nature. Nematophagous fungi and Entomogenous fungi serve to combat many nematode diseases and insect pests.

Fungi not only serve as pollution indicators but also used in bioremediation and anthropogenic pollutants. Soil fertility and humification processes are controlled by fungi. Fungi like *Fusarium, Rhizoctonia, Sclerotium, Pythium, Verticillium* and others are found to colonize and cause serious soil-borne/root-borne diseases such as wilts, root rots, damping off etc. Thus soil fungi play an important role in medicine, agriculture, industry and other biotechnological processes. Several organic acids and enzymes are produced by aspergilli, penicillia and others. Species of yeast and other fungi have been contributing, to ethanol production. Fungi like *Fusarium graminearum* are contributing to single cell protein production. The unexplored fungi are a resource for new genera and species valuable to biotechnology, agriculture, medicine, industry and others. Such biodiversity is pivotal in sustaining both managed and natural ecosystems. With increase in soil depth,the fungal population diminishes due to depletion of organic matter. Species of *Penicillium* are abundantly found in temperate climates while aspergilli dominate tropical soils. Predominance of soil fungi also changes with the climatic conditions. During summer months, aspergilli and hyphomycetes dominate while ascomycetes, penicillia and fungi-imperfecti are of predominant occurrence in winter. In monsoon mostly the rich sporulating fungi and mucorales dominate the soil fungi. The quantification of soil fungi has been conventionally difficult because of the immense potential of species and difficulties arose in culturing such fungi. PCR, 18s RNA and tRNA analyses can tackle genetic diversity of soil fungi. Morphological and biochemical data go had in hand in supporting molecular approaches. Limited information is available on the biodiversity of soil fungi, hence it is imperative to search for new forms, conserve and screen resource for new bioactive molecules, which are liable to yield valuable products, Mycobank is the recently established bioinformatics structure was initiated to link fungal names to their molecular data, pleomorphism, herbarium (germplasm) maintenance, documentation etc. (Hawksworth, 2004).

Mycorrhizal fungi

Mycorrhizae are the structures formed by the association of root in plant with fungi. Such root fungus associates were found to be normal on the root system of trees such as pine, oak, etc. The studies of the German botanist Frank (1885) in the last century had initiated world wide interest on mycorrhizae. Mycorrhizae are found associated with many plants under natural conditions. The fungi are restricted for essential nutrients and growth factors on their hosts. Such root fungus contact and dependency is of symbiotic nature.

Mycorrhizaal fungi receive carbohydrates from the host plant while plants receive fastly mobilized phosphorus and other nutrients from fungus. Seven type of mycorrhizal associations (ectomycorrhiza, ectendomycorrhiza, endomycorrhizae, arbutoid, monotropoid, ericoid and orchidaceous type) are known and saving of 50% of phosphorus ispossible, if proper soil-crop management is observed .

Arbuscular mycorrhizal fungi are the naturally occurring fungal component of soil biota in most terrestrial ecosystem. It also represents an obligate symbiotic group distinct from the rest of soil microbial biomass. AM fungal symbiosis is a dynamic process and interaction affects all physiological aspects of the host. These fungi are unique as they are partly inside and partly outside the root. The vesicles and arbuscules hyphae that are formed inside the root do not encounter competition and antagonism from soil microbes. Soil factors and host genotypes are some of the parameters that affect biological interactions in the rhizosphere. The arbuscular mycorrhizal fungi are ubiquitous in distribution and occur abundantly in phosphorus and other nutrient deficient soils. 80 – 90% of plants ranging from bryophytes to flowering plants have the dynamic association of AM fungi. AM fungi belong to Zygomycotina and cannot be cultured artificially due to their obligate symbiosis. Arbuscles are the key sites formed inside the cortical cells and are meant for nutrient exchange and remain active upto 15 days depending upon host response.

Arbuscules and vesicles are formed in cortex while spores/sporocarps and branched hypahe are formed outside the root and found associated with soil. There are nine valid genera encountered namely: *Acaulospora, Archaeospora. Entrophospora, Glomus, Gigaspora, Paraglomus, Pavispora. Sclerocystics* and *Scutellospora*. Vesicles are not reported in *Gigaspora* and *Scutellospora* but extramatrical vesicles were reported. Schubler *et al.* (2001) presented the AM fungi in phylum Glomeromycota with orders, families and genera as enlisted in Table 1.

Table 1. Classification of arbuscular mycorrhizal fungi (Glomeromycota)

Phylum : Glomeromycota

Order	Family	Genus
Glomerales	Glomeraceae	*Glomus* A
		Glomus B
Paraglomerales	Paraglomeraceae	*Paraglomus*
	Diversisporaceae	*Diversispora*
	Gigasporaceae	*Gigaspora*
	Acaulosporaceae	*Scutellospora*
		Acaulospora
		Entrophospora
Archaeosporales	Archaeosporaceae	*Archaeospora*
	Geosiphonaceae	*Geosiphon*

There are about 150 AM fungal species reported colonizing 30,000 receptive hosts. AM fungal taxonomy is more complicated and morphotaxonomic criteria viz., wall layers, ornamentation, subtending hyphae, shape, size, colour and other characters help to segregate genera and species. Presence or absence of the sporocarp is also an important criterion in identifying AM fungal genera. Molecular and biochemical parameters such as cell wall composition, protein immunogenicity, isozyme pattern, fatty and methyl ester profiles, direct sequencing of 18 sRNA gene and RAPD approach to detect polymorphism and others have been proved more useful. PCR-RELP can be used for quantitative comparison AM fungi help in soil blinding, efficient and effective mobilization of phosphorus and other nutrients, restoration of disturbed land and mine soils, establishment of forest seedlings, commercial crops, economically viable plants, medicinal and ornamental plants, aromatic plants and horticultural plants besides increasing NPK status in mycorrhizal plant and water uptake, improve plant defense against biotic and abiotic stress, increase yield and offer other benefits to host plants on inoculation with AM fungi (Manoharachary, 2004, 2010).

Root exudation and soil factors are the important ecological aspects, which can manipulate plant and microbial succession. Mycorrhizal plant exudates include many compounds useful to host plant including growth-promoting substances. Various techniques used for the isolation of AM fungal spores from the soil such as floating technique, decanting and sieving, gradient centrifugation, monoclonal and polyclonal methods are employed. Viability of AM fungi is of great significance in AM fungal application. Colonization of host plants, crop plant by AM fungi and its degree of colonization form an important factor in mycorrhizal dependency studies.

Mycorrhizal fungi play a significant role in the formation of soil-aggregates. The important role of the soil mycelium of mycorrhizal fungi in the formation of water-stable soil aggregates is well documented. Wright *et al.* (1996) discovered glomalin and named the substance after Glomales, the taxonomic order of the fungi that produce the sticky protein. The protein is a stable hydrophobic glycoprotein, which helps in binding of soil particles. As the roots grow, glomalin sloughs off into the soil where it acts as a "super glue," helping sand, silt and clay particles stick to each other. This protein is deposited on the outer hyphal walls of the extraradical mycelium and on the adjacent soil particles facilitating the aggregation of soil particles and contributes to the formation of microaggregates, a basic building block of soil structure. Glomalin stores 27 percent of total soil carbon, compared to humic acid's eight percent. It also provides nitrogen to soil and gives it the structure needed to hold water and proper aeration, movement of plant roots and stability to resist erosion.

Application of AM fungi as bio-fertilizer has been limited by the difficulties of obtaining large quantities of pure inoculum of the fungal symbiont. Resting spores are extracted from soil and are often used as inoculum. To grow AM fungi through nutrient film technique, aeroponics, root organ culture Ti/Ri – Plasmid and other techniques resulted in limited success. However, soil-less medium seems to avoid contaminants, hence soil-pot culture with a suitable host has been the most practicable method employed for multiplication of AM fungi. AM fungi which are the beneficial fungi need to be conserved through various scientific methods (Manoharachary, 2004).

Ectomycorrhiza

Ectotrophic mycorrhizae are those in which fungus completely encloses the rootlet in a sheath or mantle of issue formed of compact hypal cells and penetrates only between the cells of the root cortex. This kind of mycorrhizae is associated with many forest trees. Mycorrhizae play significant role in plant productivity and energy systems besides helping as biocontrol agents. The plants deliver photosynthate to the mycorrhizal fungi and plants receive mineral nutrients and water through mycorrhizal roots. Most of the temperatre, boreal and subarctic forest trees (*Pinus, Picea, Lanx, Abies*) amentiferous deciduour trees (*Alnus, Betula Fagus, Quercus, Populus, Solix*) tropical and semi-arid trophical plants (*Casuarina, Eucalyptus*) harbor ectomycorrhizal association, whose mycorrhizal roots function as nutrient absorbing organ in the plants (Gerdemann, 1975). Ectomycorrhizal associations are generally formed by higher basidiomycetes, ascomycetes, deuteromycetes and some members of zygomycotina.

All the mycorrhizal types share fundamental features that are associated with the interaction between symbionts features like chemotropism recognition, compatibilities and alterations during morphogenesis of the symbionts are of much significance. The events concerning between modulations of genotypes of various ecto and VA - mycorrhiza formation offers new possibilities of exploring factors controlling early events in the establishment of this symbiosis. This kind of information has become an important tool in understanding chemotropism, recognition, compatibility changes in fungal cytoskeleton, enzyme production and hormone production in the establishment of a mycorrhiza.

Most of the fungi belonging to Basidiomycotina representing *Amanita, Boletus, Caenococcum, Laccaria, Inoocybe, Pisolithus, Russula, Scleroderma, Suillus, Thelephora* etc. The possibility of a lectin-polysacharide recognition system in mycorrhizal system in mycorrhizal association is an important event in ectomycorrhizal association which needs more experimental evidences. Fungal hyphae continue to proliferate to form a mantle which consists of compact hyphae, separating living root cells from the soil and these hyphae store glycogen, proteins and lipids. The hyphae of the Hartig'snet undergo branching thus expanding the surface area for nutrient exchange.

Significance: Ectomycorrhizae help in the phosphorus uptake especially in soils of low availability. The application of ectomycorrhizal fungi may help in savings upto 50% of phosphorous, if suitable methodologies are employed to derive maximum mycorrhizal response. The other benefits made available are increased nutrient and water absorption, nutrient mobilization, increase in feeder root longevity, accumulation of nutrients and their translocation to the host tissue besides producing growth promoting substances and improving soil biomass.

Nitrogen fixing bacteria

Biological nitrogen fixation is estimated to contribute 180 x 10^6 metric tons/year globally, out of which eighty percent comes from symbiotic associations and the rest from free-living systems (Postgate, 1998). The *Fabaceac* is very important both ecologicall and agriculturally because they are the major sources of biological nitrogen fixation which is much greater than any of the free- living N-fixing organisms (Tilak *et al.*, 2005). The ability to reduce N_2 has not found in eukaryotes, but is confined to bacteria and archaea. In these groups it is widely distributed and has been demonstrated in about a hundred genera belonging to most major phylogenetic branches. The association between legumes and their symbiotic bacteria (rhizobia) is well studied.

Actinomycete genus *Frankia* form N-fixing root nodules on more than 280 species of woody plants from 8 different families (Schwintezer and Tjepkema, 1990). In temperate regions, alders (*Alnus* spp.) have great potential in various types of forestry and are documented to have considerable N_2 - fixation even at young age. *Casuarina*, another non-legume is the most important actinorhizal genus of the tropics and sub-tropics. Their ability to form symbiotic association with *Frankia* enables them to thrive in N-deficient soils and makes them desirable for agro-forestry. *Parasponia* and *Trema* spp., the only non-legumes belonging to the family *Ulmaceae* (order *Urticales*), is known to form an effective symbiosis with *Rhizobium* or *Bradyrrhzobium*. It has been suggested that these plants might have a common evolutionary origin on the basis of chloroplast sequence data, since they all form single clade within the angiosperms.

Non-symbiotic (free-living) nitrogen fixation seems to have magnitudes of agronomic significance. The importance of blue green algae (cyanobacteria) in rice cultivation is well known (Kannaiyan, 1990, Tilak and Pabbi, 1998). Nitrogen-fixing bacteria of many genera occur in high numbers in rhizosphere of a variety of plants. Some of the important nitrogen-fixing bacteria include *Achromobacter, Acetobacter, Alkaligenes, Arthrobacter, Azospirillum, Azotobacter, Azomonas, Bacillus, Beijerinckia, Clostridium, Corynebacterium, Derxia, Enterobacter, Herbaspirillum, Klebsiella, Pseudomonas, Rhodospirillum, Rhodopseudomonas, Xanthobacter* etc. Although the list is long, bacterial species of the two genera, *Azotobacter* and *Azospirillum* have been widely used and tested as inoculants to increase the crop yields. In recent years, *Acetobacter* has gained importance as an inoculant for sugarcane .The beneficial response of different crops to inoculation with various free-living diazotrophs in India has been extensively reviewed (Saxena and Tilak, 1998, Muthukumarawamy *et al.*, 2001).

Cyanobacteria (Blue-green algae)

The cyanobacteria are the largest and most diverse group of photo-synthetic bacteria. Older classification has as many as 2,000 or more species. In one recent system this has been reduced to 62 species and 24 genera. Section 19 of the current edition of *Bergey's Manual of determinative Bacteriology* describes 34 genera. Although cyanobacteria are true prokaryotes, their photosynthetic system resembles that of the eucaryotes because they have chlorophyll and a photosystem II and carry out oxygenic photosynthesis.

The ecological and agricultural importance of the blue-green algae (BGA) depends upon mainly the ability of certain species to carry out both photosynthesis and nitrogen fixation. Nitrogen-fixing BGA are either free-living or symbiotic and can be divided broadly into 3 major groups, viz., filamentous heterocystous forms, filamentous non heterocystous forms and unicellular forms. Over 300 strains are now known to fix nitrogen. Some BGA have been reported to develop symbiotic associated with eukaryotic algae, fungi, bryophytes, pteridophytes, gymnosperms and angiosperms (Tilak and Pabbi, 1998).

A tremendous degree of natural variation is observed in different attributes of cyanobacteria. All these are aquatic forms. They are present in diverse habitate such as rocks, woods, living trees, ice, hot springs, saline habitats, deserts etc. This diversity can be utilized for developing efficient strains through systematic survey of the autochthonous cyanobacteria and picking up desired organisms.

Blue-green algae have been used as biofertilizer in paddy crop in number of countries, including India (Kannaiyan, 1992). The beneficial effect on rice has been demonstrated in many areas of the country, in terms of grain yield and N-saving. A conservative estimate shows that BGA contribute 20-30 kg N/ha/season indicating a saving to that extent of the chemical fertilizer along with 10-20% increase in crop yield or without reduction in crop yield (Venkataraman and Tilak, 1990; Kannaiyan, 1990).

The nitrogen-fixing water fern *Azolla* has been used as a biofertilizer for rice crop in India, China, Vietnam, Thailand, Phillippines, Korea, Sri Lanka Bangladesh, Pakisthan, Nepal, Burma, Indonesia, Brazil and West Africa (Kannaiyan, 1992). It can contribute 40-60 kg N/ha per crop.The important factor in using *Azolla* as a biofertilzer for rice crop is its quick decomposition in soil and efficient availability of its nitrogen to rice plants (Kannaiyan, 1990). In tropical rice soils the applied *Azolla* mineralizes rapidly and its nitrogen is made available to the rice crop in a very short period.

Phosphate solubilizing microorganisms (PSM)

Phosphorus is a key nutrient limiting the plant growth. A survey of India soils reveal that 98% of them need phosphorus fertilization either in the form of chemical or biological fertilizer. Though, the P content in an average soil is 0.05% but only 0.1% of the total P present in the soil is available to the plants because of its chemical fixation and low solubility. Even after the application of chemical phosphatic fertilizer, majority of the soil P reaction products are only sparingly soluble. Under such conditions, the microorganisms offer a biological rescue system capable of solubilizing the insoluble inorganic P of soil and make it available to the plants.

Phosphate solubilizing microorganisms include bacteria fungi, actinomycetes, yeasts and blue-green algae, capable of dissolving inorganic phosphates. They can grow in medium having tricalcium, iron and aluminium phosphate, hydroxyapatite, bonemeal, rockphosphate and similar insoluble phosphate compounds as the sole phosphate source. These microbes not only assimilate P but in their presence, a large portion of soluble phosphate is released in quantities in excess of their own requirement (Gaind *et al.*, 2001).

The most efficient phosphate solubilizing bacteria (PSB) belong to genera *Bacillus* and *Pseudomonas*, though species of *Brevibacterium, Corynebacterium, Micrococcus, Sarcina* and *Achromobacter* have also been reported to be active in solubilizing insoluble phosphates. The reported *Bacillus* spp. include *B. brevis, B. cereus, B. circulans, B. firmus, B. licheniformis, B. megaterium, B. mesentricus, B. mycoides, B. polymyxa, B. pumulis, B. pulvifaciens* and *B. subtilis* isolated from the rhizosphere of legumes, cereals like rice and maize, arecanut palm, oat, jute and chilli Gaind *et al.*, 2001). Similarly, the reported *Pseudomonas* spp. include *P. striata, P. cussicola, P. fluorescens. P. punophillum, P. putida, P. syringae, P. aeruginosa, P. putrifaciens* and *P. stutzeri* isolated from root zone of plants fertilized with fresh organic matter, rhizosphere of leguminous and cereal crops, maize, desert soils etc.

The most efficient phosphate solubilizing fungi belong to geus *Aspergillus* and *Penicillium*. The common species belong to the genus *Aspergillus* include *A. niger, A. flavus, A. nidulans, A. awamori, A. carbonum, A. fumigatus, A. terreus* and *A. wentii* obtained from root nodules of legumes and rhizosphere of maize, soybean, chilli, tista soils, acidic lateritic soils and compost. A few species of *Cephalosporium, Alternaria, Cylindrocladium, Fusarium, Paecilomyces fusisporus, Penicillum digitatum, P. simplissimum, F. aurantiogriseum, Rhizoctonia* spp, *Scleroium rolfissi* are also good solubilizers of insoluble phosphate. A few species of yeasts like *Torula thermophila, Saccharomyces cerevisiae* and *Rhodotorula minta* have also been reported to solubilize inorganic phosphate. Blue-green algal members belonging to Cyanophyceae like *Anabaena* sp., *Calothrix brauni, Nostoc* sp., *Scytonema* sp., *Tolypothrix ceylonica* have also been reported as phosphate solubilizers. *In vitro* solubilization studies led to the investigation of PSM's as P releaser from native soil P pool as well as increasing availability of P from added super phosphate/rock phosphate. PSM are used as inoculant for various crops in India. The cultures used as inoculant included species of *Aspergillus, Bacillus, Escherichia, Arthrobacter* and *Pseudomonas* (Gaind *et al.*, 2001).

Phosphate solubilizing microorganisms inoculants are ecofriendly and provide yield increment with low input cost. They are known to add 30-35 kg P_2O_5 /ha (Gaind *et al.,* 2001).

The ecological distribution of different plant growth promoting and soil supporting rhizosbacteria has been reviewed by Tilak *et al.* (2005).

Applications and Biotechnology of Microorganisms

Microorganisms are omnipresent. They occur in soil, water, air, food etc. They are also present in arctic and antarctic zones, hot tropical areas, in deserts, in forest areas, on clothes etc. Microbes are present in upper horizon of atmosphere. Most of them are present on the surface as well as in body tissues of man and animals. They are friendly as well as disease causing agents.

Microorganisms in medicine

One of the land marks in the history of therapeutic microbiology is the discovery by Fleming in 1928 that the substance produced by the common fungus *Penicillium* is inhibitory for the growth of certain groups of pathogenic bacteria. The discovery of penicillin was a major breakthrough in medical microbiology and led to the search and development of a wide variety of antibiotics. The search for new antibiotics continues successfully. Even the the agony due to viral infections and tumor causing agents can be relieved by the use of antibiotics.

Microorganisms to improve soil

The discovery of bacteria involved in nitrification and nitrogen fixation marks the beginning of soil microbiology. Soil contains an abundant amount of all the elements needed for plant growth but these are present in unavailable forms. Certain specialized group of microorganisms brings about the transformation of various elements required for plant growth. Thus it was proved that the soil is not only a medium for microbial growth but also is the result of microbial activity. This recognition led to the development of soil microbiology as well as to attempts to improve soil fertility through a better understanding on the role of microbes in the mineral transformations.

Microorganisms are the effective agents in the decomposition of organic matter and the return of minerals to the soil. The life of the green plants depends on these activities. Certain bacteria like *Bacillus thuringiensis* play an important role in controlling certain insect pests and acts as a biocontrol agent.

Microorganisms are also involved in the study of geo-chemical reactions, retting of jute and coir fibres and several other branches of science and technology. They convert these products to either biomass or intermediary products that are secreted by the cells.

Microorganisms in industry

Microorganisms have been associated with man for their mutual benefit. For example, yeasts have been known to be involved with man since times immemorial and have served a steady and useful purpose. They are involved in the fermentation processes of fruit juices, wine making, brewing and bread making etc. This recognition of the ability of microorganisms to bring about desirable chemical changes in appreciable amounts led to the development of industrial microbiology and to the establishment of several industries.

Carotenoids and steroids can be obtained from fungi and algae. *Corynebacterium glutamicum* can produce large amounts of glutamic acid from sugar and ammonium salts. The development of microbial mutants has led to the possibility of large scale production of various amino acids, nucleotides and several other biochemical products. Enzymes like amylases for starch hydrolysis, proteinases for curing of leather, pectinases for producing fruit juices and several industrially important enzymes are obtained from microbial sources..

Microorganisms in dairy

Microorganisms have been associated with milk and milk products in dairy industry. Milk is a complete medium for the growth and reproduction of several types of bacteria and in particular, the lactic acid bacteria. The lactic acid bacteria have the capacity to convert milk sugar into lactic acid and this lactic acid increases the acidity of milk leading to curdling of the milk casein.

The art of cheese making developed about the same time as that of milk curdling. Cheese making also involves lactic acid bacteria and these are involved both in curdling as well as in brigning about the desirable flavours. In additioin to bacteria, some molds are also known to be important in cheese making.

Besides microorganisms which help in conversion of milk into various other desirable products, there are many others which cause the deterioration of milk and milk products, There are also many human diseases which are transmitted through milk to humans.

Microorganisms in food

Microbes have been known to play several different roles in food industry. The microbial cells may be used as food material or food supplement, or their cell components, particularly the enzymes may be used as accessory agents in food industries. Microbes could also be agents of food spoilage. A number of other processes employed in food industries are dependent on microbial fermentations for preservation or for characteristic flavour development.

Microbes in human welfare

Communities of saprophytic, parasitic and endophytic fungi living in the forest ecosystem contribute significantly to the species diversity. Recent studies have revealed that tropical fungi colonizing diversified habitats harbour diverse microfungi in abundance. It is becoming clear that both micro and macrofungi of the tropics are potential source of biotechnologically significant, industrially important and pharmaceutically valuable organic molecules *Cladosporium* (*Amorphotheca*) *resinae,* which is important in the field of petroleum product contamination. Filamentous fungi have also been used to produce different types of metabolites from alkaline media. *Cunninghamella, Mucor, Aspergillus, Botrytis, Fusarium, Paecilomyces, Sporotrichum, Verticillium* and others are reported to contain hydrocarbon assimilating species or strains. *Hansenula capsulata, Pichia lindnerii, Torulopsis glabrata* and *Candida boidinii* are selected as methylotrophic yeasts.

More than half the total production of plant residues remains unused, mainly straw and leaves and often also wastes from agricultural, forest and industrial production processes. Edible fungi serve as delicious food, spent mushroom compost can be utilized as humus fertilizer and widely grown substrates. This process has an advantage in which waste is profitable, removed and reintegrated by way of natural process into the ecosystem followed by the mobilization of cellulose or hemicellulose through mineralization. Hardly usable substrates are transformed into consumable protein rich biomass. Edible fungi represent a well characterized microbial biomass which will be generally accepted by the consumer. The consumption of edible fungi as food and drug is closely related to the history of mankind. *Lentinus elodes* and *Volvariella volvacea* belong to those species which have been cultivated for 2000 years in Asiatic countries. *Agaricus bisporus* was first mentioned in the seventeenth century in France. More recently *Agaricus bitorquis* and *Pleurotus* species have also been produced commercially. Edible fungi are cultivated worldwide under various climatic conditions. In the last two decades Taiwan has become the third largest producer of mushrooms in the world. China is the important producer of *Volvariella volvacea*. Mushrooms are considered more nutritious than many other vegetables. The nutritive value of some quality mushrooms is nearly equal to that of milk. These mushrooms contain high level of essential amino acids, less carbohydrate, more fibre, vitamins, unsaturated fatty acids with a mineral content exceeding that in fish and meat and twice that in most vegetables.

Role of fungi is well established in fermentation technology and using these industrial processes, many metabolites are commercially exploited for their antibiotic properties. Cephalosporin, penicillin and griseofulvin are derived from fungi. Yeast form an important agent in fermentation to produce ethanol from sugar or after hydrolysis from starch and cellulose. Fungi are involved in the production of foodstuffs, bread, fermented milk drink, youghurt, cheese and various yeast preparations. There are many foods used in Asian and Oriental cuisines, for example tofu, tempeh and miso whose production is dependent on fungi. *Saccharomyces, Torulopsis* and others have been found to be potential producers of glycerol. Several vitamins including Vit. B_{12}, biotin, folic acid, pantothenic acid, pyridoxine, riboflavin and thiamine are the products of fungi. *Eremothecium* and *Ashbya* are important since these fungi can make up 2.5 and 7 g/liter of Vit. B_{12}, respectively. *Blakeslea trispora,* a heterothallic fungus of mucorales group is extremely rich in mycelial β – carotene, as a potential source of carotenoids. Industrial enzymes are used in food processing and in particular amylase, glucoamylase and glucose isomerase, are exploited for the production of glucose and fructose syrups.

Fungi like *Thermomyces, Sporotrichum* and *Torula* are found to produce amylases. A number of thermophilics *viz, Acremonium cellulophilum, Aspergillus fumigatus, Chaetomium thermophile, Thermoascus aurantiacus* and others were reported to degrade cellulose and produce extracellualar enzymes. Xylanases, lipases and proteases were also found to be elaborated by thermophilic fungi. Bioactive compounds such as antibiotics including myriocin are synthesized using fungi. Important cellulolytic and lignolytic enzyme producing thermophilic fungi like *Chaetomium cellulolyticum* and *Sporotrichum pulveulentun* are the most commonly used thermophiles for upgradation of animal feed and to produce single cell protein from lignocellulosic wastes.

The disposal of municipal wastes, refuse and industrial effluents, has become a serious problem and bad odour emitted is of serious concern. Fungi like *Phanerochaete, Mucor, Sporotrichum* and others have shown their abilities in treatment.

Keratinophilic fungi possess the specialized ability to degrade hard keratine and utilize as many as unheated buildings of historic and artistic value with wall paintings such as churches and caves, provide favourable conditions for the growth of fungi. Deterioration of wall painitings by fungi is largely promoted by the dust which covers the painting. Species of *Pencillium, Aspergillus, Cladosporim, Aureobasidium, Alternaria* and *Phoma* are some of the fungal genera reported. Nearly 25 fungal species have been reported as responsible agents of deterioration of paintings in Ajanta caves. Therefore, it is recommended to use a low molecular polymer biocide at lower concentration to prevent fungal attack. Study of degradation of buildings and archaeological timber has revealed the association of *Serpula, Coniophora, Asterostroma* and other fungi.

Economic importance of algae

Chlorella is future hope of modern world as food supplement is cultivated in pilot plants. The alga is used as flour for making crackers and biscuits *Chlorella* has proteins and carbohydrates. *Chlorella* is also used in aeronautical research. Blue green alga, *Spirulina* tops the list of edible forms for single cell protein which accounts for 40-50% protein besides being rich in vitamin B_{12}. Algae serve as primary food for fishes and other aquatic animals.

Applications and biotechnology of actinomycetes

The knowledge of Streoptomyces has advanced consdiderably because of their practical importance in the production several effective antibiotics. Streptomycin, chloramphenicol, and aureomycin are produced by *S. griseus, S. venezulae and S. aureofaciens*, respectively are the most popular antibiotics among various antibiotics isolated from this genus.

Streptomycin consists of 3 groups, namely N-methyl-L-2 glucosamine, a methylpentose and an inositol derivative with 2 guanidyl residues. Streptomycin achieved its therauputical success because it is effective against a number of acid-fast and gram-negative bacteria that were not sensitive to penicillin. It is also used in certain animal and plant diseases.

Chloromycetin (chloramphenicol) was first discovered in cultures of *S. venezuelae*, but it is also produced synthetically. It is very stabe and effective against many gram-negative bacteria. sprochetes, rickettsiae, actinomycetes and large viruses.

The tetracyclines are excreted by a number of streoptomycetes including *S. aureofaciens.* They are closely related chemically, being derived from a naphthacene skeleton. The best known as chlortetracycline (aureomycin), oxytetracycline (terramycin) and tetracycline. They have spectrum of effectiveness.

The macrolide antibiotics (erythromycin, carbomycin A, picromycin etc.) are of diverse origin and have a relatively high molecular weight and are charactereised by presence of macrocyclic lactone ring.

Actinomycin was isolated as first of the antibiotics produced by streptomycetes. Actinomycin is actually a mixture of several compounds, all of them contain a phenoxazone chromophore, but all are substituted with different polypeptide chains.

Modern genetic approaches to modify microorganisms

Elucidation of gene transfer in bacteria by extra chromosomal elements has opened up the possibility of transferring forein DNA into bacteria. Genetic manipulation allows the introduction of small pieces of genetic information fro example from man into bacteria, which can then synthesize the corresponding proteins. This provides the possibility of production of hormones, antigens, antibodies and other proteins like insulin in bacterial cultures besides experimental evidences concern the transfer of nitrogen-fixing property to green plants and the cure of diseases due to biochemical defects. The recombinant DNA (rDNA) technology provides ample scope in the diagnosis and cure of hereditary defects and serious diseases. This technology is the most revolutionary technology in the 21 st century.

References

Dey R, Pal KK,. and Tilak KVBR (2005) Interaction of plant growth promoting rhizobacteria and arbuscular mycorrhizal fungi : Present status and future prospects. In: Podila G K, Varma AK (eds) Basic Research and Applications of Mycorrhizae. I K International Pvt Ltd, New Delhi, India, pp 83-110.

Frank b (1889) Ber Deut Bot Gessel 7: 322-346.

Gaind S, Mathur RS, Tilak KVBR (2000) Phosphate solubilising microorganisms. In: Yadav, Singh, Raychaudhury (eds) Recent Advances in Biofertilizer Technology. Omega Scientific Publishers, New Delhi, India, pp. 190-204.

Hawksworth DL (2004). Studies in Mycology 50 : 9-17.

Kannaiyan S. (1990) Biotechnology of Biofertilizers for Rice. Tamil Nadu Agril. Univ., Coimbatore, Tamil Nadu, India, p. 212.

Kannaiyan S (1992). *Azolla* Biofertilizer Technology for Rice Crop. Deptt. Microbiology, Tamil Nadu Agril Univ., Coimbatore, Tamil Nadu, India, p.54.

Manoharachary C (2004) Indian Phytopathol 57(1) : 1-6.

Manoharachary C, Kunwar IK, Tilak KVBR, Adholeya A (2010) Proc Nat Acad Sci India, 80(B): 1-13.

Manoharachary C, Sridhar K, Singh R, Adholeya A, Suryanarayana TS, Rawat, Johri BN (2005) Curr Sci 89(1): 58-71.

Manoharachary C, Tilak KVBR, Maheswari U, Singh D (2009) Indian Farming, November, pp. 24-30

Muthukumaraswamy R, Revathi G, Vadivelu M (2001) *Acetobacter diazotrophicus* prospects and potentialities – An overview Yadav, Singh, Raychaudhury (eds) Recent Advances in Biofertilizer Technology. Omega Scientific Publishers, New Delhi, India, pp. 126-153.

Postgate J (1998) Nitrogen Fixation Cambridge Univ. Press, 3rd ed. p.112.

Saxena AK, Tilak KVBR (1998) Free living Nitrogen Fixers : Their role in crop productivity. In : Varma A (ed) Microbes : For Health, Wealth & Sustainable Environment. Malhotra Publishing House, New Delhi, India. P.25-64.

Schubler A, Schwartzott D, Walker CA *Myco. Res* 105: 1413-1421.

Schwinyzer R, Tjepkema JD (1990) The biology of *Frankia* and actinorhizal plants. Acad Press Inc, San Diego, USA pp. 99.

Tilak KVBR, Pabbi S (1998) Microbial genetic resources. In: R.S. Paroda, Mangala Raj and P.L. Gautam PL (eds) Agrobiodiversity Conservation, Management and Use. National Academy of Agricultural Sciences and National Bureau of Plant Genetic Resources, ICAR, New Delhi, p. 85-99.

Tilak KVBR, Ranganayaki N, Pal KK, De R, Saxena AK, Nautiyal CS, Mittal S, Tripathi AK, Johri BN (2005) Curr Sci 89:136-150.

Tilak KVBR, Ranganayaki N, Manoharachari C (2006) Eur J Soil Sci 57: 67-71.

Tilak KVBR, Reddy BS (2006) Curr Sci 90:642-644.

PART-A:

INVITED LECTURES

Chapter - 1

Present scenario of viruses first isolated from naturally infected aroid food, horticultural and medicinal plants

P. Sreenivasulu, M. Padmavathi, Ch. V. Subba Reddy, K.P. Srinivas and B. Ramesh*

Department of Virology, College of Sciences, Sri Venkateswara University, Tirupati -517 502, Andhra Pradesh, India.
E-mail: pothursree@gmail.com

ABSTRACT

Ariods are grown as food crops, foliage or flowering ornamentals, aquarium or medicinal plants in different parts of the world. Many of these aroids are usually propagated through vegetative propagules like tubers, corms or stem pieces and mass multiplication by micropropagation technology. About 24 viruses representing 12 plant virus taxa have been reported to naturally infect aroid plants worldwide. Of these, 11 viruses have been first isolated from different aroids and others might have incidentally infected these plants. These viruses affect not only the growth and yield of food crops but also the growth vigor and quality of horticultural plants. Further, they have quarantine implications as they are vertically transmitted through vegetative propagules or micropropagated plants of aroids. The present scenario and future perspectives of viruses first isolated from aroid plants are discussed herein.

Key words: Aroid viruses, horticulture, medicinal plants

Introduction

The *Araceae* (aroids, arums) family is made up of nearly 100 genera and more than 1,500 species [Bown, 1988]. The major edible aroids are classified into two tribes and five genera. They are *Lasioideae* (*Cryptosperma* and *Amorphophallus*) and *Colocasioideae* (*Alocasia, Colocasia* and *Xanthosoma*) [Plucknett, 1976]. Different genera of aroids that are commonly grown as edible crops, foliage or flowering ornamentals, aquarium or medicinal plants are given in the Table 1. Edible aroids are major stable in the diet of people in the tropics and subtropics. The corms and tubers of these aroids are rich in carbohydrates which can be processed for the production of flour, beverage powder and other dried products. Aroid foliage ornamentals account for nearly one-third of the plants grown commercially in nurseries and gardens. These are grown both as outdoor and indoor ornamentals as some of these exhibit attractive natural variegation. *Zantedeschia* species are cultivated for cut flowers and as flowering ornamentals.

Table 1. Certain aroid genera of food, horticultural and medicinal importance.

Type	Plant genus (common name)
Edible plants *	*Alocasia* (Elephant's ear plant)
	Amorphophallus (konjac, elephant foot yam)
	Colocasia (taro, dasheen)
	Xanthosoma (malanga, cocoyam)
	Cryptosperma
Foliage ornamentals	*Aglaonema* (Chinese evergreen)
	Caladium
	Dieffenbachia (Dumbcane)
	Epipremnum aureum (Golden Pothos)
	Philodendron
	Spathyphyllum (Peace lily; devil's tongue)
	Scindapsus
	Alpinium
	Syngonium (Nepthytis)
	Spiranthes
Flowering ornamentals	*Anthurium* (Flamingo lily, Tailflower)
	Richardia
	Zantedeschia (calla lily, pig lily, arm lily)
Aquarium plants	*Anubias*
	Cryptocoryne (Cryptus, water trumpet)
Medicinal plants	*Pinellia*
	Typhonium
	Arisaena (dragon root, jack-in-the pulpis)

* Some species of edible aroids are also grown as ornamentals.

During the last two decades, several viruses representing different plant virus taxa have been reported to naturally infect aroid plants worldwide (Table 2). These viruses pose problems to the farmers, nursery growers and the tissue culturists in the selection of virus- free vegetative propagules and mother plants for micropropagation. Virus infections in certain aroid plants mimic the natural variegation or remain symptomless. Further, virus induced symptom expression depends on host species, virus isolates, single or mixed infection and/or growing conditions. The high temperature and plant aging often lead to masking of symptoms. As plant pathogens, viruses do have impact on the growth, quality and yield of aroid plants. Further, they have quarantine threat as they are vertically transmitted through vegetative propagules of aroids. The biological and molecular properties of some of the aroid viruses have been studied in greater detail and virus detection tests have been developed and applied in epidemiological investigations and in the formulation of management strategies.

Table. 2. Viruses reported to naturally infect aroids.

Virus (Acronym)	Genus (Family)	Reference(s)
Alfalfa mosaic virus (AMV)	*Alfamovirus* (*Bromoviridae*)	[3]
Arabis mosaic virus (ArMV)	*Nepovirus* (*Comoviridae*)	[3]
Aglaonema bacilliform virus (ABV)*#	*Badnavirus* (*Caulimoviridae*)	[4]
Bean yellow mosaic virus (BYMV)	*Potyvirus* (*Potyviridae*)	[3]
Caladium virus X (CalVX)*	*Potexvirus* (*Flexiviridae*)	[5]
Calla lily chlorotic spot virus (CCSV)*	*Tospovirus* (*Bunyaviridae*)	[6, 7]
Calla lily latent virus (CLLV)*	*Potyvirus* (*Potyviridae*)	[8]
Carnation mottle virus (CarMV)	*Carmovirus* (*Tombusviridae*)	[9]
Colocasia babone disease virus (CBDV)*	Putative *Nucleorhabdovirus* (*Rhabdoviridae*)	[10, 11, 12]
Cucumber mosaic virus (CMV)	*Cucumovirus* (*Bromoviridae*)	[13]
Dasheen mosaic virus (DsMV)*	*Potyvirus* (*Potyviridae*)	[14]
Impatiens necrotic spot virus (INSV)	*Tospovirus* (*Bunyaviridae*)	[3]
Konjac mosaic virus (KoMV)*	*Potyvirus* (*Potyviridae*)	[3, 15]
Potato virus X (PVX)	*Potexvirus* (*Flexiviridae*)	[3]
Pothos latent virus (PoLV)*	*Aureusvirus* (*Tombusviridae*)	[16, 17, 18, 19]
Soybean mosaic virus (SMV)	*Potyvirus* (*Potyviridae*)	[20]
Taro bacilliform virus (TaBV)*	*Badnavirus* (*Caulimoviridae*)	[21, 22]

Taro vein chlorosis virus (TaVCV)*	*Nucleorhabdovirus* (*Rhabdoviridae*)	[12, 23]
Tobacco rattle virus (TRV)	*Tobravirus*	[3]
Tobacco necrosis virus (TNV)	*Necrovirus* (*Tombusviridae*)	[13]
Tomato spotted wilt virus (TSWV)	*Tospovirus* (*Bunyaviridae*)	[3, 13]
Turnip mosaic virus (TuMV)	*Potyvirus* (*Potyviridae*)	[3, 24]
Zantedeschia mild mosaic virus (ZaMMV) *	*Potyvirus* (*Potyviridae*)	[25]
Zantedeschia mosaic virus (ZaMV)[a]	*Potyvirus* (*Potyviridae*)	[26, 27]

* Viruses first isolated from naturally infected aroid plants

[a] ZaMV is currently considered as an isolate of KoMV

In ICTV VIII report this virus was represented as a distinct species, but no details about this virus are available.

Among 24 viruses, 11 viruses have been first isolated from different aroids and the remaining might have incidentally infected them (Table 2). Among all the viruses, *Dasheen mosaic virus* (DsMV) is worldwide in distribution and naturally infects species of several aroid genera and a few orchids. Whereas the other viruses have relatively narrow natural host range with restricted distribution and have regional importance. Among the aroids, *Zantedeschia* species were found to be susceptible to nearly fifteen different viruses. Co-infection of aroids with DsMV and other viruses is common especially in Asia and Pacific region [Haung, 2007].

Vegetatively propagated plants are considered to be the gold mine for discovery of more and more new viruses or variants of already reported viruses. Modern virus detection and molecular characterization methods may facilitate the discovery of new viruses or virus strains from aroid plants in the near future. In this paper, the authors reviewed the present scenario and future perspectives of viruses first isolated from aroid plants.

Viruses first isolated from naturally infected aroids

Among eleven viruses first isolated from aroids, CLLV, DsMV, KoMV and ZaMMV belong to the genus *Potyvirus,* family *Potyviridae*; ABV and TaBV belong to the genus *Badnavirus,* family *Caulimoviridae*; TaVCV and CBDV to the genus *Nucleorhabdovirus,* family *Rhabdoviridae;* CalVX to the genus *Potexvirus,* PoLV to the genus *Aureusvirus,* family *Tombus viridae* and CCSV to the genus *Tospovirus,* family *Bunyaviridae* (Table 2)

I. Potyviruses

The family *Potyviridae* is the largest one among plant virus taxa [ICTV taxonomy, 2008], whose members are reported to infect agricultural and horticultural crops worldwide and some are inflicting significant crop losses. The members of the family have non-enveloped flexuous filamentous virion morphology (700-900 x 12-15 nm, bipartite virion morphology in case of bymoviruses). The host range of these viruses is narrow to moderate, and many of them are sap transmissible and a few through seed. They are vectored either by aphids, whiteflies, mites or fungi. The unifying diagnostic feature of all potyviruses is the presence of cylindrical/pinwheel inclusions in the infected tissues. The virions contain either mono- or bi-partite linear, positive sense, ssRNA genome (~ 9.5 to 11 kb) which contains VPg at 5'-end and poly (A) tail at the 3'-end. Genome is translated into single large polyprotein (340 to 368 kD) which is cleaved into 8-10 functional proteins that have different roles in pathogenesis [Urcuqui-Inchima, 2001]. The coat protein of virions consists of one polypeptide of ~ 30–40 kD. Adams *et al.* [Adams, 2005] have suggested molecular criteria for delineation of members of *Potyviridae* family into genera and species. The host range, vector transmission and genome intrinsic properties are the essential criteria. At least partial genome sequence covering 3'-UTR and complete coat protein gene may enable to find the reliable taxonomic position of the virus in this family. However, complete genome sequence information of virus decides its unambiguous taxonomic position. The *Potyviridae* family is divided into six genera viz., *Potyvirus, Rymovirus, Tritimovirus, Ipomovirus, Macluravirus* and *Bymovirus.* In addition, this family has three unassigned distinct viruses which may represent probably new genera to be created by ICTV [ICTV taxonomy, 2008]. Among these, the genus *Potyvirus* is the largest one having nearly 200 definite and tentative species. Potyviruses (DsMV, ZaMMV, CLLV and KoMV) first isolated from naturally infect aroids belong to this genus are discussed here under.

1. *Dasheen mosaic virus* (DsMV)

DsMV is the most important viral pathogen of aroid plants worldwide [Zettler, 1978]. The Florida isolate of this virus has much wider natural host range in aroids (*Aglaonema, Alocasia, Amorphophallus, Anthurium, Arisaema, Caladium, Colocasia, Cryptocoryne, Cryptospermia, Dieffenbachia, Philodendron, Spathiphyllum, Xanthosoma,* and *Zantedeschia*) and recently an isolate of this virus was reported to infect orchids *Pinellia* and *Vanilla* spp [Farreyrol, 2006]. Mixed infection in aroids involving this virus and other viruses like CalVX, CMV, ZaMMV, CLLV, KoMV or TaBV are common [Rivas, 2005; Martelli, 1998; Hu, 2009; Ahlawat, 2003]. The virus induces varied symptoms on different aroids and sometimes causes symptomless infections. In ornamental plants DsMV incited symptoms are often confused with natural variegation or mineral deficiencies. In Pacific Island countries loss of taro corm yield of approximately 20% has been reported. In mosaic affected elephant foot yam, the size and weight of tuber is decreased by 70% [Ahlawat, 2003]. In Nicaragua, DsMV was reported to cause yield losses around 25% in cocoyam [Reyes, 2006]. It also has impact on foliage aroid ornamentals by reducing growth vigour, rooting of the propagules and/or affecting the size and quality of leaves. The reports on incidence of DsMV infections based on visible symptoms are highly variable due to erratic expression of symptoms or no symptoms. Therefore, field surveys employing sensitive diagnostics need to be conducted for proper assessment of its infections. This approach has already been adopted in countries like USA, Taiwan, Japan, South Korea and China. DsMV is sap transmissible and also by aphids (*Myzus persicae, Aphis gossypii, Aphis craccivora*) in a non-persistent manner. The major means of spread in nature is through vegetative propagules of aroids. It has a narrow experimental host range and outside *Araceae* the virus infected the plants like *Tetragonia expansa, Chenopodium* spp., *Nicotiana* spp., and *Saponaria vaccaria*. The Indian isolate of DsMV has a relatively wider experimental host range [Ram, 2003]. The virus was purified and virions are non-enveloped flexuous filamentous particles measuring ~750 nm in length. The virus is moderately stable as it has a thermal inactivation point (TIP) of 55-60°C, dilution end point (DEP) of 10^{-2} and longevity *in vitro* (LIV) of 75 hr at 26°C. Like other potyviruses, this virus also induced cylindrical / pinwheel inclusions in infected tissues [Zettler, 1978].

In the SDS-PAGE analysis, purified virions of DsMV-Fl revealed upto six polypeptides ranging in mol wt from 25.0 to 45.0 kD and 29.0 kD was considered to be capsid protein [Abo-El-Nil, 1977]. But analysis of coat protein of Korean isolate of DsMV revealed multiple bands with mol wt of 42, 39 and 37 kD [Kim, 2004]. They have suggested that 39 and 37 kD proteins might be the proteolytic degraded forms of CP. Its CP gene consists of 942 nucleotides encoding 314 amino acids with deduced molecular weight of 34.7 kD.

Polyclonal antibodies to purified virions of DsMV-Fl, DsMV-FJ and DsMV-K isolates were produced [Kim, 2004; Ahlawat, 2003]. To overcome the inherent limitations of these antibodies, polyclonal antibodies (PAbs) against the bacterially expressed recombinant CP (rCP) of DsMV-Caladium isolate were also produced [Li, 1998]. They have evaluated the antisera produced against the purified virions of DsMV-Fl and antisera raised against the recombinant CP (rCP) of DsMV-Ch in enzyme-linked-immunosorbent assay (ELISA), and the latter antiserum appeared to be superior in respect of titres and background reactions Both antisera reacted with *Zucchini yellow mosaic virus, Papaya ringspot virus* type W and *Watermelon mosaic virus 2*. PAbs to rCPs of several potyviruses have been produced and found to be superior over conventionally produced antibodies (e.g., [Hema, 2003]). This recombinant DNA technology based approach can generate large amounts of stable antigen for serological testing and as an immunogen for antibodies production. DsMV-Fl antigen was reported to react positively with the antiserum of *Tobacco etch virus* (TEV) [Zettler, 1978]. The DsMV isolates from aroid plants in different countries were reported to be antigenetically related to antibodies of DsMV isolates from USA [Kim, 2004; Lima, 2004].

DsMV genome is linear, unipartite ssRNA of positive polarity of ~ 10 kb in size [Zettler, 1978; Kim, 2004]. The genomes of DMV-M13 isolate from *Zantedeschia aethiopia* from Zhejian province of China is 9991 nucleotides in length excluding poly (A) tail (DDBJ / GenBank / EMBL database accession number AJ 298033). It contains single large pen reading frame (ORF) with AUG initiation codon at 168-170 nucleotides and UAA termination codon at 9,741-9,743 nucleotides followed by 3'-UTR of 248 nucleotides. The 5'-UTR is 167 nucleotides long. The deduced polyprotein of this ORF contains 3,162 amino acids with a calculated mol wt of 362.5 kD [Chen, 2001]. Based on alignments with polyprotein sequences of other potyviruses, ten mature proteins characteristic of *Potyvirus* genus are identified (Fig. 1). The phylogenetic analysis suggests that DsMV-M13 belongs to the subgroup of potyviruses that include *Bean common mosaic virus* (BCMV), *Soybean mosaic virus* (SMV), *Zucchini yellow mosaic virus* (ZYMV) and confirmed a relationship first proposed on the basis of coat protein (CP) sequence data [Pappu, 1994a & 1994b]. Like other aphid transmissible potyviruses, DsMV-M13 also has DAG (Asp-Ala-Gly) motif near the N-terminus of CP (2,883-2,885 amino acids position). A sequence similar to the conserved potybox motif is found near the 5'-end (nucleotides 14-25). The 3'-UTR (284 nucleotides) is similar to those of other reported isolates of DsMV. Recently, Reyes *et al.* [Reyes, 2009] analysed the CP gene sequences of 10 isolates of DsMV infecting three different genotypes of cocoyam in three geographical locations of Nicaragua. These isolates showed sequence identities of 84-100% at nucleotide level and 94-100% at amino acid level. In phylogenetic tree the 10 isolates formed two distinct subgroups that correlated with geographic origin of isolates.

Chen *et al.* [Chen, 2001] have also compared the proteolytic cleavage sites of DsMV-M13 Chinese isolate with complete sequences of other members of *Potyviridae* family by multiple sequence alignment. Such analysis revealed patterns related to phylogenetic groupings.

Genetic diversity among DsMV isolates originated from different aroids and geographical regions has been investigated [Kim, 2004; Chen, 2001; Pappu, 1994a & 1994b; Reyes, 2009]. Chen *et al.* [Chen, 2001] have analysed the sequence of 3'-untranslated region (3'-UTR) and CP of four isolates of DsMV (M13 and ND Chinese isolates from *Zantedeschia* spp., TW3 from a Taiwanese dasheen cultivar) with four other published isolates (DsMV-Fl, dasheen Florida; DsMV-Ch, caladium, Florida; DsMV-TEN, dasheen Florida, DsMV-TW, aroid, Taiwan). The sizes of CP and 3'-

UTR ranged from 939 to 987 and 247-250 nucleotides, respectively. Phylogenetic dendrogram, prepared based on CP amino acid sequences of DsMV isolates and other related potyviruses, shows some grouping amongst the DsMV isolates. There is no relationship to geographical or host plant origin. Significant variation is confined to the N-terminus of CP. Pappu *et al.* [Pappu, 1994a & 1994b] have reported variations in the length of CP N-terminus and unusual amino acid repeats for American isolates and such variations and unusual amino acid repeats are also the features of isolates from China, Taiwan and Japan. Vegetative propagation of aroid plants may favor the accumulation of variations in the N-terminus of CP, the most variable region among potyvirus isolates.

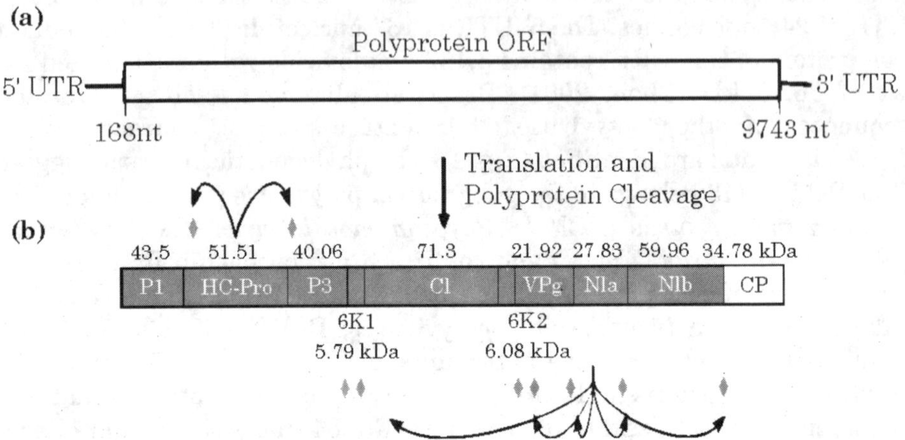

Fig. 1: Genome organization of DsMV-M13 isolate based on the full length sequence data deposited in GenBank A/C: AJ298033 [42]. (a) Polyprotein ORF with initiation and termination sites (b) Proteolytic cleavage of the encoded polyprotein into functional proteins with their deduced molecular weights.

The CP of virus isolates from Korea (DsMV-K from taro) consisted of 942 nucleotides in length and encoding 314 amino acid residues [Kim, 2004]. This isolate CP gene identity ranged from 80 to 91.5% at nucleotide level when compared to already reported virus isolates. The nucleotide sequence identity of 3'-UTR of DsMV-K isolate ranged from 86.5 to 91.8% when compared to several other isolates.

Vanilla mosaic virus (VanMV), a potyvirus naturally infecting *Vanilla tahitensis* and *V. planifolia* (orchids) in French Polynesia, Fiji, Cook Islands and Vanuatu are serologically related to DsMV [Wisler, 1987; Pearson, 1993; Wang, 1992]. The sequences covering 3'-UTR, CP and CI-6K2-NIa regions of VanMV-CI and VanMV-FP from Cook Islands and French Polynesia, respectively infecting *V. tahitensis* were compared with the sequences of several potyviruses including DsMV isolates [Farreyrol, 2006]. Both the isolates of VanMV displayed distinctive features in the N-terminal region of CPs. Identity was low to DsMV isolates across the entire CP coding region and 3'-UTR, but high across the CP core and CI-6K2-NIa regions. In accordance with ICTV criteria for species demarc-ation within the family *Potyviridae*, VanMV-CI and VanMV-FP were suggested as strains of DsMV that exclusively infect the orchid vanilla. Further, they have suggested that the isolates of VanMV may be grouped under the name DsMV-V.

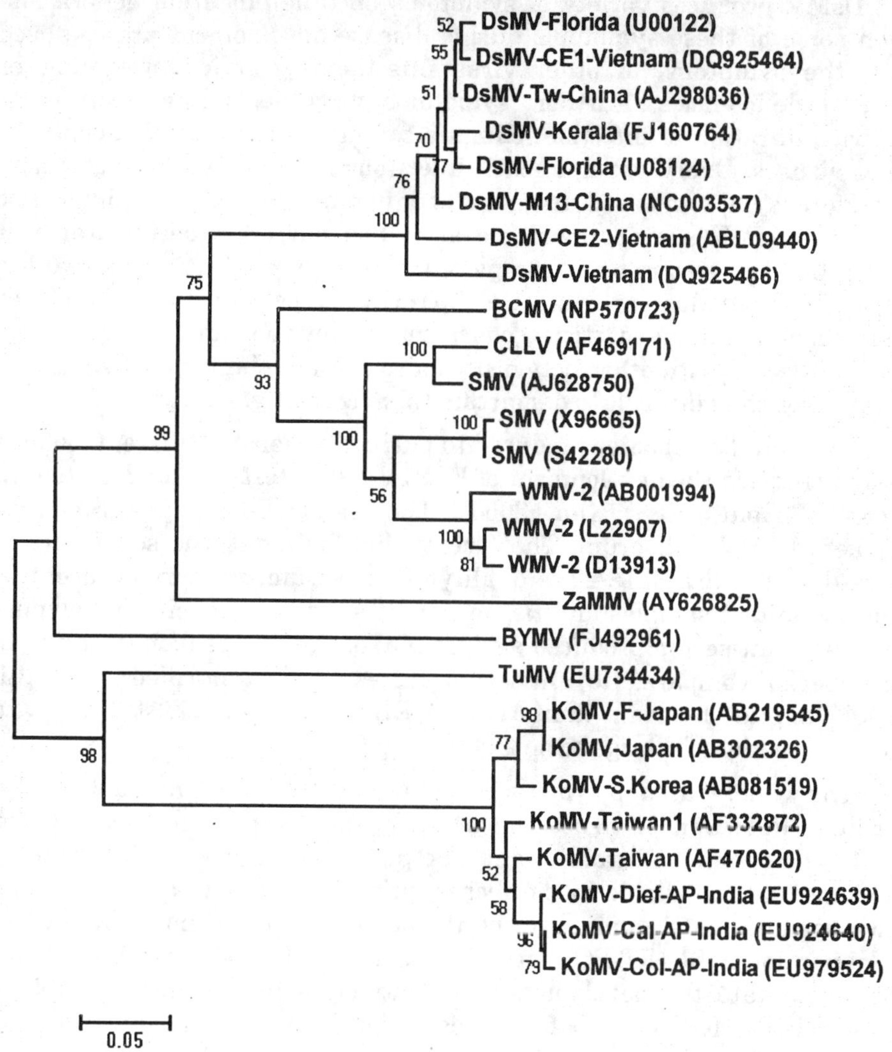

Fig. 2. Neighbor-Joining phylogenetic tree showing the relationships of CP amino acid sequences of potyvirus isolates from aroids with several other potyviruses of the genus *Potyvirus.* The acronyms of viruses: KoMV: *Konjac mosaic virus,* DsMV: *Dasheen mosaic virus,* CLLV: *Calla lily latent virus,* SMV: *Soybean mosaic virus,* TuMV: *Turnip mosaic virus,* BYMV: *Bean yellow mosaic virus,* WMV: *Watermelon mosaic virus,* BCMV: *Bean common mosaic virus,* ZaMMV: *Zantedeschia mild mosaic virus.*

DsMV was the first potyvirus isolated from aroid plants. Later other genetically distinct potyviruses viz., ZaMV, ZaMMV, KoMV and CLLV have been isolated from naturally infected aroid plants in Asia [Chen, 2006; Chang, 2006]. The phylogenetic tree based on CP gene of potyvirus isolates naturally infecting aroids and other potyviruses is given in Figure 2.

DsMV produces variety of symptoms on different aroid genera and often some of these symptoms during disease development are confused with the symptoms of other virus infections, genetic variegation or mineral deficiencies. Further, symptoms vary depending upon plant cultivar, duration of infection, climatic conditions and co-infections with other viruses. DsMV asymptomatic infections in aroids are also common. Therefore, symptom based diagnosis is only tentative. Suitable diagnostic hosts for DsMV are not available as it has narrow experimental host range. Light or electron microscopy of DsMV suspected tissues reveal the presence of cytoplasmic cylindrical / pinwheel type of inclusions and / or flexuous filamentous particles characteristic of the family *Potyviridae.* Aroids infected with other potyviruses also contain similar inclusions and virions and thus difficult to demarcate them correctly.

Polyclonal antibodies produced to purified virions of different isolates have facilitated the development of ELISA- based tests for its detection in aroids [Rodoni, 1988; Kim, 2004]. The Agdia Company (USA) also markets DsMV antiserum. The factors that influence the sensitivity of virus detection by ELISA are quality of antiserum, the virus concentration in the test samples and tissue constituents (e.g., mucilage, polyphenols). The antiserum produced against rCP of DsMV was demonstrated to be superior compared to antiserum raised against purified virus [Li, 1998]. Inspite of certain limitations of ELISA based tests, still they are found to be ideal for large scale testing.

The widely used polymerase chain reaction (PCR) based tests for sensitive detection of viruses on a small scale were also developed for DsMV detection in aroid plant samples. Based on partial genome sequence data available in databases, primers covering part of CP gene have been designed to amplify specific portions of the genome by reverse transcription-PCR (RT-PCR) in test aroid samples. The available PAbs could facilitate the development of relatively sensitive immunocapture (IC)-RT-PCR (IC-RT-PCR) for its detection. Mixed infections involving DsMV and several other viruses are common in aroids [Rivas, 2005; Kim, 2004; Hu, 2009; Gollifer, 1977]. To save time and costs, duplex or multiplex PCR tests are ideal for detection of viruses in such aroid plant samples. The genome nucleotide sequence data of viruses involved in co-infections could facilitate the designing of virus specific primers for PCR amplification. Multiplex-RT-PCR for parallel detection and discrimination of four potyviruses in calla lily is recently developed [Hu, 2009]. Such test facilitates effective screening and certification of aroid plant samples.

Generally aroids are cultivated either in fields or nurseries by planting of their vegetative propagules like corms, tubers or stem pieces. Micropropagation technology is also exploited for mass production of aroid plants. DsMV present in the source plants of vegetative propagules or in mother plants meant for micropropagation is getting vertically transmitted form one generation of plants to the next generation, and this is the major means for its survival and spread in nature. Eventhough aphids (*M. presicae, A. craccivora,* and *A. gossypii*) were reported as its vectors [Kim, 2004; Gollifer, 1977], their actual role in horizontal spread of virus in aroids is to be established. In the horticultural nurseries, it could also be spreading as a contaminant of implements and workers hands. Exchange of aroid germplasm is leading to introduction of DsMV from endemic regions into new virus-free regions due to non-rigorous testing of aroid samples at the quarantine stations. DsMV is not transmitted through the seed or by pollen. The contact spread of virus by natural rubbing of leaves is to be realized.

Disease management tactics depend upon the crop. Nelson [Nelson, 2008] recommended the following IPM (integrated pest management) principles and tactics to manage the diseases caused by DsMV in aroids.

IPM principles

- Avoid introduction or transmission of the virus during vegetative propagation.
- Prevent the introduction or spread of the disease on new hosts. Control aphids or ants where aphid transmission of DsMV is a problem.

IPM tactics

- For propagation, use cuttings or vegetative sprouts from virus-free plants only.
- Prevent mechanical/contact transmission of the virus on horticultural implements. Dip the implements in sterilizing solutions (e.g., Trisodium orthophosphate) between cuts.
- Use certified plants from tissue culture.
- Plant quarantine: intercept diseased plants at points of entry using sensitive diagnostics.
- Inspect nurseries regularly to rogue out infected plants.
- Avoid growing of taro alongside other susceptible aroids.
- Propagate aroids from seeds as DsMV is not seed transmissible.

- Control the aphids where the disease is causing significant losses.

2. *Zantedeschia mild mosaic virus* (ZaMMV)

A virus associated with mild mosaic symptoms on calla lily in Taiwan was isolated, characterized and identified as a new potyvirus, ZaMMV, based on virion morphology, antigenic relationships and partial genome sequence data [Chang, 2001]. The virions are flexuous filamentous particles measuring ~ 680 nm in length. Among 36 plant species representing 13 families sap inoculated in host range studies, the virus infected only two aroids (*Zantedeschia* spp. and *Philodendron selloum*), In ELISA, the virus failed to react with the antibodies of DsMV, TuMV and ZaMV, but positively reacted with anti-poty group specific monoclonal antibodies (Agdia Inc., USA). The sequence [1845 nt excluding 3'-poly(A)] analysis revealed that the highest amino acid sequence identity of the CP between ZaMMV and thirty other potyviruses is 55% and the highest nucleotide identity of 3'-UTR is 52%. According to the species demarcating criteria of Adams *et al.* [Adams, 2005], virus associated with mild mosaic of calla lily in Taiwan is proved to be a new species of the genus *Potyvirus,* and designated as *Zantedeschia mild mosaic virus* (ZaMMV). Phylogenetic analysis of complete CP amino acid sequence revealed that ZaMMV clustered with BCMV, *Bean common mosaic necrosis virus* (BCMNV), *Soybean mosaic virus* (SMV), DsMV and *Zucchini yellow mosaic virus* (ZYMV), as a new member of BCMV subgroup of potyviruses [Berger, 1997].

The conserved NIa-Pro cleavage site (VXXQ/A) is identified between the NIb and CP regions of ZaMMV as for DsMV. The DAG motif relating to aphid transmission is located at amino acid residues of 51 to 53 of the CP, but its aphid transmission is to be established. Interestingly, the N-terminus of the CP of ZaMMV contained 39 glutamine residues before the DAG motif. This is probably the first time that this kind of poly-glutamine sequence was discovered in plant viruses. Further studies might focus on the function(s) of this poly-glutamine tract in virus structure, biology and molecular biology.

Huang *et al.* [Haung, 2007] have developed ELISA, non-radioactive probe and IC-RT-PCR based tests for the detection of ZaMMV in aroids. ZaMMV CP gene without poly-glutamine was expressed in *E. coli* and polyclonal antibodies produced against rCP. Using these antibodies the virus is detected in field collected calla lilies and call lily flowers purchased form the market in Taiwan by indirect-ELISA (I-ELISA). Three different probes labeled with digoxigenin by PCR are evaluated for detection of ZaMMV in calla lily by dot-blot hybridization. The first probe 1175 spans entire 3' region of the NIb gene and the 5'region of CP gene. The second probe 311 covers junction region of NIb and CP genes, the sequence of which is variable among potyviruses. The probe three corresponds to 3'UTR of ZaMMV. Of these probes, probe 311 gave stronger signals and is used for the virus detection in field collected aroid samples. IC-RT-PCR is developed using produced antibodies and ZaMMV specific primers. This test confirmed the samples that gave weak signals in dot-blot hybridization and low-ELISA values. These tests have been found to be useful in distinguishing the mixed infections with ZaMMV and DsMV in Taiwan. Certification of calla lily using developed virus detection methods needs to be established to prevent its further spread through vegetative propagules and micropropagated plants.

3. *Calla lily latent virus* (CLLV)

A virus isolate (Ca-M19) capable of inducing local lesions on *Chenopodium quinoa* was first isolated from symptomless calla lily and systemic mosaic exhibiting calla lily plants co-infected with other potyviruses in Taiwan [Chen, 2006]. Potyvirus-like flexuous filamentous particles were consistently detected in Ca-M19 infected plants. In serological tests the virus did not react with DsMV, BYMV, KoMV, TurMV and ZaMMV. However, it reacted with potyvirus-specific monoclonal antibodies and several other potyviruses like BCMV, SMV, BlCMV, WMV-2 and ZYMV. Initially based on virion morphology and size and serological reactivities with potyvirus-specific monoclonal antibodies and with the antisera of other potyviruses, Ca-M19 isolated from calla lilies was identified as a member of the genus *Potyvirus*. To confirm its exact identity within this genus, 1.3 kb fragment was RT-PCR amplified from infected plants using potyvirus degenerate primers and sequenced. Comparison of generated sequences (3'-UTR and complete CP) with known potyvirus sequences in the GenBank revealed that it has greater similarity with SMV and WMV-2; 79 to 80% at amino acid level of CP and 79 to 83% at nucleotide level of 3'-UTR. These comparisons confirmed that Ca-M19 is a new species in the genus *Potyvirus* and designated as *Calla lily latent virus* (CLLV).

The CP gene of CLLV was cloned and expressed in *E. coli* system, polyclonal antibodies against the expressed recombinant CP produced and immunoassay (indirect- ELISA) was used for its detection in field collected samples [Chen, 2006]. A primer pair (M19u/M19d) for RT-PCR and DNA probe were designed for specific detection and differentiation of Ca-M19 from other viruses infecting calla lily. Multiplex-PCR for parallel detection of four potyviruses in calla lily plants is recently developed [Hu, 2009].

4. *Konjac mosaic virus* (KoMV)

KoMV (K2 isolate) that causes mosaic disease on Konjac plants (*Amorphophallus konjac*) was first isolated in Gunma Prefecture, Japan.

It infects limited number of other aroids like *Philodendron oxycardium* and *P. selloum*. But another isolate of KoMV (KoMV F), serologically distinct from DsMV, infects wider range of plants in host range studies include *A. konjac, Colocasia esculenta, P. oxycardium, P. selloum, P. verrucosum* and a few non-aroids that belong to *Chenopodiaceae* and *Amaranthaceae*. The virions of this virus are flexuous filamentous particles measuring 800-900 x 13 nm in length. It is non-persistently transmitted by aphids and vertically through the corms of konjac. The initially studied properties suggest that the virus causing mosaic of konjac in Japan is a potyvirus [Shimoyama, 1992].

Later, the genome of KoMV F isolate is sequenced. It is 9,544 nucleotides long excluding 3' poly (A) tail [Nishiguchi, 2006]. It encodes 350 kD polyprotein with 3,087 amino acid residues. Polyprotein based phylogenetic analysis shows that KoMV constitutes a branch with *Yam mosaic virus,* close to another branch with *Japanese yam mosaic virus, Turnip mosaic virus* and *Lettuce mosaic virus*. Potyvirus isolates, first isolated from aroids in other Asian countries based on partial genome sequence analyses identified as ZaMV [Sabanadzovic, 1995; Chang, 2001; Kwon, 2002] are later suggested as KoMV.

KoMV-K2 was identified as a unique potyvirus based on serological and physico-chemical properties. KoMV-K2 and F isolates differ from each other in their host range, but quite similar in symptoms on konjac plants. The partial genome sequence data of these two isolates showed that they belong to the same virus because they show 92.6 and 92.2 % identities at the amino acid level in CP and partial C-terminal NIb and 90.4% at the nucleotide level in the 3'-UTR. The sequence of KoMV showed high identity to ZaMV and Japanese hornwort mosaic virus (JHMV) at nucleotide and amino acid levels. The phylogenetic tree based on the CP data of 12 isolates (2 of KoMV, 6 of ZaMV and 4 of JHMV) indicated that KoMV is closely related to ZaMV. Kwon *et al.* [27] have reported ZaMV as a new potyvirus without comparing its properties with KoMV reported earlier than the ZaMV. Based on molecular criteria suggested by Adams *et al.* [Adams, 2005], these viruses are considered to be very similar. Among the three viruses, KoMV was the earliest virus described, and therefore it would be reasonable to state that ZaMV and JHMV are isolates of KoMV.

II. Potexviruses

Potexviruses are moderately pathogenic, causing mosaic or ringspot symptoms in a wide range of mono- and dicotyledonous plant species. Their host range is limited. They are usually mechanically sap transmissible, and in nature contact transmissible. They lack invertebrate vectors except *Potato acuba mosaic virus*. The virions are fairly stable in leaf sap. The virions of potexviruses are non-enveloped slightly flexuous filamentous particles measuring 470-580 x 13 nm. Virions contain a single linear molecule of positive sense ssRNA of about 5.9-7.0 kb. The genomic RNA is capped at 5'end and has polyadenylated tract at 3'-end. The genomes of several species have been completely sequenced and have 5-6 ORFs. The mol wt of CP is ranged from 18-27 kD. The genome expression involves 2-3 subgenomic NAs. The type species of the genus *Potexvirus* is *Potato virus X* (PVX). The criteria for demarcation of species in this genus are host range, serological and sequence of CP gene. This genus contains ~45 definitive and tentative virus species [ICTV taxonomy 2008].

1. *Caladium virus X* (CalVX)

Of the two potexviruses reported to naturally infect aroid plants (Table. 2), CalVX was first isolated from *Caladium bicolor* co-infected with DsMV in Brazil [Rivas, 2005]. The dually infected plants displayed chlorotic and necrotic ring spots. CalVX from such plants was biologically isolated by single lesion passage through sap inoculation on *Gomphrena globosa* which is not susceptible to DsMV, and subsequently it was propagated on *Nicotiana benthamiana, C. bicolor* sap inoculated with the isolated virus exhibited systemic chlorotic spots, green rings and mosaic symptoms on leaves. From dually infected plants, DsMV can be isolated using aphids as its vectors as CalVX is not aphid-borne.

CalVX is sap transmissible to 13 species that belong to *Amaranthaeae, Araceae, Chenopodiaceae, Plantaginaceae, Portulaceae* and *Solanaceae.* In addition to *Caladium* spp., the virus infected *Philodendron* spp., but not *Dieffenbachia* and *Anthurium* spp. The virus induced crystalline inclusions in the cytoplasm and nucleus of mesophyll cells of *G. globosa.*

The purified virus preparations contained slightly flexuous filamentous particles measuring 420-479 nm in length. The virus failed to react with the antisera of 17 potexviruses including PVX. The virus had a TIP of 60 to 65°C, DEP of 10^{-2} to 10^{-3} and LIV of 6 days at room temperature.

The virus capsid protein has a mol wt of 36.0 kD. Using primers specific for *Potexvirus* genus [Gibbs, 1998], 740 bp product from RNA extracts of virus infected tissues was RT-PCR amplified, cloned and sequence (GenBank accession number AY727533) analysed [Rivas, 2005]. The generated sequence shared significant identity with the replicase domain of potexviruses, ranging from 65.8 % (*Cymbidium mosaic virus*) to 60.3% (*Cassava common mosaic virus*). Adams *et al.* [Adams, 2004] suggested identity values for demarcation of species of the genus *Potexvirus* using replicase sequences of between 45.2 and 70.1% at nucleotide level and between 40.4 and 73.3% at the amino acid level. This virus is distinct from PVX, the type species of the genus *Potexvirus,* which was reported to infect *Zantedeschia* spp. [Pham, 2002]. The studied biological and molecular properties of the virus suggest that it is a distinct species in the genus *Potexvirus* and designated as Caladium virus X.

Development of duplex-RT-PCR, based on the generated genome sequences and on public databases, could be useful for parallel detection of CalVX and DsMV in single or dual infections on aroids. As the virus is sap transmissible easily from aroids to several experimental hosts and virus purification protocol developed, PAbs can be easily produced against the whole virions for development of widely used ELISAs. As potexviruses were known to be contact transmitted through implements and workers hands, proper care has to be exercised to prevent the spread of this potexvirus in aroid nurseries and gardens.

III. Plant rhabdoviruses

Plant rhabdoviruses have linear, monopartite negative sense ssRNA genome of ~12.0 kb. They are usually neither sap or seed transmitted, but vectored by aphids, leafhoppers or planthoppers. The genome has 5-6 genes encoding proteins required for successful virus replication and pathogenesis. Plant infecting rhabdoviruses are classified into two genera, *Nucleorhabdovirus* and *Cytorhabdovirus,* in the family *Rhabdoviridae,* order *Mononegavirales* [ICTV taxonomy 2008]. Members of the *Nucleorhabdovirus* replicate in the cell nucleus and bud from the nuclear membrane to accumulate in the perinuclear space, whereas members of the genus *Cytorhabdovirus* replicate and accumulate in the cytoplasm.

Two rhabdoviruses with enveloped bullet shaped virion morphology have been reported to naturally infect taro in Pacific Island countries (Fiji, the Federated States of Micronesia, Solomon islands and Vanuatu) [James, 1973; Pearson, 1999]. They are Colocasia bobone disease virus (CBDV), initially named Taro large bacilliform virus and *Taro vein chlorosis virus* (TaVCV). Based on the site of accumulation of virions in the perinuclear space of infected tissues, CBDV sand TaVCV were initially tentatively identified as putative members of the genus *Nucleorhabdovirus* [Pearson, 1999].

1. *Taro vein chlorosis virus* (TaVCV)

TaVCV infected taro plants exhibit a striking vein chlorosis symptoms, particularly at the leaf margin, which often leads to necrosis of affected tissue [Pearson, 1992]. In thin sections or sap dips, virions with bullet shaped morphology (~210 x 65 nm), characteristic of rhabdoviruses have been observed. TaVCV is serologically distinct from CBDV. It is non-sap transmissible and vectors are unknown.

The ssRNA genome of TaVCV is fully sequenced and is 12,020 nucleotides in length (GenBank accession number AY674964) [Revill, 2005]. The genome contains six open reading frames (ORFs) in the antigenomic sequence, equivalent to nucleocapsid (N) gene, phosphoprotein (P) gene, gene 3, matrix protein (M) gene, glycoprotein (G) gene and polymerase protein (L) gene that have been identified in other rhabdoviruses (Fig. 3). The putative gene products had highest similarity to the nucleorhabdovirus *Maize mosaic virus.* The TaVCV leader and trailer sequences comprised 140 and 61 nucleotides, respectively. The genome sequence analysis and clues on microscopic observation of budding of virions from nuclear membrane into perinuclear space have supported the assignment of TaVCV as a new species of the genus *Nucleorhabdovirus.* Genome variability studies identified high levels of TaVCV sequence diversity within the L gene of 20 TaVCV isolates from several Pacific island countries, maximum variability was 27.4% at the nucleotide level. Maximum variability within the N gene among 15 isolates of TaVCV at the nucleotide level was 19.3%. The observed high variability among isolates suggests that the introduction of TaVCV into the Pacific islands is not a recent occurrence [Revill, 2005].

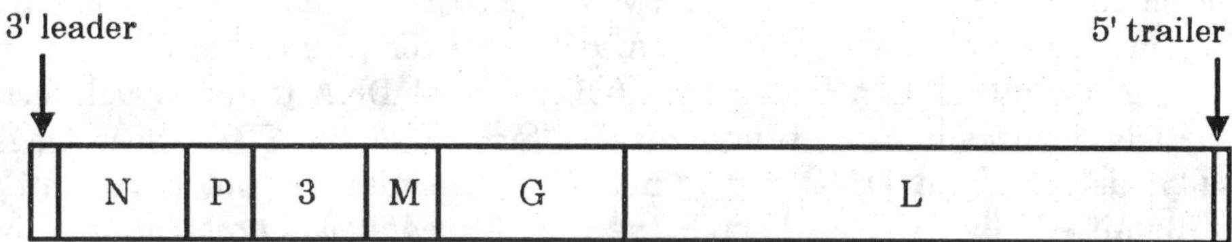

Fig. 3: Genome organization of TaVCV, in coding orientation. N, Putative nucleocapsid gene; P, putative phosphoprotein gene; 3, gene 3; M, putative matrix protein gene;G, glycoprotein gene; L, polymerase protein gene [23].

The electron microscopy of leaf sap dip preparations from suspected aroid plants gives initial clues on association of nucleorhabdoviruses, but can not suggest the exact identity of the virus. Further, this approach for routine large scale analysis of samples is not amenable as it is expensive and some times take more time. The application of immunoassays for its detection is not feasible as its purification directly from taro tissue is difficult to produce quality PAbs. The genome based tests are the option for their detection. The degenerate primers can be designed for RT-PCR amplification of the selected portion(s) of RNA genomes of nucleorhabdoviruses. Mixed infections involving nucleorhabdoviruses and other viruses (e.g., DsMV, TaBV) are common in aroids in the Pacific region. To analyze such samples either duplex- or multiplex-PCRs need to be developed for their parallel detection and differentiation. The occurrence of this virus on aroid plants other than *C. esculenta* in the Pacific Island countries is to be established. It might be perpetuating mainly through the propagules of taro as its vector is unknown. Further, this virus infection on taro and other aroids out side Pacific region has not been so far reported and needs to be checked using RT-PCR based tests.

2. Colocasia bobone disease virus (CBDV)

Alomae disease of taro in Pacific Island countries is caused by dual infection of two morphologically distinct viruses. In thin sections or sap dips, virions with large bullet shaped morphology, 300 to 335 x 50 to 55 nm characteristic of rhabdoviruses and small bacilliform particles (140 x 30 nm) were observed. The virus with large bullet shaped particles was named as Colocasia bobone disease virus (CBDV), putative member of *Nucleorhabdovirus* genus and the virus with smaller bacilliform virion morphology was identified as *Taro bacilliform virus,* member of *Badnavirus* genus [Yang, 2003a; James, 1973]. The disease in taro caused by CBDV alone is called Bobone. The taro plants infected only with CBDV are stunted and the foliage show leaf twisting and curling and remain green without tending to recover subsequently. The complex disease, Alomae, in taro is a much severe systemic necrosis disease leading finally to death of affected plants. Smaller bacilliform particle virus alone causes either milder symptoms (marginal vein chlorosis and downward leaf curling) or remains latent. Therefore, Alomae is a complex disease caused by synergistic interaction of an RNA (CBDV) and DNA (Small bacilliform particle) viruses like rice tungro disease [Azzam, 2002]. CBDV is serologically distinct from TaVCV. It is non-sap transmissible, but persistently transmitted by a planthopper vector, *Tarophagus proserpina.* The putative assignment of CBDV to the genus *Nucleorhabdovirus* is to be confirmed. For this, its genome has to be sequenced to find its exact taxonomic position.

Strict quarantine measures have to be imposed to prevent the importation of infected aroid materials into Alomae and Bobone disease-free areas. The infected plants should be rogued out and burnt. Chemical control of the *T. proserpina* is not effective when the insect populations are high.

IV. Badnaviruses

Badnaviruses are pararetroviruses that are classified in the family *Caulimoviridae,* genus *Badnavirus* [ICTV taxonomy 2008]. They have been reported to affect a wide range of tropical plant species including important crops such as citrus, banana, sugarcane, cacao, rice and yam. They have non-enveloped bacilliform virions (30 x 130 nm) encapsidating circular dsDNA genome ranging from 7.1 to 7.6 Kbp that contains 3 to 6 ORFs. Badnaviruses are vectored by either mealybugs or leafhoppers, and some are also transmitted through the seed. Badnaviruses have implications in the movement of plant germplasm. The type species is *Commelina yellow mottle virus* (ComYMV). *Taro bacilliform virus* (TaBV) and *Aglaonema bacilliform virus* (ABV) are the two distinct badnaviruses reported to naturally infect species of aroid plants [ICTV taxonomy 2008Yang 2003a &b]. The detailed information on ABV is not available.

1. *Taro bacilliform virus* (TaBV)

TaBV is reported in the taro growing areas of the South Pacific region (Samoa, Solomon Islands) [Abo-El-Nil, 1977]. Symptoms of this virus infection include chlorosis of marginal veins and downward curling of leaf margins, but often its infection appear to be latent [Yang, 2003b]. So far its natural host range is restricted to taro only. TaBV in dual infection with CBDV, a putative nucleorhabdovirus, thought to be responsible for the lethal and devastating Alomae disease of taro [James, 1973; Rodoni, 1994]. TaBV is vertically transmitted through taro corms, seed, and pollen and horizontally by mealybugs (*Pseudococcus longispinus, P. citri,* and *P. solomonensis*). Macanawani *et al.* [Macanawani, 2005] have precisely established the seed, pollen and mealybug transmission of TaBV using PCR based diagnostics.

The genome of Papua New Guinea (PNG) isolate of TaBV is completely sequenced (7,458 bp). The genome contains 4 ORFs on the plus strand that potentially encode proteins of 17, 16, 214 and 13 kD (Fig. 4; Table 3) [Yang, 2003a]. The size and organization of TaBV ORFs 1 to 3 is similar to that of most other badnaviruses, while the location of ORF 4 is similar to that of ORF 4 and ORF X of the atypical badnaviruses *Citrus yellow mosaic virus* and *Cocoa swollen shoot virus,* respectively. The phylogenetic analysis using amino acid sequence of ORF3 showed that TaBV branched most closely to *Diascoria bacilliform virus* and identified as a distinct species *Taro bacilliform virus,* genus *Badnavirus.*

Table. 3: Coding capacities of TaBV ORFs. [21].

ORF	First nucleotide	Last nucleotide*	Size (bp)	Amino acids	Coding capacity (kDa)
1	350	790	441	146	16.67
2	790	1224	435	144	15.78
3	1227	6872	5646	1881	214.34
4	1951	2283	333	110	13.1

*Including the stop codon

The sequence variability in the two regions (RT/RNaseH, C-terminal CP coding) of the genome of TaBV isolates from Pacific island countries was analysed [Yang, 2003b]. When RT/RNaseH coding region of 22 TaBV isolates from Fiji, French Polynesia, New Caledonia, PNG, Samoa, Solomon Islands and Vanuatu was examined, maximum variability at the nucleotide and amino acid level was 22.9 and 13.6%, respectively. Within the CP coding region of 13 TaBV isolates from Fiji, New Caledonia, PNG, Samoa and the Solomon Islands, the maximum variability was 30.7% at nucleotide and 19.5% at amino acid levels. Phylogenetic analysis revealed that TaBV isolates from Solomon Islands have greatest variability while those from New Caledonia and PNG have least variability. Based on the sequences of TaBV RT / RNaseH coding region, Yang *et al.* [Yang, 2003b] have developed PCR-based diagnostic test that detects all known TaBV isolates. PCR based indexing showed that TaBV is wide spread throughout the Pacific island countries.

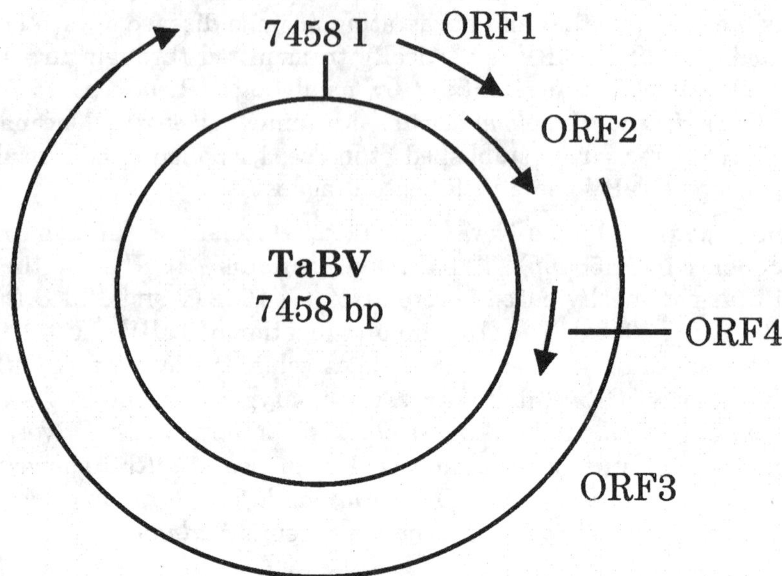

Fig. 4: Genome organization of TaBV. The numbering begins at the first nucleotide of the putative tRNA met-binding site, and the approximate position and sizes of ORFs 1–4 are indicated [21].

TaBV-like sequence was detected in all taro plants tested, including symptomless plants and those indexed as TaBV-free. Analysis of such sequences amplified from PNG and Vanuatu taro revealed that they are nearly 50% similar to the nucleotide sequence of TaBV [Yang, 2003b]. The correct identification of such sequences (integrants of TaBV or RT/RNaseH-coding region of additional badnavirus infecting taro) has to be revealed by southern and/or *in situ* hybridization studies. Diagnostic assays (e.g., multiplex-PCR) for all known taro viruses in the Pacific region need to be developed to facilitate the dissemination of virus-free propagating material throughout this region.

V. Tospoviruses

Tospoviruses are among the most damaging plant viruses. Many of the tospoviruses have wide natural host range capable of infecting both agricultural and horticultural crops besides weeds. Plant species susceptible to tospoviruses react with local and/or systemic infections. They are mechanically sap transmissible and vectored by thrips in a circulative propagative manner. They are relatively unstable and heat-inactivated at less than 50°C and are sensitive to organic solvents and detergents. The virions of tospoviruses are enveloped spherical or pleomorphic particles, 80-120 nm in diameter and display two glycoproteins (G1 and G2) on lipid envelope. The virion contains three nucleocapsids that are formed by three linear ssRNA fragments denoted L (large), M (medium) and S (small) enwrapped by the nucleocapsid (N) protein and few RNA-dependent-RNA polymerase proteins. These genomic RNAs have 5' and 3' conserved complementary sequences that facilitate the formation of non-covalently closed pseudo-circles. The M (~4,820 nucleotides) and S (12,920 nucleotides) segments are ambisense, while L (~9,000 nucleotides) is negative sense. The L RNA encodes RNA-dependent RNA polymerase, M the precursor of G1 and G2 glycoproteins and a non-structural (NSm) protein, and S RNA encodes N protein and non-structural (NSs) protein. They replicate in the cytoplasm, probably in association with endoplasmic reticulum, and form viroplasm inclusions.

Tospoviruses are classified in the genus *Tospovirus* of the family *Bunyaviridae.* The other four genera of the family infect animals. *Tomato spotted wilt virus* (TSWV) is the type species of *Tospovirus* genus. Several plant viruses have been assigned as distinct species of this genus primarily based on host range, vector transmission and amino acid sequences of the nucleocapsid protein and its serological properties. Viruses with amino acid identity lower than 90% are considered to be different species. Tospoviruses are further grouped into serogroups based on the antigenic properties of N protein [ICTV taxonomy 2008].

TSWV, INSV and CCSV are the three tospoviruses reported to naturally infect aroid plants (Table 2). TSWV is world wide in distribution and infects several agricultural and horticultural crops including aroid foliage ornamentals (Calla lily, Dieffenbachia, Anthurium). INSV has relatively restricted geographical distribution and infects several ornamental plants. It was reported to naturally infect Calla lily. Of the three tospoviruses, CCSV was first isolated from calla lily aroid plant in Taiwan [Chen, 2005; Lin, 2005].

1. *Calla lily chlorotic spot virus* (CCSV)

A tospovirus-like virus causing chlorotic spots on leaves and stems of calla lilies was first isolated from central Taiwan in the year 2000 and later it was named as Calla lily chlorotic spot virus (CCSV). The virus found to be distantly serologically related to *Watermelon silver mottle virus* (WSMoV) [Chen, 2005]. Later they considered it based on host range, thrips specificity and the preliminary results of serological tests, as a new tospovirus species. Subsequently, Lin *et al.* [2005] produced PAbs and monoclonal antibodies (MAbs) to nucleocapsid (NP) of CCSV and WSMoV to clarify the serological relationships between the two viruses. The PAbs to CCSV NP reacted stronger with the homologous antigen than with the heterologous antigen. MAbs to CCSV NP and WSMoV NP reacted specifically with the homologous, but not with heterologous, antigens in both ELISA and immunoblot analysis.

The serological cross reactions between WSMoV and CCSV using rabbit PAbs indicate that CCSV is distantly related to WSMoV and this could be the reason for false identification of CCSV as WSMoV during early field surveys [Chen, 2005]. But the MAbs produced by Lin *et al.* [Lin, 2005] able to distinguish these two viruses. CCSV is also detected by indirect ELISA in diseased melon samples in southern Taiwan (Unpublished data mentioned in [Lin, 2005]. Further, they identified mixed infections caused by CCSV and WSMoV in melon fields using MAbs.

Viruses within the genus *Tospovirus* must have similar virion morphology, genome organization and vector relationships with thrips [Van Regenmortel, 2000]. A threshold of 90% amino acid identity of tospoviral NPs is one of the important criteria for designation to the species level. To confirm the identity of CCSV, Lin *et al.* [Lin, 2005] have analysed the sequence of S RNA. It is 3,172 nucleotides in length with inverted repeats at 5'- and 3-'ends and 2 ORFs encoding the NP and non-structural (NSs) protein like other tospoviruses. But the length of this RNA differs from the S RNAs of several other tospoviruses. The NP and NSs proteins of CCSV shared low levels of amino acid identities of 20.1 to 65.1% and 19.9 to 66.1%, respectively, with those of several reported tospovirus. Phylogenetic dendrogram reveals that CCSV is a distinct member in the genus *Tospovirus.* Hence it is confirmed as a new species of the *Tospovirus* genus and belongs to the WSMoV serogroup.

Tombusviruses

Majority of members of the family *Tombusviridae* occur in temperate regions. The geographical distribution of a particular species varies from wide to restricted. The natural host range of individual virus species is relatively narrow. Members can infect either monocotyledonous or dicotyledonous plants. The experimental host range is wide and infection is often limited to roots. Diseases are characterized by mottling, crinkling, necrosis and deformation of foliage. Some virus species infections are symptomless in their natural hosts. Many virus species induce multi-vesicular bodies in the cytoplasm of infected tissues.

All species are readily transmitted by mechanical inoculation. Some may be transmitted by contact and through seeds. Viruses are often found in natural environments i.e., surface water and soil from which they can be acquired without assistance of vectors. Transmission by the chitrid fungus in the genus *Olipidium* and beetles has also been reported for members of several genera.

The virions are non-enveloped isometric particles measuring 32-35 nm in diameter. Virions sediment as one component with S_{20W} of 118-140 S. The virus particles have thermal inactivation point around 80ºC and are insensitive to organic solvents and nonionic detergents.

With the exception of the genus *Dianthovirus*, virions contain a single molecule of positive sense, linear ssRNA that contributes about 17% of the particle weight. The genome size ranges from 3.7 to 4.7 Kb, depending on the genus. Dianthovirus virions contain two genomic RNAs: RNA 1 of 3.9 Kb and RNA 2 of 1.5 Kb. The 3^1-ends are not polyadenylated. The 5^1-ends of Dianthovirus genome is protected by methylated cap.

All virions are composed of 150 copies of a single capsid protein (CP). Based on phylogenetic relationships of CPs, tombusviruses are divided into two groups. Group one possesses CPs of mol wt 37-48 kD, where as group two members have CPs with mol wt 25-29 kD.

The genomes of members of the genera *Dianthovirus* and *Avenavirus* encode three ORFs while all others encode four ORFs. The genome expression involves read through translation and subgenomic RNAs. Virus replication occurs in the cytoplasm, possibly in membranous vesicles that may be associated with cell organelles like endoplasmic reticulum, peroxisomes, mitochondria or rarely chloroplasts.

The family *Tombusviridae* is comprised of eight genera. They are: *Aureusvirus, Avenavirus, Carmovirus, Dianthovirus, Machlomavirus, Necrovirus, Panicovirus* and *Tombusvirus* [Van Regenmortel, 2000].

Pothos latent virus (PoLV)

Pothos (*Scindapsus aureus*) is an aroid foliage ornamental plant. A non-enveloped isometric virus was first isolated from symptomless hydroponically cultivated *S. aureus* plants in Italy. As this virus never induced the symptoms in the natural host, it was named as *Pothos latent virus* (PoLV). The studied properties of this virus revealed its affinities with *Tombusvirus* and *Carmovirus* genera [Sabanadzovic, 1995].

PoLV is transmissible from symptomless pothos plants to 12 out of 28 herbaceous plants that belong to five families. *Nicotiana clevelandii* and *N. benthamiana* systemically infected, whereas *Gomphrena globosa, Chenopodium amaranticolor, C. quinao, C. murale, C. foetidium, N. cervicola* and *Petunia hybrida* infected only locally with chlorotic /necrotic lesions. *Oscimum basicum* reacted with black necrotic lesions similar to those induced by tombusviruses where as *Phaseolus vulgaris* cv. Bountiful produced minute necrotic lesions. The virus was purified from systemically infected *Nicotiana* sps. and the virions are non-enveloped isometric particles measuring 30 nm in diameter. The mol wt of capsid protein of purified virions is 40.0 kD. The genome of the virus is a single ssRNA species of 4,415 nucleotides in length. The virions also contain two subgenomic RNAs (2.0 and 0.8 Kb).The PoLV genome contains four ORFs: ORF 1 encodes (read through translation) mol wt 84 kD protein with motifs of RNA polymerase, ORF 2 encodes mol wt 40 kD coat protein, the mol wt 27 kD product of ORF 3 is the movement protein and ORF 4 encoded mol wt 14 kD product is responsible for symptom severity.

The polyclonal antiserum to the purified virus was produced. It was serologically related to, but distinct from two *carmovirus* (i.e., Galingosa mosaic and Ahlum water-borne viruses) and three *Tombusvirus* species (Egg plant mottled crinkle, Sikte water-borne and Lato river viruses). The optimized ELISA test was employed to screen several aroid and non-aroid ornamentals cultivated in the same hydroponic bench along with pothos plants and none other than pothos were found to be ELISA-positive. But pothos plant samples collected outside the glasshouse were ELISA-negative. The virus successfully isolated from ELISA-positive samples.

Thin sections of ELISA positive pothos plant rootlets contained electron dense rounded bodies of 0.24 nm in diameter. Virions are scattered throughout the cytoplasm, sometimes forming discrete aggregates. Many of the virus containing cells had thinned ground cytoplasm. But in infected cells of *N. benthamiana* and *C. quinao* the ground cytoplasm was dense with virus particles often invaded the central vacuole. Virions are either scattered or regularly arranged in a crystalline lattice. Chloroplasts were rounded or variously misshapen. Mitochondria showed dialation and reduction in the number of cristae. Multivesicular bodies characteristic of tombusviruses are not observed. However, single membraned rounded to ovoid bodies about the size of mitochondrion are observed. The center of these bodies is either electron lucent or dense.

Thus, PoLV has properties somewhat intermediate between those of *Tombusvirus* and *Carmovirus* genera of the family *Tombusviridae.* Molecular analysis of PoLV genome revealed its distinctiveness from the above two genera [Rubino, 1997]. Later, Martelli *et al* [Martelli, 1998] have established a new genus *Aureusvirus* with PoLV as its type species in the family *Tombusviridae.*

While searching for causal agent of pigeonpea sterility mosaic disease (PSMD) prevalent in Indian subcontinent, a virus that has close similarities to PoLV was isolated and identified as PoLV-pigeonpea (PoLV-Pp) isolate [Lava Kumar, 2001]. But it is not the causal virus of PSMD as it was also isolated from healthy pigeonpea. Attempts to isolate a similar virus from pothos plants in India were unsuccessful. The virus is probably soil-borne and present as contaminant in pigeonpea.

Viruses infecting aroids in India

Aroid food crops like taro and elephant foot yam and foliage ornamentals are widely cultivated in India. So far, DsMV occurrence on these plants was reported [Ahlawat, 2003; Ram, 2003; Pandit, 2001] GenBank accession number from Kerala FJ160764). In addition to DsMV, a carla-like virus was suspected on elephant foot yam in Assam [Ahlawat, 2003]. The identification of these two viruses is primarily based on symptomatology, limited host range, virion morphology and antigenic relationships with the reported viruses. Recently, the genome of DsMV infecting elephant foot yam in Kerala is partially sequenced (GenBank accession no FJ160764). In Andhra Pradesh state, potyvirus isolates infecting *Colocasia*, *Caladium* and *Dieffenbachia* spp. are characterized at the genome sequence level and identified them as strains of KoMV, which was not reported so far from India (Padmavathi *et al.*, unpublished data). The partial genome sequence analysis of another potyvirus isolate from *Spathiphyllum* spp. from Andhra Pradesh revealed its distinctiveness from several other potyviruses reported to infect aroids and other potyviruses (Padmavathi *et al.* unpublished data). This indicates that careful examination of aroid plants in India may reveal the presence of more and more new viruses or new strains of already reported viruses elsewhere. Antibodies for aroid viruses occurring in India have not been produced thus far. But PCR-based methods are being used for the detection of these viruses using primers designed on the basis of sequences of virus isolates from other counties or using the potyvirid degenerate primers. We recently developed duplex-RT-PCR for two potyviruses infecting aroids in Andhra Pradesh.

A strain of *Pothos latent virus* associated with pigeonpea in India was characterized at molecular level [Lava Kumar, 2001], and its association with aroid plants in nature in India is to be established.

Future perspectives

The cultivation of aroid edible crops and horticultural plants needs to be promoted for better food security and for better climate in human habitations. Viruses infecting aroids limit their cultivation and exchange of propagules across geographical boundaries. Apparently one can not realize the impact of aroid viruses in one growing season, but over the years they may reduce the plant vigor and sometimes lead to loss of valuable aroid germplasm. To minimize the impact of viruses on the cultivation of aroids and for safe exchange of aroid materials, further research on aroid viruses may be focused on the following to complement or supplement the existing information of them and finally formulate workable management strategies.

1. During the last two decades, studies on genetic diversity among the isolates of viruses first isolated from aroids have facilitated the finding of either new viruses or new strains of already reported viruses. Research in this direction needs to be continued probably to trace and/or identify more and more new viruses or virus strains. Further, such studies may facilitate development of novel diagnostics and to understand the evolution of virus isolates.
2. Many of the viruses first isolated form aroids have narrow host range, mainly confined to aroids, their purification directly from aroids is an uphill task and thus production of quality antibodies against purified virus preparations is still a difficult task. To circumvent this, the vibrant recombinant DNA technology has already been exploited to produce quality antibodies for certain potyviruses infecting the aroids. This technology is to be applied to produce PAbs to other aroid viruses like TaBV and TaVCV that are difficult to purify. Good quality antibodies facilitate the development of immunoassays and IC-PCR for indexing and certification of aroid materials.
3. The sequence data of aroid viral genomes is available in public databases and which can be utilized to design primers for amplification of specific portions of viral genomes by PCR. Based on this approach, duplex or multiplex PCRs are already developed for parallel detection and discrimination of aroid potyviruses in Asian countries. Similar PCR-based tests need to be developed for detection of single or co-infection of DsMV and other viruses like tospo-, badna- and plant rhabdoviruses infecting aroids based on their prevalence in a given country or region.
4. In recent times, DNA microarray based methods are increasingly used for parallel detection of plant viruses on a large scale eventhough their sensitivity is comparable to ELISA [Boonham, 2007]. Presently, this novel technology costs are prohibitive for application to viruses infecting especially low value crops. In the near future this technology may be cost effective and can be applied for screening of aroid propagation materials and micropropagated plants.
5. Improved tissue culture protocols need to be developed for different aroid plant species. Such protocols and sensitive diagnostics may facilitate the mass propagation of virus-free aroids. This approach for generation of DsMV-free aroid ornamentals is partly successful; the reason could be the non-availability of highly sensitive techniques like IC-RT-PCR in those days [Zettler, 1987].

6. Aroid virus genes have not so far been exploited as transgenes to generate virus resistant transgenic aroid food crops. This approach is worthwhile to pursue for most widely occurring viruses like DsMV in the near future as one of the management strategies already successful in certain potyvirus-host combinations [Dasgupta, 2003].
7. Aroid viruses are basically perpetuating vertically through plant propagules. Aphids, mealybugs, thrips and planthoppers were reported as vectors of viruses first isolated from aroids. Their role in horizontal spread of these viruses in fields and nurseries needs to be established before putting the efforts in controlling the above insect vectors.
8. So far, TaBV is the only virus proved to be transmissible through the true seed of aroids (Taro). If other viruses are not aroid seed transmissible, it is worth while to raise the plants from seeds as virus-free stock plants. Such stock plants serve as a source for mass multiplication either by conventional or micropropagation approaches under insect vector-free conditions.
9. Management practices so far were formulated for widely occurring DsMV only in USA [Nelson, 2008]. The literature survey reveals that potyviruses are prevalent in Asian countries, and DsMV, TaVCV and TaBV are prevalent in Pacific Island countries. Workable strategies need to be formulated for the management of viruses in Asia and Pacific Island countries.

Acknowledgements

The authors are thankful to the Department of Biotechnology (DBT), Ministry of Science and Technology, Government of India for financial assistance.

References

Abo-El-Nil MM, Zettler FW, and Hiebert E (1977) Phytopathol 67:1445–1450

Adams MJ, Antoniw JF and Fauquet C (2005) Arch Virol 150:459-479

Adams MJ, Antoniw JF, Bar-Joseph M, Brunt AA, Candresse T, Fosterr GD, Martelli GP, Milne RG and Fauquet CM (2004) Arch Virol 149: 1045-1060

Ahlawat YS, Pant RP and Bhagawati KN (2003) Indian Phytopathol 56:300-301

Azzam O and Chancellor TCB (2002) Plant Dis 86: 88-100.

Berger PH, Wyatt SD, Shiel PJ, Silbernagel MJ, Drunffel K and Mink GI. (1997) Arch Virol 142: 1979-1999

Boonham N, Tomlinson J and Mumford R (2007). Ann Rev Phytopathol 45: 307–328

Bown D (1988) *Aroids: Plants of the Arum Family*. Timber Press, Portland, Oregon

Brunt AA, Crabtree K, Dallwitz MJ, Gibbs AJ and Watson L (Eds.) (1996) Viruses of Plants: Descriptions and lists from the VIDE database. CAB International. Wallingford, UK, pp, 1484

Chang, YC, Chen, Y.L. and Chung, F.C. (2001) Plant Dis 85:1289

Chen C., Chen TC, Lin YH, Yeh SD and Hsu HT (2005) Plant Dis 89: 440-445

Chen CC, Chao CH, Chen CC, Yeh SD, Tsai HT and Chang CA (2003) Plant Dis 87: 901-905

Chen CC, Hsu HT, Cheng YH, Huang CH, Liao JY, Tsai HT and Chang CA (2006) Bot Studies 47: 369-378

Chen CC, Ko WF, Jan FJ, Lin CY and Tsu HT (2002) Plant Pathol Bull 11: 242

Chen J, Chen J, Chen J and Adams MJ (2001) Arch Virol 146: 1821-1829

Dasgupta I, Malathi VG and Mukherjee SK (2003). Cur Sci 84: 341-354

Farreyrol K, Pearson MN, Grisoni M, Cohen D and Beck D (2006) Arch Virol 151: 905-919

Gibbs A, Armstrong J, Mackenze AM and Weiler GF (1998) J Virol Meth 74: 67-76

Gollifer DE, Jackson GH, Dabek AJ, Plumb RT and May YY (1977) PANS 23: 171-177

Haung CH, Hu WC, Yang TC and Chang YC (2007) Plant Pathol 56: 183-189

Hema M, Kirthi N, Sreenivasulu P and Savithri HS (2003) Arch Virol 148: 1185-1193

Hu WC, Wang CH, Lee SC, Wu CI and Chang YC (2009) Eur J Plant Pathol (In press)

Huang CH and Chang YC (2005) Arch Virol 150: 1221-1230

ICTV taxonomy 2008. www.ictvonline.org/virustaxonomy.asp

James M, Kenten RH and Woods RD (1973) J Gen Virol. 21: 145-153

Kim MK, Kwon SB, Yoon JY, Ryu KH, Heo SJ, Hong JK, Kim KH and Choi JK (2004) Plant Pathol J 20: 135-141

Kwon SB, Ha JH, Yoon JY and Ryu KH (2002) Arch Virol. 147: 2281-2289

Lava Kumar P, Jones AT, Sreenivasulu P, Fenton B and Reddy DVR (2001) Plant Dis 85: 208-215

Li RH, Zettler FW, Purcifull DF and Hiebert E (1998) Arch Virol 143: 2461-2469

Lima RCA, Lima JAA and Rubens Aguiar J (2004) Fitopatol. Brazil 29: 105

Lin YH, Chen TC, Hsu HT, Liu FL, Chu FH, Chen CC, Lin YZ and Yeh SD (2005) Phytopathol 95: 1482-1488

Macanawani AR, Ebenebe AA, Hunter D, Devitt LC, Hafner GJ and Harding RM (2005) Aust Plant Pathol 34: 73-76.

Martelli GP, Russo M, Rubino L and Sabanadzovic S (1998) Arch Virol 143: 1847-1851

Mokra V and Gotzova B (1994) Acta Horitcul 377: 361-362

Nelson SC (2008) Plant Dis PD-44: 1-9

Nishiguchi M, Yamasaki S, Lu XZ, Shimoyama A, Hanada K., Sonoda S, Shimon M, Sakai J, Mikoshiba Y and Fujisawa I (2006) Arch Virol 151:1643-1650

Pandit MK, Nath PS, Mukhopadhyay S, Devonshire BJ and Jones P (2001) Plant Pathol 50: 802

Pappu SS, Pappu HR, Lastra R and Niblett CL (1994b) Arch Virol 136: 407-413.

Pappu SS, Pappu HR, Rybicki EP and Niblett CL (1994a) J Gen Virol 75: 239-242

Pearson MN, Jackson GVH, Pone SP and Howitt RLJ (1993) Plant Pathol 42: 127–131

Pearson MN, Jackson GVH, Saelea J and Morar SG (1999) Aus Plant Pathol 28:248-253

Pham K, Langveldt SA, Lemmers MEC and Derks AFLM (2002) Acta Horitcul 568:143-148

Plucknett D L (1976) Edible Aroids: *Alocasia*, *Colocasia*, *Cyrtosperma*, and *Xanthosoma* (Araceae) In:

Simmonds NW (eds). *Evolution* of *Crop Plants*. Logman Group Limited, London, England, pp. 10-12

Ram R, Joshi A, Verma N, Kulshrestha S, Raikhy G, Hallan V, and Zaidi AA (2003) Plant Pathol 52:411

Revill P, Trinh X, Dal J and Harding R (2005) J Gen Virol 86:491-499

Reyes G, Ramsell JNE, Nyman M and Kvarnheden A (2009) Arch Virol 154: 159-162

Reyes G, Ronnberg-Wastlijung AC and Nyman M (2006) Exp Agricul 42: 301-310

Rivas EB, Duarte LML, Alexandre MAV, Galleti SR, Harakava R and Fernandes FMC (2005) J Plant Pathol 87: 109-114

Rodoni BC and Moran JR (1988) Acta Horticul 234: 281-285

Rodoni BC, Dale JL and Harding RM (1994) Papua New Guinean J Agricul, Forestry and Fisheries. 37: 14–18

Rubino L and Russo N (1997) J Gen Virol 78: 1219-1226

Sabanadzovic S, Saldarelli P, MArtelli GP, Lofortezza R, and Koenig R (1995) Eur j plant pathol 101: 171-182

Shi YH, Hong XY, Chen J, Adams MJ, Zheng HY, Lin L, Qin BX and Chen JP (2005) Arch Virol 150: 125-135.

Shimoyama J, Kameya-Iwaki M, Hanada K and Gunji T (1992) Ann Phytopathol Soc, Japan. 58: 706-712

Urcuqui-Inchima S, Haenni AL and Bernardi F (2001) Potyvirus protons: A wealth of functions. Virus Res 74: 157-175

Van Regenmortel MHV, Fauquet CM, Bishop DHL (2000) Virus taxonomy: Classification and nomenclature of viruses. Seventh report of the International Committee on Taxonomy of Viruses, Academic Press, San Diego

Wang YY and Pearson MN (1992) J Phytopathol 135: 71–76

Wisler GC, Zettler FW and Mu L (1987) Plant Dis 71: 1125–1129

Yang IC, Hafner GJ, Dale JL and Harding RM (2003a) Arch Virol 148: 937-949

Yang IC, Hafner GJ, Revill PA, Dale JL and Harding RM (2003b) Arch Virol 148:1957-1968

Zettler FW and Hartman RD (1987) Plant Dis 71: 958-963

Zettler FW, Abo-El-Nil MM and Hiebert E (1978) *Dasheen mosaic virus*. CMI/AAB Descriptions of Plant viruses, No. 191.

Chapter - 2

Cell Surface Engineering-Arming Yeast-A Review

Seshagiri Rao Gudapaty

Department of Biochemistry, Andhra University, Visakhapatnam
E-mail: drgsrao@yahoo.com

ABSTRACT

Establishment of systems to display heterologous proteins on the cell surface of microorganisms is important for many applications in microbiology and molecular biology. Some of them are separation of produced polypeptides, production of microbial biocatalysts, whole cell absorbents and live vaccines. Expression of proteins on cell surface of *Saccharomyces cerevisiae* has several advantages due to its wide use in industrial production of the above such as surface expressed proteins are covalently linked to glucan in cell wall and resistant to extraction, and as orally safe, it can be used in food and pharmaceuticals and oral vaccines. Thus arming yeast is best choice for this novel technique and the strategies and applications are discussed.

Key words: cell surface technology, agglutinin, flocculin, GPI-anchor, yeast

1. Introduction

The cell surface is functional interface between the inside and out side of the cell. Some surface proteins extend across the plasma membrane and other bond either covalently or non-covalently to cell surface components. Anchoring of surface specific proteins is combined to particular domains on the cell surface. Exploiting to display of heterologous proteins on the cell surface of microorganism is useful in separation of polypeptides, production of microbial biocatalysts, whole cell adsorbents, live vaccines and screening of novel proteins (Fig. 1). These applications are useful in biotechnology, microbiology and molecular biology [Ueda, 2000a & 2000b; Kondo, 2004].

Surface expression systems were first showed that peptides that were fused to docking proteins (pIII) of filamentous phage [Scott, 1990]. They were used for isolation of specific ligands and antibodies, and led to the development of phage display systems [Chisell, 1992]. Gram negative bacteria have been used for surface display of number of heterologous proteins [Francisco 1992 & 1993] and fused to variety of molecules like outer membrane proteins, lipoproteins, fimbrae etc. [Harrison, 1990; Newton, 1989; Hedegaard, 1989]. Gram positive bacteria are also used for surface display of proteins on the cell wall [Gunneriusson, 1996].

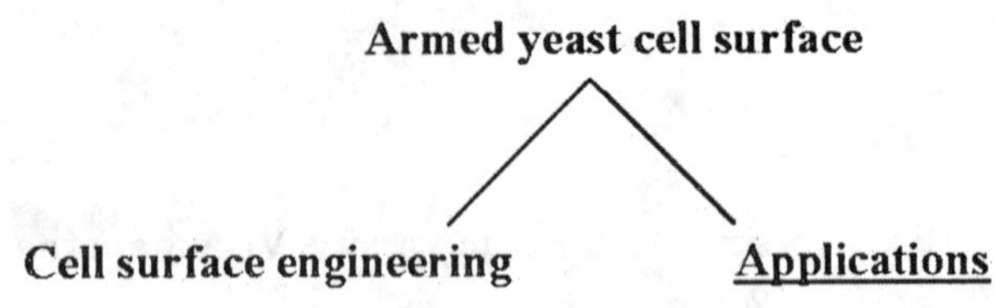

Various compartments

a) Cell recognition factor
b) Functional protein
c) Enzyme
d) Receptor
e) Antibody
f) Antigen

Cell adhesion
Molecular recognization
Immobilized catalyst
Bioconversion
Bioremediation
Signal transduction
Vaccine production
Biosensor

Fig: 1. Yeast cell surface Technology and its Applications

Expression of proteins on the cell surface of *S. cerevisiae* offers several advantages. *S. cerevisiae* is safe for oral use, so that it can be used in food and pharmaceuticals and for oral vaccines. It is widely used for industrial production of proteins, chemicals, and novel whole cell biocatalysts because surface expressed proteins linked to glucans in the cell wall.

Recent years of large scale bioconversions of biomass resources are converted to ethanol which can be used as biofuel. Production of ethanol from starchy materials needs to be improved and cost effective. As *S. cerevisiae* cannot utilize starchy materials, costly glucoamylase treatment is used and the starchy materials need to be cooked at high temperatures (140°C- 180°C). By producing cell surface expression systems to produce amylolytic enzymes on the cell surface of the *S. cerevisiae* by genetic manipulations has added advantage [Ueda, 2000a & 2000b; Schreuder, 1993 & 1996; Murai, 1997b; Matsumoto, 2002]. Moreover, the yeast can be cultivated to high density in less costly medium. Table 1 shows number of heterologous proteins that have been successfully displayed on yeast cell surface. 10^{4}-10^{5} molecules with various sizes of proteins have been displayed on each cell.

Table 1. Example of yeast cell surface display of heterologous proteins [3]

Protein and peptides	Promoter	Secretion signal Sequence	Molecular mass (kDa)
a-Agglutinin system			
Single-chain antibody	Gal1	Aga2	27–30
Single-chain T-cell receptor	Gal1	Aga2	26–28
C-terminus region of Flo1p			
Rhizopus oryzae glucoamylase	GAPDH	MFα1	62
EGFP	UPR-ICL	MFα1	27
Flocculation functional domain of Flo1p			
Rhizopus oryzae lipase	UPR-ICL	Flo1p	30
EGFP	UPR-ICL	Flo1p	27
α-Agglutinin system			
Rhizopus oryzae glucoamylase	GAPDH	Glucoamylase	62
Bacillus stearothermophilus α-amylase	GAPDH	MFα1	59
Aspergillus aculeate β-glucosidase	GAPDH	Glucoamylase	136
Trichoderma reesei endoglucanase	GAPDH	Glucoamylase	42
Rhizopus oryzae lipase	GAPDH	MFα	130
Aequorea victoria GFP	GAPDH	Glucoamylase	27
BFP	UPR-ICL	Glucoamylase	27
ECFP	PHO5	Glucoamylase	27
EYFP	MEP2	Glucoamylase	27
Apoaequorin	GAPDH	Glucoamylase	21
Hexa-His	GAPDH	Glucoamylase	0.93
ZZ	UPR-ICL	Glucoamylase	14
Fab fragment of antibody	GAPDH	Glucoamylase	50

The present article describes about the basics of molecular display of cell surface engineering, using yeasts and their uses in bioconversion, production of antibodies and proteins receptors etc. It is an example of protein engineering.

2. Principles and methods of cell surface engineering

Using genetic engineering technologies, the enzymes can be expressed by employing yeast adhesion molecule, α- agglutinin on cell surface which has glycolysyl phosphotidyl inositol (GPI) anchor [Nakamura, 2001; Hardwick, 1992; Watari, 1994] (Fig. 2A). The enzymes were immobilized and whole cells are used as cell biocatalysts by genetic immobilization. The enzyme is readily synthesized and naturally expressed on the cell surface so that tedious purification and immobilization protocols can be avoided.

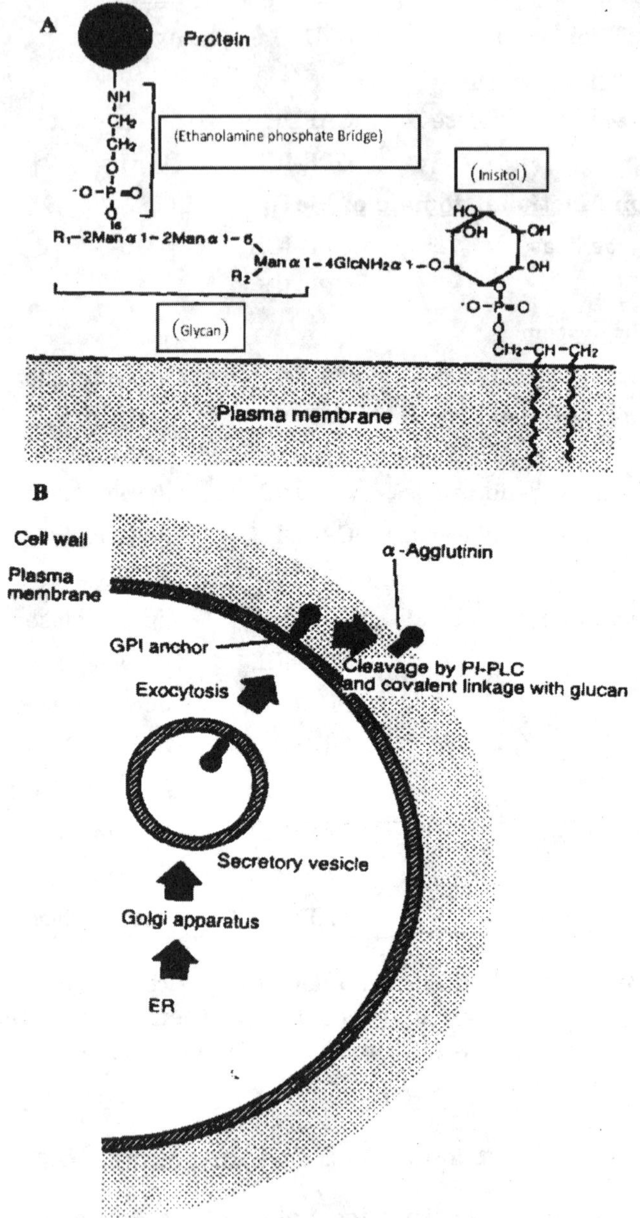

Fig. 2. GPI Anchor Protein (A) and its secretory pathway[(2)] (PI-PLC, Phosphotidylinositol-specific phospholipase C)

The yeast cell wall is rigid, about 200nm and composed of mannoproteins and β- linked glucans out side of plasma membrane with β,1-3 and β, 1-6 linked glucose [Ballou, 1982]. Two types of mannoproteins present are one attached with the cell wall with non covalent bonds and extractable with hot SDS and in other type mannoproteins are released by β, 1-3 or β,1-6 glucanase [Valentin, 1984; Fleet, 1977].

On the yeast cell surface, some of the glucanase extractable proteins like agglutinin (Ag α 1 and Ag a 1) and flocculin Flo1, Sed1, Cwp1, Tip1 and Tir1/Srp1, have GPI anchors and are essential for the viability of yeast [Watari, 1994; Ballou, 1982; Van der Vaart, 1995]. GPI anchors have been found on many eukaryotic plasma membrane proteins [Cross, 1990; Fredette, 1993] and are very well preserved. GPI moieties are covalently attached to C-termini of proteins leading to stability association with cell wall. GPI anchored proteins contain hydrophobic peptides at their C- termini. The localization of GPI anchored proteins on the cell surface occurs through secreting pathway of yeast (Fig. 2B). Different types yeast cell surface display systems include GPI anchored protein systems-agglutinin and flocculin.

2.1 Agglutinin system

Gluconase- extractable mannoproteins of *S. cerevisiae* contains the mating type specific agglutinin which is important for direct cell-cell adhesion. Mating type a and α cells express a agglutinin and α agglutinins coded by AG AI gene and AG α 1 genes respectively. Both α-agglutinin and core subunit of a agglutinin contains secretion signal regions, active region and region rich in serine and threonine and GPI anchor attachment signal (22 AA) at C- terminal end. The whole protein has 650 amino acids out of which 28 amino acids are of secretion signal sequence [Lipke, 1992]. Using recombinant DNA technology, the signal of secreted enzyme reacts with anchoring signal of α agglutinin (Fig. 3A). This results the fusion of 322 amino acid C- terminal half of α – agglutinin with GPI anchor attachment signal sequence at C-terminal end. This is used to anchor heterologous proteins on yeast surface. N-terminal end of the α– agglutinin active region is replaced by the protein encoding gene to be displayed. In case of a agglutinin, the secretion type protein Aga2P, linked core protein Aga1 P by S-S. The 69 amino acid c-terminal of Aga 2 P fusion protein and Aga 1P with in the secreting pathway are exported to the cell surface and covalently linked to the cell wall [Boder, 1997; Keike, 1999] (Fig. 3B).

2.2 Flocculin system (Flo 1P)

It is a lectin like cell wall protein of yeast coded by Flo1 and has an important role in flocculation [Teunissen, 1993]. It covers the thickness of yeast cell wall (300 nm) and has repetitive region of 1200 amino acids which makes a room for design of various anchor lengths [Watari, 1994]. It has several domains containing secretion signal, GPI anchor attachment signal, membrane anchoring domain and flocculation domain [Matsumoto, 2002; Lipke, 1992; Miki, 1982].

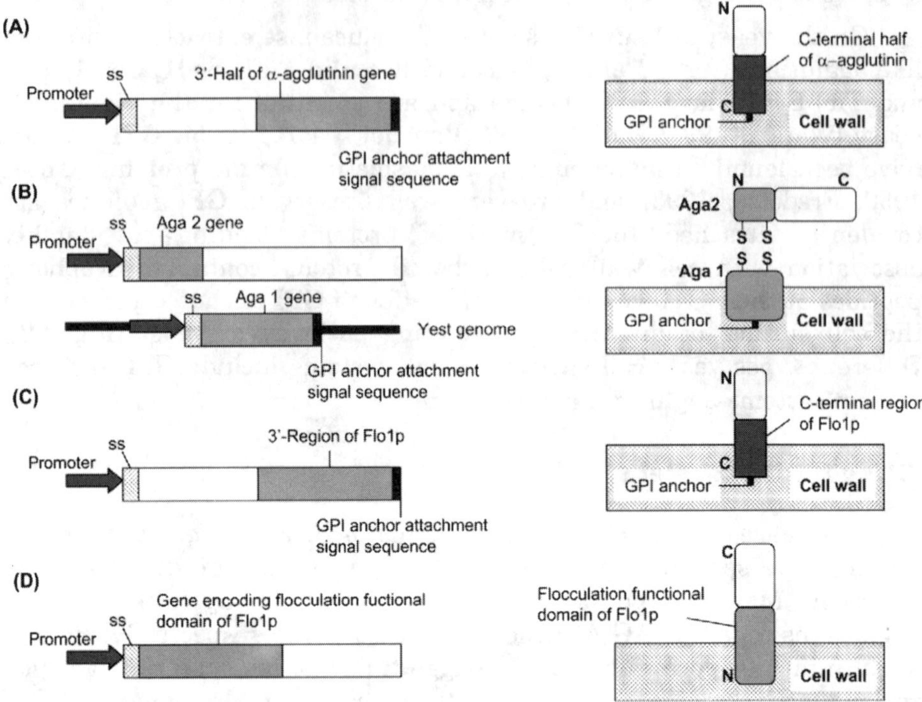

Fig. 3. Various methods of cell surface engineering

Types of cell surface display systems based flocculation are (1) C-terminal region of Flo1p having GPI attachment signal with various anchor lengths ranging from 42-1326 amino acids [Sato, 2002]. The C-terminal of the target protein is fused to anchor. The length of the anchor of Flo1p is secreted according to the purpose and nature of the target proteins (Fig.3C). (2) Other system utilizes the adhesive ability of the flocculation functional domain of Flo1 P (Fig. 3D) namely FS and FL proteins consisting of 1- 1099 and 1-1447 amino acid length respectively, with a secretion signal and insertion site for the target protein [Matsumoto, 2002].The secretion signal is bound to N-terminal of gene encoding flocculation functional domain of Flo1p.

3. Applications of cell surface engineering by armed yeast

S. cerevisiae is used in the production of ethanol from cheap starchy materials, but it lacks the amylolytic enzymes necessary for starch utilization. This is done normally by saccharification by cooking, and by treating with amylolytic enzymes like α – amylase, glucoamylase from *Rhizopus oryzae* etc., which is a costly process. To eliminate the saccharification process, modifications of yeast strains with cell surface display with these enzymes will be cost effective. Several reports [Murai, 1997 & 1998 &1999; Kondo, 2002; Shigechi, 2002 Shigechi, 2004] showed the construction with various amylolytic enzymes expressed on cell surface of yeast by engineering this with genetic manipulations. Either with one enzyme or in combination of two or more enzymes were expressed and successfully and the starch was hydrolyzed with increase the production of ethanol with the 86.5% of theoretical yield [Murai, 1999; Shigechi, 2004] (Fig. 4).

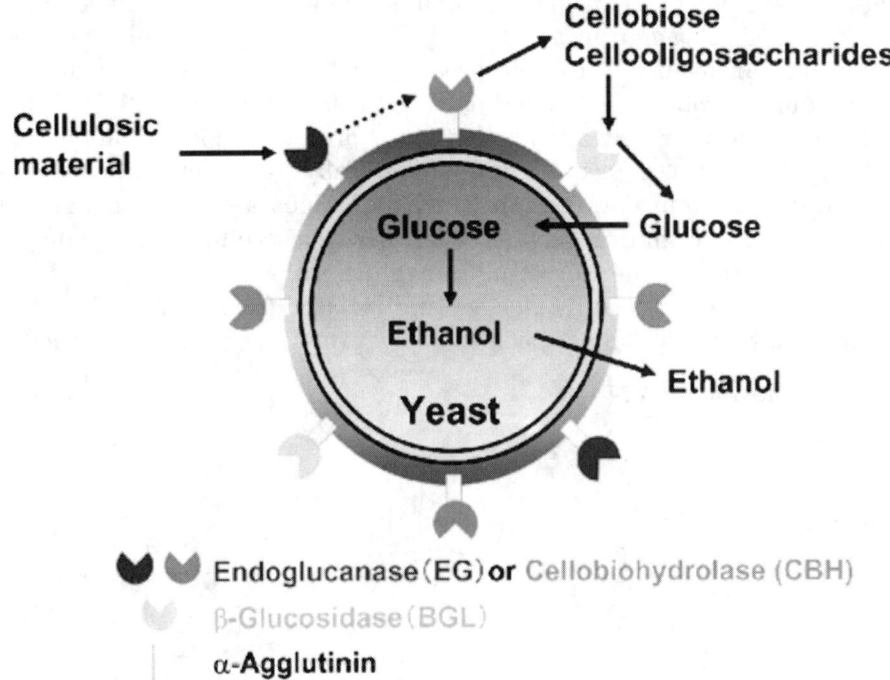

Fig. 4. Yeast strain co-displaying EG, CBH, and BGL for direct fermentation of cellulosic materials to ethanol(3)

Other important function of cell surface engineering is the expression of the cellulolytic enzyme on the cell surface yeast. As plenty of cellulosic biomass like cellulose, hemicelluloses are available for forestry residues, waste paper, agricultural and industrial wastes, utilization of these to produce inexpensive and abundant source of sugar is needed for production of ethanol. Again as yeast lacks of cellulolytic enzymes, the co-expression of the enzymes such as β – Glucosidase (BGLI), endoglucanase II (EGII) and cellobiohydrolase is successfully displayed [Murai, 1997b; Murai, 1998; Fujita, 2002 & 2004] using α-agglutinin system. Fujita *et al.* [2002] used cell surface engineering system based on α-agglutinin to co display of endogluconase II (EGII) from *Trichoderma reesei*, β-GLI from *Aspergillus aculeate* and could show a increase the ethanol production. The yield (g of ethanol/g of carbohydrate utilized) was 0.4 which is equal to 93.3% theoretical yield. The flocculent yeast expressed high ethanol production rates from soluble starch ($0.71gh^{-1}l^{-1}$) which co displayed glucoamylase from *R. oryzae* on the cell surface [Kondo, 2002]. It was shown that surface contact between starch granules and flocculent arming yeast cells is important for apparent specific activity increase of α –amylase. Starch can also be used in solid state fermentation [Khaw, 2007] by armed yeast.

Yeast cell were also engineered to display with human antigens on their surface which are able to bind specifically to antibodies to C-terminal end of agglutinins (Fig. 5 & 6). This is useful to detect monoclonal antibodies and their production from insuffient and unavailable human proteins [Chao, 2006; Tang, 2008; Shibasaki, 2007]. Introduction of selective cleavage system for a protein displayed on the cell surface of yeast was constructed successfully in order to avoid the conventional protein engineering which is tedious and time consuming. The factor Xa recognition sequence between yeast cell surface displayed protein and the anchor protein (α–agglutinin) was inserted and the GFPuv could be successfully cleaved and isolated from yeast cell surface [Michiko Kato, 2005]. This is used as a new protein purification method.

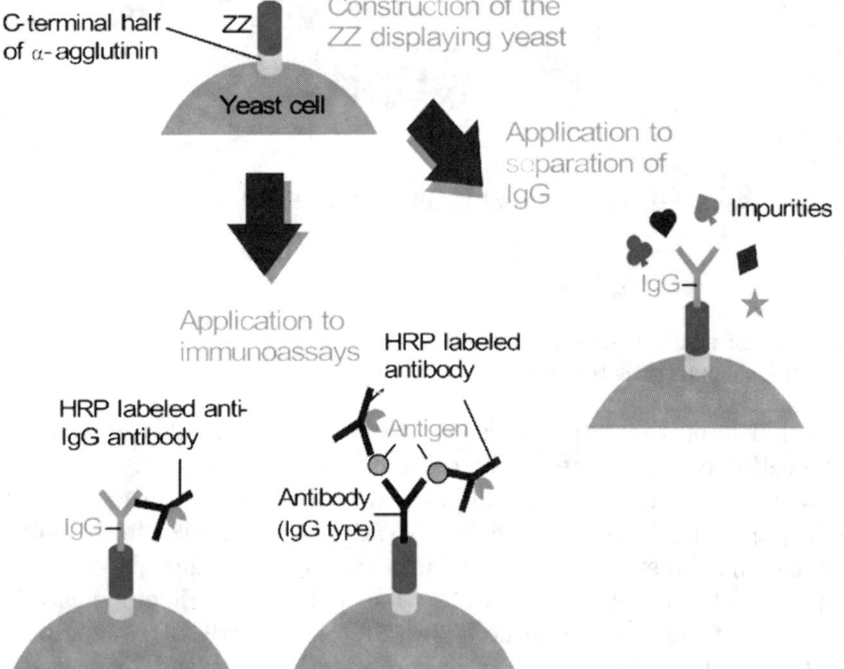

Fig. 5. ELISA to detect rabbit IgG and sandwich ELISA to detect antigen, and separation of IgG-type antibodies using yeast cells displaying ZZ [(3)]

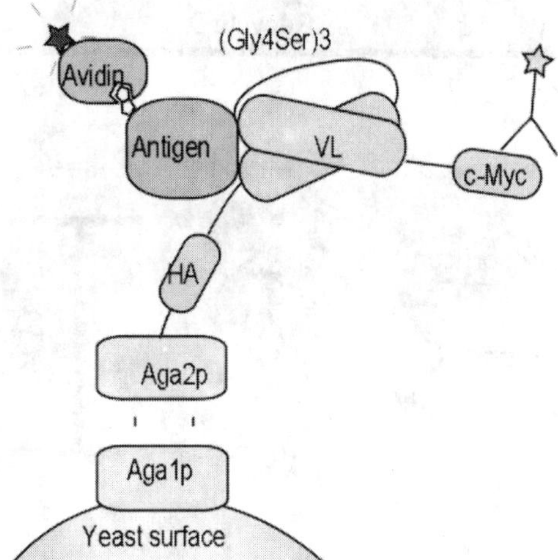

Fig. 6. Yeast surface display (40). The scFv is displayed as an Aga2 fusion protein on the surface of yeast. Expression can be detected by using fluorescent antibodies binding to the epitope tags HA and c-Myc and binding of the scFv to a biotinylated antigen can be detected using fluorescent avidin (HA, hemagglutinin; VL, variable light chain;VH, variable heavy chain (Gly4Ser) 3, flexible peptide linker).

Yeast cell surface expression of chitosanase from *Paenibacillus fukuinensis* was successfully constructed by linking C-terminal half of α-agglutinin [Takeshi, 2007]. The lipase from *Rhyzopus oryzae* (ROL) exhibits high activity suitable for food producing process and safer like yeast to use for human oral use. By using spacer mediated display and fusion with GPI anchor protein showed an increase in activity of ROL [Washida, 2001]. Using FloP1, FS & FL protein is fused with N- terminus of pro ROL protein, whose catalytic site is near C-terminus. In this system the activity of lipase reaches as much higher level than GPI display systems [Matsumoto, 2004 & 2005].

The bioremediation of heavy metal ions can be done in two ways by microbial cells (1) Binding to cell surface components (2) Gradual uptake and intracellular sequestration. The adsorbed metal ions can be recovered by simple chemical treatment such as EDTA. But intracellularly accumulated metal ions require disintegration treatment of cells. By displaying hexa- Histidine on cell surface, metal ions that are adsorbed on the cell surface is useful for bioremediation of heavy metal ions on these surface engineered yeast. *E. coli* and *Staphylococcus* species displaying hexa - His on the outer membrane adsorb Cu^{2+}, N^{2+}, Cl^{2+} and Zn^{2+} [Kotrba, 1999; Samuelson, 2000]. In case of yeast, large amount of Cd^{2+} adsorbed (Fig. 7). The hexa-His encoding gene fused with a gene coding C-terminal of α - agglutinin and showed the adsorbent of Copper ion 3 to 8 times higher [Kuroda, 2001].

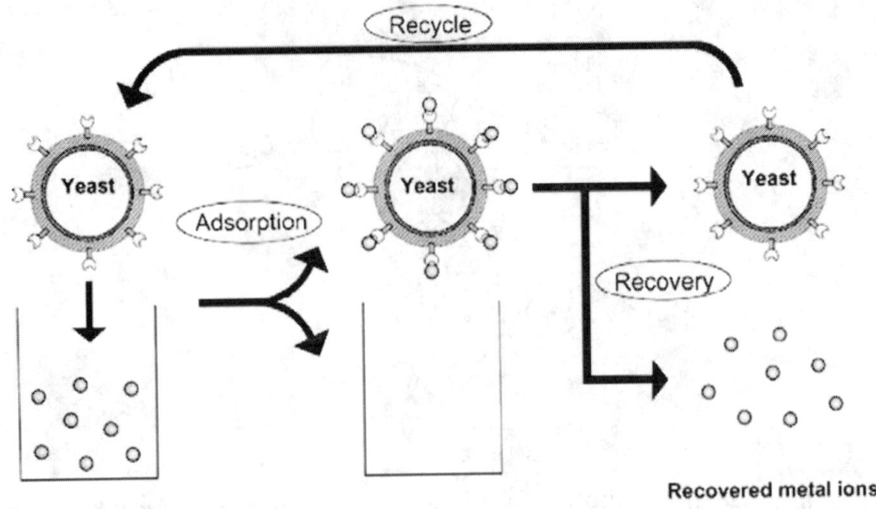

Fig. 7. Recovery of heavy metal ion adsorbed by yeast strain displaying hexa-His[3]

Yasui *et al.* (2002) constructed an arming yeast displaying a ligand binding domain of rat estrogen receptor on its cell surface can be used for screening hormone like compounds. To use as powerful and vital method for various molecular events within the cell, a green fluorescent protein (GFP) from jelly fish Victoria has been used by α – agglutinin fusion protein with GAPDH promoter on yeast cell surface as a novel visible type of reporter [Ye, 2000].

4. Conclusions

S. cerevisiae is a novel organism for cell surface engineering with new beneficial effects. They can accommodate large protein molecules not only single mutant proteins but also hetro-oligomeric multi sub units which are referred as arming yeast. This property is very well referred by proteins on to displayed cells. These displaying proteins on cell surface enhance the *S. cerevisiae* as a novel and attractive whole cell biocatalyst.

References

CE (1982) The molecular biology of the yeast *Saccharomyces*. In: Strathern, JN, Jones EW, Broach JR (eds) Metabolism and gene expression. Cold Spring Harbor Laboratory, Cold Spring Harbor, New York, pp 335–360

Boder ET, and Wittrup KD (1997) Nat Biotechnol 15:553–557

Chao G, Lau WL, Hackel BJ, Sazinsky SL, Lippow SM, Wittrup,KD(2006) Nat. Protocols.1,755-768.

Chisell D.J and McCafferty J (1992) Trends Biotechnol 10:80 -84

Cross GAM (1990) Annu Rev Immunol 8:83–110

Fleet GH, and Manners DJ (1977) J Gen Microbiol 98:315–327

Francisco JA, Earhart CF and Georgiou G (1992) Proc Natl Acad. Sci USA 89: 2713-2717.

Francisco JA (1993) Bio/Technol 11: 491-495

Fredette B, Rutishauer U and Landmesser L (1993) J Cell Biol 123:1867–1888

Fujita Y, Ito J, Ueda M, Fukuda H and Kondo A (2004) Appl Environ Microbiol 70:1207–1212.

Fujita Y *et al* (2002) Appl Environ Microbiol 68: 5136–5141

Gunneriusson E *et al* (1996) J. Bacteriol 178: 1341-1346

Hardwick KG, Boothroyd J C, Rudner A D, and Pelham HRB (1992) EMBO J 11:4187–4195

Harrison JL, Taylor IM, and O'Connor CD (1990) Res Microbiol 141:1009–1012

Hedegaard L, and Klemm P (1989) Gene 85:115–124

Keike MC *et al* (1999) Proc Natl Acad Sci USA 96:5651–5656

Khaw TS *et al* (2007) Journal of Bioscience and Bioengineering 103, 95-97

Kondo A and Ueda M (2004) Appl Microbiol Biotechnol 64: 28–40

Kondo A *et al* (2002) Appl Microbiol Biotechnol 58:291–296

Kotrba P, Doleckova L, De Lorenzo V, and Ruml T (1999) Appl Environ Microbiol 65:1092–1098

Kuroda K, Shibasaki S, Ueda M, and Tanaka A (2001) Appl Microbiol Biotechnol 57:697–701

Lipke PN, and Kurjan J (1992) Microbiol Rev 56:180–194

Matsumoto T *et al* (2002) Appl Environ Microbiol 68:4517–4522

Matsumoto T, Ito M, Fukuda H, Kondo A (2004) Appl Microbiol Biotechnol, 64: 481-485

Michiko Kato *et al* (2005) Appl Microbiol Biotechnol 69: 423–427

Miki BL, Poon NH, James AP, and Seligy VL (1982) J Bacteriol 150: 878–889

Murai T *et al* (1997b) Appl Microbiol Biotechnol 48: 499-503

Murai T, Ueda M, Kawaguchi T, Arai M, and Tanaka A (1998) Appl Environ Microbiol 64:4857–4861

Murai T *et al* (1999) Appl Microbiol Biotechnol 51:65–70

Murai T, Ueda M, Yamamura M, Atomi H, Shibasaki Y, Kamasawa N, Osumi M, Amachi T, and Tanaka A (1997). Appl Environ Microbiol 63:1362–1366

Nakamura Y *et al* (2001) Appl Microbiol Biotechnol 57:500–505

Newton SMC, Jacob CO, and Stocker, B.A.D (1989) Science 244:70–72

Samuelson P, Wernerus H, Svedberg M, and Stahl S (2000) Appl Environ Microbiol 66:1243–1248

Sato N *et al* (2002) Appl Microbiol Biotechnol 60:469–474

Schreuder MP, Brekelmans S, Van den Ende H, and Klis F M (1993) Yeast 9:399–409

Schreuder MP *et al* (1996) Trends Biotechnol 14:115–120

Scott J.K, and Smith G.P (1990) Science 249: 386-390

Shibasaki S *et al* (2007) ApplMicrobiol Biotechnol, 75,821-828

Shigechi H *et al* (2004) Appl Environ Microbiol 70: 5037–5040

Shigechi H *et al* (2002) J Mol Cat B 17:179–187

Takeshi Fukuda, Danya Isogawa, Madoka Takagi, Michiko Kato- Murai, Hisashi Kimito, Hideo Kusaoke, Mitsuyoshi Ueda and Shinichiro Suye (2007) Biosci. Biotechnol. 71, 2845-2487

Tang YQ *et al* (2008) Appl Microbiol Biotechnol .79,1019–1026

Teunissen AW *et al* (1993) Yeast 9:423–427

Ueda, M, and Tanaka,A (2000a) Biotechnol Adv 18: 121-140

Ueda M, and Tanaka A (2000b) J Biosci Bioeng 90: 125-136

Valentin E, Herrero E, Pastor FIJ, and Sentandreu R (1984) J Gen Microbiol 130:1419–1428

Van der Vaart J.M. *et al* (1995) J Bacteriol 177:3104–3110

Washida M, Takahashi S, Ueda M, and Tanaka A (2001) Appl Microbiol Biotechnol 56:681–686

Watari J *et al* (1994) Yeast 10:211–225

Yasui M *et al* (2002) Appl Microbiol Biotechnol 59:329–331

Ye K, Shibasaki S, Ueda M and Tanaka A (2000) Appl Microbiol Biotechnol 54: 90–96.

Chapter - 3

Microbes in reclamation of mined lands of Madhya Pradesh

Jamaluddin

Emeritus Scientist, CSIR, Department of Biologicial Science,
R.D. University, Jabalpur
Email:jamaluddin_125@hotmail.com

ABSTRACT

Calcium and coal mine overburden that differs in their biological, chemical and physical attributes, losses productivity and biological diversity. The number of plant species occurring was increase age of the dumps. The primary colonizes were *Blumea lacera, Eclpita proatrata* and *Melochia carchorifolia.* This was replaced by species *Arisitida adscandens, Cynodon dactylon, Euphorbia hirta, E. nummularis, Saccharum munja, Tridax procumbens and Calotropis procera.* Some of the species vary in calcite overburden dumps. Vesicular arbuscular mycorrhizal root infection of plants increased the value index which was confirmed by positive correlation. Material forest yielded highest spore population while young dumps of land 1 and 2 years gave least VAM population with dominance of few species only. The physiochemical characteristics of soil were also improved with increasing age of mine overburden dumps. The calcite rock dissolving activity of microbes also studied.

Key words: Coal mines, Reclamation, Microbes, AM fungi

1. Introduction

Mining for natural recourses generates a variety of wastes (overburden) that differs in their biological, chemical and physical attributes. Loss of productivity, biological, diversity and soil fertility as well as alteration of micro climate and environment hazards, etc. are some of the associated problems in such difficult sites, Rehabilitation and reclamation of such disturbed ecosystem is a different task as mine spoils are normally nutritionally and microbiologically poor specially AM fungi (Visser, *et al.,* 1984). The natural processes of ecological rehabilitation are controlled by factors such as soil fertility, pH, moisture temperature and soil microbial activities. Understanding of ecosystem structure including the process of primary and secondary succession is necessary for an ecosystem approach of rehabilitation processes (Thompson *et al.,* 1984). The objectives of mine overburden reclamation include re-establishment of a sustainable plant community that is functional and extremely pleasing. Therefore, study on status of AM fungi in coal mine and calcite mine areas along with AM fungi and their beneficial role for biological rejuvenation of over burden carried out.

2. Material and Methods

The study was conducted to estimate the AM fungi along with the natural regeneration in coal mine and calcium mine overburden dumps. The soil dumps of different age group 1, 2, 3 to 5 years were selected for the study. Vegetation survey was conducted by using quadrate method for each of the coal mine spoils in the month of November. The quadrate of 1m was laid out for herbs and grasses while 5m x 5m for shrub special species were used.

Mycorrhizal status was estimated by calculating root infection percentage and shone population in spoil samples. Isolation of AM fungi was subjected to wet sieving and decanting techniques given by Gerdemann and Nicolson (1963). Clearing and staining of mycorrlizal roots was made by methods of Phillip and Hayman (1970).

Soil phosphorus content of plant samples was estimated by adopting the method recommended by Jackson (1973). The plant species boosted up in nursery were planted in fresh dumps of coal mine Kusmunda; Korba was calcite dumps near Katni area. The height growth of treated plants were usual observed by usual technique.

3. Results and Discussion

It has been observed that the number of plant species improved with the age of dumps. The 5 year old dumps exhibited high floristic diversity. The 5 years reclaimed coal minc spoils were having 25 species with the high frequency of herbs. In case of calcite dumps only 12 species were noticed, in coal mine dumps primary colonizes were Blumea lacera, *Eclipla brostrata* and *Meloclina carchaorifolia*. These were replaced by species and *Avistida scandens, Cynodon dactylon, Euphorbia hirta, E. nummularis, Saccharum munja, Tridex procumbens* and *Calotropis procera*. However, in calcium mine dumps primary colonizers were, *Cynodon dactylon*, *Cyprus sp*, *Tridex procumbens*, *Calotropis* and *Butea monosperma*. The herbaceous flora was more in the planted pits of dumps which indicated the shade in the dumps.

The distribution of AM fungi in different overburden spoils of different ages showed *Glomus intraradices G. leptotichirm* and *Acaulospora sirobiculata* 2 years old dumps. *G. mossed, G. deserticola, Acaulospora scrobiculata, G. rosea, Scutellospora* sp. *Gigaspora margarita* and *Sclerocystis* sp. 3 and 5 year dumps. In calcium mine, *Glomus intraradices, Glomus mosseae, Gigaspora* sp and *Scutellospora* sp. were recorded in high frequency in 2 to 5 year old dumps. The frequency of AM fungi increased with the age of dump.

It has been pointed out that mycorrhizal formation and plant diversity in mine spoil have been increased as succession of plant species progress. This might be due to soil amendment by environmental and biological activities. Gupta and Shukla (1998) reported an increase in plant density with the increase in age of spoils which possibily due to plant succession and creation of suitable habitat for microbial activities. *Acacia nilotica, Pongamia pinnata* and *Eucalyptus* were boosted up with AM fungi in nursery and planted in the field i.e. overburden soil of coal mine. It has been observed that the plant species boosted up with AM fungi showed good growth and establishment in coal mine dumps. Similarly in calcium mine, *Jatropha curcas, Pongamia pinnata* and *Withania sominifera* were planted in the calcite dumps. Prior to plantation these spices were boosted up with AM fungi in nursery. AM inoculation helped in the calcite dumps. There was 100 percent survival of these species more has been marked increased in AM fungi l with increase in age of dumps which almost become stand at will after 5 years (Loree and Williams, 1987).

The phosphorus content of the leaves of inoculated species planted in coal mine and calcium mine increased as compared with uninoculated ones. This indicated that mine was high phosphorous acquisition by these species, Similarly Kothari *et al.* (1990) also reported high P contents of plant inoculated which AM fungi.

References

Gardemann, J.W. and Nicolson, T.H. (1963) Trans Br mycol Sc. 46: 235- 244.

Gupta O.P and Shukla, R.P. (1992) *Trop Ecol* . 82:296- 309.

Jackson, M.I.(1973) Soil chemical analysis, New Delhi, Prentice Hall, India.

Kothari, S.K. Marshner, H and Rombeld, V. (1990) *New Phytol*. 116:637- 645.

Loree, M.A.J. and William, S.E. (1987) *New Phytol*. 106 : 735-744.

Phillip J.H. and Haryman, D.S.(1970) *Trans. Br. Mycol. Soc*.55: 158-161.

Thompson, R., Vogel, W.G. and Taylor, D.D. (1984) Kenhicky Castanea 49:111-126.

Visser, S., Griffiths, C.L. Visser, S.J and Parkinson, D. (1984) *Plant and Soil.* 82:51-60.

Chapter - 4

Microbial Nanotechnology – A new revolution in Modern Microbiology

D.V.R. Sai Gopal

Professor and Head, Dept. of Virology, Guest Faculty of Nonmaterial Scienes, Dept. of Physics, Sri Venkateswara University, Tirupati – 517 502
Email: dvrsaigopal@gmail.com

Nano technology: The Future is Coming Sooner Than You Think

1. What Is Nanotechnology?

A nanometer (nm) is one billionth of a meter. The science, engineering, and technology related to the understanding and control of matter at the length scale of approximately 1 to 100 nanometers. However, Nanotechnology is not merely working with matter at the nanoscale, but also research and development of materials, devices, and systems that have novel properties and functions due to their nanoscale dimensions or components. Nanotechnology is the creation and utilization of materials. "Nanostructures"- The smallest human-made objects and they exhibit novel physical, chemical and biological properties and phenomena. Understanding Effects of nano materiala are that the Physical processes do not scale uniformly Gravity, friction, combustion, electrostatic, van derWalls, Brownian, quantum. Origins of Nanotechnology - Leucippus of Miletus, 5th century BC from the Greek was identified the nano-size materials and later Democritus of Abdera and he explained that all matter are made up of undividable particles called atoms, there is a void, which is empty space between atoms, atoms are completely solid, atoms are homogeneous, with no internal structure, atoms vary in Size, Shape and Weight. John Dalton (1803) explained that the chemical elements are made of atoms, the atoms of an element are identical in their masses, atoms of different elements have different masses, atoms only combine in small, whole number ratios such as 1:1, 1:2, 2:3 and atoms can be neither created nor destroyed. Later several events facilitated the foundations in theories of nanomaterials such as-

Ernest Rutherford (1908), Niels Bohr (1918), Einstein Pauli Bose Heisenberg proposed the theories of Von Neumann - 1932 mathematical synthesis-Gravity - Graviton - Always attracts, never repels; curves space. Gravity is the only force to which all particles are subjected – Indefinite, Strong Force - Gluon (8 kinds) - Binds quarks into nucleons and nucleons into nuclei - Limited to the atomic nucleus, Electromagnetism -Photon - Binds electrons to the nucleus; allows all physical and chemical processes – Indefinite, Weak Force - Weak bosons (3 kinds) - Causes unstable particles and nuclei to decay - Limited to the atomic nucleus. Max Knoll and Ernst Ruska (1931) discovery of electron microscopy was revolutionized both in biological and physical sciences, which it is using as one of important physical parameters in nanomaterial science. Richard P. Feynman the father of Nanoscience has stated the existence of Nanotechnology "There's Plenty of Room at the Bottom" on December 29th, 1959 (birth of Nanotechnology)- Manipulating and controlling things on a small scale and he worte the Encyclopedia Brittanica on the head of a pin- How do we write small?Information on a small scale, Better electron microscopes, The marvelous biological system, Miniaturizing the computer, Miniaturization by evaporation, Problems of lubrication, A hundred tiny hands, Rearranging the atoms, Atoms in a small world. K. Eric Drexler in 1981 reported the Development of the ability to design protein molecules will open a path to the fabrication of devices to complex atomic specifications. Later Fullerenes were discovered by Robert F. Curl Jr. Richard E. Smalley and Sir Harold W. Kroto in 1985 (C60, C70, La C82, SWNT,MWNT) and followed by IBM (1985). Nanotechnology has a long history, has a solid theoretical basis, has been looked at seriously for nearly 30 years, So, what is it? Nanotechnology could be Nano (metrology), nanoscale science (effects), nanoscale technology (fabrication), molecular nanotechnology (chemistry).

2. Aim of Nanotechnology

Nanotechnology is to learn, to exploit their properties and efficiently manufacture and employ the structures, Control of matter on the nanoscale plays an important role in scientific disciplines as diverse as Physics, Chemistry, Materials science, Biology, Medicine, Engineering and Computer simulation, Provide new types of educational opportunities to train the nanotechnologists and entrepreneurs of the future, Create the physical infrastructure to enable first-class basic research, exploration of applications, development of new industries and rapid commercialization of innovations. Within their vision of a "grand coalition" contributing to a national nanotechnology initiative, workshop participants proposed specific objectives for academe, private industry.

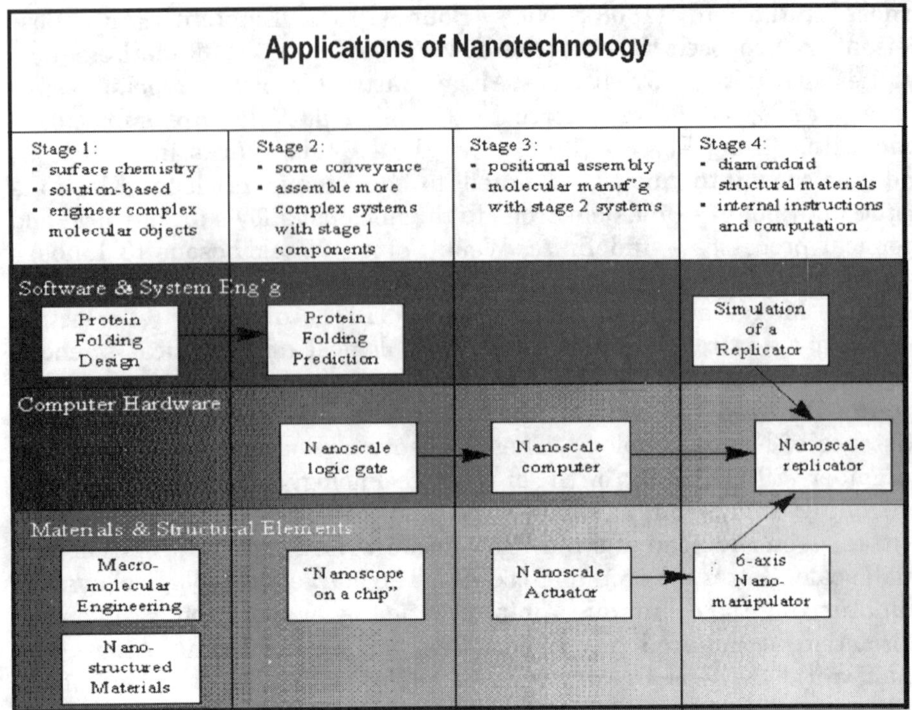

National initiative with the following objectives : Support long-term Nanoscience and Engineering research leading to fundamental discoveries of novel phenomena, processes and tools, improve institutional structures so they foster and nourish developments, encourage the type of transdisciplinary and multi institutional cooperation required in this new area. Academic application are to promote interdisciplinary work involving multiple departments, foster on-campus nanotechnology centers for greater interaction, introducing of nanoscience and engineering in existing and new courses, ereate or connect "regional coalitions" that involve industry/technology generation, ease intellectual property restrictions to improve information flow with industry, establish graduate and postdoctoral fellowships for interdisciplinary work. Definition of Nanotechnology is Nanotechnology is the popular term "for the construction and utilization of functional structures with at least onc characteristic dimension measured in nanometers". Such materials and systems can be rationally designed to exhibit novel and significantly improved physical, chemical, and biological properties, phenomena, and processes because of their size. When characteristic structural features are intermediate in extent between isolated atoms and bulk materials, in the range of about 10-9 to 10-7 m (1 to 100 nm), the objects often display physical attributes substantially different from those displayed by either atoms or bulk materials. National Nanotechnology Initiative applications are **NBIC:** Nanotechnology, Biotechnology, Information technology and Cognitive science, **GNR:** Genetics, Nanotechnology and Robotics, Power & Energy,

Creation, Storage, Transmission- cholorophyll voltaic cell. Currently known nanostructures: Carbon nanotubes, proteins, DNA, and single-electron transistors, which operates at room temperature. Rational fabrication and integration of nanoscale materials and devices herald a revolutionary age for science and technology, provided we can discover and fully utilize their underlying principles. Nanoscience R&D illustrates the challenge: What new and novel quantum properties will be enabled by nanostructures, especially at room temperatures how different from bulk behavior will be the properties of interfacial regions between contiguous nanostructures? How can new technologies exploit these properties? What are the surface reconstructions and rearrangements of atoms in nanocrystals and nanorods? Is it possible to prepare epitaxial core-shell systems in nanocrystals? Can carbon nanotubes of a single length and helicity be synthesized and purified as isolated molecular species? Is it possible to reproducibly prepare heterojunctions in one-dimensional nano-structures? What new insights in our understanding of complex polymer, supramolecular, and biological systems will come from the capability to examine single-molecule properties? How extensively can one use parallel self-assembly techniques to control the relativ arrangements of nanoscale components according to a complex, designed sequence before error rates become unacceptable? Are there processes that would lead to economic preparation of nanostructures with the control of size, shape, composition, and surface states necessary for advanced device applications? Potential applications of nanotechnology are pervasive, in the fields are Nanotechnology is fundamentally changing the way materials and devices will be produced in the future. The ability to synthesize nanoscale building blocks with precisely controlled size and composition and then to assemble those into larger structures with unique properties and functions will revolutionize segments of the materials manufacturing industry. Nanostructuring is expected to bring about lighter, stronger, and programmable materials. Reductions in life-cycle cost though lower failure rates. Innovative devices based on new principles and architectures. Use of molecular/cluster manufacturing. Applications of nanotechnology to materials and manufacturing include the Forming nanostructured metals and ceramics at exact shapes without machining, improved color printing brought about by nanometer-scale particles that have the best properties of both dyes and pigments. Nanoscale cemented and plated carbide materials and nanocoatings for cutting tools, and other electronic, chemical, and structural applications. New standards for measurements at nanoscale. Nanofabrication on a chip with high levels of complexity and functionality. Nanostructure properties: Investigate biological, chemical, electronic, magnetic, optical, and structural properties in nanostructures. Synthesis and processing: Patterning and micro-fabrication, processing with proximal probes, stamping, and embossing. Interface with bionanostructures and bio-inspired structures to multifunctional and adaptive nanostructures, to scaling approaches, to affordability at commercial scales Characterization and manipulation: Measuring and/or manipulating single macro- and supra-molecules of biological interest. Modeling and simulation: Accelerate the application of novel concepts, High performance computation to the prediction of nanostructured properties, phenomena and processes. Device and system concepts: Stimulate the innovative application of nanostructure properties to new technologies.

3. Current scientific advances

Carbon nanotubes and the use of proximal probe and lithographic schemes to fabricate individual electronic devices from these materials. Engineered individual molecules onto appropriate electrical contacts and measure transport through the molecules. Proximal probe techniques and their use to manipulate matter and thereby fabricate nanostructures. Chemical synthetic methods to prepare nanocrystals and assemble these nanocrystals into a variety of larger organized structures. Introduction of Biomolecules and Supermolecular structures into the field of Nanodevices. The isolation of Biological motors, and their incorporation into nonbiological Environments.

4. Goals for the next 10 years

Barriers and solutions-better understanding of the electronic, magnetic, and photonic interactions that occur on and are unique to this size scale. New methods to image and analyze devices and device components will be developed. It will become possible to integrate semiconductor, magnetic, and photonic nanodevices as well as molecular nanodevices into functional circuits and chips. The techniques now being developed in Biotechnology will merge with those from nanoelectronics and nanodevices. Nanodevices will have biological components. Biological systems will be probed, measured, and controlled efficiently with nanoelectronic devices and nanoprobes and sensors. There will be significant progress in nanomechanical and Nanobiomechanical systems, which will exhibit properties that are fundamentally different from their macroscopic counterparts. Iowa State scientists demonstrate first use of nanotechnology to enter plant cells AMES, Iowa. A team of Iowa State University plant scientists and materials chemists have successfully used nanotechnology to penetrate plant cell walls and simultaneously deliver a gene and a chemical that triggers its expression with controlled precision. Their breakthrough brings nanotechnology to plant biology and agricultural biotechnology, creating a powerful new tool for targeted delivery into plant cells.

5. Nanobiotechnology

Silica breaks through in plants- designer nanotubes based on mesoporous silica can now penetrate the thick cell walls of plants and deliver DNA and their activators. This opens the way to precisely manipulate gene expression in plants at the single-cell level. Mesoporous silica nanoparticles deliver DNA and chemicals into plants: Surface-functionalized silica nanoparticles can deliver and drugs into animal cells and tissues. Their use in plants is limited by the cell wall present in plant cells. A honeycomb mesoporous silica nanoparticle (MSN) system with 3-nm pores that can transport DNA and chemicals into isolated plant cells and intact leaves. The loaded MSN with the gene and its chemical inducer and capped the ends with gold nanoparticles to keep the molecules from leaching out. Uncapping the gold nanoparticles released the chemicals and triggered gene expression in the plants under controlled-release conditions. Further developments such as pore enlargement and multifunctionalization of these MSNs may offer new possibilities in target-specific delivery of proteins, nucleotides and chemicals in plant biotechnology.

6. Digital memory device based on Tobacco mosaic virus conjugated with nanoparticles

Nanostructured viruses are attractive for use as templates for ordering quantum dots to make self-assembled building blocks for next-generation electronic devices. So far, only a few types of electronic devices have been fabricated from biomolecules due to the lack of charge transport through biomolecular junctions. A novel electronic memory effect by incorporating platinum nanoparticles into tobacco mosaic virus. The memory effect is based on conductance switching, which leads to the occurrence of bistable states with an on/off ratio larger than three orders of magnitude. The mechanism of this process is attributed to charge trapping in the nanoparticles for data storage and a tunnelling process in the high conductance state. Such hybrid Bio–inorganic nanostructures show promise for applications in future nanoelectronics. Encapsulation-A single gadolinium atom encapsulated in a gadolinium fullerene, encapsulated in a single-wall carbon nanotube. Nanocomposites types- plastics, foams, aerogels, powders, membranes, coatings, films, catalysts, semi-conductors, magnets etc. Nanomaterials often have different properties than their bulk-scale counterparts - nanocrystalline copper is five times harder than ordinary copper. Nanocomposites are materials where the constituents are mixed on a nanometer scale - A nanoscale dispersion of sheet-like inorganic silicate particles in a polymer matrix is superior to either constituent in such properties as optical clarity, strength, stiffness, thermal stability, reduced permeability, and flame retardancy. Abalone shell has alternating layers of calcium carbonate and a rubbery biopolymer. It is twice as hard as and a thousand times tougher than its components.

7. Bionanoelectronics

Viruses show their good side. The Tobacco mosaic virus can be combined with metallic nanoparticles to make novel electronic memory elements. Are virus-based memory sticks just around the corner? The low conduction ('off') state in the absence of applied voltage is most probably caused by charge transfer between the virus and nanoparticles, as indicated by the semitransparent spheres for some nanoparticles. When a sufficiently high voltage is applied, this natural charge-transfer process is reversed and electrons can tunnel (yellow arrows) between the nanoparticles to produce a high conduction ('on') state.

EPILOGUE

Nanotechnology utility with living organisms-

***"The most tremendous advantage is that you can deliver several things into a living cell at the same time and release them whenever you want"* – Torney said.**

Chapter - 5

Metagenomics - Tool for an inexhaustible access to microbial communities

V.V. Lakshmi

Department of Applied Microbiology, Sri Padmavathi Mahila Visvavidyalayam, Tirupati-517502
E-mail: vedula_lak28@yahoo.co.in

ABSTRACT

Most of the microorganisms in nature are inaccessible as they are uncultivable in the laboratory. As typically <1% of bacteria can only be cultivated, it is essential to integrate traditional culture based methods with culture independent techniques and metagenomic approaches for assessing true microbial diversity. Subsequently functional genomics approaches facilitate exploring the gene pool of cultured and uncultured forms and archive environmental genetic resources utilizing high-throughput discovery approaches. These strategies involve generation and analysis of metagenomic libraries to harvest for novel biocatalyst and biomolecules of pharmaceutical importance. Combining a metagenomic approach with heterologous expression thus holds much promise for the sustainable exploitation of the biodiversity. Genome based approaches have led to revolutionizing the fields of bacterial diversity, ecology and evolution. The new tools are reshaping the way we study the environment by increasing the resolution at which microbial communities, their complexities and dynamics can be studied to reveal their functional diversity and potential. Reassembly of multiple genomes have provided insight into energy and nutrient cycling within the community, genome structure, gene function, population genetics, micro-heterogeneity, and lateral gene transfer among members of uncultured community. The application of metagenomic sequence information can facilitate the design of better culturing strategies to link genomic analysis with pure culture studies.

Key words: Metagenomics, Microbial diversity, Molecular tools, DNA micro-arrays

1. Introduction

Microorganisms represent two out of three domains of life and have vast diversity. Traditionally identification and characterization of microbial communities was limited to culturable organisms. In post genomic era, genome wide analysis provided a deeper insight into the metabolic diversity in microorganisms. The first step in this daunting task of exploration was to sequence the genomes of several adaptable bacteria that possess a variety of catabolic genes. However, it is increasingly being realized that only a fraction of microorganisms involved in biodegradation are culturable. The present culture techniques adopted are biased towards organisms that are best adapted to the laboratory conditions. This preferential growth distorts the actual microbial community composition of the environment. Hence investigations employing traditional cultivation approaches alone have become unacceptable as only tiny portion of environmental bacteria that are cultivable in laboratory are targeted (Torsvick, 2002; Daniel, 2005; Malik, *et al.*, 2008,). The biggest boom in microbial ecology started in 1995 with application of molecular tools to study the biodiversity and understand dynamics of microbial community in finer details. Culture-independent methods (CIM) emerged crucial in understanding the genetic diversity, population structure and ecological roles of the majority of habitats. CIM has revealed startling richness of the uncultured microbial world leading to renewed trend of exploring the unknown microbial territories similar to that observed in 16th-17th century, but with employing molecular tools. The initial attempt in this direction was to sequence clones from 5S rRNA cDNA library. However, most convincing scenario of vastness of microbial diversity emerged from various studies employing 16S rRNA, which were highly successful in reconstructing phylogenies and compared microbial distributions among different samples in addition to quantifying the relative abundance of each taxonomic group (Lane *et al.*, 1985; Pace *et al.*, 1985; Woese 1987; Handelsman 2004; Sanz & Kochling 2007

Genome representation of about four to seven thousand has been estimated per gm of soil using molecular tools. However, even utilizing the current r-RNA based phylogenetic approaches also only 50-200 microbial forms/gm are monitored, thus still leaving a major fraction unexplored. This scenario is mainly accounted for by the issues related to DNA extraction efficiency, primer specificity sensitivity and lower limits of detection for the available technologies, in addition to high initial amount of biomass load. These limitations destroy or distort the community structure and diversity present within both micro as well as macro habitats (Polz & Cavanaugh1998; Rappe & Giovannoni 2003). The present Gene Bank's entries of 16s RNA genes from uncultured prokaryotes have significantly out numbered those from cultured microorganisms the former being more than double. This uncultured fraction includes diverse organisms that have no affiliation or are only distantly related to the phyla of cultured ones. Thus bulk of microbiota still remains a vast untapped resource for application (Lovely 2003; Handelsman 2004 & 2005). In the past two decades integration of newer CIM techniques and improvisation of existing ones has contributed significantly to detection of several novel catabolic genes from contaminated sites especially from non-cultivable organisms (Widada *et al*. 2002, Van Hamme *et al*. 2003).

2. Metagenomic shift

Describing the phylogenetic diversity of uncultured microorganisms is only the first step in understanding biodiversity. The greater challenge lies in assigning ecological roles to the organisms as uncultured microbiota are increasingly being realized to play pivotal roles in natural environmental processes. Thus it was vital to determine function of gene at protein or metabolism levels than mere sequencing. The shift in emphasis to genome wide analysis of the currently available genome data rather than individual genes has revealed that about 30-40% of the fully sequenced organisms have no known functions assigned at biochemical level, many of which are unrelated to the known genes. Thus functional genomics has emerged as a new discipline employing major innovative technologies utilizing traditional as well as novel approaches for genome-wide searches supported by bioinformatics tools (Langer *et al*. 2006; Valenzuela *et al*. 2006; Ferrer *et al*. 2009).

Metagenomics goes a step forward and aims to transcend the limitations of the individual organism to the "meta level" by targeting community profile. Microbial community from environmental sample is mined directly by extracting data from both known and novel microbial forms. The term "metagenomics" was coined by Handelsman *et al*. 1998, while describing acid mine biofilms Thus, whereas genomics determines the complete genetic complement of a single organism only, by contrast metagenomics involves sampling the genome sequences of a community of organisms inhabiting a common environment. The definition, in application excludes studies using PCR to amplify gene cassettes or access genes of interest only as they do not provide genomic information beyond the genes being amplified. In principle, any environment is amenable to metagenomic analysis provided that nucleic acids can be extracted from sample material. Metagenomics has become the center of focus approach in last 5 years and has been applied to several niches/ environments, including oceans, soils, thermal vents, hot springs samples and Human Microbiome Project, aims to map human-associated microbial communities skin, mouth, gastrointestinal and vagina tract (Sebat *et al*. 2003; Langer *et al*. 2006; Warnecke *et al*. 2007; Hugenholtz & Tyson 2008). The approach is also termed as community genomics, ecogenomics, or environmental genomics. The process requires no prior separation of organisms from its habitat or maintaining microorganisms as either pure or mixed culture in laboratory for sequencing thus by-passing major limitations of classical approaches in microbiology. Strategies of metagenomics also circumvent several limitations due to direct cloning of large DNA as well as proper representation of finite number of clones to screen. Metagenomic studies involve massive sequencing to capture entire communities which has become possible with advances in sequencing technology like pyrosequencing. Thus access to the potential of the collective microbial genomes in a given environment simultaneously using functional sequence-based analysis employing high-throughput methods is now feasible. Areas in which metagenomics offers particular value include: medicine, alternative energy, environmental remediation, biotechnology, agriculture, biodefense, and forensics (Cowan 2005; Tringe & Rubin. 2005; Dinsdale 2008).

Metagenomics has been initiated at all three scales. Small scale target a slice of community for a selective function of interest or by employing specific techniques are initiated mainly by single-investigator or laboratory. Middle scale metagenomic projects employ multidisciplinary approach to thoroughly investigate a community of interest. The scope of these projects is diverse and is undertaken by group of collaborating investigators each targeting one facet. Large-scale project on the other hand encompass global initiatives on big scale to understand selective microbial communities in depth with detailed characterization. Metagenomics applied to community genomics of environment had the greatest impact in the last few years (Cowan 2005; Daniel 2005; Dinsdale 2008; Ferrer 2009).

3. Pipeline of the metagenomic project design can be broadly categorized into following stages

3.1 Pre-sequencing considerations

Community composition has a great influence on mode and results of metagenomic analysis. Prior knowledge of dominant population is helpful in designing subsequent strategies. The issue of assessing how much sequence data is to be obtained for a community also becomes very crucial as there is no fixed end point, unlike in complete genomes. Since species differ in gene coding densities of their genome and do not have uniform abundance in a community, it is simple to address the coverage of individual populations. To obtain greater than single sequencing coverage, the size of a moderate metagenomic library would need to be many times the size of the metagenome which can reach upto 100 Mbp with a typical Sanger sequencing. Generally a 6x to 8x coverage of microbial isolates is a common target to obtain a draft genome suitable for metagenomic project (Dinsdale 2008; Kunin *et al.* 2008; Ferrer 2009). Ultimately, objective of the study guides the sequence allocation. Recent studies have shown that even extremely low coverage of <0.01x dominant population of stratified hypersaline mat community were still sufficient to detect genetic gradients (Goldberg *et al.* 2006, Kunin *et al.* 2008).

3.2 Sampling strategies

Extraction and purification of high-quality DNA still remain main bottlenecks in metagenomic studies as there is no "one-size-fits-all" extraction method for all samples. Low-biomass of environmental sample yields insufficient DNA for library construction or pyro-sequence analysis, where microgram quantities of genomic DNA are required (Kunin *et al.* 2008) During sampling it essential to collect additional sample material and all possible collateral non-sequence data associated with an environmental sample or "meta data". Type of metadata can vary considerably depending on the sample type. Some of the common ones for environmental studies are geographical data such as global positioning system coordinates, depth/height of sample collection, data on pH, temperature, salinity of the site etc. (Venter *et al.* 2004; Kunin *et al.* 2008; Urich *et al.* 2008). This complementary information is crucial for analysis as re-sampling might not be always directly comparable. Meta data enhances the ability to interpret sequence data significantly, particularly for a comparative analysis of temporal or spatial organization. Co-localization of organisms by combining FISH with digital image analysis can provide spatial information in structured ecosystems (Lebaron *et al.* 1997; Sanz & Kochling, 2007).

3.3 Data generation

Prior to metagenomic data generation a Pre-Metagenome Community Composition Profiling (PCCP) is usually undertaken as a prescreening. This aids subsequent decision on sequence allocation and processing based on community composition of the environmental sample. This is usually followed by shotgun library preparation where in at least three different average sizes of 3, 8, and 40 kbp cloned DNA in fosmids are used. Global metagenomic projects undertaken by Joint Genome Institute (JGI) have evolved a 4: 4:1 ratio between the smaller and larger fragment sizes to produce high-quality draft assemblies (Warnecke 2007; Kunin *et al.* 2008).

Selection of sequencing technology is critical for quality as well as cost effectiveness of the project. Recently pyrosequencing has been added to the universally used Sanger's approach. Though Sanger (dye terminator) sequencing still remains the primary source of metagenomic sequence data, pyrosequencing strategies have been increasingly applied to viral and bacterial communities. Advantages of pyrosequencing over Sanger approach is a much lower per base cost and no necessity for prior cloning whereby cloning bias are avoided. The major disadvantage of pyrosequencing is its short average read lengths. This ranges from a maximum of 100 bps on the GS20 to 200 bps on the GS FLX platforms (Nulton 2006; Wommack *et al.* 2008). Any further enhancement in this capability would be an asset for metagenomics. Due to this short read length of sequence, the approaches, mainly relies on similarity searches against a reference database instead of gene calling or assembly which are not feasible in majority of instances. Combining different sequencing technologies together is being explored for producing high- quality draft assemblies. At least two stages of quality control are considered to be mandatory to get reliable data information (Edwards *et al.* 2006a&b; Goldberg 2006)

3.4 Sequence processing

This step encompasses a critical step in pre-processing of sequence reads prior to assembly, gene-prediction, and annotation. Screening by base calling of raw data coming off the sequencing machines is carried to remove cloning vector sequences followed by quality trimming to remove low-quality bases. Errors in these steps have enormous impact on downstream consequences in metagenomics. In contrast to genomics, metagenomic projects tend to retain only high quality reads and contigs as the process has low tolerance for contaminating reads, and are not easily distinguishable thereby skewing downstream analysis (Chou & Holmes 2001; Kunin *et al.* 2008). Other issues that need attention to minimize mis-assembly include non-uniform read depth repetitive regions, etc. Overlooking these aspect can greatly hamper both the quality and reliability metagenome output. Assembly process combines sequence reads into contiguous stretches of DNA based on sequence similarity, called contigs (Raes *et al.* 2007).Genome closure or finishing is not possible for most of metagenomes though it finds common place in isolated microbial genome projects. However, finishing may be considered as a viable option only for data sets pertaining to dominant populations within a metagenome (Hallam *et al.* 2006). Complete or near-complete draft-level coverage genomes have been assembled from metagenome populations present in biofilms in acid mine drainage, in activated sludge and in marine environment (Kunin *et al.* 2008 a b, c).

3.5 Gene prediction and annotation

Genes and gene fragments from a given metagenomic data set are mapped into gene families, thereby getting an estimate of their relative representation. To enhance the assignment of function to sequenced data, gene-centric trends can be adopted in metagenomics which was first described by Tringe *et al.* (2005). The method focuses on assigning roles to proteins encoded in the sequenced genes. Two different possible approaches are adopted in this mode of analysis namely unselective or targeted metagenomics. In unselective mode there is no prior selection based on either growth or PCR amplification. The data generated is a result of gene 'fishing' expedition. The approach though has advantage of providing highly valuable information of the metabolic activity of community not known earlier; it suffers from the disadvantage of possible low hit rate (Raes *et al.* 2007; Kunin *et al.* 2008 a, b). In contrast, in the second approach of targeted metagenomics, the isolated pool of DNA is deliberately 'biased' to enhance the hit rates. This approach was successful in identification of several hydrocarbon induced genes followed by capture of relevant genes, utilising expression of green florescent protein (Yun & Ryu 2005; Raes *et al.* 2007).

There are two principal modes of gene prediction - evidence-based gene calling method dependong on using homology and "ab initio" approach relying on intrinsic features of the DNA sequence to discriminate between coding and non-coding regions, allowing the identification of genes even in absence of any homologs in the available databases. (Badger *et al.* 1998; Badger & Olsen 1999). The gene prediction is generally followed by functional annotation utilizing available meta data similar to the genomic annotation. However, the fragmentary nature and lack of neighborhood context makes it more challenging in metagenomics. In genomics, gene predictions and annotations are finally checked manually to increase reliability as a part of quality control but such manual curations are not presently feasible in metagenomes which is recommended only at large contigs levels. Thus metagenomic has been successful in study single genes, pathways, organisms and communities (Tyson 2004; Tringe *et al.* 2005; Mavromatisey *et al.* 2007)

3.6 Data analysis

The collection of reads, contigs and genes resulting from metagenome pipeline have to be associated with the organisms from which they were derived in terms of taxonomic groups. This is highly desirable for meaningful interpretation of the contributing species of the ecosystem and this process is called binning or classification. Present available binning tools rarely have the resolution to discriminate between strains of the same species, so strain co-assembly is not practical. The standard processing steps in Post sequencing Community Composition (PSCC) are reevaluation of the community composition estimates directly from the metagenomic data itself. This is important for interpreting data and avoiding skewing (Kunin *et al.* 2008 a, b,c). This is followed by analyzing dominant populations by mapping of conserved phylogenetic marker genes to assess both organism identity as well as their relative abundance. If more than one dominant population is sequenced, the probable metabolic interplay among the populations is considered (Whitaker & Banfield, 2006; Raes *et al.* 2007).

3.7 Tool box of genomics/ metagenomics

Several tools are applied for the visualization and identification of polymorphism in composite population. Prominent tools used in genomics and metagenomics address microbial communities profile from diverse facets (Gilbride *et al.* 2006; Gabor *et al.* 2007). Nucleic acid based techniques like Polymerase chain reaction (PCR) and its numerous variants are applied for analysis of catabolic genes involved in the biodegradation of organic contaminants as well as for microbial community profiling (Stenuit *et al.*, 2006& 2008). Inherent limitations of extraction and purification of nucleic acids both in quality and quantity from environmental samples significantly effects analysis by PCR amplification. The presence of inhibitory substances which are co-extracted like humic acids, organic matter and clay particles significantly interferes in amplification step. Multiplex PCR technique as a single reaction have been used for the detection of mono, or dioxygenase enzyme attacking aromatic naphthalene, toluene xylene, phenol etc. thereby saving time and resources (Wilson *et al.*, 1999; Baldwin *et al.*, 2003; Kirk *et al.*, 2004; Gilbride *et al.* 2006).

Real-time PCR (RT-PCR) enables simultaneous detection and quantification of the amplified product while the reaction is still in progress as fluorescent markers are employed. With quantification limit typically of 1–2 genome copies, RT-PCR is 100- to 10,000-fold more sensitive than microarray-based methods (Inglis & Kalischuk, 2004; Eyers *et al.*, 2004b). Moreover, it does not require any time-consuming post-PCR steps for the quantification of amplicons as they are monitored in real-time and intensity of fluorescence is directly proportional to product concentration. Hence, real time PCR which allows quantification at the end of each cycle is appropriate for high-throughput mode requirements of metagenomics. Further its high analytical sensitivity for identification of specific genes in complex DNA mixture also, making it highly suitable for analysis of environmental samples (Scow & Hicks 2005; Ritalahti *et al.* 2006) Community profiling in hydrocarbon-contaminated Antarctic soil are also successfully achieved by observing real time changes.Speed, sensitivity, accuracy and advantage of possible robotic automation makes RT- PCR to occupy a prominent position among genomic and metagenomic tools (Watanabe & Hamamura 2003; Lerat *et al.*, 2005, Saleh- Lakha *et al.*, 2005; Powell *et al.*, 2006).

Genetic fingerprinting techniques provide specific pattern or profile of a given microbial community. These are based on the separation of amplicons after PCR amplification of phylogenetic markers such as rRNA (16S and 23S rRNA) along with RecA (DNA repair protein), EF-Tu, EF-G HSP70 (heat shock protein), and RpoB (RNA polymerase subunit) or functional genes using universal or specific primers. The application potentials and limitations of various finger printing techniques are reviewed (Fromin *et al.* 2002; Nocker *et al.*, 2007). Among the several genetic community profiling methods available amplified ribosomal DNA restriction analysis (ARDRA), and terminal-restriction length polymorphism (T-RFLP), ribosomal intergenic spacer analysis (RISA) along with denaturing gradient gel electrophoresis (DGGE)/temperature gradient gel electrophoresis (TGGE) are among the most popular tools used for environmental samples (Marsh *et al.* 2000; Eyers *et al.* 2004a; Forney *et al.* 2004; Fedi *et al.* 2005; Bai *et al.*, 2006). Unlike traditional gel separation methods, DGGE/ TGGE separate amplified rDNA fragments of the same length but have different base pair compositions. The number of bands produced during DGGE or TGGE is observed to be proportional to the number of dominant species in the sample. The main limitations of DGGE/TGGE is an incomplete phylogenetic specificity provided by the analyzed 500 bp fragments of 16S rRNA sequences, overestimation of a community multiple bands of the single species due to presence several copies of r RNA gene, or underestimation due to different rRNA bands having same mobility. In spite of some of these limitations, the method allows monitoring of the spatial/temporal changes in prominent microbial community structure at moderate phylogenetic resolution. This is especially prominent in the samples where microbial diversity is largely unknown (Watanabe *et al.*, 2002; Singh, 2006; Malik, 2008).

ARDRA is a simple, rapid and cost effective method to explore the bacterial community structure or screen clones. Though it is good measure for detecting structural changes of microbial community it does not provide a effective picture of phylogentic diversity in complex population (Weidner *et al.* 1996; Bai *et al.* 2006). RFLP- a modification of ARDRA utilizes fluorescent labeled PCR primers allowing quantification of terminal fragments of the digestion thereby simplifying the banding pattern still retaining reliability. This is highlighted in monitoring of several complex environmental communities as each band represented a single operational taxonomic unit or ribotype (Tajida *et al.*, 1999; Konstantinidis *et al.*, 2003, Reardon *et al.*, 2004, Macbeth *et al.*, 2004 Fahy *et al.* 2005) Despite the high resolution and sensitivity, T-RFLP being highly dependent on PCR amplification, the biases related to PCR also influence this analysis. Application of several fingerprinting techniques in combination with real-time PCR for in situ analysis of xenobiotic degrading bacterial communities and quantification of their catabolic genes showed good correlated between microbial abundance with corresponding pollutant concentration at the site (Widada *et al.*, 2002; Watanabe & Hamamura, 2003; Scow & Hicks, 2005). Some of these fingerprinting tools like RISA, T-RFLP, length heterogeneity analysis by PCR (LH-PCR), single strand conformation polymorphism (SSCP), DGGE Denaturing High Performance Liquid Chromatography (D-HPLC), have potential for high-throughput design (Barlaan *et al.*2005; Spiegelman *et al.* 2005). The term environmental gene tags are used akin to expressed sequence tags because of the fragmentary nature of the resulting data of metagenomics (Neufeld *et al.*, 2004; Ashby *et al.*, 2007; Kysela *et al.*, 2005; Yu *et al.*, 2006 van der Lelie *et al.*, 2006). Ribosomal sequence tags/ ribosomal DNA (SARST or SARD, respectively) and single-point genome signature tags (SP-GSTs) are two prominent tagging methods that have been evolved for high-throughput profiling of complex microbial communities. These techniques provide fingerprint of microbial communities in the form of concatemers of short information rich PCR-amplified tag sequences like V1, V5, V6 and other hypervariable regions of bacterial 16S rRNA genes and are useful to explore the ecological role of low abundance/novel bacteria both in qualitative and quantitative terms (Kayesala *et al.* 2005; Yu *et al.* 2006; Ashby *et al.*, 2007). Further these approaches holds great promise when combined with the habitat-specific microarrays.

4. DNA microarrays

Microarray technology is a powerful taxonomic and functional tool that is widely used at genome and transcriptome analysis including mixed microbial communities. The property of a single stranded DNA or RNA target molecule to hybridize to a complementary probe molecule that is attached to a solid support of slide/chip forms the basic principle of microarrays (Zhou, 2003). Microchips differ according to the immobilization technology used to attach the probes, the length and nature of the probes as well as synthesis and labeling of the targets. The choice between the different technologies is based on parameters such as cost, probe density, specificity, sensitivity and quantification as reviewed. (Nees & Woodworth, 2001; Zhou, 2003). Probes can either be directly synthesized *in situ* using photolithographic masks (e.g., "GeneChip" arrays), electrochemical reactions, or spotted onto the solid support after *ex situ* synthesis. As fluorescent dyes are enzymatically or chemically incorporated in the sample the readout is based on the detection of fluorescence signals. Thus microarray presents the advantage of miniaturization by spotting thousands of probes on a single slide for simultaneous gene function analysis in real time. In comparison to traditional nucleic acid hybridization techniques, microarrays provide rapid and highly sensitive detection system even up to single mismatch (Eyers *et al.*, 2004b). Three major classes of environmental microarray formats Functional gene arrays (FGA), Community genome arrays (CGA), Phylogenetic oligonucleotide arrays (POA) have been developed for microbial community analysis (Wu *et al.*, 2001; Zhou, 2003; Fantroussi *et al.*, 2003; Wu *et al.*, 2004& 2006, Rhee *et al.*, 2004 Chandler *et al.*, 2006).

Microarray approach has been successfully used to discriminate shifts in soil microbial community associated with hexachlorocyclohexane (HCH), TNT, U(VI) bioremediation and characterization of the phylogenetic and catabolic diversity of the contaminated environments (Chandler *et al.*, 2006; Garmendiaet *et al.*, 2008). Another prominent step forward in DNA arrays is to use probes produced directly from environmental DNA without any cultivations step, i.e. combining metagenomics with microarrays. Such a metagenomic array (MGA) was proposed by Sebat *et al.* (2003) for the first time to characterize microbial community of a groundwater microcosm. The MGA format is still in its early stages of development, but holds great promise for high-throughput screening of natural environments; with no prior sequence knowledge requirement for microbial communities' analysis (Gentry *et al.*, 2006). Further whole community RNA amplification (WCRA) coupling with CGA can allow monitoring functional activities in microbial communities from contaminated environments with diverse organic solvents, hydrocarbons, uranium etc. (Wu *et al.* 2006; Gao. *et al.* (2007).

To overcome the generic biases with PCR dependent approaches, new PCR-independent amplification techniques like whole genome amplification are emerging as solutions. These techniques allow access to genomic information from very low abundance microbial sources that would otherwise remain inaccessible. Multiple Displacement Amplification (MDA) using phi29 DNA polymerase and random exonuclease resistant primers without thermal cycling exhibited surprisingly uniform amplification across the genomic targets compared to PCR based whole genome amplification (Dean *et al.* 2002; Abulencia *et al.*, 2006; Binga *et al.*, 2008) The method generated larger sized products with lower error frequency holding promise for single cell genome sequencing for metagenomic studies. DNA hybridization of microbial community amplified using the ø29 DNA polymerase to CGA was successful in analysing structure of the environmental community (Wu *et al.* 2006). T7 polymerase-based linear amplification approach using fusion primers has been propo sed for mRNA based metatranscriptomics analyses Gao *et al.* (2007).

Though function-driven screening strategy have allowed discovery of several new enzymes, antibiotics with biotechnological and therapeutic applications for years now, it typically involves low-throughput screens based on the visual detection that are difficult to be adopted for metagenomic scale ((Gentry *et al.*, 2006; Kunin *et al.* 2008). However, automated colony picking, pipetting robotics, use of microtiter plates, and availability of sensitive activity assays targeting biomolecules for screening, informatics-assisted data management has aided in automation of the metagenome screens. This is particularly an advantage, as "hits" in metagenome screening are typically very low being less than 2 out of 10,000 clones. Diverse strategies are employed to enrich/select community genomes with desired traits through promoter activity rather than phenotypic expression before metagenomic library construction is tried (Lorenz & Eck, 2005).

Intracellular fluorescent biosensors or whole-cell-based biosensors are applied widely for environmental sensing and reporting of specific or classes of contaminants (Bhattacharya, 2005; Handelsman, 2005; Williamson *et al.*, 2005). For omics applications to reach its optimum potential for *in situ* studies it would be advantageous if the automated technology is made cheaper, compact and field deployable for its flexible applications (Gentry *et al.*, 2006). With the recent advances in micro-fabrication and micro-fluidics technologies, it has become possible to make the concept of 'laboratory-on-a-chip (LOC) a reality. The device being a miniaturized system assembling have all the components of sample preparation i.e., concentration, separation and purification together with hybridization/ detection/analysis steps on a single portable platform offering simple and versatile application potentials (Liu and Zhu, 2005).

5. Advantages and limitation of metagenomics approach

Strength of metagenomics lies in its potential for serendipitous discovery. Thus, it is presently applied to explore biosynthetic diversity of microorganisms from varied environments by identifing genes involved in whole biosynthetic pathways or bioremediation of important xenobiotic compounds. Discovery of a light-driven proton pump from bacterioplankton mediated by proteorhodopsin protein, was first identified in environmental DNA samples using the metagenomic approach. A more recent discovery is regarding ammonia oxidizers among archaea was thought that bacteria were solely responsible for aerobic ammonia oxidation, although their numbers often could not account for observed rates of ammonia oxidation in many habitats. Identification of an ammonia monooxygenase gene next to an archaeal marker gene encoding small subunit ribosomal RNA using bioinformatics tools implicated archaea as one of probable main ammonia oxidisers in several marine and terrestrial ecosystems which was established experimentally later (Baldwin, 2003 Daniel, 2005; Schneikeret *et al.*, 2006). Metagenomic assemblage of microorganisms has potential to answer several fundamental questions in microbial ecology as well as to assess the diversity and functionality of microorganims in situ quantitatively, in an objective way. In this bio-prospecting trend, several unique and previously unrecognized microbial forms are identified at the contaminated sites without the need of cultivation. Co-localization studies by combining FISH and digital image analysis has provided comparative analysis of temporal or spatial information in structured ecosystems to support metabolic interactions between community members as inferred from metagenomic data (Wagner *et al.*, 2006; Sanz & Kochling, 2007; Malik, 2008).

In spite of huge impact of metagenomics in different fields, there are several challenges faced with this kind of genome analysis Three notable limitations of the basic metagenomic approach include low resolution, inability to classify short metagenomic fragments and uphill task of functional verification (Cowan *et al.*, 2005, Hendelsman, 2004, 2005 Kunin, 2008a,b). The major challenge is determining the minimum size of metagenome library. Constructing metagenomic libraries from environmental sample though conceptually simple are technically very challenging. Extraction and purification of high-quality DNA for different environmental samples is still main bottleneck in metagenomics. In fact there appears to be no "one-size-fits-all" extraction method. To get more than single sequence coverage metagenomic library would need to be many times the size of the metagenome. The resolution of microbial communities by shotgun sequencing is rather low, with sufficient sequence coverage only of dominant populations to levels of permitting sequence assembly (Langer *et al.*, 2006; Goldberg *et al.*, 2006; Gabor *et al.*, 2007; Johnson & Slatkin, 2008). The large scale sequencing cost involved for metagenomes from complex environments can still prohibitive in spite of several advances in novel sequencing techniques like pyrosequencing. To address this challenge microbial community can be divided into simpler subsets that facilitate contig identification and greater genomic coverage of populations. The divide and conquer approach though difficult to achieve in cultivation approach, can be more easily addressed via metagenomic studies. The other important challenge is in identifying the source species of metagenomic fragments. Current methods to classify fragments do not perform well on sequences of less than 8 kbp. And thirdly, similar to all DNA sequence data, metagenomics data can provide information on metabolic potential for only the genes with recognizable homology with characterized proteins (Goldgerg *et al.*, 2006; Johnson & Slatkin, 2008; Cardenas & Tiedje, 2008; Shendure, 2008). Lack of reference genomes is widening gap between characterized and hypothetical proteins at an alarming rate in metagenomes data. Lack of availability of ideal phylogenetic anchors that can be globally used is still a prominent limitation. A good phylogenetic anchor needs to be equally represented in all species and serve an indictor for the diversity both in qualitative as well quantitative terms. This ideal requirement is not met in phylogenetic anchors presently available. 16S rRNA genes analysis though being the most widely used phylogenetic marker does not meet this ideal require-ment. The number of rrn operon copies vary from 1 to 15 in microorg-anism which mainly correlates with the growth rate. This leads to poor representation of slow-growing bacteria in the PCR generated 16S rRNA metagenome libraries as compared to rapidly growing counterparts. This can result in ignoring or overlooking low abundance species thereby distorting the insitu representation. Lack of reference genomes for interspecies heterogeneity/polymorphism, high level sequence conser-vation between phylogenetically unrelated genomes, confounded by low precision assembly of available softwares result in false or chimeric assemblies not represented in nature(Kysela *et al.*, 2005; Langer *et al.*, 2006, Johnson & Slatkin, 2008; Kunin *et al.*, 2008a, b,c).

Further presently with sequencing becoming more and more reachable complete genome sequences of a number of strains/individual species of some prominent organisms/ community are growing in number and diversity. Hence an array of challenges is emerging to address comparison of these multiple bacterial genomes. Integrating the information from multiple sources using comparative genome analysis for various species of a bacterial genome is attempted and is collectively described as pan-genome. These are composed of a "core genome" containing genes present in all characterized strains, and a "dispensable genome" containing genes present in two or more strains and genes unique to single strains also. The pan-genome repertoire of genome sequences of a set of related bacterial species thus turns out to be many orders larger in magnitude than any single genome. The estimated minimum number of strains/species of given organism to be characterized, ultra-high-throughout required thereof and current models designed to manage public collections of genomes in novel ways and their applicability are reviewed (Nocker *et al.*, 2007; Malik *et al.*, 2008; Tettelin *et al.*, 2008).

Metadata pertains to the metabolic information microbial population diversity, temporal spatial and physico-chemical data associated with sampling sites. For standardizing metadata, presently there are no established standards. Genomic standards consortium is involved in evolving acceptable minimal set of metadata for genomic/metagenomic projects in consultation with research communities. Further it would be significant to correlate metadata provided with deciphered ecological and environmental conditions to elucidate the factors that favor one population structure over the other (Goldbergeg *et al.*, 2006; Hallam *et al.*, 2006; Nulton, 2006; Stepanauskas & Sieracki, 2007). The diverse data generated by metagenome projects at different scales present a tremendous challenge from the data management and data mining points of view also. The mounting data is increasing at a tremendously higher rate than the increases in the laboratory computational power. More than simple storage of sequence data, the other demanding issue is 'all-versus-all' sequence comparisons required to interpret metagenomes that pushes up the computational requirements exponentially. This is forcing to adopt radical breakthrough in computing approaches like quantum computes to carry out the comparisons of metagenomic data. A pioneering effort to manage rich metadata information that include geographical location, sampling procedures and other interesting additional information regarding environment was exemplified by CAMERA project, which stands for Community Cyber infrastructure for Advanced Marine Microbial Ecology Research and Analysis. The focus of this metagenome initiative is to serve the needs of the microbial ecology research community by creating a rich, distinctive data repository dataset of global prokaryotic marine diversity of both species and genes (Overbeek 2005; Dubchak *et al.* 2006; Seshadri *et al.* 2007; Stepanauskas & Sieracki 2007).

6. Integration of -omic approaches

Sequencing the genomes is only a preliminary step in the daunting task of exploring the physiological functions of complex and highly adaptable organisms. Several of these genes generate a number of distinguishable functional entities at protein level based on differential splicing. In view of the importance of these post transcriptional modifications (PTMs) in nature, study of global protein profile during ex situ and in situ bioremediation is crucial. Further metabolomics targeting holistic analysis of the entire set of metabolites produced by cellular proteins in response to various environmental stimuli can provide critical clues to pathway analysis. Several Metabolomics studies reveal precise correlation between true metabolic fluxes analysed with proteins and mRNA abundance (Casellas *et al.* 1997; Kromer *et al.* 2004). As this approach examines all the metabolites present in a biological system there is little bias associated with the choice of metabolites to be studied. Exploring proteomics and metabolomics in gaining central positions among post-genomic techniques. Hence proteomic and metabolomics approaches have enormous potential and are powerful tool in describing biology and ecological function of an environment. These approaches and application potentials have been reviewed comprehensively (Hou *et al.* 2003; Stenuit *et al.* 2005; 2008; Dinsdale *et al.* 2008). Functional proteomics is defined as the use of proteomics to monitor and analyze the spatial and temporal properties of the molecular networks and fluxes involved in living cells. Similarly term metaproteome was used by Rodriguez-valera in 2004 to characterization entire protein complement of environmental micro biota at a given point of time. Until recently, proteomics had been mostly restricted to microbial isolates in culture due to technical difficulties in the preparation of protein extracts. Benndorf *et al.* (2007) developed a protein extraction protocol for a wide range of environmental samples, that can be coupled to the tools of protein analysis on-line linked to MS/MS via electrospray ionization source (LC-ESI–MS/MS) for automation.

Metagenomics is increasingly being viewed as a baseline technology representing the cornerstone of what can be achieved along with a range of complementary-omics technologies. Integration emergingomics technologies such as transcriptomics, proteomics, metabolomics can go long way in providing answers to fundamental questions in microbial ecology (Paul *et al.*, 2005; Malik *et al.*, 2008; Stenuit *et al.*, 2008; Zhao and Poh 2008). This outlook of 'systems biology' provides a comprehensive overview to understand the ecology and evolution of microbial ecosystems, upon which hypotheses and experimental strategies can be built up.

Over the years qualitative and quantitative techniques have proved to be invaluable tools to identify new catabolic operons and to describe environmental microbial communities. However, their application is often time-consuming and therefore limited to a small number of samples, thereby making comprehensive characterization of an ecosystem difficult. To tackle these issues, high-throughput approaches are progressively applied in environmental microbiology. They offer the advantage of miniaturization, automation and massive parallelization of time consuming steps, allowing the simultaneous "real-time" analysis. Combination of high throughput technologies like genomics, metagen-omics, and proteomics applied in novel and imaginative way along rich metadata tagging could allow us to address interesting aspects of bioremediation. Together, these approaches hold great promise for the study of microbial ecology and evolution. Synergy in currently emerging-omic approaches will hasten emergence of alternative tools to address challenging frontiers of the complex biodegradation pathways. This will deepen the understanding of microbial community structure and diversity and correlate them with vital niche-specific functions within a micro space.

References

Abulencia CB, Wyborski DL, Garcia JA, Podar M, ChenW, Chang SH, *et al.* (2006). Appl Environ Microbiol; 72(5):3291–301.

Ashby MN, Rine J, Mongodin EF, Nelson KE, Dimster-Denk D. (2007) Appl Environ Microbiol; 73(14):4532–42.

Badger, J. H., and G. J. Olsen.., A. Mironov, H. W. Mewes, and M. Gelfand. (1998). Nucleic Acids Res. 26:2941–2947.

Badger, J. H., and G. J. Olsen. (1999). Mol. Biol. Evol. 16:512–524

Bai Y, Yang D, Wang J, Xu S, Wang X, An L (2006). Res Microbiol;157:741–51.

Baldwin BR, Nakatsu CH, Nies L. (2003) Appl Environ Microbiol;69:3350–8.

Barlaan EA, Sugimori M, Furukawa S, Takeuchi K. (2005) J Microbiol Methods; 61:399–412

Benndorf D, Balcke GU, Harms H, von Bergen M (2007) ISME J 1:224-234

Bhattacharyyaa J, Reada D, Amosc S, Dooleyc S, Killhama K, Patona GI. (2005) Environ Pollut;134:485–92

Binga EK, Lasken RS, Neufeld JD. (2008) ISME J;2:233–41.

Cardenas E and Tiedje J M (2008) Curr. Opinion in Biotech, 19:544–549

Casellas M, Grifoll M, Bayona JM & Solanas AM (1997) Appl Environ Microbiol 63: 819–826.

Chandler DP, Jarrell AE, Roden ER, Golova J, Chernov B, Schipma MJ, *et al.* (2006). Appl Environ Microbiol; 72(7):4672–87

Chou H Hand Holmes M.H (2001) Bioinformatics 17:1093-1104.

Cowan D, Meyer Q, Stafford W, Muyanga S, Cameron R, Wittwer P.(2005) Trends Biotechnol;23:321–9.

Daniel R. (2005) Nat Rev Microbiol; 3:470–8.

Dean FB, Hosono S, Fang L, Wu X, Faruqi AF, Bray-Ward P, *et al.* (2002) Proc Natl Acad Sci U S A;99:5261–6.

Dinsdale EA, Edwards RA, Hall D, Angly F, Breitbart M, Brulc JM, *et al.*(2008). Nature; 452:629–33.

Dubchak, P. Hugenholtz, and N. C. Kyrpides. (2006). Bioinformatics 22:e359–e3567.

Edwards RA, Rodriguez-Brito B, Wegley L, Haynes M, Breitbart M, Peterson DM, *et al.* (2006)a. BMC Genomics;7:57.

Edwards, R. A., B. Rodriguez-Brito, L. Wegley, M. Haynes, M. Breitbart, D. M. Peterson, M.O. Saar, S. Alexander, E. C. Alexander, Jr., and F. Rohwer. (2006)b. BMC Genomics 7:57.

Fantroussi El S, Urakawa H, Bernhard AE, Kelly JJ, Noble PA, Smidt H, *et al.*(2003) Appl Environ Microbiol;69:2377–82.

Eyers L, Agathos SN, El Fantroussi S. (2004)a Environmental Microbiology. Methods and Protocols. Totowa, NJ: Humana Press;. p. 407–17.

Eyers L, George I, Schuler L, Stenuit B, Agathos SN, El Fantroussi S. (2004)b Appl Microbiol Biotechnol;66:123–30.

Fahy A, Lethbridge G, Earle R, Ball AS, Timmis KN, McGenity TJ. (2005) Environ Microbiol;7:1192–9.

Fedi S, Tremaroli V, Scala D, Perez-Jimenez JR, Fava F, Young L, *et al.* (2005) Res Microbiol;156:201–10.

Ferrer M, Beloqui A, Timmis KN, Golyshin PN (2009). J Mol Microbiol Biotechnol. 16:109-23

Forney LJ, Zhou X, Brown CJ. (2004) Curr Opin Microbiol;7:210–20

Gabor E, Klaus Liebeton, Frank Niehaus, Juergen Eck and Patrick Lorenz (2007) Biotechnol. J., 2, 201–206

Gao H, Yang ZK, Gentry TJ, Wu L, Schadt CW, Zhou J (2007). Appl Environ Microbiol 73:563–71.

Garmendia U, Heras ADL, Galva˜o TC & Lorenzo VD (2008) Microbial Biotechnology 1: 236–246.

Gentry TJ, Wickham GS, Schadt CW, He Z, Zhou J (2006). Microb Ecol; 52:159–75.

Gilbride KA, Lee DY, Beaudette LA. (2006) J Microbiol Methods 66:1–20.

Goldberg SM, Johnson J, Busam D, Feldblyum T, Ferriera S, Friedman R, Halpern A, Khouri H, *et al.*(2006) Proc Natl Acad Sci U S A, 03:11240-11245.

Hugenholtz P and. Tyson Gene W (2008) Nature 455 481-483

Hallam, S. J., K. T. Konstantinidis, N. Putnam, C. Schleper, Y. Watanabe, J. Sugahara, C. Preston, J. de la Torre, P. M. Richardson, and E. F. DeLong. (2006). Proc. Natl. Acad. Sci. USA 103:18296–18301

Handelsman J (2004) MMBR., Vol. 68, No. 4. 669–685

Handelsman J. (2005) Nat Biotechnol. 23:38–9.

Hou, B.K. *et al.* (2003) J. Chem. Inf. Comput. Sci. 43, 1051–1057

Inglis GD, Kalischuk LD. (2004) Appl Environ Microbiol;70:2296–306.

Johnson, P. L., and M. Slatkin. (2008). Mol. Biol. Evol. 25:199–206.

Kirk JL,BeaudetteLA,HartM,Moutoglis P,Klironomos JN, LeeH, *et al.* (2004). J Microbiol Methods; 58: 169–88.

Konstantinidis KT, Isaacs N, Fett J, Simpson S, Long DT, Marsh TL. (2003) Microb Ecol; 45:191–202.

Kromer, J.O., Sorgenfrei O., Klopprogge K., Heinzle E., Wittmann, C. (2004) *J. Bacteriol.*, *186*, 1769–1784.

Kunin V, Copeland A, Lapidus A, Konstantinos Mavromatis,and Hugenholtz Philip (2008a) MMBR , 72(4): 557–578

Kunin, V., S. He, F. Warnecke, S. B. Peterson, H. Garcia Martin, M. Haynes, N. Ivanova, L. L. Blackall, M. Breitbart, F. Rohwer, K. D. McMahon, and P. Hugenholtz. (2008b). Genome Res. 18: 293-297

Kunin, V., J. Raes, J. K. Harris, J. R. Spear, J. J. Walker, N. Ivanova, C. von Mering, B. M. Bebout, N. R. Pace, P. Bork, and P. Hugenholtz. (2008c). Mol. Syst. Biol. 4:198.

Kysela DT, Palacios C, Sogin ML. (2005) Environ Microbiol;7:356–64.

Lane, D. J., B. Pace, G. J. Olsen, D. A. Stahl, M. L. Sogin, and N. R. Pace. 1985. Proc. Natl. Acad. Sci. USA 82:6955–6959.

Langer M, Gabor EM, Liebeton K, Meurer G, Niehaus F, Schulze R, *et al.* (2006). Biotechnol J;1:815–21.

Lerat S, England LS, Vincent ML, Pauls KP, Swanton CJ, Klironomos JN, *et al.* (2005). J Agric Food Chem;53:1337–42.

Lebaron, P. *et al.* (1997) Appl. Environ. Microbiol. 63, 3274–3278

Liu W-T and Zhu L (2005) Environmental microbiology-on-a-chip and its future impacts TRENDS in Biotechnology Vol.23 No.4 April174-179

Lorenz P, Eck J. (2005) Nat Rev Microbiol; 3:510–6

Lovely, D.R. (2003) Nat. Rev. Microbiol. 1, 35–44

Macbeth TW, Cummings DE, Spring S, Petzke LM, Sorenson Jr KS. (2004) Appl Environ Microbiol; 70:7329–41.

Malik Seidu Beer, Michael, Megharaj M Naidu R (2008) Environment International 34 265–276

Marsh TL, Saxman P, Cole J, Tiedje J. (2000) Appl Environ Microbiol; 66:3616–20.

Mavromatis, K., N. Ivanova, K. Barry, H. Shapiro, E. Goltsman, A.C McHardy, I. Rigoutsos, A. Salamov, F. Korzeniewski, M. Land, A. Lapidus, I. Grigoriev, P. Richardson, P. Hugenholtz, and N.C. Kyrpides. (2007). Nat. Methods 4:495–500.

Nees M, Woodworth CD. (2001) Curr Cancer Drug Targets; 1:155–75.

Neufeld JD, Yu Z, LamW, MohnWW. (2004) Environ Microbiol 6:131–44.

Nocker A, Burn M, Camper AK. (2007) Microb Ecol; 54:276–89.

Nulton, R. Olson, R. Parsons, S. Rayhawk, C. A. Suttle and F. Rohwer.(2006). PLoS Biol. 4:e368.

Overbeek, R., T. Begley, R. M. Butler, J. V. Choudhuri, H. Y. Chuang, M. Cohoon, V. de Crecy-Lagard, N. Diaz, T. Disz, R. Edwards, M. Fonstein *et al.* (2005). Nucleic Acids Res. 33:5691–5702.

Pace, N.R. D.A. Stahl, D J Lane, and G.J. Olsen. (1985). ASM News 51:4–12.

Paul Debarati, Pandey,G Pandey J and. Jain RK. (2005) Trends in Biotechnology Vol. 23 No.3

Powell SM, Ferguson SH, Bowman JP, Snape I. (2006) Microb. Ecol. 52:523–32.

Polz MF, Cavanaugh CM. (1998) Appl Environ Microbiol; 64:3724–30.

Raes J, Foerstner KU, Bork P (2007) Curr Opin Microbiol. Oct; 10(5):490-8.

Rappe MS, Giovannoni SJ. (2003) Annu Rev Microbiol, 57:369-394

Reardon CL, Cummings DE, Petzke LM, Kinsall BL, Watson DB, Peyton BM, *et al.* (2004) Appl Environ Microbiol;70:6037–46.

Rhee SK, Liu X,Wu L, Chong SC,Wan X, Zhou J. (2004) Appl Environ Microbiol;70:4303–17.

Ritalahti KM, Amos BK, Sung Y, Wu Q, Koenigsberg SS, Loffler FE. (2006) Appl Environ Microbiol; 72:2765–74.

Robertson LA, Steer BA. (2004) J Microbiol Methods; 63:1–19

Saleh-Lakha S, Miller M, Campbell RG, Schneider K, Elahimanesh P, Hart MM, *et al.* (2005). J Microbiol Methods;63:1–19.

Sanz JL, Kochling T (2007). Process Biochem. 42:119–33.

Schneiker S, dos Santos VAPM, Bartels D, Bekel T, Brecht M, Buhrmester J, *et al.* (2006). Nat Biotechnol;24:997–1004. dramatic increase in organisms sequenced

Scow KM, Hicks KA. (2005) Curr Opin Biotechnol; 16:246–53.

Sebat JL, Colwell FS, Crawford RL. (2003) Appl Environ Microbiol; 69:4927–34

Seshadri, R., S. A. Kravitz, L. Smarr, P. Gilna, and M. Frazier. (2007). PLoS Biol. 5:e75.

Shendure J, Ji H. (2008) Nat Biotechnol, 26:1135-114

Singh Shailendra, Seung Hyun Kang, Ashok Mulchandani and Wilfred Chen (2008) Current Opinion in Biotechnology, 19:437–444

Spiegelman D,Whissell G, GreerCW. (2005) Can J Microbiol; 51:355–86

Stenuit B, Eyers L, El Fantroussi S, Agathos SN. (2005) Rev Environ Sci Biotechnol; 4:39–60.

Stenuit B, Eyers L, Rozenberg R, Habib-Jiwan JL, Agathos SN. (2006) Appl Environ Microbiol; 72:7945–8.

Stenuit B, Eyers L, Schuler L, Agathos S N., George I (2008) Biotechnology Advances 26 (2008) 561–575

Stepanauskas R, Sieracki ME: (2007 Proc Natl Acad Sci U S A, 104:9052-9057

Tettelin H, Riley D, Cattuto C, Medini D (2008). Curr Opin Microbiol. Oct; 11(5): 472-7.

Tiedje JM, Asuming-Brempong S, Nusslein K, Marsh TL, Flyn SJ (1999). Appl Soil Ecol; 13:109–22.

Tringe, S. G. and E. M. Rubin. (2005). Nat. Rev. Genet. 6:805–814.

Tringe, S.G., C. von Mering, A. Kobayashi, A.A. Salamov, K. Chen, H. W. Chang, M. Podar, J. M. Short, E. J. Mathur, J. C. Detter, P. Bork, P. Hugenholtz, and E. M. Rubin. (2005). Science 308:554–557.

Torsvik, V. and Ovreas, L. (2002) Curr. Opin. Microbiol. 5, 240–245

Tyson GW, Chapman J, Hugenholtz P, Allen EE, Ram RJ, Richardson PM, Solovyev VV, Rubin EM, Rokhsar DS, Banfield JF: (2004) Nature 428:37-43.

Timmis KN & Pieper DH (1999) Trends Biotechnol 17: 200–204.

Urich, T., A. Lanzen, J. Qi, D. H. Huson, C. Schleper, and S.C. Schuster. (2008). PLoS ONE 3:e2527.

Van der Lelie D, Lesaulnier C, McCorkle S, Geets J, Taghavi S, Dunn J. (2006) Appl Environ Microbiol;72:2092–101.

Van Hamme JD, Singh A, Ward OP. (2003) Microbiol Mol Biol Rev; 67:503–49.

Valenzuela Lissette, Chi An Beard, S, Orell A, Guiliani N, Shabanowitz J,. Hunt Donald F,. Jerez C A, (2006) Biotechnology Advances 24 197– 211

Venter, J.C., K. Remington, J.F. Heidelberg, A.L. Halpern, D. Rusch, J.A. Eisen, D. Wu, I. Paulsen, K.E. Nelson, W. Nelson, D.E. Fouts *et al.* (2004 Science 304:66–74.

Wagner M, Nielsen PH, Loy A, Nielsen JL, Daims H. (2006) Curr Opin Biotechnol; 17:83–91.

Warnecke, F., P. Luginbuhl, N. Ivanova, M. Ghassemian, T. H. Richardson, J. T. Stege, M. Cayouette, A. C. McHardy *et al.* (2007). Nature 450:560–565.

Watanabe, K. *et al.* (2002) Understanding the diversity in catabolic potential of microorganisms for the development of bioremediation

Watanabe K, Hamamura N (2003). Curr Opin Biotechnol; 14:289–95.

Whitaker, R. J., and J. F. Banfield. (2006). Trends Ecol. Evol. 21:508-516

Whitfield, P. D., German, A. J., Noble, P (2004). J., *Br. J. Nutr.*, *92*, 549–555.

Widada J, Nojiri H, Omori T. (2002) Appl Microbiol Biotechnol; 60:45–59.

Williamson LL, Borlee BR, Schloss PD, Guan C, Allen HK, Handelsman J. (2005) Appl Environ Microbiol; 71:6335–44.

Wilson VL, Tatford BC, Yin X, Rajki SC, Walsh MM, LaRock P. (1999) J Microbiol Methods; 39:59–78.

Woese, C. R. 1987. Microbiol. Rev. 51: 221–271.

Wommack KE, Bhavsar J, Ravel J. (2008) Appl Environ Microbiol; 74:1453

Wu L, Thompson DK, Li G, Hurt RA, Tiedje JM, Zhou J. (2001) Appl Environ Microbiol; 67:5780–90.

Wu L, Thompson DK, Liu X, Fields MW, Bagwell CE, Tiedje JM, *et al.*(2004) Environ Sci Technol;38:6775–82.

Wu L, Liu X, Schadt CW, Zhou J. (2006) Appl Environ Microbiol; 72:4931–41

Yu Z, Yu M, Morrison M. (2006) Environ Microbiol; 8:603–11.

Yun J, Ryu S. (2005) Microb Cell Fact; 4:8.

Zhao B and Poh C L (2008) *8*, 874–881

Zhou, J. (2003) Curr. Opin. Microbiol. 6, 288–294.

Chapter - 6

Microbial enzyme technology for abatement of environmental pollution in leather industry

R. Puvanakrishnan

Department of Biotechnology, Central Leather Research Institute, Chennai – 600 020
E-mail: puvanakrishnan@yahoo.com

ABSTRACT

One of the environmentally constrained operations in pretanning processes is dehairing wherein saturated lime liquors along with destructive sharpeners such as sodium sulphide are used. This generates toxic effluents, sludge forming chemicals and solid waste and they contribute high pollution load in terms of BOD, COD and total solids. Also, usage of these chemicals poses a serious threat to environment and causes health hazards. Rationalization of dehairing process by systematic use of enzymes in place of lime and sulphide is an urgent necessity. This article portrays the adaptation of latest fermentation techniques in the manufacture of enzyme formulations. In addition, it is an attempt to create awareness about the application of microbial enzyme technology for abatement of environmental pollution in leather industry.

Key words: Leather industry, environmental pollution, microbial enzyme

1. Introduction

It is well known that environmental pollution has been a major irritant to industrial development. Chemical and chemical based industries are the prime targets of environmentalists for their crusade against pollution and leather industry is not left out of the reckoning. The chemicals mainly responsible for pollution in pretanning processes are lime, sulphide and caustic soda apart from common salt and degreasing chemicals (Ahlstrom *et al.*, 2006). In fact, one third of the pollution caused by leather industries results from the wastes generated during dehairing operations. The wastes are let out into the drains which in turn empty into the main sewerage causing hazard to those who use this water. Many tanneries have been forced to close down because of their noncompliance with the standards laid down. In a short span of time, Indian Leather Industry has faced serious challenges such as German ban on pentachlorophenate, certain azo dyes, formaldehyde etc. on one hand and court order to compliance with environmental regulations on the other. The attention of tanners is focused towards revamping the processing methods, recovery systems and effluent treatment techniques to make leather processing ecofriendly. Intensive efforts have been taken up to use a viable alternative technology viz. microbial enzyme technology in pretanning processes and this could be one of the ways of solving the industrial pollution problems caused by tannery effluents.

2. Enzymatic processing - Ecofriendly

The raw skin goes through a series of chemical treatments before it turns into flattering leather. This includes soaking, liming, dehairing, deliming, degreasing and pickling. During all these steps, the chemicals used are quite toxic. Naturally, these pretanning operations make leather processing one of the worst offenders of the environment (Heidemann, 1993).

Enzyme is a biological catalyst with clear cut specificity. An important enzyme used in pretanning processes belongs to the group of proteolytic enzymes, also known as proteases which act on proteins. Although enzymes from plant, animal and microbial sources have been used, large scale use of microbial enzymes received a boost only in 1960s following the introduction of fermentation technology. The enzymes or enzymatic formulations need not be pure but must be cheap as compared to that of commercial chemicals used in leather industry (Choudhary *et al.*, 2004; Chakraborthy and Chakraborthy, 2007). Animal proteases and microbial proteases from bacteria and fungi are used in the pretanning processes of leather manufacture. The animal proteases are mixtures of trypsin, chymotrypsin and various peptidases which may contain amylase or lipase as secondary enzymes. Mainly for economic reasons, enzymes from microorganisms have come to play a significant role in recent years and enzymatic products of microbial origin are already being produced on a wide scale.

3. Application of enzymes in leather industry

For selection of microbes secreting potent proteases/lipases/tannases, conventional methods of collection can be employed. In addition, soils from forest regions, rich in biodiversity or marine soils could be collected to yield microbes with potent enzymes. To isolate microbes secreting potent proteases, the soil samples are mixed with glucose and skim milk, each of 1%, in flasks and further subjected to enrichment for three to seven days. The enriched soil samples are subjected to serial dilution to get potent microbes. Once, the selection of microbes is completed, the next stage is the selection of medium in which the microbes could be grown. Then the production of enzymes from an optimized medium is standardized.

The important stages in which microbial enzymes are used in pretanning processes are: soaking, dehairing, bating and degreasing. Soaking is the first operation in the tannery in which the skins are treated with water for making them clean and soft. Wet salted or freshly slaughtered skins do not require any chemical agent for their proper soaking. Soaking is necessary for solubilization and elimination of salts and globular proteins contained within the fibrous structure of skins (Taylor *et al.*, 1987). Soaking is carried out under alkaline conditions at low temperature in water containing sodium hypochlorite, sodium pentachlorophenate, formic acid etc. and it is accelerated by some of the non-ionic detergents and additives such as sodium sulfide or sodium tetrasulfide.

The advantages of enzymatic soaking include loosening of the scud, initiation of the opening up of the fibre structure, production of leather with less wrinkled grain and a decrease in soaking time. Enzymes from *Aspergillus parasiticus, Aspergillus flavus*, *Aspergillus oryzae* and *Bacillus subtilis* are used alone or in mixtures. Soaking is usually performed with a combination of proteolytic enzymes that are optimally active in the neutral or alkaline pH range (Kamini *et al.*, 1999).

3.1 Enzymes for dehairing

The most important operation in which enzyme is used in leather processing is dehairing. Five methods of dehairing are generally adopted viz. (i) Clipping process (ii) Scalding process (iii) Chemical process (iv) Sweating process and (v) enzymatic process (Millimet, 2003). The conventionally practiced method of dehairing is the chemical process using lime and sodium sulfide. However, the use of high concentration of lime and sodium sulphide creates an extremely alkaline environment resulting in the pulping of hair and its subsequent removal. While the efficacy of this process cannot be questioned, it has some major disadvantages. About 75% of the organic waste from a tannery comes from the pretanning yard and 70% of this waste is from hair which is rich in nitrogen. This illustrates the contribution made by the lime and sulphide process towards pollution (Jatavathu Madhavi *et al.*, 2011). Sulphide is highly toxic and has an obnoxious odour and if left untreated, it can cause major problems in the sewers. The severe alkaline condition is a health hazard for the workers.

What are the advantages of enzymatic dehairing? Enzymatic dehairing is suggested as an environmentally friendly alternative to the conventional chemical process. Hair of good quality and strength with a good saleable value is recovered. An ecologically conducive atmosphere for the workers is created. A significant nature of the enzymatic dehairing process is the time factor involved. The lime-sulfide process takes about 16h whereas the enzymatic dehairing would be completed between 8h and 20h.

Proteolytic enzymes are of great commercial importance contributing to more than 40% of the world's commercially produced enzymes. Approximately 50% of the enzymes used as industrial process aids are proteolytic enzymes. Proteolytic enzymes are more efficient in enzymatic dehairing than amylolytic enzymes. Proteolytic enzymes from plant, animal and microbial sources are used for dehairing. Microbial proteases, because of their inherent advantages, are preferred in modern practice.

Three methods of application are commonly used in the enzymatic dehairing process : (a) Paint method (b) dip method and (c) drum method. In the paint method, the enzyme solution is mixed with an inert material like kaolin, made into a thin paste, adjusted to the required pH and applied on the flesh side of skins. In the dip method of enzymatic dehairing, the skins are kept immersed in the enzyme solution at the required pH in a pit or tub. In addition, drum method is also followed.

3.2 Enzymes for bating

Another application of Microbial enzyme technology in pretanning process is known as bating. The concept of softening skins by treating them in a warm infusion of animal dung has been termed as "bating" and the product used for such process is known as a bate. The main object of bating is to remove some of the non leather forming proteinous materials such as albumins, globulins and mucoids from skins and to allow splitting of collagen fibres (Saravanabhavan *et al.,* 2003).

The principal materials which a bate contains are proteolytic enzymes, a suitable carrier for the enzyme and a deliming agent such as ammonium chloride or sulphate. Pancreatic enzymes are found to be the best for use in bating. Alternatively, microbial enzymes could be used for bating.

3.3 Enzymes for degreasing

Another major operation where enzyme is used is known as degreasing and it is the process of removal of excess natural fat from greasy skins. The presence of natural grease in certain skins, especially woolly sheep skins, results in various defects. During the degreasing operation in the pretanning process, the fat or grease is removed from the interfibrillary spaces of the skins to facilitate the even penetration of tanning materials, fat liquors etc.

Degreasing is carried out after pickling and conventionally, it is carried out by either aqueous emulsification using detergents or by solvent extraction. It is well known that organic solvents like kerosene, petrol, perchlorethylene and trichloroethylene are highly unsafe and hazardous to the workers and heavily pollute the environment. The detergents, though not hazardous, while handling and storing, cause serious pollution problems. These detergents and solvents add to the BOD (Biochemical Oxygen Demand) load of the pickling effluent and the chlorinated hydrocarbons and solvents add to the toxicity of the effluent.

Enzymatic degreasing is suggested as a viable alternative to combat the pollution problems caused by the use of solvents and detergents. Lipases catalyse the breakdown of fats and can be obtained from animal, microbial and plant sources (Thanikaivelan *et al.,* 2002, 2007). The advantages of using enzymes for degreasing are the elimination of solvents, reduction in surfactants and possible recovery of valuable byproducts.

Microbial enzymes have a role in byproducts utilization. They could be used in the treatment of fleshings from the tannery. A combination of hydrolytic enzymes viz. proteases, carbohydrases and lipases might be ideal for this use (Thanikaivelan *et al.,* 2009).

4. Conclusion

Tanneries in future will use a combination of enzymatic and chemical processes. The potential for use of microbial enzyme technology in leather processing lies mainly in areas in which pollution causing chemicals such as sodium sulfide, lime and solvents are being used. Future might witness ecolabelled leather products emerging as niche products by the use of microbial enzyme technology and the experience gained by the Indian Leather Industry in this area might greatly help to emerge as a global leader.

References

Ahlstorm L, Mathiasson L, Bharath CK, Rao JR, Nair BU (2006) JALCA 101: 388-398.

Chakraborty D, Chakraborty P (2007) Economic and Political Weekly 42(19): 1669-71.

Choudhary RB, Jana AK, Jha MK (2004) Indian Journal of Chemical Technology, 11: 659-671.

Heidemann E (1993) Fundamentals of Leather Manufacture. Eduard Roether KG. p. 21.

Jatavathu Madhavi, Jatavathu Srilakshmi, MV Raghavendra Rao and KRS Sambasiva Rao (2011) Int J Bioscience Biotechnol 3(4): 11-26.

Kamini NR, Hemachander C, Geraldine Sandana Mala J, Puvanakrishnan R (1999) Curr. Sci. 7: 80-86.

Millimet D (2003) Contemporary Economic Policy 21(3): 281-296.

Saravanabhavan S, Aravindan R, Thanikaivelan P, Rao JR, Nair BU (2003) Green Chemistry 707-714.

Tayler MM, Bailey DG, Feairheller SH (1987) J Am Leath Chem Assoc 82: 153-165.

Thanikaivelan P, Bharath CK, Saravanabhavan S, Rao JR, Chandrababu NK (2007) Clean Technology Environmental Policy 61-68.

Thanikaivelan P, Rao JR, Nair BU, Ramasami T (2002) Envirronmental Science Technology 4187-4194.

Thanikaivelan, P, Rao JR, Nair BU, Ramasami T (2009) Curr Sci. 96: 11.

Chapter - 7

Conversion of Lignocellulose into Biofuel

Rajasekhar Reddy B.*, Praveen Kumar K. and Subba Reddy, G.V.

Department of Microbiology, Sri Krishnadevaraya University,
Anantapur – 515 055
Corresponding author: rajasekhar_bontha@yahoo.co.in

ABSTRACT

In view of fast depletion of fossil fuels, alternative and renewable energy resources need to be explored. Lignocellulose has a greater potential to serve as a renewable source for production of fuels because of its continuous synthesis through photosynthetic process. Lignocellulosic biomass occurs abundantly in municipal solid wastes, plant litter, residues of agro-forest and food industries. Three components- cellulose, hemicellulose and lignin constitute lignocellulosic mass. Cellulose and hemicellulose are made with C_6 sugar and C_5 sugar, respectively. The first two components need to be separated from lignin in lignocellulose mass by chemical/biological methods. The next stage involves release of C_6 sugar and C_5 sugars from two components of lignocellulosic mass in saccharification process by enzymatic methods. Fermentation of released sugars generates ethanol production. The current status of the art of hydrolysis and fermentation technologies to produce ethanol from lignocellulose will be presented and stumbling blocks in commerciali-zation of the technology will be discussed.

Key words: Bioethanol, lignocellulose, pretreatment cellulose, fermentation

1. Introduction

In 1925, Henry Ford had quoted ethylalcohol, ethanol, as "the fuel of the future". He furthermore stated, "The fuel of the future is going to come from apples, weeds, sawdust, almost anything. There is fuel in every bit of vegetable matter that can be fermented".

Energy consumption has increased steadily over the last century as the world population has grown and more countries have become industrialized. Crude oil has been the major resource to meet the increased energy demand particular, in transport sector and this production will begin to decline before 2010 [Campbell, 1998]. Therefore, there is a great interest in development of fuels generated from renewable sources, that is to say biofuels. The term biofuel is attributed to any alternative fuel that derives from organic material, such as energy crops (wheat, sugarcane, corn, sugar beet, cassava), crop residues (e.g. rice straw, cornstover, corn cobs, rise husk) or waste biomass (food waste, live stock waste, paper waste, construction-derived wood residues). Of all the biofuels, ethanol has been trusted as an alternative fuel for the future and increasing its importance as a consequence of heightened concern for the green house effects, depleting oil reserves, and rising oil prices [Ohgren, 2007]. Fuel ethanol is mainly used as an oxygenated fuel additive. The higher octane number of the fuel mixture, when it contains ethanol, reduces the need for toxic, octane-enhancing additives such as methyl tertiary butyl ether. Due to the oxygen in ethanol molecules, there is also a reduction of carbonmonoxide emission and non-combusted hydrocarbons. It is believed that ethanol is 15% more efficient than gasoline in optimized spark- ignition engines, while it has about the same overall transport efficiency as diesel in compression-ignition engines. It is also believed that a given volume of ethanol could provide energy enough to drive about 75-80% of the distance as the same amount of gasoline, although it has the only about two-thirds of the energy content [Galbe, 2002]. This can be used as neat or blended with gasoline.

According to the Renewable Fuels Association (2003) the US annual fuel ethanol capacity was 2.9 x 10^9 US gallons in 2002, an increase of 10^9 US gallons over the production level in 2000. The bulk of the production is located in Brazil (16 billion liters produced in 2005) and the USA (10.6 billion liters in 2003) [Hamelinck, 2005]. Ethanol in both countries is mainly produced from feed stocks-corn grains and sugar cane respectively. Starch of corn grain and sucrose of sugar cane in feedstocks are filled with C_6 sugars glucose and fructose which can be conventionally fermented to ethanol by yeast or other microorganisms. In this sense, bioethanol is expected to be one of the dominant renewable biofuels in the transport sector with in the coming 20 years [Hagerdal, 2006]. However, the cost of ethanol as an energy source is relatively high compared to fossil fuels. Currently the United States, Brazil and China are in the top of countries that produce the largest quantities of fuel ethanol from food grains [Hu, 2008]. If the oil crisis continues to develop, ethanol is one of the most promising biofuels that can be used to replace gasoline for tomorrow's transportation vehicles. Until recently, the high cost of ethanol production has been a factor retarding commercial use of ethanol and requiring subsidies to promote development. Reduction of the production cost would lead to a faster commercialization of economically feasible process and would increase the competitiveness of ethanol with fossil fuels. The raw materials account for 40~70% of the total ethanol production based on sucrose or starch containing feedstocks.

Diversion of food grains for the production of ethanol using the cornstarch-based technology may not be possible in all regions of the world, especially in developing countries because of heavy pressure on food grains for feeding purposes. Alternative potential source for low cost ethanol production is to utilize non-feedstocks such as the lignocellulosic materials such as crop residues, grasses, sawdust, wood chips and solid animal waste. This is projected to replace about 30% of the petroleum currently consumed by the USA in the future.

Lignocellulosic materials continue to be investigated as a source of fermentable sugars for biofuel (cellulosic ethanol) production because of their high availability. The main lignocellulosic sources targeted to date include agricultural and forestry wastes domestic waste, short rotation herbaceous crops and wastes from conventional ethanol production crops (e.g. sugar cane bagasse, corn cobs and corn stover).

Extensive research carried out during the last two decades focused on conversion of lignocellulosic materials to ethanol [Sun, 2002]. The conversion includes two processes: release of sugars from cellulose and hemi cellulose components and fermentation of the sugars to ethanol. Current status of the art of preparation of ethanol from lignocellulosic material and stumbling block for commercialization has been presented in this report.

2. Chemical composition

Woody materials, including wood, bark and mixtures of forest residues are composed of cellulose, hemicellulose, lignin and small amounts of other species [Hu, 2008]. Cellulose is composed of anhydro-glucopyranose or glucose residue, which can be converted to glucose and provides the major source for hexose in woody biomass. Due to its crystalline nature it is hard to hydrolyze. Hemicellulose is composed of both six-carbon sugars and five carbon sugars, which include glucose, mannose, arabinose, xylose and other species. Xylose is present in large amounts. Unlike cellulose, hemicellulose as a random and amorphous structure, which makes it easily to be hydrolyzed by dilute acid or base. Lignin is the third major component in wood and comprises the glue that protects woody biomass from foreign invasion. It is mainly composed of phenolic units and represents the part of biopolymer that cannot be converted into ethanol directly or indirectly. It covers cellulose as a sheath and protects cellwall of plant from enzymatic attack by micro-organisms during infection. With the exclusion of lignin, the optimistic situation is to preserve and utilize all the carbohydrates, C_6 sugars and C_5 sugars in lignocellulose and convert them into ethanol fuel as shown in the Fig. 1.

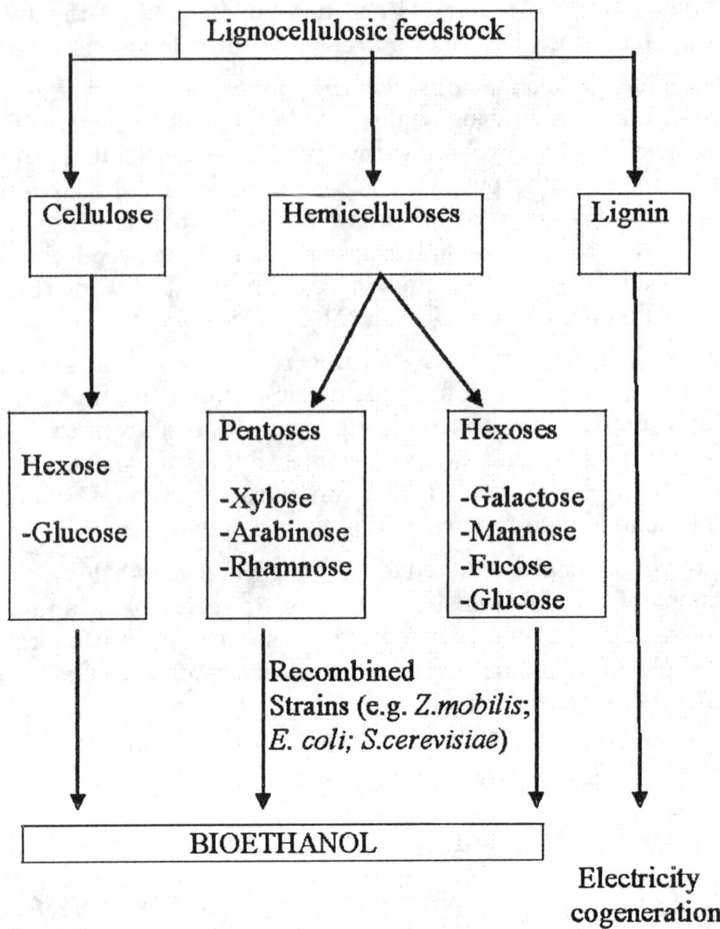

Fig. 1 Major sugars involved in the conversion of lignocellulosic Feedstock into ethanol

3. Pretreatment

The choice of pretreatment process is possibly the most important factor in the economics of the ethanol production process because it influences waste treatment, cellulose conversion rates and hydrolysis of hemicellulose sugars and their subsequent fermentation [Hu, 2008]. It heavily influences cost and performance in subsequent hydrolysis and fermentation.

The purpose of the pretreatment is to loosen the structure, to remove lignin and hemicellulose, reduce cellulose crystallinity, and increase the porosity of the materials. The pretreatment must meet the following requirements: (1) improve the formation of sugars or the ability to subsequently form sugars by enzymatic hydrolysis (2) avoid the degradation or loss of carbohydrate (3) avoid formation of by-products and (4) be cost-effective [Mosier, 2005].

In general, pretreatment can be classified into 3 categories- physical, chemical and biological.

4. Physical treatment

Uncatalysed processes generally use steam explosion or liquid hot water. Steam explosion is one of the most promising methods to make biomass more accessible to cellulase attack [Szengyel, 2000] in subsequent steps to be followed.

In steam explosion method (auto-hydrolysis) chipped biomass is treated with high pressure saturated steam and then the pressure is swiftly reduced, which makes the materials undergo an explosive decompression. It is typically initiated at a temperature of 160-260°C (corresponding pressure of 0.69-4.83MPa) for several seconds to a few minutes before the material is exposed to atmospheric pressure. The process causes hemicellulose degradation and lignin transformation due to high temperature, thus increasing the potential of cellulose hydrolysis. Steam-explosion pretreatment is economically attractive treatment.

The liquid hot water method uses compressed hot liquid water (at pressure above saturation point) to hydrolyze the hemicellulose. Xylose recovery is relatively high (88-98%), and no acid or chemical catalyst is required, which makes it environmentally attractive and economically interesting.

5. Chemical pretreatment

Common chemical pretreatment methods comprise dilute acids, alkalies, ammonia, organic solvent, sulfur dioxide, carbondioxide or other chemicals. Acids or bases promote hydrolysis and improve sugar recovery yields from cellulose by removing hemicellulose or lignin during pretreatment. Some bases can also be used for pretreatment of lignocellulosic materials and the effect of alkaline pretreatment depends on the lignin content of the materials [McMillan, 1994]. The mechanism of alkaline hydrolysis is believed to be saponification of intermolecular ester bond crosslinking xylan hemicelluloses and other components. The porosity of the lignocellulosic materials increases with the removal of crosslinks. Alkaline pretreatment uses bases like sodium hydroxide or calcium hydroxide under low temperature and pressure. Alkaline pretreatment is similar to kraft pulping, in which lignin is removed, thus improving the reactivity of the remaining polysaccharides. Meanwhile, acetyl groups and the various uronic substitutions on hemicellulose are also removed so that there is less hindrance for enzymes to access the surface of hemicellulose and cellulose. Sulfuric acid and sodium hydroxide are the most commonly used acid and base respectively [Moiser, 2005]. Lime pretreatment and ammonia pretreatment have seemed to be most attractive alkaline pretreatment, while most attention in acid pretreatment has been concentrated on the use of sulfur dioxide and sulfuric acid. There are two types of acid hydrolysis process commonly used-dilute and concentrated acid hydrolysis. Dilute acid hydrolysis has been successfully developed in pretreatment of lignocellulosic materials. It is employed for the degradation of hemicellulose leaving lignin and cellulose network in the substrate. The dilute sulfuric acid pretreatment can achieve high reaction rates and significantly improve cellulose hydrolysis. Concentrated acids such as sulfuric acid and HCl are used to treat lignocellulosic materials but these are toxic, corrosive and hazardous and require reactors that are resistant to corrosion. In addition, the conc. Acid must be recovered after hydrolysis to make the process economically feasible [Silvers, 1995]. Currently, most industrial processes are based on dilute acid hydrolysis. These methods are considered potentially cost effective and could be used on a large scale in the future [Moiser, 2005].

6. Biological pretreatment

In biological pretreatment processes, microorganisms such as brown, white and soft- rot fungi are used to degrade lignin and hemicellulose in waste materials [Ohgren, 2007]. Brown rots mainly attack cellulose, while white and soft rots attack both cellulose and lignin. White rot fungi are the cost effective basidiomycetes for biological pretreatment of lignocellulosic materials. The advantages of biological pretreatment include low energy requirement and mild environmental conditions. However, the rate of hydrolysis in most biological pretreatment processes is very low. Currently available pretreatment techniques have provided a broad range of choices that improve the conversion of carbohydrates from lignocellulosic biomass to ethanol. Some biological treatments are some times used in combination with chemical treatments [Graf, 2000].

7. Enzymatic hydrolysis

One of the most extensively investigated processes is the enzymatic hydrolysis, in which fungal cellulolytic enzymes are used to convert the pretreated cellulosic biomass in to glucose, which is then fermented to ethanol [Szengyel, 2000; Varga, 2002]. Lignocellulosics pretreated with one of the above method can be saccharified at loading rate of cellulase 10-20 FPU/g of cellulose to get fermentable sugars [Hagerdal *et al.* 2006; Itoh, 2003; Tucker, 2003]. Though bacteria and fungi can produce cellulases for the hydrolysis of lignocellulosic materials so most of research for commercial cellulase production has focused on fungi because of high enzyme titers. During the enzymatic hydrolysis, cellulose is degraded by the cellulases to reducing sugars that can be fermented by yeasts or bacteria to ethanol. Advantages of enzyme hydrolysis are high yields of sugar production under mild conditions with relatively low amounts of catalyst. Enzymes are biodegradable and thus environment friendly [Wyman, 1993]. Nonetheless the cost associated to enzymatic production is relatively high, unlike chemical or physical methods. In recent years efforts are made to cut down the cost of production due to the development of new engineered enzyme systems [Ehara, 2002]. On the other hand, a common disadvantage of the enzymatic hydrolysis is the end-product inhibition of the enzymes used to hydrolyze the cellulose and the remaining hemicellulose. This problem was reportedly minimized by performing the saccharification and fermentation processes simultaneously [Ohgren, 2006]. Different treatment methods are used in combination to improve yields of sugar production. All the above steps are carried out in different configurations.

8. Simultaneous saccharification and fermentation (SSF)

The most important process improvement made for the enzymatic hydrolysis of biomass is the introduction of SSF. This approach combines both action of the cellulase enzymes and fermentation of released sugars by microbes in one vessel and enables one-step process of sugar production on lignocellulose and fermentation into ethanol. Simultaneous saccharification of both carbon polymers- cellulose to glucose and hemicellulose to xylose and arabinose and fermentation of both C_6 and C_5 sugars in sequentially will be carried out by recombinant yeast or the organism which has the ability to utilize both C_5 and C_6 sugars [Alkasrami, 2006]. SSF gives higher ethanol yield while requiring lower amounts of enzyme because end-product inhibition from cellobiose and glucose formed during enzymatic hydrolysis is relieved by the yeast fermentation [Banat, 1998]

During pretreatment, degradation products of C_5 and C_6 sugars, primarily furfural and hydroxyl methyl furfural are formed. These components are toxic or inhibitory to the enzymatic hydrolysis and fermenting organisms and must be removed or neutralized prior to fermentation. Otherwise large amounts of fermenting microorganisms should be applied in fermentation

Both C_6 sugars and C_5 sugars are released in sugar syrup. C_6 sugars are subjected to conventional and age old fermentation to ethanol by yeast. Fermentations of C_5 sugars and C_5 sugars to ethanol can be carried simultaneously (co-fermentation) in a single and sequentially into separate reactions. The ability to ferment pentoses along with hexoses is not widespread among microorganisms [Agbogbo, 2008; Chandel, 2007]. The most promising yeasts that have the ability to use both C_5 and C_6 sugars- *Pichia stipitis, Candida shehatae* and *Pachysolan tannophilus* can be used for fermentation. However, ethanol production from sugars derived from starch and sucrose has been commercially dominated by the yeast *S. cerevisea* [Lin, 2006]. Thermotolerant yeast could be more suitable for ethanol production at industrial level. In high temperature process energy savings can be achieved through a reduction in cooling costs. Bacteria have drawn special attention from researchers because of their speed of fermentation. In general, bacteria can ferment in minutes as compared to hours for yeast. All microorganisms have limitations: either in the ability to process pentoses and hexoses, the low yields of ethanol, or the co-production of cell mass at the cost of ethanol. Furthermore, the oxygen free condition of fermentation slowly exterminates the microorganism population. Therefore, in early processes, the different sugars were fermented in different sequential reactors, and tendency towards combining reaction steps. Genetically modified organisms to make their own enzymes while also fermenting the sugars they release to ethanol with high yields is one promising approach. Also, genetic engineering and new screening technologies will bring bacteria and yeast that are capable of fermenting all five of the major biomass sugars- glucose, xylose, mannose, galactose and arabinose [Ohgren, 2006]. Mid-to long term technology will improve the fermentation efficiency of the organism (yielding more ethanol in less time), as well as its resistance, requiring less detoxification of the hydrolysate [Graf, 2000; Wooley, 1999]. All different stages in conversion of lignocelluloses to ethanol are integrated in a single reactor, consolidated bioreactor processing using genetically tailormade organisms with a purpose of minimising cost of production.

Though new technologies have greatly improved bioethanol production yet there are still a lot of problems that have to be solved for so. The major problems include maintaining a stable performance of genetically engineered yeast in commercial scale fermentation operation [Ho, 1998 and 1999], developing more efficient pretreatment technologies for lignocellulosic biomass and integrating optimal component into economic ethanol production system [Dien, 2000].

9. Economic analysis

In recent years, numerous studies aimed at quantifying the true cost of ethanol production have been undertaken by many international organizations. Bioethanol production costs are calculated by dividing the total annual costs of a system by the produced amount of fuel. According to estimates made by the National Renewable Energy Laboratory, in the US bioethanol production could cost anywhere from $0.30 to $0.38/L, depending on the technology utilized and availability of low cost feedstocks for conversion to ethanol [Wooley, 1999]. In the case of Brazil production of ethanol from sugarcane costs in a average $0.22/L from which about 60% is the raw material cost [Neves, 2007]. This could be possible because of advances made in pretreatment, enzymes and processes integration.

Brazil is one of the world leaders in bioethanol production and use as a biofuel from sugar cane by sugarcane. In the year 2005 the total ethyl alcohol production was c.a 16 billion liters. There are 243 bioethanol processing plants registered in this country and nearly 33,620 gasoline service stations authorize for sale of bioethanol. Brazilian ethanol exports have been growing significantly since 2002 and amounted to 2.6 billion liters in 2005. In the US most fuel ethanol is produced from corn by either dry-grind (67%) or wet-mill (33%) process. Since 1980 US ethanol production has risen from an average of 1 million L a day to 40 million L a day and a further doubling by 2012 has been called by federal mandates. Aiming to achieve the 28.5 billion L of renewable fuels mandated by the US government by 2012.

Concentrating efforts on the development of lignocellulose to ethanol technologies, the European Commission launched a 4-year integrated project so-called NILE (New Improvements for Lignocellulosic Ethanol), to design and investigate new solutions for an efficient and cost-effective conversion lignocellulose to bioethanol.

Of late, our country realizes importance of biofuel production to meet growing demand of energy in future. In this direction efforts are initiated by undertaking new millineum (CSIR) project on conversion of cellulose and hemicellulose into sugars and ethanol with an objective of setting up a plant for production of ethanol.

20. Conclusion

The production of ethanol from low cost and abundant cellulosic biomass provides a powerful and unique route to sustainable production of liquid transportation fuels that our society uses so extensively and offers tremendous, economic, strategic and environmental benefits.

Development of technology for effectively conducting agricultural and forestry lignocellulosic residues as well as energy crops to fermentable sugars is outstanding interest and one of the greatest challenges for society in the 21st century, envisaging improved environmental quality and sustainable energy production. For instance, a combined effect of higher hydrolysis fermentation efficiency, lower specific capital investments, increase of scale and cheaper biomass feedstocks costs could contribute with significant reduction on ethanol production costs.

References

Agbogbo FK and Coward-Kelly G (2008) Biotechnol Lett DOI 10.1007/s10529-008-9728-z.

Alkasrami A, Rudolf A, Lid 'en G and Zacchi G (2006) Enzyme Microb Technol 38: 279-286

Banat IM, Nigam P, Singh D, Marchant P and McHale AP (1998) World J Microbiol Biotechnol 14: 809-821

Campbell CJ and Laherrere JH (1998) Sci Am 3: 78-83

Chandel AK, Chan ES, Rudravaram R, Lakshmi Narasu M, Venkateswar Rao L and Ravindra P (2007) Biotechnology and Molecular Biology Review 2: pp.014-032

Dien BS, Nichols NN, O'Bryan PJ and Bothast RJ (2000) Appl Biochem Biotechnol 84/86: 181-196

Ehara K and Saka S (2002) Cellulose 9: 301-311

Galbe M and Zacchi G (2002) Applied Microbial Biotech 59: 618-628

Graf A and Koehler T (2000) Oregon Cellulose-Ethanol Study. Oregon Office of Energy. Salem OR USA. 30pp + appendices

Hagerdal BH, Galbe M, Grauslund, MFG, Liden G and Zacchi G (2006) Trends in Biotechnology 24: 549-556

Hamelinck CN, Hooijdonk GV, Faaij APC (2005) Biomass and Bioenergy 28: 384-410

Ho NWY, Chen Z and Brainard AP (1998) Appl Environ Microbiol 64: 1852-1859

Ho NWY, Chen Z, Brainard A and Sedlak M (1999) Adv Biochem Eng Biotechnol 65: 164-192

Hu G, John A, Heitmann and Orlando J Rojas (2008) BioResources 3: 270-294

Itoh H, Wada M, Honda Y, Kuwahara M and Watanabe T (2003) J Biotechnol 103: 273-280

Lin Tanaka (2006) Appl Microbiol Biotechnol 69: 627-642

McMillan JD (1994) Pretreatment of lignocellulosic biomass. In: Himmel ME, Baker JO and Overend RP (Eds). Enzymatic Conversion of Biomass for Fuels Production. American Chemical Society, Washington, DC. pp 292-324

Mosier N, Wyman C, Dale B, Elander R, Lee YY and Holtzapple M (2005) Bioresource Technology 96: 673-686

Neves MA, Kmura T, Shimizu N, Nakajima M (2007) Process Biotechnology and Molecular Biology 1-14

Ohgren K, Rudolf A, Galbe M and Zacchi G (2006) Biomass and Bioenergy 30: 863-869

Ohgren K, Vehmaanpera J, Siika-Aho M, Galbe M, Viikari L and Zacchi G (2007) Enzyme and Microbial Technology 40: 607-613

Silvers M and Zacchi G (1995) Bioresour Technol 51: 43-52

Sun Y and Cheng J (2002) Bioresource Technology 83: 1-11

Szengyel Z (2000) Ethanol from Wood – Cellulase Enzyme Production, PhD Thesis, Lund University/Chemical Engineering I, Lund, Sweden, 121pp

Tucker MP, Kim KH, Newman MM and Nguyen QA (2003) Appl Biochem Biotechnol 10: 105-108

Varga E, Szengyel and Reczey K (2002) Applied Biochemistry and Biotechnology 98: 73-88

Wooley R, Ruth M, Sheehan J, Ibsen K, Majdeski H and Galvez A (1999) Lignocellulosic biomass to ethanol-process design and economics utilizing co-current dilute acid prehydrolysis and enzymatic hydrolysis-Current and futuristic scenarios. Report No. TP-580-26517. National Reneawale Energy Laboratory. Golden Colorade USA 130pp

Wyman CE, Bain RL, Hinman ND and Stevens DJB (1993) Ethanol and methanol from cellulosic biomass. In: Johansson T, Kelly H, Reddy AKN, Williams RH, Burnham L (Eds) Renewable energy. Sources for Fuels and electricity, Island Press, Washington, pp 865-923.

Chapter - 8

Recent advances in clinical diagnostic microbiology

V. Lakshmi

Dept. of Microbiology, Nizam's Institute of Medical Sciences, Hyderabad
E-mail: lgorthi@hotmail.com

1. Introduction

The high morbidity and moratlity rates due to Infectious diseases (IDs) impose the greatest burden on health and economics of any nation. More than 95% of deaths in the developing countries are due to IDs, majority of which can be controlled with early diagnosis and appropriate therapy [David, 2004].

According to the World Health Organization (WHO), the need for accurate identification of infected patients requiring treatment, remains a major stumbling block to disease control, especially in developing countries. The burden of infectious disease in the developing world could be substantially reduced if appropriate diagnostic tests were more widely available [Rosanna, 2007]. WHO emphasises the use of rapid assays for an accurate diagnosis of symptomatic patients using cost effective methods for a definite reduction of the disease burden not only in developing nations but also globally [Rosanna, 2007].

2. Contributions of the pioneers of Clinical Micrbiology & lessons learnt [Cunha, 2004]

The history of clinical microbiology is a saga of the exploits of the great clinicians and microbiologists of the past who have worked on bacteria and their toxins, parasite lifecycles, fungi, rickettsia, Chlamydia, Mycoplasma, and protozoa.

Through the ancient times, the concept that IDs were caused by invisible living creatures had been known [Brier, 2004]. But the definitive knowledge of Microbes and the IDs caused by them emerged only in the

17th century (in 1676) when a Dutch merchant named Antoni van Leeuwenhoek observed living microbes in a drop of rainwater using a powerful magnifying lens. His discovery led to the development of the microscope, which, even to this day, is the essential and fundamental diagnostic tool of every clinical microbiologist [Elizabeth, 2009].

About a century later, Louis Pasteur reasoned that microbes might be responsible for human illnesses and confirmed the link between man & microbes through several innovative experiments. And he has been rightfully acknowledged as the Father of Microbiology [Cunha, 2004; Elizabeth, 2009].

By the 1880s, the German scientist, Robert Koch, set forth the principles of staining techniques to identify microorganisms and the methods of cultivation of bacteria in glass dishes [Cunha, 2004; Elizabeth, 2009].

A few years later the Gram's stain, invented by Hans Christian Joachim Gram, became, and continues to be, the most important and fundamental investigative tool for a presumptive and rapid diagnosis of majority of the IDs [Cunha, 2004; Elizabeth, 2009].

The first clinical microbiology labs, established in the early 1900s based on these fundamental techniques laid down by the pioneers, initiated the tracking down of several of the ID pathogens that were easily transmissible and of public health importance, like anthrax, syphilis, diphtheria, typhoid fever, malaria, tuberculosis, intestinal parasites etc. Standing on the shoulders of these scientific giants, anyone and everyone involved with IDs, can definitely benefit from the lessons of the past and be inspired by the great individuals whose insight, dedication and hard work have brought us to our current state of understanding and managing IDs [Elizabeth, 2009].

The discoveries achieved during the Golden Age of Microbiology lead to new technologies in microbiology. The present day discipline of clinical microbiology and practice of IDs is greatly influenced by the contributions of such great scientists [Elizabeth, 2009].

3. Recent advances in Clinical Microbiology

With a tremendous expansion of the knowledge about IDs, over the past 2 centuries, the clinical microbiologists and ID specialists of the 21st century are faced with far more challenging tasks that would not have been foreseen by the pioneers. They are now entering an era when priority of microbiological examinations for diagnosis of IDs substantially increases [Elizabeth, 2009].

For those involved in the diagnosis and management of infectious diseases, the last several years have been ones of wonder and enlightenment McPherson, 2006. Just when it seemed that antimicrobials and vaccines were eradicating many of the scourges of the human race, a brigade of new and old pathogens appeared or reappeared. This unprecedented and remarkable emergence and reemergence of pathogens worldwide, is mainly due to the globalization of commerce and the food supply, rapid population mobility, the HIV/AIDS pandemic and increasing microbial resistance to standard therapeutics. ID pathogens are no longer geographically restricted and have the potential for global spread with no distinctions among types of populations [David, 2004].

Clinical microbiologists and microbiology laboratories are also experiencing changes due to evolving views on 'healthcare delivery' as an economic activity, due to changes in the medical environment and the demographics of the workforce technical evolution. The physical structure of laboratories, staffing patterns, workflow, and turnaround time all will be influenced profoundly by the technical advances [Van Eldere, 2005].

The advances in diagnosis and management of IDs, made over the past few decades, have been primarily driven by the needs of specific patient populations and increasing numbers of immuno-compromized patients and the emerging / reemerging infectious agents, especially of public health importance. Several innovative and integrated methodologies are now being developed that would lead to improved and rapid diagnosis, vaccines and more effective treatments against the Ids [Cunha, 2004; Elizabeth, 2009].

4. Need for rapid diagnosis

Virtually all of the clinical impact and much of the financial impact of a clinical microbiology diagnostic test (a relatively upstream process), is determined by how the test result changes patient management. And this is a fundamental problem in traditional clinical microbiology assays that needs to be addressed critically [Ann, 1994].

To him who devotes his life to science
nothing can give happiness than making discoveries,
but his cup of joy is full only when
the results of his study find practical applications

— Louis Pasteur

Unlike other clinical laboratory tests, clinical microbiology tests, as a group, have a longer turnaround time (TAT) and the results are generally not made available in a clinically useful time. The procedures for isolating, identifying, and testing the antimicrobial susceptibility of bacterial and fungal pathogens are tedious and time consuming. The interpretation of test results is further confounded by the complexity of the testing (eg, presence of the normal or contaminating microbial flora, the nuances of antimicrobial susceptibility testing), each of which may influence clinical decision making [Charles, 2007].

Hence, increasing the clinical relevance of microbiological testing through rapid testing becomes the real challenge and a mandate. And for an immunocompromized patient, this need for a rapid result is all the more critical because infections in these patients often evolve quickly into life-threatening events [Ellen, 2006]. Thus there is a reinforced trend towards automation with minimal or negligible technical errors. Additionally, the rapidly evolving advances in pharmacotherapeutics are demanding the diagnostic laboratories for more evidence based results [Fre´ de´ ric Pene, 2009].

The fundamental advantage of the recent and advanced technology, particulalry using molecular technologies, is that it allows for delivery of "real-time" results in a clinically meaningful time. The technology thus has a definite clinical impact and is revolutionizing ID practices [Ian, 2005]. Clumsy, contamination-prone techniques have largely been replaced with automated, real-time detection technology performed in closed systems. Extensions of the well-known techniques (eg, microscopy, culture methods and immunoassay), as well as from more unfamiliar arenas like the molecular methods (eg, genomics, proteomics, multivariate index arrays and multimodal aggregated panels of biomarkers) are finding a significant place in the clinical microbiology laboratories. The petri dish and traditional tissue stains have been supplanted by nucleic acid amplification technology and in situ oligonucleotide hybridization for "growing" and "seeing" some microorganisms [Ian, 2005; Raquel, 2006].

Diseases due to uncultivable bacteria could represent emerging infectious diseases. However, the growing importance of these pathogens remains ill understood and undefined [Bhattacharya, 2002]. Non-culture based approaches, especially molecular genetic methods are evolving as the most important tool in our understanding of these enigmatic pathogens [Ian, 2005; Raquel, 2006; Bhattacharya, 2002].

The emerging technology that helps in a ready access to nucleic acid sequences and the conceptual advances that allow inference of evolutionary relationships from certain sequences have brought about the identification and detection of novel and previously uncharacterized microorganisms [Elizabeth, 2009; Bhattacharya, 2002].

Laboratory based, traditional clinical microbiology is now being rapidly replaced by advanced techniques, that are highly specific and sensitive and most importantly, facilitate rapid and sometimes point of care (POC) testing to detect the pathogens or their products Rosanna, 2007; Christopher, 2001; Point of Care Diagnsotic Testing]. The challenge of developing POC tests are expected to give laboratory-quality test results but should be simple to perform, requiring minimal training of users and field applicable [Rosanna, 2007; Christopher, 2001; James, 2007; Point of Care Diagnsotic Testing, Thomas, 2003]. All the chemistry must occur in small volumes and must be packaged into a self-contained device that is portable and stable at room temperature. The cost can be lowered by multiplex testing for several infections within a syndrome. The ultimate goal is to have a 'lab on a chip' that fulfils all the criteria of an ideal test [Technologies Magazine, 2004; The TDR Diagnostics Evaluation Expert Panel, 2006].

5. Challenges of advanced techniques

5.1 Statistical methods of evaluation performance parameters of new techniques [Les Irwig, 2002; Rosanna, 2007]

The performance parameters of sensitivity, specificity, predictive values, precision, and accuracy are the traditional standards by which laboratorians have compared and contrasted clinical laboratory tests and methods. During the recent past, new statistical approaches, such as likelihood ratios, discrepant analysis, consensus- and discriminate-function analysis and receiver operating characteristic curves have been used to define performance parameters of laboratory tests. Just as new diagnostic tests should undergo rigorous study and evaluation in controlled clinical trials, new statistical methods developed to define performance parameters should also undergo rigorous evaluation. At the same time, it also might be useful to define the performance parameters of older microbiology tests using newer statistical approaches. And all Statistical approaches, like the diagnostic tests, should be validated before being widely used (2,21).

When selecting between different tests or methods, it should be remembered that no single performance characteristic defines the laboratory or clinical utility of a test. Furthermore, the primary considerations in selecting a test should be the appropriateness of the test or method for the health care setting in which it will be used and the clinical purpose for which it will be ordered [Rosanna, 2007; Les Irwig, 2002; Elizabeth, 2007; John, 2007; Michael, 2002; Carol, 2007].

For eaxmple

- Screening tests should have a high sensitivity and relatively low specificity, whereas the opposite is true for confirmatory tests.
- Predictive values vary according to the prevalence of the disease or condition in the population being tested, so one test may yield very different predictive values when used to test different patient populations.
- Discrepant analysis is a statistical approach that has been used widely in clinical microbiology during the past few years for the evaluation of NAA tests.

For these reasons and others, such as test complexity, a test that is appropriate for one health care setting may be inappropriate for another.

5.2 Turn Around Time (TAT) [Charles, 2007]

Test results that are not available within a reasonable period of time are unlikely to be clinically relevant or to be used appropriately. There is a need to

- define a reasonable test TAT for each method, for each type of health care professional who uses the results, and for different health care organizations.
- determine which measure of test TAT should be used and to develop a system to monitor and improve test TAT.
- determine what can be done to decrease test TAT to acceptable levels, and how much it will cost to achieve the shorter TAT.
- answer the question of whether one can satisfy clinicians' demands for shorter TAT, or if the laboratory should even try to do so. Satisfying physician's expectations for improved test TAT is a complex process for which there is no single solution.
- It is reasonable to try and achieve the shortest possible test TAT, but not at an unreasonable cost or if it means using less accurate diagnostic methods.
- In some situations use of a less accurate method is acceptable if it means a significantly shorter test TAT. eg. stat Gram's stain of CSF.

5.3 Approach to automation [Neil, 2007; Ann, 1994]

A clinical laboratory considering automation for the first time or as a change or extension of some existing automation should begin with an evaluation of needs. Failure to properly evaluate the needs of the laboratory and to understand the current state and processes in the laboratory are two key reasons why automation projects are sometimes not successful.

The laboratory should undertake to complete a map of the current workflow from the arrival of patient specimens through completion of testing, reporting of results, and disposal of expired specimens. A good rule of thumb in guiding decisions about automation is whether the proposed system or component handles 80% of the workload, which includes specimen containers and handling situations. If these can be standardized and automated, the laboratory may achieve a significant improvement in its processes with corresponding improvements in quality and a reduction of labor and turnaround time, which should be sufficient to justify the investment in automation and the planning and evaluation time involved.

6. Appropriate vs Inappropriate testing [Charles, 2007]

Laboratory medicine lacks the tools necessary to define appropriate test use; nonetheless, existing laboratory test characteristics, although inadequate, provide a common starting point for developing definitions of appropriate test use.

It is easy to recognize that the use of some tests is appropriate than tests that are inappropriate for clinical use. The challenge for tests in the latter category is not in recognizing them, but rather to decrease or eliminate their use by physicians. Unfortunately, the tests in this category is increasing especially with the introduction of the molecular assays for diagnosing IDs. To overcome these issues, the clinical microbiologist has to have the support of the infectious diseases service and educate potential users who are affected by the changes including the technologists who will implement the changes.

7. Cost effectiveness of clinical microbiology tests
[Van Eldere, 2005]

The real challenge for clinical microbiology will be to use new techniques to improve its cost-effectiveness and impact on infectious disease management.

Appropriate test use depends on the use of the most cost-effective test for a given purpose. Unfortunately, the term cost-effective is poorly defined. Defining cost effectiveness is perhaps even more challenging than defining clinical relevance, partly because the cost effectiveness of a test depends on its clinical relevance, but also because many other variables must be considered. A better, albeit descriptive, definition is "the least expensive method that yields clinically relevant test results in a timely manner and that does not increase costs elsewhere in the health care system"

For many clinical microbiology tests the alternative methods differ not only by cost but also by their performance parameters and test TAT. A related issue is determining the point at which the costs required to achieve incremental gains in test performance parameters or test TAT make the method no longer cost-effective. The clinical microbiology laboratories can aim to increase the cost effectiveness of tests and should make every attempt to perform only clinically relevant testing. Tests that are not clinically relevant cannot be cost effective, streamline the work-up of common bacterial and fungal pathogens and make use of the best available technologies.

8. Technological innovations [Elizabeth, 2009]

Independent of any challenge or issues, technological innovations will happen and the drivers are readily apparent. Diagnosticians have taken great leaps forward in their level of overall sophistication and familiarity with the new technology. The change has already set in and will continue to do so

- Beyond traditional light microscopes, an array of instruments enables researchers to analyze microbes and devise ways to combat them. Further improvements in magnification and new techniques for staining specimens made it possible to identify and see inside disease-causing microbes.
- Electron microscopes, for example, magnify objects hundreds of thousands of times-enough to see and understand the internal mileu of the viruses.
- Flourescence microscopy offers rapid screening for Acid fast bacilli, Fungal hyphae, autoantibodies etc.
- Other advanced techniques, let researchers take microbes apart to understand their inner workings.
- Chromogenic agars to cultivate and identify the bacteria directly without the need for further biochemical tests are now being used routinely in clinical Microbiology laboratory [Hara, 2005]
- Rapid immunochromatography test devices for infectious diseases are being increasinlgy used in many small diagnostic laboratories in place of the elaborate and technically demanding ELISA [Chris Baylis].
- Automated instrumentation and expert systems to rapidly identify and test and analyse for the antimicrobial spectrum of bacteria and yeasts are available [Ann Robinson, 1994; Van Eldere, 2005].
- Point of care tests especially using Stat-PCR would probably become the order of the day [Michael, 2002; Reuters Health Information 2007].
- Unique and specific biomarkers for each infectious disease, that could be detected by rapid methods, are already being identified and would be the targets for the next generation of assays based on proteomics (protein biomarker detection) [Charles, 2008; Elizabeth, 2009].
- Single organism detection will evolve to multiple organism detection by microarrays [Plamen, 2008].
- Real – Time Multiplex molecular assays like Multiplex PCR with Microarrays, Multiplex PCR with solid or liquid microarray dete-

ction of amplicons, Synthesized arrays, Printed arrays etc would soon be introduced for routine diagnosis of infectious diseases [Charles, 2008; Plamen, 2008; Raquel, 2006; Elizabeth, 2009].

- Integration of molecular methods with proteomics and the revolutionizing technology of Mass spectrometry (eg. Multiplex Analysis via PCR-MS), diagnosis of IDs will be even more precise [MALDI; Daniel, 2005; Zhang, 2008; Della-Latta].
- Sequence-based identification methods will continue to evolve like the DNA Sequencing, FISH technology which give specific results in a matter of few hours and thus have definite and exiting diagnostic and therapeutic advantages [Micheal].
- Bioinformatics and understanding the basis of Sequence-Based technology for molecular characterization of the infectious agents will gain further importance [European Congress of Clinical Microbiology and Infectious Diseases Barcelona].
- Genotyping and molecular epidemiology, molecular testing of drug resistance and as mentioned earlier, proteomics will eventually impact microbiology laboratories [Elizabeth, 2009].
- DNA sequencing and micro array technologies developed under the auspices of the human genome project are making steady inroads into ID practice [Elizabeth, 2009].
- Phylogenetic analysis and identification of bacteria, fungi and viruses by direct DNA sequencing are quickly entering the main stream [Elizabeth, 2009].

The trends are clear toward earlier, more actionable information. And within the next few years, Clinical Microbiology laboratories will move rapidly towards advanced and automated techniques to support more targeted interventions in patients being managed in cancer centers, transplant centers, children's hospitals, HIV clinics, and other foci of immunocompromized patients [Elizabeth, 2009].

Innovative thinking is needed on the part of laboratorians to keep clinical microbiology at the center of evidenced based medical practice. The future is full of opportunities cleverly disguised as challenges, from the impacts of consumerism to the exciting innovations in companion diagnostics.

For example: the Foundation for Innovative New Diagnostics (FIND) [David, 2004; Chris Baylis] was created by the Bill and Melinda Gates Foundation during the diagnostic-development activities for Tuberculosis, in 2003, with an aim to respond specifically to the need for better diagnostics in the developing world by harvesting the best biotechnologies that are available, and developing and validating new tests for use in patient management, disease control, and surveillance.

9. The future of clinical microbiology

[FIND Symposium; Lim Daniel, 2005]

The unprecedented opportunities for innovations in diagnostics development are are a result of the convergence of advances in two important areas:

1. The potential for the discovery of novel diagnostic targets or biomarkers through genomics, and an increased understanding of pathogen virulence factors and host gene expression using microarrays and Proteomics [Charles, 2008; Plamen, 2008].
2. Rapid technological advances in areas such as material science, nanotechnology and microfluidics [Lim, 2005].

This integration has been fuelled by the recent events of bioterrorism and the urgent need for rapid detection of potential bioweapons [Wolfgang, 2001; Lance, 2001].

- Nanotechnology has allowed the miniaturization of complex reaction processes into small, self-contained packages [Lim, 2005].
- Microelectromechanical systems (MEMS) have been exploited to develop personal computers and cell phones that operate from a microchip made from a silicon wafer. Biomedical applications of MEMS are called bioMEMS or microfluidics.
- Such technology allows specimen processing, biochemical reactions and detection of the resulting product in a flow-through format with on-chip control of thermo-pneumatic pumps, micro-heaters and temperature sensors, miniaturized fluorescence detectors, sample/analyte concentrators and filters.
- Some of the novel sensor technologies, such as surface plasmon resonance (SPR) and mirror resonance technology, allow for the analysis of three-dimensional biomolecular interactions in real-time,without the need for labels or purification. Changes in the SPR signal are monitored through a reduction in the reflected light intensity with regard to both angle and wavelength. These sensors are now being exploited as remote sensing devices for the agents of bioterrorism.
- Fibre-optic technology that is based on long period grating (LPG) and optical-fibre sensors, is a breakthrough for telecommunications.
- With specially designed affinity coatings to ligands, such as monoclonal antibodies, that can bind target molecules and can be adapted for multiplex diagnostics detection. Such technology is useful for screening combinatorial-chemistry libraries, exami-

ning large arrays of candidate drugs and discovering new pharmaceuticals.

- As wavelength is a direct measurement of the bound mass and is a characteristic of the individual fibre profile and grating period, this can be adapted for use in diagnostics where the difference in wavelengths between the control and the test sample is directly proportional to the concentration of bound analyte. This type of technology provides accurate analyte quantification to parts-pertrillion levels within 5 minutes. The technology can be developed into a compact, lightweight biochemical detection platform that is suitable for both centralized (laboratory) and decentralized (point-of-care) clinical settings.
- Thus integrating sophisticated reagents and detection systems into easy-to-use formats that deliver patient empowering point-of-care information for the detection of ID pathogens, is a goal that now seems achievable. The convergence of knowledge of pathogen–host interactions with the remarkable advances made in biotechnology during the past few decades should offer a great opportunity for the development of 'leapfrog' technologies that overcome many of the diagnostic problems faced by patients and healthcare workers in developing countries [Elizabeth, 2009].

We are faced with unique challenges as the future of clinical microbiology unfolds, first within the content of the immuno-compromized host and then within the context of the general population. What was once seen in science fiction TV shows is becoming our reality [Sintchenko, 2001].

It is not difficult to foresee a future in which

- clinicians not only rely on patients' vital signs and traditional laboratory diagnostics but also rely on patients' DNA profiles to determine predisposition to infection or predict adverse effects of therapy [Sintchenko, 2001; Michelle, 2007].
- the costs of rapid diagnosis are small in comparison to the costs of waiting for results, both in terms of humanity and in terms of dollars [Hara, 2005].
- Near-point-of-care molecular testing is a reality, taking us into the realm of same-day testing of microbes and antibiotic susceptibility [Hara, 2005].

Information technology [Donald, 2002; Ruben, 2009; Lawrence, 2002] is a critical element of any automation solution. Numerous options, large and small, are now available from a variety of vendors to meet needs ranging from the automation of single tasks (eg, sorting) to several functions (preanalytical processing) to TLA systems with interfaced analyzers. The cost and volume thresholds for laboratories to implement automation have steadily dropped. And over the next 5 to 10 years, nearly every mediumsized or larger clinical laboratory and even many smaller laboratories will probably implement some form of automation.

Web technologies in a clinical laboratory would further provide a rich set of tools to enhance the value of the laboratory's contribution to patient care.

10. Conclusions

The primary objective of any clinical microbiologist is to provide rapid and specific diagnosis of an infectious disease that would have a favorable impact on patient management and outcomes [Elizabeth, 2009; Sintchenko, 2001].

The greatest challenge for the future clinical microbiologists, while paying tributes to pioneers like Louis Pasteur, need to familiarize with the powerful advancing diagnostic options and aim to integrate the different and exiting technologies of material science with the science of Biology [Elizabeth, 2009].

The present day Clinical Microbiologists are well positioned, should they accept these challenges of the rapidly evolving new technology and take advantage of the several exiting opportunities. Through a combination of talents and expertise in medical care and scientific innovation, together with critical management skills, teams of dedicated laboratory medicine professionals can continue to improve health care and increase the "value proposition" inherent in the practice of Clinical Microbiology and IDs [Sintchenko, 2001].

"Failing to prepare
is
preparing to fail"

The need of the hour is "A back to School" training of the future clinical microbiologists in all the advanced technologies as a preparedness to face the challenges of IDs and their early recognition, under the changing environments.

There is only one science and the application of science.
These two activities are linked as the fruits of the tree.

— Louis Pasteur

However, it should be borne in mind that much can be learned by studying the past to better understand the infectious disease challenges of the present. Medical history teaches us where we came from, where we stand in medicine at the present time, and in what direction we are marching. It is the compass that guides us into the future.

With all the new knowledge and information derived from advanced technology every 21st Century clinical Microbiology Laboratory, being targeted and affected by the booming new technology and emerging infectious diseases, should face the real challenge to most efficiently and effectively use the available techniques (traditional and advanced) to provide an accurate result that would help the patient in the most cost effective and clinically relevant manner. As advanced technologies become more and more available and evolve from batch mode to on-demand formats, they will have an ever-greater impact on patient treatment and management. Such changes will undoubtedly continue and lead the field of diagnostic microbiology inevitably to a truly modern discipline As real-time molecular diagnostic technology improves, so should the delivery of real-time patient results [Elizabeth, 2009; Van, 2005].

References

Ann Robinson (1994). Clinical MicrbiologyReviews, 7/2: 185-199.

Bhattacharya S, Vijayalakshmi N, Parija SC (2002). IJMM; 20/4: 174 – 177.

Brier B (2004) Infect Dis Clin N Am: 18: 17–27.

Carol A. Rauch, MD, James H. Nichols (2007). Clin Lab Med 27; 45–858

Charles D. Hawker (2007). Clin Lab Med, 27: 749–770.

Charles Y.C, Anatoly U, Taral G, Silvi R, Shigeo Y, David S, Carolyn W, Lawrence WD, David W (2008). J Pediatr; 153:76-83.

Christopher P (2001). BMJ; 322; 1285-1288.

Chromogenic & Fluorogenic Microbiological Methods. Lecture Presentation by Dr Chris Baylis

Cunha B.A. (2004). Infect Dis Clin N Am., 18: xi–xv.

Daniel V. Lim, Joyce M. Simpson, Elizabeth A. Kearns and Marianne F. Kramer. (2005) Clinical Microbiology Reviews. 18/4: 583–607.

David Mabey, Rosanna W. Peeling, Andrew Ustianowski and Mark D. Perkins (2004). Nature reviews, Microbiology 2: 231 – 240.

Della-Latta, P., Whittier, S., Wu, F. Impact of rapid identification of *C. albicans* and *C. glabrata* directly from blood cultures using PNA FISH technology on selection of antifungal therapy. Abstract number: P 1382.

Donald P. Connelly (2002). Clin Lab Med; 22: 529–545.

Elizabeth A. Wagar (2007). Clin Lab Med; 27: 909 – 930.

Elizabeth MM, Donna MW. Future trends in diagnosis of Infections in the immunocompromised population. In Diagnostic Microbiology of the Immunocompromised Host. Ed. Randall T H, Karen CC, Yi-Wei Tang, Donna MW (2009) ASM Press, Washington DC, USA; pp 443 – 462.

Ellen Jo Baron. (2006) Clinical Infectious Diseases, 43: 1318–1323.

Evaluation of diagnostic tests for infectious diseases: general principles. The TDR Diagnostics Evaluation Expert Panel. (2006) Nature Reviews/Microbiology; S 21 – 33

Fre´de´ric Pene, Emilie Courtine, Alain Cariou, Jean-Paul Mira.Toward theragnostics. (2009) Crit Care Med, 37/1(Suppl.): S 50 – S 58.

Hara MO (2005). Clinical Microbiology Reviews, 147– 62 Vol. 18, No. 1.

Ian R. Poxton (2005). Med Princ Pract, 14 (suppl 1): 20–26.

James H. Nichols (2007). Point of Care Testing Clin Lab Med 27: 893–908.

John L. Carey, Teri A. Oldaker (2007). Clin Lab Med; 27 :687 - 707.

Lance R. Peterson, John D. Hamilton, Ellen Jo Baron, Lucy S. Tompkins, J. Michael Miller, Catherine M. Wilfert, Fred C. Tenover and Richard B. Thomson (2001). J Clinical Infectious Diseases; 32:605–11.

Lawrence Drew W (2002). Clin Lab Med 22 xi.

Les Irwig, Patrick Bossuyt, Paul Glasziou (2002) BMJ; 324;669-671.

Lim Daniel V; Simpson Joyce M; Kearns Elizabeth A; Kramer Marianne F (2005). Clinical microbiology reviews;18(4):583-607.

Lucia M. Berte (2007) Clin Lab Med 27;771– 790

McPherson & Pincus: Henry's Clinical Diagnosis and Management by Laboratory Methods, 21st ed. (2006) W. B. Saunders Company.

Michael D.D. McNeely (2002) Clin Lab Med 22 515–528.

Michael L. Wilson (2002) Clin Lab Med 22: 491– 503

Micheal AZ, Maria Chan, James CE, Andria FG, Ira M Lubin.CLSI MM9 – A Nucleic acid sequencing methods in diagnsotic laboratory medicine: approved Guideline, vol 24 / 40.

Michelle Beaver. (2007) Infection Control Today, Electronic version. 10/29.

Molecular Diagnostics Expanding rapid nucleic acid testing (2004) IVD Technologies Magazine.

Nanostructures and Nanobiosensors for Lab-on- a- chip. Centre for Biomems & Nanobiosystems, University of Cincinnati (April 2008) 18th European Congress of Clinical Microbiology and Infectious Diseases Barcelona, Spain, 19–22.

Neil Benn, Fabrice Turlais, Victoria Clark, Mike Jones, Stephen Clulow (2007). Clin Lab Med; 27: 29–39.

Plamen A. Demirev1 and Catherine Fenselau (2008) J. Mass Spectrom; 43: 1441–1457

Portfolio of new diagnostic technologies 36th IUATLD World Congress of Lung Health, Mark Perkins, FIND Symposium.

Raquel Cuchacovich (2006). Infect Dis Clin N Am, 20 : 735–758.

Reuters Health Information 2007. Microchip Test Quickly Spots Pathogenic, Drug-Resistant Bacteria http://www.medscape.com/viewarticle/562424.

Rosanna W. Peeling, Peter G. Smith and Patrick M. M. Bossuyt (2007). Nature reviews, Microbiology: S1 – 6.

Ruben Amarasingham, Laura Plantinga, Marie Diener-West, Darrell J (2009). Arch Intern Med. 169 (2): 108 – 114.

Sintchenko V, G L Gilbert. (2001) J Clin Pathol; *54*: 441-442.

The University of Bristol, School of Chemistry mass spectrometry resource Matrix-assisted Laser Desorption/Ionisation (MALDI).

Thomas L. Hearn (2003). Rapid Point of Service Testing – A Public Health Perspective. Keystone POCC AACC Meeting.

Van Eldere, J (2005) Clin Microbiol Infect; 11 (Suppl. 1): 15–18.

Wolfgang F. Klietmann and Kathryn L. Ruoff (2001). Clin Microbiol Rev; 14(2): 364–381.

Zhang C, O. Crasta, S. Cammer, R. Will, R. Kenyon, D. Sullivan, Q. Yu, W. Sun, R. Jha, D. Liu, T. Xue, Y. Zhang, M. Moore, P. McGarvey, H. Huang, Y. Chen, J. Zhang, R. Mazumder, C. Wu and B. Sobral. (2008). An emerging cyberinfrastructure for biodefense pathogen and pathogen-host data Nucleic Acids Research, 36, D884– D891, Database issue.

Chapter - 9

Development of consolidated bioprocess (CBP) for bioethanol production using lignocellulosic substrates

L. Venkateswar Rao, Chand Pasha, J. Sridevi and A. Vimala Rodhe*

Department of Microbiology, University College of Science, Osmania University, Hyderbad.
*E-mail: vrlinga@gmail.com

ABSTRACT

The world is seeking alternatives for highly priced and highly in-demand fossil fuels like petroleum. Ethanol has been made since ancient times by yeast fermentation. All beverage ethanol and most of the industrial ethanol is still made by fermentation. Zymase, an enzyme from yeast, converts simple sugars into ethanol and carbon dioxide. In India ethanol is made from sugarcane molasses but as the demand for alternative fuels is increasing there is a need for exploring other renewable substrates. Currently much attention is being focused on the use of lignocellulosic biomass, as it is an ideal renewable resource for the production of fuel ethanol via fermentation. Lignocellulosic substrates (*Prosopis juliflora, Lantana, Typha, Saccharum, Rice straw, wheat straw, Sorghum straw and corn cobs*) can serve as an excellent, cheap source for bioethanol production as they are available in large quantities with great diversity. The complex structure of these lignocellulosic substrates makes it necessary to carryout various pretreatment steps involving chemicals/ enzymes for them to be utilized by the fermenting organism (*Saccharomyces cerevisiae*). Selection of suitable cellulase producing organism which is capable of utilizing lignocellulosic substrates and their development aids in cost reduction of bioethanol production. Use of cellulases for the enzymatic hydrolysis of cellulosic substrates can be preferred over acid hydrolysis as the latter results in the accumulation of fermentation inhibitors making it necessary to carry out detoxification steps adding to the process cost. Use of developed or adapted or genetically engineered yeasts fermenting both hexoses and pentoses is an important aspect of bioethanol production. Adoption of a new technology for large-scale ethanol production from lignocellulose might result in economic and environmental benefits. Unfortunately, several factors currently prohibit the commercial production of ethanol from lignocel-lulosic materials. The main problem is the production cost for cellulase enzymes, which is an important component of the bioconversion process, remains high enough to be a significant proportion of the total costs for ethanol production. Enzyme production cost estimates range as high as 25 to 50% of the total ethanol production costs which significantly limit the economic viability of this process. The second factor is the strain utilized for ethanol production. *Saccharomyces cerevisiae* is an organism of choice for ethanol production, but it is being limited due to short range of its substrate utilization and inability to tolerate inhibitors. The third factor in bioethanol production is the substrate and its pretreatment cost. Even though research is going on since many years on pretreatment, no successful process is developed. If above stated three problems are solved ethanol production costs may be reduced, with the potential to make ethanol (from lignocellulose) a viable supplemental fuel source in the light of current political, economic and environmental issues. The present review focuses on various aspects involved in the development of Consolidated Bioprocess (CBP) for bioethanol production using lignocellulosic substrates.

Key words: Bioethanol, Lignocellulosic materials, Thermotolerant yeast.

1. Introduction

Energy is the lifeline of modern societies. India having 17% of the world's population has just 0.8% of the world's known oil and natural gas resources. Meeting our energy requirements in a sustainable manner continues to be a major challenge. Transport sector alone consumes 50 percent of petroleum products. India currently imports 70% of crude oil, which accounts for over $ 40 billion per year and adds $ 425 million for every $ 1.00 dollar increase in the price of crude oil. Crude oil is touching $70 a barrel and there is no sign of relief in the foreseeable future. Thus India is becoming vulnerable in the matter of energy security. Developing and promoting biofuels will mitigate this vulnerability and other environmental effects arising from the use of fossil fuels such as coal and petroleum.

Ethyl alcohol or ethanol is a well known substance that has a wide range of applications. It can be used in the chemical industry as a solvent or agent for chemical reactions and uses include the manufacture of industrial products like detergents, paint, varnish and cosmetic products. It is also utilized as a substrate to produce ethyl esters (Biodiesel) and to manufacture industrial-grade diamonds.

Although the referred uses have been the traditional driving forces that fostered the international trade of ethanol, its expanded use as an automotive fuel has created a new market segment that is viewed as the fastest growing and most important market for the product in the future (Pimentel, 2003).

2. Bioethanol as fuel

Bioethanol is an alcohol made from sugar, starch, cellulose and products containing sugars. It is fermented, distilled and used as a substitute or partial substitute for petrol. Bioethanol is a clean-burning, non-petroleum liquid fuel. Countries' dependence on imported oil, environmental issues and employment in rural areas (Farrell *et al.*, 2006) are reasons for the consideration of the replacement of fossil fuels by bioethanol. Ethanol can be used as a transportation fuel as well as a gasoline fuel extender and an octane enhancer. Ethanol can be mixed with gasoline in 10% (E10), 20% (E20) and 22% (E22) blends without engine modifications (Bullen *et al.*, 2006) but higher-level blends (such as 85% or 95%) require some engine modification. (Von Blottnitz and Curran 2006). Ethanol has an octane number of 113, allowing a higher compression ratio in the gasoline engine. Ethanol serves as an octane-booster in itself and in the form of ethyl tertiary butyl ether (ETBE), thus it can replace the toxic additive benzene. Using ETBE blended at 15-16% level with gasoline, the carbon dioxide, nitrogen oxides and some aromatic compounds emissions diminish, but aldehyde emission increases at a low level (<1%) (Niven *et al.,* 2005). As a fuel additive, ethanol provides oxygen to the fuel and thus improves fuel combustion and reduces tailpipe emission of carbon dioxide and unburned hydrocarbons. It gives further advantages to the fuel because of its high heat of vaporization, low flame temperature, greater gas volume change, high specific energy, lower volumetric energy content, which means shorter distance travel compared to gasoline (Gautam and Martin *et al.,* 2000; Ragauskas *et al.,* 2006).

In fossil fuels, a fair amount of hydrocarbons may end up as 'half-burnt' products - chief among these are the polyaromatic hydrocarbons, which are implicated in carcinogenesis (causing cancer). Where as ethanol ensures the complete combustion of other compounds in a fuel mix. A key barrier to ethanol's expanded role in fuel consumption is its price differential with gasoline. Since a major part of the total production cost is the cost of feedstock, reducing feedstock costs could lead to lower costs of ethanol production (Wyman 1996; Andersson *et al.*, 2006). The way to reduce the cost of the feedstock is to switch from the traditional starch-based feedstocks (Ericsson *et al.*, 2004; Wright, 2006) to cellulosic biomass. For this reason, there is a great deal of interest in producing ethanol from cellulosic feedstock. However, low-priced feedstock requires costly preparation, which makes it necessary to develop effective and relatively inexpensive pretreatment procedures. The high cost of the commercial cellulase enzymes induces researchers to develop more affordable solutions, such as the production of cheaper own cellulase (Katzen *et al.*, 1995) from the same cellulosic biomass substrate that is used for ethanol production.

3. World ethanol production and trade.

Major findings:

- World's ethanol production will pass 20 Bln gallons in 2012.
- Ethanol production is expected to grow in 2008 - 2012.
- U.S. and Brazil are leading the world in production of ethanol.
- Emergence of new ethanol producers in Asia and Latin America.
- Cuba has the capacity to manufacture as much as 3.2 billion gallons of ethanol annually from its sugar crop.

Many countries are taking an increased interest in ethanol as an alternative fuel, with the USA and Brazil currently leading the way. In 2005, Brazil produced 4.8 million gallons of ethanol and the USA produced 3.9 billion gallons, making these countries the two largest producers in the world, accounting for over 90 per cent of the total world production. There are a limited number of studies on ethanol markets, as the industry has experienced a boom only in recent years (Gallagher *et al.*, 2006). The recent increase in crude oil and gasoline prices has also opened a new market for ethanol as a fuel extender (Eidman 2006). Asia is the largest potential producer of bioethanol from crop residues and waste crops that could produce up to 291 GL/year of bioethanol (Kim and Dale, 2004). The demand for ethanol is approximately divided into 68% for fuel, 21% for industries and 11% for beverages. Approximately 9% of the ethanol is produced synthetically and consequently, fermentation is responsible for 91% of global ethanol production (Wheeler *et al.*, 1991).

Table 1. World ethanol production forecast 2008 - 2012 by a country, (Millions of Gallons)

	2008	2009	2010	2011	2012
Brazil	4,988	5,238	5,489	5,739	5,990
U.S.	6,198	6,858	7,518	8,178	8,838
China	1,075	1,101	1,128	1,154	1,181
India	531	551	571	591	611
France	285	301	317	333	349
Spain	163	184	206	227	249
Germany	319	381	444	506	569
Canada	230	276	322	368	414

Indonesia	76	84	92	100	108
Italy	50	53	55	58	60
ROW	2,302	2,548	2,794	3,040	3,286
World	**16,215**	**17,574**	**18,934**	**20,293**	**21,653**

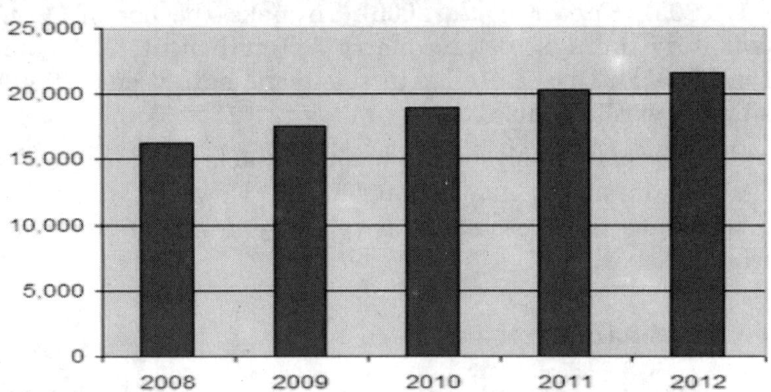

Fig. 1

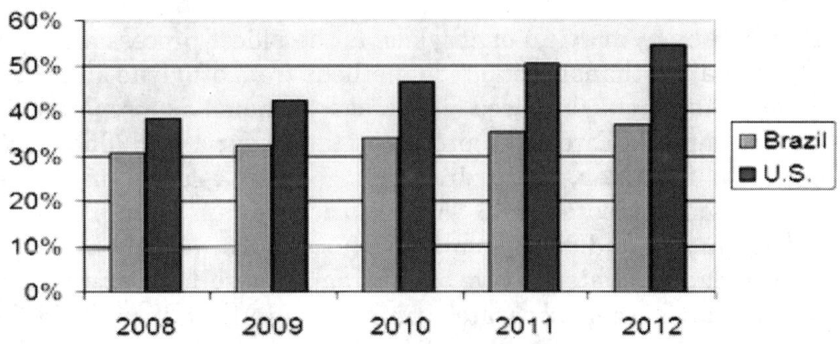

Fig: 2.

Courtesy: World's ethanol production forecast 2008-2012

4. Indian strategies of biofuel program.

India is not self-sufficient in petroleum and has to import about two-thirds of its requirement. The prices have crossed US$ 70 per barrel and the expenditure on crude purchase is in the range of Rs.1800 billion per year, impacting in a big way, the country's foreign exchange reserves (Chandel *et al.*, 2006b). A blend of ethanol in fossil fuel will, make a big difference both in our import as well as in environmental management.

The Government of India through its Planning Commission has initiated a national program to cultivate vast areas of waste/degraded lands by plantation of oil-bearing and bio energy trees. After successful field trials, the government made it mandatory to blend 5% biofuel in fossil fuel for the supply in 9 states (1. Andhra Pradesh 2. Goa. 3. Gujarat 4. Haryana 5. Karnataka 6. Maharashtra 7. Punjab 8. Tamilnadu and 9. Uttar Pradesh) and four contiguous union territories (1. Daman and Diu 2. Dadra Nagar Haveli 3. Chandigarh 4. Pondicherry) with effect from January 1, 2003. (Kapoor *et al.*, 2006). To make the use of 5% biofuel blend mandatory throughout the country 363 million liters of ethanol is required in 2003/04. Due to decline in sugarcane output only 196 million liters of ethanol was available.

According to many analysts, the major obstacle in widespread use of ethanol as fuel is simply a question of cost. However, there is great potential to reduce these costs considerably by developing economically viable technologies.

5. Ethanol production

Ethanol may be produced synthetically or by direct fermentation of sugars or from other carbohydrates such as starch and cellulose that can be converted to sugar.

5.1. Synthetic ethanol

a.1. Indirect hydration of ethylene is the oldest process among the two major ethanol production methods from ethylene and started more than one hundred years ago. Ethanol is prepared from ethylene in a three-step process using sulfuric acid (John 1969). In the first step, the hydrocarbon feedstock containing 35-95% ethylene is exposed to 95-98% sulfuric acid in a column reactor to form mono- and diethyl sulfate. This is subsequently hydrolyzed with enough water to give 50-60% aqueous sulfuric acid solution. Ethanol is then separated from the dilute sulfuric acid in a stripper column. The last step of this process is to concentrate the sulfuric acid and recycle to the process.

a.2. Direct hydration of ethylene was commercialized in 1947. In this process, an ethylene-rich gas is combined with water and passed through a fixed-bed catalyst reactor, in which ethanol is formed according to the following reaction (Nelson and Courter 1954). Ethanol is then recovered in a distillation system. Both direct and indirect hydration of ethylene gives rise to undesirable by-products such as diethyl ether, which reduce the quality of ethanol. Other processes that are used to make ethanol synthetically are not commercially important.

5.2 Ethanol production by fermentation.

Ethanol can be produced by fermentation from different kinds of raw materials. The raw materials are classified into three categories: Saccharine, starchy and cellulosic materials (Table 1). Some sugars can be converted directly to ethanol, whereas starch and cellulose must first be hydrolyzed to sugar before conversion to ethanol.

Table 2. Major resources for fermentative production of ethanol

Saccharines	Starchy Materials	Cellulosic Materials
Sugarcane	Grains	Wood
Sugar beet	Potatoes	Agricultural residue
Molasses	Root crops	Municipal solid waste
Fruits/juices		Waste paper

6. Raw materials

The factors affecting the choice of raw materials are as follows (Cowling and Kirk 1976; Rao and Pasha 2006):

1. Materials should be inexpensive, should have consistent quality and should be available throughout the year.
2. Ease of handling, along with associated transport and storage costs.
3. Sterilization requirements and any potential denaturation problem.
4. The quantity of target product and its rate of formation should be high.
5. The levels and range of impurities should be less.
6. Overall health and safety implications.

Most ethanol conversion systems encountered in literature, are based on a single feedstock. But considering the hydrolysis fermentation process, it is possible to use multiple feedstock types. This may aid in achieving the desirable large scale production in future.

Most of the polymeric raw materials are available at prices lower than refined sugars. However, transportation costs of the raw materials make it necessary to use locally available raw materials. Consequently, each country should develop ethanol production based on the available raw material in that country (Butterworth, 2006).

6.1. Saccharine materials

Fermentation involves microorganisms that use the fermentable sugars and produce ethanol and other byproducts. These microorganisms can typically use the 6-carbon sugars, one of the most common being glucose. Therefore, biomass materials containing high levels of glucose or precursors to glucose are the easiest to be converted to ethanol. Feed stocks that are rich in sugars (saccharine) include sugarcane (also molasses) sugar beet, sweet sorghum and various fruits. However, these materials are all in the human food chain and are too expensive to use for fuel ethanol production.

6.2 Starchy materials

Another potential ethanol feedstock is starch which is made up of long chains of glucose molecules. Some of the starchy materials commonly used around the world for ethanol production includes, potato, sweet potato, cassava and cereal grains like sorghum, maize, barley and wheat Starchy materials require a reaction with water (hydrolysis) to break down the starch into fermentable sugars (saccharification) in presence of Specific enzymes (amylase) which can cleave the chemical bonds.

6.3 Lignocellulosic materials

Lignocellulosics are abundant sources of carbohydrate, continuously replenished by photosynthetic reduction of carbon dioxide by sunlight energy (Parikka, 2004). They are the most promising feedstock for the production of energy, food and chemicals (Solomon *et al.*, 1999; Kim and Dale, 2004). Agricultural wastes and in fact all lignocellulosics can be converted into products that are of commercial interest such as ethanol, glucose xylitol, vinegar and single cell protein (Hood, 2004; Das and Singh, 2004). Lignocellulosics are formed by three main polymeric constituents, cellulose, hemicelluloses and lignin (Lynd *et al.*, 2005) in various proportions (Table 2).

6.4 Cellulose

It is the most abundant cell wall component forming a supporting matrix for the cell membrane. It is a regular linear homopolymer of repeating D-glucose units in ß1-4 linkage (Sluiter *et al.*, 2004).

6.5 Hemicellulose

The second major structural component, hemicellulose has an intermediate degree of complexity and is a heteropolymer of pentose and hexose residues. Typically, hemicelluloses in softwood are glucomannans, whereas those in hardwoods are mainly xylans. The principal pentose sugar in hemicellulose is xylose (i.e., b-D-xylopyranose as found in xylan). Other sugars present are galactose, arabinose, rhamnose and methylglucuronic acid units

6.6 Lignin

Lignin is a three-dimensional network built up of dimethoxylated (syringyl, S), monomethoxylated (guaiacyl, G) and non-methoxylated (p-hydroxyphenyl, H) phenylpropanoid units. Lignin is highly resistant to chemical and biological degradation and confers mechanical resistance to wood. The utilization of lignocelluloses requires pre-treating the biomass to open the lignocellulosic polymers, to allow their hydrolysis by either enzymes or acids.

Table 3. Composition (%) of agricultural and wood residues.

Substrates	Cellulose	Hemi Cellulose	Lignin
Bagasse	33	30	29
Barley straw	40	20	15
Corn cobs	42	39	14
Corn stakes	35	15	19
Groundnut shells	38	36	16
Oat straw	41	16	11
Pine	41	10	27
Rice straw	32	24	13
Rice hulls	36	15	19
Saw dust	55	14	21
Sorghum straw	33	18	15
Wheat straw	30	24	18
Hard wood stems	40-55	34-40	18-25
Prosopis stem	48.5	21.7	26.10
Lantana wood	45.1	17.0	27.25
Soft wood stems	45-50	25-35	25-35
Grasses	25-40	35-50	10-30
Paper	85-99	0	0-15
Sorted refuse	60	20	20
Leaves	15-20	80-85	0
Cotton seed hairs	80-95	5-20	0
Newspaper	40-55	25-40	18-30
Waste papers from chemical pulps	60-70	10-20	5-10
Primary wastewater solids	8-15	NA	24-29
Solid cattle manure	1.6-4.7	1.4-3.3	2.7-5.7
Switch grass	45	31.4	12.0

Source: Cheung and Andrson (1997), Boopathy (1998) and Dewes and Hunsche (1998).

7. Pretreatment of lignocelluloses

A major obstacle in the effective utilization of lignocellulose is the chemically unreactive nature of cellulose. The lignin-hemicellulose matrix, cellulose crystallinity and its low surface area make the lignocellulose very resistant to enzymes (Alex *et al.*, 2006). Making cellulose accessible to the enzyme is an essential factor in order to increase the rate of hydrolysis (Philippids and Smith 1995). Therefore, pretreatment is necessary before enzymatic hydrolysis (Wyman *et al.*, 2005). Enzymatic hydrolysis has been recognized to be an attractive method for hydrolysis of cellulose and hemicelluloses contained in pretreated biomass suspension (Himmel *et al.*, 1996). The methods of lignocellulose pretreatment can be classified into three groups: chemical, physico-chemical and biological pretreatment.

8. Enzymatic hydrolysis of cellulose

Enzymatic hydrolysis of cellulose is carried out by highly specific cellulase enzymes (Suurnakki *et al.*, 2000). Different kinds of "cellulases" may be used to cleave the cellulose.

9. Screening of cellulase-producing microorganisms

Bacteria belonging to *Clostridium, Cellulomonas, Bacillus, Thermomonospora, Ruminococcus, Bacteriodes, Erwinia, Acetovibrio, Microbispora* and *Streptomyces* can produce cellulases (Bisaria 1989). *Cellulomonas fimi* and *Thermomonospora fusca* have been extensively studied for cellulase production. Although many cellulolytic bacteria, particularly the cellulolytic anaerobes such as *Clostridium thermocellum* and *Bacteroides cellulosolvens* produce cellulases with high specific activity, they do not produce high enzyme titers (Vladimir *et al.*, 2005). Because the anaerobes have a very low growth rate and require anaerobic growth conditions, most research for commercial cellulase production has focused on fungi (Duff and Murray 1996).

Fungi that have been reported to produce cellulases include *Sclerotium rolfsii, P. chrysosporium and species* of *Trichoderma, Aspergillus, Schizophyllum* and *Penicillium* (de Vries *et al.*, 2001; Mach and Zeilinger 2003; Krogh *et al.*, 2004). Fungi of the genera *Trichoderma* and *Aspergillus* are thought to be potent cellulase producers and crude enzymes produced by these microorganisms are commercially available for agricultural use.

10. Improving enzymatic hydrolysis

The factors that affect the enzymatic hydrolysis of cellulose include substrates, cellulase activity and reaction conditions (temperature, pH and other parameters). To improve the yield and rate of the enzymatic hydrolysis, research is focused on optimizing the hydrolysis process and enhancing cellulase activity (McCarthy *et al.*, 2004).

10.1. Substrates

Subtrate concentration is one of the main factors that affect the yield and initial rate of enzymatic hydrolysis of cellulose. Low substrate levels, result in increase of the yield and reaction rate of the hydrolysis (Cheung and Anderson 1997). However, high substrate concentration can cause substrate inhibition, which substantially lowers the rate of the hydrolysis and the extent of substrate inhibition depends on the ratio of total substrate to total enzyme (Penner and Liaw 1994).

10.2. Cellulase loading

Increasing the dosage of cellulases in the process, to a certain extent, can enhance the yield and rate of the hydrolysis, but would significantly increase the cost of the process. Cellulase dosage of 10 FPU/g cellulose is often used in laboratory studies because it provides a hydrolysis profile with high levels of glucose yield in a reasonable time (48-72 h) for a reasonable enzyme cost (Gregg and Saddler 1996).

Cellulases can be recovered from the liquid supernatant or the solid residues. Enzyme recycling can effectively lower the enzyme cost (Movagharnejad, 2005). The efficiency of cellulose hydrolysis decreases gradually with each recycling step.

10.3 End-product inhibition

Cellulase activity is inhibited by cellobiose and to a lesser extent by glucose. Several methods have been developed to reduce the inhibition, including the use of high concentrations of enzymes, the supplementation of ß-glucosidases during hydrolysis and the removal of sugars during hydrolysis by ultra filtration or simultaneous saccharification and fermentation (SiSF). The production of enzymes is influenced by induction and catabolite repression (de Groot *et al.*, 2003). 2-Deoxy-D-glucose (DG), a toxic glucose analogue has frequently been employed to isolate glucose-deregulated mutants (Haq *et al.*, 2001). Hence development of 2 deoxy glucose resistant mutants will result in increased cellulase production (Spiridonov and Wilson 2000).

10.4 Formation of inhibitors in hydrolysis

In addition to the sugars, several inhibitors are released in the hydrolysis process. During acid hydrolysis of lignocellulosics, aliphatic acids (acetic, formic and levulinic acid), furan derivatives and phenolic compounds are formed in addition to the sugars. Furfural and 5-hydroxymethyl furfural (HMF) are the most important furans, formed by decomposition of pentoses and hexoses respectively (Zhao *et al.*, 2005). Acetic acid is formed by hydrolysis of the acetyl groups in the hemicellulose as a consequence of deacetylation of acetylated pentosan (Lawford and Rousseau, 1993). These compounds are known to inhibit ethanol fermentation performance synergistically (Keating *et al.*, 2006). Therefore, in order to improve the fermentability and quality of the hydrolyzates, it is necessary to remove the inhibitory compounds before fermentation.

11. Removal of inhibitors

Hydrolysis of lignocellulosic materials always goes together with the formation of by-products that inhibit the fermentation process. Removal of inhibitors can be done either by reducing the formation of inhibitors or by the process of detoxification.

11.1 Reducing the formation of inhibitors

Modification of the hydrolysis process in order to minimize the formation of hydrolysis by-products is of interest in order to obtain a higher sugar yield by decreasing the formation of by-products and achieve a better fermentability of the hydrolyzate. Among the methods used for hydrolysis of the lignocellulosic substrates, the enzymatic hydrolysis is probably the best as lower concentrations of inhibitors are formed. However, it is important to note that enzymatic hydrolysis alone is not best in terms of process economy at present stage of development (Qureshi and Manderson 1995).

11.2 Detoxification

In order to remove the inhibitors and increase the hydrolyzate fermentability, several detoxification treatments, including chemical, physical and biological methods have been used. These methods include neutralization, over liming, ion exchange resins, adsorption into activated charcoal, Tin oxides and treatments with enzymes such as peroxidase and laccase (Palmqvist and Hahn-Hagerdal, 2000). Since detoxification increases the cost of the process (Sivers *et al.*, 1994) it is important either to overcome detoxification step or to develop cheaper and efficient methods.

b.1 Overliming: The most economical method of detoxification involves treatment of hydrolyzates with solid calcium hydroxide called as "overliming" (Rana-tunga *et al.*, 2000). Overliming is one of the classical chemical detoxifi-cation methods (Leonard and Hajny, 1945). In this method, CaO or $Ca(OH)_2$ is added to the hydrolyzate to adjust the pH to 9 or 10. Furan derivatives make transient complexes with calcium ions and this complex will be removed in solids (Purwadi *et al.*, 2004). Extraction with organic solvents such as ether or ethyl acetate has also been applied to remove most of the inhibitors, such as phenolics, weak acids and furans (Palmqvist 1998).

b.2 Activated charcoal treatment: Activated charcoal treatment is an efficient and economic method for reduction in the amounts of phenolic compounds (Parajo and Dominguez 1996), acetic acid, other aromatic compounds, furfural and hydroxy methyl furfural (Mussato and Roberto 2001) normally found in hemi-cellulosic hydrolyzates.

b.3 Biological detoxification methods: Biological detoxification is another method to enhance the fermen-tability of hydrolyzates. Lignin is degraded by several fungal enzymes like lignin peroxidase, Mn-dependent peroxidase and laccase or mono-phenol oxidase (Lee, 1997). Jonsson *et al.* (1998) and more recently Chandel *et al.*, 2006a, applied laccase to remove phenolic monomers and phenolic acids from willow hydrolyzates treated with steam and SO_2. Treatment with the peroxidase and laccase enzymes, obtained from the lignolytic fungus *Trametes versicolor*, has shown to increase the ethanol productivity in a hemicellulose hydrolyzate of willow (Jonsson *et al.*, 1998). *Debraryomyces* an other lignolytic fungus helps in detoxification by removing monomeric and polymeric phenolics (Pasha and Reddy 2005).

12. Microorganisms for ethanol production

Ethanol fermentation is a biological process in which fermentable sugars are converted to ethanol and CO_2 by microorganisms. The microorganism is the key factor in the conversion of sugars to ethanol. There are several important characteristics of the microorganism in an industrial "hydrolyzate-to-ethanol" process. Ideally, the microorganism should give a high ethanol yield, have a high ethanol tolerance, and be resistant to inhibitors of hydrolyzates, with no oxygen requirement and should utilize a broad range of substrate. It is also desirable to have a strain with high rate of sugar consumption, minimal nutrient require-ment, high salt tolerance, high shear tolerance, thermotolerance, safety for humans and no spore formation (Picataggio and Zhang 1996). Strain development studies are being carried out to meet the above criteria.

Although many bacteria, fungi and yeast are employed in ethanol fermentation, the most commonly used microbe is yeast. Bioconversion of various agro residues into ethanol is also carried out by different yeasts (Nigam 2001; Krishna *et al.*, 2001; Ali *et al.*, 2004; Sharma *et al.*, 2004;

Palmarola 2005, Chandel *et al.*, 2006a). *Saccharomyces* is also generally recognized as safe (GRAS) organism as a food additive for human consumption and is therefore ideal for producing alcoholic beverages and for leavening bread. *Saccharomyces cerevisiae*, which can produce ethanol to give concentration as high as 18 percent in the fermentation broth, is the preferred organism for most ethanol fermentation. Theoretically, 100 grams of glucose produces 51.11 g of ethanol and 48.8 g of carbon dioxide. However, in practice, the microorganisms use some of the glucose for growth and hence the actual yield is always less than 100%. *Saccharomyces cerevisiae* has several desirable characteristics for fermentation of hydrolyzates, which are widely cited in the literature. But the main disadvantage is its limited substrate utilization range, specially its inability to ferment xylose. This is particularly disadvantageous when fermenting hardwood hydrolyzates. Still, *S. cerevisiae* is probably the first choice for most hydrolyzates, especially from softwoods, where glucose and mannose constitute the largest part of the sugars. Some identified xylose fermenting yeast strains are *Pachysolen tannophilus, Pichia stipitis, Candida shehatae* and the most widely applied recombinant bacterial strains are *E. coli, Klebsiella oxytocoa* and *Zymomonas mobilis* (Krishnan *et al.*, 1997; Himmel *et al.*, 1997).

Zymomonas mobilis has several appealing properties like it is used as a biocatalyst for ethanol production. It shows a homoethanol fermentation pathway and tolerates up to 120 g/l ethanol. *Z. mobilis* is a GRAS organism and has simple nutritional requirements. Despite its advantages as an ethanologen, it is not well suited for all the biomass conversions because it can ferment only glucose, fructose and sucrose (Swings and Deley 1977).

Genetically engineered microorganism that can convert pentose and hexose sugars simultaneously to ethanol can greatly improve ethanol production efficiency and reduce the production cost (Sun and Cheng 2002). Both yeasts (such as *Saccharomyces* and *Pichia* species) and bacteria (*E. coli, Klebsiella* and *Zymomonas*) have been genetically engineered to ferment glucose, xylose and arabinose sugars (Ingram 1999; Nigam 2001). Though recombinant technologies have greatly improved bioethanol production, still there are lots of problems, which are to be solved. One of the major problems include maintaining a stable performance of recombinant yeast in commercial scale fermentation operations (Chan *et al.*, 2004), and developing more efficient technologies for lignocellulosic biomass (conversion) and integrating best components into an economic ethanol production system.

13. Thermotolerant yeasts

Thermotolerant yeasts are known for their growth below 20°C and up to 40°C (Kiransree *et al.*, 2000; Limtong, *et al.*, 2005; Rao and Pasha

2006). Thermotolerance is defined as increase in cellular resistance to hyperthermia occurring after either a brief exposure to 42-45°C followed by recovery at 37°C (acute thermotolerance) or continuous exposure to longer periods at non-lethal temperatures of 38-40°C (chronic thermo tolerance). A shift from 23 to 30°C do not result in any major alterations in the pattern of protein synthesis (Lepock 2003) but a shift to 36°C (mild heat shock) does. It is assumed that the response might play a major role in the protection of cells from thermal injury. Indeed, a rapid shift in cultivation temperature of *S. cerevisiae* from 23 to 36°C leads to protection from death due to a subsequent extreme (52°C) heat treatment.

13.1 Application

The idea of using thermotolerant yeast attracted many ethanol producers and researchers over the years.In any fermentation process it is necessary to reduce energy requirement and production costs. This can be achieved by the use of thermo tolerant yeasts, which help in decreasing or minimizing cooling, distillation and mixing costs and by theoretically increasing growth and productivity rate. They also reduce the risk of contamination and also make the product recovery very easy. They are especially attractive in tropical countries like India where ambient temperature ranges from 30 to 45°C. They have faster fermentation rates and can also be used in leavening process in baking industry. In addition they can be used for efficient simultaneous saccharification and fermentation of cellulose by cellulases because the temperature optimum of cellulose enzymes (40-45°C) is close to the fermentation temperature of thermotolerant yeasts (Bollok *et al.*, 2000). Higher substrate solubility, saccharification and fermentation rates, anaerobic conditions by low oxygen solubility, continuous ethanol removal make this method efficient. In this regard thermotolerant yeasts would be advantageous in many warmer regions of the world where cooling costs are more expensive. Still few draw backs of thermotolerant yeasts like lower ethanol and substrate tolerance and increased media evaporation makes their utilization limited.

13.2 Isolation and improvement

To find the suitability of thermotolerant yeast strains for ethanol production many researchers focused n their isolation and selection. There are several reports in the literature on thermotolerant *Kluyveromyces* yeast strains (Banat and Marchant 1995; Singh *et al.*, 1998). But very few reports describe the growth and ethanol fermentation of *Saccharomyces cerevisiae* at elevated temperatures (Banat *et al.*, 1992; Kiransree *et al.*, 2000). A Japanese group at Kumamoto University (Kida *et al.*, 1992) obtained a thermotolerant flocculating strain of *Saccharomyces cerevisiae* KF-7 through protoplast fusion of two yeast strains. A further investigation on its stability under repeated-batch fermentation and on improvement of its osmotolerance (Morimura *et al.*, 1997) has been reported. Barron *et al.* (1995) and Banath *et al.* (1996) are well known for their work on thermotolerant yeast strain, *Kluyveromyces marxianus* IMB3, capable of producing 10 g/l of ethanol during growth at 45°C on a medium containing milled paper and commercial cellulase by simultaneous saccharification and fermentation (SiSF). Recently bio-ethanol production using lignocellulosics and thermotolerant yeasts has gained a large attention (Ryabova *et al.*, 2003; Ballesteros *et al.*, 2004; Kadar *et al.*, 2004). But there are very few reports on the selection of yeasts that are able to grow and ferment biomass at higher temperatures at commercial levels (Banat *et al.*, 2000). Enrichment and isolation program for new ethanol-producing thermotolerant yeasts strains from the fermentation tanks of brewing or distilleries (D'Amore *et al.*, 1989; Banat *et al.*, 1992), hot soil region (Banat *et al.*, 1992; Kiransree *et al.*, 2000) and hot springs (Ueno *et al.*, 2001) and waste compost (Balakumar *et al.*, 2001; Fong Jiunn 2006) samples resulted in the selection of several thermotolerant yeasts. But often they are not suitable to use directly at high temperature fermentations (Abdel-Fattah *et al.*, 2000). Hence a lot of efforts are required to improve the strains' thermotolerance and fermentation at high temperatures using UV (Sridhar *et al.*, 2002; Rajoka *et al.*, 2005), chemical mutagenesis (Balakumar *et al.*, 2001; Ueno *et al.*, 2002), adaptations (Balakumar *et al.*, 2001; Matmati *et al.*, 2002) protoplast fusion (Ezeronye and Okerentugba 2001) and recombinant DNA technology (Becker and Boles 2003). It is necessary to focus on strain improvement, immobilization (Riordan *et al.*, 1996) and media optimization (Gough *et al.*, 1996). But only few researchers attempted to understand the mechanism of thermotolerance in yeast (Piper 1993).

In our laboratory after a mass-screening program of isolating thermotolerant yeasts for ethanol production, several thermotolerant yeast strains were isolated from the soil samples collected within hot regions of Kothagudem thermal power station, A.P., India, where the temperature in summer months are more than 45°C. Out of these isolates, VS_1, VS_2, VS_3 & VS_4 were found to be suitable for ethanol production at 40°C. Among these isolates VS_3 strain was found to have good thermotolerance, osmotolerance and ethanol tolerance compared to VS_1, VS_2 & VS_4. Experiments were conducted to find out the suitability of VS_3 strain for ethanol production from various carbon sources using Solid Substrate Fermentation, Submerged Fermentation and Immobilization (Kiran Sree *et al*, 1999& 2000). Later strain improvement studies were carried out to further improve the thermotolerance, osmotolerance and ethanol tolerance of VS_1 and VS_3 strains using adaptation, physical (UV) (Sridhar *et al*, 2002), chemical (1-methyl-3-nitro-1-nitrosoguanidine-MNNG) mutagenesis and protoplast fusion.

14. Fuel Grade Ethanol Recovery

Fuel grade ethanol can be recovered by two stage distillation process. In the first stage 95% ethanol is obtained. The remaining 5% water is removed by azeotropic distillation or by using molecular sieves which gives 99% ethanol.

Molecular sieves

Molecular sieve dehydration technology utilizes microporous particles such as alumino-silicates, possessing a very precise pore size. The pores make it possible to separate small molecules from large ones through selective adsorption. For example, ethanol dehydration is accomplished with molecular sieves having a pore diameter of 3Å, which entraps water molecules, which have a diameter of 2.5Å. Ethanol molecules, which have a diameter of 4Å, cannot enter and therefore, flow around the material. However, process equipment size restrictions kept liquid phase molecular sieves from making significant inroads into the ethanol industry.

15. Economic feasibility of biofuels

The import cost of crude oil during the year 2004-2005 was around Rs. 1,21,500 crores (Subramanian *et al.*, 2005). Historically, the projected cost of bioethanol has dropped from US$1.22 per liter to about US$ 0.31 per liter because of continuous improvement in pretreatment, enzyme application and fermentation (Wyman, 1999). Further economic analysis of the bioethanol process yielded a projected cost of as low as US$ 0.20 per liter by 2015, if enzymatic processing and biomass improvement targets are met (Wooley *et al.*, 2001). The process used for ethanol production, using cellulolytic enzymes makes the process cost effective. Recently the cost of enzymes was reduced by a combination of protein engineering and process development. The use of novel, tailored cocktails of enzymes with higher specific activities are required for further cost reductions (Gray *et al.*, 2006) World leaders in enzyme production, Genencore, Novozyme and logen are actively working towards reducing cost of cellulase production for bioethanol production using new and improved enzymes developed by protein engineering and directed evolution.

In India ethanol is mainly produced from molasses but other crops like sweet sorghum, which require less input than sugarcane can be utilized for the purpose. Looking into production cost economics, ethanol from damaged food-grains involves lowest cost but the availability is insufficient to meet the demand. The next lowest cost option is corn but it is used in the preparation of corn flour and in other food products for human consumption thus none are available for ethanol production in India. Therefore the most promising alternative is agro-residues which are available in plenty in India, unfortunately the production of ethanol from agro residues is nil. Fuel ethanol production will remain a significant industry and become a potentially self-sustainable agricultural based system for the 22nd century, if the utilization of lignocellulosic becomes a commercial reality. R&D work for converting lignocellulosic biomass into ethanol is continuing in various laboratories and more research in this area needs to be encouraged for reducing the cost of ethanol production. (Chandel *et al.*, 2006b)

16. Future perspectives

The use of biofuels is hampered by scarce production and high cost. If blending of biofuel in gasoline is increased from 5%; it creates further hike of price due to high demand. It is proposed that a limited subsidy should be given for an initial period. This would help in bringing awareness quickly. Established biotech companies in India like- Biocon India Ltd. Bangalore, Advance Biochemicals, Pune and Reliance Life Sciences Ltd. should come forward to develop new technologies for production of biofuels. The present need is the deregulation of feedstock and its pricing, simplification of licensing for biofuel production, which will make biofuel industry strong and vibrant.

References

Abdel-Fattah, W.R., Fadil, M., Nigam, P. and Banat, I.M. (2000) *Biotechnol & Bioeng* 68, 531-535.

Alex, B., Mikhail, B., Neil, G., John, K., Vera, M., Satoshi, K., and Jack, S. (2006) *J Biotechnol* 125,198-209.

Ali, M., Nancy, D., Daniel, S., Yat-Chen,Ch., Christina, E., and Zhang, M. (2004) *Biotech Lett* 26, 321-325.

Andersson, E., Harvey, S., and Berntsson, T. (2006) Energy efficient upgrading of biofuel integrated with a pulp mill. *Energy* 31, 1384-1394

Balakumar, S., Arasaratnam,V. and Balasubramaniam, K. (2001) *W. J. Microbiol & Biotechnol* 17, 739 – 746.

Ballesteros, M., Oliva, J.M., Negro, M.J., Manzanares, P. and Ballesteros, I. (2004) *Proc Biochem* 39, 1843-1848.

Banat, I.M. and Marchant, R. (1995) *W.J. Microbiol and Biotechnol* 11, 304–306.

Banat, I.M., Nigam, P. and Marchant, R. (1992) *W. J. Microbiol. Biotechnol* 8, 259–263.

Banat, I.M., Singh, D., Marchant, R. (1996) *Acta Biotechnol* 16, 215-223.

Banat, I.M., Singh, D., Nigam, P. and Marchant, R. (2000) Rec Advan in Food Sci 1, 11-55.

Barron, N., Marchant, R., McHale, L. and McHale, A.P. (1995) *Appl. Microbiol Biotechnol* 43, 518-520.

Becker, J. and Boles, E. (2003) *Appl Environ Microbiol* 69, 4144-4150

Bisaria, V.S. and Mishra, S. (1989) *Crit Rev Biotechnol* 9, 61–103.

Bollok, M., Reczey, K. and Zacchi, G. (2000) *Appl Biochem Biotechnol* 84, 69–80.

Boopathy, R., Gurgas, M., Ullian, J. and Manning, J.F. (1998) *Curr. Microbiol* 37, 127–131.

Bullen, R.A., Arnot, T.C., Lakeman, J.B. and Walsh, F.C. (2006) *Biosens & Bioelectro* 21, 2015-2045.

Butterworth,B. (2006) *Refocus* 7,60-61.

Chan, A. W., R. Hoffman, and B. McInnis. (2004) *Ecol and Soci* 9,1-17.

Chandel, A.K., Kapoor, R.K., Lakshmi Narasu, M., Viswadevan, V., Saravana Kumaran, S.G., Rudravaram, R., Venkateswar Rao, L., Tripathi, K.K., Lal, B. and Kuhad, R.C. (2006b) Int. J. Global Energy Issues. In Press

Chandel, A.K., Kapoor, R.K., Singh, A. and Kuhad R.C. (2006a) *Biores. Technol.* In press.

Cheung, S.W. and Anderson, B.C. (1997) *Bioresour Technol* 59, 81-96.

Coral, G., Colak, O., Unaldi, M. N. and Arikan, B. (2003) *Turk J Biol* 27,1-5.

Cowling, E.B. and Kirk,T.K. (1976) *Biotechnol & Bioeng Sympo* 6, 95–123.

Das, H. and Singh,S.K. (2004) *Crit Rev Food Sci Nutr* 44, 77–89.

De Groot, M.J., Van, D.E., Vondervoort, P.J., De, Vries, R.P., Vankuyk, P.A., Ruijter, G.J. and Visser, J. (2003) *Microbiol* 149,1183-1191.

De Vries, R.P. and J. Visser.(2001) *Microbiol Mol Biol Rev* 65, 497–522

Dewes, T. and Hunsche, E. (1998) *Biological Agriculture and Horticulture* 16, 251-258.

Duff, S.J.B. and Murray, W.D. (1996) *Biores Technol* 55, 1-33.

Eidman,V.R.(2006) The Economics of Energy Production in Agriculture. "Energy in Agriculture: Managing the Risk,"conference presentation, June 27-28, 2006, Kansas City, Missouri.

Ericsson, K., Nilsson. and Lars, J. (2004) *Biomass Bioenergy* 26, 205–220.

Ezeronye, O.U. and Okerentugba, P.O. (2001) *Lett Appl Microbiol* 32,190-193.

Farrell, A.E., Plevin, R.J., Turner, B.T., Jones, A.D., O'Hare, M. and Kammen, D.M. (2006) *Science* 311, 506–508.

Fong Jiunn, C.N. (2006) *Extremophiles* 10, 363-372.

Gallagher, P., Schamel, G., Shapouri.and Brubaker, H. (2006) *Agribusiness* 22, 109-134.

Gallo, B.J. (1979) International Symposium on Genetics of Industrial Microorganisms, 3rd, Madison, WI, 1978. Abstract No. 35.

Gautam, M. and Martin, D.W. (2000) *J Power Energy* 214, 165–182.

Gmitter, F.G. Jr., Ling, X.B. and Deng, X.X. (1990) *Theoret and Appl Geneti* 80, 785-790.

Gough, S., Flynn, O., Hack, D.J. and Marchant, R. (1996) *Appl Microbiol Biotechnol* 46, 187–190.

Gray, K.A., Zhao, L.and Emptage, M. (2006) *Curr Opin Chem Biol* 10,141-146.

Gregg, D.J. and Saddler, J.N. (1996) *Biotechnol Bioeng* 51, 375 – 383.

Haq, I., Khurshid, S., Ali, S., Ashraf, A., Qadeer, M.A., Rajoka, M.I. (2001) Mutation of *Aspergillus niger* for hyper-production of citric acid following fermentation of blackstrap molasses. *World J. Microbiol & Biotechnol* 17, 35-37.

Himmel, M. E., William, S. A., Baker, J. O., Nieves, R. A. and Steven, R. T. (1996) Cellulases: structure, function and applications. In: Handbook on Bioethanol: Production and Utilization. Wyman, C.E., Ed., Taylor and Francis, Washington. pp. 143– 161.

Himmel, M. E., Adney, W. S., Baker, J. O., Elander, R., McMillan, J. D., Nieves, R.A., Sheehan, J. J., Thomas, S. R., Vinzant, T. B. and Zhang, M. (1997) Advanced Bioethanol Production Technologies. A perspective. –In: Fuels and Chemicals from Biomass. B.D. Chap. 1, pp. 2-45.

Hood, E.E. (2004) Bioindustrial and biopharmaceutical products from plants. In New directions for a diverse planet. Proceedings of the 4th International Crop Science Congress, Brisbane, Australia, Sept 26–Oct 1; ISBN 1 920842 2-29

Ingram, L.O., Aldrich, H.C., Borges, A.C.C., Causey, T.B., Martinez, A. Morales, F., Saleh, A., Unverwood, S.A., Yomano, L.P., York, S.W., Zaldivar, J. and Zhou, S.D. (1999) *Biotechnol. Prog* 15,855-866.

John, J.A. (1969) Hydration, in Miller SA (ed.), Ethylene and industrial derivatives. Ernest Benn, London, 690-801.

Jonsson, L.J., Palmqvist, E., Nilvebrant, N.O. and Hahn-Hägerdal, B. (1998) *Appl Microbiol Biotechnol* 49,691–697.

Kadar, Z.S., Szengyel, Z.S. and Reczey, K. (2004). *Indust Crop & Products* 20, 103-110.

Kapoor, R.K., Chandel, A.K., Kuhar, S., Gupta, R. and Kuhad, R.C. (2006) Bioethanol from crop residues, production forecasting and economics: An Indian perspective, In: R.C. Kuhad and A. Singh editors, Lignocellulose Biotechnology: Current and Future Prospects, New Delhi I.K.International pp.32-45.

Katzen, R. and Monceaux, D.A. (1995) *Appl. Biochem. Biotechnol* 51, 585-592.

Keating, J.D., Panganiban, C. and Mansfield, S.D. (2006) *Biotechnol & Bioeng* 93, 1196 – 1206

Kim, S. and Holtzapple, M.T. (2006) *Bioresour Technol* 97, 583–591.

Kim, S. and Dale, B.E.(2004) *Biomass Bioenerg* 26, 361–375.

Kiran Sree, N., Sridhar, M., Suresh, K., Banat, I.M. and Venkateswar Rao, L. (2000) *Bioresour. Technol* 72, 43-46.

Kiran Sree, N., Sridhar, M., Suresh, K., Rao, L.V. and Pandey, A. (1999) *Proc Biochem* 34, 115 – 119.

Krantz, M., Nordlander, B., Valadi, H., Johansson, M., Gustafsson, L. and Hohmann, S. (2004) *Eukar Cell* 3,1381-1390.

Krishna, S.H., Reddy, T.J. and Chowdary, G.V. (2001) *Bioresour Technol* 77, 193-196.

Krishnan, M.S., Xia, Y., Ho, N.W.Y. and Tsao, G.T. (1997) Fuel ethanol production from lignocellulosic sugars studies using genetically engineered

Saccharomyces yeast. –In: *Fuels and Chemicals from Biomass. B.D. Saha, J. Woodward (eds.)* ACS Symp. Ser.

Krogh, K.B.R., Mørkeberg, A., Jorgensen, H., Frisvad, J.C. and Olsson, L. (2004) *Appl Biochem Biotechnol* 113, 389–401.

Lawford, H.G. and Rousseau, J.D. (1993)*Appl. Biochem. Biotechnol* 39, 301-322.

Lee, J. (1997) *J. Biotechnol* 56, 1-24 Limtong, *et al.*, 2005

Leonard, R.H. and Hajny, G.J. (1945) *Ind Eng Chem* 37,390–395.

Lepock, J.R. (2003) *Int. J. Hyperthermia* 19, 252–266.

Lynd, L.R., van Zyl, W.H., McBride, J.E. and Laser, M. (2005) *Curr Opin Biotechnol* 16, 577–583

Mach, R.L. and Zeilinger, S.(2003) Appl Microbiol Biotechnol 60, 515–522.

Matmati, N., Morpurgo, G., Babudri, N. and Marini, A. (2002) *J .Basic. Microbiol* 42, 345-54.

McCarthy, J.K., Uzelac, A., Davis, D.F. and Eveleigh, D.E. (2004) *J Biol Chem* 279,11495–11502

Morimura, S., Ling, Z.Ya.and Kida, K. (1997) *J. Ferment & bioeng* 83, 271-274..

Movagharnejad, K. (2005) *Biochem Eng J* 24, 217-223

Mussato, S.I. and Roberto, I.C. (2001) *Biotechnol Lett* 23, 1681–1684

Nelson, C.R. and Courter, M.L (1954) *Chem. Eng. Progr* 50, 526-532.

Nigam, J.N. (2001) *J Appl Microbiol* 90,208– 215.

Niven R (2005). *Renew Sustain Ener Rev* 9,535–555.

Palmarola-Adrados, B., Galbe, M. and Zacchi, G. (2005) *J Chem Technol Biotechnol* 80, 85-91 Piper 1993

Palmqvist, E. and Hahn-Hagerdal, B. (2000) Bioresour.Technol 74, 25-33

Palmqvist, E., Grage, H., Meinander, N.Q. and Hahn-Hägerdal, B. (1999) *Biotechnol Bioeng* 63,46–55.

Parajo, J.C., Dominguez, H., and Dominguez, J.M. (1997) Enzyme Microb Technol 21,18–24.

Parikka, M. (2004) Biomass Bioenergy 27, 613–620.

Pasha, C. and Reddy, G. (2005) Food Chem 89, 449 – 453.

Pasha.C., Aruna, A., Maqsood, A.M. and Rao, L.V. (2005) *J Appl Microbiol* 98, 318–323.

Penner, M.H. and Liaw, E.T. (1994) Kinetic consequences of high substrate to enzyme ratios in *Trichoderma* cellulase based saccharification systems. In enzymatic conversion of biomass for fuels production (M.E. Himmel, J.O. Baker and R.P. Overend, eds.).

Philippidis, P.G. and Smith, K.T. (1995) *Appl. Biochem. Biotechnol* 51, 117-123.

Picataggio, S.K. and Zhang, M. (1996) Biocatalyst development for bioethanol production from hydrolysates, In Wyman CH (ed.), Handbook on bioethanol: production and utilization. Taylor & Francis, Washington DC, 163-178.

Pimentel, D. (2003) *Nat Resour Res* 12, 127–134.

Piper, P W. (1993) FEMS Microbiol. Rev 11, 339-356.

Purwadi, R., Niklasson, C. and Taherzadeh, M.J. (2004) *J. Biotechnol* 114, 187-198.

Qureshi, N. and Manderson, G.J. (1995) Bioconversion of renewable resources into ethanol: an economic evaluation of selected hydrolysis, fermentation, and membrane technologies. *Energy Sourc* 17, 241-265

Ragauskas, A.J. Williams, C.K. Davison, B.H. Britovsek, G. Cairney, J. Eckert, C.A. Frederick Jr. W.J., Hallett, J.P. Leak D.J. and Liotta C.L. (2006) *Science* 311, 484–489.

Rajoka, M.I., Ferhan, M. and Khalid, A.M. (2005) *Lett in Appl Microbiol* 40: 316-321

Ranatunga, T.D., Jervis, J., Helm, R.F., McMillan, J.D. and Wooley, R.J. (2000) *Enzyme. Microb Technol* 27, 240–247.

Rao, L.V. and Pasha, C. (2006) Thermotolerant yeasts for bioethanol production using lignocellulosic substrates. In "Yeast" Book chapter edited by T. Satyanarayana and B.N Joshi published by J.K International, New Delhi. 2006.

Riordan, C., Love, G., Barron, N., Nigam, P., Machant, R., McHale, L. and McHale, A.P. (1996) *Bioresour Technol* 55, 171-173.

Ryabova, O.B., Chmil,O.M. and Sibirny, A.A. (2003) *FEMS Yeast Res* 4, 157-164.

Sharma, I.K. (1981) *J. Econ. Taxon. Bot* 2,245–248.

Sharma, S. (2004) Lantana-whose weed any way! developing strategic directions for integrated utilization and control, In: Workshop on *Lantana camara*: Problems and prospects (Volume of abstracts), organized by IIT, Delhi, HESCO, Dehra Dun and Department of Science and Technology (DST), Govt. of India at Dehra Dun from Feb. 10–11.

Singh, D., Banat, I.M., Nigam, P. and Marchant, R. (1998) Biotechnol. Lett 20, 753 -755.

Sivers, V.M., Zacchi, G., Olsson, L. and Hahn-Hagerdal, B. (1994) *Biotechnol Prog* 10: 555-560.

Sluiter, A., Hames, B., Ruiz, R., Scarlata, C.,Sluiter, J. and Templeton,D. (2004) Determination of structural carbohydrates and lignin in biomass. NREL, Golden, Co.

Solomon, B.O., Amigun, B., Betiku, E., Ojumu, T.V. and Layokun, S.K. (1999) *JNSChE* 16, 61–68.

Spiridonov, N.A. and Wilson, D.B. (2000) *J Bacteriol* 182, 252–255.

Sreenath, H.K. and Jeffries, T.W. (2000) *Biores Technol* 72, 253-260

Sridhar, M., Kiran Sree, N. and Venkateswar Rao, L. (2002) *Biores Technol* 83, 199 – 202

Subramanian, K.A., Singal, S.K., Saxena, M. and Singhal, S. (2005) *Biomass Bioenergy* 29,65-72.

Sun, Y. and Cheng. J. (2002) *Biores Technol* 83, 1-11.

Suurnakki, A. M., Tenkanen, ML., Niku-Paavola, L, and Viikari, M. Siika-aho. and Buchert, J. (2000) *Cellulose* 7, 189–209

Swings, J. and DeLey, J. (1977) *Bacteriol Rev* 41, 1–46.

Ueno, R., Urano, N. and Kimura, S. (2002) *Fisheries Sci* 68, 571–578.

Vladimir, V., Schantz, N.Z. and Schwarz, W.H. (2005) *FEMS Microbiol Lett* 249, 353-358.

Von Blottnitz, H. and Curran, M.A. (2006) *J. Cleaner Product* 34, 2654-2661.

Wheeler, K., Jansheka, R H. and Sakuma, Y. (1991) Ethyl alcohol, In Chemical Economics Handbook. SRI International, USA.

Wooley, R., Ruth, M., Sheehan, J. and Ibsen, K. (2001) Lignocellulose biomass to ethanol process design and economic utilizing co-current dilute acid prehydrolysate and enzymatic hydrolysis current and futuristic cases. *NREL/TP*, 580-26157.

Wright, L. (2006) *Biomass Bioenergy* 30, 706-714.

Wyman, C.E. (1996) Ethanol production from lignocellulosic biomass: overview. – In: Handbook on bioethanol: production and utilization. C.E. Wyman (ed.) Taylor and Francis, Washington, DC, Chapter 1, 1-18.

Wyman, C.E., Dale, B.E., Elander, R.T., Holtzapple, M., Ladisch, M.R. and Lee, Y.Y. (2005) *Bioresour. Technol* 96, 1959–1966.

Wyman, C.E. (1999) *Ann Rev Energy Environ* 24, 189-226.

Zhao, J., Wang, M., Yang, Z. and Yang, Z. (2005) *Enzyme & Microbial Technol* 37, 246-253.

Chapter - 10

Microbial lipases- An Overview

S. Ram Reddy and Pallavi Pogaku

Department of Microbiology, Kakatiya University, Warangal, A.P
E-mail: sanditi_ramreddy@yahoo.com

ABSTRACT

Lipases (triacylglycerol acylhydrolase, EC 3.1. 1.3) are part of the family of hydrolases that act on carboxylic ester bonds. Their physiological role is to hydrolyze triglycerides to diglycerides monoglycerides, fatty acids and glycerol. Bacterial lipases are of interest because they are generally more stable than those from other organisms, especially against high temperatures and severe operational conditions and/or storage conditions. They usually exhibit good chemioselectivity, regioselectivity and enantioselectivity. Lipases exhibit broad substrate specificity and can be found with optimum activities over a wide range of temperatures. These interesting properties make lipases the most versatile biocatalysts. Bacterial lipases are mostly inducible enzymes, requiring some form of oil, fatty acid, fatty acid alcohol or fatty acid ester for induction. Lipases are usually secreted out in the culture medium during late logarithmic or stationary phases. A few researchers have performed systematic medium optimization and fermentation studies for lipase production. The cost of microbial lipase production depends on the enzymatic yield which is defined by the cost of the amount of enzyme produced, the cost of downstream processing and the stability of the enzyme at post-harvest. Recently, the production of microbial lipases in organic solvent is gaining much attention. In view of the advantage of whole cells immobilization technique, studies have been made on lipase production in this direction. Lipases are mostly induced in presence of fats or oil in the culture medium. Tween 80 was shown in several reports to be an effective substrate for lipase production. Forty seven different bacterial lipases are classified into six families on the basis of amino acid sequence homology. The purification efficiency is estimated by the determination of the enzyme activity yield and the purification factor. As a general rule, four steps of chromatography are needed to reach high purity for lipases. Lipases have become an integral part of the modern food industry. The use of enzymes to improve the traditional chemical processes of food manufacture has been developed in the past few years. Their applications are also found in the detergent, food, leather, textile, oil and fat, cosmetic, paper and pharmaceutical industries. Because of their excellent capability for specific regioselective reactions in a variety of organic solvents with broad substrate recognition, lipases have emerged as an important biocatalyst in biomedical applications. Lipases were generally added to the detergents primarily in combination with proteases and cellulases. Leather processing involves the removal of subcutaneous fat, de-hairing and stuffing. Tanning processes are usually performed in an alkaline environment, so alkalophilic microbes ought to be better for exploration.

Key words: Lipases, bacteria, fermentation, applications

1. Introduction

The annual market for industrial enzymes has reached around US$1.6 billion in 2000 (Demain, 2000). Twelve major suppliers and four hundred minor producers assure the world demand of enzymes. Europe alone produces around 60% of the total world supply of industrial enzymes which are mainly (at least 75%) hydrolytic including proteases, carbohydrases and lipases (Sharma *et al.,* 2001). Lipases (triacylglycerol acylhydrolase, EC 3.1. 1.3) are part of the family of hydrolases that act on carboxylic ester bonds. Their physiological role is to hydrolyze triglycerides to diglycerides, monoglycerides, fatty acids and glycerol. Lipase producing bacteria were reported nearly a century ago by the microbiologist (Jaeger and Eggert, 2002), but it is not until the 1950s that the first works on fungal lipases were reported (Ghosh *et al.,* 1996). Lipases catalyze esterification, interesterification, acidolysis, alcoholysis and aminolysis in addition to the hydrolytic activity on triglycerides. As hydrolases, they do not require cofactors. They usually exhibit good chemioselectivity, regioselectivity and enantioselectivity. Finally, lipases possess broad substrate specificity and can be found with optimum activities over a wide range of temperatures. These interesting properties make lipases as the most versatile biocatalysts. Their applications are found in the detergent, food, leather, textile, oil and fat, cosmetic, paper and pharmaceutical industries (Kademi *et al.,* 2004). Commercial use of lipases in various industries is a billion-dollar business (Jaeger and Eggert, 2002; Pandey *et al.,* 1999a). In the detergent industry, about 1000 tons of lipases are sold every year (Godfrey and West, 1996).

A typical reaction catalyzed by lipases is shown in Fig 1.1.

The interest in lipases arises due to the ability of these enzymes to catalyze the hydrolysis as well as synthesis of fatty acid esters. Lipases act on a variety of substrates including natural oils, synthetic triglycerides and esters of fatty acids. Many lipases are resistant to solvents and hence find use in the synthesis of chiral drugs. They show varied substrate and positional specificities and find use in various industries like, food, chemical, pharmaceutical, cosmetic, leather and detergent.

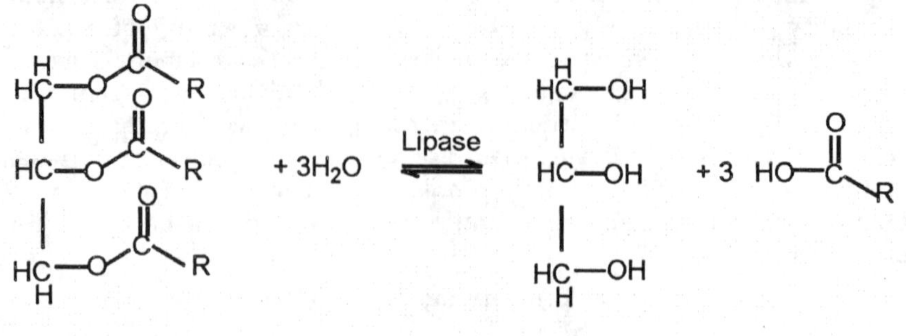

Triacylglycerol + $3H_2O$ $\overset{\text{Lipase}}{\rightleftharpoons}$ Glycerol + 3 Fatty acid

Fig. 1. Hydrolysis of triacylglycerol by lipase

2. Sources of lipases

Lipases (E.C. 3.1.1.3) are produced by several organisms namely, bacteria, fungi, archea and eukarya, as well as by animals and plants (Olson *et al.*, 1994). Most of the enzymes for commercial application are obtained from the microorganisms, including bacteria, fungi and yeast. Enzymes from these sources have wide application in different industries such as food, pharmaceuticals, leather, detergents, textiles, paper and pulp, waste management, etc. Food industry is the largest consumer of enzymes and approximately 45% of bulk share goes to it. Detergents are the second stake holder in enzymes consumption. Among all the industrial enzymes, hydrolytic enzymes account for 85%. The market size was approximately US$ 1.6 billions in 2002 and has witnessed ~12% annual growth over a period of last one decade. It is assumed that the market continues to grow fast and reaches US$ billions by next decade.

Table 1. Some commercially available microbial lipases and their sources

Source	Trade name	Supplier
Bacteria		
Achromobacter sp.	Lipase ALC, Lipase ALG	Meito Sangyo Co.
Alcaligenes sp.	Lipase PLC, Lipase PLG, Lipase QLC, Lipase QLG	Meito Sangyo Co.
Burkholderia cepacia	Lipase SL	Meito Sangyo Co.
Chromobacterium viscosum	Lipase CV	Genzyme
		Calbiochem

Pseudomonas cepacia	Lipase PS "Amano"	Amano
Pseudomonas fluorescens	Lipase AK "Amano"	Arnano
Pseudomonas stutzeri	Lipase TL	Meito Sangyo Co.
Thermus aquaticus		Fluka
Thermus flavus		Fluka
Thermus thermophilus		Fluka
Fungi		
Aspergillus niger	Lipase A "Amand' 6	Amano
	Lypolyve AN	Lyven
Penicillium camembertii	Lipase G "Arnano" 50	Amano
Penicillium roquefortii	Lipomod™338P - L338P	Biocatalysts Ltd
Mucor japanicus	Lipase M "Amano" 10	Amano
Mucor rhiehei	Piccnate	Gist-Brocades
Rhizomucor rhehei	Palatase®*	Novozymes
Rhizopus arrhizus		Fluka
Rhizopus delernar		Fluka
Rhizopus oryzae	Lipase F-AP15	Amano
	Lipase L036P - L036P	Biocatalysts Ltd
	Lipase F-DS	Amano
Rhizopus oryzae	Lipornod™ 627P - L627P	Biocatalysts Ltd
	LipopanC®) F	Novozymes
	Novol inteir (with pailease)	Novozymes
	Greasex ®, NovoCor ® AD	Novozymes
Rhizopus niveus	Newlase F	Amano
	Lipozyme « TL IM	Novozymes
Rhizopus sp.	Lipase UL	Meito Sangyo Co.
Thermomyces lanuginosus	Lipolase®, Lipolase® Ultra	Novozymes
	Lipo Prime™, Lipex ®	
Penicillium sp./ *Aspergillus* sp.	Lipomod™ 621 P - 1,621	Biocatalysts Ltd
Yeasts		
Candida antarctica	Novozym® 435	Novozymes
	Noopazyme®	Novozymes
Candida cylindracea	Lipomod™ 34P - L034P	Biocatalysts Ltd
	Lypolyve CC	Lyven

	Lipase MY	Meito Sangyo Co.
Candida lipolytica		Fluka
Candida rugosa	Lipase AY "Amano" 30	Amano
	Resinase®	Novozymes
	Lipase AYS "Amano"	Amano
	Lipase II	Genzyme
		Seikagaku, Calzyme
Candida utilis		Fluka
Mammalian		
Porcine pancreas	Lipomod™ 224P - L224P	Biocatalysts Ltd
	Lipase 30, Lipase-CE	SPL
		Seikagaku, Calzyme, Worthington
Vegetal		
Wheat germ		Sigma
Mixtures		
Candida cylindracea + porcine pancreas	Lipomod™ 29P - L029P	Biocatalysts Ltd

An important and fast growing segment in the enzyme industry is of therapeutic enzymes, which are used to prepare health care products, e.g. penicillin acylase, or could be used directly also in the treatment of disorders, e.g. asparaginase. Some examples of pharmaceutical products prepared by the use of therapeutic enzymes include semi synthetic penicillins, cephalosporin, steroids, chiral compounds etc. Table 1 shows some of the therapeutic enzymes produced commercially through microbial sources.

3. Distribution of lipase producers

The presence of lipases in bacteria had been observed as early as 1901 A.D. for *Bacillus prodigiosus, Bacillus pyocyneus*, and *Bacillus fluorescens* (Jaeger *et al.,* 1994) which represent some of today's best studied lipase producers, now named, *Serratia marcescens*, *Pseudomonas aeruginosa* and *Pseudomonas fluorescens* respectively. Subsequently, a variety of bacteria, fungi and actinomycetes has been reported to produce lipases (Sztajer *et al.,* 1988, Rapp and Backhaus, 1992). Lipases also represent important virulence factor of many plant and animal pathogens. Lipases with novel properties have been discovered from microorganisms isolated from antarctic ocean (Feller *et al.,* 1990), hot springs (Gowland *et al.*, 1987, Lee *et al.*, 1999), compost heaps (Gowland *et al.*, 1987, Rathi *et al.*, 2000) and highly salty or sugary environments (Elwan *et al.*, 1985, Ghanem *et al.*, 2000). Lipase producers have been reported to grow at varied pH and temperatures. The fungi are usually reported to require acidic pH for growth and lipase production (Arima *et al.*, 1972, Pokorny *et al.*, 1994). Many bacteria are found to prefer neutral pH but there are reports of alkalophilic bacteria also (Gao *et al.*, 2000 Ghanem *et al.*, 2000). Psychrophilic and thermophilic organisms, as well as organisms having different oxygen demand (aerobic, microaerophilic and anaerobic) are reported to produce lipases.

4. Production of lipases

Bacterial lipases are mostly inducible enzymes, requiring some form of oil, fatty acid, fatty acid alcohol or fatty acid esters for induction. However, there are a few reports of constitutive lipase production by bacteria (Elwan *et al.*, 1983, Gao *et al.*, 2000). Lipases are usually secreted out in the culture medium; although there are a few reports of intra-cellular accumulation of lipases (Mourey, 1981, Lee and Lee, 1989) as well as cell bound lipases (Large *et al.*, 1999). The onset of lipase production is organism-specific, but, in general, it is released during late logarithmic or stationary phases (Matselis and Roussis, 1992, Makhzoum *et al.*, 1995, Pallavi *et al.*, 2010).

The cultivation period from 36 to 168 h has been reported for the different lipase producing organisms. Fast growing organisms were normally found to secrete the lipase within 12-24 h (Stuer *et al.*, 1986, Lee and Lee, 1989, Chartrain *et al.*, 1993, Lee and Rhee, 1993, Lin *et al.*, 1996, Imamura and Kitaura, 2000). The production of lipases from bacteria is heavily patented although the published work is rather scanty (Nakanishi and Ikeda, 1986, Inoue *et al.*, 1987, Holmes, 1990, Ishida *et al.*, 1995, Ishikawa *et al.*, 1995, Lawler and Smith, 2000). A few researchers have performed systematic medium optimization and fermentation studies for lipase production. The organisms are normally grown in a complex nutrient medium containing a carbon source (usually oil), a nitrogen source (organic/inorganic), phosphorus source (sodium or potassium phosphate) and mineral salts, supplemented with micronutrients ($MgSO_4$ or $CaCl_2$) (Pallavi and Reddy, 2010). A pH range between 8.0-10.0 has been used for lipase production by alkalophillic bacteria (Pallavi *et al.*, 2010)

A variety of non-conventional carbon sources like beef tallow, wool-scour effluent, whey, n-hexadecane and n-paraffins and Tweens have been used for lipase production (Gowland *et al.*, 1987, Brahimi-Horn 1991, Gilbert *et al.*, 1991a, Chen *et al.*, 1998, Kosugi and Suzuki 1998, Fonchy *et al.*, 1999). Several researchers have studied the effects of various poly-saccharides and detergents on lipase production. Winkler and Stuckmann (1979) have reported that the exolipase production by Serratia marcescens was enhanced by polysaccharides like glycogen, hyaluronate, laminarin, pectin B and gum arabic while the exolipase production by *Pseudomonas aeruginosa* was enhanced by treatment with glycogen (Schulte *et al.,* 1982), hyaluronate (Jaeger and Winkler, 1984) and alginate (Wingender and Winkler, 1984). A "detachment hypothesis" has been proposed by Winkler and Stuckmann (1979) suggesting the interaction of polysaccharides with cell surface structures as probable mechanism of this phenomenon. They have suggested the presence of hypothetical binding sites for newly synthesized lipases. The polysaccharides were suggested to interact with these lipase molecules by either competi-tion for these sites or by changing the conformation of exolipase. Lin *et al.* (1995) have reported that addition of Triton X-100 increased the lipase activity by 50-fold. Table 2 summarizes the carbon and nitrogen sources used and assay methods for production of lipases by different microorganisms.

Table 2. Carbon and nitrogen sources used and assay methods for production of lipases (Ashok Pandey *et al.,* 2005)

Organism	C-source	N source	Lipase U/ml	Assay	Reference
Bacillus sp.	Tween 80	NH_4NO_3	4	Titrimetry using olive oil	Gowland *et al.* 1987
Bacillus sp. THL 027	Rice bran oil	Yeast extract, peptone	8.3	Titrimetry using olive oil	Dharmsthiti and Luchai 1999
Bacillus licheniformis	Maltose	Soybean mela	2.8	Titrimetry using tributyrin	Sztajer and Maliszewska 1988a
Flavobacterium	Tween 80	Yest extract, peptone	0.2	Spectrophotometry using pNPP	Labuschagne *et al.* 1997
Pseudomonas fluorescens	Starch	Soybean meal	1.4	Titrimetry using tributyrin	Sztajer and Maliszewska 1988a
Pseudomonas aeruginosa EF2	Tween 80	KNO_3	8.3	Titrimetry using olive oil	Gilber *et al.* 1991a
Pseudomonas sp.	Ground soybean, soluble starch	Corn steep liquor	87.5	Titrimetry using olive oil	Gao *et al.* 2000
Pseudomonas nitroreduscens	Soybean meal, soluble starch	$(NH_4)_2SO_4$, urea	500	Titrimetry using olive oil	Watanable *et al.* 1977

Pseudomonas fragi	Soybean meal, soluble starch	$(NH_4)_2SO_4$, urea	30	Titrimetry using olive oil	Watanable *et al.* 1977
Pseudomonas aeruginosa YS-7	Soybean oil	-	85	Spectrophotometry using pNP laurate	Shabtai and Daya-Mishne 1992
Pseudomonas aeruginosa MB 5001	Peptonized milk	Beef extract, peptone	100	Titrymetry using olive oil	Chartrain *et al.* 1993
Pseudomonas pseudoalcaligenes	Soymeal	Peptone, yeast extract	100	Spectrophotometry using pNP laurate	Lin *et al.* 1996.
Pseudomonas sp.	Castrol oil	Polypeptone	105	Titrimetry using olive oil	Yamamato and Fujiwara 1988

4.1. Production process

The cost of microbial lipase production depends on the enzymatic yield which is defined by the cost of the raw materials used, the production process, the cost of downstream processing and the stability of the enzyme post-harvest (Kanwar *et al.,* 2002). Lipase productions have been performed mostly in submerged, batch mode of fermentations

Submerged fermentation

Submerged fermentation (SmF) is the most commonly used for lipase production. Lipase production in batch processes is mostly growth-associated and occurs during the early exponential rate. High lipase yield is obtained when cell growth reaches stationary phase (Lin *et al.,* 2001) where some essential substrates become limiting. Nakashima *et al.* (1989) developed a computer-assisted fed-batch culture for controlling the specific feed rate of meat extract into the culture medium to study the intracellular lipase production by *Rhizopus chinensis*. The productivity of the fed-batch culture of *Pseudomonas fluorescens* was eight time that of the batch method while the apparent yield has also increased by 1.6-fold using the controlled feeding of olive oil (Suzuki *et al.,* 1988).

Recently, the production of microbial lipases in organic solvent is gaining much attention. This technique offers many advantages such as higher solubility of the substrates, easy product recovery into the organic phase, possibility to isolate unwanted inhibitory substrates and/or products by using solvents-water two-phase systems and finally reaction equilibrium may be shifted towards synthesis by continuously removing the products (Hun *et al.,* 2003).

Immobilization

In view of the advantage of whole cell immobilization technique, studies have been made on lipase production as well. Immobilized cells of *R. arrhizus* on polyurethane foam were viable and produced lipase (70-80 U/l) for 120h period (Elibol and Ozer, 2000a). Immobilized cells in polyacrylamide could be reused three times with only 10% loss at each consecutive use (Hemachander *et al.,* 2001). Polypropylene powders (Accurel MP 1000) were used as adsorbents for organic solution containing *n*-hexadecane and olive oil to immobilize *Acinetobacter radioresistens* cells (Liu and Tsai 2003). Production of lipase in a fermentor using a mutant strain of *Corynebacterium* species and partial purification and immobilization was also carried out and concluded that crude lipase was purified by salting art method followed by dialysis and immobilized using calcium alginate gel matrix followed by glutaraldehyde cross linking purification process increased specific activity of enzyme from 2.76 to 114.7 IU/mg (Roy *et al.,* 2004).

Solid-state fermentation

Solid-state fermentation (SSF) offers many advantages over submerged fermentation (SmF) for the production of various enzymes, including lipases (Pandey 1992, Pandey *et al.,* 2000). High lipase productions were obtained by cultivation of *Rhizopus* sp. (Christen *et al.,* 1995, Ul-Haq *et al.,* 2002), *Aspergillus* sp. (Olama and EI-Sabaeny 1993, Mahadik *et al.,* 2002) and *Penicillium* sp. (Riviera-Muñoz *et al.,* 1991, Chen *et al.,* 1999, Miranda *et al.,* 1999). SSF was carried out using inert support (amberlite, a polymeric resin) which was supplemented with culture medium to study the lipase production by *R. delemar* (Christen *et al.,* 1995). Recently, cheap agricultural by-products have been gaining a great interest as suitable substrates in SSF: rice bran (Rao *et al.,* 1993a, 1993b, Bhushan *et al.,* 1994), wheat bran (Mahadik *et al.,* 2002), almond meal (Ul-Haq *et al.,* 2002), babassu oil cake (Chen *et al.,* 1999), gingelly oil cake, coconut oil cake (Benjamin and Pandey, 1997) and olive oil cake (Cordova *et al.,* 1998, Kademi *et al.,* 2003a). Comparing the potential of nylon sponge as inert support to food and agroindustrial wastes (Barley bran and triturated nut) for lipase production by *Yarrowia lipolytica,* Dominguez *et al.* (2003) found that lipase activities obtained on wastes were 5-fold higher than those on nylon sponge. Production of lipase by *Bacillus megaterium* AKG-1 using wheat brain in solid substrate fermentation using spectrophotometric assay, Anurag Sekhan *et al.* (2004) concluded that with wheat bran as substrate, maximum lipase yield (18.6 IUg-1 wet weight) was obtained after 24 hours of incubation and on further incubation resulted in drop in lipase activity.

4.2 Optimal conditions for production

Lipases are enzymes of secondary metabolism with the function to initiate the metabolism of lipids when they become available in the extracellular environment. Lipids, insoluble in water, must then be hydrolyzed to more polar compounds, which are furthermore absorbed as nutrients by microbial cells. The production of lipases, which are mostly extracellular, is therefore often regulated in response to the presence of lipids in growth medium (Haas *et al.,* 1999). Microbial lipase fermentations are affected by different factors such as medium composition (carbon and nitrogen sources), temperature, pH, aeration, agitation, size and age of inoculums. Each factor can have various effects on lipase production depending on culture conditions and microorganism under examination (Haas and Bailey 1993). Sangiliyandi and Gunasekaran (1996) and Pallavi *et al.* (2010) isolated lipase producing *Bacillus licheniformis* and *Staphylococcus* sp. from an oil mill refinery effluent, oil contaminated soils and observed that maximum production lipase at incubation period of 48 hours.

Lipases are mostly induced in presence of fats or oil in the culture medium. On the contrary, *Pseudomonas* sp. produced lipases constitutively and presence of organic, natural oil/fats or fatty acid esters had no significant effect on the production (Gao *et al.,* 2000). *Mucor* sp. lipase was also produced constitutively in the absence of lipids in batch culture, but the regulatory system of lipase synthesis seemed to be stimulated in presence of palm oil on solid medium (Abbas *et al.,* 2002). Triglycerides such as tributyrin (5 g/1) were used to induce the lipase production by *Kluyveromyces marxianus* (Deive *et al.,* 2003). Lipase C (acting on sn-2 position of the triglyceride) was produced when *Geotrichum* sp. F040 I B was grown in a medium containing tributyrin while lipase A acting on sn-1, 3 position was not. Tributyrin increased the production of extracellular and cell-bound enzyme (Ota *et al.,* 2000). *Cryptococcus* sp. S-2 lipase activities were influenced by the presence of triolein (Kamini *et al.,* 2000). Olive oil is the most used lipid substrate to induce lipase production by bacteria (Rathi *et al.,* 2002), yeasts (Corzo and Revah, 1999, Muralidhar *et al.,* 2001) and fungi (Omar *et al.,* 1987, Gulati *et al.,* 2000, Lima *et al.,* 2003). The production of alkaline lipase by *Acinetobacter radioresistens* was stimulated when a combination of olive oil and *n*-hexadecane was added to the culture medium (Liu and Tsai, 2003). The best volumetric ratio of *n*-hexadecane:olive oil was 7.0 for a maximum lipase activity of 21.86 U/ml at 24 h. Using GYP media, Sarkar *et al.* (1998) showed that olive oil at a concentration of 7% (v/v) resulted in an increase in lipase production by *Pseudomonas* sp., but at higher concentration the activity ceased drastically. The inhibition of lipase synthesis at higher olive oil concentrations could be due to poorer oxygen transfer into the medium which can alter fungal metabolism and consequently the production of lipases (Elibol and Ozer, 2000b).

In some conditions, fatty acids also stimulated the lipase production as reported for *A. calcoaceticus* BD413 (Mahler *et al.,* 2000), *A. radioresistens* (Liu and Tsai, 2003), *C. rugosa* (Dalmau *et al.,* 2000), *R. oligosporus* (Nahas, 1988) and *G. candidum* (Shimada *et al.,* 1992). However, the presence of short chain fatty acids inhibited the lipase production by *Pseudomonas* sp. (Gao *et al.,* 2000). Other oils from vegetable or animal sources such as corn oil (Corzo and Revah, 1999, Ozer and Eiihol, 2000, Lima *et al.,* 2003), sesame oil (Maia *et al.,* 2001), rape oil (Hiol *et al.,* 1999), castor oil (Omar *et al.,* 1987), whale oil (Omar *et al.,* 1987) and tung oils (Omar *et al.,* 1987) are less expensive than olive oil and also have positive effect on enzyme production and in some cases higher lipase production than on olive oil can be obtained (Maia *et al.,* 2001). These oils can be used to replace olive oil and consequently can help to reduce production cost.

Tween 80 was shown in several reports to be an effective substrate for lipase production (Bhatnagar and Johri, 1988, Muralidhar *et al.,*2001, Gopinath *et al.,* 2003). *Yarrowia lipolytica* lipase was enhanced with the presence of Tween 80 between 0.5 and 2 g/l in the culture medium (Corzo and Revah, 1999). Addition of Tween 80 to the culture medium gave similar extracellular enzyme activities and lower cell-bound lipase of *C. rugosa* compared to media containing other lipidic compounds. This result indicated that enzyme biosynthesis was followed by an almost complete secretion (Dalmau *et al.,* 2000). It seemed that Tween 80, in addition to its induction effect on lipase biosynthesis, increased cell permeability, and consequently cell-bound lipase decreased significantly.

Use of fats or oils to induce lipase production also has some disadvantages such as limitation of nutrient and energy availability due to low solubility of carbon source, difficulties in maintaining lipid in an emulsified state in culture, and the fact that lipid causes turbidity in the growth medium, which hinders optical or spectral assessment of the organism's growth (Haas *et al.,* 1999). Haas and Bailey (1993) suggested that glycerol, a highly soluble final product of triglyceride hydrolysis, might be the real inducer molecule for lipase production. In glycerol medium, the maximum lipase level was achieved much earlier during growth and the enzyme activity remains stable at its peak value for a considerable period of time (compared to glucose and olive oil). The use of glycerol as main carbon source rather than glucose or lipid supported mycelial growth and lipase production of *R. oryzae* (formerly *Rhizopus delemar*) (Haas *et al.,* 1999). Gum arabic used to emulsify lipid substrates can enhance enzyme production by improving the availability of the substrates (Schmidt-Dannert *et al.,* 1994, Gulati *et al.,* 1999, Mahler *et al.,* 2000). It is believed that Gum arabic helps in releasing the enzyme bound to the membrane into the medium (Schmidt Dannert *et al.,* 1994). Various sugars, including fructose (Kamini *et al.,* 2000), lactose (Kamini *et al.,* 2000), maltose (Fadiloglu and Erkmen, 2002), galactose (Gopinath *et al.,* 2003) and glucose (Elibol and Ozer, 2000a), can have a stimulating effect on enzymatic production. However, in some cases, high levels of glucose cause catabolite repression on enzyme production (Elibol and Ozer, 2000a).

Various nitrogen sources were tested in the study of lipase production. A minimal culture medium can be used with an organic nitrogen source such as urea (Deive *et al.,* 2003), corn steep liquor (Gao *et al.,* 2000, Burkert *et al.,* 2003), or an inorganic nitrogen source such as sodium nitrate (Kanwar *et al.,* 2002, Gopinath *et al.,* 2003), ammonium sulphate and ammonium nitrate. Also, a rich medium containing peptone (Fadiloglu and Erkmen, 2002; Benjamin and Pandey, 1997), tryptone or yeast extract can be used. Organic nitrogen sources can be costly and a burden in downstream processing. Lawrence *et al.* (1967) investigated the production and characterization of lipases from a *Micrococcus* and *Pseudomonas* and observed that effect of different media upon growth and production of lipase and concluded that growth and lipase production by the *Micrococcus* was unaffected by the concentration of peptone between 0.5 and 2% but growth and initial lipase production by the *Pseudomonas* was considerably stimulated by an increase of peptone. Noor *et al.* (2006) reported the S5 lipase, which is an organic solvent tolerant enzyme isolated from *Pseudomonas* sp. strain S5 which can degrade BTEX (Benzene, Toluene, Ethyl Benzene and Xylene) and concluded that maximum lipase production was exhibited when peptone was utilized as the sole nitrogen source and Tween 60 enhanced lipase production.

Aeration has a highly variable effect on the production of lipases. Some microorganisms produce great amounts of lipases when the culture medium is gently aerated, while significant aeration is required for other microorganisms. Oxygen concentration is often used as a criterion for determining whether fermentation is oxygen-dependent or oxygen-independent. However, oxygen transfer rate (OTR) appears to be the critical factor in the production of lipases (Chen *et al.,* 1999, Elibol and Ozer, 2000b). Improving OTR either by aeration, agitation or perfluorocarbon (PFC) inclusion resulted in an increase in lipase production of *Rhizopus arrhizus* (Elibol and Ozer, 2000b).

Ogino *et al.* (1999) reported the growth of *P. aeruginosa* in presence of organic solvents. The strain produced a solvent-stable enzyme which is also active in presence of various solvents. The strain grew in media containing solvent of log *P* of 7.6 and 6.6, *n*-tetradecane and *n*-dodecane respectively as well as in absence of solvent. *Bacillus sphaericus* 205y was also able to grow in presence of solvents up to 75% and also could produce a solvent-stable lipase (Hun *et al.,* 2003). The enzyme activity was increased by the presence of *n*-hexane and *p*-xylene for the lipase of *Pseudomonas* sp. G6 (Kanwar *et al.,* 2002). When n-hexadecane was used as sole carbon source in a mineral medium, growth and lipase production were stimulated.

Advantages to work at higher temperatures have already been mentioned (higher diffusion rates, increasing mass transfer effects, increased solubility of lipid substrates in water and reduced risk of contamination) (Abdel-Fattah, 2002). Contamination would end up with less biomass, unwanted end-product and enzyme degradation by proteolysis (Johri *et al.,* 1985, Gabelman, 1994). Most lipases show their optimal activity between 25°C-40°C. Abdou (2003) reported that purified lipase from *Serratia marcescens* had optimum activity at 37°C but retained 90% of its total activity at 50°C Lipase can be active in extreme pH conditions, as low as 2 for *Kluyveromyces marxianus* CECT 1018 lipase (Deive *et al.,* 2003) and as high as 11 for *Pseudomonas* sp. (Gao *et al.,* 2000), *B. thermocatenulatus* lipase (Schmidt-Dannert, 1994) and *H. lanuginosa* S-38 lipase (Liu *et al.,* 1972).

Zhang Zhen-qian and Guan Chun-yun (2009) reported the study of screening of rape seed soil sample in Victoria blue solution solid medium, with olive oil as sole carbon source. The lipase of *Enterobacter agglomerans* which has the highest activity was identified and the optimisation of carbon source, nitrogen source were carried out and concluded that optimum conditions are: initial pH 7.0, agitation at 30°C for 48 hours, 2% lactose as ideal carbon source, 1.5% beef extract as nitrogen source.

Bayoumi *et al.* (2007) screened 153 aerobic thermophilic bacterial isolates from desert soil samples collected from different localities in Egypt. *B. licheniformis* B42 and *Geobacillus stearothermophilus* B78 were selected as the most potent bacterial isolates for lipase production from slaughter house wastes as cheap raw substrate and concluded that thermoalkalophilic lipase production by *B. licheniformes* B42 and *G. stearothermophilus* B78 the optimum conditions were: pH 9.9, temperature 50°C–55°C and carbon sources fructose and glucose. Immanuel *et al.* (2008) investigated lipase production by milk isolate *Serratia rubidaea* under submerged culture, and concluded that the effect of carbon source on lipase production indicated that starch was suitable substrate to maximum lipase production and effect of lipids and surfactants showed that gingelly oil and Tween 20 were ideal for lipase production. The molecular mass of partially purified enzyme was observed to be 54kDa.

5. Classification of bacterial lipases

Arpigny and Jaeger (1999) have classified 47 different bacterial lipases into six families on the basis of amino acid sequence homology.

They have also compared the data available from three-dimensional structures of some enzymes and biochemical properties of different lipases reported by different researchers. A search of available data banks (GenBank, Swiss Protein Sequence Database, The Protein Information Resource, The Protein Research Foundation Database, and The Brookhaven Protein Databank) revealed 217 entries of lipolytic enzymes from bacteria. Many of the entries turned out to be redundant and finally, 47 different lipases were identified and grouped into six families based on amino acid sequence homology (Table 3).

Table 3. Classification of lipases and their characteristics.

Family	Sub-family	Enzyme-producing species	Accession No.	Similarity (%)	Properties
I	1	*Pseudomonas aeruginosaa*	D50587	100	True lipases
		Pseudomonas fluorescens C9	AF031226	95	
		Vibrio cholerae	X16945	57	
		Acinetobacter calcoaceticus	X80800	43	
		Pseudomonas fragi	X14033	40	
		Pseudomonas wisconsinensis	U88907	39	
		Proteus vulgaris	U33845	38	
	2	*Burkholderia glumaea*	X70354	35	
		Chromobacterium viscosuma	Q05489	35	
		Burkholderia cepacia	M58494	33	
		Pseudomonas luteola	AF050153	33	
	3	*Pseudomonas fluorescens SIKW1*	D11455	14	
		Serratia marcescens	D13253	15	
	4	*Bacillus subtilis*	M74010	16	
		Bacillus pumilus	A34992	13	
	5	*Bacillus stearothermophilus*	U78785	15	
		Bacillus thermocatenulatus	X95309	14	
		Staphylococcus hyicus	X02844	15	Phospholipase
		Staphylococcus aureus	M12715	14	
		Staphylococcus epidermidis	AF090142	13	
	6	*Propionibacterium acnes*	X99255	14	
		Streptomyces cinnamoneus	U80063	14	
II		*Pseudomonas aeruginosa*	AF005091	100	o.m.-bound esterase
		Aeromonas hydrophila	P10480	31	Acyltransferase

	Salmonella typhimurium	AF047014	17	o.m.-bound esterase
	Photorhabdus luminescens	P40601	17	
	Streptomyces scabiesa	M57297	15	
III	*Streptomyces exfoliatus*	M86351	100	Extracellular lipase
	Streptomyces albus	U03114	82	
IV	*Moraxella* sp.	X53053	33	Lipase 1
	Moraxella sp.	X53868	100	Lipase 2
	Archaeoglobus fulgidus	AE000985	28	Carboxy-lesterase
	Alicyclobacillus acidocaldarius	X62835	25	
	Pseudomonas sp. B11-1	AF034088	24	
	Alcaligenes eutrophus	L36817	24	
	Escherichia coli	AE000153	20	Esterase
V	*Moraxella* sp.	X53869	100	Lipase 3
	Psychrobacter immobilis	X67712	88	
	Pseudomonas oleovorans	M58445	34	PHA-depolymerase
	Haemophilus influenzae	U32704	34	Putative esterase
	Sulfolobus acidocaldarius	AF071233	25	Esterase
	Acetobacter pasteurianus	AB013096	15	Esterase
VI	*Pseudomonas fluorescens*	S79600	100	Esterases
	Synechocystis sp.	D90904	24	
	Spirulina platensis	S70419	22	
	Rickettsia prowazekii	Y11778	16	
	Chlamydia trachomatis	AE001287	15	

Family-I

Family I comprises a total of 22 members subgrouped into 6 subfamilies. Subfamilies I.1 and I.2 extend the *Pseudomonas* group I and II lipases encoded in an operon together with their cognate intramolecular chaperones, which have been designated Lif (lipase-specific foldases) (Jaeger *et al.*, 1994). These lipases are secreted via the type II pathway, whereas those belonging to subfamily I.3 use the type I secretion pathway. The *Bacillus* lipases grouped in subfamily I.4 are the smallest lipases known, with a molecular mass of 19.6 kDa. They seem to be well suited for biotechnological applications as is *B. thermocatenulatus* lipase belonging to subfamily I.5 (Schmidt *et al.,* 1997). Lipases from *Staphylococcus hyicus* and *S. aureus* belong to the best characterized lipases originating from gram-positive bacteria (Simons *et al.,* 1997).

Family II (The GDSL family)

The enzymes grouped in family II did not exhibit the conventional penta-peptide GlyXaa-Ser-Xaa-Gly but rather displayed a Gly-Asp-Ser-(Leu) [GDS(L)] motif containing the active-site serine residue. In these proteins, this important residue was found to lie much closer to the N-terminus than in other lipolytic enzymes.

Family III

This family of lipases was identified primarily by Cruz *et al.* (1994) and mentioned by Wei *et al.* (1998) who solved the 3D structure of the *Streptomyces exfoliatus* (M11) lipase. This enzyme displays the canonical fold of α/β -hydrolases and contains a typical catalytic triad.

Family IV (The hormone-sensitive lipase (HSL) family)

A number of bacterial enzymes (family IV) displayed a striking amino acid sequence similarity to the mammalian HSL. They were included in family IV.

Family V

Like proteins in the HSL family, enzymes grouped in family V originated from mesophilic bacteria (*Pseudomonas oleovorans, Haemophilus influenzae, Acetobacter pasteurianus*) as well as from cold-adapted (*Moraxella* sp., *Psychrobacter immobilis*) or heat-adapted (*Sulfolobus acidocaldarius*) organisms.

Family VI

With a molecular mass in the range 23-26 kDa, the enzymes in this family were among the smallest esterases known.

6. Purification and characterisation

The importance of lipases has been widely recognized in numerous fields of applications (Houde *et al.,* 2004). The source and nature of enzyme, the eventual application and degree of purity are needed to determine 'the purification schemes. This subject has been recently and extensively reviewed (Palekar *et al.,* 2000, Saxena *et al.,* 2003). Enzyme purity is evaluated after each step with the measurement of overall activity (U) and specific activity (U/mg). The purification efficiency is estimated by the determination of the enzyme activity yield and the purification factor. As a general rule, four steps of chromatography are needed to reach high purity for lipase (Palekar *et al.,* 2000).

Pre-purification steps include separation of microorganism from the medium supernatant by either centrifugation (Schmidt-Dannert *et al.,* 1994) or filtration or both (Kamini *et al.,* 2000, Abdou, 2003). Enzymes are concentrated by precipitation from a large supernatant volume with ammonium sulphate (Patkar *et al.,* 1993, Talon *et al.,* 1995, Abbas *et al.,* 2002), acetone, ethanol (Rua *et al.,* 1993), acids or by ultrafiltration (Snellman *et al.,* 2002). These steps involve the treatment of large supernatant volumes and precipitation techniques usually end up with higher enzyme yields (Aires-Barros *et al.,* 1994). Iizumi *et al.* (1990) studied the purification and characterization of a thermostable lipase from newly isolated *Pseudomonas sp.* KWI-56 and reported that enzyme was purified 13.9 fold by acetone precipitation and gel filtration by HPLC with 2.9% recovery. The thermostability was very high, such that more than 96% of initial activity remained after incubation at 60^{0}C for 24 hours.

Ion exchange chromatography using mostly weak anion exchanger DEAE-sepharose was applied to purify lipase from *Staphylococcus warneri* (Talon *et al.,* 1995) and *C. antarctica* (Patkar *et al.,* 1993), DEAE-Sephacel for *C. rugosa* (formerly *C. cylindracea*) (Rua *et al.,* 1993), DEAE-Sephadex A-50 for lipase from *G. candidum* (Gopinath *et al.,* 2003), CM-Cellulose (weak cation) followed by DEAE cellulose for *Serratia marcescens* (Abdou, 2003), and strong anion exchanger such as Mono Q for *Acinetobacter* sp. RAG-I lipase (Snellman *et al.,*2002), Q-Sepharose for lipase from *B. thermocatenulatus* DSM 730 (Schmidt-Dannert, 1994) and for *M. hiemalis f. hiemalis* lipase (Hiol *et al.,* 1999).

Gel filtration is widely used for lipase purification even if it is known to have a low capacity (Saxena *et al.,* 2003). Gel filtration has been used for the lipases from *Bacillus* sp. THL 027 (Dharmsthiti and Luchai 1999), *Serratia marcescens* (Abdou, 2003), *G. candidum* (Gopinath *et al.,* 2003), *M. hiemalis f hiemalis* (Hiol *et al.,* 1999), *Mucor* sp. (Abbas *et al.,* 2002) and *R. oryzae* (Hiol *et al.,* 2000). Using aqueous two-phase system, Gulati *et al.* (2000) purified 11-fold the *A. terreus* lipase without any loss in activity. Purification and partial characterization of a lipase from *Antrodia cinnamomea* was made by ammonium sulphate precipitation and DEAE – Sepharose chromatography. The yield and purification factor were 33.7% and 17.2 folds, respectively. The molecular weight of the purified enzyme was estimated to be 60kDa and 54kDa by SDS-PAGE (Chen-Hang Shu *et al.,* 2005 and Pallavi Pogaku and Ram Reddy, 2009b).

Stability of the purified lipases

Bacterial lipases are of interest because they are generally more stable than those from other organisms, especially against high temperatures and severe operational conditions and/or storage conditions (Dharmsthiti and Luchai, 1999). *Bacillus* THLO27, a thermophilic strain isolated from oil-contaminated area close to a seashore restaurant, was found to grow with up to 3% NaCI and its lipase was active with up to 5% NaCl (Dharmashiti and Luchai, 1999).

Some factors affecting substrate specificity are the state of the substrate interface, lipid distribution, effect of metal ions, etc (Abbas *et al.*, 2002). Broad substrate specificity has been reported for *B. stearothermophilus* lipase for soluble and insoluble substrates (Kambourova *et al.*, 2003). The highest activity was against tributyrin, and triacylglycerols were preferred over monoacylglycerols. This lipase expressed also activity towards soluble esters. Tributyrin, tricaprylin, tripalmitin and triolein were hydrolyzed by lipases from *Cryptococcus* sp. (Kamini *et al.*, 2000) and *Mucor hiemalis sp. hiemalis* (Hiol *et al.*, 1999). Lipase from *Mucor* sp. hydrolysed vinyl esters and triglycerides (Abbas *et al.*, 2002). Oils such as palm oil, sunflower oil, rape seed oil and olive oil were hydrolysed by lipases produced by *R. oryzae* (Hiol *et al.*, 2000), *Cryptococcus* (Kamini *ct al.*, 2000) and *Mucor* sp. (Abbas *et al.*, 2002).

Good solvent stability in methanol, ethanol, isopropanol, acetone, glycerol, sorbitol and *n*-hexane has also been reported for lipase from *B. stearothermophilus* (Kambourova *et al.*, 2003). Lipase isolated from *B. sphaericus* 205y was stable and showed an increased activity in presence of *p*-xylene and *n*-hexane (Hun 2003). Enzyme activity of lipase from *Cryptococcus* sp. S-2 was also enhanced with dimethylsulfoxide and diethyl ether. The enzyme was stable in all organic solvents tested (Kamini *et al.*, 2000). *P. aeruginosa* LST-03 produced a lipase stable in the presence of cyclohexane (Ogino *et al.*, 1999). High stability of lipase from *R. oryzae* in hexane, heptane, cyclohexane and isooctane was also observed (Hiol *et al.*, 2000).

7. Biotechnological applications

Lipases possess the unique feature of acting at an interface between the aqueous and nonaqueous (*i.e.* organic) phase; this feature distinguishes them from esterases. Lipase activity generally depends on the available surface area. Lipases are the most versatile biocatalysts and they bring about a range of bioconversion reactions such as hydrolysis, interesterification, esterification, alcoholysis, acidolysis and aminolysis. Lipases act under extremely mild conditions (Sharon *et al.*, 1999). They can be used in a variety of organic solvents and often show selectivity for a specific reaction type.

Lipases are an excellent alternative to many classical organic techniques in the selective transformation of complex molecules. They possess many features that favor their use as an excellent biocatalyst. They impart specificity to a reaction, in which a chemical process is typically more non-specific. In addition, the use of enzyme can decrease side reactions and simplify post-reaction separation problems (Pandey *et al.*, 1999). Lipase-catalyzed processes are reported to offer cost-effecti-veness too, in comparison with traditional downstream processing in which energy consumption and toxic by-products might often pose problem (Jansen, *et al.*, 1996). The alkaline thermophilic lipases find application in detergent industry. Many fatty food stains and human serum contain triglycerides which are hydrolyzed by lipases to produce fatty acids, monoglycerides and diglycerides, which are easier to remove than unhydrolyzed triglycerides (Fuji *et al.*, 1986).

Lipases in food industry

Lipases have become an integral part of the modern food industry. The use of enzymes to improve the traditional chemical processes of food manufacture has been developed in the past few years. Yoneda *et al.* (1996) have patented a process on Pseudomonas lipase, which was claimed to be useful in, for example, food processing and oil manufacture. Alcoholysis of cod liver oil for the production of omega- 3 polyunsaturated fatty acids was investigated by using Pseudomonas lipase (Zuyi and Ward, 1993). A few bacteria produce flavour esters and find use in cheese industry. The production of flavour esters by lipases of Staphylococcus warneri and Staphylococcus xylosus has also been described by Talon *et al.* (1996). Synthesis of fatty acid esters by a recombinant Staphylococcus epidermidis lipase has been described by Chang *et al.* (2001). Chromobacterium viscosum lipase was shown to have good potential for the instant generation of aroma and flavour compounds and could be stored at least for one month. In this case, the lipase activity was immediately regenerated on dehydration (Carlile *et al.*, 1996).

Biomedical applications of lipases

Because of their excellent capability for specific regioselective reactions in a variety of organic solvents with broad substrate recognition, lipases have emerged as an important biocatalyst in biomedical applications. Recently, Parmar *et al.* (1996) have reviewed a variety of substrates accepted by hydrolytic enzymes, including lipases, to produce compounds in high enantiomeric excess, which can be used as chiral building blocks for the synthesis of compounds of pharmaceutical interest. Conventional chemical synthesis of drugs containing a chiral center generally yields equal mixtures of enantiomers. During the past decade, many studies have shown that racemic drugs usually have the desired therapeutic activity residing mainly in one of the enantiomers and the other enantiomer might interact with different receptor sites, which can cause unwanted side effects (Pandey *et al.,* 1999). Akita *et al.* (1997) have performed enzymic hydrolysis in organic solvents for the kinetic resolution of water-insoluble acetyloxy esters using immobilized Pseudomonas sp. lipase to produce chiral intermediates for the synthesis the antibiotic (-)-indolmycin. A method was developed by Jimenez *et al.* (1997) to synthesize methyl (R)- and (S)-2-Tetradecyloxirane carboxy-late through sequential kinetic resolution catalysed by *Pseudomonas sp.* lipase. Both the enantiomers are a potent anti-diabetic and antioxidant agent.

Lipases in pesticide industry

A variety of pesticides (insecticides, herbicides, fungicides or their precursors) made with the applications of lipases are currently in the use (Pandey *et al.,* 1999). The most important application of lipases has been in the organic synthesis of pesticides for the production of optically active compound (Reddy, 1992). Generally, these compounds were produced through the resolution of racemic mixtures of alcohol or carboxylic esters; stereospecific synthesis reactions were also employed. Akita *et al.* (1995) described a highly stereospecific synthesis of the versatile chiral synthon possessing two stereogenic centres, which was subsequently converted into a homochiral intermediate for the synthesis of the biologically active potent pesticide nikkomycin-B. Mitsuda *et al.* (1990) have reported use of Achromobacter lipase for enantioselective hydrolysis of the acetic acid ester of racemic a-cyano-3-phenoxybenzyl alcohol (CPBA) for the production of (S)-CPBA, an active insecticidal stereoisomer.

Lipases as detergents

Lipases were generally added to the detergents primarily in combination with proteases and cellulases. However, other enzymes such as amylases, peroxidases and oxidases are also reported to be added in detergent preparations (Kottwitz *et al.,* 1994). Removal of oil/fatty deposits by lipase is attractive owing to its suitability under milder washing conditions. To be a suitable additive in detergents, lipases should be both thermostable as well as alkalophilic and capable of functioning in the presence of the various components of washing powder formulations (Jaeger *et al.,* 1994).

Pseudomonas lipase preparations have been used for preparation of washing powder formulations. *Pseudomonas medocina* (Lumafast*) and *Pseudomonas alcaligenes* (Lipomax*) lipases have been manufactured by Genencor international USA, as detergent additive (Jaeger *et al.,* 1994, Reetz and Jaeger, 1998). The Novo group has reported a highly alkaline, positionally non-specific lipase, from a strain of *Streptomyces* sp. that was useful in laundry and dish-washing detergents as well as industrial cleaners (Pandey *et al.,* 1999). Several lipase-producing organisms and their manufacturing processes are patented for preparation of detergent lipases (Yokoe and Mase, 1988, Mukoyama and Umehara, 1989, Holmes, 1993, Ishida *et al.*, 1995, Lawler and Smith, 2000).

Bioprocessing of leather

Leather processing involves the removal of subcutaneous fat, de-hairing and stuffing. Tanning processes are usually performed in an alkaline environment, so alkalophilic microbes ought to be better for exploration. Many *Bacillus* sp. strains, which grew successfully under highly alkaline conditions, were found to be useful in leather processing (Haalck *et al.,* 1992).

Lipases in environmental management

Lipases have been used for the degradation of wastewater contaminants such as olive oil from oil mills (Vitolo *et al.,* 1998). The treatment process involved the cultivation of lipase-producing microbial strains in the effluents. Wakelin and Forster (1997) have investigated the microbial treatment of waste from fast-food restaurants for the removal of fats, oils and greases. They cultivated pure and mixed microbial flora known to produce lipases and other enzymes. Acinetobacter sp. was the most effective of the pure cultures, typically degrading 60-65% of the fatty material.

8. Conclusions

Determination and comparison of specific activities and substrate specificities of different lipases are absolutely required to investigate the physiological functions of lipases as well as to judge their usefulness for biotechnological applications. However, serious doubts remain as to the comparability of results published by different laboratories. As outlined above, no standard substrate or assay system is available to determine specific lipase activities or to distinguish lipases from esterases. For this reason, a recent study that compares the enzyme activities of 11 bacterial lipases obtained from both gram-positive and gram-negative bacteria is very important (Simons *et al.,* 1998). All tested lipases showed high activity towards the short-chain triglyceride tributyrylglycerol and the long-chain trioctanoylglycerol, with lipases from *P. aeruginosa* and *Proteus vulgaris* being the most active. Phospholipids like 1,2-dioctanoyl-sn-glycero-3- phosphocholine were either poor substrates (for lipases from *S. aureus, S. epidermidis, S. warneri, Acinetobacter calcoaceticus, Arthrobacter* sp., *P. aeruginosa, Burkholderia cepacia*, and *P. vulgaris*) or not hydrolyzed at all (by lipases from *B. thermocatenulatus and S.*

marcescens). The only exception was *S. hyicus* lipase, which readily hydrolyzed phospholipids with high specific activity (Simons *et al.*, 1998). In summary, it is obvious that many lipase genes are being found by accident and the corresponding proteins remain largely uncharacterized. Novel lipase-producing microorga-nisms have been described (Berger *et al.*, 1995; Jarvis and Thiele, 1977; Svetlitshnyi *et al.*, 1996), but the cognate lipase genes and proteins still need to be isolated and characterised

References

Abbas H, Hiol A, Deyris V and Comeau L (2002). *Enzyme. Microbiol. Technol.* 31: 968-975.

Abdel-Fattah YR (2002). *Biotechnol Lett.* 24: 1217-1222.

Abdou AM (2003). *J.Dairy. Sci.* 86: 127-132

Aires-Barros MR, Taipa MA and Cabral JMS (1994). Isolation and purification of lipases, In- Lipases: their structures, biochemistry and applications. Wooley P, Peterson SB (eds), Cambridge University Press, Cambridge. 243-270.

Akita H, Chen CY and Uchida KA (1995). *Tetrahedr. Asymm.* 6: 2131-2134.

Akita H, Umezawa I and Matsukura H (1997). *Chem. Pharm. Bull.* 45: 272-278. CA 126: 251049.

Anurag Sekhan, Nebtu Dahiya, Ram P, Tewari and Gurinder S. Hoondal (2004). *Ind. J. Microbiol* 44(3): 219-220.

Arima K, Liu W H and Beppu T (1972). *Agric. Biol. Chem.* 36(11): 1913-1917.

Arpigny J L and Jaeger KE (1999). *Biochem. J.* 343: 177-183.

Ashok Pandey, Colon Webb, Carlos Ricardo Soccol and Christian Larroche (2005). Enzyme technology, Asiatech Publishers Inc. New Delhi, India.

Bayoumi RA, louboudey SS EI, Sidlay NM and Abd-El-Rahman MA (2007). *J. Appl. Research.* 3(12): 1752-1765, INSI Net publication.

Benjamin S and Pandey A (1997). *Acta Biotechnol.* 17:241-251

Berger JL, Lee BH and Lacroix C (1995). *Appl. Microbiol. Biotechnol.* 44:81–87.

Bhatnagar S and Johri BN (1988). Proceeding of International Symposium in Budapest. 195-209.

Bhushan B, Dosanjh NS, Kumar K and Hoondal CS (1994). *Biotechnol Lett.* 16:841-842

Brahimi-Horn MC, Mickelson CA, Gall AM, Guglielmino MG and Sparrow LG(1991). *Enzym. Microb. Technol.* 13(9): 740-746.

Carlile K, Rees GD, Robinson BH, Steer TD and Svensson M (1996). *J. Chem. Soc. Faraday Trans.* 92: 4701-4708. CA 126:141283.

Chang R.C., Chou S.J. and Shaw J.F. (2001). *J. Agric. Food Chem.* 49(5): 2619-2622.

Chartrain M, Katz L, Marcin C, Thien M, Smith S, Fisher E, Goklen K, Salmon P, Brix T, Price K and Greasham R (1993). *Enzym. Microb. Technol.* 15:575-580.

Chen J-Y, Wen C-M and Chen T-L (1999). *Biotechnol. Bioeng.* 62: 311-316

Chen S-J, Cheng C -Y and Chen T -L (1998). *J. Ferment. Bioeng.* 86(3): 308-312.

Cordova J, Nemmaoui M, Ismaili-Alaoui M, Morin A, Roussos S. Raimbault M and Benjilali B (1998). *J. Molecular Catalysis B-Enzymatic.* 5(1-4):75-78

Corzo G and Revah S (1999). 70:173-180.

Corzo G and Revah S (1999). *Biores. Technol.* 70:173-180.

Cruz H, Perez C, Wellington E, Castro C and Servin-Gonzalez L (1994). *Gene.* 144:141-142

Dalmau E, Montesinos JL, Lotti M and Casas C (2000). *Enzyme. Microbiol Technol.* 26:657-663.

Deive FJ, Costas M and Longo MA (2003). *Biotechnol Lett.*25: 1403-1406.

Demain AL (2000). *Trends. Biotechnol.*18: 26-31.

Dharmsthiti S and Luchai S (1999). *FEMS Microbiol Lett.* 179:241-246.

Dominguez A, Costas M, Longo MA and Sanroman A (2003). *Biotechnol Lett.* 25: 1225-1229.

Elibol M and Ozer D (2000a). *Process Biochem.* 36: 219–223.

Elibol M and Ozer D (2000b). *Process Biochem.* 36: 325-329.

Elwan SH, El-Hoseiny MM, Ammar MS and Mostafa SA (1983). *G. Bacteriol., Virol. Immunol.* 76(7-12): 187-199. CA 103: 19500.

Elwan SH, El-Hoseiny MM, Ammar MS and Mostafa SA (1985). *J. Microbiol.* 20(1): 103-114. A 105: 187316

Fadiologlu S and Erkmen O (2002). *J. Eng. Environ. Sci.* 26: 249-254.

Feller G, Thiry M, Arpigny JL, Mergeay M and Gerday C (1990). *FEMS Microbiol. Lett.* 66: 239-244.

Fonchy E, Morin A, Rodrigue N, Müller B and Chalier P (1999). *Appl. Environ. Micrbiol.* 65(7): 3114-3120

Fuji T, Tatara T and Minagwa M (1986). *J. Am. Oil Chem. Soc.* 63(6):796-799

Gabelman A (1994). In- Bioprocess production of flavour, fragrance, and color ingredients, Wiley J & Sons (eds), Wiley-lnterscience Publication. New York.

Gao X-G, Cao S-G and Zhang K-C (2000). *Enzym. Microb. Technol.* 27:74-82.

Ghanem EH, Al-Sayed HA and Saleh KM (2000). *J. Microbiol. Biotechnol.* 16(5): 459-464.

Ghosh PK, Saxena TK, Gupta R, Yadav RP and Davidson S (1996). *Sci. Prog.*79:119-157.

Gilbert EJ, Drozd JW and Jones CW (1991). *J. Gen. Micobiol.* 137: 2215-2221.

Godfrey T and West S (eds) (1996). Industrial Enzymology. The application of Enzymes in Industry, New York, Stockton Press.

Gowland P, Martin K and Trichur KS (1987). *FEMS Microbiol. Lett.* 48:339-343.

Gulati R, Saxena RK and Gupta R (2000). *Process Biochem.* 36:149-155.

Haalck L, Hedrich HC, Hassink J and Spener F (1992). *Prog. Biotechnol.* 8:505-512.

Haas MJ and Bailey DC (1993). *Food Biotechnol.* 7(l):49-73

Haas MJ, Bailey DG, Baker W, Berka TR, Cichowicz DJ, Derewenda ZS, Genuario RR, Joerger RD, Klein RR, Scott K and Woolf DJ (1999). *J. Sci. Technol* of *fats, oils and waxes.* 101(10): 364-370

Hemachander C, Bose N and Puvanakrishnan R (2001). *Process Biochem.* 36: 629-633.

Hiol A, Jonzo MD, Druet D and Comeau L (1999). *Enzyme Microbiol. Technol.* 25:80-87.

Hiol A, Jonzo MD, Rugani N, Druet D, Sarda L and Comeau L-C (2000). *Enzyme. Microbiol Technol.* 26: 421-430.

Holmes PE (1990). PCT Int. Appl. WO 9010695 A1. CA 114: 205549.

Houde A, Kademi A and Leblanc D (2004). *Appl Biochem. Biotechnol.*

Hun CJ, Rahman RNZA, Salleh AB and Basri M (2003). *Biochem. Eng. J.* 15:147-151.

Iizumi T, Nakamura K and Fukase T (1990). Agric. Biol. Chem. 54:1253-1258

Imamura S and Kitaura S (2000). *J. Biochem.* 127:419-425.

Immanuel Grasion, Palanichamy Eakkiraj, Austin Jebadhas, Palanisoumy Iyapparaj and Arunahalam Palaversam (2008). *Food. Technol. Biotechnol.* 46(1):60-65.

Inoue S, Kimura Y, Takei A and Fushimi K (1987 Jpn. Kokai Tokkyo Koho JP 62210987 A2. CA 108:73818

Ishida R, Suzuki M, Kotsuka T and Sakimoto K (1995). PCT Int. Appl WO 9506720 A1. CA 122:309812.

Ishikawa H, Ogino H and Myamoto K (1995). Jpn. Kokai Tokkyo Koho JP 07227276 A2. CA 123:312336.

Jaeger KE and Eggert T (2002). *Curr. Opin. Biotechnol.* 13(4):390-397

Jaeger KE and Winkler UK (1984). *FEMS Microbiol. Lett.* 21:33-38.

Jaeger KE, Ransac S, Dijkstra B, Colson C, Heuvel M and Misset O (1994). *FEMS Microbiol. Reviews* 15:29-63.

Jansen AEM, Vaidya AM and Halling PJ (1996). *Enz. Microb. Technol.* 18: 340-346.

Jarvis GM and Thicle JH (1997). *J. Microbiol. Methods* 29:41-47.

Jimencz O, Bosch MP and Guerrero A (1997). *J. Org. Chem.* 62:3496-3499.

Johri BN, Jam S and Chouhan S (1985). Proceedings of Indian Academy of Sciences (Plant Sciences), 94(2-3):175-196.

Kademi A, Lee B and Houde A (2004). In-Concise Encylopedia of Bioresource Technology, Pandey A (ed), Haworth Press, Binghamton, NY, USA: 433-442.

Kademi A, Moriri A, lsmaili—Alaoui M, Roussos S, Baratti J and Houde A (2003a). Growth kinetics and esterase/lipase production by thetmophilic fungi in solid state fermentation on olive cake, In-New Horizons in Biotechnology, Pandey A, Roussos S, Augur C, Soccol CR (eds.), Kluwer Academic Publishers

Kambourova M, Kirilova N, Mandeva R and Derekova A (2003). *J. Mol. Catalysis B: Enzymatic.* 22: 307-313.

Kamini NR, Fujii T, Kurosu T and Iefuji H (2000). *Process Biochem.* 36 : 317-324.

Kanwar L, Kumar Gogol B and Goswami P (2002). *Biores. Techno.* 84 : 207-211

Kosugi Y and Suzuki H (1998). *Biotechnol. Bioeng.* 31: 349-356.

Kottwitz B, Upadek H and Carrer G (1994). *Chim. Oggi* 12: 21-24.

Large KP, Mirjalili N, Peacock MOLM, Zormpaidis V, Walsh M, Cavanagh ME, Leadlay PF and Ison AP (1999). *Enzym. Microb. Technol.* 25(7): 569-575.

Lawler D and Smith S (2000). U. S. US 6162634 A. CA 134:41190

Lawrence RC, Fryer TF and Reiter B (1967). *J. Gen. Microbiol.* 48 : 401-418.

Lee DW, Koh YS, Kim KJ, Kim BC, Choi KHJ, Kim DS, Suhartono MT and Ryun YR (1999). *FEMS Micibiol. Lett.* 179: 393-400.

Lee SY and Lee B H (1989). *Biotechnol. Appl. Biochem.* 11 : 552-563.

Lee SY and Rhee JS (1993). *Enzym. Microb. Technol.* 15(7): 617-623.

Lee YK and Choo C L (1989). *Biotechnol. Bioeng.* 33(2): 183-90.

Lima VMG, Krieger N, Sarquis MIM, Mitchell DA, Ramos LP and Fontana J (2003). *Food. Technol. Biotechnol.* 41(2) : 105-110.

Lin SF, Chiou CM and Tsai YC (1995). *Biotechnol. Lett.* 17: 959-962.

Lin SF, Chiou CM, Yeh CM and Tsai YC (1996). *Appl. Environ. Microbiol.* 62(3): 1093-1095.

Lin YC, Wu JY and Chen TL (2001). *Biotechnol . Bioeng.* 76(3): 214-218.

Liu IL and Tsai SW (2003). *Appl. Biochem. Biotechnol.* 104 : 129-140

Mahadik ND, Puntambekar US, Bastawde KB, Khire JM and Gokhale DV (2002). *Proc. Biochem.* 38 : 715-721.

Mahler GF, Kok RG, Cordenons A, Hellingwerf KJ and Nudel BC (2000). *J. Ind. Microbiol. Biotechnol.* 24 : 25-30.

Maia MMD, Heasley A, Camargo de MMM, Melo EHM, Morais MA Jr, Ledingham WM and Lima Filho JL (2001). *Biores. Technol.* 76: 23-27.

Makhzoum A, Knapp JS and Owusu RK (1995). *Fd. Microbiol.* 12(4): 277-290.

Matselis E and Roussis IG (1992). *Fd. Sci. Technol.* 25 (5): 433-437.

Miranda OA, Salgueiro AA, Pimentel MCB, Lima Filho JL, Melo EHM and Duran N (1999). *Biores. Technol.* 69: 145-147.

Mitsuda S, Yamamoto H, Umemura T, Hirohara H and Nabeshima S (1990). *Agric. Biol. Chem.* 54: 2907-2912.

Mourey A (1981). *Rev. Fr. Corps Gras*, 28(2): 55-8 (French). CA 94:188319.

Mukoyama K and Umehara K (1989). U.S. US 5227300 A 13. CA 119:220517

Muralidhar RV, Chirumamila RR, Niarchant R and Nigam P (2001). *Biochem. Eng. J.* 9: 17-23

Nahas E (1988). *J. Gen. Microbiol.* 134 : 227-233.

Nakanishi K and Ikeda M (1986) Jpn. Kokai Tokkyo Koho JP 61058586 A2. CA 105:41284.

Ogino H, Miyamoto K, Yasuda M, Ishuni K and Ishikawa H (1999). *Biochem. Eng. J.* 4 : 1-6.

Olama ZA and EI-Sabaeny AH (1993). *Microbiol. SEM.* 9: 134-141.

Olson GJ, Woese CR and Overbeek R (1994). *J. Bacteriol.* 176: 1-6.

Omar IC, Nishio N and Nagal S (1987). *Agri. Biol. Chem.* 51 : 2145-2151.

Ota Y, Sawamoto T and Hasuo M (2000). *Biosci. Biotechnol. Biochem.* 64 : 2497-2499

Ozer D and Elibol M (2000). *J. Food Sci. Technol.* 37 : 661-664.

Palekar AA, Vasudevan P and Yan S (2000). *Biocat. Biotrans.* 18: 177-200.

Pallavi Pogaku and Ram Reddy S (2009a). Prod. Socio- economic development : Challenges before women scientists, technologist and engineers.NIT (Orissa): 415-420

Pallavi Pogaku and Ram Reddy S (2009b). Prod. Biotechnology in life sciences current trends and developments. 92-99

Pallavi Pogaku and Ram Reddy S (2010). *J. Microb. World.* 12(2): 168-173.

Pallavi Pogaku, Suresh A, Srinivas P and Ram Reddy S (2010). *Afr. J. Biotechnol.* 9(6): 882-886.

Pandey A, Benjamin S, Soccol CR, Nigam P, Krieger N and Soccol VT (1999). *Biotechnol. Appl .Biochem.* 29: 119 -131.

Pandey A, Soeeol CR and Mitchell D (2000). *Process Biochem.* 35 : 1153-1169.

Parmar VS, Bisht KS, Singh A and Jha A (1996). *Proc. Ind. Acad. Sci.* 108: 575-583.

Patkar SA, Bjorking F, Zundel M, Schulein M, Svendsen A, Heldt-Hansen HP and Gormsen E (1993). *Ind. J. Chem.* 32B : 76-80.

Pokorny D, Friedrich J and Cimerman A (1994). *Biotechnol. Lett.* 16(4): 363-366.

Rao PV (1993). *Process Biochem* 28 : 385-389.

Rao PV, Jayaraman K and Lakshmanan CM (1993a). *Process Biochem.* 28 : 385-389.

Rapp P and Backhaus S (1992). *Enzym. Microb. Technol.* 14: 938-943.

Rathi P, Bradoo S, Saxena RK and Gupta R (2000). *Biotechnol. Lett.* 22: 495-498.

Rathi P, Goswami VK, Sahai V and Gupta R (2002). *J. Appl. Microbiol.* 93 : 930-936.

Reddy DR (1992). In: Industrial Biotechnol. edited by Malik V. S. and Sridhar P., 387-404, Oxford and IBH Publishing Co., New Delhi.

Reetz MT and Jaeger KE (1998). *Chem. Phys. Lipids* 93: 3-14.

Rivera-Munoz C, Tinoco-Valencia JR, Sanchez S and Farres A (1991). *Biotechnol. Lett* 13: 277-280

Roy Nityananda, Lalitagauri Roy and Parimal Chattupadhyay (2004). *Ind. J. Exp. Biol.* 42 : 202-207.

Rua ML, Diaz-Niaurino T, Fernandez VM, Otero C and Ballesteros A (1993). *Biochem. Biophys. Acta.* 1156 : 181-189

Sangiliyandi G and Gunasekaran P (1996). *Ind. J. Microbiol.* 36 : 109-110.

Saxena RK, Sheoran A, Gin B and Davidson WS (2003). *J. Microbiol. Meth.* 52: 1-18.

Schmidt Dannert C, Rua ML, Wahl S, Schmid RD (1997). *Biochem Soc. Trans.* 25 : 178-82.

Schmidt-Dannert C, Sztajer H, Stochlein W, Menge U and Schmid RD (1994). *Biochem. Biophy. Acta.* 1214: 43-53.

Schulte G, Bohne L and Winkler UK (1982). *Can J. Microbiol.* 28: 636-642.

Sharma R, Chisti Y and Banerjee UC, (2001). *Biotechnol Adv.* 19: 627-662.

Sharon C, Nakazato M, Ogawa H I and Kato Y (1999). *Ind. J. Exp. Biol.* 37: 481-486.

Shimada Y, Sugihara A, Nagao T and Tominaga Y (1992). *J. Ferment. Bioeng.* 74(2): 77-80.

Simons JWF A, Van Kampean MD, Riel S, Gotz F, Egmond MR and Verheij H M (1998). *J. Biochem.* 253(3): 675-683

Simons J-WFA, Boots J-WP, Kats MP, Slotboom AJ, Egmond MR, Verheij HM (1997). *Biochem.*36:14539–14550.

Snellman EA, Sullivan ER and Colwell RR (2002). *Eur. J. Biochem.* 269: 5771-5779

Stuer W, Jaeger KE. and Winkler U K (1986). *J. Bacteriol.* 168: 1070-1074.

Suzuki T, Mushiga Y, Yamane T and Shimizu S (1988). *Appl. Microbiol. Biotechnol.* 27 : 417-422.

Svetlitshnyi V, Rainey F, Wiegel J (1996). *Int. J. Syst. Bacteriol.* 46:1131–1137.

Sztajer H, Maliszewska I and Wieczorek J (1988). *Enzyme Microb. Technol.* 10(8): 492-497.

Talon R, Dublet N, Montel M C and Cantonnet M (1995). *Curr Microbiol.* 30(1): 11-16.

Talon R, Montel M C and Berdague JL (1996). *Enzym Microb. Technol.* 19(8): 620-622.

UL-Haq I, Idrees S and Rajoka MI (2002). *Pro. Biochem.*37 : 637-641

Vitolo S, Petarca L and Bresci B (1998). *Biores. Technol.* 67: 129-137.

Wakelin N G and Forster C F (1997). *Biores. Techno*l. 59(1): 37-43.

Wei Y, Swenson L, Castro C, Derewenda U, Minor W, Arai H, Aoki J, Inoue K, Servin-Gonzalez L and Derewenda Z S (1998). *Structure* 6: 511-519.

Wingender J and Winkler U K (1984). FEMS *Microbiol. Lett.* 21: 63-69.

Winkler FK and Stuckmann M (1979). *J. Bacteriol.* 138 : 663-670.

Yokoe M and Mase T (1988). Jpn. Kokai Tokkyo Koho JP 63039579 A2. CA 109: 72058.

Yoneda T, Takada H, Ohno K and Sasuga J (1996). Pat. WO 9627002. CA 125: 269257.

Zhang-Zhen-quian and Guan Chun-Yu (2009). *Afr. J. Biotechnol.* 8(7): 1273-1279.

Zuyi L and Ward O P (1993). *Biotechnol. Lett.* 15: 393-398.

Chapter 11

Lactic acid bacteria for Health and Wealth

Gopal Reddy, Kiran Kumar, K., Chaitanya, K. and Hari Krishna, N.

Department of Microbiology, UCS, Osmania University, Hyderabad – 7.
E-mail : gopalred@hotmail.com

ABSTRACT

Man lives in the community of microbial population. Microbes are present in and around man. Most of them are useful and some are harmful. It is necessary to propagate the useful ones and control the harmful ones. The type of microbes that associate with human body depends on the environment of living. The first association with microbes starts immediately after birth. The first colonizers of new born baby are lactic acid bacteria (LAB). These are useful to man for good health and help in giving wealth if properly managed. LAB are used as probiotics and for several metabolic products. The important metabolic product from LAB is lactic acid (LA) which exists in stereo isomeric forms like L(+) and D(-). Only L(+) lactic acid is useful to man as energy food. LA is produced in several fermented foods by LAB and some fungi. It gives taste, serves as energy and preservative in foods. LA is used as industrial feed stock for several synthetic products. Demand for LA is increasing due to its use in synthesis of polylactate as biodegradable plastic. LA is produced by chemical synthesis and microbial fermentation of which the former gives racemic mixture and latter gives stereo specific molecule depending on the nature of the fermenting organism. LA is the oldest molecule of microbial fermentation depending on the nature of the fermenting organism and is still in high demand. It is produced by fermentation with molasses sugar as substrate by saccharolytic LAB. Increasing molasses costs lead to search for low cost alternative substrates for LA fermentation. In recent times, amylolytic LA fermentation is gaining prominence due to its low cost starchy substrates. It requires two step process of chemical hydrolysis followed by LA fermentation by saccharolytic organisms. Approaches are made for efficient amylolytic LA fermenting bacteria in single step fermentation. We have isolated several amylolytic LAB and one of such isolates identified as *Lactobacillus amylophilus* GV6 is studied extensively. It is projected as a potential industrial strain for large scale LA production using low cost starchy substrates.

Key words : Lactobacillus, Probiotics, Lactic acid, Health.

1. Human microbiota

The human microbiome or human microbiota is the assemblage of microorganisms that reside on the surface and in deep layers of skin, in the saliva and oral mucosa, in the conjunctiva, and in the gastrointestinal tracts. They include bacteria, fungi, and archaea. Some of these organisms perform tasks that are useful for the human host. However, the majority have no known beneficial or harmful effect. Those that are expected to be present, and that under normal circumstances do not cause disease, but instead participate in maintaining health, are deemed members of the normal flora [Reuter, 2001] or microbiota. (wikipedia). If you ask the average person how microbes (or germs) impact their lives, they would immediately think of disease. This is not a silly view, as pathogenic microorganisms have greatly affected human populations throughout our existence. Until about 1930, microbes were the major cause of death in humans, with infectious disease infant mortality rates above 50%. From today's perspective this is a horrendous statistic; over half of all infants did not make it to adulthood! With the advent of antibiotics, vaccines and better water sanitation, humanity has reduced the impact of pathogenic microbes, but they will always remain an important social concern. However, apparently, harmless and beneficial bacteria far outnumber the harmful varieties.

Populations of microbes (such as bacteria and yeasts) inhabit the skin and mucosa. Their role forms part of normal, healthy human physiology. However, if microbe numbers grow beyond their typical ranges (often due to a compromised immune system) or if microbes populate atypical areas of the body (such as through poor hygiene or injury), disease can result. It is estimated that 500 to 1000 species of bacteria live in the human gut [Sears, 2005] and a roughly similar number on the skin [Grice, 2009; Pappas, 2009]. Bacterial cells are much smaller than human cells, and there are atleast ten times as many bacteria as human cells in the body (approximately 10^{14} versus 10^{13}) [Savage, 2009; Berg, 1996]. Though members of the flora are found on all surfaces exposed to the environment (on the skin and eyes, in the mouth, nose, small intestine), the vast majority of bacteria live in the large intestine. Many of the bacteria in the digestive tract, collectively referred to as the gut flora, are able to break down certain nutrients such as carbohydrates that humans otherwise could not digest. The majority of these commensal bacteria are anaerobes, meaning they survive in an environment with no oxygen. Normal bacterial flora can act as opportunistic pathogens at times of lowered immunity [Reuter, 2001].

Escherichia coli is a bacterium that lives in the colon; it is an extensively studied model organism and probably the best-understood bacterium [Lee, 2003]. Certain mutated strains of these gut bacteria do cause disease; an example is *E. coli* O157:H7. A number of types of bacteria, such as *Actinomyces viscosus* and *A. naeslundii*, live in the mouth, where they are part of a sticky substance called plaque. If this is not removed by brushing, it hardens into calculus (also called tartar). The same bacteria also secrete acids that dissolve tooth enamel, causing tooth decay.

Human microflora consists of 400 different species of bacteria. Some of these are beneficial and others are potentially harmful. A balance between the two is vital for human health and wellbeing. One way of maintaining a balance between the beneficial and harmful bacteria in our intestines is to eat the types of food that contain beneficial bacteria. Beneficial bacteria that can be introduced into the digestive system through food are called probiotics.

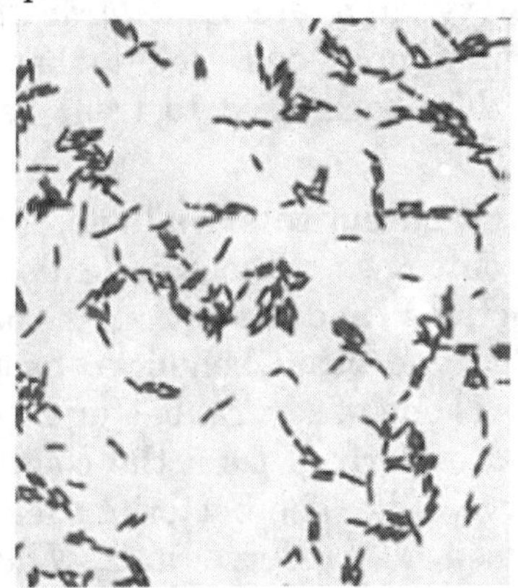

Fig. 1. Morphology of Lactobacillus under light microscope.

2. *Lactobacillus* species

Lactobacillus is a genus of Gram-positive (Fig. 1) facultative anaerobic or microaerophilic rod-shaped bacteria [Makarova, 2009] with an incident rate of 20-60. They form over a kilogram of our body weight and are a major part of the lactic acid bacteria group, named as such because; most of its members convert lactose and other sugars to lactic acid. Lactobacillus encompasses a significant number of different species that are largely diverse. Several members of the genus have had their genome sequenced [Chandra Iyer, 2009]. These lactic acid bacteria are used to preserve food and feed. In humans, they are commonly present in the vagina [Dicks, 2000] and gastrointestinal tract, hence referred to as microflora, or gut flora, where they are symbiotic and make up a small portion of the gut flora and are usually benign. Many species are prominent in decaying plant material. The production of lactic acid makes its environment acidic, which inhibits the growth of some harmful bacteria and fungus from invading the body. These bacteria break down food remains that have not been digested earlier in the digestive system. The 'gut flora' also produces vitamin K, which is essential for normal blood clotting.

Most commercially-promoted fermented milk products with probiotic properties contain *Lactobacillus* or *Bifidobacterium*. Natural Yogurts and Yakult, a fermented milk product, are examples of foods which contain probiotics. Particular strain of lactobacillus, bifidobacterium, and streptococcus promote good health in man and animals and prevent taking harsh antibiotics and explore the benefits of probiotics. Population of *Lactobacillus acidophilus* are small organisms that stop the growth of harmful bacteria and increase resistance to infections. The active culture can be found in some yogurts and made as therapeutic formulations in the form of supplemental liquid, powder or tablet form. *L. acidophilus* is part of the natural flora of the gastrointestinal and genitourinary tracts. Other lactobacilli found in the vagina are *L. delbruekii* and *L. gasseri* and disturbance of the vaginal flora can lead to bacterial vaginosis (BV). Hence, *L. acidophilus* has been used to treat bacterial vaginosis and yeast infections for decades.

Compared to the pharmaceuticals available, the use of probiotics can help to achieve desired outcomes without any side effects. It is advisable to prevent yeast infections and maintain a healthy population of *Lactobacillus acidophilus* and other beneficial bacteria for good health. *Lactobacillus bulgaricus* is one of several bacteria used to produce yogurt. The name *L. bulgaricus* is derived from the country Bulgaria where it was first used to preserve milk. The bacterium feeds on milk to produce lactic acid which is used to preserve milk. *Lactobacillus bulgaricus* naturally exists in the human gastrointestinal tract. The bacterium is helpful to people suffering from lactose intolerance which occurs in individuals who lack the enzyme to break down lactose to simple sugars. *Lactobacillus bulgaricus* and other beneficial bacteria are used to help in lactose intolerance.

Lactobacillus plantarum is a natural inhabitant of the human gastrointestinal tract. The first sample of *Lactobacillus plantarum* was isolated from human saliva. This strain has one of the largest genome known among lactic acid bacteria. Lactobacillus plantarum significantly reduces intestinal gas and its effects last for months after treatment.

Lactobacillus casei is found in fermented milk and has beneficial properties for human health. The human digestive tract consists of *Lactobacillus casei*, a natural flora that prevents an overpopulation of ingested lactic acid bacteria from gaining residence in the gastrointestinal tract. Live *Lactobacillus casei* reduces diarrhea and helps modifies microflora in the body. *Lactobacillus casei* produces DL-lactic acid and amylase that complements the growth of *Lactobacillus acidophilus*.

Deficient growth of *Lactobacillus acidophilus*, *Lactobacillus bulgaricus*, *Lactobacillus plantarum*, and *Lactobacillus casei* could be due to the following:

- Antibiotics
- Poor diet
- Chlorinated water
- Fluoride water
- Birth control pill
- Stress

3. *Lactobacillus* species as probiotics

Lactobacillus and Bifidobacteria cultures are increasingly used as probiotics in pharmaceuticals and in foods. The selection of strains is performed often for technological rather than for microecological reasons. Detailed reports about species and strains composition of these micro-organisms in the intestinal microflora of man are rare. Investigations are performed with samples originating from infants and adults, taken from faeces and from upper sections of the intestinal tract including mouth and stomach, and from caecum and colon. Post mortem cases as well as test subjects under physiological conditions were analyzed using an automatic capsule system sampling at defined times in different parts of the intestinal tract. The fate of selected strains after oral intake was studied, too. Furthermore, influences of the microflora originating from food were considered. The identification of autochthonous (indigenous) and allochthonous (transient) species could be achieved with descriptions of new species in the genera Lactobacillus and Bifidobacterium. *L. gasseri* and *L. reuteri* [Amelia, 2010] proved to be predominant autochthonous *Lactobacillus* species in infants as well as in adults. Both species were occasionally present even in the stomach. This was also the case with an anaerobic lactic acid bacterium, previously named *Catenabacterium catenaforme*, later classified as *L. ruminis*, a non-motile variant of this species. The bifidobacterial microflora differed in composition between infants and adults and in different stages of the host's life. Up to 5 species or special strains of bifidobacteria could be present in different, individually fixed, combinations. Species typical for infants were *B. bifidum*, *B. infantis*, *B. breve*, and *B. parvulorum*. Typical for adults were 4 different variants of *B. adolescentis*, *B. bifidum* and *B. longum* could often be found in both groups, but in lower numbers. *B. longum* showed some oxygen tolerance whereas *B. bifidum* and *B. adolescentis* required strict anaerobic and fastidious conditions for cultivation. The autochthonous *Lactobacillus* and *Bifidobacterium* [Reuter, 2001] microflora in man will remain stable life-long. With lactobacilli, however, some successions may be caused by transient species derived from food or from the oral cavity, thus giving the impression of an altered microflora. Nevertheless *L. gasseri*, *L. reuteri*, *L. ruminis*, and to some degree, *L. salivarius*, may be present as autochthonous species all the time. With bifidobacteria, a decreasing tendency in counts and in multiple composition in elderly people exists. Furthermore, this microflora is also influenced by consumption habits, which are probably caused by geographical circumstances.

3.1 Food production

Some *Lactobacillus* species are used for the production of yogurt, cheese, sauerkraut, pickles, beer, wine, cider, kimchi, chocolate, and other fermented foods, as well as animal feeds, such as silage. Sourdough bread is made using a "starter culture," which is a symbiotic culture of yeast and lactic acid bacteria growing in a water and flour medium. Lactobacilli, especially *L. casei* and *L. brevis*, are some of the most common beer spoilage organisms. The species operate by lowering the pH of the fermenting substance by creating the lactic acid, neutralising it to the desired extent.

3.2 Probiotics and biotherapeutics

Some *Lactobacillus* spp. and other lactic acid bacteria may possess potential therapeutic properties including anti-inflammatory and anti-cancer activities, as well as other features of interest. A study by researchers from the Beth Israel Deaconess Medical Center and the UCLA in 2009 demonstrated the protective effects of some strains of these bacteria for anti-tumor and anti-cancer effects [Chen, 2009]. Dietary administration alleviated the risks of certain types of cancers and suppressed colonic tumor incidence, volume and multiplicity induced by various carcinogens in mice. For a few strains, oral administration effectively reduced DNA adduct formation, ameliorated DNA damage and prevented putative preneoplastic lesions such as aberrant crypt foci induced by chemical carcinogens in the gastrointestinal tract. Reports also indicated that some cultures administered to animals inhibited liver, colon, bladder, and mammary tumors, highlighting potential systemic effects of probiotics with anti-neoplastic activities [Chandra Iyer, 2009].

Lactobacilli are also used to restore particular physiological balance such as in the vaginal ecosystem. Their role is [Makarova, 2006] to physically protect the vaginal epithelium by building a thick layer separating the epithelium from pathogens, [Dicks, 2000] to physiologically keep the balance of the vaginal ecosystem in maintaining the pH at ~4.5 and [11], generating hydrogen peroxide against pathogens. *Lactobacillus* is highly tolerant to low pH and can easily maintain low pH and protect the vaginal ecosystem from Gram negative and Gram positive bacteria. Atleast one *Lactobacillus* species in the colon may play a role in defense against flu viruses. Antibiotics that kill *Lactobacillus* seem to impair production of interleukin-1 beta (IL-1 beta), which is necessary to combat influenza and some other viruses [Chen, 2009; Ichinohe, 2011]. *Lactobacillus* thus has quite good importance in the areas of promoting health and protecting humans and animals against pathogens, viruses and carcinogens.

4. Lactic acid

Apart from contributing to health lactobacilli are important micro-organisms in context of its contribution to the fermentation industry for the production of the first biotechnologically produced enigmatic chemical, lactic acid (2-hydroxypropanoic acid), the most widely occurring natural organic acid. It has numerous applications in the food, pharmaceutical, leather and textile industries and also as a chemical feedstock. The increasing applications of lactic acid in newer avenues like biodegradable plastics, medical care, agriculture and also as green solvents, has led to constant increase in its demand [Datta, 1993; Litchfield, 1996]. Lactic acid can be produced either by chemical synthesis or by carbohydrate fermentation technology. Chemical synthesis gives racemic lactic acid, whereas, fermentation technology can produce desired stereoisomers such as L(+) or D(-) or a racemic mixture (DL) of lactic acid using homolactic organisms [Datta, 1993; Litchfield, 1996]. The product stereospecificity is influenced by the intended applications of lactic acid. The racemic polymer and L(+) lactate polymer differ from each other in their physical properties [Kulakarni, 1971]. The presence of L(+) lactic acid with higher optical purity in the lactate polymer confers higher melting point and crystallinity. Human beings and animals have only L-lactate dehydrogenase (L-LDH) which degrades compounds derived from L(+) lactic acid, hence D(-) lactic acid accumulates in the blood resulting in health problems like acidurea and/or bone decalcification. Therefore, only the L(+) isomer of lactic acid is used in human consumables as food preservatives or taste-enhancing additives etc. Hence, in recent years, fermentative production of L(+)lactic acid has been receiving considerable interest from industry, for human consumption or making biodegradable polylactic acid (PLA) plastics and coatings [Litchfield, 1996; Seeley, 1992].

5. Substrates for LA fermentation

Refined sugars such as glucose and sucrose are the most commonly used substrates for commercial production of lactic acid by fermentation processes, however, they are expensive. Most abundantly available, cost-effective raw material such as starch is being used to some extent for lactic acid production. However, this is a two-step process that includes saccharification by acid or base or microbial amylase, followed by *Lactobacillus* fermentation. Hence, an alternative strategy that involves direct conversion of starch to lactic acid using both amylolytic and acid producing organisms would considerably reduce the cost of production, as the liquefaction and saccharification processes will be eliminated. In the recent past some work has been carried out to develop production strains using starch as fermentation substrate. Though the capacity of lactic acid bacteria (LAB) to convert starch into lactic acid is known, it is not a frequent characteristic Giraud, 1991]. Starch-degrading amylolytic lactic acid bacteria (ALAB) have been reported. These include *L. amylophilus* [Nakamura, 1979], *L. amylovorus* [Nakamura, 1981], *L. plantarum* A6, Lactobacillus strains LEM220, 207 and 202 [Champ, 1983], Leuconostoc strains [Lindgren, 1984], *L. acidophilus* [Kandler, 1986], *L. cellobiosus* [Sen, 1984], *L. plantarum* [Olympia, 1995], L. fermentum strains Ogi E1 and MW2 [Agati, 1998], *L. amylolyticus* strains [Bohak, 1998] and *L.manhitovorans* [Morlon-Guyot, 1998]. Among the most representative species, *L. amylophilus* and *L. amylovorus* were studied for their ability to produce lactic acid from starch [Zhang, 1991; Mercier, 1992; Yumoto, 1995; Litchfield, 1996]. *Lactobacillus amylophilus* produces only L(+) lactic acid and *L. amylovorus* produces DL lactic acid from starch. All wild strains of *L. amylophilus* reported so far showed about 90% starch to L(+) lactic acid conversion efficiency from pure starches at low substrate concentrations. However, at high starch concentrations, the lactic acid production was low [Zhang, 1991; Mercier, 1992; Yumoto, 1995; Litchfield, 1996] For an economically viable process, it is necessary to isolate or develop amylolytic Lactobacillus organisms with high starch conversion efficiency to L(+) lactic acid. In this direction, *L. amylophilus* strain GV6 was isolated and evaluated for its ability to produce high yields of L(+) lactic acid at high substrate concentrations with corn starch and soluble starch [Vishnu, 2000 & 2002]. The organism also showed a higher amylolytic efficiency than previously reported strains of *L amylophilus* [Vishnu, 2002]. The lactic acid production from various crude starchy substrates further reduces the production cost. *Lactobacillus amylophilus* GV6 degraded pure and crude starchy substrates efficiently indicating the presence of efficient amylolytic enzymes in culture broth (Vishnu, 2002). It was observed that strain GV6 is efficient at producing lactic acid at 1–10% substrate concentrations of pure starches and crude starchy substrates (flours). Strain GV6 produced high yields of lactic acid with pure starches such as soluble starch, corn starch and potato starch, yielding 92–96% at low substrate concentrations in 2 days and 78–89% at high substrate concentrations in 4–6 days (Table 1).

The strain GV6 also produced high yields of lactic acid with crude starchy substrates such as wheat flour, sorghum flour, cassava flour, rice flour and barley flour, yielding 90–93% at low substrate concentration in

2 days and 80% or more at high substrate concentrations in 6–7 days (Table 2).

Table 1. Fermentative production of lactic acid by *L. amylophilus* GV6 with various pure starches as substrates [33].

Substrate Conc (g/L)	Fermentation Time(days)	Substrate Degraded (g/L)	Lactic acid produced(g/L)	Lactic acid yield(g/g)[a]
Soluble starch				
10	2	10.0	9.6	0.96
40	3	38.5	35.6	0.93
70	3	67.5	61.0	0.90
100	4	83.3	75.7	0.89
Corn starch				
10	2	9.4	8.8	0.94
40	3	37.6	32.6	0.89
70	4	58.1	49.1	0.85
100	5	50.1	39.2	0.78
Potato starch				
10	2	9.5	8.7	0.92
40	3	35.9	32.0	0.89
70	4	46.5	40.0	0.85
100	6	47.0	38.1	0.81

CD at 5% - confidence difference was arrived at 5% level of significance for comparision of any combination of two mean values. [a] g/g – gram lactic acid per gram substrate utilized.

Pure starches may give more accessibility to amylolytic enzymes, resulting in higher yields of lactic acid. However, the lactic acid yields by *L. amylophilus* GV6 were comparable when low cost crude starchy substrates were used. Fermentative production of lactic acid was also checked with cheap byproducts like wheat bran, Wheat bran is an underexploited cheap byproduct obtained during the milling process of wheat.

The ability of *Lactobacillus amylophilus* GV6 to hydrolyze raw starch in wheat bran to produce L (+) lactic acid was studied in solid state fermentation (SSF), opening a novel method for the utilization of these agricultural byproducts. L (+) Lactic acid production by *L. amylophilus* has been reported by various groups from processed and unprocessed starch but there are no reports available from natural starches in SSF. All amylolytic wild strains reported so far have high yield efficiencies at low substrate concentrations whereas at high substrate concentrations lactic acid production was low. *L. amylophilus* GV6 showed high yield efficiency at high substrate concentration in SSF [Naveena, 2005].

Expensive pure starch could be replaced with cheaper underutilized abundantly available byproduct like wheat bran. The replacement of pure starch with wheat bran will bring down the cost of substrate 100 times and will make the whole process more economical in terms of substrate costs. When compared to sucrose and hydrolyzed starch, the substrate costs will be still lowered to many folds.

Table 2. Fermetative production of lactic acid by *L. amylophilus* GV6 with various crude starches as substrates.

Substrate Conc (g/L)	Starch Content (g/L).	Fermentation Time (days)	Substrate Degraded (g/L)	Lactic acid produced (g/L)	Lactic acid yield (g/g)[a]
Sorghum Flour					
20	13.6	2	12.6	11.6	0.92
60	40.8	4	33.0	29.4	0.89
100	68.0	6	35.3	28.7	0.81
CD at 5%	-	-	1.7	3.5	0.03
Cassava flour					
20	16.5	2	15.2	14.0	0.92
60	49.4	4	38.1	33.6	0.88
100	82.0	6	37.0	30.3	0.82
CD at 5%	-	-	1.6	3.7	0.03
Wheat flour					
20	13.8	2	12.6	11.7	0.93
60	41.4	4	33.2	29.9	0.90
100	69.0	6	35.8	29.0	0.81
CD at 5%	-	-	2.6	3.3	0.01
Rice flour					
20	15.6	2	14.3	13.0	0.91
60	46.8	4	35.9	30.9	0.86
100	78.0	6	36.9	29.5	0.80
CD at 5%	-	-	2.0	2.6	0.01
Barley flour					
20	13.8	2	12.6	11.3	0.90
60	41.4	4	33.0	27.3	0.86
100	69.0	7	35.3	25.5	0.81
CD at 5%	-	-	2.6	3.2	0.01

CD at 5% - confidence difference was arrived at 5% level of significance for comparision of any combination of two mean values. [a] g/g – gram lactic acid per gram substrate (starch) utilized.

6. Low cost nitrogenous substrates:

Research on the development of an efficient and economical process for the lactic acid production in our laboratory involved the screening of media components with a view to replace costly ingredients like peptone and yeast extract by renewable low cost materials in submerged fermentation of which red lentil and bakers yeast cells completely replaced commercial peptone and yeast extract [Mohammad, 2005]. L(+) Lactic acid fermentation was studied using *Lactobacillus amylophilus* GV6 under the influence of inexpensive nitrogen sources (red lentil-RL, and Baker's yeast cells-YC) and starch as carbon substrate by response surface methodology (RSM). Central composite rotatable design (CCRD) was employed to determine maximum lactic acid production at optimum values for process variables RL, YC and incubation period (IP) and a satisfactory fit model was realized. Lactic acid production was significantly affected by RL and IP interactions as well as by independent variables RL and YC. Maximum lactic acid production of 13.5 g/15.2 g starch was obtained with RL 0.8%, YC 1% and IP of 48 h, with 92% lactic acid yield efficiency (g lactic acid produced/g substrate utilized) and 40% increase (from 50 g to 92 g/100 g starch utilized) in lactic acid production. This is the first report on response optimization in direct fermentation of starch to lactic acid using inexpensive nitrogen sources substituting peptone and yeast extract in anaerobic submerged fermentation by amylolytic lactic acid bacteria.

The laboratory studies made on amylolytic lactic acid fermentation by *Lactobacillus amylophilus* using a low cost carbon and nitrogen substrates can be taken up for pilot scale assessment. This leads to an initation for development of production technology for stereospecific L(+) lactic acid as a low cost high volume fermentation product.

7. Conclusions

Lactic acid bacteria are the first colonizers of human body. They help to promote good health in different ways inside and outside the human body. The concept of probiotic therapy is gaining more prominence in heathcare using lactic acid bacteria. Lactic acid as an industrial product has numerous applications. Production of stereospecific lactic acid as a low cost chemical is more important. Amylolytic lactic acid bacteria are very important for development of low cost technology to produce lactic acid as a high value-low cost microbiological product. *Lactobacillus amylophilus* can be a potential industrial strain for lactic acid fermentation.

References

Agati V, Guyot JP, Morlon-Guyot J, Talamond P & Hounhouigan DJ (1998) Jour of Appl Microbiol 85 : 512–520.

Amelia Bentrup (2010) *Lactobacillus reuteri* Good for Health, Swedish study http://www.naturalnews.com/030498_Lactobacillus_reuteri_good_health.html

Berg *et al.* (1996) Trnds in Microbiol 4 (11): 430–435.

Bohak I *et al* (1998) Syst and Appl Microbiol 21: 360–364.

Champ M, Szylit O, Raimbault P & Abdelkader N (1983) Jour of Appl Bacteriol 55: 487–493.

Chandra Iyer and James Versalovic (2009) Lactobacillus Molecular Biology: From Genomics to Probiotics chap. 8 *Lactic Acid Bacteria: Probiotics With Anti-Cancer Activities.*

Chen X, Fruehauf J, Goldsmith J D, Xu H, Katchar K K, Koon H W, Zhao D, Kokkotou E. G. *et al.* (2009Gastroenterology 137 (3): 914.

Datta RS, Tsai P, Patrick B, Moon SH & Frank JR (1993) Technological and economic potential of poly lactic acid and lactic acid derivatives. Intl Cong on Chem from Biotechnol, pp. 1–18. Germany.

Dicks LMT, M Silvester, PA Lawson, MD Collins. (2000) Intl J of Syst and Evol Microbiol 50 (3): 1253–1258.

Giraud E, Brauman A, Keleke S, Lelong B & Raimbault M. (1991Appl Microbiol and Biotechnol 36, 379–383.

Grice *et al.* (2009) Science 324(5931):1190-2.

Ichinohe T, Pang I K, Kumamoto Y, Peaper D R, Ho J H, Murray T. S, Iwasaki A (2011) Proc of the Natl Acad of Sci 108 (13): 5354–5359.

Kandler, O. & Weiss, N (1986) In Berg's Man of Syst Bacteriol 2 : pp.1209 -1234.

Kulakarni R K, Moore E G, Hegyell A F & Leonard F (1971) Jour of Biomed Mtrls Res 5:169–181.

Lee PS, Lee KH (2003) Biotechnol Bioeng 84 (7): 801–814.

Lindgren, S. & Refai, O (1984) Jour of Appl Bacteriol 57: 221–228.

Litchfield, J.H. (1996) Adv in Appl Microbiol 42: 45–95.

Makarova *et al.* (Oct 2006) Proc of Natl Acad of Sci (U S A) 103 (42): 15611–15616.

Mercier P, Yerushalmi L, Rouleau D, Dochain D (1992) Jour of Chem Technol and Biotech. 55: 111–121

Mohammad Altaf, Basa Janakiram Naveena and Gopal Reddy (2005) Food Technol. Biotechnol. 43 (3) 235–239 (2005).

Morlon-Guyot J, Guyot JP, Pot B, Jacobe de Haut I. & Raimbault M (1998) Intl Jour of Syst Bacteriol 48 : 1101–1109.

Nakamura LK & Crowell CD (1979) Dev in Ind Microbiol 20 : 531–540.

Nakamura LK (1981) Int Jour of Syst Bacteriol 31: 56–63.

Naveena BJ, Altaf Md, Bhadriah K, Reddy G (2005) Biores Technol 96 : 485–490.

Olympia M, Fukuda H, Ono H, Kaneko Y. & Takan, M (1995) Jour of Ferm and Bioeng 80 : 124–130.

Pappas S (2009) Your Body is a habitat for bacteria. Sci Now Daily News.

Reuter G (2001) Curr Issues Intest Microbiol 2(2):43-53.

Savage D C (1977) Ann Rev of Microbiol 31:107–133.

Sears CL (2005) Anaerobe 11: 247–251.

Seeley RS (1992) Chem Business. pp. 28–28.

Sen S & Chakrabarthy SL (1984) Jour of Ferm Technol 6, 407–413.

Vishnu C, Seenayya G & Reddy G (2000) Bioproc Engg 23 : 155–158.

Vishnu C, Seenayya G, and Gopal Reddy (2002) World Jour. of Microbiol and Biotechnol 18: 428-433.

Yumoto I & Ikeda K (1995) Biotechnol Lett 17 : 543–546.

Zhang DX & Cheryan M (1991) Biotechnol Lett 13: 733–738.

Chapter - 12

Microbes – Wheels of Organic Farming

M.N. Sreenivasa

Department of Microbiology, University of Agricultural Sciences,
Dharwad- 580 005
E-mail: sreenivasamn@gmail.com

1. Introduction

Increasing consciousness about conservation of environment as well as health hazards associated with agrochemicals and consumers preference to safe and hazard free food are the major factors that lead to the growing interest in organic agriculture.

Organic farming offers the possibility of sustaining crop yields and maintenance of soil health. It avoids or largely excludes the use of synthetic fertilizers, pesticides etc and solely depends on the use of crop residues, animal manures, biological inoculants, off-farm organic wastes, crop rotation etc., to maintain soil health and productivity. Use of organic manures enhance the bulk density of soil and there by reduces resistance to penetration by plant roots. Organic production methods are supportive of environment. Organic agriculture encourages a balanced host/ predator relationship through augmentation of beneficial insect population, biological and cultural pest control.

2. Importance of Organic Agriculture

1. The demand for organic food is steadily increasing both in the developed and developing countries with an annual average growth of 20-25 per cent.
2. Decline in productivity of soil.
3. Indiscriminate use of pesticides affects human and animal health, biodiversity, wildlife etc. and cause environmental pollution.
4. High cost of inputs in conventional agriculture.

5. Conventional agriculture is not sustainable.
6. Deficiency of micronutrients.
7. Global warming due to rise in carbon pool in atmosphere.
8. Growth rate of agriculture production (1.5%) is much below the population growth rate (2.0%). Our country to be economically strong should dependent on agriculture and allied enterprices.

3. Microbial role in Organic Farming

Microorganisms are playing a key role in organic farming in terms of organic matter decomposition, bio-fertilizers, bio-pesticide etc.

At UAS, Dharwad, the results of several experiments conducted revealed *Phanerochaete chrysosporium* to be an efficient lignocellulolytic fungus for decomposing several agroresidues including sugarcane trash, redgram stalk, cotton stalk etc. This is know to produce lignin degrading enzymes like lignin peroxidase, manganese peroxidase, glyoxal oxidase and reduce C:N ration. At the same time, scientists have worked out the possibility of enrichment of decomposed material by inoculating free living nitrogen fixers like *Azotobacter* and p-solubilizers. Such attempts have given fruitful results in enrichment of vermicompost with higher amount of N and P contents.

In continuation to decomposition and enrichment of agro residues, the results of several interaction studies clearly showed that coinoculation of *Trichoderma harzianum,* a cellulolytic fungus and P- solubilisers viz *Pseudomonas striata, Bacillus polymyxa* along with the application of rock phosphate resulted in better decomposition of cotton stalks and nutrient uptake by the test crop.

Table 1. Population of microflora before sowing and after harvest

Treatments	Total Bacteria (No. x 106/ g soil)		Fungi (No. x 103/ g soil)		Actinomycetes (No. x 102/ g soil)	
	I	AH	I	AH	I	AH
T1: 100% org	55.3*	67.2	56.7	69.0	46.7	58.8
T2: 75: 25 org: inorg	55.3	64.6	56.7	64.0	46.7	58.2
T3: 50:50 org: inorg	62.7	71.6	57.0	74.0	42.7	63.6
T4: 100% inorg	57.3	74.0	58.3	81.4	42.0	61.2
SEm±	-	3.33	-	2.65	-	3.99
CD at P=0.05	-	NS	-	8.17	-	NS
CV (%)	-	10.7	-	8.20	-	14.8

* Mean of three replications.

The cropping system influences soil microflora. The dynamics of soil microflora and soil enzymes was studied is a long term integrated nutrient management experiment in a fixed site in two cropping systems at UAS, Dharwad. Both soil microflora and soil enzymes were highest in groundnut-sorghum cropping system as compared to sorghum- safflower cropping system. The organic C, available N, available P, soil microflora and enzyme activities increased significantly with the application of organic fertilizers along with inorganic. Positive correlation was observed in soil microflora, soil enzymes and nutrient availability. These results highlight the significance of application of organic matter in maintenance of soil fertility and bio-dynamics (Tables 1, 2, 3 and 4). Several efforts to enrich vermicompost with beneficial microflora gave fruitful results (Table 5).

Table 2. Population of FNF and PSB

Treatments	Free living N2-fixers (No. x 103/g soil)		PSB (No. x 103/ g soil) with zone		PSB (No. x 103/ g soil) without zone	
	I	AH	I	AH	I	AH
T1: 100% org	32.0	49.4	3	4	30.7	43.4
T2: 75: 25 org: inorg	32.0	45.2	4	4	30.7	59.2
T3: 50:50 org: inorg	32.3	56.0	6	8	31.0	55.0
T4: 100% inorg	31.3	43.4	5	8	31.3	51.0
SEm±	-	2.65	-	-	-	2.21
CD at P=0.05	-	8.71	-	-	-	6.81
CV (%)	-	12.2	-	-	-	9.50

Table 3. Dehydrogenase activity, soil respiratory activity and grain yield of sunflower

Treatments	Dehydrogenase activity (Ml of H2 released /g air dried soil)		Soil respiratory activity (wt of CO2 (g)/100 g soil/ 2 hrs)		Seed yield (q/ac)
	I	AH	I	AH	
T1: 100% org	4.2	15.0	0.11	0.14	5.43
T2: 75: 25 org: inorg	4.2	16.1	0.11	0.22	5.68
T3: 50:50 org: inorg	4.3	16.4	0.11	0.21	5.75
T4: 100% inorg	4.3	16.1	0.11	0.19	5.81
SEm±	-	0.24	-	0.10	
CD at P=0.05	-	0.73	-	NS	
CV (%)	-	3.3	-	12.6	

Table 4. Soil Biological properties (Bengal gram) RARS, Raichur for the year 2005-06

Treatments	Total Bacteria (Nox10^6/ g soil)		Total Fungi (Nox10^3/ g soil)		Total Actino-mycetes (Nox10^2/ g soil)		Freeliving N_2 fixers (Nox10^3/ g soil)		PSM (Nox10^3 / g soil)		Dehydrogenize Activity (micro liters of H_2 released/g of soil)		Respiratory activity (wt. of CO_2 (g)/100 g soil/2 hrs.)	
	I	AH	I	AH	I	AH	I	AH	I	AH	I	AH	I	AH
T_1 75% OM	63	84	56	78	44	56	31	43	28 (3)	36 (6)	3.98	4.36	0.11	0.16
T_2 100% OM	63	96	58	89	47	61	31	48	32 (2)	41 (6)	4.25	4.96	0.11	0.18
T_3 50%+50%	66	118	58	86	50	68	37	52	33 (3)	42 (8)	4.63	5.11	0.14	0.24
T_4 100% IOM	49	94	46	68	29	54	22	31	25 (2)	38 (5)	2.93	3.73	0.08	0.16
T_5 100% IOM + OM	53	112	52	92	41	71	33	43	31 (3)	40 (6)	3.90	4.30	0.10	0.18
SEM ±	4.10		2.90		2.40		2.10		2.60		0.21		0.01	
CD at P = 0.05	12.40		8.60		7.60		6.30		7.80		0.63		0.03	

Table 5. Population of *Azospirillum* and P-Solubilizers and N & P contents in vermi compost

Treat-ments	Population of *Azospirillum* (MPN No x 10^5/g)				Population of PSM (MPN No x 10^7/g)				N-content (%)	P-content (%)
	Period in months				Period in months					
	I	II	III	Mean	I	II	III	Mean		
AZO	32	54	84	57.0	13	16	21	16.7	0.64	0.68
PSB	14	18	21	17.7	39(3)	54(5)	61(9)	51.3	0.56	0.98
AZO+ PSB	33	70	95	66.0	41(4)	63(8)	69(11)	57.7	0.79	1.15
Mean	26.3	47.3	66.3	46.9	31	44.3	50.3	42		
SEM±	0.91	1.62	1.74		1.24	1.36	1.89		0.03	0.04
CDat P=0.05	2.70	4.80	5.30		3.60	4.10	5.7		0.09	0.13

x 10 ^{5}/g of Lignite & *Pseudomonas striata* 36 x 10 ^{7}/g of lignite.

N - Content in UIC : 0.56% P-content in UIC : 0.68 %

Of late, lot of importance is given to protect the environment from pollution. Hence several scientists are exploiting microbial technologies for production of bioethanol from agro residues. Bioethanol can be blended with petrol up to 20%. At present, ethanol is blended up to 5% with diesel for transportation. Several microbial cultures viz., *Trichoderma reesei* and *Pachysolen tannophilus* NCIM – 3445 showed their efficiency in terms of ethanol yield. Such studies certainly help to protect environment from pollution.

Another potential area which uses microbial technology in organic farming is biocontrol of pathogens and crop pests by which usage of chemical insecticides can be reduced and inturn environmental pollution can be avoided. Several microorganisms are known to improve plant growth directly through nutrient mobilization and production of plant growth hormones and indirectly through suppression of plant pathogens or by inducing systemic resistance in plants. The soil microorganisms having these multiple beneficial traits are referred to as plant growth promoting rhizobacteria (PGPR). Recently they are also referred to as plant health promoting rhizobacteria (PHPR). Fluorescent pseudomonades play a key role in biocontrol of plant pathogens as they have rapid growth, simple nutritional requirement and ability to utilize diverse organic substrates. They are known to produce highly potent broad spectrum antifungal molecules against various phytopathogens. They are reported to produce antibiotics, siderophores, HCN etc. They also compete for space and nutrients with soil borne plant pathogens. Fluorescent pseudomonads are promising bioinoculants for agricultural system to increase productivity as they are cost effective and ecofriendly. Anther fungus, *Trichoderma harzianum* is also being used extensively. Some of the microbial pesticides NPV, GV, *Bacillus thuringiensis, B popilleae, Metarrhizium anisopliae, Beavaria bassiana, Nomuraea rileyi, Verticillium lacanii etc.,* are being used under field conditions for biological control of crop pests. All these organisms can grow well and show activity at an optimum RH (Minimum 75%-85%) and temperature 28-30°. The moisture and temperature play a key role on the growth of micro organisms. The environmental conditions should also be favourable for their action after spraying these microbial pesticides on to the crops. Heavy rainfall/ higher temperature after spraying will result in loss of microbial pesticides due to washing or inactivation of enzymes. Hence it is better to spray these microbial pesticides during late evening or early hours of the day for efficient activity.

Organic farming should be taken as means of enhancing soil fertility, soil health and productivity and to achieve sustainability. Of late, many farmers committed suicide due to high cost of inputs in conventional agriculture by which they could not repay their debts. Organic farming definitely will be the answer for all these socio-economic problems as well as soil fertility and productivity problems. Organic farming is locally available and renewable resource based system, hence easily affordable to even small and marginal farmers and keeps farmers away from debt trap.

We need to create awareness among farmers so as to reduce environmental pollution and poisoning of food.

References

Chhonkar, P.K. 2003. Journal of the *Indian Society* of *Soil Science*, Vol. 51, No. 4, pp. 365-377.

Dwivedi Vandana 2005. Organic farming: Policy initiatives, Paper presented at the *National Seminar on National Policy on Promoting Organic Farming*, 10-11 March, 2005. Pp.58-61.

GOI 2001. The report of the working group on organic and biodynamic farming, *Planning Commission, Government* of *India*. pp:1-25.

lomte, M.H. and Pendke, M.S. 2004. Organic Farming Inventory. Proceedings of the State level Seminar on Present and Future Prospects of Organic Farming, *Marathwada Agricultural University, Parbhani*, pp. 1-133.

Nagaraj Naik and M. N. Sreenivasa, 2008. *Karnataka Journal* of *Agricultural Sciences*. 22: 231-232.

Nagaraj Naik, Bhat S.N., Sreenivasa M.N and J.H. Kulkarani, 2008. *Proceedings* of *National Seminar on Urban development*. NESA, Bangalore.

Nekar, M.M., Babalad, H.B., Bhat, S.N and M.N. Sreenivasa, 2009. *Journal* of *Oilseeds Research.* Vol 26 : 390-392.

Rajendran, T.P., Venugopalan, MY and Tarhalkar, P.P., 2000. *Technical Bulletin* No.1/2000, Nagpur,.p.39.

Ramesh, P, Mohan Singh and Subba Rao, A. 2005. *Current Science*, 88. pp.561-568.

Sreenivasa, M. N, 2006 *Vatika IAHS* 4:19-20.

Sreenivasa, M. N. and A. Naganagoud, 2005 Enrichment of vermicompost using beneficial microbes. *Proc. 46th AMI conf.* 8-10 December, 2005 Osmania University Hyderabad.

Sreenivasa, M. N. and V. P. Nagalikar, 2005, Rhizosphere microflora and enzyme activities in response to application of organic amendments. *Proc. 46th AMI conf.* 8-10 December, 2005 Osmania University Hyderabad.

Venkateswarlu, B., Srinivasa Rao, Ch., Ramesh, G., Venkateswarlu, S., and Katyal, J.C. 2005. Effect of long term incorporation of legume biomass on soil quality under rainfed conditions. Extended abstract of a paper presented at the *International Conference on Soil Water an Environment: Issues and Strategies*, 28th January to 1st Feb., 2005, New Delhi, pp. 118.

Chapter - 13

The industrial applications of lignin and lignolytic enzymes

M.A. Singara Charya

Department of Microbiology, Kakatiya University, Warangal - 506 009
E-mail : mascharya@yahoo.co.in

ABSTRACT

Lignin is the most abundant aromatic compound on earth, and is second only to cellulose in its contribution to living terrestrial biomass. It is formed by the oxidative coupling of several related phenylpropanoid precursors: coniferyl alcohol, sinapyl alcohol and p-hydroxycinnamyl alcohol. Apart from its effective role in carbon sequestration, lignin determines the rigidity, strength and resistance to the plant structure. In the industrial applications of lignin, it was used a green alternative to plastic, called arboform, used for the manufacture of toys, loudspeakers boxes, car parts, golf tees, ballpoint pens etc. Its importance is also established as emulsifier in animal feed and effective and economical adhesive. The lingo-sulfonates are used on unimproved road surfaces as soil stabilization and dust control agents. Wood-rotting basidiomycetous fungi that cause white rot in wood are the most efficient lignin degraders in the nature. They secrete three important types of lignolytice enzymes, laccases, lignin peroxidase, and manganese peroxidase. The potential applications of lignolytic enzymes are well documented in bio-pulping and bio-bleaching process in paper and rayon industries. Their role in food, fodder, textile, industries and degradation of many xenobiotics is also reported. The role of lignolytic enzymes in pharmaceutical industries, synthesis of biocatalysts, biosensors for the detection of phenolic compou- nds in food industries and in the field of nanobiotechnology is also documented.

Key words: Lignin, lignolytic enzymes, white rot fungi, laccase, lignin peroxidase, manganese peroxidase, biobleaching,

1. Introduction

The term 'lignin' is introduced by de Candole in 1819. Lignin is the second most abundant organic compound on earth after cellulose. The major structural elements of wood are cellulose, hemi-cellulose and lignin. Lignin is a water impermeable aromatic polymer found in all higher plants and serves as a barrier against microbial attack [Ramesh Chander and Ajay Singh, 2007]. The wood consists of three major components: Cellulose (40-60%), Hemi cellulose (10-30%), Lignin (15-30%). Lignin content varies in different plants *viz.* in woody angiosperms it is 18-25%, in gymnosperms it is 25-35%, while in monocotyledons it ranged between 10-30%. There are three major groups of lignins, those are, Guaiacyl lignin - composed of coniferyl alcohols with small amounts of coumaryl and sinapyl alcohol derived units (conifers, lycopods, ferns, horsetail), Guaiacyl-syringyl lignin - composed of equal amounts of coniferyl alcohol and sinapyl units (dicotyledons) and Guaiacyl-syringyl-p-hydroxyphenyl lignin - composed of equal amounts of p-coumaryl alcohol, coniferyl alcohol and sinapyl alcohol (grasses).

The plants synthesize vast quantities of insoluble, aromatic macromolecules, including tannins and most importantly for lignin. These aromatic macromolecules are biodegradable, but their biodegradation is often slow. Lignin is tough, rigid polymer, provides structural support and is main structural component of secondary wall, only found in vascular plants (Fig. 1). Lignin is synthesized (Fig. 2) by one-electron oxidation of the precursors p-coumaryl alcohol, coniferyl alcohol and sinapyl alcohol generating phenoxy radicals which then undergo polymerization. Lignin is synthesized by plants from carbon-dioxide by way of shikimic acid. Shikimic acid is converted to L-phenyl alanine, which is the primary precursor of all three cinnamyl alcohols (p-coumaryl alcohol, coniferyl alcohol and sinapyl alcohol). The cinnamyl alcohols are polymerized enzymatically by phenol oxidases, enzymes that oxidize the aromatic alcohols forming free radicals that spontaneously and irreversibly polymerise by complex phenol coupling reactions. The polymerization process result in a formation of a highly branched, structurally complex molecule comprised of polyphenol proponoid units inter connected by varied types of covalent linkages. The subunits are nine joined together with a variety of bond types, mainly carbon-carbon and ether bonds the α- aryl-ether bonds being the most abundant [Adlere, 1977]. Also heterogeneous ring structures such as dibenzodioxocin ring occur [Brunow *et al.*, 1998a]. Softwood (gymnosperm) lignin comprises mainly of coniferyl alcohol subunits, which makes its structure more dense and resistant towards microbial degradation than hardwood (angiosperm) lignin consisting of both guaiacyl and syringyl subunits [Faix *et al.*, 1985; Blanchette, 1995; Burlat *et al.*, 1997]. In addition, softwood has higher lignin content (25-33% of dry weight) than hardwood (20-25% of dry weight) [Sjöström and Westermark, 1998]. The proportion of major linkages in lignin is showed in Table 1. A structural model of spruce lignin polymer according to Brunow *et al.* [1998a] (Fig. 3).

Table 1. Proportion of major linkages in lignin

Linkage type	Softwood (spruce) [%]	Hardwood (birch) [%]
β-O-4-Aryl ether	46	60
α-O-4-Aryl ether	6–8	6–8
4-O-5-Diaryl ether	3.5–4	6.5
β-5-Phenylcoumaran	9–12	6
5-5-Biphenyl	9.5–11	4.5
β-1-(1,2-Diarylpropane)	7	7
β-β-(Resinol)	2	3
Others	13	5

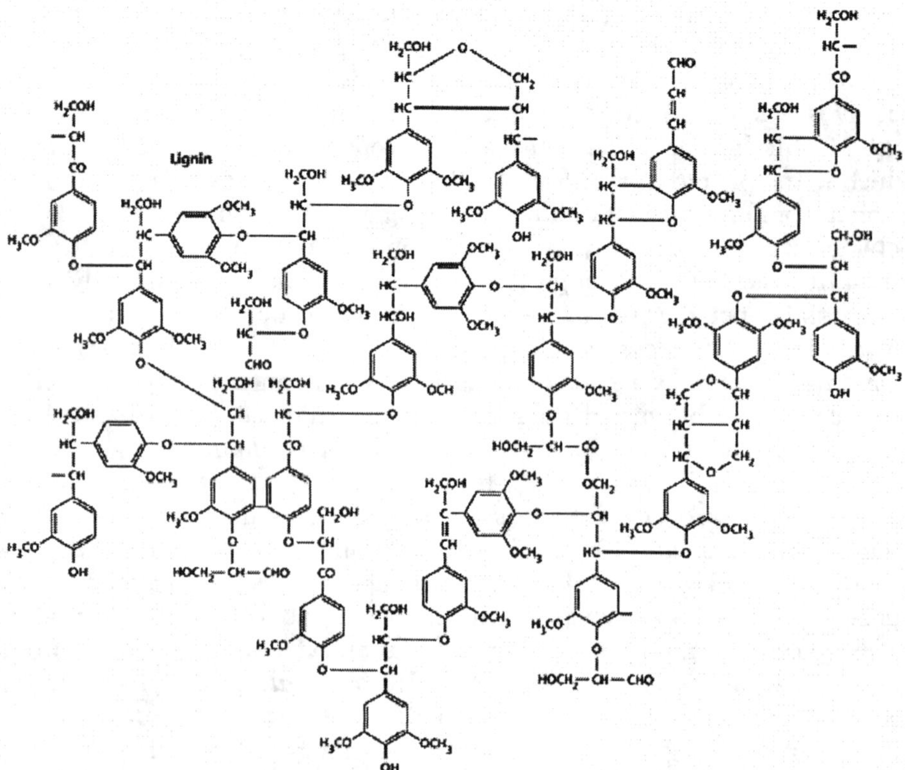

Fig. 1. Schematic diagram of lignin

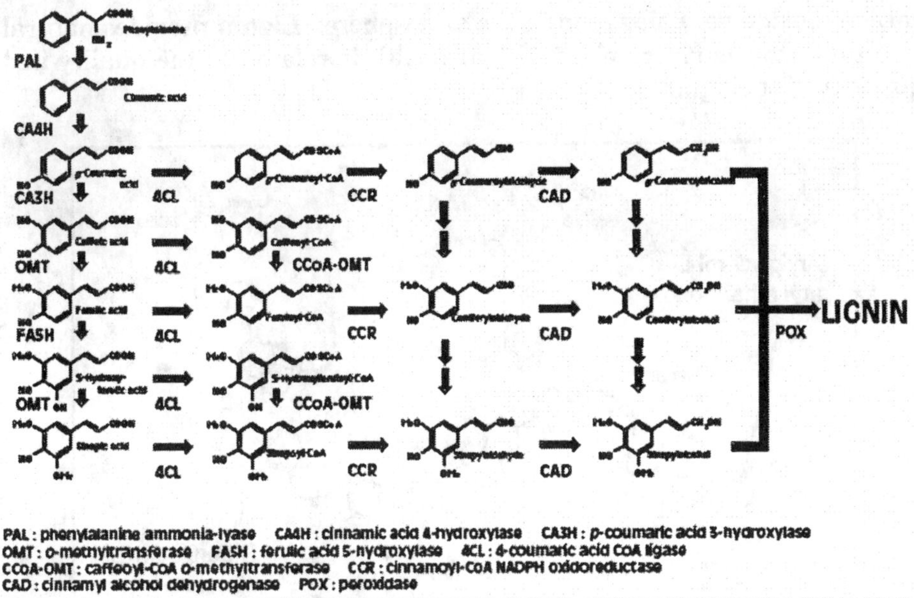

PAL : phenylalanine ammonia-lyase CA4H : cinnamic acid 4-hydroxylase CA3H : *p*-coumaric acid 3-hydroxylase
OMT : o-methyltransferase FASH : ferulic acid 5-hydroxylase 4CL : 4-coumaric acid CoA ligase
CCoA-OMT : caffeoyl-CoA o-methyltransferase CCR : cinnamoyl-CoA NADPH oxidoreductase
CAD : cinnamyl alcohol dehydrogenase POX : peroxidase

Fig. 2. Pathway of lignin biosynthesis

2. Applications of lignin

2.1 Role in disease resistance in plants

Lignin, a major component of cell walls of vascular plants is considered the first line defense against successful penetration of invasive pathogens. Lignification renders the cell wall more resistant to mechanical pressure applied during penetration by fungal appresoria as well as more water resistant and thus less accessible to cell wall degrading enzymes [Vance *et al.*, 1980; Nicholson, 1992; Zeyen *et al.* 2002]. Lignification is essential for the structural integrity of plant cell walls and is crucial for plant development. The production/secretion of variety of phenol compounds, group of phytoalexins and diversified Pathogenesis Related (PR) proteins by the diseased plants was also considered that their role is essential in the disease management of plants. Apart from these, the production of gums, resins and mucilaginous secretions of the plants during infections or wounds were essentially produced by lignin. Obviously, there is a close interaction between disease management and lignin production.

2.2 Ecological advantages of lignin

Huge amounts of free carbon-dioxide is sequestered and immobilized in the form of lignin, so that the quantity of green house gas is reduced. Slow degradation of lignin in the soil is beneficial because of the release of carbon-dioxide very slowly in to the atmosphere. Lignin provides natural resistance to the entire plant. Plant health is related to the quality and quantity of the lignin.

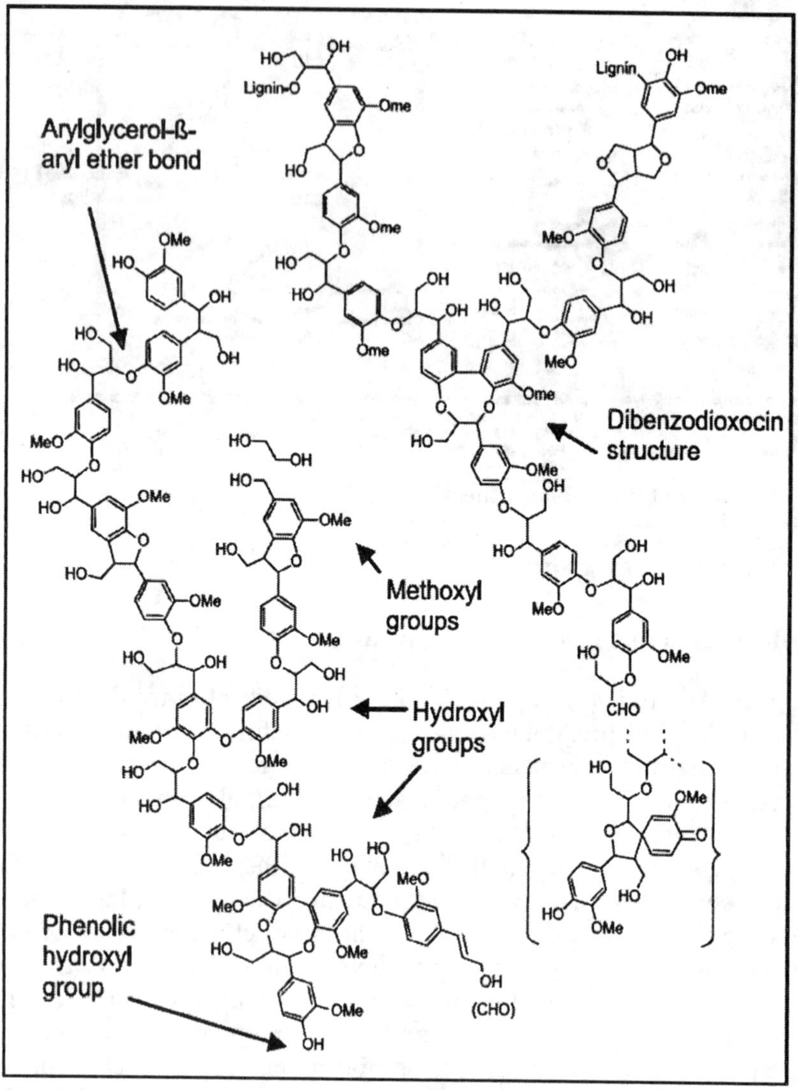

Fig. 3. A structural model of spruce lignin polymer according to Brunow *et al.* (1998). The figure is reproduced by the permission of the American Chemical Society.

2.3 Industrial applications of lignin

Lignin which is now considered as a waste product during pulp and paper processing find many applications. Apart from its important application as fire wood in many parts of the world, it is the basis for household furniture, doors, windows, etc. The ancient people have great wisdom in selecting lignin for this purpose viewing its strong resistance towards microbial attack. The durability, stability and complexity of chemical composition made it number one choice for these house hold articles. In industrial sector, lignin can be used as an emulsifying, sequestering, binding or dispersal agent. The first commercial use of lignin by Marathan Corporation in Rothschild, Wisconcin, USA in 1927 as leather tanning agent.

The company with the name to Lingo Tech., USA used as polyurethane foam. In 1998, a German company, Tecnaro developed a process for turning lignin into Arboform, which behaves identical to plastic for injective molding. Therefore, it can be used in place of plastic for several applications. When the item is discarded, it can be burned just like wood. This will be a green alternative to plastics. This arboform is used for the manufacture of toys, loud speaker boxes, car parts, golf tees, ball point pens, etc. The bio-plastic dubbed arboform, derived from wood pulp based lignin can be mixed with hemp, flax or wood fibers and other additives such as wax to create a strong, non-toxic alternative to petroleum based plastics.

Lignin can be used as animal feed serving as emulsifiers in animal feed and as a raw material in the production of vanillin, which is widely used as an ingredient in food flavours, in pharmaceutical and fragrance in perfumes and odour masking products. Lignin also can be employed as adhesive, with its mucilaginous and gummy properties. It is an effective and economical adhesive acting as binding agent or 'glue' in pellets or compressed materials. The binding properties of lignin also make the product a useful component of the making of charcoal briquettes, ceramics, linoleum paste, play wood and particle boards. Lignin degradation scheme proposed by Leisola and Garcia [1989] is showed in Fig. 4. Major thermochemical lignin conversion process and their potential useful products is showed in Fig. 5.

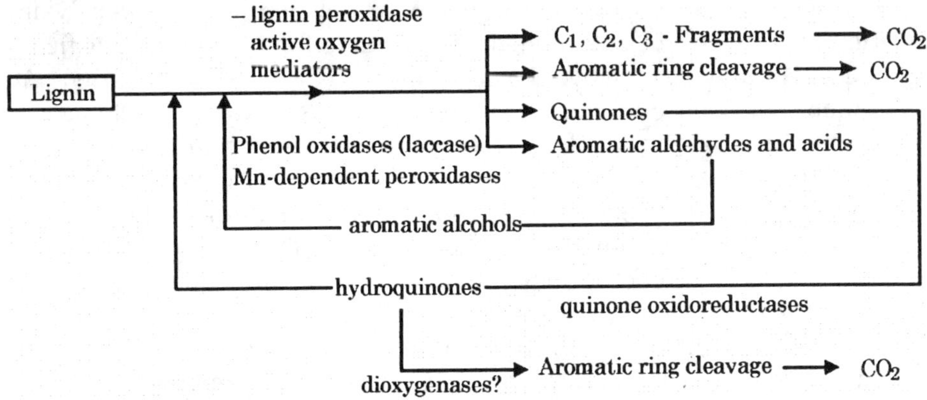

Fig. 4: Lignin degradation scheme as proposed by Leisola and Garcia (1989)

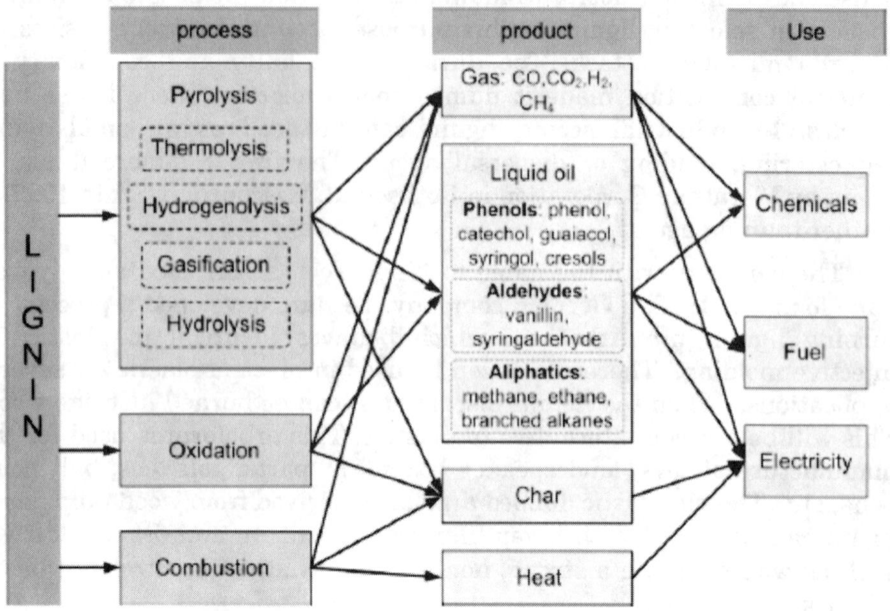

Fig 5: Major thermochemical lignin conversion process and their potential products

2.4 Future of lignin utilization

Lignin is available in huge amounts as a by-product of the pulping process, and this will further increase as the lignocellulosic ethanol process gets commercialized. Although the prospects of macromolecular lignin being utilized as raw materials in many industrial processes as a substitute for phenol in phenol-formaldehyde resins and fillers in polymers [Setua, 2000], carbon fibers, binders, polyurethane foams, epoxy resins and biodispersants [Lora and Glasser, 2002] have been discussed, the most prominent lignin utilization so far appears to be its depolymerization into lower-molecular weight compounds. Lignin can be gasified to produce H_2 and CO (syngas) which in turn can be synthesized into methanol and various other chemicals through known technologies. Alternatively, pyrolysis oils can be obtained from lignin in the way similar to biomass pyrolysis. However, pyrolysis oil produced from lignin pyrolysis is highly oxygenated and unstable during storage and, hence, requires further upgrading [Chen *et al.*, 2001]. Thus, fuel production does not appear to be a much promising route for the value addition of lignin unless some efficient and direct methods are developed. Instead, production of high-value bulk and fine chemicals such as acids, alcohols, phenols and aldehydes show better prospects of commercialization.

Lignin is probably the only viable renewable source for the production of aromatic compounds [Zakzeski *et al.*, 2010]. The first stage in chemical production includes the cleavage of primary linkages to result in monomeric aromatics with various substituted functional groups. The selective cleavage of these functionalities by oxidation then can produce specific value-added compounds including aromatic aldehydes and acids. Partial or complete removal of the functionalities by reduction will provide simple aromatics, including benzene, toluene, and phenol. Lignin processed in practical biorefineries may have been generated from a variety of biomass feedstock, isolated using different techniques, and contaminated with impurities to various extents. The development of specific pathways and catalysts applicable to such situations is crucial for the realization of an efficient biorefinery converting lignin to value added chemicals. A lignocellulosic biorefinery utilizing carbohydrates to produce fuel ethanol and lignin to produce fine chemicals offers a good prospect as an alternative to petroleum refineries. With wide varieties of fine and bulk chemicals possible from lignin, lignocellulosic biorefineries will have a diversified product range, which in turn will make them more competent and sustainable.

3. Estimation of lignin

Lignin content was estimated by the method [Chesson, 1978]. One gram of the sample was refluxed with 150 ml boiling 80% (v/v) ethanol in a refluxing flask for 5 min. The solution was allowed to cool and the contents were filtered. The residue was again refluxed for two hours with 150 ml water at 100°C. Soluble sugars and hot water soluble pectic substances present in the sample were removed by ethanol and water respectively. The residue obtained, after refluxing with water was refluxed for two hours with 150 ml of 0.5% (w/v) ammonium oxalate at 100°C. During this process hot oxalate soluble pectic substances were removed. Again the residue was refluxed for two hours with 150 ml 0.5M sulphuric acid at 100°C, cooled and filtered. The filtrate containing hemicellulose was treated with 10 ml of 72% (v/v) sulphuric acid at room temperature for 4 hours, then diluted to 0.1M sulphuric acid and refluxed at 100°C for a further period of 2 hours. The contents of the flask were cooled and filtered. The filtrate contained cellulose and the residue left behind was lignin.

4. Lignin degrading Enzymes

Enzymes involved in lignin degradation (Table 2) are laccase (EC 1.10.3.2; benzenediol:oxygen oxidoreductase), lignin peroxidase (E.C. 1.11.1.14; Diarylpropane; oxygen, hydrogen peroxide, oxido-reducase), manganese peroxidase (E.C. 1.11.13; Manganese (II) hydrogen peroxide, oxido-reductase), versatile peroxidase (EC 1.11.1.16), and H_2O_2-forming enzymes such as glyoxal oxidase (GLOX) and aryl alcohol oxidase (AAO, EC 1.1.3.7), cellobiose dehydrogenase which are produced by lignin-degrading white-rot and litter-decomposing fungi in different combinations [Hatakka, 1991; Hatakka,1994]. Lignin-modifying enzymes oxidize substructures of lignin, are extracellular and typically they are present as multiple isoforms with similar function.

4.1 Laccases

Laccases are blue copper-containing oxidases that catalyze one-electron oxidations of aromatic amines and phenolic compounds such as phenolic substructures of lignin. The terminal electron acceptor in the catalytic reaction is molecular oxygen, which is reduced to water [Thurston, 1994]. Laccases are glycoproteins and those of white-rot fungi generally have molecular weight between 60-80 kDa and p*I* 3-6 [Hatakka, 2001; Baldrian, 2006]. Although laccase has been extensively studied for decades the catalytic mechanism is not fully understood. Laccase catalyses the formation of phenoxyl radicals and their unspecific reactions leading finally to C-hydroxyl oxidation to ketone, alkyl-aryl–cleavage, demethoxylation and C-C cleavage in phenolic lignin substructures, as well as polymerization reactions [Youn *et al.* 1995]. Fungal laccase has been suggested to participate in morphogenesis, fungal plant pathogen interaction, stress defense and detoxification of byproducts of lignin degradation [Thurston, 1994; Baldrian, 2006].

Laccase assay: The laccase activity was determined using 2, 2'-azino-bis (3- ethylbenzthiazoline-6-sulfonic acid) (ABTS) (Sigma) as the substrate (Rasera *et al*., 2009). The laccase reaction mixture contained 0.5 ml of 0.45 mM ABTS, 1.2 ml of 0.1 M phosphate buffer (pH 6.0) and 0.5 ml of filtrate to give a final reaction volume of 2.2 ml. The oxidation of the substrate (ABTS) was monitored by the increase in the absorbance at 420 nm using Shimadzu UV-1800 spectrophotometer (Tokyo, Japan) over 90 s at 30°C (± 2°C), using ε = 3.6 × 104 cm–1 M–1. Enzymatic activity was expressed as 1 U = 1 μmol of ABTS oxidized per min at 25°C (± 1).

Mode of action of the laccase enzyme : Laccases contain 4 copper atoms termed Cu T1 (where the reducing substrate place) and trinuclear copper cluster T2/T3 (where oxygen binds and is reduced to water). As a one-electron substrate oxidation is coupled to the four-electron reduction of oxygen the reaction mechanism cannot be entirely straightforward. Laccase can be thought to operate as a battery, storing electrons from individual oxidation reactions in order to reduce molecular oxygen. Hence, the oxidation of four reducing substrate molecules is necessary for the complete reduction of molecular oxygen to water. In general terms, substrate oxidation by laccase is a one-electron reaction generating a free radical. The initial product is typically unstable and may undergo a second enzyme-catalysed oxidation or otherwise a non-enzymatic reaction such as hydration, dis-proportionation or polymerisation. The bonds of the natural substrate, lignin, that are cleaved by laccase include, Cα-oxidation, Cα-Cβ cleavage and aryl-alkyl cleavage (Fig. 6a). Laccases are similar to other phenol-oxidising enzymes, which preferably polymerise lignin by coupling of the phenoxy radicals produced from oxidation of lignin phenolic groups [Bourbonnais *et al.* 1995]. Due to this specificity for phenolic subunits in lignin and its restricted access to lignin in the fibre wall, laccase has a limited effect on pulp bleaching unless redox mediators (eg. 2,2′-azino-bis(3-ethybenzthi-zoline-6- sulfonic acid (ABTS)) will be introduced in the reaction (Fig.6b) [Bourbonnais and Paice, 1996]. Catalytic cycle of a laccase-mediator oxidation system [Banci *et al*: 1999] showed in Fig. 7.

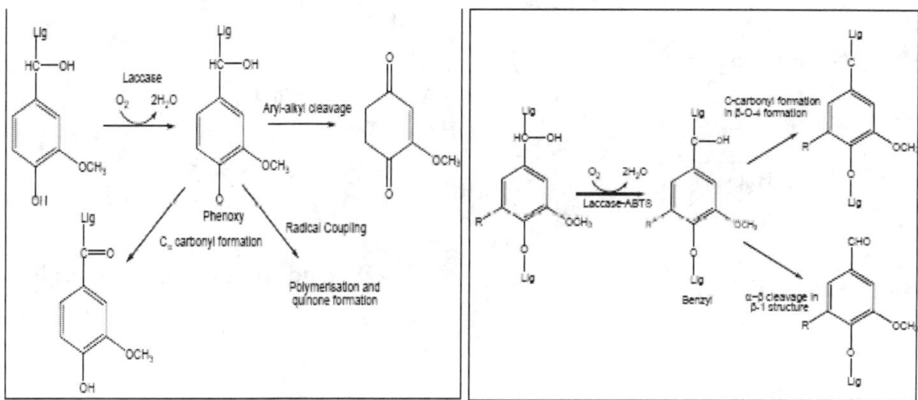

Fig. 6: Oxidation of (a) phenolic subunits of lignin by laccase and (b) non-phenolic lignin model compounds by a laccase mediator system

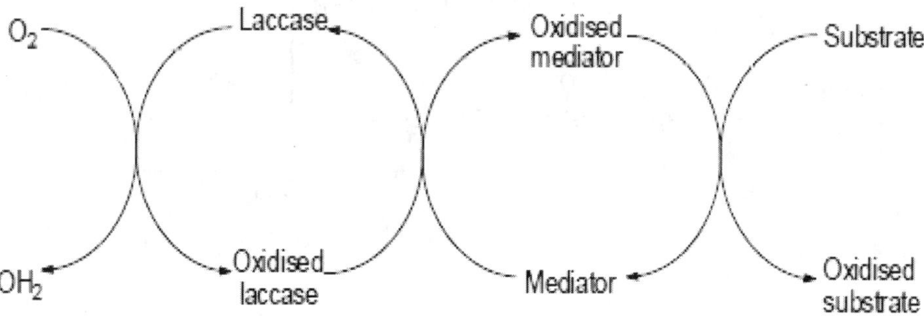

Fig. 7. Catalytic cycle of a laccase-mediator oxidation system

4.2 Peroxidases

The role of peroxidase in lignin degradation (Fig. 8 & 9) was brought into light from experiments where enzyme catalase was found to inhibit lignin degradation [Faison and Kirk, 1983]. Lignin-degrading peroxidases, i.e. Manganese peroxidase (MnP), Lignin peroxidase (LiP) and Versatyl peroxidase (VP), are structurally related hemecontaining glycosylated peroxidases.

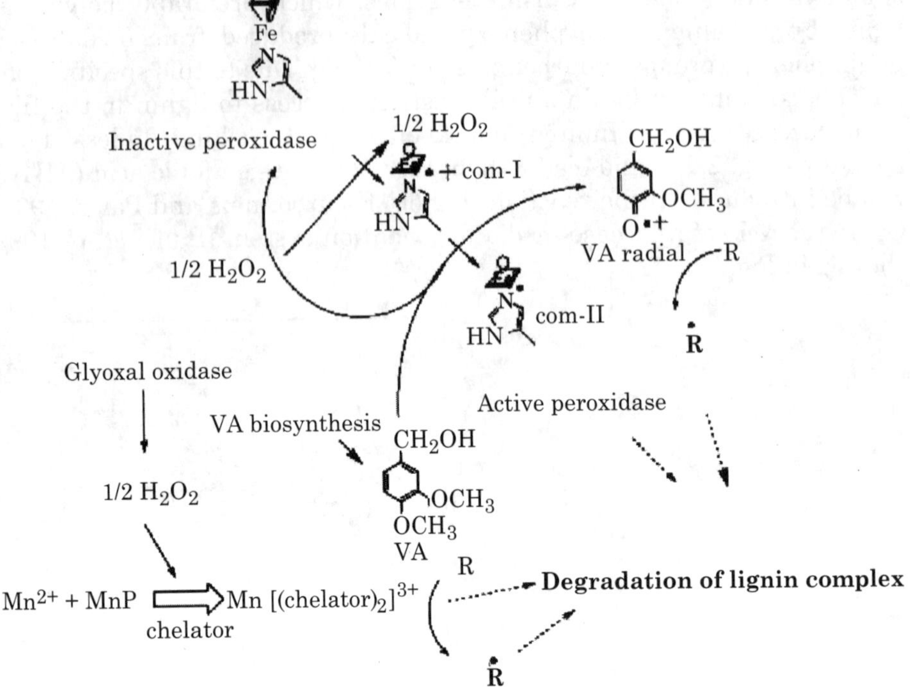

Fig 8. Reaction mechanism of lignin peroxidase and manganese peroxidase for lignin degradation

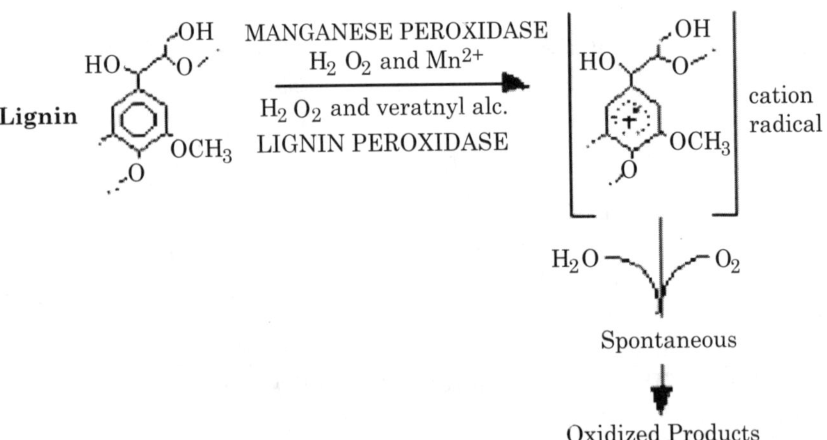

Fig. 9. Single electron oxidation of an aromatic ring in lignin by extracellular peroxidases and subsequent oxidation cation radical by oxygen.

4.3 Lignin peroxidase

Lignin peroxidase (LiP) also called ligninase is a hemeprotein that is produced by many white-rot fungi [Kirk and Farrel, 1987] and that is approximately 37,000 daltons in size. It bears structural and mechanical similarity to the more familiar plant peroxidases and utilizes hydrogen peroxide and organic peroxides to oxidize a variety of substrates [Tien *et al.*, 1986]. They differ from the other oxidoreductases in that they have very low pH optima and much higher redox potentials. Lignin peroxidases can act on a wide range of substrates with their reactivity being largely determined by the redox potential of the substrate. They can catalyze the oxidation of substrates with a reduction potential greater than or equal to 7.55 eV [Hammel *et al.*, 1986]. In their work, Haemmerli *et al.* [1986] showed that lignin peroxidases oxidize lignin monomers, dimers and trimers as well as polycyclic aromatic compounds such as benzopyrene. It can catalyse oxidation of both phenolic and nonphenolic aromatic compound as substrates; the phenolic compounds are oxidized to yield products similar to those produced by classical peroxidases, while the oxidation of the nonphenolic methoxybenzenes is unique to the lignin peroxidases [Kersten *et al.*, 1985]. H_2O_2 is required for the catalytic reaction carried out by LiP. These reactions are neither substrate-specific nor stereo-specific. In a LiP-catalyzed reaction, the heme-containing enzyme is first oxidized by H_2O_2 to generate the two-electron oxidized intermediate Compound I [Koduri and Tien, 1995]. Compound I is then reduced by one electron donated by a substrate molecule (such as veratryl alcohol or H_2O_2) to give a 1-electron oxidized enzyme intermediate, Compound II, and a free radical product. The so formed radical intermediate can then act on lignin or other substances. The catalytic cycle is subsequently completed by the one-electron reduction of Compound II by a second substrate molecule [Renganathan and Gold, 1986]. In the absence of a reducing substrate, Compound II can undergo a series of reactions with H_2O_2 to form Compound III, oxyperoxidase [Renganathan, 1986; Cai and Tien, 1992]. Compound III decays either spontaneously or in the presence of H_2O_2 and a reducing substrate to give ferric LiP and O_2.

In addition, a number of studies have proposed an important role of veratryl alcohol (VA), a fungal secondary metabolite, in stimulating LiP-mediated oxidation of a large number of resistant substrates, such as small aromatic compounds, lignin and proteins. Some authors propose that VA acts as a redox mediator and facilitates the turnover of the enzyme and protects it from H_2O_2 inactivation [Dejong *et al.* 1994]. Therefore, LiP catalyzes veratryl alcohol to veratryl aldehyde, with the enzyme bound veratryl alcohol cation radical forming an unstable intermediate [Khindaria *et al.* 1995]. This highly unstable intermediate is stabilized by the acidic microenvironment in the active site of the enzyme, which prevents deprotonation of the radical and its subsequent reaction with the oxygen. This radical can act as a redox mediator but not as a diffusible oxidant for LiP-catalyzed lignin degradation [Khindaria *et al.*, 1996]. LiP oxidizes a variety of substrates, including veratryl alcohol, and both phenolic and non-phenolic substructures of lignin in a hydrogen peroxide dependent reaction to form a phenoxyl or aryl-cation radicals. The cation radical formation in non-phenolic lignin structures leads to unspecific reactions including cleavage of C-C-bond, demethoxylation, ring opening and depolymerization [Kirk and Farrel, 1987]. It has been suggested that veratryl alcohol could act as a redox-mediator in reactions catalyzed by LiP, although the half-life of the radical is too short to be involved in oxidative reactions far from place of origin [Hatakka, 2001]. However, LiP has been demonstrated to oxidize Mn^{2+} to Mn^{3+} via radical reactions involving veratryl alcohol, oxalate and O_2 [Popp *et al.* 1990].

Lignin peroxidase (LiP) assay: Lignin peroxidase (LiP) (EC 1.11.1.14) activity was determined spectrophotometrically at 310 nm through the oxidation of veratryl alcohol to veratryl aldehyde (molar absorptivity, ε310 = 9300 M-1 cm^{-1}). The reaction mixture contained 300 µL veratryl alcohol (8 mM), 600 µl 0.3M citrate/0.4M phosphate buffer (pH 4.5), 60 µL culture supernatant, and 1890 µL distilled water. The mixture was incubated for 2 min at 30°C and the reaction was initiated by addition of 150 µL H_2O_2 (5 mM). The absorbance was immediately measured in one-minute intervals after addition of H_2O_2. One unit (U) of LiP activity was defined as activity of an enzyme that catalyzes the conversion of 1 µ mole of veratryl alcohol per minute (Takamiya *et al.*, 2008).

4.4 Manganese peroxidase

MnP has been found from most of the lignin-degrading wood and litter inhabiting fungi studied so far [Hatakka, 2001; Khindaria *et al.* 1996; Steffen, 2003]. MnP oxidizes Mn^{2+} to Mn^{3+} in a hydrogen peroxide dependent reaction [Gold *et al.*, 2000; Hofrichter, 2002]. To stabilize the formed Mn^{3+} an organic chelator, e.g. a dicarboxylic acid that is produced by the fungus, is needed [Gold *et al.*, 2000; Wariishi *et al.*, 1992]. The oxid-ative reaction performed by chelated Mn^{3+} yields radical formation from various phenolic substrates, carboxylic acids, and unsaturated lipids. In phenolic substructures of lignin this leads to the formation of phenoxyl radical intermediates and thereafter to several reactions such as demethoxylation, quinone formation and C-C-cleavage [Hofrichter, 2002].

4.5 Manganese dependent peroxidase

Manganese dependent peroxidase (MnP) is heme-containing enzyme and oxidizes Mn (II) to Mn (III) that chelates with organic acids that form the principal substrates of the enzyme. This Mn III-organic acid complex can freely diffuse away from the active site of the enzyme [Mester and Field, 1998] and can oxidize secondary substrates such as phenolic compounds, lignin model compounds [Wariishi *et al.* 1988; Wariishi *et al.* 1989; Tuor *et al.* 1992], high molecular weight chlorolignin [Lackner *et al.* 1991], chlorophenols [Grabski *et al.* 1998] etc. MnP can utilize both hydrogen peroxide and various organic hydroperoxides, such as syringic acid, ferulic acid, sinapic acid etc. [Erikkson *et al.* 1990].

Table 2. Enzymes involved in the degradation process of lignocelluloses

Name of enzyme	**Reaction**
Endo-1,4-β-glucanase (3.2.1.4)	Endohydrolysis of 1,4- β -D-glucosidic linkages of cellulose
Exo-1,4- β –glucanase (3.2.1.91)	Hydrolysis of 1,4-β-D-glucosidic linkages releasing cellobiose from cellulose
β –Glucosidase (3.2.1.21)	Hydrolysis of terminal non-reducing β -D- glucose residues with release of β-D-glucose
Endo-1,4-β-xylanase (3.2.1.8)	Endohydrolysis of 1,4-β-D-xylosodic linkages of hemicellulose with xylans
β-xylosidase (3.2.1.37)	Hydrolysis of 1,4-β-D-xylans to remove successive D-xylose residues
Ligninase (1.11.1.14)	Oxidation in alkyl side chains, C-C cleavage in side chain of lignin, cleavage of aromatic ring, oxidation of benzyl alcohol to aldhehydes
α-L-Arabinofuranosidase (3.2.1.55)	Hydrolysis of terminal non-reducing α-L-arabinofuranosidase residues in α-L-arabinosidase
Acetylesterase (3.1.1.6)	An acetic ester + H_2O = an alcohol + acetate
α-Glucuronidase (3.2.1.139)	α-D-glucuronosidase + H_2O = an alcohol + D-glucuronate
β-Mannosidase (3.2.1.25)	Hydrolysis of terminal β-D-mannose residues in β-D-mannosides
α-Galactosidase (3.2.1.22)	Hydrolysis of terminal α-D-galactose residues in α-D-galactosides
Mn-dependent peroxide (1.11.1.13)	Dependent on H_2O_2 and Mn^{2+}
Laccase (1.10.3.2)	4-Bezendiol + O_2 = 4-benzosemiquinione + 2 H_2O
Horseradish peroxide (1.11.1.7)	Donor + H_2O_2 = oxidized donor + 2 H_2O
Superoxide dismutase (1.15.1.1)	$O_2^{\cdot-} + O_2^{\cdot-} + 2H^+ = O_2 + H_2O_2$
Glyoxalate oxidase (1.2.3.5)	Glyoxylate + H_2O + O_2 = oxalate + H_2O_2
Glucose-1-oxidase (1.1.3.4)	β-D-Glucose + O_2 = D-glucono-1,5-lactone + H_2O_2
Aryl alcohol oxidase (1.1.3.7)	An aromatic primary alcohol + O_2 = aromatic aldehyde + H_2O_2
Pyranose-2-oxidase (1.1.3.10)	D-Glucose + O_2 = 2-dehydro-D-glucose + H_2O_2
Cellobiose:quinine oxidoreductase (1.1.5.1)	Cellobiose + quinine = cellobiono-1.5-lactone + phenol
Cellobiose dehydrogenase (1.1.99.18)	Cellobiose + quinine = cellobiono-1.5-lactone + reduced acceptor

Table adopted and modified from Leonowicz et al. (1999)

MnP is a glycoprotein with varying degree of carbohydrate content in different isoenzymes [Pease *et al.* 1989]. In *Phanerochaete chrysosporium* MnP, the mature protein has a calculated molecular weight of 37,711 dalton and contains 358 amino acids which is preceded by a leader peptide of 24 amino acid residues [Pease *et al.* 1989]. Based on their structural studies Sundaramoorthy *et al.* [1994] proposed a three dimensional structure of MnP containing a single binding site for Mn (II) on the surface of the protein. This binding site is acidic in character and is comprised of three carboxylate ligands, a heme propionate oxygen and two water molecules. Youngs *et al.* [2000] showed that MnP contains two Calcium ions that are thought to provide thermal stability to the enzyme.

4.6 Manganese independent peroxidase (MiP)

Ligninolytic cultures of several white rot fungi have been found to produce peroxidases different from MnP and LiP which are sometime referred to as MiP which was identified as distinct from MnP by its independence of Mn^{2+} ions. The role of these enzymes in lignin degradation is not well elucidated. Comparative studies done on the different kinds of peroxidases showed that the ligninolytic activity of MiP responded differently towards inhibitors than of LiP and MnP except of cysteine, to which the three activities were equally sensitive but differed in the mode of inhibition. MiP activity was detected in *Trametes versicolor*, *P. chrysosporium* and *Coriolopsis polyzona* [Waldner *et al.,* 1988; Nerud *et al.,* 1991].

Manganese peroxidase (MnP) assay: Activity of manganese peroxidase (MnP) (EC 1.11.1.13) was measured by using guaiacol as a substrate. The increase in absorbance at 465 nm due to oxidation of guaiacol ($\varepsilon 465$= 12,100 M^{-1} cm^{-1}) was measured. The reaction mixture contained 300 µL sodium succinate buffer (0.5 M, pH 4.5 at 27ºC), 300 µL guaiacol (4 mM), 600 µL manganese sulphate (1mM), 300 µL culture supernatant and 1200 µL distilled water. The mixture was incubated for 2 min at 30ºC and the reaction was initiated by addition of 300 µL hydrogen peroxide (1mM). The absorbance was measured in 1 min intervals after addition of hydrogen peroxide. One unit of MnP activity was defined as activity of an enzyme that catalyzes the conversion of 1µ mole of guaiacol per minute (Atalla *et al.,* 2010).

4.7 Hydrogen peroxide producing enzymes

It is thought that for the generation of oxygen radicals, laccase requires presence of H_2O_2 extracellularly. Different white-rot fungi have different mechanisms for generation of hydrogen peroxide. Aryl-alcohol oxidase (AAO) and glyoxal oxidase (GOX) are implicated to be the main enzymes involved in production of H_2O_2 extracellularly. Generation of

H_2O_2 by AAO may involve a catalytic system. Benzyl alcohols are oxidized to benzaldehydes and to lesser extent benzaldehydes to benzoic acid by AAO under production of H_2O_2 during the reaction. Aryl-alcohol dehydrogenase (AAD) then reduces benzaldehyde and benzoic acid to regenerate the benzyl alcohols [Guillen *et al.* 1992; Guillen *et al.* 1994]. Initially, there is a two-electron reduction of the aldehyde by AAD and the subsequent electron transfer from veratryl alcohol to oxygen catalyzed by AAO. AAD requires high cellular NADPH as its reducing power [Mulheim *et al.* 1991]. Thus, the continuous production of H_2O_2 requires a large drain of the reducing equivalents. However, this is not a problem for the white-rot fungi since there is excessive carbon in nature which can readily be used as a source of reducing equivalents for lignin degradation.

The other enzyme capable of generating H_2O_2 is glyoxal oxidase, a copper metalloenzyme, produced by wood-rot fungi and forms an essential component of the extracellular lignin degradation pathways [Whittaker *et al.* 1999]. It contains a free radical-coupled copper active site [Whittaker *et al.* 1996]. It uses simple aldehyde, α- hydroxycarbonyl, or α-dicarbonyl compounds as substrates to produce H_2O_2. The activity of glyoxal oxidase is expressed during secondary metabolism, when the ligninases are also expressed. Work of Hammel *et al.* [1994] demonstrated release of 0.8- 0.9 equivalent of glycolaldehyde (a glyoxal oxidase substrate) in *P. chrysosporium*, after each Cα-Cβ cleavage of the arylglycerol β-aryl ether lignin model 1-(3, 4-dimethoxyphenyl)- 2-phenoxypropane-1,3-diol. Further, in the same work oxidation of glycolaldehyde by GOX (glyoxal oxidase) then yielded 0.9 and 2.8 molecules of oxalate and H_2O_2, respectively per reaction. The cleavage of arylglycerol α- aryl ether structures by ligninolytic enzymes eventually recycles H_2O_2, useful for subsequent cleavage reactions. Methoxyl groups in lignin are attacked by alcohol oxidases to form methanol which is further oxidized to aldehydes thereby producing H_2O_2. Other oxidases that have been postulated as likely possible donors of H_2O_2 are glucose 1- oxidase [Kelley and Reddy, 1986a, 1986b] and fatty acyl-coenzyme A oxidase [Greene and Gould, 1984].

Glyoxal oxidase assay: Glyoxal oxidase activity was determined as described previously, using methylglyoxal as the oxidase substrate and with phenol red or 2,29-azinobis(3-ethylbenz-thiazoline-6-sulfonic acid) as the substrate for horseradish peroxidase in the coupled-enzyme assay (Philip *et al.* 1995).

4.8 Aryl alcohol oxidase assay (AAO)

The standard AAO activity was measured spectrophotometrically [Patricia, 2005] by monitoring the oxidation of veratryl alcohol to veratraldehyde at 310 nm. The reaction mixture contained 8 mM substrate in airsaturated 100 mM sodium phosphate buffer, pH 6. One unit of AAO activity is defined as the amount of enzyme that converts 1 μmol of alcohol to aldehyde per min at 24°C.

4.9 Cellobiose dehydrogenase

Cellobiose dehydrogenase (CDH) is involved in lignin degradation and carries out one electron reductions of a range of substrates such as quinones, semiquinones, organic free radicals etc., by simultaneously oxidizing cellobiose to cellobiono lactone [Archibald *et al.* 1997]. CDH is known to be present in different fungi including the white rot *T. versicolor*, *P. chrysosporium* etc. and has been given various names like cellobiose oxidoreductase (CBO) and cellobiose:quinone oxidoreductase (CBQ). Although the biological function of CDH has not been fully understood, a mechanism that involves hydroxyl radical generation has been suggested. It is implicated that the radical can degrade and modify cellulose, hemicellulose and lignin.

Cellobiose dehydrogenase assay: CDH activity was determined by the cellobiose-dependent reduction of 3,5-di-*tert*-butylbenzoquinone (1,2) (TBBQ) at 420 nm (Watermark, 1974). The assay mixture contained Na acetate (100 mM; pH 4.5), ethanol (20%, vol/vol), cellobiose (2 mM), and TBBQ (0.33 mM) in a total volume of 1.5 ml. To prevent interference caused by laccase-mediated reoxidation of the hydroquinone formed from TBBQ by CDH, the reaction was made anoxic by bubbling N2 through the mixture for 3 min. One unit of enzyme was defined as the amount that reduced 1 mmol of TBBQ min21 ml21.

4.10 Versatile peroxidase

Versatile peroxidase is able to oxidize both veratryl alcohol and Mn^{2+} in hydrogen peroxide dependent catalytic cycle, which means that it possesses the catalytic activity of both MnP and LiP [Mester and Field, 1998]. The molecular weight of lignin-degrading peroxidases varies from 35 - 48 kDa for LiP to 38 - 62 kDa for MnP. The isoelectric points of LiPs and MnPs are generally between 3 and 4 for wood-inhabiting fungi [Hatakka, 2001; Hofrichter, 2002]. However, even neutral MnPs have been found from litter-decomposing fungi. Lignin-degrading peroxidases are globular proteins formed by 11-12 -helices in two domains [Martinez, 2002]. Between the two domains situates the central cavity with heme bound by two histidine residues conserved in all lignin-degrading peroxidases [Conesa *et al.* 2002]. Four to five disulfide bridges and two structural Ca^{2+} cations stabilize the structure of the protein and the active site [Martinez, 2002; Conesa, 2002]. Three acidic amino acids involved in binding of Mn^{2+} are conserved in MnPs and VPs but are absent from LiPs [Conesa *et al.*, 2002].

Versatile peroxidase assay : Manganese-dependent activity of versatile peroxidase (VP) was measured at 270 nm by monitoring the formation of Mn^{3+}-malonate complex in 50 mM sodium malonate buffer at pH 4.5 in the presence of 19 mM H_2O_2 (Wariishi *et al.*, 1992). The VP activity was expressed in U/ml. One unit of versatile peroxidase was defined as the amount of enzyme required to form 1 μmol Mn^{3+}-malonate complex per minute ($\varepsilon = 1.159 \times 10^4 M^{-1} cm^{-1}$).

5. Wood rotting fungi

5.1 White Rot Fungi

White rot breaks down all major wood components (cellulose, hemicelluloses and lignin) more or less simultaneously, and commonly causes rotted wood to feel moist, soft and spongy, or stringy and to appear white bleached. Wood affected by white rot normally does not crack across the grain and will only shrink and collapse when severely degraded. The strength of the infested wood decreases gradually until it becomes spongy to the touch and stringy when broken. Lignin degradation process by white-rot fungi has been termed "enzymatic combustion", because the causative agents are oxidative extracellular enzymes [Kirk and Farrel, 1987]. The White-rot fungi extensively decompose the important structural components of wood, including both cellulose and lignin and converts in to carbon-dioxide and water. White rot fungi oxidatively degrade lignin by demethylation and/or hydroxylation of aromatic rings to produce diphenolic structures followed by rapid oxygenolytic cleavage of the polyhydroxylated aromatic nuclei yielding aliphatic products. Biodegradation of lignocellulosic paddy straw by white rot fungi and white rot fungal usage in biopulping and biobleaching was reported [Vijaya *et al.*, 2005a, 2005b]. Lignolytic enzymes were screened for their lignolytic activity from different white rot fungal species [Vinay Sagar *et al.*, 2008, 2010]. White rot fungal screening was reported for its highest lignolytic enzyme activities and identified on molecular level [Metuku *et al.*, 2011]. Decolorization reports were there on rayon effluents and bleaching of kraft pulp by *Stereum ostrea* [Rama Krishna *et al.*, 2011]. The biological delignification of lignocellulosic feed stocks, *Prosopis juliflora* and *Lantana camara* was carried out with *Pycnoporus cinnabarinus,* a white rot fungus, at different scales under solid-state fermentation (SSF) and the fungal treated substrates were evaluated for their acid and enzymatic saccharification [Rishi *et al.*, 2010]. Lignolytic enzymes produced by white rot fungi is showed in Table 3.

5.2 Brown Rot Fungi

Brown rot primarily decays the cellulose and hemicelluloses in wood, leaving a brown residue of lignin, the substance which holds the cells together. Wood affected by brown rot is usually dry and fragile, readily crumbles into cubes because of longitudinal and transverse cracks (tending to crack across the grain). Infected wood may be greatly weakened, even before any external evidence of decay can be seen. Brown rot is generally more serious than white rot. Old infestations of brown rot which have dried out will turn to powder when crushed. They are often labeled as "dry rot." This common term is deceiving, because dry wood will not rot! Actually, wood kept dry will never decay. Enzymatic profile was screened from brown rot fungi in broth media supplemented with agricultural wastes [Hima Bindu *et al*. 2010, 2011].

Table 3. Ligninolytic enzymes produced by white rot fungi

Enzyme	E.C. No	Catalyzed reactions	Fungi	References
Laccase	1.10.3.2	Phenol oxidation	*Trametes versicolor*	Vinay Sagar *et al*. 2008; Yaropolov *et al*. 1994
Lignin peroxidase	1.11.1.14	Phenol polymerization	*Phanerochaete chrysosporium*	Haglund, 1999; Erden *et al*. 2009
Manganese peroxidase	1.11.1.13	Phenol oxidation; Oxidize Mn^{2+} to Mn3+	*Phanerochaete chrysosporium*	Hofrichter, 2002
Cellobiose-quinone oxireductase	1.1.5.1	Quinone reduction; Celobiose degradation	*Phanerochaete chrysosporium*	Soares, 1998
Aryl alcohol oxidase	1.1.3.7	H_2O_2 production	*Pleurotus saborcaju*	Martinez *et al*. 2009
Glyoxal oxidase	1.2.3.5	H_2O_2 production	*Phanerochaete chrysosporium*	Martinez *et al*. 2009
Manganese independent peroxidase	1.11.1.7	Activity on aromatic substrates	*Phanerochaete chrysosporium*	Wyatt, 1995; Ruiz-Duenas, 2009
Versatile peroxidase	1.11.1.16	Oxidizes Mn^{2+}; High redox-potential aromatic compouds	*Pleurotus sp.*	Ruiz-Duenas, 2009
Cellobiose dehydrogenase	1.1.99.18	Lignin degradation; Unite the hydrolytic and oxidative systems; Dispose manganese for MnP through precipitate reduction from manganese oxide (MnO_2)	*Phanerochaete chrysosporium*	Kersten, 2007; Carvalho *et al*. 2009

5.3 Chemistry of lignin degradation by white rot fungi

Lignin decayed by white rot fungi contain less carbon, methoxyl and hydrogen than corresponding sound (non-decayed) lignin. White rotted lignin contain more oxygen, more carbonyl groups and more carboxyl groups than sound lignin. The microbiologically introduced carbonyl groups are largely conjugated to aromatic rings, as are about a third of the newly introduced carboxyl groups. The remainder of the new carboxyls are conjugated to alpha, beta unsaturated aliphatic structures. Degraded lignins can be up of 25% deficient in methoxyl groups. White rot fungi oxidatively degrade lignin by demethylation and hydroxylation of aromatic rings to produce diphenolic structures. Rapid oxygenolytic cleavage of the polyhydroxylated aromatic nucleus, yielding aliphatic products. These reactions are catalysed by extracellular or cell surface bound oxygenases. Certain white rot fungi attack the aromatic rings of lignin faster than the side chains. White rotted lignins contain increased quantities of aromatic carboxyl groups as compared to sound lignins. The new vanillic acid moieties can arise by fungal oxidation of terminal phenyl propane units that occur originally in the natural polymer, or those resulting from fungal degradation of the macromolecule.

$$\text{Lignin} \xrightarrow[\text{Alkaline nitrobenzene oxidation}]{} \text{Vanillic acid}$$

6. Biological applications of lignolytic enzymes (Table 4)

6.1 Paper and pulp industry

Among the applications of lignolytic enzymes in different industries, the pulp and paper industry, which is the sixth largest energy consumer in the industrial sector, is using these enzymes extensively for energy conservation. This industrial sector investing 24.5% of its total manufacturing cost towards energy in three major processing steps i.e. pulping, bleaching and paper production [Shukla *et al.*, 2005]. The use of fungus/ fungus secreted lignolytic enzymes prior to pulping offered an attractive opportunity for mechanical wood pulp facilities. This technology saved an estimated 30% of the energy consumed in refining the mechanical pulp. The technology also improved the paper strength, reduced the pith content, and also reduced the emissions of volatile organic compounds. The application of lignolytic enzymes saved 25% of chlorine containing bleaching chemicals. This bio-bleaching technology not only reduced the consumption of bleaching chemicals, but also reduced absorbable organic halogen, improved pulp and paper quality and improved brightness and at the same time reduced the effluent toxicity and pollution loads.

The lignolytic enzymes are at present in the market for paper and pulp, textile, food, and other industries. Being specific, energy saving, and biodegradable, these group of enzymes fit well with the development of highly efficient, sustainable and eco-friendly industries.

Table 4. Biological applications of lignolytic enzymes

Enzyme	Applications	References
Laccase	Spore resistance- Rhizomorph formation- Pathogenesis- Fruit bodies formation- Pigments synthesis- Lignin degradation	Mayer, 2002; Minussi *et al.* 2007
Lignin Peroxidase	Biodegradation of lignin- Defense of fungi against pathogens	Score *et al.* 1997; Trejo-Hernandez *et al.* 2009
Manganese Peroxidase	Degradation of lignin- Intere-specific fungal interactions	Score *et al.* 1997; Trejo-Hernandez *et al.* 2009
Food Industry		
Laccase	Phenolic removal from the food and beverage- Ascorbic acid determination- Sugar beet pectin gelation	Minussi *et al.* 2007; Ghindilis, 2000; Selinheimo *et al.* 2006
Lignin peroxidase	Source of natural aromatics- Production of vanillin	Lesage-Meesen *et al.* 1996; Lomascolo *et al.* 1999; Barbosa *et al.* 2008
Manganese peroxidase	Production of natural aromatic flavours	Lomascolo *et al.* 1999; Barbosa *et al.* 2008; Zorn *et al.* 2003
Pulp and paper industry		
Laccase	Depolymerization of lignin- Delignify wood pulps Bleaching of kraft pulps	Bourbonnais *et al.* 1997; Chandra, 2002; Widsten, 2008
Lignin peroxidase	Decolouriment of kraft pulp Mill effluents	Ferrer *et al.* 1991; Sigoillot *et al.* 2005
Manganese peroxidase	Kraft pulp bleaching	Michel *et al.* 1991; Mailaja *et al.* 2007; Wasenberg *et al.* 2003
Textile industry		
Laccase	Textile dye degradation and bleaching	MC Kay, 1979; Cripps *et al.* 1990; Cooper, 1993; Riu *et al.* 1998; Shin, 2004; Kunamneni *et al.* 2008; Gomes *et al.* 2009
Lignin peroxidase		
Manganese peroxidase		
Bioremediation		
Laccase	Biodegradation of xenobiotics- Polycyclic aromatic hydrocarbons (PAHs) degradation	Pointing, 2001; Bamforth, 2005; Rodriguez, 2006; Anastasi *et al.* 2009

Lignin peroxidase	Degradation of azo, hetero-cyclic, reactive and polymeric dyes- Mineralizationof environ-mental contaminants Xenobiotic and pesticides degradation	Gomes *et al.* 2009; Bumpus, 1987; Abraham *et al.* 2002; Wen *et al.* 2002
Manganese peroxidase	PAH's degradation Synthetic dyes- Bleach from paper prod-ucing plants DDT, PCB, TNT	Koller *et al.* 2000; Robles-Hernandez *et al.* 2008
Organic synthesis, Medical, Pharmaceutical, Cosmetics and Nanotechnology Applications		
Laccase	Polymers production- Coupling of phenols and steroids- Medical agents Carbon-nitrogen bonds construction- Complex natural products synthesis- Personal hygienic products- Biosensors and bioreporters	Milstein *et al.* 1989; Bauer *et al.* 1999; Xu, 2005; Kunamneni *et al.* 2008; Mikolasch *et al.* 2007; Christenson *et al.* 2004
Lignin peroxidase	Functional compounds- synth-esis Cosmetics and dermato-logical for skin- Bioelectro-catalytic activity at atomic resolution	Barbosa *et al.* 2008; Christenson *et al.* 2004; Belinky *et al.* 2005
Manganese peroxidase	Acrylamide polymerization Polymer styrene degradation Direct electron transfer (DET)	Iwahara *et al.* 2000; Lee *et al.* 2006

6.2 Textile industry

Lignolytic group of enzymes are used in commercial textile applications to improve the whiteness in conventional bleaching of cotton. Potential benefits of the application include chemicals, energy and water saving. These enzymes are used *in situ* to convert dye precursors for better, more efficient fabric dyeing [Kunamneni *et al.* 2008]. Laccase group of enzymes find potential applications for cleansing such as cloth and dishwashing. They are also applied in cleansing formulations to eliminate the odour on fabrics, including cloth, sofa surface and curtain, or in a detergent to eliminate the odour generated during cloth washing [Lantto *et al.* 2004].

6.3 Food industry

Many substrates of lignolytic enzymes such as carbohydrates, unsaturated fatty acids, phenols and thiol containing proteins are important components of various food and beverages. Their modification by laccases may lead to new functionality, quality improvement and cost reduction [Kirk *et al.*, 2002]. These are also used as oxygen scavengers for better food packing. The flavour quality of vegetable oils can be improved with lignolytic enzymes by eliminating dissolved oxygen. Laccases can also deoxygenate food items derived partly or entirely from extracts of plant materials. Treatment with lignolytic enzymes enhanced the colour of tea based products. Various enzymatic treatments have been proposed forfruit juice stabilization, among which it can be found the use of laccase group of enzymes [Alper and Acar, 2004]. The enzymes are added to the dough used for producing baked products, to exert an oxidizing effect on the dough constituents and to improve the strength of gluten structures in dough and /or baked products.

Wine stabilization is one of the main applications of the lignolytic enzymes in the food industry as alternative to physical-chemical adsorbents. Musts and wines are complex mixtures of different chemical compounds such as ethanol, organic acids (aroma), salts and phenolic compounds (colour and taste) (Barbosa *et al.* 2008). These enzymes are also used to improve storage life of beer. Haze formation in beer is a persistent problem in the brewing industry. These laccase group of enzymes remove polyphenols and reduce the problems in the brewing industries.

6.4 Organic synthesis

Increasing interest has been focused on the application of lignolytic enzymes as a new biocatalysts in organic synthesis. These enzymes have provided an environmentally benign process of polymer production without the use of hydrogen peroxide (Kobayashi and Higashimura, 2003; Mita *et al.* 2003). Lignolytic enzymes catalysed cross linking reactions of new urushiol analogues for the preparation of 'artificial urushi' polymeric films was demonstrated (Ikeda *et al.* 2001). These enzymes are also known to polymerize various amino and penolic compounds. To improve the production of fuel ethanol from renewable raw materials, lignolytic enzymes from Trametes versicolor was expressed in Saccharomyces cerevisiae to increase its resistance to phenolic fermentation inhibitor in lignocellulose hydrolyzates (Larsson *et al.* 2001). Lignolytic enzymes also used to synthesize various functional organic compounds including polymers with specific mechanical/electrical/optical properties, textile dyes, pigments, flavour agents and pesticides.

6.5 Bioremediation

Lignolytic enzymes have many possible applications in bioremediation. They have been applied to degrade various substances such as undesirable contaminants, by products or discarded materials. They are also used to degrade plastic waste having olefin units and polyurethanes. These enzymes are also been used to eliminate odour emitter from places such as garbage disposal sites, live stock farms or pulp mills (Maximo and Costa, 2004). They also used for decolorizing dye home effluents and waste waters from olive oil mills, pulpmills etc. removing colour phenolic compounds. Another potential environmental application for lignolytic enzymes is the bioremediation of contaminated soils as these are able to oxidize toxic organic pollutantssuch as various xenobiotics, PAHs, chlorophenols and other contaminants. They are used to reduce the concentration synthetic heterocyclic compounds such as, halogenated organic pesticides in the soil (Gomes *et al.* 2009).

6.6 Pharmaceutical industry

Many products generated by lignolytic enzymes are antimicrobial, detoxifying or active personal care agents. Due to their specificity and bio-based nature these enzymes can be used in the synthesis of complex medical compounds as anesthetics, anti-inflammatory, antibiotics, sedatives etc. (Nicotra *et al.*, 2004). One potential application of lignolytic enzymes based in situ is the generation of iodine, a reagent widely used as disinfectant and also they have been reported to possess significant HIV-1 reverse transcriptase inhibitory activity (Wang and Ng, 2004).

A novel application field of lignolytic enzymes is in cosmetics. Laccase based hair dyes could be less irritant and easier to handle than current hair dyes. Recently, cosmetic and dermotological preparation containing protein for skin lightening have also been developed (Martinez *et al.*, 2009). These enzymes are used as deodorants for personal hygiene products including tooth paste, mouth wash, detergent, soap and diapers and protein engineered laccase also used to reduce allergenicity (Tachihara *et al.* 2006).

6.7 New lignolytic enzyme based biocatalysts

Enzymatic catalysis in organic solvents has opened a new field of biotechnological application of enzymes. The ability to use enzyme in non-aqueous solvents greatly expanded the potential scope of economic impact of biocatalysis. When biological catalysts are placed in the unnatural environment they exhibit a number of removable novel properties such as altered stereo-selectivity, enhanced stability and increased rigidity (Desantis and Jones, 1999). The investigation on molecular evolution of lignolytic enzymes was carried out for organic cosolvent resistance. The white rot fungal laccases expressed in Saccahr-omyces cerevisiae was engineered by *in vitro* evolution in the presence of increasing concentration of acetonitrile and ethanol. The turnover rates of mutant enzymes at high concentration of organic solvents were several fold improved (Alcalde *et al.*, 2005).

6.8 Nanobiotechnology

The lignolytic enzymes can be used as biosensors or bioreporters. A number of biosensors containing laccase and lignin peroxidase have been developed for immunoassay and for determination of glucose, aromatic amines and phenolic compounds (Gomes *et al*., 2004). Lignolytic enzymes covalently conjugated to a bio-binding molecule and used as a reporter for immuno-chemical, histo-chemical, cytochemical or nucleic acid detection assays (Xu, 1999). The bioreporter applications are of interest for the high-sensitivity diagnostic field. In addition to biosensors, these enzymes could be immobilized on the cathode of biological cells that could provide power for small transmitter systems (Palmore, 2004). Fuel cells are very attractive energy sources, particularly at micro-, mini-, portable- or mobile scale. Those have higher energy conversion/usuage efficiency and lower pollution effect than any of the existing/emerging energy sources (Marcia *et al.* 2011). Lignolytic enzymes also appliedas biocatalysts for the electrode reactions. These enzymes based miniature biological fuelcell is of particular interest for many medical applications calling for a power source implanted in human body (Bamiere *et al.* 2004).

Now-a-days the use of environment friendly processes are becoming more and more popular in almost all industries and therefore, biotechnological processes are coming to the forefront of research. An application of biotechnology in industrial sector offers great potential for reducing capital and energy costs, improving properties and reducing the environmental impact on production processes.

In this context the contemporary research is oriented to the detection of lignolytic enzymes in complex substrates. Once a fungus has been shown to degrade lignin in experiments with radio-labeled synthetic lignins, the question arises as to which lignolytic enzymes the organism is expressing. Here, two approaches were identified, one of these is to infiltrate a high molecular weight lignin model compound in to the lignocellulosic substrate, and then to determine by subsequent product analysis whether fungus cleaves it in the same way that purified lignolytic enzyme does. Another approach currently under development, and useful when the gene for the enzyme of interest has been sequenced, is to isolate fungal RNA from the substrate and use reverse transcription/polymerase chain reaction technique to determine whether the gene for the enzyme is being expressed. These new research tools are helpful to improve our knowledge about the degradative mechanisms in the diversified microorganisms by producing versatile peroxidase group of enzymes.

Thus, this group of enzymes are highly versatile in nature and they find applications in a wide variety of industries. The biotechnological significance of these enzymes has led to a drastic increase in the demand for these enzymes in recent times. Lignolytic enzymes are promising to replace the conventional chemical processes of several industries. In this way, there is a broad field of investigation that is almost entirely open to new findings and it is quite reasonable to propose that many new applications will be fond in the near future.

Acknowledgements

The Author is thankful to the Head, Department of Microbiology, Kakatiya University,Warangal for providing necessary facilities.

References

Abraham WR, Nogales B, Golyshin PN, Pieper DH, Timmis KN (2002) Current Opinion Microbiol 5(3):246-253

Adlere (1977) Wood Sci Technol 11:169-218

Alcalde M, Bulter T, Zummarrga M, Garcia AH, Mencia M, Plou FJ, Ballesteros A (2005) J Biotechnol Screen 10:624-631

Alper N, Acar J (2004) Nahrung 48:184-187

Anastasi A, Coppola T, Prigione V, Varese G (2009) Role of laccases and peroxidases. J Hazardous Mat 65(1-3):1229-1233

Archibald FS, Bourbonnais R, Jurasek L, Paice MG, Reid ID (1997) J Biotechnol 53:215-236

Atalla M, Mabrouk, Zeinab H, Kheiralla, Eman R, Hamed, Amani A Youssry, Abeer A (2010) Abd El Aty Agric Biol J N Am 1(4):591-599

Baldrian P (2006) FEMS Microbiol. Rev.30:215-242

Bamforth SM, Singleton I (2005) J Chem Technol Biotechnol 80(7):726-736

Bamiere F, Ferry F, Rochefort D, Leech D. (2004) Electrochem Commun 6: 237-244

Banci L, Ciofi-Baffoni S, Tien M (1999) Biochemistry 38:3205

Barbosa ES, Perrone D, Vendramini ALA, Leite SGF (2008) BioRes 3(4):1042-1050

Bauer CG, Kuhn A, Gajovic N, Skorobogatko O, Holt PJ, Bruce NC, Makower A, Lowe CR, Scheller FW (1999) Fresenius' J Anal Chem 364(1-2)179-183

Belinky P, Lasser H, Dosoretz C (2005) Methods of producing lignin peroxidase and its use in skin and hair lightening. Patent IPC Class: AA61K896F1 (Rakuto Bio Technologies Ltd.). Origin: Arlington; VA US, USPC Class: 42462

Blanchette RA, (1995) Canadian J Botany 73:S999–S1010

Bourbonnais R and M.G. Paice (1996) Tappi Journal 79:199

Bourbonnais R, Paice M, Reid I, Lanthier P, Yaguchi M (1995) Appl Environl Microbioly 61:1876

Bourbonnais R, Paice MG, Freiermuth B, Bodie E, Borneman S (1997) Appl Environ Microbiol 63(12)4627-4632

Brunow G, Kilpeläinen I, Sipilä J, Syrjänen K, Karhunen P, Setälä H, Rummakko P (1998a) American chemical society, ACS symposium series 697, Washington DC, pp.131-147

Bumpus JA, Aust SD (1987) Appl Environ Microbiol 53(9):2001-2008

Burlat V, Ambert K, Ruel K, Joseleau JP (1997) Plant Physiol Biochem 35: 645–654

Cai DY, Tien M (1992) J Biol Chem 267:11149-11155

Carvalho W, Canilha L, Ferraz A, Milagres A (2009) Química Nova, vol. 32(8) 1-5

Chandra RP, Ragauskas AJ (2002) Enz Microb Technol 30(7):855-861

Chen CA, Pakdel H, Roy C (2001) Bioresour Technol 79 :277

Chesson A (1978) J AppL Bacteriol 45:219-330

Christenson A, Dimcheva N, Ferapontova EE, Gorton L, Ruzgas T, Stoica L, Shleev S, Yaropolov AI, Haltrich D, Thorneley RNF, Aust SD (2004) Electroanal 16(13-14)1074-1092

Conesa A, Punt PJ, Hondel CAMJJ (2002) J Biotechnol 93:143-158

Cooper P (1993) *J Soc Dyers Col* 109(3):97-101

Cripps C, Bumpus JA, Aust SD (1990) Appl Environ Microbiol 56(4):1114-1118

Dejong E, Field JA, Debont JAM (1994) FEMS Microbiol Rev 13:153-188

Desantis G, Jones J (1999) Curr Opin Biotechnol 10:324-330

Erden E, Ucar CM, Gezer T, Pazarlioglu NK (2009) Braz J Microbiol 40(2):346-353.

Eriksson KEL, Blanchette RA, Ander P (1990) Biodegradation of Lignin, In: Microbial and Enzymatic Degradation of Wood and Wood Components (Timell T.E. ed.), Springer-Verlag, New York, pp. 225-399

Faison BD, Kirk TK (1983) Appl Environ Microbiol 46:1140-1145

Faix O, Mozuch MD, Kirk TK (1985) Holzforschung 39:203–208

Ferrer I, Dezotti M, Duran N (1991) Biotechnol Lett 13(8)577-582

Ghindilis AL (2000) Biochem Soc Trans 28(2):84-89

Gold MH, Youngs HL, Sollewijn Gelpke MD (2000) Manganese peroxidase. In: Sigel, A., Sigel, H., editors. Metal Ions in Biological Systems, Marcel Dekker Inc, New York, pp. 559-586

Gomes E, Aguiar AP, Carvalho CC, Bonfa MRB, DA Silva R, Boscolo M (2009) Braz J Microbiol 40(1)31-39

Gomes SASS, Nogueira JMF, Rebelo MJF (2004) Biosens Bioeloctr. 20: 1211-1216

Grabski AC, Grimek HJ, Burgess RR (1998) Biotechnol Bioeng 60:204-215

Greene RV, Gould JM (1984) Biochem Biophys Res Commun 118:437-443

Guillen F, Martinez AT, Martinez MJ (1992) European J Biochem 209:603-611

Guillen F., Martinez A.T., Martinez M.J., and Evans C.S. (1994) Appl Microbiol Biotechnol 41:465-470

Haemmerli SD, Lcisola MSA, Sanglard D, Fiechter A (1986) J Biol Chem 261: 6900-6903

Haglund C (1999) Biodegradation of xenobiotic compounds by the white-rot fungus Trametes trogii. Molecular Biotechnology Programme, Uppsala University School of Engineering, pp. 30

Hammel KE, Kalyanaraman B, Kirk TK (1986) Journal of Biological Chemistry 261:6948-6952

Hammel KE, Mozuch MD, Jensen KA, Kersten PJ (1994) Biochem 33:13349-13354

Hatakka A (1994). FEMS Microbiol. Rev., 13: 125-135.

Hatakka A. (2001) Biodegradation of Lignin. Biopolymers.Lignin.Humic Substances and Coal. A multivolume handbook. Wiley. Vol 1 (Steinbűchel A., ed) Chapter 5

Hofrichter M (2002) Enz Microb Technol 30(4):454-466

Ikeda R, Tanaka H, Uyama H, Kobayashi H (2001) 74:1067-1073

Iwahara K, Hirata M, Honda Y, Watanabe T, Kuwahara M (2000) Biotechnol Lett 22(17):1355-1361

Kelley RL, Reddy CA, (1986b) J Bacteriol 166:269-274

Kelley RL, Reddy CA, (1986a) Archives in Microbiol 144:248-253

Kersten P, Cullen D (2007) Fungal Genetics Biol 44(2):77-87

Kersten PJ, Tien M, Kalyanaraman B, Kirk TK (1985) J Biol Chem 260:2609-2612

Khindaria A, Yamazaki I, Aust SD (1995) Biochem 34:16860-16869

Khindaria A, Yamazaki I, Aust SD (1996) Biochem 35: 6418-6424

Kirk O, Borchert TV, Fuglasang CC (2002) Curr Opin Biotechnol 13:345-351

Kirk TK, Farrell RL (1987) Ann Rev Microbiol 41:465-505

Kobayashi S, Higashimura H (2003) Prog Polym Sci 28:1015-1048

Koduri RS, Tien M (1995) J Biol Chem 270: 22254-22258

Koller G, Moeder M, Czihal K (2000) Chemosph 41(12):1827-1834

Kunamneni A, Ghazi I, Camarero S, Ballesteros A, Plou FJ, Alcalde M (2008) Proc Biochem 43(2): 169-178

Lackner R, Srebotnik E, Messner K (1991) Biochem Biophys Res Commun 178:1092-1098

Lantto R, Schonberg C, Buchert J, Heine E (2004) Textile Res 74:713-717

Larsson S, Cassland P, Jonsson LJ (2001) Appl Environ Microbiol 67:1163-1170

Lee JW, Lee SM, Hong EJ, Jeung EB, Kang HY, Kim MK, Choi IG (2006) The J Microbiol 44(2):177-184

Leisola MSA, Garcia S (1989) Lignin degradation mechanism in Enzyme systems for lignocellulose degradation, Galway, Ireland, Elsevier publication pp 89-99

Leonowicz A, Matuszewska A, Luterek J, Ziegenhagen D, Wojtas-Wasilewska M, Cho NS, Hofrichter M, Rogalski J (1999) Fungal Genetics and Biology 27:175-185

Lesage-Meesen L, Delattre M, Haon M, Thibault JF, Ceccaldi BC, Brunerie P, Asther M (1996) J Biotechnol 50(2-3):107-113

Lomascolo A, Stentelaire C, Asther M, Lesage-Meessen L (1999) Trends Biotechnol 17(7):282-289

Lora JH, Glasser WG (2002) J Polym Environ 10 :39

Maijala P, Mettala A, Kleen M, Westin C, Poppius-Levin K, Herranen K, Lehto JH, Reponen P, Maentausta O, Hatakka A (2007) Treatment of softwood chips with enzymes may reduce refining energy consumption and increase surface charge of fibers. In: 10th International Congress on Biotechnology in the Pulp and Paper Industry. Madison Wisconsin, Book of Abstracts. pp.65

Marcia JMM, Admir CS, Helene CTR (2011) Env Biotech 13 (6): 717-730

Martinez AJ, Ruiz-Duenas FJ, Martinez MJ, Rio JC, Gutierrez A, (2009) Curr Opin Biotechnol 20(3): 348-357

Martinez AT (2002) Enz Microbiol Technol 30: 425-444

Martinez AT, Ruiz-Duenas FJ, Martinez MJ, Rio JC, Guttierrez A (2009) Current Opinion in Biotechnol 20(3):348-357

Maximo C, Costa FM (2004) Process Biochem 39:1475-1479

Mayer AM, Staples RC (2002) Phytochem 60(6)551-565

MC Kay G (1979) J American Dyestuff Rep 86(4):29-36

Mester T, Field JA (1998) J Biol Chem 273:15412-15417

Metuku RP, Burra S, SVSSSL Hima Bindu N, Pabba S, Singara Charya MA (2011) J Cell Tissue Res 11(1): 2557-2562

Michel MC, Jr FC, Dass BS, Grulke EA, Reddy ACG (1991) Appl Environl Microbiol 57(8)2368-2375.

Mikolasch A, Schauer F (2009) Appl Microbiol Biotechnol 82(4):605-624

Milstein O, Nicklas B, Huttermann A (1989) Appl Microbiol Biotechnol 31(1) 70-74

Minussi RC, Miranda MA, Silva JA, Ferreira CV, Aoyama H, Marangoni S, Rotilio D, Pastore GM, Duran N (2007) African J Biotechnol 6(10)1248-1254

Mita N, Jawaki SI, Hirishi U, Kobayashi S (2003) Macromol Biosci 3:253-257

Mulheim A, Waldner R, Sanglard D, Reiser J, Schoemaker HE, Leisola MSA (1991) Eur J Biochem 195:369-375

Nerud F, Zouchova Z, Misurcova Z (1991) Biotechnol Let 13:657-660

Nicholson RL, Hammerschmidt R (1992) Ann Rev Phytopathol 30:369–389

Nicotra S, Cramarossa MR, Mucci A, Paynoni VM, Riva S, Forti L (2004) Tetradydra 60:595-600

Palmore GTR (2004) Trends Biotechnol 22: 99-100

Patricia Ferreira, Milagros Medina, Francisco Guillen, Marıa Jesus Martinez, Willem JH Van Berkel, Angel T Martinez (2005) Biochem J 389:731–738

Pease EA, Andrawis A, Tien M (1989) J Biol Chem 264:13531-13535

Philip J Kersten,Christian Witek, Amber Vanden Wymelenberg, Daniel Cullen (1995) J Bacteriol 6106–6110

Pointing SB (2001) Appl Microbiol Biotechnol 57(1-2)20-33

Popp JL, Kalyanamaran B, Kirk TK (1990) Biochem 29:10475-10480

Rama Krishna G, SVSSSL Hima Bindu N, Singara Charya MA (2011) Int J Pharm BioSci 2(3):277-287

Ramesh Chander K, Ajay Singh (2007) Lignocellulose biotechnology: future prospects, I.K.International

Rasera K, Ferla J, Dillon AJP, Riveiros R, Zeni M (2009) Desalin., 245: 657-661.

Renganathan V, Gold MH (1986) Biochem 25:1626-1631

Rishi Gupta, Girija Mehta, Yogender Pal Khasa, Ramesh Chander Kuhad (2010) Biodegrad 22(7):797-804

Riu J, Schonsee I, Barcelo D (1998) J Mass Spec 33(7):653-663

Robles-Hernandez L, Gonzales-Franco AC, Crawford DL, Chun WWC (2008) Tecnociencia Chihuahua 2(1):32-39

Rodriguez S, Toca JL (2006) Biotechnol Adv 24(5):500-513

Ruiz-Duenas FJ, Martinez AT (2009) Microb Biotechnol 2(2):164-177

Score AJ, Palfreyman JW, White NA (1997) Int Biodeter Biodegrad 39(2-3):225-233

Selinheimo E, Kruus K, Buchert J, Hopia A, Autio K (2006) Effects of laccase, xylanase and their combination on the rheological properties of wheat doughs. J Cereal Sci 43(2):152-159

Setua DK *et al.* (2000) Polym Compos 21:988

Shin KS (2004) The JMicrobiol 42(1)37-41

Shukla OP, Rai UN, Subramanyam SV (2005) Enviro News Archives 10(2):16-20

Sigoillot C, Camarero S, Vidal T, Record E, Asther M, Perez-Boada M, Martinez MJ, Sigoillot JC, Asther M, Colom JF, Martinez ÁT (2005) J Biotechnol 115(4)333-343

Sjöström E, Westermark U (1998) Chemical composition of wood and pulps: Basic components and their distribution. In: Sjöström E., Alén R., editors. Analytical methods in wood chemistry, pulping, and papermaking. Springer-Verlag, Berlin, Germany, pp 1-35.

Soares SCHL (1998) Estudos mecaníticos da degradação de efluentes de indústrias de papel e celulose por fungos basidiomicetos degradadores de madeira. Universidade Estadual de Campinas. Instituto de Química, Campinas, São Paulo, pp. 122

Steffen KT (2003) Degradation of recalcitrant biopolymers and polycyclic aromatic hydrocarbons by litter-decomposing basidiomycetous fungi. Dissertationes Biocentri Viikki Universitatis Helsingiensis, 23/2003. Ph.D. Thesis, Department of Applied Chemistry and Microbiology, University of Helsinki, Helsinki, P. 68

Sundaramoorthy M, Kishi K, Gold MH, Poulos TL (1994) J Biol Chem 269:32759-32767

Hima Bindu SVSSSL N, Singara Charya MA (2010) Nat Acad Sci Lett 33: (7&8) 235-239

Hima Bindu SVSSSL N, Krishna Bolla, RamPrasad Metuku, Samatha Burra, Shiva Krishna Pabba, Singara Charya Maringanti (2011) J Rec Adv Appl Sci 26:19-26

Tachihara T, Miyoshi H, Komai TJP (2006) JP325441A2

Takamiya M, Magan N, Warner PJ (2008) J Hazardous Mat 154:33–37

Thurston CF (1994) Microbiology 140: 19-26

Tien M, Kirk TK, Bull C, Fee JA (1986) J Biol Chem 261:1687-1693

Trejo-Hernandez MR, Lopez-Nybgyua A, Quintero Ramirez R (2001) Proc Biochem 36(7) 635-639

Tuor U, Wariishi H, Schoemaker HE, Gold MH (1992) Biochem 31:4986-4995

Vance CP, Kirk TK, Sherwood RT (1980) *Ann Rev Phytopathol* 18:259–288

Vijaya CH, Singara Charya MA (2005a) Env. Sci .7: 77-83

Vijaya CH, Vijaya Laxmi M, Singara Charya MA (2005b) Natl J Life Sci 2:133-138

Vinaya Sagar Goud J, Jaya Prakash Goud M, Singara Charya MA (2008) Asian J Microbiol Biotech Env Sci 10:765-766

Vinaya Sagar GJ, Kandukuri Vasu, Singara Charya MA (2010) Geobios 37:25-28

Waldner R, Leisola MSA, Fiechter A (1988) Appl Microb Biotechnol 29:400-407

Wang HX, Ng TB (2004) Biochem Biophy Res Commun 315:450-454

Wariishi H, Akileswaran L, Gold MH (1988) Biochem 27:5365-5370

Wariishi H, Valli K, Gold MH (1989) Biochem 28:6017-6023

Wariishi H, Valli K, Gold MH (1992) J Biol Chem 267:23688-23695

Wasenberg D, Kryriakides I, Agathos SN (2003) Biotechnol Adv 22(1-2)161-187

Wen X, Jia Y, Li J (2009) Chemosph 75(8) :1003-1007

Westermark U, Eriksson KE (1974) Acta Chem.Scand Ser B 28:209–214

Whittaker MM, Kersten PJ, Cullen D, Whittaker JW (1999) J Biol Chem 274: 36226-36232

Whittaker MM, Kersten PJ, Nakamura N, SandersLoehr J, Schweizer ES, Whittaker JW (1996) J Biol Chem 271:681-687

Widsten P, Kandelbauer A (2008) Enz Microb Technol 42(4)293-307

Wyatt AM, Broda P (1995) Microbiol 141:2811-2822

Xu F (1999) The encyclopedia of bioprocessing technology. John Wiley & Sons NY 1545-1554

Xu F (2005) Ind Biotechnol 1(1):38-50

Yaropolov AI, Skorobogatko OV, Vartanov SS, Varfolomeyev SD (1994) Appl Biochem Biotechnol 49:3 257-280

Youn HD, Hah YC, Kang SO (1995) FEMS Microbiol. Lett., 132: 183-188

Youngs HL, Moenne-Loccoz P, Loehr TM, Gold MH (2000) Biochemistry 39:9994-10000

Zakzeski J, Bruijnincx PCA, Jongerius AL, Weckhuysen BM (2010) Chem Rev 110:3552

Zeyen RJ, Kruger WM, Lyngkjær MF, Carver TLW (2002) Physiol Mol Plant Pathol 61:315- 323

Zorn H, Langhoff S, Sscheibner M, Nimtz M, Berger RG (2003) Biol Chem 384(7):1049-1056.

Chapter 14

Helicobacter pylori: Biology and Diseases

C.E. Prasad*[*]*, Lakshmi Kiran Chelluri*[**]*, Subhash Pawar and S. Ram Reddy

Department of Microbiology, Kakatiya University, Warangal -506 009
[*]Department of Respiratory Medicine, Gandhi Medical College, Hyderabad
[**]Department of Stem Cell Transplant Biology, Global Hospital, Hyderabad
E-mail: sanditi_ramreddy@yahoo.com

ABSTRACT

Helicobacter pylori is the causal agent of human gastritis and stomach ulcers. At least half of the world population is infected by the bacterium, making it the most widespread infection in the world. The World Health Organization (WHO) has declared the bacteria to be a class I carcinogen. Majority of people with *H. pylori* however has asymptomatic fever, less than 10% of individuals colonized with *H. pylori* develop peptic ulcer disease, gastric cancer mucosa associated lymph tissue lymphoma (MALT). *H. pylori* consists of a large diversity of strains and the genomes of three strains completely sequenced consist of nearly 1-7 million base pairs and with 1,550 genes. The survival of *H. pylori* in the acidic stomach 5 dependant on urease and it would die without the enzyme. Two related mechanisms by which *H. pylori* promote cancer are under investigation. One mechanism involves the enhanced production of free radicals and an increased rate of host cell mutation. The other proposed mechanism is a perigenetic pathway that involves enhancement of the transformed host cell. Recent studies suggest an increased *H. pylori* prevalence in patients with various extra gastrointestinal disorders including skin, cardiovascular, rheumatic and liver diseases. *H. pylori* is contagious, however the exact route of transmission is not known. Though a number of diagnostic methods are developed the most reliable method is biopsy check during endoscopy with a rapid urease test, histological examination and microbial culture. The standard first line therapy includes one week triple therapy consisting of the antibiotic clarithromycin and a proton pump inhibitor such as omeprazole.

Key words: *Helicobacter pylori*, gastritis, peptic ulcers, urease, extra gastrointestinal disorders, triple therapy.

1. Introduction

Helicobacter pylori was first discovered in the stomach of patients with gastritis and stomach ulcers nearly 25 years ago by Dr Barry J. Marshall and Dr J. Robin Warren of Perth, Western Australia. At that time (1982-83), the conventional thinking was that no bacterium can survive in the human stomach as the stomach produces extensive amounts of acid which is similar in strength to the acid found in a car battery. Marshall and Warren literally rewrote the text books with reference to what causes gastritis and gastric ulcers (Blaser *et al.*, 2004). In recognition of their very important discovery, they were awarded the 2005 Nobel Prize for Medicine and Physiology (Blaser *et al.*, 2006). In their original paper, Warren and Marshall contended that most stomach ulcers and gastritis were caused by infection of this bacterium and not by stress and spicy food as had been assumed before (Marshall *et al.*, 1984; Akhter *et al.*, 2007). To demonstrate that *H. pylori* caused gastritis and was not merely a bystander, Marshall drank a broth culture of *H. pylori* (Bonnevie *et al.*, 1977). He became ill several days later with nausea and vomiting. An endoscopy ten days after inoculation revealed signs of gastritis and the presence of *H. pylori* (Calam *et al.*, 2001). These results suggested that *H. pylori* was the causative agent of gastritis. Marshall and Warren went on to show that antibiotics are effective in the treatment of many cases of gastritis (Bhasin *et al.*, 2000). In 1987, the Sydney gastroenterologist Thomas Borody invented the first triple therapy for the treatment of duodenal ulcers. In 1994, the National Institute of Health (USA) published an opinion stating that most recurrent duodenal and gastric ulcers were caused by *H. pylori* and recommended that antibiotics be included in the treatment regimen (Faisal *et al.*, 1990).

2. *Helicobacter pylori*

Helicobacter pylori is a spiral shaped bacterium that infects over 3000 million people world wide, making it one of the most common bacterial infections (Everhart *et al.*, 2000). Since this discovery, the World Health Organization (WHO) has declared the bacteria to be a Class 1 carcinogen (Graham *et al.*, 2005). It invades the mucosal lining of the stomach and is the cause of up to 95% of duodenal and upto 75% gastric ulcers and has also been associated with gastric cancer and lymphoma (Parsonnet *et al.*, 1994; Parsonnet *et al.*, 2004; Tsuji *et al.*, 2003). Despite intense investigations into the spread of *H. pylori*, the precise mode of transmission remains unclear (Eidt *et al.*, 1994). The presence of *H.pylori* in the stomach induces a chronic, active, inflammation in almost everyone infected (Liu *et al.*, 2006; Lundergardh *et al.*, 1994). Majority of people with *H. pylori* however has a symptomatic fever, less than 10% of individuals colonized with *H. pylori* develop peptic ulcer disease, gastric cancer mucosa associated lymph tissue lymphoma (MALT) (Fischbach *et al.,* 2004).

Recent studies indicate that *H. pylori* might be related to non ulcer dyspepsia (Viskum *et al.*, 1974). Chronic inflammation and increased immune response have been observed in a variety of cardiovascular disorders, including COPD and bronchitis (Roussos *et al.*, 2003, 2006). Moreover, active pulmonary tuberculosis (TB) is frequent among patients with partial gastrectomy for peptic ulcer patients (Realdi *et al.*, 1999). And it has increased two to three fold compared to findings in ulcer free controls (Fillipou *et al.*, 2002; Tsang *et al.*, 1998). Based on these observations, many recent studies have evaluated the relation between various respiratory disorders and *H. pylori* infection (Roussos *et al*., 2006).

Morphology : *H. pylori* is a helix-shaped Gram-negative bacterium, about 3 micrometres long with a diameter of about 0.5 micrometres (Enroth *et al.*, 1999). It is microaerophilic *i.e.* it requires oxygen but at lower concent-ration than is found in the atmosphere (OToole *et al.*, 2000). It contains a hydrogenase which can be used to obtain energy by oxidizing molecular hydrogen (H_2) that is produced by intestinal bacteria. It produces oxidase, catalase, and urease (Ottemann *et al.*, 2002). It is capable of forming biofilms (Stark *et al.*, 1999), and can convert from spiral to a possibly viable but nonculturable coccoid form (Chan *et al.*, 1994). Both of them are likely to favor its survival and be factors in the epidemiology of the bacterium. The coccoid form can adhere to gastric epithelial cells (Lieu *et al.*, 2006).

H. pylori possess five major outer membrane protein (OMP) families (Kusters *et al.,* 2006). The largest family includes known and putative adhesions (Appelmelk *et al.*, 1999). The other four families include porins, iron transporters, flagellum-associated proteins (Dunn *et al.*, 1993; Dunn *et al.*, 1997), and the proteins of unknown function. Like other typical Gram-negative bacteria, the outer membrane of *H. pylori* consists of phospholipids and lipopolysaccharide (LPS) (Alm *et al.*, 2000). The O antigen of LPS may be fucosylated and mimic Lewis blood group antigens (Boren *et al.*, 1993), found on the gastric epithelium (Kusters *et al.*, 2006). The outer membrane also contains cholesterol glucosides, which are found in few other bacteria (Kusters *et al.*, 2006). *H. pylori* has 4–6 flagella. All gastric and enterohepatic Helicobacter species are highly motile due to flagella (Josenhans *et al.*, 2000). The characteristic sheathed flagellar filaments of helicobacters are composed of two copolymerized flagellins, FlaA and FlaB (Rust *et al.*, 2008; Akhiani *et al.*, 2002).

Genome

H. pylori consists of a large diversity of strains, and the genomes of three have been completely sequenced (Tomb *et al.*, 1997; Everhart *et al.*, 2002). The genome of the strain "26695" consists of about 1.7 million base pairs, with some 1,550 genes (Alm *et al.*, 2000). The two sequenced strains show large genetic differences, with up to 6% of the nucleotides differing. Study of the *H.pylori* genome is centered on attempts to understand pathogenesis, the ability of this organism to cause disease (Alm *et al.*, 2000). Approximately 29% of the loci are in the pathogenesis category of the genome database (Appelmelk *et al.*, 1999).

Both sequenced strains have an approximately 40 kb-long Cag pathogenicity island, a common gene sequence believed responsible for pathogenesis that contains over 40 genes (Boneca *et al.*, 2003).This pathogenicity island is usually absent from *H. pylori* strains isolated from humans who are carriers of *H. pylori* but remain asymptomatic (Baldwin *et al.,* 2007). The *cagA* gene codes for one of the major *H. pylori* virulence proteins (Krausse *et al.*, 2001). Bacterial strains that have the *cagA* gene are associated with an ability to cause ulcers (Blaser *et al.*, 1995). The *cagA* gene codes for a relatively long 1186 amino acid protein (Kauser *et al.*, 2005). The *cag* pathogenicity island has about 30 genes, part of which code for a complex type IV secretion system (Backert *et al.*, 2008). The low GC content of the *cag* PAI relative to the rest of the *Helicobacter* genome suggests that the island was acquired by horizontal transfer from another bacterial species (Tomb *et al.*, 1997; Thjodleifsson *et al.*, 2007).

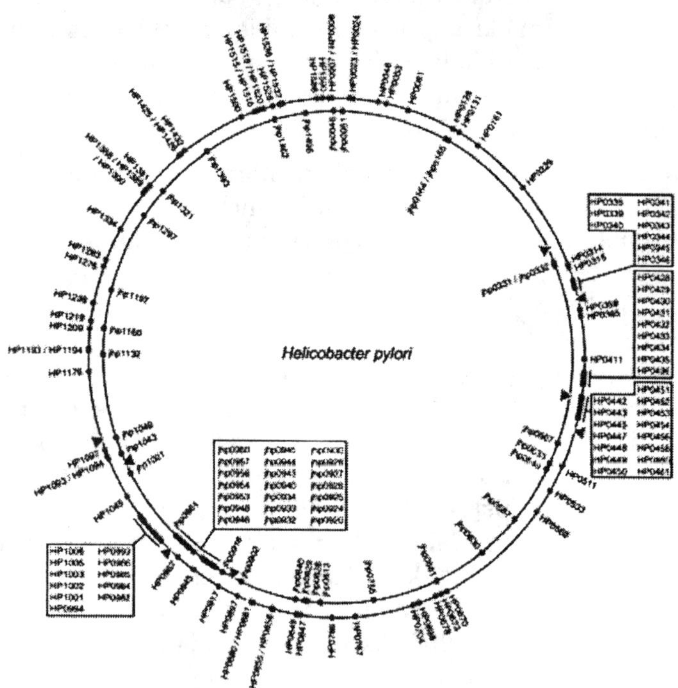

Cytotoxin-associated gene antigen (CagA)

One of the most intensely studied proteins of *H. pylori* has been CagA, yet its function has not been determined (Aviles *et al.*, 2004). The

CagA protein may be delivered from attached bacteria into the host cytoplasm by an *H. pylori* specific type IV secretion system encoded by the cag pathogenicity island cag PAI (Cover *et al.*, 1995), where it is phosphorylated on tyrosine residues (Stein *et al.*, 2000). It may then be able to influence eukaryotic signal transduction pathways and cytoskeletal plasticity (Asahi *et al.*, 2000; Segal *et al.*, 1999). The cag PAI has biological significance and may provide an explanation for the observation that *H. pylori* (Blaser *et al.*, 1996), which produces CagA is more interactive with the host than those that do not (Vander *et al.*, 1997).

Vacuolating cytotoxin A (VacA)

Epidemiological and experimental evidence has clearly demonstrated the association between VacA production and the occurrence of tissue damage. The best known VacA effect is its ability to induce cytoplasmatic vacuoles in various eukaryotic cell lines by altering intracellular membrane traffic at the late endosomal and prelysosomal (Aviles *et al.*, 2004; Cover *et al.*, 1994), level. It is suggested that expression of *H. pylori* virulence is triggered by contact with the host cell (Talley *et al.*, 1997). As seen for other pathogenic bacteria, the effector molecules of virulence are actively secreted on the host cell surface even translocated across the mammalian host cell membrane. Intracytoplasmatic vacuoles (V) in gastric epithelial cells induced *in vivo* by *H. pylori*. Ultrathin section. Bar, 0.5 μm. Vacuoles appear as round structures of up to several micrometers in diameter a size that can be reached only upon fusion of several smaller compartments (Smoot *et al.*, 1997). The toxin could promote endosome-endosome fusion either directly or through stimulation of a fusogenic protein (Talley *et al.*, 1997). However, it is known that ammonia is also able to cause cell vacuolation. The specific mechanisms by which the VacA causes these effects are not well understood, but recent studies indicate that VacA interacts directly with a cytoplasmatic target molecule (Pai *et al.*, 1999).

Taxonomy

Kingdom	:	Bacteria
Phylum	:	Proteobacteria
Class	:	Epsilon proteobacteria
Order	:	Campylobacteria
Family	:	Helicobacteriaceae
Genus	:	Helicobacter
Species	:	*H. pylori*

Apart from *H. pylori*, several other species of *Helicobacter* are reported (Goodwin *et al.*, 1989), however, *H. pylori* is the most widely known species of the *Helicobacter* genus. It is pathogenic to humans.

Gastric Helicobacter species

Gastric *H. pylori* species have adap-ted to the in hospitable conditions found at the gastric mucosal surface, and it is currently thought that the stomach of all mammals can be colonized by members of the genus *Helicobacter* (Svanes *et al.*, 1999). Urease is thought to allow short term survival in the highly acidic gastric lumen whereas motility is thought to allow rapid movement towards the more neutral part of the gastric mucosa (Faisal *et al.*, 1990). This may explain why both factors are prerequisites for colonization of the gastric mucosa (Nakamura *et al.*, 1998). Gastric flagellar motility then assists in penetration into the viscous mucus layer, where the more neutral conditions allow the growth of the gastric Helicobacter species (Goodwin *et al.*, 1989).

3. Symptoms

Most people who are infected with *H.pylori* have no symptoms and do not know that they are infected (Blaser *et al.*, 1996). A number of these bacteria may just live harmlessly in the lining of the stomach and duodenum region of humans and has been found to be associated with many disorders in the gastric area including chronic superficial gastrititis, chronic active gastritis, peptic ulcer disease, and even gastric cancer (Blaser *et al.*, 1996). When first infected, the patient develops chronic superficial gastritis. This is an infection of the mucous layer only, with no effect on the epithelial cells. If it is left untreated, this condition will probably span the rest of the patient's life (Blaser *et al.*, 1996). This condition can, however, develop into chronic active gastritis, which is an inflammation of the stomach's epithelial cells below the mucous layer. Because this affects the cells, the condition of chronic active gastritis can develop into gastric cancer (Blaser *et al.*, 1996). The bacteria do not actually invade the cells but rather release chemicals that these tissues absorb. It is these chemicals that cause inflammation of the tissues. Few patients, however, actually develop chronic active gastritis, peptic ulcer disease and gastric cancer from *H.pylori* infection (Blaser *et al.*, 1996).

4. Pathogenicity

To colonize the stomach, *H. pylori* must survive the acidic pH of the lumen (Akada *et al.*, 2000), and burrow into the mucus to reach its niche, close to the stomach's epithelial cell layer (Alamuri *et al.*, 2004). The bacterium has flagella and moves through the stomach lumen and drills into the mucoid lining of the stomach (Ottemann *et al.*, 2002). Many bacteria can be found deep in the mucus, which is continuously secreted by mucous cells and removed on the luminal side (Alm *et al.*, 1999). To avoid being carried into the lumen, *H. pylori* senses the pH gradient within the mucus layer by chemotaxis and swims away from the acidic contents of the lumen towards the more neutral pH environment of the epithelial cell surface (Schreiber *et al.*, 2004). *H. pylori* are also found on the inner surface of the stomach epithelial cells and occasionally inside epithelial cells (Petersen *et al.*, 2003). It produces adhesins which bind to membrane-associated lipids and carbohydrates and help it adhere to epithelial cells (Warren *et al.*, 1983).

H. pylori produce large amounts of the enzyme urease, molecules of which are localized inside and outside of the bacterium (Calam *et al.*, 2001). Urease breaks down urea which is normally secreted into the stomach to carbon dioxide and ammonia, ammonia is converted into the ammonium ion by taking hydrogen from water upon its breakdown into hydrogen and hydroxyl ions (Backert *et al.*, 2008). Hydroxyl ions then react with carbon dioxide, producing bicarbonate which neutralizes gastric acid (Werdmuller *et al.*, 1997). The survival of *H. pylori* in the acidic stomach is dependent on urease, and it would eventually die without the enzyme (Mobley *et al.*, 1991). The ammonia that is produced is toxic to the epithelial cells, and, along with the other products of *H. pylori* including protease, vacuolating cytotoxin A (VacA) (Kusters *et al.*, 2006). Certain phospholipases also damage those cells (Smoot *et al.*, 1997).

Colonization of the stomach by *H. pylori* results in chronic gastritis, an inflammation of the stomach lining (Ernst *et al.*, 2000). The severity of the inflammation is likely to underlie *H. pylori*-related diseases (Shiotani *et al.*, 2002). Duodenal and stomach ulcers result when the consequences of inflammation allow the acid and pepsin in the stomach lumen to overwhelm the mechanisms that protect the stomach and duodenal mucosa from these caustic substances (Leontiades *et al.*, 1999). The type of ulcer that develops depends on the location of chronic gastritis, which occurs at the site of *H. pylori* colonization (Dixon, 2000). The acidity within the stomach lumen affects the colonization pattern of *H. pylori* and therefore ultimately determines whether a duodenal gastric ulcer will form (Cave *et al.*, 1996; Caygill *et al.*, 1991).

In people producing large amounts of acid, *H. pylori* colonizes the antrum of the stomach to avoid the acid secreting parietal cells located in the corpus, the main body of the stomach (Kusters *et al.*, 2006). In these patients *H. pylori* can also colonize the corpus of the stomach, where the acid-secreting parietal cells are located (Santucci *et al.*, 2000).

However, chronic inflammation induced by the bacteria causes further reduction of acid production and, eventually, atrophy of the stomach lining, which may lead to gastric ulcer and increases the risk for stomach cancer. About 50-70% of *H. pylori* strains in Western countries carry the *cag* pathogenicity island *cag* PAI (Peek *et al.*, 2006). Western patients infected with strains carrying the *cag* PAI have a stronger inflammatory response in the stomach and are at a greater risk of developing peptic ulcers and stomach cancer than those infected with strains lacking the island. (Viala *et al.*, 2004).

Two related mechanisms by which *H. pylori* could promote cancer are under investigation. One mechanism involves the enhanced production of free radicals near *H. pylori* and an increased rate of host cell mutation (Xue *et al.*, 2001). The other proposed mechanism has been called a perigenetic pathway (Tsuji *et al.*, 2003), that involves enhancement of the transformed host cell phenotype by means of alterations in cell proteins such as adhesion proteins. It has been proposed that *H. pylori* induce inflammation and locally high levels of TNF-α and interleukin 6 (Stemmler *et al.*, 2005). According to the proposed perigenetic mechanism inflammation associated signaling molecules such as TNF-α can alter gastric epithelial cell adhesion and lead to the dispersion and migration of mutated epithelial cells without the need for additional mutations in tumor suppressor genes such as genes that code for cell adhesion proteins (Suganuma *et al.*, 2008).

5. *Helicobacter pylori* associated disease

Peptic ulcer disease is a common clinical ailment, once thought to be caused by over secretion of acid and pepsin (Schubert *et al.*, 2008), an enzyme of the stomach that promotes digestion by breaking down proteins (Linz *et al.*, 2007). Researchers latter found, however that although the injury caused acid and pepsin is necessary for the formation of ulcer's, acid secretion levels of the majority of patients with gastric duodenal ulcers are normal (Akhter *et al.*, 2007). An ulcer is now known to be the result of an imbalance between aggressive and defensive mechanisms in the stomach to infection by *H. pylori* (Beales *et al.*, 1995). Humans are the only known host of *Helicobacter pylori*.

Recent studies suggest an increased *H. pylori* prevalence in patients with various extra gastrointestinal disorders, including skin, cardiovascular, rheumatic and liver diseases (Gasbarrini *et al.*, 2004; Carloni *et al.*, 2000). Table 1 summarizes those extra digestive pathologies, characterized by a high prevalence of *H. pylori* infection (Realdi *et al.*, 1999; Roussos *et al.*, 2002). At present, there is no definite proof of a causal relationship between *H. pylori* and these diseases (Roussos *et al.*, 2006). The observed associations might be explained by a potential etiopathogenetic role of *H. pylori* infection in these disorders (Crabtree *et al.*, 1998). It is well known that *H. pylori* colonization of the gastric mucosa stimulates the release of a variety of pro inflammatory cytokines, including interleukin IL-1, IL-8 and tumor necrosis factor-α. Moreover, a cross molecular mimicry between bacterial and host antigen exists in *H. pylori* infected patients (Langman *et al.*, 1976). Therefore, *H. pylori* might have a pathogenic role in diseases characterized by abnormal activation of inflammatory mediators and induction of autoimmunity (Carloni *et al.*, 2000; Bamford *et al.*, 1997).

Vascular diseases association with *H. pylori* infection (Parsonnet *et al.*, 1994, Houghton *et al.*, 2005). (Kanbay *et al*, 2005). Skin diseases association with *H. pylori* infection (Goodman *et al*, 2006). Autoimmune diseases association *H. pylori* infection (Kanbay *et al.*, 2005) Hepatic biliary and other diseases association *H. pylori* infection (Solnick *et al.*, 2001).

Table 1. Extra digestive diseases associated with *H. pylori* infection

Vascular diseases
Ischaemic heart disease
Primary Raynaud's phenomenon
Primary headache
Skin diseases
Idiopathic chronic urticaria
Rosacea
Alopecia areata
Autoimmune diseases
Sjogren's syndrome
Autoimmune thyroiditis
Autoimmune thrombocytopenia
Schoenlein-Henoch purpura
Hepatic biliary and other diseases
Liver cirrhosis
Growth retardation
Chronic idiopathic sideropenia
Sudden infant death
Diabetes mellitus

Chronic inflammation and increased immune response have been observed in a variety of respiratory disorders, including chronic obstructive pulmonary disease (COPD) and bronchiectasis (Huang *et al.*, 1997; Nelson *et al.*, 2000; Silva *et al.*, 1989) (Table 2). Moreover, active pulmonary tuberculosis (TB) is frequent among patients with partial gastrectomy for peptic ulcer disease (Lundegardh *et al.*, 1994). The prevalence of chronic bronchitis in peptic ulcer patients is increased two to three fold compared with findings in ulcer-free controls (Langman *et al.*, 1976). Based on these observations, many recent studies have evaluated the relation between various respiratory disorders and *H. pylori* infection

(Kanbay *et al.*, 2006). Table 2 summarizes those respiratory diseases whose association with *H. pylori* infection has been studied (Roussos *et al.*, 2003, 2006), (Viskum *et al.*, 1974; Bonnevie *et al.*, 1977; Svanes *et al.*, 1999; Lundegardh *et al.*, 1994; Gocyk *et al.*, 2000; Philippou *et al.*, 2004; Rosenstock *et al.*, 2000; Keatings *et al.*, 1996; Tsang *et al.*, 2000; Sanaka *et al.*, 1998).

Table 2. Respiratory diseases implicated to be associated with *H. pylori* infection

Pulmonary tuberculosis
Chronic obstructive pulmonery disease (COPD)
Bronchial asthma
Emphysema
Bronchiectasis
Active bronchitis
Lung cancer

6. Epidemiology

At least half the world's population is infected by the bacterium, making it the most widespread infection in the world (Pounder *et al.*, 1995). Actual infection rates vary from nation to nation (Graham *et al.*, 1991). The third world has much higher infection rates than the Western Europe North America, Australasia (Goodman *et al.*, 2001), where rates are estimated to be around 25% (Pounder *et al.*, 1995). Infections are usually acquired in early childhood in all countries (Kusters *et al.*, 2006). However, the infection rate of children in developing nations is higher than in industrialized nations, probably due to poor sanitary conditions (Alberg *et al.*, 2005). In developed nations it is currently uncommon to find infected children, but the percentage of infected people increases with age, with about 50% infected for those over the age of 60 compared with around 10% between 18 and 30 years (Pounder *et al.*, 1995; Kikuchi *et al.*, 2005). The higher prevalence among the elderly reflects higher infection rates when they were children rather than infection at later ages (Kusters *et al.*, 2006; Cave *et al.*, 1996; Malaty *et al.*, 2007). Prevalence appears to be higher in African-American and Hispanic populations, although this is likely related to socioeconomic rather than racial factors (Smoak *et al.*, 1994; Everhart *et al.*, 2000). The lower rate of infection in the West is largely attributed to higher hygiene standards and widespread use of antibiotics (Graham *et al.*, 2005). Despite high rates of infection in certain areas of the world, the overall frequency of *H. pylori* infection is declining (Malaty, 2007). However, antibiotic resistance is appearing in *H. pylori* (Graham *et al.*, 1991). There are already many metronidazole- and clarithromycin-resistant strains in most parts of the world (Megraud, 2004).

7. Transmission

H. pylori is contagious, although the exact route of transmission is not known (Megraud., 1995; Cave., 1996). Person to person transmission by either the oral oral and fecal oral route is most likely (Brown, 2000; Ogihara *et al.*, 2000). Consistent with these transmission routes (Mitchell *et al.*, 2002). The bacteria have been isolated from feces, saliva and dental plaque of some infected people (Brown, 2000). Transmission occurs mainly within families in developed nations, yet can also be acquired from the community in developing countries. *H. pylori* may also be transmitted orally by means of fecal matter through the ingestion of waste-tainted water, so a hygienic environment could help decrease the risk of *H. pylori* infection (Brown, 2000).

8. Prognosis

H. pylori colonizes the stomach and induces chronic gastritis, a long-lasting inflammation of the stomach. The bacterium persists in the stomach for decades in most people. Most individuals infected by *H. pylori* will never experience clinical symptoms despite having chronic gastritis. Approximately 10-20% of those colonized by *H. pylori* will ultimately develop gastric and duodenal ulcers. *H. pylori* infection is also associated with a 1-2% lifetime risk of stomach cancer and a less than 1% risk of gastric MALT lymphoma (Wundisch *et al.*, 2005). It is widely believed that in the absence of treatment, *H. pylori* infection once established in its gastric niche persists for life (Brown *et al.*, 2000). In the elderly, however, it is likely infection can disappear as the stomach's mucosa becomes increasingly atrophic and inhospitable to colonization (Wundisch *et al.*, 2005). The proportion of acute infections that persist is not known, but several studies that followed the natural history in populations have reported apparent spontaneous elimination (Goodman *et al.*, 2005; Goodman *et al.*, 2001).

While *H. pylori* has been disappearing from the stomach of humans, the incidence of the related disorders acid reflux disease (Fallone *et al.*, 2000). Barrett's esophagus, and esophageal cancer have been raising dramatically (Blaser, 2005). In 1996, Blaser advanced the hypothesis that *H. pylori* have a beneficial effect by regulating the acidity of the stomach contents. It lowers the impact of regurgitation of gastric acid into the esophagus, (Blaser and Atherton, 2004) and Blaser (2005). The hypothesis is not universally accepted as several randomized controlled trials failed to demonstrate worsening of acid reflux disease symptoms following eradication of *H. pylori* (Graham *et al.*, 2007; Delaney *et al.*,

2005). Nevertheless, Blaser has refined his view to assert that H. pylori is a member of the normal flora of the stomach (Blaser, 2006). He postulates that the changes in gastric physiology caused by the loss of *H. pylori* account for the recent increase in incidence of several diseases, including type 2 diabetes, obesity and asthma (Blaser *et al.*, 2008). His group has recently demonstrated that *H. pylori* colonization is associated with a lower incidence of childhood asthma (Chen *et al.*, 2007).

9. Diagnosis

Diagnosis of infection is usually made by checking for dyspeptic symptoms and by tests which can indicate *H. pylori* infection (Glupczynski *et al.*, 1998). One can test noninvasively for *H. pylori* infection with a blood antibody test, stool antigen test, or with the carbon urea breath test in which the patient drinks ^{14}C- and ^{13}C-labelled urea, which the bacterium metabolizes, producing labelled carbon dioxide that can be detected in the breath (Karnes *et al.*, 1991; Stenstrom *et al.*, 2008). However the most reliable method for detecting *H. pylori* infection is a biopsy check during endoscopy with a rapid urease test, histological examination, and microbial culture (Graham *et al.*, 2002). There is also a urine ELISA test with 96% sensitivity and 79% specificity (Zoller *et al.*, 1993). None of the test methods is completely fail safe. Even biopsy is dependent on the location of the biopsy (Kuiers *et al.*, 1995). Blood antibody tests, for example, range from 76% to 84% sensitivity. Some drugs can affect *H. pylori* urease activity and give false negatives with the urea-based tests (Logan *et al.*, 2001).

Culture : *H. pylori* can be cultured only when a specimen cantaining the pathogen has been obtained and, in this case, that means obtaining a biopsy specimen by endoscopy methods (Akosy *et al.*, 2003). Culture varies but in general they involve the biopsy specimen and culturing them on a variety of specialized agar plates at elevated temperatures for at least seven days (Cutler *et al.*, 1995). Culture is generally regarded as the gold standard for detecting a bacterium (Suerbaum *et al.*, 2002). For *H. pylori*, however, the success of the technique depends on local technique and access to facilities and can be regarded as going no more than 60-90% sensitive, though being 100% specific the cost of each test is high (Perez-Perez et al., 2002).

Histology: Histological examination of tissue biopsy samples usually four, taken from different part of the stomach lining permits detection of the bacterium together with evaluation of tissue damage (Vinette *et al.*, 2004). Most infection can be detected with haemotoxylin and eosin (H and E) stain of gastric tissue but special stains like Geimsa can be used if H and E results are not conclusive (Allerberger *et al.*, 1996). The sensitivity of Wright Giemsa and Brown Hopps stains has been shown to be 100%

compared with H and E (Dooley *et al.*, 1989). Though histological detecting of the organism has generally been considered to have lower sensitivity at 80 -95% with 100% specificity (Aksoy *et al.*, 2003).

Antibody measurement: Patients infected with *H. pylori* have immunoglobulin antibodies to the organism (Stenstrom *et al.*, 2008). Tests for the detection of antibodies to *H. pylori* circulating in blood have excellent sensitivity and specificity of above 95% and are cheap and simple compared with invasive techniques. Due to the presence of different strains of *H. pylori* antigen for antibody manufacture is generally prepared by using preparations from several different stains (Allerberger *et al.*, 1996). Antibody assays in blood have measured Ig G and Ig A antibodies which have been shown to be specific for *H. pylori* and not other gram negative organisms (Perez-Perez *et al.*, 2002), where both IgG, IgA assays have been compared with other testing methods like culture and histology IgG assays tend to have slightly higher sensitivity and specificity hence to anti-IgG methods tend to be favored (Crabtree *et al.*, 1991). The commercially available assays are of two sorts either microtitre plate assays for use in a laboratory near patient (Sorberg *et al.*, 1997). Both types of assay usually have a cut off value set with control sera so that they differentiate patients with *H. pylori* infection from those who do not rather than quantity the concentration of circulating anti-*H. pylori* immunoglobulin (Raymond *et al.*, 1996). Laboratory based micro titer assays (Crabtree *et al.*, 1991), and near patient testing devices, perform equally well compared with standard techniques.Antibody tests have high sensitivity and specificity when compared with other methods (Peterson *et al.*, 2002). They also have the advantage of being non invasive and near patient versions are available which can produce results from a small finger prick of blood within a few minutes (Portorreal *et al.*, 2002). At least one commentator has suggested that for *H. pylori* infection, serology testing may be the gold standard (Blaser *et al.*, 1990).

10. Prevention

Rising antimicrobial resistance increases the need for a prevention strategy for the bacteria. There have been extensive vaccine studies in mouse models, which have shown promising results. Researchers are studying different adjuvants, antigens, and routes of immunization to ascertain the most appropriate system of immune protection, with most of the research only recently moving from animal to human trials. An intramuscular vaccine against *H. pylori* infection is undergoing Phase I clinical trials and has shown an antibody response against the bacterium (Moayyedi *et al.*, 2005). Its clinical usefulness requires further study (Malfertheiner *et al.*, 2007). A Japanese study published April 1, 2009 in the journal cancer prevention research found that eating as little as 2.5 ounces of broccoli sprouts daily for two months reduces the number of colonies of *H. pylori* bacteria in the stomach by 40% in mice and humans (Sanaka *et al.*, 1999). This treatment also seems to help by enhancing the protection of the gastric mucosa against *H. pylori*, but is relatively ineffective on related gastric cancers (Rauws *et al.*, 1990).

11. Treatment

Once *H. pylori* is detected in patients with a peptic ulcer the normal procedure is to eradicate it and allow the ulcer to heal (Armstrong *et al.*, 1987). The standard first line therapy is a one week triple therapy consisting of the antibiotic clarithromycin, and a proton pump inhibitor such as omeprazole (Nakamura *et al.*, 2005). Variations of the triple therapy have been developed over the years, such as using a different proton pump inhibitor, as with pantoprazole rabeprazole and replacing amoxicillin with metronidazole for people who are allergic to penicillin (Malfertheiner *et al.*, 2007). Such a therapy has revolutionized the treatment of peptic ulcers and has made a cure to the disease possible. Previously the only option was symptom control using antacids, H_2-antagonists or proton pump inhibitors alone (Rauws *et al.*, 1990; Graham *et al.*, 1991).

An increasing number of infected individuals are found to harbour antibiotic resistant bacteria (Huang *et al.*, 2002). This results in initial treatment failure and requires additional rounds of antibiotic therapy (Bamford *et al.*, 1999). Alternative strategies such as quadruple therapy, which add a bismuth colloid (Stenstrom *et al.*, 2008), for the treatment of clarithromycin resistant strains of *H. pylori* (Bhasin *et al.*, 2000). The use of levofloxacin as part of the therapy has been suggested some practitioners of functional medicine use herbal formulas to treat *H. pylori* infection and claim a great rate of success (Bhasin *et al.*, 2000). However, there are no peer reviewed clinical studies that provide evidence of the effectiveness of herbal formulas though many remedies are offered (Bamford *et al.*, 1999, Stenstrom *et al.*, 2008, Nakamura *et al.*, 2005, Hentschell *et al.*, 1993).

References

Akada J.K., Shirai M., Takeuchi H., Tsuda M and T Nakazawa (2000). *Mol. Microbiol.* 36 :1071-1084.

Akhiani A. A., Pappo J., Kabok Z., Schon K., Gao W L. E Franzen and N. Lycke (2002). *J. Immunol.* 169 : 6977-6984.

Akhter Y., Ahmed I., Devi S.M and Ahmed N (2007). Infect Agent Cancer 2: 2.

Suerbaum S, Josenhans C (2007) *Nat. Rev. Microbiol.* 5: 441–452.

Aksoy, D.Y., Ayber M., Ozaslan E., Kav, T., Engin, D., Ercis, S., Altinok, G., Hascelik, G., Uzunalimoglu, B., Arslan, S (2003). *Hepato gastroenterology* 50: 1047-1049.

Alamuri, P., and Maier, R.J (2004). *Mol. Microbiol.* 53 : 1397-1406.

Alberg, A.J., Brock, M.V. and Samet, J.M (2005). *J. Clin. Oncol.* 23 : 3175-3185.

Allerberger, F., Oberhuber, G., Wrba, F., Puspok, A., Dejaco, C., Dierich, M.P. (1996). *Hepato gastroenterology* 43: 1653-1659.

Alm, R.A., Bina, J., Andrews, B.M., Doig, P., Hancock, R.E. and Trust, T.J. (2000). *Infect. Immun.* 68 : 4155-4168.

Alm, R.A (1999) *Nature.* 397: 176-180.

Appelmelk, B.J (1999). *Infect. Immun.* 67 : 5361-5366.

Armstrong, J.A., Wee, S.H., Goodwin, C.S. and Wilson, D.H. (1987). *J. Med. Microbiol.* 24 : 343-350.

Asahi M (2000) *J. Exp. Med.;* 191: 593–602.

Aviles-Jimenez, F., Letley, D.P., Gonzalez-Valencia, G., Salama, N., Torres, J. and Atherton, J.C. (2004). *J. Bacteriol.* 186 : 5182-5185.

Backert, S. and Selbach, M. (2008). *Cell. Microbiol.* 10 (8): 1573–81.

Baldwin DN, Shepherd B, Kraemer P, *et al.* (February 2007). *Infect. Immun.* 75 (2): 1005–16.

Bamford KB, Andersen L(1997 Curr. Opin. *Gastroenterol.* 13 :25–30.

Bamford, J.T.M., Tilden, R.L., Blankush, J.L. and Gangeness, D.E. (1999) *Arch. Dermatol.* 135 : 659-663.

Beales, I.L.P., Davey, N.J., Pusey, C.D., Lechler, R.I. and Calam, J. (1995 *Lancet* 346: 381-382.

Bhasin, D.K., Sharma, B.C. and Ray P, *et al.* (2000 *Helicobacter.* 5(2) : 84-87.

Bizzozero, G. (1893) Archiv. für. Mikroskopische. Anatomie. 42: 82–152.

Blaser, M.J. (1990). *J. Inrect. Dis.* 161:626–633.

Blaser MJ, et al. (1995) Cancer. Res. 55:2111–2115.

Blaser, M.J., and Crabtree, J.E. (1996). *Am. J. Clin. Pathol.* 106 : 565-567.

Blaser, M.J. and Atherton, J.C. (2004). *J. Clin. Invest.* 113(3): 321–33.

Blaser, MJ (2005) *Gastroenterology* 128: 1512–1515.

Blaser, M.J. (2006) *EMBO reports* 7(10): 956–60..

Blaser, M.J., Chen, Y. and Reibman, J. (2008) *Gut* 57(5): 561–7.

Bonnevie O. (1977) *Gastroenterology* 73: 1000-1004

Boren, T., Falk, P., Roth, K.A., Larson, G. and Normark, S. (1993) *Science* 262:1892-1895.

Brown, L.M. (2000) *Epidemiol. Rev.* 22(2): 283–297.

Calam, J. and Baron, J.H. (2001) *BMJ* 323 : 980-982.

Carloni E *et al* (2000) *Dig. Liver. Dis.* 32; S214–216.

Cave, D.R. (1996) *Am. J. Med.* 100(5A): 12S–17S; *discussion* 17S–18S.

Caygill, C.P., Knowles, R.L. and Hall, R. (1991) *Eur. J. Cancer. Prev.* 1: 35-37

Chan, W.Y *et al* (1994) *Am. J. Clin. Pathol.* 102(4) : 503–507.

Chen, Y. and Blaser, M.J. (2008 *J. Infect. Dis.* 198(4): 553–560.

Cover, T.L *et al.* (1995). *J. Clin. Microbiol.* 33: 1496–1500.

Cover, T.L *et al* (1994) J. Biol. Chem. 269: 10566–10573.

Crabtree, J.E. (1998) *Dig. Dis. Sci.* 43(Suppl 9): 46–55.

Crabtree, J.E., Shallcross, T.M., Heatley, R.V. and Wyatt, J.I. (1991) *J. Clin. Pathol.* 44: 326-328.

Cutler, A.F *et al* (1995) *Gastroenterology* 109: 136-141.

Delaney, B. and McColl, K. (2005) *Aliment. Pharmacol. Ther.* 22 (Suppl 1) : 32–40..

Dixon, M.F. (2000) *Baillieres Best Pract. Res. Clin. Gastroenterol.* 14(1): 27–40.

Dooley CP *et al.* (1989) *N. Engl. J. Med.* 321: 1562–1566.

Dunn, B.E. (1993). Edited by C.S. Goodwin and B.W. Worsely. CRC Press. Proteins, antigens, and typing methods for *Helicobacter pylori*. In *Helicobacter pylori*: *Biology and Clinical Practice* 181-207.

Dunn, B.E., Cohen, H. and Blaser, M.J. (1997) *Clin Microbiol Rev* 10: 720–741.

Eidt, S., Stolte, M. and Fischer, R. (1994) *J. Clin. Pathol.* 47 : 436-439.

Enroth, H *et al* (1999) *Helicobacter* 4:7-16.

Ernst, P.B. and Gold, B.D. (2000) *Annu. Rev. Microbiol.* 54: 615-640.

Everhart, J.E *et al* (2000) *J. Infect. Dis.* 181(4): 1359–63.

Everhart JE, Kruszon-Moran D, Perez-Perez G (2002) *Clin. Diagn. Lab Immunol.* 9: 412–416.

Faisal, M.A., Russell, R.M., Samloff, I.M. and Holt, P.R. (1990) *Gastroenterology* 99:1543-1544.

Fallone, C.A *et al* (2000) *Am. J. Gastroenterol.* 95 : 659-669.

Filippou, *et al* (2002) *J. Clin. Gastroenterol.* 34 : 189-190.

Fischbach, W *et al* (2004) *Gut* 53 : 34-37.

Gasbarrini, A *et al* . (2004) *Helicobacter.* 9(Suppl. 1): 57-66.

Glupczynski, Y. (1998) An overview. *Br. Med. Bull.* 54: 175–186

Gocyk, W *et al* (2000) *Med. Sci. Monit.* 6: 1085-1092.

Goodwin, C.S *et al* (1989) *Int. J. Syst. Bacteriol.* 39 : 397-405.

Goodman, K.J. and Cockburn, M. (2001) *Epidemiology* 12(2): 266–71

Goodman KJ *et al.* (December 2005) *Int. J. Epidemiol.* 34 (6): 1348–55.

Goodman KJ, Joyce SL, Ismond KP. (2006) *Curr. Gastroenterol. Rep.* 8:458–64.

Graham D.Y *et al* (1991) *Dig. Dis. Sci.* 36:1084-1088.

Graham, K.S. and Graham, D.Y. (2002) Contemporary Diagnosis and management of H. Pylori – Associated Gastronitestinal Diseases. *Newton, Handbook in Health Care,* pp 87-100.

Graham, D.Y. and Shiotani, A. (2005) *Gut* 54(6):735-8

Graham, D.Y., Yamaoka, Y. and Malaty, H.M. (2007) *Helicobacter.* 12 (Suppl 2): 64–8.

Hentschell, E *et al* (1993) *N. Engl. J. Med.* 328:308-312.

Houghton J, Wang TC: (2005) *Gastroenterology* 128:1567–1578.

Huang, J.Q., Sridhar, S. and Hunt. R.H. (2002) *Lancet* 359 :14-22.

Huang, S.L., Su, C.H. and Chang, S.C. (1997) *Am. J. Respir. Crit. Care Med.* 156 : 1436-1439.

Josenhans, C., Eaton, K.A., Thevenot, T. and Suerbaum, S. (2000) *Infect. Immun.* 68(8): 4598–603.

Kanbay, M., Gur, G., Akcay, S. and Yilmaz, U. (2005) *Respir. Med.* 99:1213-1216.

Kanbay, M, Kanbay, A. and Boyacioglu, S. (2006) *Respir. Med.* 5

Karnes, *et al*. (1991) *Gastroenterology* 101:167-174

Kauser, Fet al. (2005) *J. Clin. Microbiol.* 43:1538–1545.

Keatings, V.M *et al* (1996) *Am. J. Respir. Crit. Care Med.* 153 : 530-534.

Kikuchi, S. and Dope, M.P. (2005) *Helicobacter.* 10(suppl 1) : 1-4.

Krausse, R *et al* (2001) *Infection* 29: 154-158.

Kusters, J.G., van Vliet, A.H. and Kuipers, E.J. (2006) *Respiration* 63:236–240.

Kuipers EJ, Perez-Perez GI, Meuwissen SG, Blaser MJ (1995) *J. Natl. Cancer. Inst.* 87: 1777–1780.

Langman, M.J. and Cooke, A.R. (1976) *Lancet* 1: 680-683.

Leontiades, G.I., Sharma, V.K. and Howden. C.W. (1999) *Arch. Intern. Med.* 159: 925-940.

Linz, B., Balloux, F., Moodley, Y., Manica, A. and Liu, H. *et al*. (2007) *Nature* 445: 915–918.

Liu, Z.F *et al* (2006) *J. Med. Microbiol.* 55 (Pt 8): 1009–15.

Logan, R.P. and Walker, M.M. (2001) *BMJ* 323 (7318): 920–2.

Lundegardh, G., Helmick, C., Zack, M. and Adami, H.O. (1994) Dig. Dis. Sci. 39:340-346.

Malfertheiner P, Megraud F, O'Morain C, *et al.* (June 2007) *Gut* 56 (6): 772–81.

Malaty, H.M. (2007) *Best Pract. Res. Clin. Gastroenterol.* 21(2): 205–14.

Marshall, B.J. and Warren, J.R. (1984) Lancet I 1311-1315.

Megraud F (September 2004) *Gut* 53 (9): 1374–84.

Megraud F (1995) *Aliment. Pharmacol. Ther.* 9 Suppl 2: 85–91.

Mitchell, H. and Megraud, F. (2002) Helicobacter. 7(Suppl 1): 8-16

Moayyedi, P *et al* (2005) Cochrane Database Syst. Rev. 25

Mobley, H.L.T., Hu, L.T. and Foxall, P.A. (1991) Scand. J. Gastroent. 26 (Suppl. 187): 39-46.

Nakamura, H (1998) Infect. Immun. 66 : 4832-4837.

Nakamura, S *et al* (2005) Cancer 104 : 532-540.

Nelson, S., Summer, W.R. and Mason, C.M. (2000) Semin. Respir. Infect. 15: 24-31.

Ogihara, A *et al* (2000) J. Gastroenterol. Hepatol. 15: 271-276 .

O'Toole, P.W., Lane, M.C. and Porwollik. S. (2000) Microbes Infect. 2:1207-1214.

Ottemann, K.M. and Lowenthal, A.C. (2002) *Infect. Immun.* 70 (4): 1984–90.

Pai R., Cover T. L., Tarnawski A. (1999) *Biochem. Biophys. Res. Comm.*; 262: 245–250.

Parsonnet, J., and Isaacson. P.G. (2004) N. Engl. J. Med. 350 : 213-215.

Parsonnet, J *et al* (1994) N. Engl. J. Med. 330:1267-1271.

Peek, R.M. and Crabtree, J.E. (2006) *J. Pathol.* 208(2): 233–48.

Perez-Perez GI et al. (2002) Gut 50: 295–298.

Petersen, A.M. and Krogfelt, K.A. (2003) *FEMS Immunol. Med. Microbiol.* 36(3): 117-26.

Peterson, W.L. and Graham, D.Y. (2002) Helicobacter pylori. Feldman M Friedman LS Sleisenger MH eds. Sleisenger and Fordtran's gastrointestinal and liver disease: pathophysiology, diagnosis, management, WB Saunders Philadelphia 7th ed 2002:732-746..

Philippou, N (2004) World J. Gastroenterol. 10: 3342-3344

Portorreal, A. and Kawakami, E. (2002) Arq. Gastroenterol. 39: 198-203.

Pounder, R.E. and Ng, D. (1995) *Aliment. Pharmacol. Ther.* 9(Suppl 2): 33–9.

Rauws, E.A.J. and Tytgat, G.N.J. (1990) Lancet 335: 1233-1235.

Raymond, J *et al* (1996) Eur. J. Clin. Microboil. Infect. Dis. 15: 415-417..

Realdi, G., Dore, M.P. and Fastame, L. (1999) Dig. Dis. Sci., 44: 229-236.

Rosenstock *et al* (2000) Scand J. Public. Health 28: 32-40

Roussos, A *et al* (2002) J. Gastroenterol. 37: 332-335.

Roussos, A., Philippou, N. and Gourgoulianis, K.I. (2003) World J. Gastroenterol. 9 : 5-8.

Roussos, A., Philppou, N., Mantzaris, G.J. and Gourgoulianis, K.I. (2006) Respiration. 73(5): 708 – 14

Rust, M., Schweinitzer, T. and Josenhans, C. (2008) "Helicobacter flagella, Motility and Chemotaxis". *'Helicobacter pylori: Molecular Genetics and Cellular Biology* (Yamaoka Y, ed.), Caister Academic Press.

Sanaka, M *et al* (1998) J. Clin. Gastroenterol. 27: 331-334.

Sanaka, M., Kuyuma, Y., Yamanaka, M. and Iwasaki, M. (1999) Am. J. Gastronicterol. 94: 1984.

Santucci, A., Trabalzini, L., Bovalini, L., Neri, P. and Martelli, P. (2000) Electrophoresis 21:37123-37171.

Schreiber, S., Konradt, M. and Groll C, *et al* (2004 *Proc. Natl. Acad. Sci. U.S.A.* 101(14): 5024–9.

Schubert, M.L. and Peura, D.A. (2008) *Gastroenterology* 134(7): 1842–60.

Segal ED, Cha J, Lo J, Falkow S, Tompkins LS (1999) Proc. Natl Acad. Sci. U S A 96: 14559–14564.

Shiotani, A. and Graham, D.Y. (2002) *Med. Clin. North Am.* 86(6): 1447–66.

Silva, J.R., Jones, J.A., Cole, P. and Poulter, L. (1989) Thorax 44: 668-673.

Smoak, B.L., Kelley, P.W. and Taylor, D.N. (1994) *Am. J. Epidemiol.* 139(5): 513–9.

Smoot, D.T. (1997) *Gastroenterology* 113(6 Suppl): S31–4; discussion S50.

Solnick, J.V., and Schauer. D.B. (2001) Clin. Microbiol. Rev. 14:59-97.

Stark, R.M., Gerwig, G.J. and Pitman, R.S., *et al* (1999)" *Lett. Appl. Microbiol.* 28(2): 121–6.

Stein M., Rappuoli R., Covacci A. (2000) *Proc. Natl. Acad. Sci. USA.*; 97: 1263–1268

Stemmler, S *et al* (2005) Genes Immun. 6:225-230.

Stenström, B., Mendis, A. and Marshall, B. (2008) *Aust. Fam. Physician* 37(8): 608–12.

Suganuma, M., Yamaguchi, K. and Ono, Y. *et al* (2008) *Int. J. Cancer* 123(1): 117–22.

Svanes, C., Lie, S.A., Lie, R.T., Soreide, O. and Svanes, K. (1999) Scand J. Gastroenterol. 34: 18-24

Talley N. J., and Hunt R.H. (1997) Gastroenterology 113: S67-S77.

Thjodleifsson Bet al. (2007) Scand J. Infect. Dis. 39: 683–689.

Tomb J.F., White O. and Kerlavage A.R. *et al* (1997) *Nature* 388(6642): 539–47

Tsang K.W *et al* (1998) Am. J. Resp. Crit. Care Med., 158. 1047-1051.

Tsang K.W *et al* (2000) Respir. Med. 94:756-759.

Tsuji S., Kawai N., Tsujii M., Kawano, S and Hori, M (2003) *Aliment. Pharmacol. Ther.* 18 (Suppl) 1: 82–9.

vander Hulst R *et al* (1997) J. Infect. Dis. 176 : 196-200.

Viala J., Chaput C and Boneca I.G. *et al* (2004) *Nat. Immunol.* 5(11): 1166–74.

Vinette K.M., Gibney K.M., Proujansky R and Fawcett P.T (2004) BMC Microbiol. 4: 5.

Viskum K (1974) Scand. J. Respir. Dis. 55: 284-290.

Warren J. R and B. J Marshall (1983) Lancet I:1273-1275.

Wundisch T *et al* (2005) J. Clin. Oncol. 23: 8018-8024.

Xue F.B., Xu Y.Y., Wan Y., Pan B.R., Ren J. and Fan D.M (2001) World J. Gastroenterol. 7: 801-804.

Zoller L., Sobe D. Faulde M (1993) Kin. Lab 39: 45-54.

Chapter - 15

Current trends on the role of plant growth promoting rhizobacteria in sustainable agriculture – Indian scenario

Kolluru V.B.R. Tilak

Department of Botany, Osmania University, Hyderabad 500 007, India
E-mail: tilakkvbr@gmail.com

ABSTRACT

In recent years, root inhabiting (rhizosphere) bacteria (rhizobacteria) are gaining importance. The beneficial effects of plant growth promoting rhizobacteria (PGPR) has been variously attributed to their ability to improve crop production, protection against phytopathogens,soil health and environmental quality. However, there is a need for isolation, identification and improvement of potent PGPRs and their deployment as the promising bio-inoculants. This article provides a brief account of PGPR's from Indian work point of view on isolation of rhizosphere competent bacteria, methods to improve root colonization by introduced bacteria, mechanisms involved to enhance plant-micorbial interactions that enable inoculants to successfully colonize host rhizosphere and promote growth of host plants and the present and possible future trends in bioinoculant technology .

Key words: PGPR, Agriculture, Current trends, Indian scenario

1. Introduction

Soil microorganisms are important in soil processes that determine plant productivity. Our knowledge of soil microorganisms is limited in part by the inability to study them. Torsvik *et al.* (1990) estimated that 1 g of soil contains 4000 different bacterial "genomic units" as evident from DNA-DNA reassociation. The estimate suggest that about 5000 bacterial species have been described. Approximately 1% of the soil bacterial population can be cultured by standard laboratory practices. It is not certain that only this 1% represent the entire bacterial population (Kirk *et al.*, 2004). Colonization of plant root system is the very first step in nearly all interactions between plants and soil borne microbes. Region of contact between root and soil has been termed "rhizosphere" by Hiltner (1904). Plant roots release enormous amounts of exudates that may represent upto 10-20% of the photosynthates and this leads to significant stimulation of the microbial density and activity. Broadly, the three separate, but interacting components in the rhizosphere are rhizosphere (soil), rhizoplane, and the root itself. Rhizosphere is the zone of soil influenced by roots through the release of exudates that affect activity of root-colonizing rhizobacteria. The rhizoplane is the root surface, including the strongly adhering soil particles. Lastly the root itself forms a part of the system, because certain endophytes, colonize root tissues. Microbial colonization of the rhizoplane and/or root tissues is root colonization, whereas the colonization of the adjacent volume of soil under the influence of the root is rhizosphere colonization. The rhizosphere effect is the most immediate influence of plants on soil systems. However, since the microbial inoculations would mainly be applied to soils before the plant is grown, the strains should also survive in the soil and show optimal saprophytic potential. To meet such requirements, sufficient knowledge is needed about the bacterial traits that affect soil and rhizosphere colonization by microbes (Nautiyal 2000; Tilak *et al.* 2005; Nautiyal 2006; Manoharachary *et al.* 2006; Dey *et al.* 2010).

Inoculation of plants with beneficial bacteria can be traced back to centuries. Although bacteria were not known to exist until von Leeuwenhoek in 1683 discovered microscopic 'animals'and their utilization as plant growth stimulants in agriculture also has a lone history. Theophrastus (372–287 BC) suggested the mixing of different soils as a means of 'remedying defects and adding heart to the soil' (Vessey 2003). From their experience, farmers knew that mixing of soil from a previous legume crop with soils for growing nonlegumes improved the yield most often. By the end of 19th century, the practice of mixing "naturally inoculated" soil with seeds became a recommended practice of legume inoculation in the USA. A decade later, the first patent "Nitragin" was registered for plant inoculation with *Rhizobium* sp. (Nobbe and Hiltner 1896). Eventually, the practice of legume inoculation with rhizobia became more common. For almost 100 years, *Rhizobium* inoculants have been produced globally. Inoculation with nonsymbiotic, associative rhizosphere bacteria, like *Azotobacter* was used on a large scale in Russia (Cooper, 1959). *Bacillus megaterium* for phosphate solubilization was used in the 1930s on large scale in Eastern Europe. In the late 1970's, *Azospirillum* was found to enhance nonlegume plant growth and biocontrol agents, mainly of the *Pseudomonas fluorescens* and *P. putida* groups began to be intensively investigated in 1980's (Bashan 1998). In recent years, various rhizobacteria, *Aeromonas veronii*, *Azotobacter* sp., *Azoarcus* sp., Cyanobacteria (predominantly of the genera *Anabaena* and *Nostoc*), *Alcaligenes, Burkholderia, Comamonas acidovorans*, *Enterobacter*, *Erwinia*, *Flavobacterium*, *rhizobia* (including the *Allorhizobium*, *Azorhizobium*, *Bradyrhizobium*, *Mesorhizobium*, *Rhizobium*, and *Sinorhizobium*), *Gluconacetobacter diazotrophicus*, *Herbaspirillum seroepdicae*, *Serratia*, *Variovorax paradoxus* and *Xanthomonas maltophilia*, as well as many unidentified rhizobacteria proved beneficial for plant growth. Therefore, their use as biofertilizers or biological control agents (bioinoculants) for agriculture improvement has been the focus of many research groups in India and abroad (Saxena and Tilak, 1994,2005; Subba Rao 2000; Sindhu and Dadarwal 2001; Goel *et al.* 2001; Pal *et al.* 2004; Tilak *et al.* 2006, 2011; Tilak 2007; Vogeti *et al.* 2011 a,b). Seed inoculation with plant growth promoting rhizobacteria (PGPR): *Proteus vulgaris*, *Klebsiella planticola*, *Bacillus subtilis* improved the yield and nutrient uptake in mustard, sunflower and teff (Angaw *et al.*, 2011a). Similarly beneficial effect on nodulation and nitrogen-fixation due to combined inoculation of *Rhizobium* and PGPR's inpigeonpea (Tilak *et al.*, 2006); mungbean, soybean, lentil and chickpea was also reported recently (Angaw *et al.*, 2011b).

The immediate response to soil bioinoculants varies considerably depending on the bacteria, plant species, soil type, bioinoculant density and environmental conditions. In general, shortly after introduction into the soil, the bacterial population declines progressively, but they must compete with the often better-adapted native microflora and withstand predation by protozoans. The major role of bioinoculant formulation is to provide a more suitable microenvironment that prevents rapid decline of bacteria introduced in the soil. Bioinoculants have to be designed to provide a dependable source of beneficial bacteria that survive in the soil and are available to the plant. A better understanding of interactions between rhizobacteria and other microorganisms is obligatory for sustainable management of soil fertility and crop production. Colonization of the rhizosphere by rhizobacteria results in modifications of plant growth and development. Mechanisms involved in growth promotion by plant growth-promoting rhizobacteria can be divided into indirect and direct effects (Figure 1). Direct effects include enhanced provision of nutrients and the production of phytohormones (Nautiyal and Tilak, 2006; Pal *et al.*, 2006) while the indirect effects involve aspects of biological control: ie. production (Anith *et al.*, 1998 a,b; Pal *et al.*, 2000; 2003; Dey *et al.*, 2004; Nautiyal, 2000; Tilak *et al.*, 2005; Nautiyal, 2006; Pal *et al.*, 2008; Nautiyal and Tilak, 2009).

2. Isolation of native rhizobacteria

During the last two decades, there were several attempts to isolate bacteria from the rhizotic (rhizosphere, rhizoplane and endorhizosphere) zones of different crop plants (Ladha *et al.*, 1983; Glick *et al.*, 1995; Malik *et al.*, 1999; Sindhu *et al.*, 2000; Tilak and Reddy, 2006). A rapid and novel procedure for the isolation of PGPR was described by Glick *et al.* (1995). This entails screening soil bacteria for the ability to utilize 1-aminocyclopropane-1-carboxylate (ACC) as a sole N source, a trait that is a consequence of the presence of the enzyme ACC deaminase. This trait appears to be limited to soil bacteria capable of stimulating plant growth. The method appears to be rapid and should replace the otherwise slow and tedious processes of testing individual strains that promote plant growth (Glick *et al.* 1995).

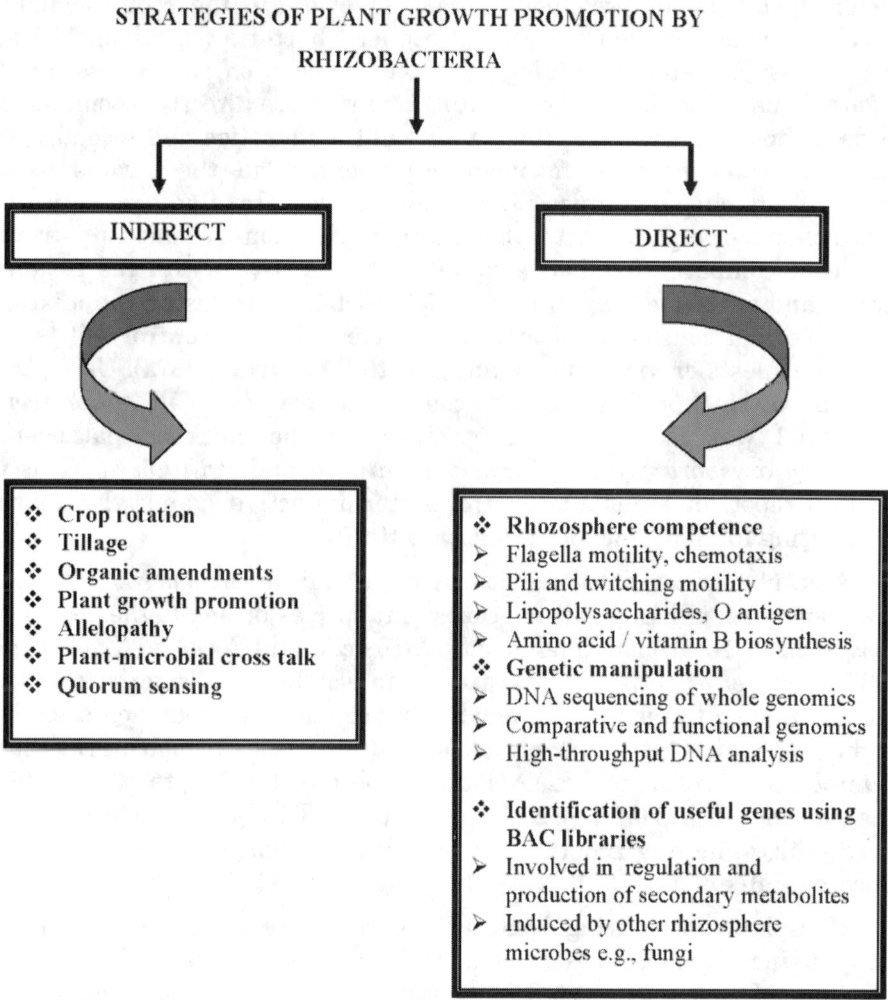

Fig. 1. Indirect and direct strategies of plant growth promotion by rhizobacteria

A rapid and novel procedure for the isolation of PGPR was the sceening of soil bacteria for their ability to utilize the compound 1-aminocyclopropaane-1-carboxylate (ACC) as a sole N source, a trait that is a consequence of the presence of activity of enzyme ACC deaminase. This trait appears to be limited to soil bacteria that are capable of stimulating the plant growth. This method appears to be rapid and should replace the otherwise slow and tedious process of testing individual strains for their ability to promote plant growth.

The isolation and development of plant beneficial bacterial strains applicable to a variety of crops, soils and locations will depend upon the development of improved and rapid detection and screening procedures that more rapidly screen and identify beneficial strains (Nautiyal, 1997a). A few methods have been developed for selecting rhizosphere competent bacterial strains for greenhouse assay to evaluate the root colonizing capability of the native chickpea rhizospheric bacteria (Nautiyal, 1997a). In this assay, system screening time was reduced on two counts. First, spontaneous rifampicin resistant (Rif^r) strains were directly inoculated to seeds without any check for the stability of the mutation and secondly, no attempts were made to taxonomically identify all the strains being screened for chickpea rhizosphere competence. The findings indicated that isogenic or equally rhizospheric competitive non-isogenic Rif^r strains should be compared for their survival and competition with the isogenic parent and with under specific ecological niche, before using the mixture for stable and consistent biological seed treatment to control soil borne pathogens/pests or to promote plant growth (Nautiyal, 1997a). Using this method, a chickpea rhizosphere-competent bacterium *P. fluorescens* NBRI1303 was isolated for suppression of the chickpea pathogenic *Fusarium oxysporum* f. sp. *ciceri, R. bataticola* and *Pythium* sp. This is the first report of a single biocontrol bacterium active against three most devastating fungi of chickpea (Nautiyal, 1997b).

A number of diazotrophic bacteria isolated from various root fractions of kallar grass, a highly salt-tolerant plant species belong to the genera of *Azospirillum, Azoarcus, Enterobacter, Klebsiella* and *Zoogloea* (Zafar *et al.* 1987; Malik *et al.* 1997). In an attempt to isolate rhizobacteria from the rhizotic zones of greengram using seven selective and four non-selective media, the dominant genera belonged to *Pseudomonas, Bacillus, Enterobacter, Proteus* and *Klebsiella* were obtained by Gupta *et al.* (1998). The presence of *Kurthia* sp. a Gram+ve, urea hydrolysing bacteria in the wheat rhizosphere was noticed using nutrient agar supplemented with glucose (Malik *et al.* 1999).

Phosphate-solubilizing bacteria are routinely screened by a plate assay using Pikovskaya (PVK) agar (Pikovskaya 1948). The relative efficiency of isolates is tested by selecting those capable of producing a halo/clear zone on plates owing to the production of organic acids into the ambient medium (Katznelson *et al.* 1962; Gaind *et al.* 2000). However, the reliability of this halo-based technique is questioned, as many isolates did not produce any visible halo/zone on agar plates but could solubilize various types of insoluble inorganic phosphates in liquid medium (Gupta *et al.* 1994). Therefore, a novel defined microbiological growth medium,

National Botanical Research Institute's phosphate growth medium (NBRIP), is more efficient than PVK, for screening phosphate-solubilizing microorganisms (Nautiyal 1999). It was suggested that soil microbes should be screened in NBRIP broth assay for the identification of the most efficient phosphate solubilizers (Nautiyal 1999). It is envisaged that usage of this formulation based upon qualitative analysis will facilitate quick screening of phosphate solubilizing bacteria. The results indicate that the formulation can also be used as the quality control test for the expeditious screening of commercial bioinoculant preparations, based on phosphate solubilizers (Mehta and Nautiyal 2001).

The glucose dehydrogenase gene, the enzyme in the oxidation pathway contributes sufficiently to normal phosphate solubilization ability in several bacteria. It is possible to enhance further the biofertilizer potential of PGPR by incorporating the genes involvd in mineral phosphate solublization without affecting their ability to fix nitrogen or produce phytohormones for dual benefit to agricultural crops. Glucose dehdrogenase from Gram negative bacteria can be engineered to improve their ability to use different substrates, function at higher temperatures and EDTA tolerance etc. through site-directed mutagenesis (Sashidhar and Podile, 2010; Sashidhar *et al.*, 2010).

3. PGPR- microbe interactions

In recent past indiscriminate use of excessive fertilizers and pesticides for maximum crop production has resulted in nutrient sufficient rhizosphere environment leading to less defensive root environment for pathogens. Such environment has also resulted in changed rhizosphere activity by which the associative plant growth promoting bacteria remaining free in the rhizosphere and the pathogens by virtue of their host dependence interact more freely with the host roots thereby increasing the incidence of pathogenicity. In the past one decade, scientists are looking back at the rhizosphere system and have identified plant growth promoting rhizobacteria (PGPR), plant health promoting rhizobacteria (PHPR) as well as nodule promoting rhizobacteria (NPR) which have capability of promoting plant growth (Kloepper, 1980; Nautiyal and Tilak, 2009). An interaction of PGPR with mycorrhizal fungi has been reviewed by Dey *et al.* (2005).

A large array of bacteria including species of Pseudomonas, Azospirillum, Azotobacter, Bacillus, Klebsiella, Enterobacter, Alkali-genes, Arthrobacter and Serratia have been shown to enhance plant growth.

There are several mechanisms by which PGPR may stimulate growth including production of growth regulating substances, phytohormones, suppression of plant pathogens through antibiosis, siderophore production, nitrogen fixation, mineralisation of organic phosphorus, production of ammonia and phyto alexins/flavonoid–like compounds, enhancement of mineral uptake, production of cyanide and other metabolities, ability to colonise root surface rapidly, production of lytic enzymes such as chitinase and ability to utilize diverse substrates for own growth found in root exudates. (Glick, 2000; Gupta *et al.*, 2000).

Existence of PGPR around roots of legumes and non-legumes is well established. In leguminous crops, inoculation with *Rhizobium, Bradyrhizobium* and PGPR affects symbiotic nitrogen flixation by enhancing the colonisation and nodulation of legume roots by rhizobia/ bradyrhizobia. (Gupta *et al.*, 1998 a,b,c).

Seed inoculation with certain PGPR like *Pseudomonas, Enterobacter, Bacillus subtilis, Klebsiella planticola, Proteus vulgaris* etc, resulted in improved grain yields in sunflower (Malik *et al.*, 1996) and also in different cereals and legumes under field conditions (Tilak 2011 a,b). Dey *et al.* (2000) reported beneficial effects of groundnut with different strains of fluorescent pseudomonads on pod yield. Combined inoculation of PGPR's and *Rhizobium* enhanced the nodule occupancy and nitrogen fixation in pigeonpea (Tilak *et al.* 2006). Suppression of root-borne diseases of maize like soft rot, collar rot by fluorescent *Pseudomonas* sp. has also been reported by Pal *et al.* (2001). Combined inoculation of *Pseudomonas aeruginosa* and AM fungus (*Glomus fasciculatum*) enhanced the growth, nutrient uptake and yield of sweet potato (Vogeti *et al.*, 2010, 2011).

In view of their growth promotary role in plant productivity and possible application as biocontrol agents, current research efforts with pseudomands are directed to understand their *in situ* behaviour in the rhizosphere environment for sustenance, competition and regulation of iron transport.

4. Genetic manipulation of plants and bacteria

Many processes in the rhizosphere influence plant disease and plant nutrition by affecting the dynamics of microbial population and communities. The action of both plant-beneficial and plant-deleterious soil-borne microbes depends on their ability to establish in the rhizosp-here. Therefore, knowledge of bacterial growth conditions in the rhizosphere is important for understanding rhizosphere colonization. Two approaches have been adopted to determine the basis of root colonization. The first focuses on bacterial traits strongly suspected to be involved in the colonization process (e.g. motility) and the colonization ability of genetically engineered strains of PGPR compared with wild-type strains. In a second approach, there are mutagenesis strategies to isolate new bacterial traits involved in colonization. Analyses of non-pathogenic plant -rhizobacteria interactions involved studies of bacterial traits expected to lead to successful colonization. Once rhizobacteria reach a root and get tightly bound to it, the third determining step for a successful colonization of the root system, is their ability to sustain and grow on the root. It is generally considered that the ability of microbes to compete for nutrients available in seed and root exudates is critical for the colonization process. Root exudates may be recognized, utilized, and processed by rhizobacteria as nutrients, signals or toxins An indirect evidence of the strong implication of root exudates in plant–microbe interaction is the identification of new colonization genes encoding bacterial cell surface and secreted proteins involved in transports or signal transduction.

In general, a proper characterisation of target soils and rhizosphere as habitats for introduced microbes, as well as adequate strategies are the key to success on enhanced plant-microbial interactions (Pal *et al.*, 2011). In consistency of PGPR's could be due to inefficiency of strains to clonize host roots (Dutta and Podile, 2010). As climatic conditions, soils, plants, and microbes are all variable and/or diverse, there is no general rule for how introductions into soils can be optimized. However, it is clear that since soil generally represents a hostile environment for introduction of an alien microbe and since the microbes in the soil are subjected to a range of adverse abiotic and biotic conditions, the rhizosphere competence of the introduced microbe depends to a large extent on how conducive to its survival and functioning the target environment is or can be made, in terms of either natural or induced ecological selectivity (through genetic engineering) or available protective niches. It is anticipated that knowledge of key ecological and molecular characteristics that determine bacterial survival and adaptation in soil may also lead to methods for genetic manipulations directed to improve the survival of bacteria introduced into soils. This in turn is critical for the application of beneficial microorganisms as inoculants to support plant growth (Pal *et al.*, 2011). Although the inability of commercial inocula to compete in the rhizosphere has been well documented, factors involved in colonization, remain elusive. Hardly anything is known about the ecological and molecular basis of rhizosphere colonization, despite its importance in plant-microbe interactions. Therefore, an understanding of the interactions between the biocontrol agent and complex rhizosphere microbial community in which the agent must function, is essential (Podile and Kishore, 2006; Das *et al.*, 2010; Neeraja *et al.*, 2010). From this understanding, one may develop models to predict the conditions under which a biocontrol strategy will succeed or fail. However, developing predictive models for biocontrol is a complex endeavour because of the multi - component nature of the system. Biocontrol often requires the introduction and establishment of biocontrol agent in the microbial community on or near the plant surface. Once introduced to the rhizosphere, various mechanisms account for the successful application, or failure of agent to control phytopathogenic fungi involving complex interactions between the plant rhizosphere microflora, phytopathogenic fungi, the biocontrol agent, and various biotic and abiotic factors encountered in soil, plant released compounds, induction of systemic resistance, niche exclusion, and environment stress, e.g., salt, pH, temperature, moisture, chemical fertilizers, pesticides and fungicides, etc. It is envisaged that an understanding of the mechanisms underlying these interactions may permit us to offer a logical basis for selection and manipulation of more effective biocontrol agents.

Engineering of the rhizosphere is another emerging field in which little is known, both at ecological and molecular level. The interest in engineering the rhizosphere has several reasons as plants might resist soil- borne pathogens more effectively, be better hosts to beneficial microbes, remediate toxic waste, or attract communities of soil microbes that enhance plant health. The ecology of rhizosphere is a relatively new research area. Consequently, there is little understanding of how environmental factors will affect bacterial colonization and persistence on roots and the resulting effect on plant. However, it is known that the bioinoculants introduced to the rhizosphere, are involved in complex biological interactions with the host plant and the surrounding rhizosphere microbes. The introduced bacteria are nourished by the root exudates and are thus dependent on the host plant. At the same time, the introduced bacteria may affect the host by inducing physiological changes. Interactions with indigenous rhizosphere microbes may be neutral, antagonistic (e.g., competition for nutrients, production of antibiotic compounds, parasitism, or predation), or synergistic (i.e., the promotion of *Rhizobium* induced nodulation of legumes). These microbial interactions are greatly influenced by many environmental parameters, including, temperature, moisture, fertilizer regimen, and soil type (Pal *et al.* 2011).

Nautiyal and Dion (1990) were the first to report on the capability to catabolize opines by pseudomonads, as carbon and nitrogen. Subsequently, impact of the novel substrate mannopine (MOPby creating nutritional bias on rhizobacterial colonization, was examined using mannopine utilization (Mut) as the model system Opines are excellent substrates for various soil microorganisms outside *Agrobacterium*.

5. PGPR as bioinoculants

There is an enormous volume of literature on the application of bacteria for improvement of plant performance (Saxena and Tilak 1994; Malik *et al.* 1996,1999; Saxena *et al.* 2000; Gupta *et al.* 2003; Shende *et al.* 2009; Tilak *et al.* 2010; Angaw *et al.* 2011 a,b), but few bacteria like *Azotobacter* and *Azospirillum* have been developed as commercial products. The organisms under most scrutinyfor potential use in agriculture are bacteria belonging to the genera *Pseudomonas* and *Bacillus species* (Nautiyal *et al.*, 2002; Nautiyal *et al.*, 2006; Tilak *et al.*, 2006;

Tilak and Reddy, 2006; Podile and Kishore, 2006). Positive response of plant growth promoting chitinolytic Paenibacillus elgii to tobacco root exudates was reported by (Das *et al.*, 2010).

During the last two decades, several examples of rhizobacteria capable of providing substantial disease control in the field have been reported (Saxena *et al.*, 2000; Nautiyal *et al.*, 2002, 2006). Many bacterial genera have shown their potential for biocontrol both under *in vitro* and *in vivo* conditions.

The usefulness of *Bacillus* as a source of antagonist for many plant pathogens is well known. Several potent strains from different species of *Bacillus* have been tested on a wide variety of plant species for their ability to control several diseases. *Bacillus* has ecological advantages because it produces endospores that are tolerant to extreme environmental conditions such as heat and desiccation (Nautiyal *et al.*, 2006).

Fluorescent pseudomonads have revolutionized the field of biological control of soil-borne plant pathogenic fungi. During the last one decade, they have emerged as the largest potentially most promising group of PGPR involved in the biocontrol of plant diseases. They have received the most attention for several compelling reasons, such as their colonization ability in roots, simple nutritional requirements, particularly their ability to use many carbon sources that exude from roots and to compete with the indigenous microflora in the rhizosphere. Apart from these qualities, pseudomonads are amenable to genetic manipulation. These characteristics make them ideal as a promising bioinoculant. There are numerous examples of biocontrol of several devastating fungal plant pathogens of important crops by fluorescent pseuomonads and has been reviewed from time to time (Nautiyal 1997c; Anith *et al.* 1998; Pal *et al.* 2000; Saxena *et al.*, 2000; Nautiyal 2006).

6. Conclusions

There are problems preventing commercial use of bioinoculants. Improvement of the biocontrol mechanisms of these bacteria by ecological or genetic means is an important approach for enhancing their performance as bioinoculants. However, inconsistency of the bioinoculants demonstrates that there is still a considerable need for extensive studies on rhizobacterial populations to understand the different criteria, which influence the composition of the rhizobacterial flora.

Recent advances in our understanding of the ecology and molecular biology of the systems responsible for bioinoculants are opening the ways to strain improvement The new tools of recombinant DNA technology, mathematical modeling, and computer technology combined with a continuation of the more classical approaches such as crop rotation, various tilling strategies, addition of organic amendments etc. to harness the power of microbes to improve soil, plant, human and the environment health.

Acknowledgement

The author is grateful to the National Academy of Sciences, India for providing him the financial support as Senior Scientist, Platinum Jubilee fellowship.

References

Angaw Tsigie, Tilak, KVBR, Saxena AK (2011a) Biol Fertil.Soils 47:971-974.

Angaw Tsigie, Tilak, KVBR, Saxena AK (2011b) J Indian Bot Soc 90(1&2):103-108.

Anith KN, Tilak KVBR and Khanuja SP (1998) Curr Sci 77:671-677.

Anith KN, Tilak KVBR, Khanuja SPS and Saxena AK (1998) World J Microbiol Biotech 14: 939-941.

Bashan Y (1998) Biotechnology Advances 16:729-770.

Cooper R (1959) Soils and Fertilizers 22: 327.

Das NS, Dutta S, Kondareddy A, Neeraja C, Sarma VSRNP, Srinivas V, Appa Rao P (2010) J Plant Growth Regul, Springer.

Dey R, Pal KK, Bhatt DM, Chauhan SM (2004) *Microbiol Res* 15 9: 371-394.

Dey R Pal KK, Cauhan SM, Bhatt DM (2000) International Arachis Newsletter 20 : 77-79.

Dey R, Pal KK, and Tilak KVBR (2005) Interaction of plant growth promoting rhizobacteria and arbuscular mycorrhizal fungi : Present status and future prospects. In: Podila G K, Varma AK (eds) Basic Research and Applications of Mycorrhizae . I K International Pvt Ltd, New Delhi, India, pp. 83-110.

Dey R, Pal KK, Tilak, KVBR (2011) Proc Nat Acad Sci India (In press).

Dutta S, Podile, AR (2010) *Critical Rev.Microbiol.* 36 : 232-244.

Gaind S, Mathur RS, Tilak KVBR (2000) Phosphate solubilising microorganisms. In: Yadav, Singh, Raychaudhury (eds) Recent Advances in Biofertilizer Technology. Omega Scientific Publishers, New Delhi, India, pp.190-204 .

Glick BR (2003) Plant growth promoting bacteria. In Glick BR, Pasternak JJ (eds) Molecular Biotechnology –Principles and Applications of Recombinant DNA. ASM Press, Washington DC, USA, pp 436-454.

Glick BR, Karaturovic DM, Newell PC (1995) Can J Microbiol 41 : 533-536.

Goel AK, Sindhu SS, Dadarwal KR (2001) Application of plant growth promoting rhizobacteria as inoculants of cereals and legumes. In: Yadav, Singh, Raychaudhury (eds) Recent Advances in Biofertilizer Technology. Omega Scientific Publishers, New Delhi, India, pp 207-256 .

Gupta A, Gopal M, Tilak KVBR (2000) Indian J Expt Biol 387 : 856-862.

Gupta A, Saxena AK, Gopal M, Tilak KVBR (1998) J Sci Indus Res 57: 726-731.

Gupta A, Saxena AK, Murali G, Tilak KVBR (2003) Trop Agric 80: 28-35.

Gupta R, Singal R, Shankar A, Kuhad RC, Saxena RK (1994) J Gen Appl Microbiol 40:255–260.

Hiltner L (1904) Arb Dtsch Landwirtsch Ges Berl 98:59–78.

Johri BN, Rao ChVS, Reeta Goel (1997) Fluorescent pseudomonads in plant disease management. In Dadarwal KR (ed) Biotechnological Approaches in Soil Microorganisms for Sustainable Crop Production. Scientific Publishers, Jodhpur, India, pp 193-221.

Katznelson H, Peterson E, Rouatt JW (1962) Can J Bot 40:1181–1186.

Kirk JL, Beaudette A, Hart M, Moutoglis P, Kironomos JN, Lee H, Trevors JT (2004) J Microbiol Method 58: 169-188.

Kloepper JW, Leong J, Teintze M, Schroth MN (1980) Curr Microbiol 4 : 317-330.

Ladha JK, Barraguis WL, Watanabe I (1983) Can J Microbiol 29 :1301-1308.

Malik KA, Bilal R, Mehnaz S, Rasul G, Mirza MS, Ali S (1997) Pl Soil 194: 37-49.

Malik RDS, Kavimandam SK, Tilak KVBR (1996) Indian Farming 45 : 15-16.

Malik RDS, Kavimandam SK, Tilak KVBR (1999) Indian J Exp Biol 37: 92-93.

Manoharachari C, Kunwar IK, Sarath Babu K (2006) Proc Nat Acad Sci India, 76(B)IV: 321-342.

Mehta S, Nautiyal CS (2001) Curr Microbiol 43:51-56.

Nautiyal CS (1997a) Curr Microbiol 34:12-17.

Nautiyal CS(1997b) Curr Microbiol 35:52-58.

Nautiyal CS (1997c) FEMS Microbiol Ecol 23:145-158.

Nautiyal CS (1999) FEMS Microbiol Lett 170:265–270.

Nautiyal CS (2000) Biocontrol of plant diseases for agricultural sustainability. In: Upadhyay RK, Mukherji, KG,. Chamola BP (eds.) Biocontrol potential and its exploration in sustainable agriculture, Kluwer Academic/Plenum Publishers, USA, pp 9-23.

Nautiyal CS (2006). Biological Control of Plant Diseases by Natural and Genetically Engineered Fluorescent *Pseudomonas* spp. In Ray RC, Owen PW (eds) Microbial Biotechnology in Horticulture. Enfield (NH) USA: Science Publishers, Enfield (NH) USA, pp 125-162.

Nautiyal CS, Dion P (1990) Appl Environ Microbiol 56: 2576-2579.

Nautiyal CS, Johri JK and Singh HB (2002) Can J Microbiol 48:588-601.

Nautiyal CS, Mehta S and Singh HB (2006) J Microbiol Biotechnol 16:184-192.

Nautiyal CS, Tilak KVBR (2009) Agriculturally important rhizobacteria as bioinoculants for enhancing plantgrowth and soil health In: Khachatourians GG, Arora DK, Rajendran TP, Srivastava AK.(eds.) Agriculturally Important Microorganisms Vol 11 Academic World International, New Delhi, India, pp 77.

Neeraja C, Kondareddy A, Purushotham P, Suma K, Sarma PVSRN, Moerschbacher BM, Appa rao Podile (2010) Critical Reviews Biotechnology pp 1-11.

Nobbe F, Hiltner L (1896) Inoculation of the soil for cultivating leguminous plants. U.S. Patcnt 570 813.

Pal KK, Dalsania TL, Manesh T, Gondaliya S, Ghorai S, Dey R (2008) Biochemical and molecular characterization of some salt tolerant plant growth promoting groundnut rhizobacteria. 49th Annual Conference of AMI, New Delhi.

Pal KK, Dey R, Bhatt DM (2004) *J. Oilseeds Res,* 21: 284-287.

Pal K, McSpadden Gardener B (2006). Biological control of plant pathogens. The Plant Health Instructor DOI: 10.1094/PHI-A-2006-1117-02.

Pal KK, Rinku Dey, Bhatt DM, Chauhan S (1999) International Arachis Newsletter 19: 51-53.

Pal KK, Tilak KVBR, Saxena AK, Dey R, Singh CS (2000a) Microbiol Res 155: 233-242.

Pal KK, Tilak KVBR, Saxena AK, Dey R, Singh CS (2000b) Ind J Microbiol 40: 21-24.

Pal KK, Tilak KVBR, Saxena AK, Dey R, Singh CS (2001) Microbiol Res 156: 209-223.

Pikovskaya RI (1948) Mikrobiologiya 17:362–370.

Podile AR, Kishore GK(2006) Plant growth-promoting rhizobacteria. In: Gnanamanickam SS. (ed), Plant Associated Bacteria, Springer, Netherlands, pp 195-230.

Sashidhar B, Krishna Kishore I, Lalitha G, Kondareddy A, Gopinath K, Appa Rao P (2010) J Molecular Microbiol Biotechnol 18:109-119.

Sashidhar B, Podile AR (2010) J Appl Microbiol 109: 1-12.

Saxena AK, Pal KK, Tilak KVBR (2000) Bacterial biocontrol agents and their role in plant disease management. In: Upadhyay RK, Mukerji KG, Chamola BP(eds) Biocontrol Potential and its Exploitation in Sustainable Agriculture.: Kluwer Academic/Plenum, New York, pp.25-37.

Saxena AK, Tilak KVBR (1994) Indian. J Microbiol 34: 91-106.

Shende R, Saxena AK, Lata, Chauhan SVS, Tilak KVBR (2010) Proc Nat Acad Sci, India 80: 155-159

Sindhu SS, Dadarwal KR (2000) Indian J Microbiol 40:211–246

Subba Rao NS (2000) Biofertilizers for agriculture and forestry. Oxford & IBH Publ.Co., New Delhi, India.

Tilak KVBR (2007) J.Eco-Friendly Agric 2(1):1-19.

Tilak KVBR, Manoharachary C (2011). Diversity and bioprospecting of plant growth promotin soil microorganisms (PGPSM)-Aspects and prospects. In Vimala Y, Trivedi PC, Govil CM (eds), Advances in Botany. Pointer Publishers, Jaipur, India, pp. 1-16.

Tilak KVBR, Pal KK, Dey R (2010) Microbes for Sustainable Agriculture. IK International Publ., New Delhi, India. pp.200.

Tilak KVBR, Ranganayaki N, Manoharachari C (2006) Eur J Soil Sci 57:67-71.

Tilak KVBR, Ranganayaki N, Pal KK, De R, Saxena AK, Nautiyal CS, Mittal S, Tripathi AK, Johri BN (2005) Curr Sci 89:136-150.

Tilak KVBR, Reddy BS (2006) Curr Sci 90:642-644.

Torsvik V, Salte K, Soerheim R, Goksoeyr J (1990) Appl Environ Microbiol 53: 1519-1524.

Vessey JK(2003) Pl Soil 255: (2) 571-586.

Vogeti S, Brunda Devi K, Tilak KVBR, Bhadraiah B. (2009). Indian Phytopath. 62: 258-260.

Vogeti S, Brunda Devi K, Tilak KVBR Bhadraiah B (2011) Proc Nat Acad Sci India 81(IV).

Zafar Y, Malik KA, Niemann EG (1987) J Appl Microbiol 3 : 45-56.

Chapter - 16

Green Fungus for the Benefit of Green Plants

Nagamani, A.* and Sarojini Chakravarthy, K.

Mycology and Plant Pathology Laboratory, Department of Botany, Post Graduate College of Science, Osmania University, Saifabad, Hyderabad – 500 004

*Corresponding author: mani1159@rediffmail.com

ABSTRACT

A saprophytic fungus which produces green conidia on dead organic matter is identified as beneficial organism to the plants. It was characterized by Persoon in 1794 as the genus *Trichoderma*. Now, more than 110 species with very high genetic diversity were recognized and numerous species were identified as biocontrol agents against several plant pathogens. The diversity of species has been worked in many countries along with its telomorph "*Hypocrea*". The anamorphs were paired with the corresponding telomorphs in recent years. Around 35 species were reported from India. However, in India the research studies on *Trichoderma* species diversity is limited. The identification of species is based on morphological and molecular analysis. Among the species, *T. harzianum* is most widely used as biocontrol agent. The other species applied as biocontrol agents belong to *T. atroviride, T. asperellum, T. hamatum, T. longibrachiatum, T. reesei, T. virens, T. stromaticum*. The plant pathogens like *Rhizoctonia, Sclerotium, Fusarium, Pythium, Phytophthora* were controlled by various species of *Trichoderma*. The production of hydrolytic enzymes and more than 100 metabolites from *Trichoderma* influence the survival of the plant pathogens. The plant growth substances produced by *Trichoderma* spp. enhance the growth and development of plants and induces resistance against plant pathogens. The rhizosphere competence' ability of this genus has facilitated it to grow in association of roots. The synergistic effect of all beneficial properties of the genus *Trichoderma* popularized it to use as biofungicide and biofertilizer in annual and perennial crops. In addition to its ability as biocontrol agent, it is also used in bioremediation. The detail benefits of fungus to the plants are discussed in this review.

Key words: Trichoderma, Biocontrol, Biofungicide

1. Introduction

Trichoderma is a fungal genus, first described and identified by its anamorphic state isolated chiefly from soil (Persoon 1794). It was later envisaged into four genera. *Trichoderma* spp. are ubiquitous colonizers of cellulosic materials and can thus often be found wherever decaying plant material is available (Kubicek *et al.*, 2008; Jaklitsch, 2009). *Trichoderma* Pers. ex Fr., a genus was placed earlier under Deuteromycotina, Hyphomycetes, Phialosporae, Hyphales, Dematiaceae. After recognizing the teleomorph as Hypocrea Fr., now it belongs to Hypocreales, Ascomycotina (Rifai and Webster 1966a,b; Webster 1964). *Trichoderma* is readily identified by a unique morphology that includes rapid growth rate in culture, production of numerous spores that are varying shades of green, rarely white conidial balls, on repetitively branched, however with a complex conidiophore structure. The reverse side of the colonies is often uncoloured, buff and yellow shades. The colour of the colony is mostly in shades of green (Fig. 1). Many species produce prodigious quantities of thick walled chlamydospores in submerged mycelia (Gams and Bisset 1998). The colonies of few species emit coconut order (Rifai, 1969).

Most of the *Trichoderma* species are morphologically very similar and were considered for many years as a single species: *T. viride* (Bisby 1939). The identification and characterization of *Trichoderma* was worked out into a monograph (Rifai, 1969), in which this genus was classified into nine species aggregates. Later, several biological species were recognized under each aggregate. Bisset (1984, 1991 a,b,c and 1992) elevated Rifai's species aggregates to species level and recognized two to several species within each of five sections of the genus. Due to the plasticity of phenetic characters and subtle differences among the species, it is difficult to apply these morphological criteria like colony, conidiophore, phialides, conidia (Fig. 1) and ascocarp, ascus and ascospore to identify different species. As new species were discovered, a consolidated taxonomical scheme was needed.

Several articles were published to decide whether these artificial approaches have a molecular or physiological basis (Bisset *et al.*, 2003; Kuhls *et al.*, 1997; Overton *et al.*, 2006; Samuels 1996; Samuels *et al.*, 1998, 2006). DNA methods brought additional valuable criteria to the taxonomy of *Trichoderma* which are being used today for studies that include identification (Hermosa *et al.*, 2001; Lübeck *et al.*, 2000) and phylogenetic classification (Kullnig-Gradinger *et al.*, 2002; Lieckfeldt and Seifert 2000). The image analysis of HPLC chromatogram is also was used for the identification of strains (Thrane *et al.*, 2001).

The taxonomy and phylogeny of several *Trichoderma/ Hypocrea* species have been confirmed by a combination of morphological, physiological and genetic approaches (Druzhinina and Kubicek 2005; Gams

2006; Jaklitsch 2009; Kullnig *et al.*, 2002, Lieckfeldt *et al.*, 2001; Overton *et al.*, 2006). Their studies revealed the molecular data obtained from analyses of gene sequences is essential to distinguish phenetically cryptic species and to establish phylogenetic relationships. On the other hand, phenetic data from morphological and metabolic studies validated conclusions derived from the phylogenetic analyses, and also contributed useful biological information about the species. However, none of these markers was correlated with the antagonistic characteristics assessed.

The molecular approaches with multiloci databases comparing the ITS 1&2 and transition elongation factor (tef) assumed the major role in species recognition (Dodd *et al.*, 2000; Kopchinskiy *et al.*, 2005). The identification of species, a list of correctly identified strains of *Trichoderma* and their GenBank numbers for sequences of translation-elongation factor EF-1α and internal transcribed spacer rDNA was reviewed (Samuels 2006a). In the recent literature, *Trichoderma* is divided into 5 sections with several clades; in each section such as Section 1-Pachybasium, Section 2-Longibrachiatum, Section 3-Trichoderma, Section-4 Hypocreanum and Section-5 lone lineages (Bisset 1984, 1991 a, b, c and 1992; Samuels *et al.*, 1998, 2006b).

Recently several species were added to the list (Samuels *et al.*, 2000). The studies to define species in *Trichoderma* were exemplified while describing new species from Asia (Bisset *et al.*, 2003). The first 100 *Trichoderma* species were characterized by molecular data (Druzhinina *et al.*, 2006). Accordingly, researchers followed several biochemical and molecular methodologies together with traditional taxonomy and identified more than 110 species to date. Till to date only thirty five species were reported from India (Bhagat and Pan, 2010; Bhramaramba and Nagamani, 2010; Nagamani *et al.*, 2002, 2006; Nagamani and Chakravarthy, 2009; Rahel and Nagamani, 2010). The taxonomic and ecological aspects of different species from India have been reviewed (Nagamani *et al.*, 2002, 2006). Mostly the species were generally typified with the asexual stage as it is most common in nature and culture.

Most isolates of the genus *Trichoderma* that were found to act as mycoparasites of many economically important aerial and soil-borne plant pathogens, have been classified as *T. harzianum* Rifai (Gams and Meyer, 1998). Due to the fact that the species *"harzianum"* is generally considered as a group made of mycoparasitic and biocontrol strains, and there is large morphological plasticity that results in character overlaps with other species, the identification of the species may be difficult. Several authors have reported a large genetic variability among *T. harzianum* isolates (Bowen *et al.*, 1996; Gómez *et al.*, 1997; Grondona *et al.*, 1997; Muthumeenakshi *et al.*, 1994). In fact, it has been demonstrated that at least four distinct species are present within the biocontrol. *T. harzianum* aggregate: *T. harzianum, T. atroviride, T. longibrachiatum* and *T. asperellum* (Hermosa *et al.*, 2000). Coevolution of organisms antagonistic to pathogens results in many *Trichoderma* strains being inactive against fungi other than those against which they were originally selected (Ihrmark *et al.*, 2010). This is strongly advantageous in that they are less likely to act against non-target organisms, but it does mean that a new selection process must take place for each crop/pathogen combination (Grondona *et al.*, 1997).

Some species within the clades are able to antagonize phytopathogenic fungi (Kevei and Nagy 2003). The competent substrate colonization especially rhizosphere; antibiosis and/or mycoparasitism, inducing plant disease resistance are the main mechanisms operate here. This potentiality of *Trichoderma* species as biocontrol agents of plant diseases was first recognized in the early 1930's (Weindling 1932). This antagonistic ability is the base for effective use of different *Trichoderma* strains as an alternative to the chemical control against a wide range of fungal plant pathogens (Howell 2003, Kaewchai *et al.*, 2009; Papavizas, 1985; Samuels, 1996). This has culminated in the commercial production of various *Trichoderma* species for the disease management and growth enhancement of various crops in India (Mohiddin *et al.*, 2010) and worldwide (Nagamani and Chakravarthy 2009, Harman *et al.*, 2010). A large variety of applications has been developed owing to various synergistic actions of *Trichoderma* strains (Benitez *et al.*, 2004). Even though, the paradigm of *Trichoderma* as a biocontrol agent is most common, it also has been used for many other purposes which have beneficial effects on green plants. The biodegradable capacity is a source of useful enzymes in different industrial sectors (Harman and Kubicek 1998). The various benefits of this green fungus on green plants include:

1. Plant disease control
2. Plant disease resistance induction
3. Abiotic stress tolerance- drought resistance, salinity resistance
4. Plant growth promotion and productivity- increased shoot and root growth; nutrient uptake; photosynthetic rate, germination percentage; biofertilizers
5. Biodegradation and bioremediation.

Each benefit of this green fungus on green plants is reviewed below.

2. Plant disease control

The application of chemical fungicides is the most common method (Knight *et al.*, 1997) for control of plant pathogens. However, chemical fungicides have a negative effect on environment and human health (Nannipieri 1994) and resulting in developing resistance in plant pathogenic fungi (Benitez *et al.*, 2004). To counteract all these harmful effects, the biofungicides have become a potential alternative (Kulkarni *et al.*, 2007). *Trichoderma* species are potent mycoparasites and are used for biological control of fungal diseases on agricultural crops.

T. harzianum* (NFCCI 2241)** ***T. asperellum ***T. virens*** ***T. longibrachiatum***

***T. reesei* (NFCCI 2242): Colony on PDA and parasitizing the mycelium of *Rhizoctonia solani* by coiling around the hyphae (Chakravarthy, 2008)**

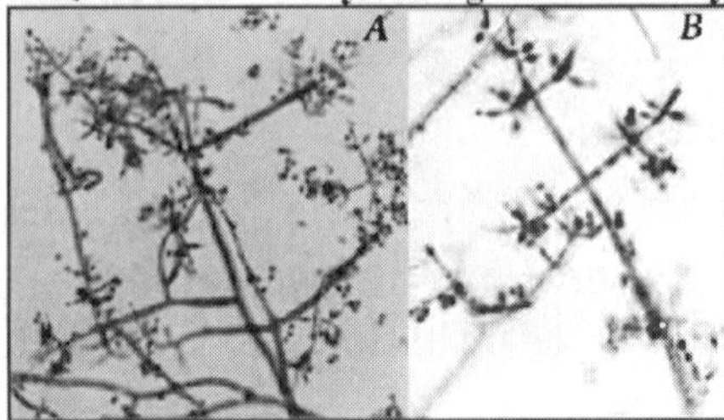

Photomicrograph of *T. hazianum* (Chakravarthy, 2008)

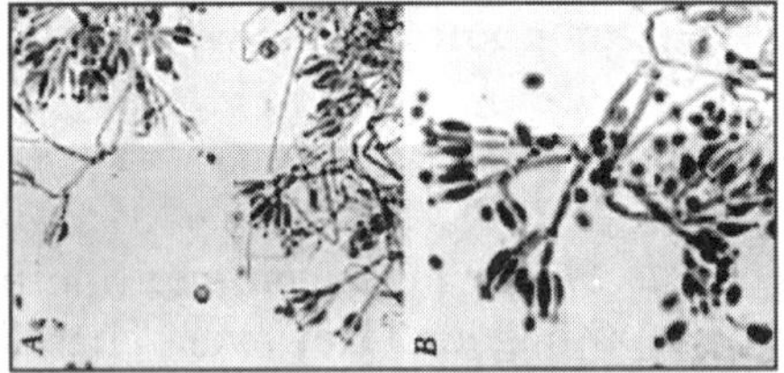

Photomicrograph of *T. virens*(Chakravarthy, 2008)

Fig. 1. *Trichoderma* species on PDA plates

The biocontrol fungi have much greater potential than bacteria to grow and spread in the rhizosphere through possession of hyphal growth and wide host range (Whipps 2001). Among a variety of fungal species and isolates examined as bioagents, the *Trichoderma* species lucidly dominated. This domination is remarkably proved because of the various mechanisms operate (Howell 2003, Vinale *et al.*, 2008a) such as mycoparasitism and antibiosis (Weindling 1932, Dennis and Webster 1971a).

Mycoparasitism is the foremost identified mechanism which involves very complex process. A range of plant pathogenic fungi are known to be attacked by the *Trichoderma* through mycoparasitism (Howell 2003). It follows sequential events such as (i). Recognition (ii). Attack and (iii). Subsequent penetration of the host.

(i) Recognition: the *Trichoderma* senses the pathogen when chitinases degrade pathogen cell wall and attack begins.

(ii) Attack: *Trichoderma* attaches to the pathogen that binds to the pathogen lectins then *Trichoderma* coils around the pathogen.

(iii) Penetration: Now the *Trichoderma* produces "Cell Wall Degrading (CWD) Enzymes" and peptaibols which facilitate the hyphae to enter the lumen of the parasitized fungus and starts assimilation.

The biochemical evidence for the coiling of the pathogen was detected and the gene involved in controlling this mechanism is G-α-gene (*tga1*) (Omero *et al.*, 1999). The genes involved in mycoparasitism were induced by the action of CWDEs and molecular biology of mycoparasitic interactions between pathogen and antagonist has been reviewed (Vinale *et al.*, 2008a).

The production of various chemical constituents is one of the foremost action of *Trichoderma* in plant disease control is generalized as antibiosis. The antifungal compounds at this juncture are identified as CWD enzymes and secondary metabolites. *Trichoderma* species releases hydrolytic enzymes at a constitutive level, which induces the host (plant pathogen) to release the molecules which are in turn sensed by the *Trichoderma* to parasitize the host (Harman *et al.*, 2004a, Woo and Lorito 2007). Cell wall degrading enzymes (CWDEs) showed antifungal activity towards broad spectrum of fungal pathogens that played a substantial role in biocontrol (Vinale *et al.*, 2008a). The antifungal CWDEs of *Trichoderma* spp. comprise of cellulases, chitinases, glucanases and proteases. The antifungal activity of purified endochitinase and chitobiosidase (Chitinolytic enzymes) produced by *Trichoderma harzianum* was studied (Lorito *et al.*, 1993, 1996a). Chitinases are divided into 1, 4-ß-acetylglucosaminidases (GlcNAcases), endochitinases and exochitinases. In fact Chitinases and *β*-1,3-glucanases are the best-studied antifungal proteins and whole-genome sequencing reveal between 20 and 36 different genes encoding chitinases, the hydrolytic enzymes that are involved in the mycoparasitic attack. Many GlcNAcases and their genes from *T. harzianum*, *T. virens* and *T. atroviride* have been described (Harman *et al.*, 2004b; Kim *et al.*, 2002). Chit36 (chitinase gene) secreted from *T. harzianum* inhibited the spore germination of *Botrytis cinerea* and the growth of both *Sclerotium rolfsii* and *Fusarium oxysporum* (Viterbo *et al.*, 2001). *T. virens* expressing Chit42 enzyme enhanced the biocontrol activity against *Rhizoctonia solani* than the wild type in cotton seedlings (Howell 2003). Endo- and exo-chitinases produced by *T. koningii* destroyed the cell wall integrity of *Sclerotium cepivorum* infecting onion roots (Metcalf and Wilson 2001). Many ß-1,3-glucanases have been isolated from *T. harzianum*, *T. atroviride* and *T. virens*. The enzyme from *T. harzianum* inhibited the growth of *B. cinerea*, *R. solani* and *Phytophthora citrophthora* (Benitez *et al.*, 1998b). Glucanases alone or in synergism with chitinases reduced the growth of *Gibberella fujikuroi* (De La Cruz and Llobell 1999). Cellulases (ß-1,4-glucanases), comprising cellobiohydrolases, endoglucanases (*egl1, egl2*) and ß-glucosidases, have not been widely studied earlier for biocontrol purposes, although cellulose is abundant in oomycetes (Bartnicki-García 1968). *T. harzianum* produces many cellulases which have a greater biocontrol activity against *Pythium ultimum* on cucumber seedlings (Migheli *et al.*, 1998). Biocontrol of *B. cinerea* by *T. harzianum* is due to the production of proteases which inactivated the hydrolytic enzymes produced by the pathogen on bean leaves (Elad and Kapat 1999). Moreover, purified mixes of CWDEs with different lytic activities showed improved antifungal effects against various plant pathogens, sometimes comparable to those obtained by using synthetic pesticides alone (Vinale *et al.*, 2008a).

Chitinase encoding genes are the most used to improve plant defense against fungal pathogens. These enzymes are capable of degrading the linear homopolymer of β-1,4-N-acetyl-D-glucosamine, the main component of the cell wall of most phytopathogenic fungi, and have shown strong inhibitory activity *in vitro* on germination and hyphal growth (Lorito *et al.*, 1996b). Although hydrolytic enzymes are produced by plants but they are not sufficient to confer resistance. Very few plants have resistance to multiple pathogens. On the other hand the plants transformed with chitinase and *β*-1, 3-glucanases together have been observed to inhibit fungal growth and have shown resistance to multiple pathogens. There are many evidences to show that fungal chitinases alone has increased the resistance of transformed plants against pathogenic fungus. The rice plant transformed with an endochitinase gene (*ech*42) from the biocontrol fungus *T. atroviride* increased the resistance to sheath blight caused by *Rhizoctonia solani* and rice blast caused by *Magnaporthe grisea* (Liu *et al.*, 2004). The genes encoding for chitinase genes from *Trichoderma* species are utilized to bring out transgenic plants.

Trichoderma spp. produce a plethora of secondary metabolites with biological activity (Ghisalberti and Sivasithamparam 1991; Sivasithamparam and Ghisalberti 1998). Secondary metabolites include antibiotics, which are natural products able to inhibit microbial growth. Antibiotic production is often well correlated with biocontrol ability, and the application of purified antibiotics was found to show effects on the host pathogen similar to those obtained by using the corresponding living microbe. Ghisalberti *et al.* (1990) demonstrated that the biocontrol efficacy of *T. harzianum* isolates against *Gaeumannomyces graminis* var. *tritici* is related to the production of pyrone-like antibiotics. The production of secondary metabolites by *Trichoderma* spp. is strain dependent and includes antifungal substances belonging to a variety of classes of chemical compounds. Ghisalberti and Sivasithamparam (1991) classified metabolites into three categories: (i) volatile antibiotics, i.e. 6-pentyl-α-pyrone (6PP); (ii) water-soluble compounds and (iii) peptaibols. But earlier Dennis and Webster (1971 b,c) classified them into volatile and non-volatile. The insecticide tolerant strains of *T. reesei* and *T. longibrachiatum* when tested in vitro inhibited the growth of *R. solani* by producing both volatile and non-volatile metabolites (Chakravarthy and Nagamani, 2007, Chakravarthy *et al.*, 2011).

The activity of metabolites of *Trichoderma* manipulates the physiology of the host through various mechanisms. *Trichoderma* strains which induce phytoalexins have more biocontrol efficiency than non-inducers in controlling plant disease. *T. virens* strains that induced high levels of phytoalexins showed improved biocontrol efficacy than non-induced strains of *T. virens* (Hanson and Howell, 2004). This clearly indicates the plant responded resistance was controlled by the action of secondary metabolites of *Trichoderma*, thus, plant disease control has been achieved. Low molecular weight and volatile metabolites enhance the status of one organism by affecting the physiology of competitor organisms. In contrast, high molecular weight metabolites are polar may exert their activity on direct interactions by contact between *Trichoderma* species and their antagonist (Cardoza 2005; Kubicek and Harman 1998). However, the synergism of metabolites and enzymes produced by *Trichoderma* also appears to be important in plant disease control (Lorito *et al* 2006b; Reino *et al.*, 2008). A detailed account on the secondary metabolites of biocontrol agent *Trichoderma* was reviewed by Reino *et al.*, (2008) and Vinale *et al.* (2008b).

Competition between the *Trichoderma* and plant pathogens is other means which affects the growth and infection of the plant pathogen. Competition for carbon, nitrogen and other growth factors, together with competition for space or specific infection sites modify environmental conditions those may be used by the BCA to starve the plant pathogens. *Trichoderma* has a strong capacity to mobilize and take up soil nutrients, thus making it more efficient and competitive than any other soil microbes (Benitez *et al.*, 2004). Sivan and Chet (1989) demonstrated that competition for nutrients is the major mechanism used by *T. harzianum* to *Fusarium oxysporum* f. sp. *melonis*. The pathogen like *Botrytis cinerea* is sensitive to lack of nutrients is easily inhibited by *Trichoderma* where the latter fungus is highly efficient in utilizing the nutrients at faster rate than the former fungus.

The strains of *T. harzianum* T-22, and *T. virens* as well as other rhizosphere competent biocontrol agents provide long-term protection or other advantages to plants from a single application at the beginning of the season. *Trichoderma* isolates establish themselves on roots, grow with the developing root system, and remain functional for at least the life of an annual crop (Harman 2000). Strains within this genus includes a wide spectrum of evolutionary solutions that range from very effective soil colonizers with high biodegradation potential to non-strict plant symbionts that colonize the rhizosphere. The rhizosphere competence gives long-term protection at a substantial distance from the infection court, thus eliminating the pathogen causing disease. In addition, it has reported recently that the fungi are generally confined to the roots; the changes in expression are substantially greater in the shoots than in the roots (Shoresh *et al.*, 2010). Moreover, the *Trichoderma* strains able to modify external pH in rhizosphere and adapt its own metabolism to the surrounding growth conditions. This would consequently reduce the pathogenicity of phytopathogens, because most of its factors could not be effective in causing the disease (Benitez *et al.*, 2004).

3. Plant disease resistance induction

The capacity to induce resistance against a broad range of plant diseases which are caused by various classes of plant pathogens namely *Rhizoctonia solani*, *Colletotrichum lindemuthianum*, *C. graminicola*, *C. arbiculare*, *Botrytis cinerea*, *Magnopothe griseae, Pseudomonas syringae* pv. *lachrymans*, *Xanthomonas campestris* pv. *phaseoli*, *X. oryzae* pv. *oryzae*, and *Phytophthora capsici* in a variety of plants like cotton, bean, tomato, pepper, tobacco, cucumber, maize, and rice is wide spread in this genus, the *Trichoderma*. Metcalf and Wilson (2001) demonstrated that some *Trichoderma* strains can colonize only local sites on roots but rhizosphere competence strains colonize entire root surface for several weeks or months (Harman 2000). The root colonization by *Trichoderma* strains results in increased levels of defense plant enzymes (Harman *et al.*, 2004a). Furthermore, Vargas and others (2010) suggested that the presence of a sucrose-dependent network in the fungal cells regulates the symbiotic association between host and BCA. This benefits the plant to acquire resistance against the plant pathogens. Thus, the *Trichoderma* plays an eminent role in induction of resistance in the host plants. This phenomenon is supported by the findings of Yedidia *et al.* (1999). In their study the plant response was noticed by an increase in peroxidase and chitinase activities, and deposition of callose-enriched wall appositions on the inner cell walls. Howell *et al.* (2006) supported these finding while working with cotton seedlings treated with *Trichoderma* species. Marra *et al.* (2006) reported that the signaling induces systemic effects in plants, thus the fungus (*Trichoderma*) reprogramme plant gene expression resulting in alteration of plant responses to their environment.

Several benefits pertaining to induced plant resistance are conferred and reviewed recently (Harman 2011a). These benefits include priming and defense; increased photosynthesis and carbohydrate metabolism and, abiotic stress response (Mastouri *et al.*, 2010; Shoresh *et al.*, 2010). Among other benefits, increased plant growth especially roots under stress (Harman 2000); systemic resistance to diseases hypersensitive response (HR), induced systemic resistance (ISR) and systemic acquired resistance (SAR) (Lorito *et al.*, 2010); enhancement of seed vigor (Mastouri *et al.*, 2010; Shoresh *et al.*, 2010) and improved nitrogen use efficiency (NUE) (Harman and Mastouri 2010) are reported. Altogether, *Trichoderma* enhanced plant growth to certain extent because respiratory systems and carbohydrate metabolisms are upregulated and thereby increase energy and sugar supply to the growth of the plant. This benefits the plant by an increase in photosynthesis; thus, induce growth response (Shoresh *et al.*, 2010).

The xylanase produced by *Trichoderma* strains induced ethylene production and plant defense responses in tobacco (Lotan and Fluhr 1990). In a different study a group of proteins and peptides produced by the strains of *T. virens* induced terpenoid phytoalexin biosynthesis and peroxidase activity which was shown to have elicitor activity in cotton (Hanson and Howell 2004). The bioactive molecules like these are produced during the multiple interactions between *Trichoderma*, plant pathogens and host roots function as inducers of antagonistic gene expression cascade in *Trichoderma* and also function as elicitors of plant defense mechanisms (Harman *et al.*, 2004a). These reactions are expressed as HR, ISR and SAR against several plant pathogens (Dana *et al.*, 2001) as well as hostile abiotic conditions (Harman *et al.*, 2004b). The studies on Barley, Tobacco, Potato and Strawberry plants treated with *T. viride* or *T. harzianum* showed tolerance against the pathogens like *Fusarium* sp. (McIntyre *et al.*, 2004), *Alternaria alternata*, *A. solani*, *B. cinerea*, *R. solani* or *Colletotrichum* sp. (Howell, 2003) by expressing above mentioned mechanisms. The details of interactions of *Trichodema*-plant-pathogen interactions were available elsewhere (Vinale *et al.*, 2008a). The saprophytic strains of *Trichoderma* can provide plant growth promotion in absence of plant pathogens (Whipps 1997).

4. Abiotic stress tolerance

The abiotic stress tolerance against drought, pH, salinity and temperature is increased as a part of systemic resistance because the fungi (*Trichoderma*) improve the redox status of the plant. As a result, symptoms of biotic and abiotic stress can be limited (Shoresh *et al.*, 2010). The *Trichoderma* strains enhance the activity of pathways to convert glutathione and ascorbate to reduced forms (Harman 2011a). Seeds treated with *T. harzianum* resisted water deficit condition and enhanced deep rooting (Harman 2000) and cocoa seedlings showed continuous growth in drought condition (Bae *et al.*, 2009). *Trichoderma* inoculated plants had higher chlorophyll content and photosynthetic efficiency under drought stress (Shoresh *et al.*, 2010). *Trichoderma* treated plants showed the salt induced inhibition of leaf metabolic activity, as a consequence the effect of salinity was reduced (Yildirim *et al.*, 2006). Tomato seedlings were much less damaged by the temperature extremes in the presence of *T. harzianum* (Shoresh *et al.*, 2010). However, the mechanisms and elicitors for the enhanced resistance against abiotic factors is still unknown (Shoresh and Harman 2008). But the enhancement of these pathways in chloroplast is expected to increase photosynthetic efficiency which in turn induces nitrogen use efficiency (NUE) in plants. In extensive field trails of cereals it was found that the amount of nitrogen fertilizer could be reduced by 40-50% in the presence of *Trichoderma* with no reduction in yield. Hence, Harman (2011b) opined that the advantage of *Trichoderma* species that benefits the plants could be obtained across a wide variety of crops without extensive modification of varieties. The reduced vigor of maize seeds due to oxidative stress was restored vigor with the treatment of *T. harzianum* T22 (Bjorkman *et al.*, 1998).

Culture filtrates of the four *Trichoderma* species *T. pseudokoningii*, *T. harzianum*, *T. reesei* and *T. viride*, extracted with different organic solvents were found to be toxic against two problematic weeds of wheat, *Phalaris minor* L. and *Rumex dentatus* L. It is concluded that the culture filtrates of *Trichoderma* species have herbicidal potential (Javaid and Ali 2011).

5. Plant growth promotion and productivity

The potential of *T. harzianum* as growth inducer of various floriculture and horticulture crops was demonstrated by Chang *et al.* (1986). Plant seeds treated with the spores of *Trichoderma* increased the rate of seed germination (Benitez *et al.*, 1998a). Later, various researchers reported a similar type of benefit while using various strains of *Trichoderma*. *T. harzianum* 1295-27 was shown to solubilize phosphate and micronutrients that may be made available to provide plant growth those are unavailable to plants in certain soils (Altomare *et al.*, 1999). Benitez *et al.* (1998a) concluded that *Trichoderma* produces growth factor that enhances the rate of seed germination, plant growth, and yield. The findings of Zimand *et al.* (1996) clearly explained the cause for this benefit to plants. *T. harzianum* T 39 inhibited the conidial germination of *Botrytis cinerea,* reduced the germ tube elongation and the activity of pathogen secreted enzymes which are responsible for pathogenicity. In response to reduced activity of enzymes of the pathogen, the host accumulates oligogalacturonides which can elicit the host defense weaponry, thus checking the disease development in bean. In addition, the activity of BCA could reduce the effect of concentration of inhibitory substances present in soil those hinder the plant growth (Windham *et al.*, 1986). The investigations (Benitez *et al.*, 2004; Harman *et al.*, 2004b) revealed that *Trichoderma* spp. produce organic acids like gluconic, citric and fumaric acids. These acids decrease soil pH and thus allow the solubilization of phosphates, micronutrients and mineral cations like iron, manganese and magnesium. In turn, these are useful for plant metabolism which causes a generalized increase in the uptake of many elements including arsenic, cobalt, cadmium, chromium, nickel, lead, boron, zinc, aluminum and sodium (Harman *et al.*, 2004b; Yedidia *et al.*, 2001). Although, the *Trichoderma* colonize the roots, its effect was identified in the shoot growth, where the fungus is absent. This differential expression in shoots is probably due to translocated signals from the roots colonized by the *Trichoderma*. The treatment with *T. hamatum* or *T. koningii* increased the productivity of lettuce, tomato and pepper plants upto 300% (Vinale *et al.*, 2004) and with *T. longibrachiatum* (M1Ps) or *T. reesei* (NFCCI 2242) upto 25% grain yield in rice plants (Chakravarthy 2008). Cutler *et al.* (1989) reported that the compounds produced by antagonistic fungal strains were able to act as growth promoting factors. This hypothesis suggested that these secondary metabolites of *Trichoderma* may act as auxin like compounds. Thus, the growth promotion probably due to production of plant hormones or increased uptake of plant nutrients of the plant (Chet *et al.*, 1993, Altomare *et al.*, 1999), control of one or more pathogens (Baker *et al.*, 1986; Hoitink *et al.*, 2006, Harman *et al.*, 2004a,b) or strengthening the plant defense mechanisms (Zimand *et al.*, 1996; Shoresh and Harman 2008; Woo and Lorito, 2007).

A detailed study conducted by Harman and group (2004c) using T22 on maize indicated that *Trichoderma* species are resistant to toxic metabolites produced by deleterious microflora and produced different enzymes that are capable of degrading them in the root zone. Thus, T22 directly increased root growth, controlled deleterious non-pathogenic root microflora and their toxic metabolites.

The multifaceted effect of *Trichoderma* strains as root growth and crop productivity enhancer, resistance to abiotic stresses and uptake and use of nutrients through root colonization has received an importance to use it as biofertilizer. Haque and others (2010) reported that a combination of 50% fertilizer and 50% biofertilizer (*T. harzianum*) increased the seed production by 5.34% per plant.

6. Bioremediation and biodegradation

Unjudicial use of pharmaceuticals, pesticides, and petrochemicals deliberately is causing an effect on the life of plants, animals and man equally in nature, because they persist as such or added as toxic metabolites. In the same way the soils are also affected by these factors, as well as by bad management, mining or industrial pollution, and global problems like desertification and climate change. The uses of naturally occurring microorganisms or genetically engineered microorganisms are employed to transform or detoxify the pollutants. These pollutants bioacc-umulate in environment, and they ultimately affect the biodiversity. In this context the development of healthy and fertile soils is of fundamental importance for agricultural areas.

The novel character of *Trichoderma* producing varied enzymes has an impact on the pollutants. These enzymes are capable of degrading polycyclic aromatic hydrocarbons (PAHs), polyphenolics, ferrocyanides, metallocyanides and cyanides. The additional use of these enzymes produced by the rhizosphere competent *Trichoderma* help to increase the fertility of soil by biodegrading soil pollutants mentioned above. Thus, it helps the plants by ameliorating the soil pollution (Harman *et al.*, 2004b).

Trichoderma spp. (1) degrade cyanide by the activity of characterized enzymes, (2) take up and degrade metallocyanides, (3) are resistant to high levels of a variety of toxicants, and (4) some strains are highly rhizosphere competent and form stable, long lasting plant microbe communities. So, it is found that wheat plants are protected from cyanide toxicity when the roots are colonized by *Trichoderma* strains when applied (Harman *et al.*, 2004b). A few studies proved that it is possible to recover degraded soils through biological processes, including soils polluted by copper and gold mining activities (Anonymous 2010). *Trichoderma viride* removed the organic carbon and polyphenolic compounds by growing on olive oil mill wastewaters, which made the wastewaters to be treated easily for further use (D'Urso *et al.*, 2008). The most effective removal of arsenic was observed in the *Trichoderma* sp., sterile mycelial strain, *Neocosmospora* sp. and *Rhizopus* sp. fungal strains. These fungal strains can be effectively used for the bioremediation of arsenic-contaminated agricultural soils (Srivastava *et al.*, 2011).

Trichoderma strains play an important role in the bioremediation of soil contaminated with pesticides and herbicides. They have the ability to degrade a wide range of insecticides: organochlorines, organophosphates and carbonates.

Trichoderma harzianum is effective in endosulfan degradation along with other microfungi and algae (Prasad, 2011). The fungus *Trichoderma harzianum* 2023, used as a biocontrol agent, is resistant to many pesticides; when grown in liquid culture, this species actively transforms pentachlorophenol into pentachloroanisole (Rigot and Matsumura 2002). Several studies on the isolation of bacterial coculture and mixed cultures capable of degrading endosulfan reported the degradation of β-endosulfan by *Aspergillus niger*. *Trichoderma harzianum*, *Phanerochaete chrysosporium*, and *Mucor thermohyalospora* MTCC 1384 have also been examined for endosulfan degradation (Katayama and Matsumura, 1993).

Trichoderma were the most tolerant among the filamentous fungi from estuarine sediments collected from two contaminated sites and the soils contaminated mainly by polycyclic aromatic hydrocarbon (PAHs) of industrial pollutants (da Silva *et al.*, 2003). *Trichoderma viride,* has the ability to degrade PAHs. (Ch´avez-Gomez *et al.*, 2003). *Trichoderma sp.* were the best in the removal of ammonia (84%) from domestic sewage (Thanh and Simard, 1973).

Various wastepaper materials, a major component of solid waste, were treated with the cellulase enzyme from *Trichoderma viride*, thus bioconverting their cellulose component into fermentable sugars (van Wyk and Mohulatsi, 2003).

Trichoderma viride strain in an airlift bioreactor exhibited an extraordinary capability to remove high concentrations of the highly toxic and soluble hexavalent chromium (Morales-Barrera *et al.*, 2006). Harman *et al.*, (2004b) concluded that the uses of fungi (*Trichoderma*) and their associations with plants appear to be highly promising new tools for the treatment or amelioration of soil and water pollution.

Biodegradation of crop residues by fungi is dependent on the plant and fungus species. Residues reduction varied markedly with plant species. However, there was a relatively good relation between residues biodegradation and fungi lignocellulolytic enzyme activities. The deuteromycetes grew fast and produce more hemicellulolytic and cellulolytic enzymes in easily degradable plant residues like wheat, barley, and pea (Sinegani *et al.*, 2005). Hence, the fungi like *Trichoderma* and *Aspergillus* are known to play a key role in biodegradation of agricultural wastes and decomposition of straw (Helal, 2005) and also enhancing the availability of good quality manure and biofuels (Goyal and Sindhu, 2011).

Traditionally straw incorporation is considered an important strategy to improve soil quality and reduce dependence on mineral fertilizers (Tejada *et al.*, 2008). Cellulolytic fungi applications such as *Aspergillus, Chaetomium*, and *Trichoderma* (Bowen and Harper, 1990) have shown promising results. Fontaine *et al.* (2004) showed that exogenous cellulase supplementation accelerated decomposition of cellulose in soil. Therefore, using exogenous cellulase may be a potential means to accelerate straw decomposition and increase soil fertility (Han and He, 2010). Even timber wastes were inoculated with the combinations of different fungi, where *T. reesei* and *Phanerochete chrysosporium* combination was found to be superior in obtaining quality product of vermicompost with higher agronomic value (Kumar and Shweta, 2011). *Trichoderma koningii* in a consortium of fungi effectively degraded the lignocellulosic substances in horticultural plant residues thus increased the humic substances which are added to the ecosystem to increase the soil fertility (Goyal and Sindhu, 2011). By treating the rice straw with the fungal consortium of *A. niger* (F44) and *T. viride* (F26), the C/N ratio was reduced to 19.5 from an initial value of 29.3 in three weeks of the biodegradation process, thus showing the potential of this method for use in large-scale composting of rice straw (Kausar *et al.*, 2010).

Straw is usually either burnt in the field causing environmental hazards such as respiratory diseases or disposed in a way that does not benefit the farmers to its maximum extent. *Trichoderma* species were often cited as high cellulose-decomposers (Tengerdy and Szakacs, 2003; Zayed and Abdel-Motaal, 2005). Biological treatment of agricultural residues is a new method for the improvement of residue digestibility (Jalk *et al.*, 1998). These scientific studies proved the application of *Trichoderma* species alone or in combination with other fungi is beneficial for the plant growth and productivity by degrading cellulose, hemicelluloses, lignin, and total carbon of agricultural by-products and residues. In turn, this process is having an additional advantage in utilizing the agricultural wastes as organic manure and compost.

7. Conclusions

The use of *Trichoderma* species as biological control agents has been investigated for over 70 years but it is only relatively recently that strains have become available commercially. Many *Trichoderma* strains, mainly *T. harzianum*, *T. viride* and *T. virens* (formerly *Gliocladium virens*), have been identified as having potential applications in biological control. A partial list of genera of plant pathogenic fungi affected by *Trichoderma* includes: *Armillaria, Botrytis, Chondrostereum, Colletotrichum, Dematophora, Diaporthe, Endothia, Fulvia, Fusarium, Fusicladium, Helminthosporium, Macrophomina, Monilia, Nectria, Phoma, Phytophthora, Plasmopara, Pseudoperonospora, Pythium, Rhizoctonia, Rhizopus, Sclerotinia, Sclerotium, Venturia, Verticillium,* and wood rot fungi (Lumsden *et al.*, 1993; Monte, 2001; Nagamani and Chakravorthy, 2009). *T. harzianum, T. longibrachiatum, T. koningii, T. hamatum, T. asperellum, T. viride* are applied as biocontrol agents against plant pathogenic fungi (Deacon, 2006; Hjeljord and Tronsmo, 1998) for a wide variety of crops and climates. The main antagonist used in disease control in agriculture is the fungus *Trichoderma harzianum* Rifai, a low cost biocontrol agent that can establishes itself in different pathosystems, has moderate effects on soil balance and does not harm beneficial organims that contribute towards pathogen's control. *T. harzianum* is a safe and effective biocontrol agent in both natural and controlled environments that does not accumulate in the food chain and to which it has not been described resistance (Harman *et al.*, 2004a).

Biocontrol agents are widely regarded by the general public as "natural" and therefore, non-threatening products, although risk assessments must clearly be carried out on their effects on non-target organisms. Moreover, knowledge concerning the behaviour of such antagonists is essential for their effective use. *Trichoderma* exists in nearly all agricultural soils (Hagn *et al.*, 2003) and in other environments and is very common in diverse habitats (Samuels, 2006a).

Trichoderma has attracted special attention and is considered as wonder fungal organism because different species act as biological control agents of plant diseases; producer of cellulolytic, hemicellulolytic, chitinolytic and proteolytic enzymes which have various agricultural and industrial applicayions. It is a bioagent in biodegrading chlorophenolic compounds, PAHs, cyanides, arsenic recognized as a valuable organism in soil bioremediation. It's efficiency in decomposing of crop and horticultural residues in fact plant organic matter is utilized to increase the soil fertility. In addition, it also identified as enhancer of plant growth (Windham *et al.*, 1985) and amplifier of resistance against plant pathogens (Chet *et al.*, 2006). For these reasons *Trichoderma* is finding a place in various fields of industrial and agricultural applications.

*Trichoderma reesei (=Hypocrea jecorina*is) is an economically important producer of cellulases and hemicellulases and is used for heterologous protein expression (Kubicek and Penttilä 1998). Its harmful effects are few, some species are known as opportunistic pathogen of human beings (Richter *et al.*, 1999; Kredics *et al.*, 2003; Kuhls *et al.*, 1999) and cause of disease in commercially produced mushrooms (Samuels *et al.*, 2002). This biocontrol agent has no harmful effects on humans, wild life and other beneficial organisms. *Trichoderma* spp. has proved effective and selective enough against most of the fungal diseases. Today, more than 50 different *Trichoderma*-based agriculture products can be found as registered in many countries in five continents, and are sold and applied to protect and improve yield of vegetables, ornamentals and fruit trees (Woo *et al.*, 2006). After several decades of research, the *Trichoderma* has been proved safe and never been documented having an adverse reaction on humans and livestock (Anonymous, 2005).

Trichoderma spp. are successful colonizers of their habitats, which is reflected both by their efficient utilization of the substrate at hand as well as their secretion capacity for antibiotic metabolites and enzymes (Schuster and Schmoll, 2010). They are able to deal with such different environments as the rich and diversified habitat of a tropical rain forest as well as with the dark and sterile setting of a biotechnological fermentor shake flask. Under all these conditions, they respond to their environment by regulation of growth, conidiation, enzyme production, and hence adjust their lifestyle to current conditions, which can be exploited for the benefit of mankind.

References

Altomare C, Norvell WA, Bjorkman T. and Harman GE. 1999. App Envrn Microbiol 65: 2926-2933.

Anonymous 2005. Fertilizer recommendation guide-2005, Bangladesh Agricultural Research Council (BARC). Soil publication No. 45. Farmgate, Dhaka-1215, Bangladesh. 260 p.

Anonymous 2010. *Trichoderma* spp. and its potential in soil remediation, DRYNET.

Bae H, Sicher RC, Kim MS, Kim SH, Strem MD *et al.* 2009. J Exp Bot 60: 3279-3295.

Baker R, Elad Y, and Sneh B. 1986. Physical, biological and host factors in iron competition in soils. In: Iron Siderophores and Plant Diseases. ed. Swinburne TR, Plenum Press, New York. pp 77-84.

Bartnicki-García S. 1968. Annu Rev Microbiol 22:87-108.

Benı´tez T, Delgado-Jarana J, Rinco´n AM, Rey M, and Limo´n MC. 1998a. Biofungicides *Trichoderma* as a biocontrol agent against phytopathogenic fungi. In: Recent research developments in microbiology. Ed : Pandalai SG, Trivandrum: Research Signpost; 1998. pp. 129–150.

Benítez T., Limón, J., Delgado-Jarana, J., and Rey, M. 1998b. Glucanolytic and other enzymes and their control. Pages 101-127 in: *Trichoderma* and *Gliocladium*, Vol. 2. Eds. Harman GE and Kubicek CP. Taylor and Francis, London.

Benitez T, Rincon AM, Limon MC, and Codon AC. 2004. Int. Microbiol 7: 55-59.

Bhagat S and Pan S. 2010. J Mycol Pl Pathol 40: 145-157.

Bhramaraba S and Nagamani A. 2010. Antagonistic efficacy of *Trichoderma* on the bakanae pathogen of rice. Abstract No. 34. National Seminar on Microbial diversity - Exploration and Bioprospecting, Dept. of Microbiology, Kakatiya Univ., AP. India 27th and 28th March 2010.

Bisby GR. 1939. Trans Br Mycol Soc 23: 149-168.

Bisset J. 1984. Can J Bot 62: 924-931.

Bisset J. 1991a. Can J Bot 69: 2357-2372.

Bisset J. 1991b. Can J Bot 69: 2373-2417.

Bisset J. 1991c. Can J Bot 69: 2418-2420.

Bisset J. 1992. Can J Bot 70: 639-641.

Bisset J, Szakacs G, Nolan AN, Druzhinina I, Gradinger C and Kubicek CP. 2003. Can J Bot 81: 570-586.

Bjorkman Y, Blanchard LM and Harman GE. 1998. J Am Soc Hortic Sci 123: 35-40.

Bowen RM and Harper SHT. 1990. Soil Biol Biochem 22: 393–399.

Bowen JK, Franicevic SC, Crowhurst RN, Templeton MD and Stewart A. 1996. New Zealand Journal of Crop and Horticultural Science 24: 207–17.

Cardoza RE. 2005. Secondary metabolites produced by *Trichoderma* and their importance in the biocontrol process. In: Mellado-Duran E, Barredo JL (eds) Microorganisms for industrial enzymes and biocontrol. Res Signpost, India, p 207.

Ch´avez-Gomez B, Quintero R, Esparza-Garcia F, Mesta, Howard AM, Zavala FJ, Hernandez-Rodriguez C, Gill´en T, Poggi-Varaldo HM, Cortes-Barrera J and Rodriguez-Vasquez R. 2003. Bioresource Technol. 89: 177–183.

Chakravarthy SK. 2008. Biological control of sheath blight pathogen of rice by the species of *Trichoderma* Pers. ex. Fr. Ph D Thesis submitted to Osmania Univerisity, Hyderabad. pp. 197.

Chakravarthy SK and Nagamani A. 2007. J Mycol Pl Pathol 37: 82-86.

Chakravarthy SK, Nagamani A, Rahel RK and Bhramaramba S. 2011. Adv Environ Biol 5: 2632-2638.

Chang YC, Chang YC and Baker Y. 1986. Pl Dis 70: 145-148.

Chet I, Barak Z and Oppenheim A. 1993. Genetic engineering of microorganisms for improved biocontrol activity; PP 628 in Biotechnology in Plant Disease Control Chet J (ed), Wiley Liss. Inc. New York.

Chet I, Viterbo A, Brotman Y and Lousky T. 2006. Enhancement of plant disease resistance by the biocontrol agent *Trichoderma*. Life Sciences. URL: http://www.weizmann.ac.il/

Cutler HG, Himmetsbach DS, Arrendale RF, Cole PD and Cox RH. 1989. Koninginin A Agri Biolog Chem 53: 2605–2611.

D'Urso A, Gapes D and Bravil M. 2008. Bioremediation of olive oil mill wastewaters by fungal (*Trichoderma viride*, strain 8/90) sequencing batch reactor. www.aidic.it/IBIC2008/webpapers/12D'Urso.pdf

da Silva M, Umbuzeiro GA, Pfenning LH, Canhos VP and Esposito E. 2003. Soil Sediment Conta 12: 345-356.

Dana MM, Limon MC, Mejias R, Mach RL, Benitez T, Pintor-Toro JA and Kubicek CP. 2001. Curr Genet 38: 335-342.

De La Cruz J and Llobell A 1999. Eur J Biochem 265: 145-151.

Deacon J. 2006. Fungal interactions: mechanisms and practical exploitation. In Fungal biology. 4th Edition. Blackwell publishing ltd. pp 240.

Dennis C and Webster J. 1971a. Trans Brit Mycol Soc 57: 363-369.

Dennis C and Webster J. 1971b. Trans Brit Mycol Soc 57: 25-39.

Dennis C and Webster J. 1971c. Trans Brit Mycol Soc 57: 41-48.

Dodd SL, Crowhurst RN, Rodrigo AC, Samuels GJ, Hill RA and Stewart A. 2000. Mycol Res 104: 23-34.

Druzhinina I and Kubicek CP. 2005. J Zhejiang Univ Sci 6B: 100–112.

Druzhinina IS, Kopchinskiy AG, Komoń M, Bissett J, Szakacs G and Kubicek CP. 2005. Fungal Gen Biol 42: 813-828.

Druzhinina IS, Kopchinskiy AG and Kubicek CP. 2006. Mycosci 47:55–64.

Elad Y and Kapat A. 1999. Eur J Pl Pathol 105: 177-189.

Fontaine S, Bardoux G, Benest D, Verdier B, Mariotti A, and Abbadie L. 2004. Soil Sci Soc America J 68: 125–131.

Gams W. 2006. *Hypocrea* and *Trichoderma* studies marking the 90th birthday of Joan M Dingley (Ed). Stu Mycol 56: 1-177.

Gams W and Bisset J. 1998. Morphology and identification of *Trichoderma*. In: *Trichoderma* and *Gliocladium* Vol I: Basic Biology, Taxonomy and Genetics. Kubicek CP and Harman GE (eds). Taylor and Francis, London, UK. pp 3-34.

Gams W and Meyer M. 1998. Mycologia 90: 04–915.

Ghisalberti EL and Sivasithamparam K. 1991. Soil Biol Biochem 23:1011–20.

Ghisalberti EL, Narbey MJ, Dewan MM and Sivasithamparam K. 1990. Pl Soil 121: 287–291.

Gómez I, Chet I and Herrera-Estrella A. 1997. Mol Gen Genet 256: 127-135.

Goyal S and Sindhu SS. 2011. Microbiol J 1: 126-138.

Grondona *et al* 1991. Appl Environ Microbiol 63:3189-3198.

Hagn A, Pritsch K, Schloter M and Munch JC. 2003. Biol Fertil Soils 38: 236-244.

Han W and He M. 2010. Bioresource Technol 101s: 3724–3731.

Hanson LE and Howell CR. 2004. Phtopathology 94: 171-176.

Haque MM, Haque MA, Ilias GNM and Molla AH. 2010. The Agriculturists 8: 66-73.

Harman GE. 2000. Pl Dis 84: 377-393.

Harman GE. 2011a. New Phytologist 189: 647-649.

Harman GE. 2011b Phytoparasitica 39: 103-108.

Harman GE and Kubicek CP. 1998. *Trichoderma* and *Gliocladium*. Vol. 2. Enzymes, Biological Control and Commercial Applications. London, UK: Taylor and Francis.

Harman GE and Mastouri F. 2010. Enhancing nitrogen use efficiency in wheat using *Trichoderma* seed inoculants, Vol. 7. St. Paul, MN, USA: Int Soc Pl-Microbe Interactions 1–4.

Harman GE, Howell CR, Viterbo A, Chet I and Lorito M. 2004a. Nature Microbiol Rev 2: 43-56.

Harman GE, Lorito M and Lynch JM. 2004b. Adv Appl Microbiol 56: 313-330.

Harman GE, Petzoldt R, Comis A and Chet J. 2004c. Phytopathol 94: 147-153.

Harman GE, Obregon MA, Samuels GJ and Lorito M. 2010. Changing models of biocontrol in the developing and developed world. Pl Dis 94: 928-939.

Helal GA. 2005. Mycobiol 33: 150-157.

Hermosa MR, Grondona I, Iturriaga EA, Diaz-Minguez JM, Castro C, Monte E and Garcia-Acha I. 2000. Appl Environ Microbiol 66:1890-1898.

Hermosa MR, Grondona I, Dı´az-Mı´nguez JM, Iturriaga EA and Monte E. 2001. Curr Gen 38: 343–350.

Hjeljord L and Tronsmo A. *Trichoderma* and *Gliocladium* in biological control: an overview. In: Kubicek CP and Harman GE (eds), *Trichoderma* and *Gliocladium*, Vol. 2. Enzymes, Biological Control and Commercial Applications. London, UK, Taylor and Francis Ltd.

Hoitink HAJ, Madden LV and Dorrance AE. 2006. Phytopathol 96: 186-189.

Howell CR. 2003. Pl Dis 87: 4-10.

Howell CR. 2006. Phytopathol 96: 178–180.

Ihrmark K, Asmail N, Ubhayasekera W, Melin P, Stenlid J and Karlsson M. 2010. Evol Bioinform Online. 156:1-26.

Jaklitsch WM. 2009. Stud Mycol 63: 1-91.

Jalk D, Nerud R and Siroka P. 1998 Folia Microbiol 43: 687-689.

Javaid A and Ali S. 2011. Herbicidal activity of culture filtrates of *Trichoderma* spp. against two problematic weeds of wheat. Natural Product Res 25: 730-740.

Kaewchai S, Soytong K and Hyde KD. 2009. Mycofungicides and fungal biofertilizers. Fungal Div 38: 25-50.

Katayama A and Matsumura F. 1993. Environ Toxicol Chem 12: 1059–1065.

Kausar H, Sariah M, Mohd Saud H, Zahangir Alam M and Razi Ismail M. 2010. Int Biodet Biodegra 64: 594-600.

Kevei F and Nagy E. 2003. Food Technol Biotechnol 41: 37–42.

Kim DJ, Bae JM, Uribe P, Kenerley CM and Cook DR. 2002. Curr Genet 40: 374-384.

Knight SC, Anthony VM, Brady AM, Greenland AJ, Heaney SP, Murray DC, Powell KA, Schulz MA, Spinks CA, Worthington PA and Youle D. 1997Ann Rev Phytopathol 35: 349–72.

Kopchinskiy AG, Komoń M, Kubicek CP and Druzhinina IS. 2005. Mycol Res 109: 658-660.

Kredics L, Antal Z, Doczi I, Manczinger L, Kevei F and Nagy E. 2003. Acta Microbiologica et Immunologica Hungarica 50: 105-117.

Kubicek, CP and Harman GE.1998. *Trichoderma* and *Gliocladium,* Vol I & II. Taylor and Francis Ltd., London, U.K.

Kubicek CP and Penttila ME. 1998. Regulation of production of plant polysaccharide degrading enzymes by *Trichoderma*. In: Harman GE and Kubicek CP (eds). *Trichoderma* and *Gliocladium*, Vol. 2: Enzymes, Biological Control and Commercial Applications. London: Taylor and Francis 49-71.

Kubicek CP, Komon-Zelazowska M and Druzhinina IS. 2008. J Zhejiang Univ Sci B 9: 753-763.

Kuhls K, Lieckfeldt E, Samuels GJ, Meyer W, Kubicek CP and Börner T. 1997. Mycologia 89: 442-460.

Kuhls K, Lieckfeldt E, Borner T and Gueho E. 1999. Medical Mycol 37: 25-33.

Kulkarni M, Chaudhari R and Chaudhari A. 2007. Novel tension-active microbial compounds for biocontrol applications. In: General concepts in Integrated Pest and Disease Management (eds: A. Ciancio and K.G. Mukerji). Springer: 295-304.

Kullnig-Gradinger CM, Szakács G and Kubicek CP. 2002. Mycol Res 106: 757-767.

Kumar R and Shweta. 2011. Bioresou Technol 102: 1475-80.

Lieckfeldt E and Seifert KA. 2000. Stud Mycol 45: 35-44.

Lieckfeldt E, Kullnig C, Samuels GJ and Borner T. 2001. Mycol Res 105: 313-322.

Liu M, Sun X, Zhu J, Xu T, Harman GE and Lorito M. 2004. J Zhejiang Univ Sci 5: 133-136.

Lorito M, Harman GE, Hayes CK, Broadway RM, Tronsmo A, Woo SL and Di Pietro A. 1993. Phytopathol. 83: 302-307.

Lorito M, Farkas V, Rebuffat S, Bodo B, Kubicek CP. 1996a. J Bacteriol 178: 6382–6385.

Lorito M, Woo SL, D'Ambrosio M, Harman GE, Hayes CK, Kubicek CP and Scala F. 1996b. Molecular Pl–Microbe Interaction 9: 206-213.

Lorito M, Woo SL, Harman GE and Monte E. 2010. Ann Rev Phytopathol 48: 395–417.

Lotan T and Fluhr R. 1990. Pl Physiology 93: 811-817.

Lübeck M, Poulsen SK, Lübeck PS, Jensen DF and Thrane U. 2000. FEMS Microbiol Lett 185: 129-134.

Lumsden RD, Lewis JA and Fravel DR. 1993. Formulation and delivery of biocontrol agents for use against soilborne plant pathogens. In: Hall FR and Barry JW (eds). Biorational Pest Control Agents: Formulation and Delivery. American Chemical Society, Washington, pp. 166-188.

Marra R, Ambrosino P, Carbone V, Vinale F, Woo SL, Ruocco M, Ciliento R, Lanzuise S, Ferraioli S, Soriente I, Gigante S, Turra D, Foglisno V, Scala F and Lorito M. 2006. Curr Gen 50: 307-321.

Mastouri F, Bjorkman T and Harman GE. 2010. Phytopathol 100: 1213–1221.

McIntyre M, Nielsen J, Arnau J, van der Brink H, Hansen K and Madrid S. 2004. Proceedings of the 7th European Conference on Fungal Genetics. Copenhagen, Denmark.

Metcalf DA and Wilson CR. 2001. Pl Pathol 50: 249-257.

Migheli Q, Gonzalez-Candelas L, Dealessi L, Camponogara A and Ramon-Vidal D. 1998. Phytopathol 88: 673-677.

Mohiddin FA, Khan MR, Khan SM and Bhat BH. 2010. Pl Pathol J 9: 92-102.

Monte E. 2001. Intl Microbiol 4: 1-4.

Morales-Barrera L and Cristiani-Urbina E. 2006. Enzyme Microbial Technol 40: 107-113.

Muthumeenakshi S, Mills PR, Brown AE and Seaby DA. 1994. Microbiol 140: 769–777.

Nagamani A, Manoharachary C, Agarwal DK and Chowdhary TN. 2002. Monographic contribution on *Trichoderma*. Associate publishers, New Delhi, 47 pp

Nagamani A, Kunwar IK and Manoharachary C. 2006. Handbook of soil fungi. I K Int Pvt Ltd, New Delhi. pp 409-475.

Nannipieri P. 1994. The potential use of soil enzymes as indicators of productivity, sustainability and pollution. In: Soil Biota-Management in Sustainable

Farming Systems, ed. Pankhrust CE, Doube BM, Gupta VVSR, and Grace PR. 238-244 pp. Australia: CSIRO.

Omero C, Inbar J, Rocha-Ramirez V, Herrer-Estrella A, Chet I and Horwitz BA. 1999. G protein activators and cAMP promote mycoparasitic behavior in *Trichoderma harzianum*. Mycol Res 103: 1637-1642.

Overton BE, Stewart EL and Geiser DM. 2006. Stud Mycol 56: 39-65.

Papavizas GC. 1985. Ann Rev Phytopathol 23: 23-54.

Persoon CH. 1794. RÖmer's Neues Mag. Bot. 1: 81-128. (original not seen).

Prasad MNV. 2011. A State-of-the-Art report on Bioremediation, its Applications to Contaminated Sites in India. Publication of Ministry of environment and forests, Govt. of India. pp 1-88.

Rahel RK and Nagamani A. 2011. Mechanisms of biocontrol gaents in controlling the collar rot of *Mentha arvensis* L. Abstract No. 17. National seminar on Recent trends in Microbial technology. Dept. of Microbiology. Kakatiya Univ. A.P., India 26th and 27th Match 2011.

Reino JL, Guerrero RF, Hernandez-Galan R and Collado IS. 2008. Secondary metabolites from species of the biocontrol agent *Trichoderma*. Phytochem Rev 8: 89-123.

Richter S, Cormican MG, Pfaller MA, Lee CK, Gingrich R, Rinaldi MG and Sutton DA. 1999. J Clinical Microbiol 37: 1154-1160.

Rifai MA .1969. Mycological Papers. No 1-116. CMI, Kew, England.

Rifai MA and Webster J. 1966a. Trans Brit Mycol Soc 49: 289-296.

Rigot J and Matsumura F. 2002. J Environ Sci Health. Part B-Pesticides, Food Contaminants, and Agricultural Wastes 37, 201–210.

Samuels GJ. 1996. Mycol Res 100: 923-935.

Samuels GJ. 2006a. Phytopatho 96: 195-206.

Samuel GJ. 2006b. ISTH subcommission on *Trichoderma* and *Hypocrea*. http//www.ISTH.info

Samuels GJ, Petrini O, Kuhls K, Lieckfeldt E and Kubicek CP. 1998 Stud Mycol 41: 1-54.

Samuels GJ, Pardo-Schultheiss R, Hebbar P, Lumsden RD, Bastos CN, Costa JC and Bezerra JL. 2000. Mycol Res 104: 760-764.

Samuels GJ, Dodd SL, Gams W, Castlebury LA and Petrini O. 2002. Mycologia 94: 146-170.

Samuels GJ, Dodd SL, Lu BS, Petrini O, Schroers HJ and Druzhinina IS. 2006. Stud Mycol 56: 67-133.

Schuster A and Schmoll M. 2010. Appl Microbiol Biotechnol 87: 787-799.

Shoresh M and Harman GE. 2008. Plant Signaling and Behavior 3: 737-739.

Shoresh M, Harman GE and Mastouri F. 2010. Ann Rev Phytopathol 48: 21-43.

Sinegani AS, Emtiazi G, Hajrasuliha S and Shariatmadari H. 2005. African J Biotechnol 4: 1058-1061.

Sivan A and Chet I. 1989. Degradation of fungal cell walls by lytic enzymes of *Trichoderma harzianum*. J Gen Microbiol. 135: 675–82.

Sivasithamparam K and Ghisalberti EL. 1998. Secondary metabolism in *Trichoderma* and *Gliocladium*. In *Trichoderma* and *Gliocladium*. Vol I Kubicek CP and Harman GE (Eds). Taylor and Francis, London, pp 139-191.

Srivastava PK, Vaish A, Dwivedi S, Chakrabarty D, Singh N and Tripathi RD. 2011. Sci Total Environ. 15: 2430-2442.

Tejada M, Gonzalez JL, Garc´ıa-Mart´ınez AM and Parrado J. 2008 Biores Technol 99: 4949–4957

Tengerdy RP and Szakacs G. 2003. Biochem Engg J 13: 169-179.

Thanh NC and Simar RE. 1973. Mycopathol Mycol Applica 51: 223-232.

Thrane U, Poulsen SB, Nirenberg HI and Lieckfeldst E. 2001. FEMS Microbiol Letts 203: 249-255.

van Wyk JPH and Mohulatsi M. 2003. Biores Technol 86: 21-23.

Vargas WA, Crutcher FK and Kenerley CM. 2010. New Phytologist 189: 777–789.

Vinale F, D'Ambrosio G, Adadi K, Scala F, Marra R, Turra D, Woo SL and Lorito M. 2004. J Zhejiang Univ Sci 30: 2-8.

Vinale F, Sivasithamparam K, Ghisalberti EL, Marra R, Woo SL and Lorito M. 2008a. Soil Biol Biochem 40: 1-10.

Vinale F, Sivasithamparam K, Ghisalberti EL, Marra R, Barbetti MJ, Li H, Woo SL and Lorito M. 2008b. Physiological Mol Pl Pathol 72: 80-86.

Viterbo A, Haran S, Friesem D, Ramot O and Chet I. 2001. FEMS Microbiol Lttrs 200: 164-174.

Webster J. 1964. Trans Brit Mycol Soc 47: 75-96.

Weindling R. 1932. Phytopathol 24: 837-845.

Whipps JW. 1997. Adv Botanical Res 26: 1-134.

Whipps JM. 2001. J Experimental Bot 52: 487-511.

Windham MT, Elad Y and Baker R. 1986. Phytopathol 76: 518-521.

Woo SL and Lorito M. 2007. Exploiting the interactions between fungal antagonists, pathogens and the plant for biocontrol. In: Vurro M, Gressel J (Eds.), Novel Biotechnologies for biocontrol agent enhancement and management. IOS, Springer Press, Amsterdam, The Netherlands, pp. 107-130.

Woo SL, Scala F, Ruocco M and Lorito M. 2006. Phytopathol 96: 181-185.

Yedidia I, Benhamou N, Chet I. 1999. Appl Environ Microbiol 65: 1061–1070.

Yedidia I, Srivastava AK, Kapulnik Y and Chet I. 2001. Pl Soil 235: 235-242.

Yildrim E, Taylor AG and Spittler TD. 2006. Sci Hortic 111: 1-6.

Zayed G. and Abdel-Motaal H. 2005. Biores Technol 96: 929-935.

Zimand G, Elad Y and Chet I. 1996. Phytopathol 86: 1255–60.

Chapter - 17

Sclerotium rolfsii- A potential mycoherbicide for *Parthenium histerophorus*

A.K. Pandey and Rekha Shukla

Department of Biological Sciences, Rani Durgavathi University, Jabalpur - 482 001
Email : Akprd7@yahoo.com

ABSTRACT

Parthenium hysterophorus L. is an exotic, obnoxious and pernicious weed posing serious threat to human livestock and our biodiversity. The universal occurrence and the detrimental effects of the weed necessitate its eradication from crop/forest lands and other location of public uses. It is very difficult to control through conventional methods due to one or other reason. During the recent past, several fungal pathogens as biocontrol agents are well established and have been patented and few of them have been successfully commercialized as mycoherbicides. *Sclerotium rolfsii* a soil borne, necrotrophic pathogen an isolate of it incites severe collar rot disease in *Parthenium.* It has shown very strong mycoherbicidal potential under natural condition, but broad host range of the species creates doubt about its suitability as mycoherbicide. Secretion of oxalic acid and endopolygacturonase concomitantly with rapid mycelial growth appears to be the key requirement for establishing infection. There has been a rapid increase in using molecular techniques for the identification of fungi at genus, species and subspecies level. Recent advances in molecular techniques have enabled inter and intraspecific grouping of fungi with few morphological differences. Suitable mass production and formulation strategies are required to evaluate the bioherbicidal potential of a *Sclerotium rolfsii* in the field conditions. For mass production solid substrate fermentation (SSF) liquid /submerged fermentation (SmF) and biphasic fermentation have been recommended by many workers.

Key words: *Parthenium hysterophorus, Sclerotium rolfsii*, Mycoherbicide

1. Biology of target weed: Parthenium

Carrot weed, *Parthenium hysterophorus* L. (Family *Asteraceae)* is an alien, annual herbaceous plant supposed to have originated in North East Mexico and was accidentally introduced in India during 1956. It was first time observed and identified by Rao (Chandra and Vartak, 1970) and since then has spread like wild fire and is present in almost every corner of India (Oudhia, 2000, a,b,c,d; 2001 a and b; Oudhia and Tripathi 2000). In a couple of years it has covered waste lands, railway yards, marshy patchy, unused cultivable lands, grasslands, roadsides along the canals and other areas (Oudhia and Tripathi, 2000). Serious infestation in economically important agricultural crops like sugarcane, sunflower, maize, wheat, chickpea, onion, garlic, sorghum etc. and orchards have also observed. It is also known by certain other names like fever few, congress weed, star weed, carrot, white cap, white top, bitter weed, false ragweed and also by different vernacular names like Chatak chandni, broom brush, oradi, Gajari, phandriphuli, Nakshathra gida, Vyori barina and safed topi (Evans, 1997).

1.1 Ill effects of Parthenium

Parthenium is Poisonous, pernicious and aggressive weed posing a serious threat to human being and live stocks. It causes severe human and animal health problems. Sesquiterpene lactones derived from plants of Asteraceae are responsible for contact dermatitis (Hosmani, 1995). The active principle is parthenin. In addition, chemical inhibitors like caffeic acid and p-coumaric acid are also responsible for skin allergies like irritation, contact dermiatitis phytophotodermatitis, itching and pigmentation of the skin (Tower and Suba Rao, 1992; Rao *et al.*, 1977). Respiratory allergies and air borne contact dermatitis are the other common complaints due to Parthenium conact (Rajashekhar, 2005). It is also responsible for the bitter disease in live stock. For the first time cases were reported in Texas during 1930 in USA and during 1973 in India (Mahadevappa, 1997). In India this weed causes yield losses up to 40% in several crops and is reported to reduce forage production up to 90-95%. Its invasion in forest areas and fallow lands has posed a threat to the plant biodiversity of medicinal plants in hilly region Jeevan Rao and Shantharam (1997).

1.2 Management of Parthenium

The universal occurrence and the detrimental effects of the weed necessitate its eradication from crop/forest lands and other location of public uses. Several methods have been derived to combat this. Management strategies comprise of following methods:

1.3 Physical methods

Hand weeding is still commonly used procedure throughout the world as means of weed control. Undoubtedly, it is an effective means of control but has certain serious short coming. It is tedious and unrewarding social activity. Eradication of weeds through cutting and burning have also been levied but within a few days of operation new plant sprout from the left over roots and stems (Kaur and Sharma, 1986). Health hazards are also one of the major draw back of the method. Manual removal is not cost effective and can be advocated only in limited situation. It is feasible only in high value crops. Moreover, the relief from this method is temporary and needs to be repeated. Mechanical removal with the help of tractor, plough etc is possible only up to a certain extent and only in open fields.

1.4 Chemical methods

Chemical herbicides are relatively new tools in man's fight against weeds. 2-4D a chemical herbicide has revolutionized the agricultural weed management. Several pre and post emergence herbicides have been recommended to control Pathenium in India (Chinnusamt *et al.* 2005; Kandaswamy, 2005). Recommended dose of these chemicals kill the standing Pathenium population and to some extent suppress immediate germination of the seeds deposited in the soil. Chemical control has supremacy over other measures on the score of the quick relief, time save and cost effectiveness. Research on chemical control was tried with promising results during the seventies and eighties. Widespread uses of chemical pesticides have led inevitably greater risks to users and consumer. Significant health and environmental hazards have arisen at specific non-agricultural sites as our requirements for synthetic chemical pesticide increased.

Transportation and disposal of toxic materials and waste products are an increasing risk to the country as our dependence upon chemical pesticides increase. As a result of outcries mentioned above, the discovery and development of an alternative to chemical pesticide is an essential agenda for agricultural scientist faced with daunting task to protect crops and livestock on a large scale from a wide range of ever shifting spectrum of agriculture. Biological control of weeds especially with fungal pathogens has fascinated the scientist world over as an effective and ecofriendly approach for many weed problems.

1.5 Biological control

During the last few years much emphasis has been laid on controlling Parthenium through biological agents like insects, pathogens and by crating competition that result in survival of the fittest. Biological weed control agents are organism that suppress or kill weedy plant without significantly injuring desirable plants (Templeton, 1979; Templeton and Freeman, 1987; Freeman, 1987; Trujillo, 1981; Evans, 2001). Efforts made by the scientists have reached to a point where following strategies have been clearly defined.

1.6 Mycoherbicide

Mycoherbicide is a component of a broad term "bioherbicide". According to Watson (1989) bioherbicide are the living entities used deliberately to suppress the growth or reduced the population of a weed speicies. The concept of mycoherbicide was first introduced by Daniel *et al.* (1970) who demonstrated that an endemic pathogen might be rendered completely destructive to its weed host by applying a massive dose of inoculums at a particularly susceptible stage of weed growth. Use of the pathogen in a product form and an application technique similar to the chemical tactic are the salient features distinguishing the mycoherbicides from classical agents.

Significant work on mycoherbicides to control *Parthenium hysterophorus, Lantana camara, Xanthium strumarium* etc. prominent weed of M.P. and Chattisgarh have also carried out by a group of worker at Jabalpur. Morethan 25 fungi associated with different parts of *Pathenium hysterophorus* have been isolated by Rajak *et al.* (1990). These include several necrotrophic pathogens and many of them viz. *Colletotrichum gloeosporioides* f.sp. *parthenii, C. dematium, Sclerotium rolfsii, Fusarium oxysporum, F. solani, Myrothecium roridum, Alternaria alternate and A. macrospore* have significant potential and satisfy the conditions required for mycoherbicidal agents for Parthenium (Pandey, 1998; Gayatri, 1998).

Herbicidal potential of these fungal metabolites have been also recorded against *Lantana camara* (Saxena, 2000; 2001). Saxena (2001) obtained very high activity in all the 3 fractions separated from CFCF for *A. alternata* (FGCC #25) against *L. camara*. Microbial toxins evaluated for promising herbicidal pontential have been enlisted in Table.1.

2. *Sclerotium rolfsii*: The biological agent

The genus *Sclerotium* (Class: *Deuteromycetes*; Sub Class: *Hyphomycetidae*; Order: *Agonomycetales*; Family: *Agonomycetaceae*) was described by Saccardo (1911) to include fungi which formed differential sclerotia and that had sterile mycelia. Many diverse fungi are included in this genus and they generally form small, tan to dark brown to black spherical sclerotia which are initially differentiated into a rind, cortex and medulla (Singleton *et al.*, 1992). *S. rolfsii* is a soil borne plant pathogen of world wide importance with a very extensive host range including more than 500 plant species (Cillieirs *et al.*, 2000; Harlton *et al.*, 1995; Punja, 1988;

1985). Most *S. rolfsii* disease has been reported on dicotyledonous host, but with sexual monocotyledonous species also being infected. It is rampant in subtropical and tropical region than in temperature regions (Yorinoni, 1994). Growth of *S. rolfsii* is optimal at 27-30°C and sclerotia do not survive at temperature below 0°C. The larger number of sclerotia produced by *S. rolfsii* and their ability to persist in the soil for several years, as well as prolific growth rate of the fungus (2-3 cm/dm culture) make it a well suited facultative parasite and a pathogen of major importance throughout the world. The first confirmed report of losses due to the pathogen was made by Rolfs in 1892 on tomato (*Lycoperiscon esculentum*) in Florida in the USA (Aycock, 1966).

2.1 Host tissue infection and pathogenicity

The extent of mycelia growth from sclerotia in soil is influenced by the form of germination (eruptive or hyphal) and by the presence of volatile compound, decaying plant tissues and soluble nutrients (Punja *et al.* 1984). The maximum distance from the host surface over which mycelium form an erruptively germinating sclerotia can grow and infect without an exogenous nutrient source (competence distance) is a function of many factors such as fungal isolates, soil moisture texture and depth (Punja, 1985). Mycelium from erruptively germinating sclerotia can infect host tissue without requiring an exogenous food base of non-living organic matter provided they are located with in the competence volume. For hyphally germinating sclerotia to infect, a food base must indeed be provided because mycelia growth is sparse (Punja & Grogan, 1981).

2.2 Mechanism of infection

A sequence of events takes place between the initial mycelia contact of host and the maceration and death of the host tissue (Bateman and Beer, 1965; Punja *et al.* 1985). Sclerotium is a very advance pathogen i.e. death of host tissues occurs prior to hyphal penetration due to the production of large quantity of oxalic acid and polygalacturonases (Bateman, 1972) which act in concert (Bateman & Beer, 1965; Punja *et al.*, 1985). Oxalic acid sequesters calcium from the cell walls to form calcium oxalate and lowers the tissue pH to the optimum for endopolygalacturonases and cellulase activity (Punja *et al.*, 1985). Pathogen also produces cellulases that play a secondary role in tissue destruction and disease development (Wood, 1960; Bateman, 1968; 1969, 1970). Fungal hyphae grow through tissues both inter and intracellularly. They contain numerous micro bodies (Hansseler *et al.*, 1978) presumed site of oxalate synthesis. Pectic substances in host cell walls are depleted in advance of growing hyphae and peroxide is released. The end result is the typical water soaked, necrotic and macerated appearance of tissues infected by *S. rolfsii.*

Secretion of oxalic acid and endopolygalacturonase concomitantly with rapid mycelia growth appears to be the key requirement for establishing infection (Punja *et al.*, 1985).

2.3 Epidemiology

The spatial pattern of inoculums of *S. rolfsii* in naturally infected fields is clustered or clumped. Disease incidence due to *S. rolfsii* may increase following periods of temperature and moisture fluctuation, cycles of drying and wetting have been reported to stimulate germination of sclerotia. The presence of an organic substrate for mycelia growth may enhance disease severity (Mac Carter and Kays, 1984; Javed and Coley Smith, 1973; Milthorpe, 1941). Temperature and moist conditions appear to enhance disease development regardless of whether organic matter is present. Initial infection occurs at soil level, where sclerotia are most likely to be stimulated to germinate by drying and remoistening. Several factors like temperature moisture and biotic factors are known to influence survival of sclerotia. Temperature above 50ºC for extended period are lethal to sclerotia which survive better in moist soil near the surface rather than buried while mycelium survives better in sandy soils than in fine textured soils (Chattopadhyay and Mustafie, 1977). Oxalic acid produced by this fungus might be playing a key role in ligin bio-degradation through stimulation of lignin degrading enzymatic activities. Oxalate might be induces foliar wilting during fungal infection by manipulating guard cells.

2.4 Production of oxalic acid

Sclerotium rolfsii synthesizes and secretes millimolar concentrations of oxalic acid (ethanedioic acid) which is mainly responsible for its pathogenicity. Major pathway for oxalate biosynthesis involves the cleavage of isocitrate to succinate and glyoxylate by isocitrate lyase (Schilling and Coworkers, 2000) which is present in the extract of hyphae synthesizes glyoxylate dehydrogenase. Glyoxalate dehydrogenase is the major enzyme involved in oxalate biosynthesis by *S. rolfsii* (Maxwell and Bateman, 1968).

2.5 Sexuality and genetics

The teleomorph or the sexual state is described as *Athelia rolfsii* by Curzi and Tu and Kimbrogh (1978). The sexual state as not occur commonly in nature. It forms on media that are poor nutrients in older culture grown at low light intensity and following the growth of mycelium from a rich to a poor medium. Although reports of the occurrence of the basidial state are infrequent basidiospores can infect host tissue under greenhouse conditions.

2.6 Variability and vegetative compatibility

The term vegetative compatibility refers to the ability of individual strains of the same fungal species to undergo mutual asexual hyphal anastomosis which resulted in viable fused cells containing nuclei of both parental strains in a common cytoplasm (Singleton *et al.*, 1992). Compatible pairings are characterized by formation of anastomosis bridge formed from growth of lateral branches followed by fusions and the individual mycelia cannot be discovered therefore belong to the same vegetative compatibility group (VCG). The occurrence of the same VCG on different host species or from widely different geographic areas reflects the wide host range or spread of the pathogen (Cilliers *et al.* 2000). Also the fact that isolates from within a given geographical area the diverse suggest that genetic changes have occurred within sub-population. Vegetative compatibility groups have been used to estimate genetic variability or relatedness with in *many fungal species such as Fusarium field population* (Chulze, 2000) Cryphonectria parasitica (Murill) Barr (Powell, 1995). *Verticillium dehliae* Kleb (Dobinson *et al.*, 1998). *Cryph-onectria cubensis* (Van Heerdeen and Wingfeild, 2001). Rhizoctonia (Adam, 1998) and *S. rolfsii* (Punja *et al.*, 1985, Cilliers *et al.*, 2000, Harlton *et al.*, 1995). Incompatible isolates can be characterized by formation of a zone of antagonism (or barrage zone) as a result of disruption and plasmolysis of the hyphae of both isolates which results in thinning out of the mycelium in the zone of interaction.

2.7 Molecular characterization

There has been a rapid increase in using molecular techniques for the identification of fungi at genus, species and subspecies level. Recent advances in molecular techniques have enabled inter and intraspecific grouping of fungi with few morphological differences (Bruns *et al*. 1992). Multilocus modular markers, including restriction fragment length polymorphism (RFLPs) (Elias *et al.*, 1993) and random amplified polymorphic DNAs (RAPDs) based on polymerase chain reaction (PCR) have also proven to be useful for characterizing some populations of fungal pathogens. Another PCR based method, amplified fragment length polymorphism (AFLP) has been used to characterize fungal populations (Gonzalez *et al*. 1998). DNA amplification technique through PCR can be used not only to multiply specific regions of the DNA but also to study fungal population biology (White *et al.*, 1990). The internal transcribed spacer (ITS) region of the ribosomal DNA of fungi has been used to detect intra specific variation in fungal isolates. Harlton *et al*. (1995) used RFLP of ITS region of rDNA and reported the presence of genetic variation in *S. rolfsii*. High DNA sequence similarity was among isolates belonging to same vegetative compatibility groups in Rhizoctonia isolates, but only little similarity when different AGs were compared to each other (Adams, 1998). ITS DNA sequence has been employed to help identify some 60 fungal isolates including *S.rolfsii* associated with cotton diseases in the field (Lyon and Becerra – Lopez Lavalle, 2000). Nalim *et al.* (1995) used molecular markers & detected that all *S. rolfsii* isolates within in an MCG produced identical patterns and that some MCGs shared the same pattern for ITS region.

2.8 Biochemical characterization

Morphological identical organism some time behaves differently and it is very difficult to categorize them. Therefore, analysis of extra-cellular, intracellular isozyme and total protein pattern proved to be very useful for categorization of morphological identical organisms (Paterson, 1989). They determine the diversity of protein in ecologically variant isolates and it is definitely related to alleles or loci (Sweetinghan *et al.*, 1986). Biochemical diversity is interrelated to genetic diversity because it reveals the diversity of particular alleles or loci. Isozyme analysis is a versatile and inexpensive technique which can be used to determine the account of genetic variation among organisms. The interpretation of banding pattern interns of specific alleles allows the determination of ratio of alleles the determination of ratio of alleles expressed and common among individuals. These ratios are an excellent mean of determining phylogenetic relationship among organism.

Isozyme analysis is frequently used by many workers to solve the taxonomic problems of very closely related isolates of the same species. This technique is commonly used to make on the separation or combination of species, subspecies, varieties and inter sterility groups have also been identified.

Isozyme analysis studies are also used by several other workers for discrimination of fungi (Sweetinghan *et al.*, 1986). Cruickshank (1990) clearly differentiated various isolates of Rhizoctonia by pectic zymography. Cruickshank and Pitt (1987) have recommended the extra cellular isozyme pattern for characterization of *Penicillium* spp. Upadhyay *et al.* (2003) differentiated ecologically different *Agaricus bisporus* isolates on the basis of amylase and protease alleles into two categories.

2.9 Phytopathogenic potential

The pathogen is known to incite severe infection primarily in the stem of plants. However, it may infect any part of the plant under unfavorable conditions including roots, fruits, seeds, leaves and flowers. An indigenous strain of *S. rolfsii* is reported to cause collar rot disease of the *Parthenium* (Pandey *et al.*, 1990). When the fungus attacks plant that have already developed some hard tissues it doesn't invade throughout the stem, but grow into the cortex and slowly or quickly grides the stem which eventually dies. First sign of infection in mature plant is the formation dark brown lesions on the stem at or first beneath the soil line. The disease is characterized by progressive yellowing and wilting to leaves which is followed by shriveling of collar regions of the stem. Infection extents rapidly and attains a height of more than 10 cm. The pathogen produce abundant white fluffy cottony mycelium which covers the entire infected areas and subsequently several mustard red like dark reddish brown sclerotia form on the surface of mycelial tuff. Lesions and eventually dissimilar growth gradually girdle tissues of stem of mature plant and sclerotial formation can also frequently observed on seed and leaves of the weed.

2.10 Mass production

Suitable mass production and formulation strategies are required to evaluate the bioherbicidal potential of a *S. rolfsii* in the field conditions. For mass production solid substrate fermentation (SSF) liquid/submerged fermentation(SmF) and biphasic fermentation have been recommended by many workers (Singh *et al.*, 2000; Han and Yuan, 2003). A brief account of these is as follows:

Solid substrate fermentation

Solid substrate fermentation is defined as the growth of a micro-organism in a layer of moist solid substrate, with no free-moving water present, but with enough moisture present for the growth and meta-bolism of the microorganism. There are several inherent problem associated with solid substrate fermentations. Among these are high labor costs, difficulties in maintaining sterility, lack of control on fermen-tation conditions and recovery of the spores from the substrate (Churchill, 1982). These problems are in large part responsible for the development of a highly developed submerged culture fermentation industry (Gray, 1986). Many fungi fails to produce spores in submerged culture, solid substrate fermentation may sometimes offer the only method of spore production. This technique has slightly modified by the workers where the agent first mass produce in liquid conditions and then transferred to solid substrate. Solid substrate provides both nutrient and base for growth and sporulation (Singh, 2002).

Liquid fermentation

From the practical and economic standpoint, fungi that sporulate in liquid culture are favoured over those that require additional steps to induce sporulation, and this factor may prove to be an essential requirement for the commercial development of a fungus as a mycoherbicide (Bowers, 1982, 1986).

For early evaluations and in small-scale experiments, such as virulence and host range experiments, ample inoculum can usually be produced in shake flasks. However, with shake flasks it is difficult to control many of the fungal growth parameters that influence mycelial growth or sporulation, such as pH, temperature, agitation and aeration.

2.11 Formulation

Formulation is the blending of active ingredients, such as fungal spores, with inert carriers, such as diluents and surfactants in order to alter the physical characteristics to a more desirable form. They may include diluting to a common potency, enhancing stability and/or biological activity, improving mixing and spray ability, and possibility integrating the mycoherbicide into a pest management system (Daigle *et al.*, 1990). Generally following two types of formulations have been recommended.

Solid-based formulations

Pathogens that infect at or below the soil line are probably delivered best in a solid or granular formulation. Granular formulation are better suited for use as preplant or preemergence mycoherbicides because (1) the granules provide a buffer from environment extremes; (2) the granules can serve as a food base for the fungus, resulting in a longer period of persistence; (3) the granules are less likely to be washed away from the treated areas than are spores.

Sprayable or liquid formulation

A suspension of a bioherbicide agent in water would be the simplest spray formulation because water is inexpensive, readily available, easy to handle, and necessary to maintain life for all plant pathogens. Many weeds are covered with a waxy cuticle that prevents water from spreading evenly, and this prevents an equal distribution when aqueous materials, such as mycoherbicides, are applied.

Surfactants aid in wetting the plants and in dispersing the fungal spores throughout the spray mix. Because spores of mycoherbicedes are finite units, it is of paramount importance that the surface area be covered with the material as evenly and equally as possible. This formulation did not remove the usual obstacle in making a foliar-applied mycoherbicide effective: a long dew period greater than 8h (Auld, 1993; Walker, 1981a). Because foliar-applied mycoheribicides are exposed to rain, abrasion, ultraviolet rays and dessication, invert emulsion formulation have been developed to protect the pathogen until infection of the target weed. Quimby *et al.* (1998) developed the first invert emulsion formulation for mycoherbicides. Invert emulsion are water-in-oil emulsion where water is the dispersed phase and oil is the continuous phase. The oil phase consisted of a paraffinic oil, paraffin wax, soybean oil, and soybean lecithin as an emulsifier. The water phase contained humectants and surfactants. The oil phase and water phase were mixed in ratios of 5:6 to 2:1, respectively. Water evaporated slowly from the very viscous formulations. Paraffin wax is an effective adjuvant for reducing the rate of water evaporation from invert emulsion spray droplets. These viscous formulations were sprayed with a special air assist nozzle. Various liquid formulations tried for many agents are listed in Table.2.

Table 2. Liquid based formulations of mycoherbicide

Weed host	Pathogen	Formulating reagents
Velvet leaf	*Fusarium lateritium*	Water + Tween-20 surfact ant (0.02%) Experimental formulation - water
	Colletotrichum coccodes	Water+Sorbitol (0.075%)
Northern jointvetch	*C.gleosporioides* f. sp. *aeschynomene*	Comercial formulation component A; dried spores, component B; dehydrating agent+surfactant
Spurred anoda	*Alternaria macrospore*	Water+nonoxynol surfacta- nt (0.02%; Sucrose (5% w/v)
Giant ragweed	*Protomyces gravidus*	Water
Field bindweed	*Phomopsis convolvulus*	Water+Gelatin (0.1%)
Jimsonweed	*Alternaria cassiae*	Water+nonxynol surfactant (0.04%)
Florida beggarweed	*Colletotrichum truncatum*	
Siclepod	*Alternaria cassiae*	Water + nonoxynol surfac- tant (0.04%), Paraffin wax, mineral oil, Soybean oil, Corn syrup, lecithin
Common pusslane	*Dichotomophtohora portulaceaceae*	Water + Tween-20 surfactant (0.02%)
Hemp serbania	*Colletotrichum truncatum*	Water + nonoxynol surfact- ant (0.04%), Paraffin wax, mineral oil, Soybean oil, Corn syrup, lecithin

Eastern Black nightshade	*Colletotrichum cocoides*	Water + Tween-20 surfactant (0.02%)
Strangler vine	*Phytophthora palmivora (De Vine)*	Commerical formulation Chlamydospores in water
Horse purslane	*Gibbago trianthemae*	Water + Tween-20 surfactant (0.02%)
Water hyacinth	*Cercospora rodamanii*	Mycelial fragment + wettable powder
Malva pusilla	*Colletotrichum gloeosporioides* f. sp. *malvae*	Spores+silica gel

2.12 Field testing

Field evaluation is one of the most important components of a complete evaluation of the efficacies of pathogen as biological control agents for weeds. Although an organism may perform effectively in green house or growth chamber experiments. The use of the fungus in the field is often characterized by visible activity against the target weed (Singh *et al.*, 1995).

Mycoherbicidal control of weeds including Parthenium initiated at Jabalpur, yielded many pathogenic fungi including *Sclerotium rolfsii.* They incite severe infection and causes significant damage to the weed and thus have considerable potential as mycoherbicide (Mishra *et al.*, 1995, 1996). The pathogen was found to be highly effective against Pathenium at field trial using mycelial preparation with organic matter (Mishra, 1993). *Fusarium oxysproum* and *F.solani* tested against Parthenium successfully reduced the population by morethan 65% and also colonized on seeds. *Alternaria alternate* (FGCC#25) was tested in small scale plots provided significant control of *Lantana camara*. *Sclerotium rolfsii* Sacc. is a necrotrophic fungus responsible for severe color rot infection in many weeds including *Parthenium hysterophorus* (Pandey *et al.* 1998; Shukla & Pandey, 2006). Studies conducted by earlier workers revealed that the pathogen have significant mycoherbicidal potential and satisfied more or less all the requirement necessary for its development as mycoherbicide against this weed. Despite of its natural infection in natural condition broad host range of the species create doubt about its suitability as mycoherbicide. However, there are several reports where different isolates have shown significant variation in their morphology, biochemical and molecular characteristics (Sharma *et al.*, 2000; Cilliers *et al.*, 2000; Adandonon, 2004).

References

Adams, GC (1998) *Adv. Pl. Pathol.* 6: 535-552.

Adondonon, A (2004).Damping off and stem rot of cowpea in Benin, caused by *Sclerotium rolfsii.* Ph.D. Thesis. University of Pretoria, Pretoria.

Auld, B (1993) *Crop Protect.* 12:477-479.

Bateman, DF (1968) *Net. J. Plant Pathol.* 75:67-80.

Bateman, DF (1969) *Phytopathol.* 59:37-42.

Bateman, DF (1970) *Phytopathol.* 60:1846-1847.

Bateman, DF and Beer, SV (1965) *Phytopathol.* 55:204-221.

Bateman, DF (1972) Physiol. Pl. Pathol. 2:175-184.

Bowers, RC (1986) Weed Sci. 34:24-25.

Bowers, RC (1982) Commercialization of microbial weed control agents. In: Biological control of weeds with plant pathogen (Charudattan, R and Walker H.L. eds.) John Wiley, New York, Pp.157-173.

Bruns, TD, White TJ and Taylor JW (1992) *Ann. Rev. Eco. Sys.* 22:525-564.

Chandra, MEA, Vartak, VD (1970) *India PANS*, 16:212-213.

Chattopadhyay, SB and Mustafie, TP (1997). *Current Sci.* 46: 226-228.

Chinnusamy, C, Prabhakaran, NK, Shanmugasundaram, R., Senthil, A and Prabhu Kumar, G (2005). Post emergence herbicidal control of *Parthenium hysterophorus* under non cropped situation. Proc. 2nd Int. Conference on Parthenium Management, 3-5 Dec, University of Agriculture Sciences, Bangalore, India, Pp.147-149.

Chulze, SN, Ramirez, ML, Torres, A and Leslie, JF. *Appl. Environ. Microbiol.* 66(12): 5312-5315.

Churchil, BW (1982) Mass production of microorganisms for biological control. In: Biological control of weeds with plant pathogen. (R. Charudattan and H.L. Walker.eds.), Wiley, New York, Pp.139-156.

Cilliers, AJ, Herselman, L and Pretorius, ZA (2000) *African. Phytopathol.* 90:1026-1031.

Cruickshank, RH, and Pitt, JJ (1990) Isozyme variation in *Aspergillus flavus* and closely related species. In: Modern concepts in Penicillium and Aspergillus classification (Samson, RA and Pitt, JJ. eds.) Plenum Press, New York, Pp. 259-265.

Cruisckhank, RH, and Pitt, JJ (1987) *Mycologia* 79(4): 614-620.

Dobinson, KF, Patterson, NA, White, GJ, and Grant, S (1998) *Mycol.Res.* 102 (9):1089-1095.

Evans, HC (1997) Biocontrol News Information 18(3):89-98.

Evans, HC, Greaves, MP and Watson, AK (2001). Fungal Bioncontrol agents of weeds. In: Fungi as bioncontrol agents (Butt, TMG, Jackson, W and Magon, N. eds.) CABI Publishing, Wallingford Oxon, UK.Pp.169-192.

Freeman, TE (1987). Microbial herbicides. In: Advances in Agricultural Microbiology (Subba Rao, NS, ed.) Oxford and IBH Publishers Ltd. New Delhi, Pp. 419-428.

Gayathri, S (1998) Evaluation of some indigenous strains of *Colletotrichum dematium* for the management of *Parthenium hysterophorus* L. Ph.D. Thesis, R.D. University, Jabalpur.

Gonzalez, M, Rodriguez, MEZ, Jacob, JL, Hernandez, F, Acosta, J, Martinez, O and Simpson, J (1998) *Phytopathol.* 88: 292-299.

Gray, PP (1986). Mass production of microorganism, fermentation techniques and equipment. Workshop Proceedings: Potential for mycoherbicides in Australia. Agric. Research & Veterinary Centre, Orange. New South Wales.Australia Pp.4-7.

Han, Jian-Rang, Yuan, Jing Ming (2003) *J. Industrial Microbiol & Biotechnol.*

Hanssler,G, Maxwell, DP and Maxwell, MD (1978) *Phytopathol.* 92:157-167.

Harlton, CE, Levesque, CA and Punja, ZK (1995) *Phytopathol.* 85:1269-1281.

Javed, ZUR and Coley-Smith, JR (1973) *Trans. Br. Myco. Soc.* 60:441-451.

Jeevan Rao and Shantaram, MV (1997). Parthenium menace in soil treated with urban solid wastes In:Proceedings first Int. Conference on Parthenium management, Oct UAS, Dharwad, Karnataka, Pp. 20-21.

Kandasamy, OS (2005). Parthenium weed: Status and prospects of chemical control in India. In: Proceedings of International Conference on Parthenium Management, 5-7 Dec. University of Agriculture Sciences, Bangalore, India, Pp. 134-146.

Mac Carter, SM and Kays, SJ (1984) *Plant Dis.* 68: 299-302.

Mahadevappa, M (1997). Ecology, distribution, menace and management of Parthenium In: Proceedings of I[st] International Conference on Parthenium Management (Mahadevappa, M and Patil, VS. eds.) 6-8 Oct 1997, University of Agricultural Sciences, Dharwad, India, 1-12.

Maxwell, DP and Bateman, DF (1968) *Phytopathol.*68: 1635-1642.

Milthorpe, FL (1941) *Proc. Linnean. Soc.* N.S. Wales. 66: 65-75.

Mishra, J, Pandey, AK, Hasija SK (1995) *Ind. Phytopathol.* 48: 476-79.

Mishra, J, Pandey, AK, Hasija, SK. (1996a) *J. Pathol. Res.* 9(1): 19-24.

Nalim, FA, Starr, JL, Woodward, KG, Segner, S and Keller NP (1995) Phytopathol. 85: 1507-1512.

Oudhia, P (2000a) *Insect Environment* 5(4):168

Oudhia, P (2000b) *Int. Rice Res. Notes* (IRRN) 25(1):34.

Oudhia, P (2000c) *Crop Res.* 20(3):560-562.

Oudhia, P (2000d) *Crop Res.* 20(3): 563-566.

Oudhia, P and Tripathi, RS (2000) *Mahamaya Research on crops* 11(1):111-115.

Pandey, AK, Mishra, Jyothi and Hasija, SK (1998) *J. Mycol. Pl. Pathol.* 28:284-287.

Pandey, AK, Hasija, SK, and Rajak, RC (1990) *Nat. Acad. Sci. Letters* 13:369-370.

Paterson, RM, Bridge, PD, Crosswait,MJ and Hawksworth, DL (1989) *J. Gen. Microbiol.* 135:2979-2991.

Powell, WA (1995) *Mycologia* 87:738-741.

Punja, ZK (1985) *Phytopathol.* 23:97-127.

Punja, ZK (1998). *Sclerotium rolfsii,* a pathogen of many plant species. In: Advances in Plant pathology (ed. G.S. Sidhu) Academic Press. San Diego. Ca. 523-534.

Punja, ZK and Grogan, RG (1981) *Phytopathol.* 71:1099-1103.

Punja, ZK, Jenkins, S.F. and Grogan, RG (1984) *Phytopathol.* 74: 1290-1295.

Punja, ZK, Smith, VL, Campbell, CL, and Jenkins, SF. (1985) *Plant Dis.* 69:469-474.

Quimby, PC, Fulgham, FE, Jr., Boyette,CD, and Connick, WJ (1998). An invert emulsion replaces dew in biocontrol of sickle pod- a preliminary study. In: Pesticides, Formulation and application system. American Society for testing and Materials. Philadelphia. (Hovde, DA and Beestman, GB. Eds.) Pp. 264-270.

Rajak, RC, Farkya, S, Hasija, SK and Pandey, AK (1990) *Proc. Nat. Acad. India* 60:165-168.

Rajshekar, N (2005) Personal communication, Dept. of Skin and J.J.M. Medical College Devangere.

Saccardo, PA (1911) *Ann. Mycol.* 9:249-257.

Saxena, S, Pandey, AK, Rajak, RC (2000) *Indian Phytopathol.* 53(a):490-493.

Saxena, S, Pandey, AK and Rajak, RC (2001) Potential of an idiolite from Alternaria alternate as biorational agrochemicals In: Frontiers in Fungal Biotechnology and Plant pathogen Relationships (Manoharachary, C, Bagyanarayana, G, Bhadraiah, B, Satyaprasad, K, Reddy, BN and Nagamani, A. eds.) Allied Publishers, Hyderabad, Pp. 22-28.

Saxena, S, Pandey, AK (2001) *App. Microbiol. Biotechnol.*55:395-403.

Schilling, BM, Henning, A and Rao, U (2000) *Bioprocess Engineering* 22:51-55.

Singh, J (2002). Management of Parthenium by developed mycoherbicide from *Colletotrichum* sp. Ph.D thesis, RD.VV. Jabalpur.

Singh, SR, Singh, HV, Singh, Y (2000) Efficacy of *Trichoderma* sp. against *Rhizoctonia solani* and reduction of substates for their mass production. In: Proc of *Phytopathol Soc.*Golden Jublee 1:393-394.

Singh, US, Kohmoto,K and Singh, RP (1995). Pathonenesis and host specificity in plant disease: Histopathological, Biochemical, genetic and molecular bases, Pergmon Elsevier, Amsterdam.

Singleton, LL, Mihali, JD and Rush, CM (1992) *American Phytopathol. Soc. Minnesota.*USA

Shukla, R and Pandey, AK (2006) *Ann. Pl. Protect. Sc.* 14(1):202-205.

Swwetingham, MW, Cruisckhank, RH and Wong, DH (1986). Trans Br. Mycol. Soc. 86:305-311.

Templeton, GE, Te Beest, DO and Smith Jr. RJ (1979) *Ann Rev. Phytopathol.* 17:301-310.

Towers, GHN and Suba Rao, PV (1992). Impact of the pantropical weed, *Parthenium hysterophorus* L. on human affairs. In: Proceedings of the 1st International weed control congress Melbourne, Australia (ed. Richardson, RG) Weed Sci.Society of Victoria 134-138.

Tu, CC and Kimbrough, JW (1978) Bot. Gaz. (Crawfordsvilla) 139:4540466.

Upadhyay, MK, Pandey, AK and Rajak, RC (2003) *J. Basic and Appl. Mycol.* 2 (I & II): 77-80.

Van Heerdeen, SW and Wingfield, MJ (2001) *Mycol. Res.* 105:94-99.

Waler, HL (1981a) Weed Sci. 33:212-215.

White, TJ, Lee, T, Bruns, S and Taylor, J (1990). Amplification and direct sequencing of ribosomal RNA genes for Phylogenetics. In: PCR protocols (Innis, M.A, Gelfland, DH, Sninsky, JJ and White TJ. Eds.) Academic Press, New York.

Wood, RKS (1960) *Ann. Rev. Plant Physiol.* 11: 299-322.

Yorinori, JT (1994). Fungal disease In: Brazilian Agricultural Research Enterprise Tropical Soybean : Improvement and Production FAO, Rome.

Chapter - 18

Recent Developments in the Analysis of Mycotoxins Associated with Foods and Feeds

S.M. Reddy and S. Girisham

Department of Microbiology, Kakatiya University, Warangal
E-mail: sivasrigirisham@gmail.com

ABSTRACT

Mycotoxins, natural secondary metabolites of low molecular weight and exhibiting great structural diversity are elaborated by fungi on wide range of agricultural commodities under varying climatic conditions. Although mycotoxicoses caused by direct consumption of contaminated food and feed stuffs poses the greatest risk of animals and humans, the entry of mycotoxins or their metabolites in to food chain by carry over in to milk, animal tissues and egg should not be under estimated. Requirement of application of regulatory limits has prompted the development of vast number of analytical methods for detection and quantification of mycotoxins in food and feed at lowest concentration possible. The chemical diversity, varying concentrations, wide range of agricultural commodities, foods and biological samples, poses a great challenge to analytical chemist

The design of sampling procedures for different mycotoxins has been an international concern. Current methods include an extraction step, a cleanup step to reduce or eliminate unwanted co-extracted matrice components and a separation step with suitable specific detection ability. Modern cleanup techniques such as immmunoaffinity column (IAC) or solid phase extraction (SPE) methods help to simplify protocol, improve selectivity and performance characteristics. Screening of samples contaminated with many mycotoxins is frequently performed by TLC which yields qualitative and semi- quantitative results. ELISA are often used for rapid screening.

Biological detection methods are based on the toxic effects of mycotoxins used death, pathological lesions, changed biochemical events and immunological events or immuunotoxic effects as indices for toxicity. But these are not specific and their sensitivity is generally low and used only for screening for general toxicity. Physicochemical assay methods such as

TLC, HPLC, GC and HPLC-MS_MS are highly sensitive but are laborious, require expensive instrumentation and need cleanup of the samples. Thus, these are of limited use for making routine quantitative screening for mycotoxins in agricultural commodities.

To overcome the difficulties with biological and physicochemical methods, new immunological methods such as Radioimmunoassay (RIA), florescence polarization immunoassay (FPIA), enzyme immunoassay (EIA), direct competitive assay (dc-ELISA), indirect competitive assay (idc-ELISA) or double antibody assay, Immmunoaffinity chromatography (IAC), combination with HPLC or TLC, anti–idiotype antibody based immunoassay and biosensors proved to be highly sensitive for detection of mycotoxins.

The established state of the art chromatography based methods for mycotoxin analysis are increasingly being complemented by biosensors and infrared screening techniques that are fast and cost effective. The development of kits and simple tests requiring no instruments and improve detection capacity and measurement for detection of minute amounts of multiple mycotoxins is the need of the hour and important step towards safer foods and feeds.

Key words: Mycotoxins, Aspergillus, Penicillium, Fusarium, Fumonisins, Aflatoxin, Trichothecenes and Zearalenone, HPLC and TLC.

1. Introduction

Mycotoxins are natural, secondary metabolites produced by fungi on agricultural commodities in the field and during storage under a wide range of climatic conditions. About 250 different filamentous fungal species, mainly belonging to *Aspergillus*, *Penicillium* and *Fusarium*, have been identified. Several hundred different mycotoxins have been discovered so far, exhibiting great structural diversity, which results in different chemical and physicochemical properties. Aflatoxins and Ochratoxins (produced mainly by *Aspergillus* sp.), fumonisins, trichothecenes and zearalenone (produced by *Fusarium* sp.), patulin (produced mainly by *Penicillium* sp.), and ergot alkaloids (produced in the sclerotia of *Claviceps* sp.) received special attention due to their frequent occurrence and their severe effects on animal and human health (D'Mello and MacDonald, 1997). Mycotoxins are potent toxins and have a wide range of actions on animals and humans including cytotoxic-, nephrotoxic and neurotoxic, carcinogenic, mutagenic, immunosuppressive and estrogenic effects. Although mycotoxicoses caused by direct consumption of contaminated food and feed stuffs poses the greatest risk to animals and humans, the entry of mycotoxins or their metabolites into the food chain by "carryover" into milk, animal tissue or eggs should not be underestimated. National and international institutions and organizations, such as the European Commission (EC), the US Food and Drug Administration (FDA), the World Health Organization (WHO) and the Food and

Agriculture organization (FAO) of the United Nations, have recognized the potential health risks to animals and humans posed by food- and feed-borne mycotoxin intoxication and addressed this problem by adopting regulatory limits for major mycotoxin classes and selected individual mycotoxins. The FAO has compiled comprehensive worldwide regulations and directives regarding mycotoxins in food and feed as of December 2003 (FAO, 2004). The Joint Expert Committee on Food Additives (JECFA), a scientific advisory body of FAO and WHO, provides mechanisms for assessing the toxicity of food additives, veterinary drug residues and contaminants, has recently evaluated the hazards related to several mycotoxins, including fumonisins B1, B2 and B3, ochratoxin A, deoxynivalenol, T-2 toxin, HT-2 toxin, and aflatoxin M1 (WHO, 2002). The report explains the nature of each toxin, including its absorption and excretion as well as toxicological effects and it includes general considerations of analytical methods, sampling, associated intake issues and control mechanisms.The EC has set maximum levels for some mycotoxins, including aflatoxins, ochratoxin A, patulin, deoxynivalenol and zearalenone, T-2 and HT-2 toxins in different foods. The requirement to apply these regulatory limits has prompted the development of a vast number of analytical methods for the detection and quantification of mycotoxins in food and feed. Pittet (2005) and Kraska *et al.* (2008) have discussed recent developments in mycotoxins analysis of foods and feeds.

Moulds of phylogenitically unrelated are reported to elaborate same mycotoxins. Similarly some of these moulds elaborate more than one mycotoxins which are chemically unrelated .Thus the situation is complex and remained unexplainable in terms of biochemical makeup The elaboration of these mycotoxins in nature is equally complex and the conditions which lead to their elaboration still could not be understood precisely .The degree of their elaboration qualitatively and quantitatively could not be determined precisely. The natural sample which appears to be normal may be with sufficient amount of mycotoxins which might lead to health hazards. This is mainly because of most of these toxins are highly toxic and their concentrations in nano grams may cause irrepairable health hazards. Thus, there is a need to have precise methods of analytical detection and quantification. Further, they should not only be precise but also to be simple and economical in operation at the field level.

In view of its basic need, large numbers of techniques are in the process of development. However, many of them proved to be set with one or other drawbacks.

2. Methods of analysis

Different methods of mycotoxins analysis are available and the one(s) chosen by a laboratory will vary. Some laboratories may use more than one method, depending upon its circumstances. Cost of mycotoxin analysis depends partly upon the method used to detect their presence. When comparing cost between laboratories, it is important to know if the methods used by the laboratories are similar or different. Some methods of mycotoxin analysis are able to detect several different mycotoxins with one test, while others detect only one mycotoxin. Some are able to determine the concentration of mycotoxins present, while others cannot. Different methods adapted for the detection and estimation of different mycotoxins are précised in Fig. 1. Some laboratories may screen samples for the presence of mycotoxins, using a relatively fast, easy and cheap method of analysis. If that analysis finds a mycotoxin to be present, then the sample is re-analyzed using another method that confirms the presence of the mycotoxins and determine the amount present.

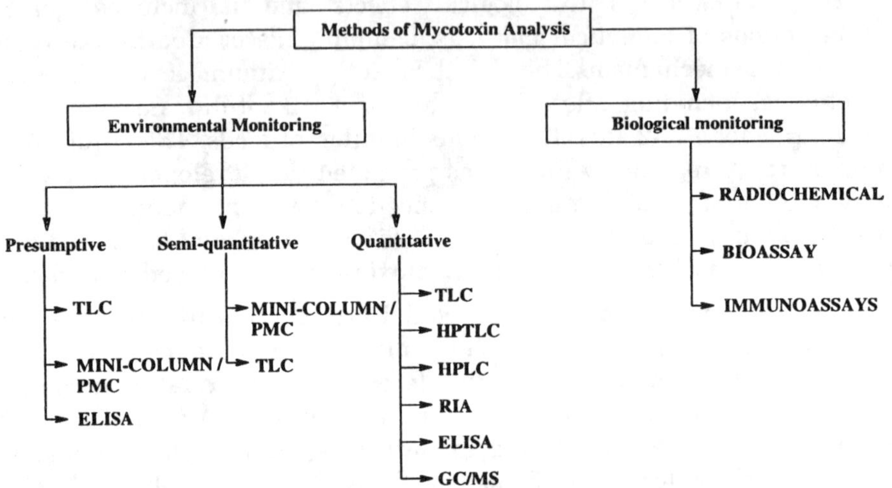

Fig. 1. Schematic presentation of different methods of mycotoxin analysis

The methods of mycotoxins analysis may be broadly grouped in to

a. **Physico-chemical analysis**- instrumental methods which involve sophisticated instruments and are highly sensitive.

b. **Biological analysis**- Detection of mycotoxins employing biological organisms is most widely used. These are useful in tracing source of known and unknown mycotoxins in food. However, their use in the surveillance of food and food stuffs for mycotoxins is of limited use.

c. **Immunoassays**- Biochemical separation techniques are promising and developing fast and expected to play an important role in mycotoxin detection methodology in the near future.

Two methods of analysis commonly used to detect aflatoxins at grain elevators are black light and enzyme-linked immunosorbent assay (ELISA). The black light test for aflatoxin is based on the glow (fluorescence) of aflatoxin under ultraviolet light, commonly called black light. Research has demonstrated that results obtained using this method is rarely falsely negative, but often falsely positive. In other words, if glowers are observed in an appropriately collected sample, aflatoxin may be present but one cannot say that with certainty. Other chemicals present in the grain may also cause it to glow. If no glowers are found, then it is very likely that aflatoxin is not present in significant concentration.

Failure to detect mycotoxins in the sample provided for analysis may be the result of sample selection such results do not prove mycotoxins are not present in other parts of the feed from which the sample was collected.

If glowers are found to be present, the concentration of aflatoxin present in a sample cannot be reliably determined by counting the number of glowers present under black light. Samples judged to be positive for aflatoxins by black light should be analyzed by a more reliable method to confirm the presence of aflatoxin and determine the concentration present.

Enzyme-linked immunosorbent test kits are available for aflatoxin, fumonisin, vomitoxin and zearalenone. The kits test for only one mycotoxin, so a separate kit is necessary for each. They are based upon the interaction between the mycotoxin and antibodies produced against the mycotoxin. Antibodies against the mycotoxin are bound to the sides and bottom of a plastic well. Mycotoxins in a sample of *grain* are disso-lved into a solution, which is placed in the plastic well. If the mycotoxin for which the kit is designed to detect its presence in the solution, the mycotoxin binds to the antibody in the well. After a series of solutions are added and removed from the well, the presence of the mycotoxin is detected based on a color change in the final step. Recently number of enzyme linked immunosorbent assays (ELISA) have been developed (Bullerman and Gourama, 2001) and are commercially available. Some ELISA kits are designed to determine the concentration of mycotoxin in the grain, others are designed to detect the presence of the mycotoxin above a specified concentration, but cannot be used to determine how much above that concentration is present.

More sophisticated methods of mycotoxin analysis may be used in laboratories for several reasons. They can detect mycotoxins in many different kinds of feeds, detect them more reliably and allow the concentrations of mycotoxins present to be more precisely determined. Analytical services using such methods usually cost more and take longer time than ELISA tests. Each method has advantages and disadvantages (Table 1) One has to use discretion for employing different techniques based on precision required. Bioassay methods also proved to be more useful and precise in detection of different mycotoxins (Table 2).

Table 1. Advantages and disadvantages of different methods for detection of mycotoxins in foods/feeds.

Technique	Detection level (µg/g)	*Advantages*	Disadvantages
TLC	10	Versatile, Compact, Rapid	Tedious
HPLC	1	High precision, Good recovery, High sensitivity, Automated	Expensive
Minicolumn	5	Fast, Easy to handle, Good for routine analysis	Low precision
RIA	5	High specific, More sensitive, Fast	Expensive

Table 2. Detection of mycotoxins by bioassay

Technique	Type of assay	Mycotoxin
Microbial Assays		
	Bacillus megaterium	MT
	B. stearothermophilus	MT
	B. thuringensis	MT
	B. subtilis	AF
	Streptomyces and Nocardia spp.	AF
	Lactic acid bacteria	AF
	Rumen bacteria	AF
	Salmonella typhinarium	MT
	Staphylococcus aureus	CPA
Cytological assays:		
	Tracehal organ cultures	MT
	Calf kidney cells	AF
	Cultured embryonic lung cells	AF
	Cultured liver cells	AF
	Cultured mouse cells	AF

Immuno assays:		
	Immunochemical assays	MT
	Radioimmunoassay	AF
Invertebrate assays:		
	Brine shrimp	MT
	Mollusc eggs	AF
	Planaria	AF
	Insects	MT
	Paramecia	AF
	Amoeba	AF
Vertebrate assays:		
	Chick embryo	MT
	Day-old chick	MT
	Ducklings	MT
	Rabbit skin test	TR
	Mouse water refusal	*TRE*
	Fish	MT
	Mice	MT
	Bull Spermatozoas	AF
	Amphibians	MT
	Rats	MT

AF = Aflatoxins; MT = Multitoxins; TR = Trichothecenes

It is debatable whether culturing and identification of molds present in feed provides useful information for the assessment of mycotoxin contamination. The presence or absence of molds do not necessarily indicate the mycotoxin contamination.

The identities of molds found in feeds may be of some value in assessing the possibility of mycotoxin contamination. Identification of the genus and species of a mold present in feed is of most important. However, it is difficult to determine if a particular mold present in feed is capable of producing mycotoxins. Molds capable of producing mycotoxins are called toxigenic. Even if the genus and species of the mould are identified, it may not be possible to determine if that particular mold belongs to a toxigenic strain. If only the genus is identified, it is even more difficult to predict its toxigenicity. If an identified mold is believed to be toxigenic, it is not possible to determine if the mold has or is actively producing mycotoxins in the food. The presence *or* absence of mycotoxins can be determined only by chemical analysis for mycotoxins.

2.1. Physico - chemical methods

The physico chemical analysis is are based on the differences in physicochemical properties of different mycotoxins. In general the physicochemical assays comprise of six important steps (Table 3).

1. Sampling and sample preparation, 2. Extraction, 3. Cleanup, 4. Separation, 5. Detection and Quantization and 6. Confirmation

Table 3. Basic steps in Physico-chemical analysis of mycotoxins

Step	Description	Purpose
Sampling	Probe of automatic sampler	Representative sample
Sample preparation	Grinding, mixing, sub sampling	Representative sample
Extraction	Shaker or blender	Separate the toxin from compounds insoluble extraction solution
Clean-up	Liquid-liquid partitioning (Seperatory funnel)	Separate the toxin from groups of compounds of sample extract
	Thin layer chromatography (TLC)	Separate the toxin from remaining compounds.
Final separation	Gas-liquid chromatography (GLC)	The sample extract might interfere with toxin
	Liquid-Liquid chromatography (LC), Minicolumn chromatography	
Detection and quantitation	Fluorescence on TLC plate Fluorescence in Solution UV absorption in solution GLC flame detector	Detection and measurement of response
Confirmation	TLC separation and detection of derivative of mycotoxin, Biological test, Mass spectrometry	Identification of chemical compounds

2.1.1 Sampling and sample preparation

Sampling plays a crucial part in the precision and determination of mycotoxins level due to the very heterogeneous distribution of the toxins in agricultural commodities and products intended for human and animal consumption .The distribution of mycotoxins in the sample material is an important factor to be considered in establishing regulatory sampling criteria. It has been recognized by many international and national authorities and organizations such as FAO (2004), EC (2006) and FDA (2007). EC has laid down the method of sampling and analysis for official control of levels of various mycotoxins in food stuffs, repealing all former directives and amendments on this subject. It includes methods of sampling and analysis of different mycotoxins in cereals, dried fruits, fruit juices, must and wine, ground nuts and nuts, spices, milk, coffee and their products including baby foods. Mycotoxins in liquid food are distributed homogenously but in food grains it is concentrated in so called "hot spots". Mycotoxins specially produced by *Aspergillus*, aflatoxins are distributed very heterogenously with large particle size such as dried figs or groundnuts.

The number of contaminated particles may be very low but contamination with in a particle can be very high. To obtain the same representativeness for batches of food products with the large particle size, the weight of the incremental sample taken has to be larger than in cases of batches with smaller particle size. The number of incremental samples to be taken from different places of a lot depending on the weight of the entire lot. This may result in rather large aggregate samples up to 30kg in the case of aflatoxins determination in dried figs, groundnut and nuts. However, handling and sample preparations of large quantities pose difficulties in implementing such sampling plans especially with regard to the validation process of sampling methods. Therefore, EC regulation stipulates the division of aggregate samples intended for human consumption in to up to three laboratories samples of less than 10 kg for homogenization and analysis. This is not required for the products intended for further sorting or processing before human consumption or for use as an ingredients.

The grain inspections, packers and stockyards administration (GIPSA) of USDA developed general guidelines for grain, rice and hops. The FDA is constantly updating their investigative operation manual (IOM) that describes general procedures for field investigators and inspectors and includes information on sampling and sampling schedules for various occasions.

The minimum total sample size ranges from 4.5 kg to 34 kg depending on the heterogenecity expected of the sample type. Kay (2001) elegantly discussed different approaches for sampling grain in different countries and provided an interpretation of major differences between the methods in terms of lot size, tolerance, sampling techniques and sampling size.

A properly drawn sample in the range of 2 to 40 kg is too large for analysis. Therefore, it must be sub sampled to obtain analytical samples in the range of 25 to100 g (most often 50 g) the mycotoxin contamination may still be heterogeneous. However, this error can be greatly minimized by thoroughly grinding and blending the samples. The objective is to homogenize the mycotoxin within the sample to the greatest extent possible. In case of aflatoxin contaminated peanuts, for examples, grinding the sample for 6-7 min in a vertex, aflatoxins are fairly uniformly distributed.

Each step in the process is of vital importance as it may lead to wrong reporting. The sampling is not only random from a sufficiently large size but also large number so that, it is a representative sample. Similarly extraction method of mycotoxins is important in selection of suitable solvent and conditions under which it is extracted and probable contaminant in this extract.

Out of several available analytical procedures the methods of choice should depend on nature of mycotoxin, degree of quantitation required (Screening very accurate quantitation) the cost that can be incurred and the time that can be allotted. In general there are three steps involved in most of the analyses. These include:

2.1.2. Extraction

The organic solvents most commonly used for extraction of mycotoxins are: chloroform, acetonitrile, methanol and acetone. These solvents are mixed with a given ratio of a more polar solvent (water, dilute acid, aqueous solution of salts) to aid the breaking of weak electrostatic bonds which bind some mycotoxins to other substrate molecules (e.g. proteins).

The ground sample or preferably aqueous slurry, is either shaken with the extraction solvent for 30-45 minutes or blended at high speed for about 3 minutes. It should be noted that an explosion proof blender is recommended for use with inflammable solvents such as acetone and methanol. In blending care must be taken to ensure that the entire sample is continuously washed with the extraction solvent. If large number of samples are to be analyzed, the shaking protocol may be preferable because modern shakers are capable of holding upto 20 samples concurrently. Accelerated solvent extraction (ASE), microwave associated extraction (MAE) may help to speed up and automate the extraction process and offer a robust and time saving alternative to classical solvent extraction techniques.

Once the solid sample has been shaken or blended with the extraction solvent, the liquid is separated from the solids either by filtration or centrifugation. The extract is then either cleaned up further to isolate the toxins or applied directly to the determinative step in the procedure.

2.1.3. Clean-Up

Mycotoxins are such a diverse group of chemical compounds that it is difficult to find a simple procedure which specifically removes non-mycotoxin interfering compounds, whilst leaving the mycotoxins in the extract. For this reason it is difficult to find a good method for screening a wide-range of mycotoxins simultaneously. It is possible, however, to devise procedures which remove interfering non-mycotoxin compounds from the extract of a particular commodity and leave a particular mycotoxin or group of mycotoxins in the extract. This process not only removes sample impurities but also facilitate to concentrate mycotoxin prior to determinative step. However, many screening methods including ELISA require no clean up other than dilution of the extract and/or filtration before analysis. Some of the common clean-up techniques which have been used are:

***(i) Defatting*:** This is carried out prior to the toxin extraction step and uses petroleum ether or hexane to extract lipids from the sample using a Soxhlet extractor. This step is only required when the subsequent clean-up step is not capable of removing the lipids.

***(ii) Column–chromatography*:** This technique has very wide applications, and is used in a number of regulatory or officially approved methods. A glass column is packed with one or more adsorbent materials and the crude extract is added to the top of the column. The column is then eluted with a series of solvents or solvent mixtures which are designed to first wash off interfering compounds and then elute the desired mycotoxins, whilst other interfering compounds remain strongly bound on the column.

A miniature column, called a mini-column is used in many rapid aflatoxin assay methods to remove interfering compounds and to qualitatively detect aflatoxin down to a few parts per billion (ppb)

***(iii) Precipitation*:** This is an extremely useful technique whereby certain chemicals, sometimes in colloidal form are added to the crude extract and these adsorb certain pigments, proteins and other interfering compounds onto their surface. The complex so formed precipitates out of solution and can be filtered off, leaving a cleaned-up solution. Useful precipitating agents include cupric carbonate, ammonium sulphate, lead acetate and ferric gel.

***(iv) Liquid-liquid partition*:** It is commonly used, often in conjunction with one or the other clean-up procedures, to provide additional clean-up and also to transfer toxins from one solvent system to another, whilst at the same time effecting a considerable increase in concentration of the toxins. The partition is carried out in a separating funnel which contains the two immiscible solvents. The funnel is shaken for a few minutes to allow the dissolved compounds, including the mycotoxins, to partition between the two phases. The solvents are selected so that the mycotoxins are preferentially partitioned into one of the solvents. Care must be taken in the choice of solvents in order to minimize the risk of emulsion formation.

(v) Thin layer chromatography (TLC): Although this technique is used mainly for detection and quantification of toxins, it is also useful for separating the mycotoxins from interfering compounds in the extract. Sometimes, it is necessary to try a range of developing solvents in order to separate the toxins of interest from the interfering compounds. If this proves impossible, or too time consuming, then two dimensional TLC may be employed.

(vi) Other clean-up procedures: The clean-up procedures outlined above are those employed in officially approved methods for aflatoxin analysis and they are also used in analytical methods for many other mycotoxins. Other clean-up procedures such as reverse dialysis, solid phase extraction (SPE), immunity columns and base extraction are also used for the detection of different mycotoxins.

There is a strong trend towards the use of immune affinity columns (IAC). In mycotoxin analysis as a cleanup and enrichment technique for sample extracts or liquid samples, IACs contain immobilized antibodies that exclusively retain a certain mycotoxin or mycotoxin class. Due to their high specificity, IACs produce cleaner matrix components and excellent signal to raise ratios compared to less selective super critical fluid extraction (SPE) sorbent materials. Grochowaski (2004) has developed method based on isotope dilution procedure with the semipermiable membrane cleanup standards.

Solid- phase extraction columns (SPE) are usually porous silica, the surface of which has been modified to provide selective absorption of either the analyte or impurities. In some cases the analyte is retained on the columns, while impurities pass through and are washed off. In other cases, the SPE columns are designed to trap impurities and permit the analyte to pass through. The advantage of latter type of column is that only extraction solvent is needed to purify the sample, eliminating the need for additional solvent for elution.

(vii) Work-Up: After the clean-up step, the extract must be "worked-up" in order to prepare it for the detection and/or quantification step. Following clean-up the extract is often dissolved in a large volume of aqueous solvent, so it must be transferred into a small volume (l0-50ml) of a volatile solvent, such as chloroform. The chloroform solution often requires drying and this is achieved by passing it through a bed of anhydrous sodium sulphate. The solvent is evaporated off, as described below to near dryness. Care must be taken at this stage as some mycotoxins can break-down if the dry extract is heated at 100°C. To avoid this, the evaporation is best carried out using a rotary evaporator at 30°C to reduce the volume to a few ml, followed by evaporation to dryness in an inert atmosphere using a sample concentrator. If such items of equipment are not available, then evaporation should be carried out using a steam-bath, preferably under a stream of nitrogen. Great care should be taken to ensure that flasks are removed from the steam-bath just before they go dry and that vials are removed on the point of dryness. The extracts are now ready for detection and quantification.

2.1.4. Separation

Separation of mycotoxins of interest involves minicolumn chromatography, TLC, HPLC and GC. Minicolumn chromatography is useful for screening and quick decision needed to whether accept or reject a food lot. TLC the most reliable and relatively simple is a good separation technique. Two dimensional TLC provides good separation resulting in low limits of detection. HPLC is an alternative as it offers automatic ultimate separation and quantitation.

2.1.5. Detection and quantification

The final step in an analytical protocol involves determining whether toxin is present, using atleast one detection technology. Usually we make a distinction between reference (confirmatory) methods, which allow detecting, identifying and quantifying mycotoxins in various matrices and rapid screening methods which are aimed essentially at detecting the presence of a mycotoxin or group of toxins.

The high demand for rapid screening methods for mycotoxin analysis reflects the need for fast and cost effective on site determination of the level of mycotoxin contamination in foods and feeds. Immunological based screening methods have shown great potential and are increasingly applied in routine analysis and monitoring of mycotoxins.

Future trends in screening methods include the further development of fast and simple tests requiring no instrumentation and improved detection capability for simultaneous measurement of multiple mycotoxins.

It is fortunate that the aflatoxins and many of the other important mycotoxins are fluorescent under ultraviolet light. This enables them to be detected at very low levels (parts per billion) and this method of detection is used in the majority of analytical methods for mycotoxins.

Detection or "qualitative assay" is usually by TLC or mini-column and only requires the use of a qualitative (not accurately determined) standard. Such a standard is also sufficient to enable a semi-quantitative assessment to be made by TLC, using a "dilution-to extinction" principle. Fully quantitative determinations, which require the use of a standard of known concentration, can be carried out by a variety of techniques including TLC, high performance liquid chromatography (HPLC), high performance thin layer chromatography (HPTLC) and the fluorotoxin-meter (FTM). Though HPLC and GLC techniques have excellent sensitivity and reproducibility and can be adapted to automation, they do have certain disadvantages such as

(i) Samples require to be highly processed in relation to clean-up to achieve sensitive detection limits.

(ii) Only one sample can be analyzed at any given time.

(iii) Capital output is high

(iv) Running costs are high

(v) Highly trained persons are normally required to operate the equipment and interpret the results.

(vi) They are time consuming.

Analytical methods which are simple and can be handled by inexperienced operator and institutions where good laboratory facilities are not available (Table 4). Kokkonen and Jostoi (2009) have developed automated solvent extraction step followed by filtering, concentrating and analysis of the crude extract with liquid chromatographic tanden mass spectrometric (LC-MS/MS) determination of nearly 31 mycotoxins produced by main mycotoxin producing fungi. The performance was good for semi quantitative detection varying between 1 and 1,250 $\mu g/kg^{-1}$. This is a simple and rapid technique for determination of several co-existing mycotoxins and is ideal for detection levels of relevant to animal welfare. However, LC-MS methodologies involve the most expensive apparatuses and require the services of high level professionals. The cost of the complete system could be prohibitive.

Table 4. Advantages, disadvantages and economics of mycotoxins in foods and feeds

Service	Precision of the Method	Advantages	Disadvantages	Cost (Approxiamate)
Mold culture and identification	Microbiological culturing; microscopic or biochemical identification tests.	Determines presence of molds, but not necessarily toxigenic strains.	Does not determine if molds found to be present. It may take a relatively long time to obtain results	Rs. 500-1000
Chemical analysis	Enzyme-Linked Immuno-sorbent Assay (ELISA)	Rapid, relatively inexpensive	Only detects one mycotoxin at a time. Proof of mycotoxin presence usually requires additional testing.	Rs. 500-1000

Thin layer chromatography (TLC)	Can detect more than one kind of mycotoxin	Slower than ELISA. Proof of mycotoxin presence may require additional testing. May not work well for feeds other than grains.	Rs. 200-500
High performance liquid chromatography (HPLC)	Easier to provide quantitative results. May be able to detect more than one kind of mycotoxin.	May be more expensive. Proof of mycotoxin presence may require additional testing	
Gas chromatography / mass spectroscopy (GC/MS)	Can identify the mycotoxin without additional testing. May be able to detect more than one kind of mycotoxin.	More expensive	Rs. 1000

2.1.6. Confirmation

It is essential that confirmatory tests be carried out if a mycotoxin is thought to have been detected, especially when a new commodity or new source of commodity is being analyzed. Failure to do this could easily lead to false-positive results (i.e. an interfering compound is mistaken for a known mycotoxin). Detection and confirmation of different mycotoxins by observation under visible light, UV light and chemical methods are précised in Table 5.

Confirmatory tests will help to confirm sample that have been determined to contain mycotoxins based on rapid screening tests and other to quantitate the amount of toxin present. This usually involve a chromatographic technique such as TLC, GC, HPLC or liquid chromatography with mass spectroscopy (LC-MS) or tandem mass spectrometry (LC-MS/MS) to further separate mycotoxins from the extract impurities. HPLC of extracts of foods and feeds is the most prevalent and sensitive method for identification and quantalization of mycotoxins. Often, the mycotoxins extracted from the field samples undergo cleanup using commercial immunoaffinity columns before their analysis by HPLC. The columns are available for all important mycotoxins such as AFS, Ochratoxin A, T2 toxin, DON, Citrinin, fumonisins, Zearalenone, Patulin and Moniliformin. Multiplex columns are also available for these mycotoxins which can detect more than one mycotoxin produced by single fungus (Logrieco *et al.*, 2007; Gonzalez Pereyra *et al.*, 2008)

Most confirmatory tests involve the formation of a derivative which has different properties (e.g. colour of fluorescence or polarity), than the presumptive mycotoxin. When detection and/or quantification is by TLC, then derivatives can be formed "on the plate" either by spotting a reagent on the plate before development or by spraying on a reagent after development. An example of the former approach is the use of trifluoro-acetic acid (TF A) to form the hemi-acetal derivatives of aflatoxin B,G and M which are identified, if present, as slower running fluorescent spots. A commonly used spray reagent for the aflatoxins is 50% sulphuric acid which reacts with the blue and green-blue fluorescent aflatoxins to give yellow fluorescent derivatives (Table 5).

Table 5. Detection and confirmation of different mycotoxins

Name of the toxin	Solvent system	Spray reagent	Detection		References
			UV	Visible	
Aflatoxins	C : A (95:5)	13	B g	--	Stack and Pohland (1975)
Ochratoxin A	T : Ea : F (6 : 3 : 1)	8	B b	--	Gimeno (1979)
Patulin	T : Ea : F (6 : 3 : 1)	10	--	Y	Subramanian (1982)
Sterigma-tocystin	C : M : A (1:1:1)	8	Y	--	Adye and Mateless (1964)
Citrinin	T : Ea : F (6 : 3 : 1)	1, 2, 3	--	y, by, lb	Gorst-Allman and Steyn (1979)
Cyclopiaz-onic acid	T : Ea : F (6 : 3 : 1)	1,2,3,11	--	b, rb, br, p	Rathinavelu and Shanmuga-sundaram (1984)
Ochratoxin A	T : Ea : F (6 : 3 : 1)	2, 3, 12	-, -, b, b	y, pb, -	Piskorska-Pliszczynska and Jusckiewicz (1977)
Pnitrem A	C;M (97:3)	1,2,3	---	p,p,g	Gorst-Allman and Steyn (1979)
Deoxyniva-lenol (DON)	T : Ea : F (6 : 3 : 1)	4,7,8	Ch, bl	Y	Ramakrishna and Bhatt (1987)
Diacetoxy Scirpenol (DAS)	C: M (97: 3)	6,9	Bg	-,br	Kamimura *et al.* (1981) Ramakrishna and Bhatt (1987)
Fusarinone –X	C: M C:M (97: 3)	8	Bl	-	Kamimura *et al.* (1981) Ramakrishna and Bhatt (1987)

HT-2 toxin	C: M (97: 3)	6	bg,-	-	Kamimura *et al.* (1981) Ramakrishna and Bhatt (1987)
Nivalenol (NIV)	E: T :F (5:4:1)	4,7,8	-,Chl, bl	y,-,-	Kamimura *et al.* (1981) Ramakrishna and Bhatt (1987)
T-2 toxin	C: M (97: 3)	6,9	bg,-	-,P	Kamimura *et al.* (1981) Ramakrishna and Bhatt (1987)
Zearal-enone	C: M (97: 3)	1,2,3,4,5, 7,8	br, Ch, bl	br,do,ip,- -	Pathre *et al.* (1979)
Fumon-isins	W:M (3:1)	4,11	-	Br	Kamimura *et al.* (1981)
Verru-carins	C:M (98:2)	14	---	P	Rao *et al.*, 1985
Roridins	C:M (98:2)	14	---	P	Rao *et al.*, 1985
Satra-toxins	C:M (98:2)	14	---	P	Rao *et al.*, 1985

Solvent systems: C = Chloroform; M= Methanol; W-Water;
Detection colours: bl = Bluc; Ch= Charring; Y = Yellow; bg = blue green; br = Brown; P = Pink

Spary reagents : 1= Ce (SO_4) 2 1% 6N H_2 SO_4; 2=2,4-DNP; 3 = $FeCl_3$ 3% in ethanol ; 4= P- anisal dehyde; 5=50% H2SO4 in methanol; 6 = 20% H_2SO_4; 7 = H_2SO_4; 8=20% $AICI_3$; 9= Chromatropic acid;10= 20% Phenyl hydrogen hydrochloride: 11 = $FeCl_3$ 3% in ethanol;

12 = 1% p-dimethylaminobenzaldehyde in n-butanol and HCl fumes: 13 = Ammonia fumes;14 = Ploroglucinol-HCl.

Some of the precautions required in rapid mycotoxins analytical methodology are

1. Minimum sample preparation prior to analysis.
2. Good sensitivity and specificity for the appropriate mycotoxins (s) under test.
3. Applicability for wide range of natural substrates.
4. Easy adaptability to infield situations using untrained personnel or atleast using the minimum of laboratory equipment and technical assistance.
5. Laboratory oriented prototype to assess the full capabilities of the system for toxin determination.
6. Availability of final results within a short period of time from sample preparation (5-30 min).

2.2 Immuno Assays

Intensive studies to develop new sensitive, specific and simple methods have led to many improved and innovative analytical methods for mycotoxins analysis in the last a few years (Trukeress and Pohland, 2001 and De Vries *et al.*, 2002). Simplified and efficient clean up procedures such as solid phase extraction cartridges have been established. Better analysis and quality control has been established by using certified reference materials (Boenke, 1997). More sensitive TLC, HPLC and GC techniques are now available. Sensitive and versatile high resolution MS and GC/tandem MS/MS are coming to the market. MS methods have also been incorporated into HPLC systems. New Chemical methods including capillary electrophoresis, fluorescence polarization immunoassay and biosensors are emerging and have gained application for mycotoxin analysis (Chu, 2000; Maragos *et al.*, 2001; Maragos and Plattner, 2002; Nasir and Jolly, 2002). Recent progress in nanotechnology has lead to new innovative approaches for monitoring of mycotoxins in grains as well as new diagnostic tools for mycotoxicosis.

Immunoassays are gaining wide acceptance as an analytical tool for mycotoxins. Antibodies against almost all of the mycotoxins are now available (Chu, 2000). New immunoassay systems are emerging (Maragos and Plattner, 2002). Immunoscreening methods have been widely accepted as a simple approach to screening for mycotoxins in several commodities. Many immunoscreening kits which require less than 15 min to complete are commercially available. These assays eliminate many of the above steps. Automation of mycotoxin analysis has become a reality by combining these newer techniques. Instead of using antibodies, molecularly imprinted polymers are being used for solid phase extraction of mycotoxins such as ochratoxin A (Jodlbaur *et al.*, 2002), rather than analysis of toxin. PCR methods based on primers of key enzymes involved in the biosynthesis of mycotoxins have been introduced for the determination of toxigenic fungi present in food (Schnerr *et al.*, 2002).

Although several biological methods are available, most of these methods are nonspecific and relatively insensitive. To overcome the difficulties encountered with the chemical and biological methods, new immunochemical methods have been developed (Chu, 1996). Many immunoassay kits for mycotoxin analysis are also commercially available (Beier and Stanker, 1996). The application of mycotoxin immunoassay is not limited to foods and feeds, but also used as a sensitive approach to monitor mycotoxins in body fluids and tissues or organs of humans and animals that have been exposed to the mycotoxins (Pestka, 1994).

Immunochemical methods are based on the specific interaction between antibodies and the toxin. Like most other natural products, mycotoxins must first be conjugated to a protein/polypetide carrier before subsequent use in immunization for antibody production (Chu, 1995). Newer innovative approaches for the coupling of mycotoxin to marcomolecule have been developed in recent years (Chu, 1996). With the advances in hybridoma technology monoclonal antibodies against many mycotoxins were also made. Antibodies against almost all the important mycotoxins have been made available. Many types of immunoassays, including radioimmunoassay (RIA) and enzyme-linked immunosorbent assays (ELISA), as well as several novel immunochemical screening tests, have been developed. Most of these methods are very sensitive, specific and simple to operate. Specific antibodies have also been used as immunohistochemical reagents and to arm affinity columns that are used as a cleanup tool for analysis of mycotoxins by other methods. Antibodies against different mycotoxins are précised in Table 6.

Table 6. Antibodies against Mycotoxins

Mycotoxins	Type of antibodies	References
AAL toxin	Pab	Szurdoki *et al.* (1996)
Aflatoxins: B1, B2, G1, G2	mAb, pAb	Hsu and Chu (1992)
Aflatoxin metabolities: B2a, Q1, M_1aflatoxical, DNA adducts	mAb, pab	Yu and Chu (1998)
Citrinin	pAb	Usleber *et al.* (1996)
Cyclopiazonic acid	mAb, pAb	Haung and Chu (1993)
Ergot alkaloids	mAb, pAb	Sheltry (1996)
Fusarochromanone	pAb	Yu *et al.* (1990)
Fumonisin	mAb, pAb	Yu and Chu (1999)
Kojic acid	pab	
Ochratoxin A	mAb, pAb	Clarke *et al.* (1994)
Patulin	pAb	Mc Elory and Weiss (1993)
Paxilline related	mAb, pab	Garthwaite *et al.* (1994)
PR-toxin	pab	
Rubratoxin B	pab	
Secalonic acid	pab	Brigys *et al.* (1994)
Sporidesmin	mAb, pAb	Li *et al.* (1996)
Sterigmatocystin	PAb	Schmitt *et al.* (1996)
Trichothecenes: DAS, DON, FX, DOVE, AeDON, NIV, Rgridin A, T-2 toxin, and	mab, pAb	Wang and Chu (1991); Barna-vetro *et al.* (1994)

T-2 toxin metabolites (HT-2, T-2-tetraacetate 3'-OH-T-2, dep-T-2)		
Versicolorin A	pAb	Reynolds and Pestka (1991)
Zearalenone	mAb, pAb	Berna Vertro *et al.* (1994).

2.2.1 Radioimmunoassay (RIA)

The RIA procedure involves incubation of specific antibody simultaneously with a solution of unknown sample or known standard and a constant amount of labeled toxin. After separation of the free and bound toxin, the radioactivity in those fractions is then determined. The toxin concentration of the unknown sample is determined by comparing the results to a standard curve that is established by plotting the ratio of radio activities in the bound fraction and free fraction versus log concentration of unlabeled standard toxin.

2.2.2 Florescence Polarization Immunoassay

Demands to increase speed and lower costs have lead to continue improvements in these immuno assays (Maragoes, 2009). Since many mycotoxins are of low molecular weight (below 1KDa) immunoassays for their detection have generally been constructed in competitive heterogeneous formats which required separation of bound and free labels (tracer) are not suitable for low molecular substances. On the other hand fluorescence polarization immunoassays will be most convenient as it does not require separation of bound and free (labels). The benefits of application of FPIA to mycotoxin analysis include the elimination of steps to separate bound and free labels giving the potential for very rapid immunoassays which take from 2-15 min to complete. Important assay parameters such as speed and sensitivity are dependent upon not only the quality of the antibody and tracer combination can yield rapid and sensitive mycotoxin assays. Apart from using most sensitive antibody and tracer combination available, matrix can be controlled using a number of established procedures ranging from simple dilution to matrix-matched calibration. Further, several existing FPIA instruments are field portable.

2.2.3 Enzyme Immunoassay

Enzyme immunoassay (EIA) is a general term for the immunoassays involving use of an enzyme as a marker for the detection of immunocomplex formation. The general principle of EIA is similar to that of RIA, there is an amplification system present in this assay, and thus, EIA is more sensitive. Since no radioactive substances are used, this assay avoids the problems encountered in handling radioactivity. Enzyme labelling can be done by conjugation of enzyme to Ag or Ab via the periodate oxidation and subsequent reductive alkylation method or by cross-linking using glutaraldehyde. Some of the methods used in conjugation of hapten to proteins can also be used. Although horseradish perodxidase (HRP) and alkaline phosphatase are the two enzymes most commonly used, others, such as glucose-6-phosphate dehydrogenase coupled with oxido-reductase and luciferase, glucose oxidase, beta-galactosidase, and urease have also been used. ELISA is further divided into two major types. One type is competitive ELISA (c-ELISA), which can be used for the analysis of both hapten and macromolecule, while the other is noncompetitive sandwich ELISA, which is used for divalent and multivalent Ag. c-ELISA is used most frequently for mycotoxin analysis. Bodea *et al.* (2004) have developed a flow injection immunoassay (F1-1a) method for detection of aflatoxins in milk. Thirumala Devi *et al.* (2000) have discussed elegantly advances made in the immunological detection of different mycotoxins. Stark (2010) discussed various aspects of ELISA in the detection and quantization of different mycotoxins.

(i) Direct competitive ELISA (dc-ELISA)

In this assay, specific *antibodies* against mycotoxins are coated on the ELISA plate. The sample or mycotoxin standard solution is generally incubated simultaneously with enzyme-conjugate or incubated separately in two steps. The amount of enzyme bound to the plate is then determined by incubation with a chromogenic substrate solution. The resulting color/fluorescence, which is inversely proportional to the mycotoxin concentration present in the sample, is then measured instrumentally or by visual comparison with the standards. In this assay, the mycotoxin-enzyme conjugate (marker) and free mycotoxin compete for the same binding site on the solid-phase antibody. Although HRP is most commonly used as the enzyme for conjugation, other enzymes such as alkaline phosphatase and beta-galactosidase also have been used. Excluding the time for sample preparation, dc-ELISA itself generally can be completed in 0.5-2 hr. In general, dc-ELISA is approximately 10-100 times more sensitive than RIA when purified standard is used and as little as 2.5 pg of pure mycotoxin can be measured (Table 7).

Since a cleanup step is usually not necessary, many samples can be analyzed within a relatively short period. Dc-ELISA can detect 0.05-50 ppb of mycotoxins in foods and feeds. Like RIA, the sensitivity of ELISA is improved when a cleanup treatment is included in the assay protocol.

Table 7. Sensitivity of Direct Competitive ELISA for Selected Mycotoxins

Mycotoxins	Foods/feeds	Standard range (ng/assay)	Detection limits (μg/kg) or (μg/l)	References
AAL	C			Yu and Chu (1998)
AFB/Afs	C, Wh, P, Pb	0.0025-1	1-10 (1)[b]	Barna Vetro *et al.*, (1994)
AFM,	M, Ch	0.0012-1	0.10 (0.01)	Biancardi (1997)
CPA	C, P, MF	0.1-0.6	(0.05-0.1)	Yu and Chu (1998)
DON	C, Wh	0.012-125	1000 (10)	Sinha and Savard (1997)
3-AcDON	B	0.0002-0.4	16	Sinha *et al.* (1995)
15-Ac-DON	Wh	0.002-0.025	50-100	
Fm	C, MF, M, BR	0.0005-50	10-500	Scott and Trucksess (1997)
H-Fm	C	0.1-15	(5-10)	
OA	Wh, B, K	0.025-0.5	30 (1-2)	Tapai *et al.* (1997)
T-2	C, Wh	0.0025-0.2	2.5-50 (1)	
Type A TCTC	C	0.0025-0.2	50-100	
ZE	C	0.02-2.5	50 (10)	Tanaka *et al.* (1995)

(ii) Indirect Competitive ELISA (idc-ELISA or Double-Antibody ELISA)

In the indirect competitive ELISA (idc-ELISA), a mycotoxin-protein (or polypeptide) conjugate is first prepared and then coated to the microtiter plate before assay. The plate is then incubated with specific rabbit (or other type) antibody in the presence or absence of the homologous mycotoxin. The amount of antibody bound to the plate coated with mycotoxin-protein conjugate is then determined by reaction with goat anti-rabbit (or anti-other type) IgG-enzyme complex (which is commercially available) and by subsequent reaction with the substrate. Thus, toxin in the samples and toxin in the solid phase compete for the same binding site with the specific antibody in the solution. The idc-ELISA has also been widely used for analysis of a number of mycotoxins (Table 8) with a sensitivity that is comparable to or slightly better than that of the direct ELISA in some cases. This type of ELISA requires fewer antibodies (100 times less) and does not require preparation of a toxin-enzyme conjugate. However, it takes more analytical time (2 hr). To optimize the assay, selection of secondary antibody-enzyme conjugate in the idc-ELISA is important.

Table 8. Senistitivity of Indirent Competitive ELISA for selected Mycotoxins

Mycotoxins	Foods/Feeds	Standard range (ng/assay)	Detection limits(ug/kg)or(ug/L)
Idc –ELISA			
Aflatoxins	C,P, Peanut Buffer (Pb), F,Rs,S	0.0002-1	0.25-5
AFM	Milk (M)	0.0001-0.005	0.005
Diocetoxy scirpenol	C, Wheat (Wh)	0.005-5	0.005
Dioxynivalenol	C, Wh	0.010-100	300(1)h
3-Ac-DON	Rye (R)	0.005-1	1000(10)
Fumonisins	Corn (C)	1-100	1
Nivalenol	B	0.05-5	200(50)
Ochratoxin A	B,K, mixed feed (M F), Wh	0.004-10	(30)
Sabatoxin	S,.U,W	0.0001-5	0.06-50(1)
T-2	M,S,U	0.002-0.2	5(0.2-1)
IIT-2	Urine (U)	0.005-0.5	(0.5)
Zearalenone	C,W,M,K,F	0.05-2.5	1-60
M-dc-ELISA			
AFB	B,P,Pb,C	0.01-1.5	10
T-2	Barley (B)	0.05-5	(50)

2.2.4 Immunoscreening methods

By shortening the incubation time and adjusting the antibody and enzyme concentrations in the dc- or modified idc-ELISA assay system, it is possible to do a quick screening test toxin levels in less than 30 min (e.g., 20 ppb) (Schneider *et al.*, 1991). Based on the same principle as the dc-ELISA, several other types of immunoscreening tests with sensitivity similar to ELISA were also developed rather than coating the antibody onto the microtiter plate, the antibody is immobilized on a paper disk or other membrane (Usleber *et al.*, 1993) which is used directly as a strip (Smith *et al.*, 1994) or mounted either on a plastic card (card screening test), a plastic strip as dipstick(Edrington *et al.*, 1997), in a plastic cup (Cup test and Cite), or in a syringe (Cite probe and Idexx probe) (Sarr *et al.*, 1995). Antibody has also been coated on polystyrene beads (Mayura *et al.*, 1998). The reaction is carried out on the wetted membrane disk. Thus, after reaction, the absence of color (or decrease in color), generally blue, at the sample spot indicates the presence of toxin in the sample. The reaction is generally very rapid and takes less than 10-15 min to complete. Like the above formats, dipstick-type enzyme immunoassay has been developed for quick screening of *Penicillium islandicum* in rice grain and 3-acetyl-DON (Chu, 1995) and T-2 toxin (De Saeger and Van Peteghem, 1999) in wheat. Abouzied and Pestka (1994) immobilized different mAbs against AFB_1, FmB_1 and Ze.

Another screening test is the immunoaffinity method, which was originally designed for mycotoxins such as AF, OA, and ZE that fluoresce (Trucksess *et al.*, 1994). In this assay, sample extracts diluted in phosphate buffer are applied to the affinity a column in which specific antibody was covalently bound to the solid matrix. After washing to remove the unbound materials, specific mycotoxin is then eluted from the column with the appropriate solvent system and then subjected to other chemical analyses. For mycotoxins with native fluorescence such as AF, OA and ZE, the toxin level is then equate could be directly determined fluorometrically or be determined after derivatization to enhance the fluorescence (Hansen, 1990). For fumonisin and DON screening, it is necessary to introduce a fluorophore to the materials eluted from the immunoaffinity column (IAC). The sensitivity to the IAC screening tests for AFBI, FmBI, OA, and ZE is 2 ppb, 1 ppm, 5 ppb and 0.2 ppm, respectively.

The application of various immune screening tests to mycotoxins has been summarized by Pestka (1994) and Chu (1996), and most of the screening tests are commercially available as kits. All of the rapid-screening test kits permit monitoring of mycotoxins semi quantitatively and have been found to be effective in screening mycotoxins in the field by FSIGS (Pestka, 1994).

2.2.5 Complementation of *immunoassays with chemical*

(i) Immunoaffinity Chromatography

With the availability of antibodies against various mycotoxins, immunoaffinity columns (IACs) were made by conjugating antibodies to a solid-phase matrix. These columns are then used either in a screening test as discussed above or as a cleanup column for subsequent chemical analysis (Chu, 1996). IAC was first used in the RIA and later for recovery of AFM from urine and milk samples (Wu *et al.*, 1983) for subsequent analysis. Although earlier applications of this technique were primarily aimed at biological fluids (Gropman *et al.*, 1992), the IAC has gained wide application as a cleanup tool for a number of mycotoxins and is not limited to fluid samples (Scott and Truckses, 1997). IACs for a number of mycotoxins are also commercially available (Table 9). It is apparent that once the contaminants are removed by specific IACs, the solution can be directly subjected to liquid chromatographic (LC) quantitation either off-line in an automated system, or by fluorometry. IAC not only serves as a cleanup tool but also concentrates the mycotoxin from a large amount of sample. Thus, it lowers the detection limits sometimes as low as parts-per-trillion (PPT) of mycotoxins can be measured.

Lateral flow immunochromatography (LTF), also called rapid one-step assay of mycotoxins has become the most popular method for rapid identification of mycotoxins and rough estimation of their concentration. LTF is a simple comprising a loading pad, zone of colored particles coated with a mouse monoclonal antimycotoxin antibody, a zone nitrocellulose membrane, a test line that contains immobilized mycotoxin, a positive control line that contains a secondary antimouse antibody and absorbent pad. The sensitivity of LTF is very high and is comparable to those of sophisticated methodologies such as LC-MS-MS and Surface Plasmon Resonance (SPR).

Table 9. Immunoaffinity Chromatography of Mycotoxins

Mycotoxins	Commodities	Analysis	Selected ref.
AFM	M[a]	F1 h	Fremy and Chu.(1989)
AFB	F, P	TLC	Candlish *et al.* (1988)
AFB	Cff,IP,nuts, figs, etc	HPLC/PsCD	Barmark and Larsen (1994)
	C,P,Ct,F, Pb, BR	F1/Br; HPLC/ PCD	Groopman *et al.* (1988)
AFL	S, U	HPLC	Kussak *et al.* (1998)
AFQ 1	U	HPLC	Kussak *et al.* (1994)
AF-adducts	T, S, U	HPLC,ELISA	Groopman *et al.* (1992)
AF-albumin	S	HPLC,ELISA	Sabbioni *et al.* (1998)
CPA	C, P, feed	ELISA	Yu *et al,* (1998)
DON	C, W	HPLC	Yu *et al.* (1999)
FmB, HFms	C, starch	HPLC; LC/MS	Ware *et al.* (1994)
OA	Cff, T, W, C, So	HPLC	Nakajima *et al.* (1994)
OA	Cff, C, So	CE	Corneli *et al.* (1998)
ZE	M, U, C	Elisa,HPLC,MS	Azcona *et al.* (1990)

(ii) Combination with HPLC and TLC

With the availability of sensitive ELISA methods, this technique has proved effective as a post column monitoring system for HPLC (Kramer *et al.*, 1994). This is especially useful for the analysis of compounds with no specific absorption, such as TCTCs and Fms. For example, in the analysis of various type A TCTC mycotoxins, the sample extract with no cleanup treatment was first subjected to HPLC with a C-18 reverse-phase column. Individual fractions eluted from the column were analyzed by ELISA using "generic" antibodies against group A TCTCs. This approach cannot only identify each individual group A TCTC, but can also determine its concentration quantitatively. As little as 2 ng of T-2 toxin and related TCTCs as well as their metabolites can be monitored by this method. A combination of HPLC and ELISA technology proved to be an efficient, sensitive, and specific method for the analysis of TCTC (Park and Chu, 1996) and other mycotoxins (Chu, 1991).

(iii) Other newly developed immunochemical methods

(a) Anti-idiotype antibody-based immunoassays

The development of immunochemical methods for mycotoxin detection has led to a great demand for specific antibodies and related immunochemical reagents for the assay. An alternate approach to preparing immunochemical reagents is through generating anti-idio-type (anti-ID) antibodies (Nisonoff, 1991). Anti-ID (Ab2) for large molecules has been well developed and have been applied to clinical diagnosis and immunotherapy. Recent studies have succeeded in generating Ab2 against a number of small-molecular-weight haptens including mycotoxins Af_s (Krska *et al.*, 1997), FmB1 (Chu, *et al.*, 1995), and T-2 toxin (Chanh *et al.*, 1992).

(b) Development of Biosensors

Although development of An/Ab-based biosensors for mycotoxin detection was initiated in the late 1980s, application of this technology has emerged only as promising in recent years. Nevertheless, some of the principles and methodology are useful for designing the biosensors. In the so-called "hit-and-run" assay (Warden *et al.*, 1990) for T-2 toxin column was equilibrated with a fluorescein isothiocyanate (FITC)-labeled Fab fragment of IgG (anti-T-2 toxin). Samples containing T-2 toxin were injected into the column. The FITC-Fab that eluted together with the samples containing T-2 toxin was then determined in a standard flow through fluorometer whereas these methods were not as sensitive as the regular ELISAs. Several biosensors with some modifications such as line resolved chromatography (Bacigaupo *et al.*, 1994), fiber optic immunosensor (Strachen *et al.*, 1997) miniatured ELISA (Yacoub-Geroge *et al.*, 1996) and surface plasman resonance biosensor (Mullett *et al.*, 1998) can be used for rapid screening of mycotoxins. Maragos (2009) elegantly discussed the importance of biosensors in the detection of mycotoxins and felt their increasing role in rapid high-throughput detection when mycotoxins are in very low concentration.

Technology is not only low cost but also reliable and rapid and various types of assays have already been devised for several of the major groups of mycotoxins. The limit of detection based on fluorescent antibodies against DON, AFB, and Ochratoxin A are at ppb level (Ngundi *et al.*, 2006) and they are regenerable. Prieto-Siman *et al.* (2007) gave clear description and explanation of Physico-chemical principles of new biosensor devices for detection of different mycotoxins. A biosensor array based on surface Plasmon Resonance (SPR) has made it possible to assay several mycotoxins simultaneously (Vander Gaag *et al.*, 2003). Electronic noses are being developed to identify volatile biomarker compounds emitted from food grains facilitating the detection of different mycotoxins (Karlshoj *et al.*, 2007).

It is apparent that immunoassay techniques have gained wide acceptance as analytical tools for mycotoxins. Antibodies against almost all the important mycotoxins are currently available. Sensitive, simple and specific immunoassays have been established for the analysis of various mycotoxins in foods, feeds, animal tissues and body fluids.

(c) Immunoassay for Mycotoxin-Producing Fungi

Other than mycotoxins, antibodies against specific fungi (Lu *et al.*, 1995) as well as several enzymes involved in the biosynthesis of afla-toxin (Lu *et al;* 1997) and trichethecenes (Park and Chu, 1996) have also been generated. Five mAbs against *Aspergillus flavus were* recently produced by Candlish *et al.* (1997) and those mAbs were used in the immunoassays for identification of *A. flavus*.

A real-time polymerase chain reaction (PCR) assays are being used for detection of mycotoxigenic fungi in foods and feeds. Niessen (2007) described in detail PCR methods for the detection of fungi that produce mycotoxins such as aflatoxins, T2 toxin, DON, fumonisins and patulin in different agricultural commodities.

3. Conclusions

The established state of the art chromatography based methods for mycotoxin analysis are increasingly being complemented by number of new devices screening methods including lateral flow devices (LFDs), biosensors and IR screening techniques that are fast and cost effective. Nevertheless these techniques will have to compete with both classical confirmatory methods and MTP-ELISA which are now widely used for mycotoxin screening. Mycotoxin detection is a major problem in developing countries where contaminated food commodities may readily reach food stores and home. The development of sophisticated kits for the detection of minute amounts of mycotoxins is the most important step towards safer foods and feeds in these countries.

Acknowledgement

Thanks are due to Head, Department of Microbiology, Kakatiya University for providing necessary facilities.

References

Abouzied M M, Pestka J J (1994) J AOAC Int. 77: 495-501.

Adye J, Mateless R C (1964) Biochem. Biophy. Acta. **86**:418.

Azcona J I, Abouzied M M, Pestka J J (1990) J. Food **Prot. 53,** 577-580.

Bacigaupo MA, Ius A, Meroni G, Dovis M, Petruzzelli E (1994) Analyst 119: 2813-2815.

Badea *et al* (2004) Analytica Chimica Acta 520 (2004) 141–148

Barmark A L, Larsson K (1994) J. AOAC Internl 77: 46-53.

Barna Vetro T, Gyongyosi A, Solti L (1994) Appl. Envir. Microb. **60:** 729-731

Beier R C, Stanker L H (1996) American Chemical Society: Washington, DC, 528 pp.

Biancardi A (1997) Indust. Aliment. **36:** 870-876.

Boenke A(1997) Food Chem. **60**: 255-262

Brigys I R, Towers N R, Molan P C (1994) J. Agri. Food Chem. **42 :** 2769-2777.

Bullerman L B, Ryu D and Jackson L S (2001) Kluwer Academic/Plenum publishers, New York, pp. 195-204

Candlish A A G, Bowo M S W, Smith J E (1997) Biotecnol. Tech. **11:** 21–24

Candlish A A, Haynes C A, Stimson W H (1988) J. Food Sci. Technol. 23:479.

Chanh T C, Kennedy R C,Hewetson J F (1992) Int. J. Clin. Lab. Res. R: 28-35.

Chu F S (1991) Pub. NewYork pp. 87-136.

Chu F S (1995) In Analyzing Food for Nutrition Habelling and Hazardous contaminants. (Eds. I J Jeon and W G Ikins) Marcel Dekker New York, pp. 283-332.

Chu F S (1996) In Fumonisims in food. (eds L Jackson, J W Deveries and L B Bullerman) Plenum Press, New York. Pp. 123-134.

Chu FS (2000) In Food borne disease Handbook, Volume 3: Plant toxicant (Eds YH Hui, RA Smith and DG Spoerke), 2nd Edition, Marcel Dekker, New York. Pp. 683-713.

Clarke JR, Marquardt RR, Frohlich AA, Pitura RJ (1994) J. Food Protec. **57:** 991-995.

Corneli S, Maragos CM (1998) J. Agric. Food. Chem. 46: 3162-3165.

D Mello H J P F, Placinta C M, Macdonald A M C (1999) Animal Feed Science and Technology **80**: 183-205.

De Saeger S, Vanpeteghem C (1999) Appl. Envir. Microb. **62:** 1880-1884.

De Vries J W, Truckses M W, Jakkram L S (2002) Mycotoxins and Food safety: Advances in Experimental Medicine and Biology vol 504.(eds) Kluwer Academic/Planum Publishers, New York.

Edrington TS, Kubena LF, Harvey RB, Rotting Laus GE (1997) Poultry Sci. **79**: 1205-1211.

EC, European Commission (2006) The Rapid System for Food and Feed (RASFF) Annual Report 2005. Health and Consumer Protection Directorate-General, Office for Official Publication of the European Communities, Luxembourg.

FAO, Food and Agriculture Organization (2004) Food and Nutrition paper 81 Food and Agriculture Organization of the United Nations, Rome Italy Accessed 10 July 2007.

FDA (2007) Investigative Operation Manual. Washington (DC): Food and Drug Administration. Accessed 10 July

Fremy J M, Chu F S (1989). Immunochemical methods of analysis for aflatoxin MI. Van: H.P. van Egmond (Ed.) Mycotoxin in Dairy Products pp 97-126. Elsevier Science, London.

Garthwaite I, Sprosen J, Briggs L, Collin R, Jowers N R (1994) Food Agric. Immunol. **6:** 123-129.

Gimeno A (1979) J. Assoc.off. Anal.Chem.**63:** 579-585.

Gonzalez Pereyra M L, AlonsonV A, Sager R, Morlaco M B, Magnoli C E, Astoreca A L, Rosa C A, Chiacchiera S M, Dalcero A M, Cavaglien L R (2008) J. Appl. Microbiol 204: 2034-2042

Gorst-Allman Ch P, Steyn P S (1979) J.Chromatogr. **175**:325-331

Grochowaski A (2004) Presented at the Scientific Conference –Veterinary Food Hygiene–Zearalenone Mycotoxins–Mycotoxins detoxicants, Olsztyn, Poland, 1-12

Groopman J D, Donahue K F (1988) J. Assoc. off. Anal. Chem. 71: 861

Groopman J D, Haster J A, Trudel L J, Pikul A, Donahue P R, Wogun G N (1992) Cancer Res. 52: 267-274

Hansen T J (1990) J. Food Protect. 53: 75-77

Haung X, Chu F S (1993) J. Agric. Food Chem. 41: 329-333.

Hsu K H, Chu F S (1992) Food Agric. Immunol. **4:** 83-91.

Jodlbaur J N, Maier M, Lindner W (2002) J. Chromatogra. **945**: 45-63.

Kamimura H, Nishijima M, Yasuda K, Saito A, Nagayama I, Oshiyam H, Naoi H (1981) J. Assoc. Off. Anal.Chem.**64**:1067-1073.

Karlshoj K, Nielsenpr, Larsen T (2007) J. Food Sci. 72: M: 287-292.

Kay S (2001) European Commission Directorate General JRC, Institute for Health and Consumer Protection Accessed 10 July 2007

Kokkonen M, Jostoi M N (2009) Food Anal. Methods 2:128-140

Kramer J, Seitz P, Baltes H (1994) IEEE. J. Quantum Electronics Vol. 30. No. 11, pp. 2726-2730.

Krska R, patricia Schubert Ullrich, Alexander Molinelli, Michael Sulyok, Susan Macdonald, Colin Craws (2008) Food Addi and Contamin 25: 152-163.

Krska R, Schuhmacher R, Weingartner J, Grasser Bacer M (1997) Cereal Res. Commun. 25: 327-329

Kussak A, Andersson B and Andersson K (1994) J. Chromatogra. Biomed. Appl. 656: 329-334.

Kussak A, Andersson B, Andersson K, Nilsson C A (1998) Chemosphere 36: 1841-1848.

Li S Z, Chen P Y, Marquardt R R, Han Z K, Clarke J R (1996) J. Agric. Food Chem. **44:** 372-375.

Logrieco A, Moretti A, Perrone G, Mule G (2007) Int. J. Food. Microbiol 229: 22-26

Lu P, Marquardt R R, Kierekjar zuk D (1995) Lett. Appl. Microb. 20: 41-45.

Lu Z W R, Dantzer E C, Hopnans V, Prisk J E, Cunnick J E, Murphy P A, Hendrich S (1997) J. Agric. Food Chem. 45: 803-809.

Maragoes Ch (2009) Toxins 1; 196-207.

Maragos C M (2009) World Mycotoxin J.2, 221-238.

Maragos C M, Jolley M E, Plattner R D, Nasir M S (2001) J. Agric. Food Chem. 49: 596-602.

Maragos C M,Plattner R D (2002) J. Agri. and Food Chem. **50:** 1827-1832.

Mayura K, Abdel M A, Mc Kenzie K S, Sarr A B, Edwards J F, Naguib K, Phillips T D (1998) Toxicol. Sci. 41: 175-182.

Mc Elory, L J,Weiss C M (1993) Can. J. Microb. **39:** 861-863.

Mullett W, Lai E P C,Yeung T M (1998) Anal.Biochem 25: 161-167.

Nakajima M, Tsubouchi H, Miyabe M,Ueno Y (1997) Food and Agricultural Immunology 9, 77-83.

Nasir M S and Jolley M E (2002) J. Agric. Food Chem. **50:** 3116-3121.

Ngundi M M, Qadri S A, Wallace E V, Moore M H, Lassman M E, Shriver- lake L C, Ligler F S, Taitt C R (2006) Environ. Sci. Technol. 40: 2352-2356.

Niessen (2007) Int. J. Food Microbiol 229: 38-46.

Nisonoff A (1991) J. Immunol. 147: 2429-2438.

Park J, Chu F S (1996) J. AOAC. 79: 365-371.

Pathre S V, Mirocha C J, Fenton S W (1979) J. Assoc. Off. Anal. Chern. 62, 1268-1278

Pestka J J (1994) Food Agric. Immunol 6; 219-234

Piskorska-Pliszczynska, Jusckiewicz T (1977) Arch.Inst. Pasteur Tunis. **3-4**:279-288.

Pittet, A, (2005) Mitt.Lebensm.Hyg. 96:424-444.

Prieto-Siman B, Noguee T, Campas M (2007) Trends Analyt Chem 26: 689-702

Ramakrishna Y, Bhatt R V (1987) Curr. Sci.**56**:524-526.

Rao G V, Rao P S, Girisham S, Reddy S M (1985) Curr Sci. 54: 507-509.

Rathinavelu A, Shanmugasundaram E R B (1984) J. Assoc. off. Anal.Chem. **67**:38-40.

Reynolds G, Pestka J J (1991) J. Food Protect. **54:** 105-108

Sabbioni G and Sepai O (1998) Marcel Decker, New York. Pp. 183-226.

Sarr A B, Mayura K, Kubena L F, Harvey R B, Phillips T D (1995) Toxicol Lett. **75**: 145-151.

Schanerr AW, Hooker D C, Baute T S,Tamburic I L (2002) Plant Disease **86**: 1123

Schmitt K, Martlbauer E E, Usleber R, Gessler Lepschy J, Abranson D (1996) In Immuno assays for residue analysis: Food safety. (eds. R C Beier and L H Stanker) Acs symposium series book no. 621. American chemical Society: Washington, DC. pp. 314-321.

Schnieder E, Dietrich R, Martlbauer E, Usleber E,Terphan G (1991) Food. Agric. Immunol. **3:** 185-193.

Scott P M, Trucksess M W (1997) J. Assoc. off. Anal. Chem. **80:** 941-949.

Scott P M,Trucksess M W (1997) J. AOAC. **80:** 941-949.

Shelby R A (1996) American Chemical Society: Washington DC. Pp. 231-242

Sinha R C, Savard M E (1997) Can. J. Plant. Pathol. **19:** 8-12.

Sinha R C, Savard M E, Lau R (1995) J. Agric. Food Chem. **43:** 1740-1744.

Smith J E, Solomons G M (1994) Mycotoxins in Human Nutrition and Health. Agro Industrial Research Divison, European Commission Directorate-General XII. Pp. 1-300.

Stack M E, Pohland A E (1975) J. Assoc. off. Anal.Chem. **58**:110-113.

Stark A A (2010) Molecular Mechanism of Detection of Aflatoxins and Other Mycotoxins. In Mycotoxins in Food, Feed and Bioweapons (eds) M Rai and A Varma Springer-Verlag Berlin Heidelberg Cha 2, 21-37.

Strachen N J C, P G John, I G Millar (1997) Food Agric. Immunol 3: 177-183.

Subramanian (1982). J. Assoc. off. Anal.Chem.**65**:5-7.

Szurdoki F, Trousdale E, Gee S J, Ward B, Hammock B D,Gilchrist D (1996) In Immunoassays for Residue Analysis: food safety. (eds. Beier RC and Stanker LH) Acs symposium series book no. 621. American Chemical society: Washington, DC pp 330-340.

Tanaka T, Teshima R, Ikebachi H, Sawada J, Ichinoe M (1995) J. Agric. Food Chem. **43:** 946-950.

Tapai K, Teren J,Mesterhazy (1997) Cereal Res. Commun. **3:** 307-308.

Thirumala Devi K, Mayo M A, Reddy G, Reddy S V, Delfurse P Reddy D V (2000) J. Agric. Food Chem. 48: 5079-5082.

Trucksers M W, Pohland (2001) J. AOAC. 84.

Trucksess M W, Stack M E, Nesheim S, Albert R H, Romer T R (1994) J . AOAC Inter. 6, 1512–1521.

Usleber E, Martlbauer E, Abramson D (1996) In Immunoassays for Residue analysis: Food safety. (Beier RC and Stanker LH, eds.) ACS symposium series Book no. 621. American chemical society: Washington, DC pp 322-329.

Usleber E, Schnider E, Martlbauer E,Terplan G (1993) J. Agric. Food Chem. **41:** 2019-2023.

Vander Gaag B, Spalk S, Dietrich H, Stigter E, Boonzaaijer G, Vanosanbruggen T, Koopalk (2003) Food Control. **24:** 252-254.

Wang C R and Chu F S (1991) Appl. Envir. Microb. **57:** 1026-1030.

Warden B A, Sentissi A, Ehrat M, Cecchini D J, Allam K, Giese R W (1990). Chromatographic enzyme immuno assay for T-2 toxin. J. Immunol. Methods 131: 77.

Ware G M, Umrigar P P, Carman A S, Kuan S S (1994) Anal. Lett. **27:** 693-715.

World Health Organization (2002) Fifth – Sixth report of the Joint FAO/WHO Expert Committee on Food Additives, WHO Technical Report Series 906, World Health organization, Geneva, Switzerland p 70

Wu S G, Yang, Sun T (1983) J. Clon. Oncol. **5:** 81-84

Yacoub-Geroge E, Wolf H, Koch S, Woias P (1996) Sensors Acturators. B.B. Chem. **34:** 429-434.

Yu Fing-Yih and Chu F S (1998) J AOAC. **81:** 749-756.

Yu Fing-Yih and Chu F S (1999) Food. Agric. Immunol. **11:** 297-306.

Yu J, Wei R D, Mirocha C J and Chu F S (1990) Food Agric. Immunol. **2:** 55-64.

Yu W, Chu F S (1999) J. Agric. Food Chem. **47:** 4815-4820.

Chapter - 19

Chitin, Chitosan and Chitooligosaccharides: A Treasure in Store

Appa Rao Podile, Pallinti Purushotham, Subha Narayan Das, P.V.S.R.N. Sarma and Jogi Madhu Prakash*

Department of Plant Sciences, School of Life Sciences, University of Hyderabad, Hyderabad – 500 046, India
E-mail: podilerao@yahoo.com; arpsl@uohyd.ernet.in

ABSTRACT

Chitin is the second most abundant, next only to cellulose, polysaccharide in nature. The exploitation of chitinous materials seems to be an infinite treasure. The oligosaccharides of chitin/chitosan, prepared by hydrolyzing chitin/chitosan with chitinase/chitosanase, have various potential applications in the fields of food, agricultural, and pharmaceutical industries. Numerous studies on chitin have focused on its physico-chemical properties. In this paper, we describe various aspects of chitin research including sources, structure, the types of chitinases, their mode of action with special emphasis on transglycosylation for the synthesis of chitooligosaccharides as well as biotechnological applications of chitin and chitooligosaccharides. We project, through this review, the potential wealth in chitin, chitosan and chitooligosaccharides for wide range of applications with a special emphasis on their use in agriculture.

Key words: Chitin, Chitosan, Chitinases, Chitooligosaccharides, Transglycosylation.

1. Chitin is the second most abundant polysaccharide

Chitin, poly β-(1, 4)-*N*-acetyl-D-glucosamine, is a natural polysaccharide derivative of cellulose synthesized by an enormous number of living organisms. Considering the amount of chitin produced annually, it is the most abundant polymer after cellulose. Chitin occurs in nature as ordered crystalline micro-fibrils forming structural components in the exoskeleton of arthropods or in the cell walls of fungi. It was also projected that the chitin serves many functions in a number of other living organisms in the lower plant and animal kingdoms, where reinforcement and strength are required. The annual production of cellulose and chitin is around one trillion and 100 billion tons, respectively [Tharanathan, 2003; Kim, 2006] which makes them an almost unlimited source of raw material for environmentally friendly and biocompatible products, as well as biofuels. Despite the widespread occurrence of chitin, up to now, the main commercial sources of chitin have been crab and shrimp shells. In industrial processing, chitin is extracted from crustaceans by acid treatment to dissolve calcium carbonate followed by alkaline extraction to solubilize proteins. In addition, a decolorization step is often added to remove leftover pigments and obtain a colourless product.

2. Chitin exits at least in two different forms

Depending on its source, chitin is extracted as two allomorphs, namely α- and β-chitin [Rinaudo, 2006], which can be differentiated by infrared and solid-state NMR spectroscopy together with X-ray diffraction. A third allomorph γ-chitin has also been described, but from a detailed analysis, it seems that it is a variant of α family [Atkins, 1985]. The structures of α- and β forms differ only in that the piles of chains are arranged alternately antiparallel in α-chitin, whereas they are all parallel in β-chitin. The γ-structure shows that two chains run in one direction and another chain runs in the opposite direction. Among the chitin variants, α-chitin is the most abundant biopolymer in the nature; it occurs in fungal and yeast cell walls, in krill, in lobster and crab tendons and shells, and in shrimp shells, as well as in insect cuticle. It is also found in or produced by various marine living organisms. According to the crystalline structure, α-chitin has strong intersheet and intrasheet hydrogen bonding, and β-chitin has weak hydrogen bonding by intra-sheets. Therefore, in contrast to α-chitin, β-chitin is characterized by a weak intermolecular force, and it has been confirmed to exhibit higher reactivity and affinity for solvents than α-chitin [Kurtia, 1994].

3. Chitosan is the soluble deacetylated form of chitin

Chitin becomes soluble in aqueous acidic media and is called chitosan when the degree of deacetylation (DDA) of chitin reaches about 50% (depending on the origin of the polymer) [Rinaudo, 2006]. The solubilisation occurs by protonation of the –NH2 function on the C-2 position of the D-glucosamine repeat unit, whereby the polysaccharide is converted to a polyelectrolyte in acidic media. Chitosan was classified according to the fraction of *N*-acetylated residues (F_A) or degree of *N*-acetylation (DA), the degree of polymerization (DP) or the molecular weight (MW), the molecular weight distribution (PD, for PolyDispersity), and the pattern of *N*-acetylation (P_A) or sequence. Chitosan exhibits a variety of interesting physicochemical and biological properties. Chitosan is the only pseudo-natural cationic polymer and thus, in combination with its non-toxicity, biocompatibility and biodegradability, finds in numerous applications in agriculture, cosmetics, water treatment and medicine [Neeraja, 2010].

4. Chitin hydrolysis by chitinases is widespread across biological systems

The breakdown of chitin is catalyzed by chitinases (EC 3.2.1.14) which hydrolyze chitin to *N*-acetyl glucosamine, chitobiose or smaller chitooligosaccharide (CHOS) fragments. They are found in a wide range of organisms, including bacteria, fungi, viruses, plants, and animals. The presence of chitinases in such organisms is closely associated with the physiological roles of their substrates [Bhattacharya, 2007]. For instance, bacteria produce chitinases so that they can use chitin as a source of carbon and nitrogen for growth, whereas chitinases in yeasts and other fungi are important for autolysis, nutritional and morphogenetic functions. Plant chitinases play a role as defensive agents against patho-genic fungi and some parasites by disrupting their cell walls, whereas viral chitinases are involved in the pathogenesis of host cells. Animal chitinases are involved in dietary uptake processes. Human chitinases are particularly associated with anti-inflammatory effects against T-helper-2-driven diseases, such as allergic asthma. Chitinases from bacteria and fungi are extremely important for maintaining a balance between the large amount of carbon and nitrogen trapped in the biomass as insoluble chitin in nature [Li, 2006].

5. Chitinases are grouped in to two different families

Based on amino acid sequence similarity, chitinases are classified into families 18 and 19 of glycoside hydrolases (GH) [Henrissat and Davies, (Henrissat, 1997); described in the Carbohydrate Active enZYme (CAZy) database, http://www.cazy.org/]. The members of the two different families differ in their amino acid sequences, three-dimensional structures, and molecular mechanisms of catalytic reactions [Aronson, 2003]. Family 18 chitinases have catalytic domains of triosephosphate isomerase [TIM barrel $(\beta/\alpha)_8$] fold with a conserved DxDxE motif [Vaaje-Kolstad, 2004] on the β4-strand and catalyze the hydrolytic reaction by substrate-assisted mechanism [Terwisscha, 1996; Van Aalten, 2001], whereas family 19 chitinases comprise of high percentage of α-helices with two lobes in the catalytic domain and adopt the single displacement catalytic mechanism [Brameld, 1998; Hoell, 2006]. A complete survey of *Trichoderma* chitinases suggested that family 18 chitinases are further classifiable into subgroups A (bacterial / fungal), B (plant /fungal) and C (killer toxin-like chitinases) [Seidl, 2005]. Family 18 chitinases are represented in most living organisms, whereas family 19 enzymes are mostly found in plants, where they contribute to defence against chitinous pathogens.

6. Multiple chitinases and chitin binding proteins are involved in chitinolysis

Chitinases can be divided into two major categories: exochitinases and endochitinases [Dahiya, 2006]. Exochitinases can be further divided into two subcategories: chitobiosidases, which cleave diacetylchitobiose units from the non-reducing end of the chitin chain, ß-(1,4)-*N*-acetyl-glucosaminidases (NAGase), which cleaves *N*-acetylglucosamine (NAG) oligomers, generating NAG monomers. Endochitinases cleave glycosidic linkages randomly at internal sites along the chitin chain, eventually providing a variety of low molecular mass NAG oligomers such as diacetylchitobioses and chitotrioses [Dahiya, 2006].

Efficient chitin degradation also depends on the action of a family 33 chitin binding proteins (CBPs). The CBPs bind to the insoluble crystalline chitin, leading to structural changes and increased accessibility of substrate. The function of family 33 CBPs was first demonstrated for *Serratia marcescens* CBP21 [Vaaje-Kolstad, 2005 & 2005]. *S. marcescens* CBP21 was designated as "chitin oxidohydrolase" as it acts on the surface of crystalline chitin, to introduce chain breaks and generates oxidized chain ends, promoting further degradation by chitinases [Vaaje-Kolstad, 2010].

7. Chitinases exhibit at least two different modes of catalysis

The hydrolysis of the glycosidic linkage of chitinous substrates like chitin and chitosan is a nucleophilic substitution at the anomeric carbon, and can lead to either retention or inversion of the anomeric configuration. Both hydrolysis reactions take place through general acid catalysis, and require a pair of carboxylic acids at the enzyme's active site. One carboxylic acid is acting as a proton donor, facilitating leaving group departure, and the other acts as a base (inverting mechanism) or as a nucleophile (retaining mechanism). In both the mechanisms, the position of the proton donor is within hydrogen-bonding distance of the glycosidic oxygen. The inverting mechanism (also called the single displacement mechanism) is a "one-step" reaction, where the protonation of the glycosidic oxygen occurs simultaneously with a nucleophilic attack on the anomeric carbon by an activated water molecule. This water molecule is located between a carboxylic group and the anomeric carbon and it is activated by the carboxylic group that acts as a base. Since the water molecule approaches the anomeric carbon from the side of the cata-lytic base, this mechanism leads to inversion of the anomeric configuration.

Chitosanases belonging to families GH46, GH75 and GH80 and family 19 chitinases use the inverting mechanism [Davies,1995; Fukamizo, 1995; Iseli, 1995; Adachi, 2004; Cheng, 2006; Cantarel, 2009]. The retaining mechanism (also referred to as the double displacement mechanism) is a two-step reaction, where the first step involves the protonation of the glycosidic oxygen (by the catalytic acid) and a congruent nucleophilic attack on the anomeric carbon atom by the nucleophile (the second carboxylic acid). This attack leads to breakage of the glycosidic linkage and the formation of a covalent linkage between the anomeric carbon and the catalytic nucleophile [Vocadlo, 2001]. Subsequently, this intermediate is hydrolyzed by a water molecule that approaches the anomeric carbon from a position close to that of the original glycosidic oxygen, leading to retention of the anomeric carbon configuration. Family 18 chitinases use a special variant of the double displacement mechanism, referred to as the substrate-assisted double displacement mechanism. Here, the carbonyl oxygen atom from the *N*-acetylgroup of the sugar bound in subsite-1 act as the nucleophile, leading to formation of an oxazolinium ion intermediate. Because of this involvement of the *N*-acetyl group in catalysis, productive substrate-binding of chitosan and chitosan oligomers to family 18 chitinases requires that a NAG is bound in the -1 subsite [Terwisscha, 1995; Tews, 1997; Sørbotten, 2005; Brameld KA, 1998; Van Aalten, 2001].

Chitinases and chitosanases can have endo- or exo-activity, where both the endo- and exo-mode of action can be combined with processivity. Processive enzymes will not release the substrate after one cleavage, but remain associated with the substrate so that a new cleavage can take place as the polymer substrate slides through the substrate-binding cleft. Processivity is difficult to analyze when insoluble substrates such as chitin is degraded, but can be studied using chitosan as substrate [Sørbotten, 2005; Horn, 2006; Eijsink, 2008]. Whereas processivity generally is considered to be favourable for the hydrolysis of crystalline substrates, processivity has been shown to reduce enzyme efficiency towards soluble and more accessible polymeric substrates such as chitosan [Horn, 2006; Zakariassen, 2009]. Thus, for the industrial production of CHOS, the use of non-processive enzyme variants may be beneficial at least in some cases.

8. A few of the family 18 chitinases are transglycosylases (TGases)

Some of the GHs, besides hydrolysis of glycosidic bonds, are capable of transglycosylation (TG) to form new glycosidic bonds between donor and acceptor saccharides [Eneyskaya, 1997; Fujita, 2007; Umekawa, 2008]. In retaining GHs, the TG reaction occurs through a double-displacement mechanism [Ly, 1999]. In the first step the glycosidic oxygen is protonated by a catalytic acid, and the anomeric carbon is a target for a nucleophilic attack from the catalytic base, leading to cleavage of the glycosidic bond and to formation of a glycosyl-enzyme intermediate. In the second step, the intermediate decomposes with different possible outcomes. Hydrolysis is attained by attack of a water molecule on the glycosyl-enzyme intermediate that is assisted by the conjugated base of the catalytic acid residue and that leads to release of a hydrolyzed sugar. Alternatively, TG may occur if the water molecule is outcompeted by another acceptor such as a carbohydrate or an alcohol (Fig. 1).

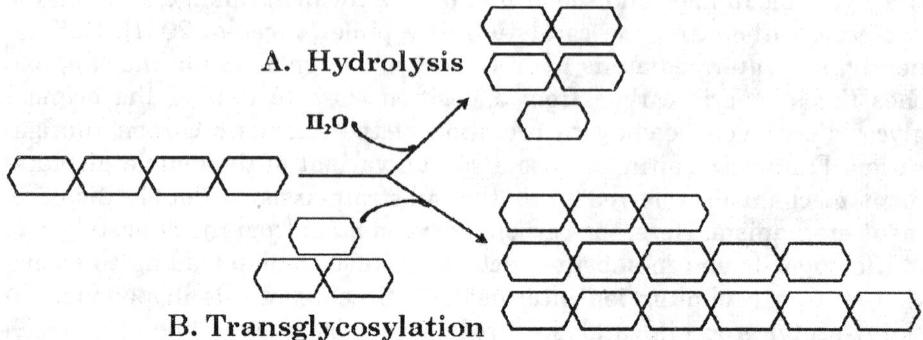

Fig. 1. Transglycosylation mechanism of glycoside hydrolases. Two possible mechanisms of glycoside hydrolases and transglycosidase. (A) Hydrolysis occurs when the glycosyl-enzyme intermediate is broken down by water, (B) Transglycosylation occurs when a sugar moiety acts as the glycosyl acceptor. *N*-acetyl glucosamine.

Since TG is a kinetically controlled reaction, efficient TG requires an enzyme with an active site architecture that disfavours correct positioning of the hydrolytic water molecule and/or favours binding of incoming carbohydrate molecules, through strong interactions in the aglycon subsites [Williams, 2000]. Family 18 chitinases are retaining GHs that are special in the sense that the nucleophilic attack is carried out by the acetamido group of the sugar bound to the -1 subsite. Consequently, the intermediate is not a covalent glycosyl-enzyme intermediate, but a noncovalent oxazolinium ion. The positioning and nucleophilicity of the acetamido group are heavily affected by residues on the enzyme, in particular by a conserved aspartate and tyrosine [Van Aalten, 2001; Bokma, 2002; Synstad, 2004]. Aspartate is part of the fully conserved diagnostic DxDxE motif that includes the catalytic acid (Glutamate) and another aspartate at a relatively buried position in the core of the [$(\beta/\alpha)_8$ (TIM-barrel) fold], which is considered the catalytic domain in family 18 chitinases.

TG effect of GH 18 chitinases was observed in wide range of organisms. These include chitinases from bacteria like *Coccidioides immitis* [Fukamizo, 2001], *Serratia marcescens* [Aronson, 2006], *Vibrio harveyi* [Suginta, 2005], *Bacillus circulans* [Sasaki, 2002] and from *Bacillus* sp.

(Wako Pure Chemical Inc.), Fungi like *Aspergillus fumigatus* [Lu, 2009], *Mucor hiemalis* [Yamanoi, 2004], *Trichoderma reesei* [Usui, 1990], the cycad *Cycas revoluta* [Taira, 2009], and plants like *Nicotiana tabacum* [Ohnuma, 2011], and *Arabidopsis thaliana* [Ohnuma, 2011b] also have TG GHs. The efficacy of TG is not constant for all the TG chitinases. It varies from enzyme to enzyme (Table 1).

Table 1. Details of representative examples of transglycosylating chitinases from different biological sources

Source	Maximum length of TG products obtained	Duration of formation * (min)	Substrate concentration (mM)	References
CrChi-A	 (DP9)	30	1.0[a]	Taira *et al* [49]
NtChiV	 (DP8)	10	5.0[a]	Ohnuma *et al.* [50]
AtChiC	 (DP8)	10	4.6[a]	Ohnuma *et al.* [51]
AcMNPV chitinase	 (DP8)	10	4.6 - 5.0[a]	Fukamizo *et al.* [105]
Sp ChiD	 (DP13)	45	3.5[a]	Purushotham and Podile (un published)
	 (DP8)		1.0[b]	
Chitinase A	 (DP8)	30	1.0[c]	Suginta *et al.* [44]
	 (DP6)		1.0[d]	

⬭ *N*-acetyl glucosamine

* Further details not available

[a]DP6 as substrate, [b]DP5 as substrate, [c]DP4 as substrate, [d]DP3 as substrate.

TG appears to be a common phenomenon in GHs and can in principle be controlled by favoring binding and correct positioning of a sugar to the intermediate. Recently Zakariassen *et al.* [Zakariassen, 2006] explained the enzymatic features that determine TG activity by investigation through site-directed mutagenesis on two family 18 chitinases, ChiA and ChiB from *S. marcescens*, with inherently little TG activity. Mutation of the middle Asp in the diagnostic DxDxE motif, which interacts with the catalytic Glu during the catalytic cycle, yielded the strongly TG mutants ChiA-D313N and ChiB-D142N, respectively. Correlations between a lower K_m and increased TG have been observed for ß-mannanase Man5A from blue mussel *Mytilus edulis* [Larsson, 2006], endo-ß-*N*-acetylglucosaminidase from *Mucor hiemalis* [Umekawa, 2008] and chitinase A from *Vibrio carchariae* [Suginta, 2005].

The TG activity of these chitinases implicates a great potential for the synthesis of CHOS and polysaccharides of chitin origin. Hypertransglycosylating chitinases may find applications for synthetic purposes. Because of an increased demand for bioactive carbohydrate oligomers, it is pertinent to have access to several technologies yielding oligomeric products of desired isomerism in high yields. An example is inhibition of the human acidic mammalian chitinase (AMCase), a family 18 Chitinase that is a therapeutic target in asthma amelioration. CHOS with specific number and sequences of acetylated and deacetylated units have been shown to inhibit family 18 chitinases due preferences of acetylated sugar moieties in individual subsite yielding nonproductive binding [Cederkvist, 2008]. Relatively short CHOS are easily produced by enzymatic hydrolysis of chitosan [Cederkvist, 2008]. A partially deacetylated soluble form of chitin can be used to produce longer CHOS of particular length and sequence of acetylated and deacetylated units. Partially deacetylated CHOS are hydrolyzed slower than fully acetylated, which is most beneficial when using TGases. Moreover, TG technology is complemen-tary to glycosynthase technology, which requires the use of mutated glycosyl hydrolases that are no longer hydrolytically active [Perugino, 2004; Shaikh, 2008].

9. Chitinases are inhibited by wide varieties of compounds

In addition to general enzyme inhibitors, such as organic compounds and oxidizing/reducing agents, a number of reports are available on the natural chitinase inhibitors. Allosamidin, an antibiotic produced by *Streptomyces* sp., is a known specific inhibitor of chitinases from insects, yeast, fungi, and human serum. Allosamidin is a pseudotrisaccharide that mimics the oxazolinium reaction intermediate. It is similar to NAG but lacks a pyranose ring oxygen and contains an oxazoline ring in which the methyl group is substituted by dimethylamine. Allosamidin is a competitive inhibitor and exerts its inhibitory effect by acting as a nonhydrolyzable analog of the oxazolinium ion intermediate [Tews, 1997]. Allosamidin inhibited the *Bacillus* sp. BG-11 chitinase by 50 and 70% at a concentration of 30 and 50 µg/ml of enzyme solution, respectively, with an IC_{50} value of 40 µM [Bhushan, 1998]. Psammaplin A, a brominated tryrosine-derived compound, was found to be a noncompetitive inhibitor of chitinase B from *S. marcescens*, a family 18 chitinase. Crystallographic studies suggest that a disordered Psammaplin A molecule is bound near the active site [Tabudravu, 2002]. Omura *et al.* [Omura, 2000] isolated a modified cyclic peptide named argifin from *Gliocladium* sp. FTD 0668. The IC_{50} value of argifin against *Lucilia cuprina* chitinase was 3.7 mM. Another chitinase inhibitor, argadin (modified cyclic pentapeptide), was isolated from *Clonostachys* sp. FO-7314 by Arai *et al.* [Arai, 2000]. The IC_{50} of argadin against *L. cuprina* chitinase was 34 nM at 20°C and 150 nM at 37°C. Cocrystallization of *S. marcescens* chitinase with argadin/ argifin revealed ordered binding of these cyclic peptides in the active site. A new chitinase inhibitor, CI-4, was isolated from *Pseudomonas* sp. IZ08 by Izumida *et al.* [Izumida, 1996].

Inhibition of family 18 chitinases is emerging as a target for pest and fungal control as well as asthma and inflammatory therapy. To this regard, it is desirable to have access to non-toxic inhibitors that are easy to produce, and have high efficiency and specificity. CHOS that are partially *N*-acetylated have the potential to fulfill these requirements. The primary cause of the inhibitory effect of CHOS comes from the fact that family 18 chitinases have an absolute preference for an *N*-acetylglucosamine in the -1 subsite. Therefore, any CHOS that preferably binds to a family 18 chitinase in such a way that a D (Deacetylation) ends up in the crucial -1 subsite will act as an inhibitor [Cederkvist, 2008].

10. Chitin and its derivatives have potential applications

The enormous amount of renewable resource like chitin and chitosan require large scale disposal and recycling [Das, 2001]. Unused renewable polysaccharides could become precious bioresources, if they are processed/ converted by modern biotechnologies to make value-added products [Zhao, 2010]. Chitin and its derivatives have high economic value owing to their versatile biological activities. CHOS are useful in various fields because they have antimicrobial activity, antitumor activity, immuno-enhancing effects, and protective effects against infection with some pathogens [Neeraja, 2010]. NAG and chitobiose, the major end products from the chitinases, also have pharmacological applications. NAG possesses anti-inflammatory effects and was used to treat ulcerative colitis, gastrointestinal inflammations, and as a nutritional substrate for pediatric chronic inflammatory bowel disease [Salvatore, 2000]. Chitobiose serves as a building block for transgly-cosylation reaction to generate longer CHOS [Kobayashi,1997]. The derivative of NAG, glucosamine, helps regeneration of joint cartilage and was used extensively to treat osteoarthritis [Huskisson, 2008]. The CHOS are of interest to the food, agriculture, and medicine-related industries in light of the diverse applications for these molecules. Use of oligochitosan as a broad-spectrum vaccine against plant diseases [Yin, 2010] highlights the need to produce specific products for agriculture. Smaller CHOS with degree of polymerization (DP) ranging from 2 to 6 were obtained from chitin by chemical or enzymatic depolymerisation, or metabolic engineering [Austin, 1981; Inaba, 1984; Samain, 2007]. Most biological activities require CHOS ≥ 4 [Hamel, 2010], but the synthesis of oligomers with ≥ 6 DP has been a daunting task. The TG activity of chitinases has been actively investigated for synthesis of higher chain CHOS.

CHOS are useful to improve the crop health. The capacity to produce CHOS on a large scale would allow the use of synthetic CHOS of desired length and sequence to induce immunity in plants. CHOS elicit a variety of defence reactions in plants and also in mammals. Despite their biological interest and potential agronomical usefulness, CHOS with well-defined structure remain poorly accessible. In the future, with more in-depth studies, we believe there will be more useful bioactive compounds generated by TGases that beneficial to human health and the environment.

10.1 Chitin/chitosan

The key property of chitin-derived products for use in various biomedical applications is the immuno-modulating effect [Aam, 2010; Muzzarelli, 2010]. Various mechanisms of immuno-enhancement activity of chitin and its derivatives have been reported. Chitin and chitosan affected C3 and C5 components of complement system and concluded that complement system is activated by chitin and chitosan through the alternative pathway. The intensity of complement and macrophage activation of chitin is less than chitosan; therefore, chitin is more biocompatible [Minami, 1998].

Hemostasis through blood coagulation is an important step for wound healing. The main cellular components in blood coagulation are platelets. Okamoto *et al.* [2003] reported that chitin is an effective agent for hemostasis maintenance through aggregating platelets, and suggested that the effect of chitin and chitosan is due to both physical and chemical properties of these biopolymers, especially their amino groups [Okamoto, 2003].

Chitin and chitosan are among the candidates to battle obesity and hypercholesterolemia, as they could reduce the amount of cholesterol in rats [Razdan, 1994]. Chitin derivatives such as *N*-succinyl-chitosan [Kamiyama, 1999], carboxymethyl chitin [Dev, 2010], chitosan hydrogel [Ishihara, 2006], and hydroxyethyl chitin [Zhao, 2006] possess essential features for drug delivery carrier. Chitin and its derivatives have antioxidant properties [Ngo, 2009], hence preventive effects on various diseases. Chitin that was chemically modified to obtain aminoethyl-chitin, showed antioxidant activity against free radicals such as 1,1-diphenyl-2-picrylhydrazyl (DPPH), hydroxyl, superoxide, and peroxyl [Je, 2006].

The wide usage of heavy metals in industry continues to cause serious and widespread health problems. It is important to remove these metals from the environment. Chitin phosphate could absorb uranium in the presence of sodium carbonate solution [Sakaguchi, 1981]. Chitin and chitosan have been shown to have copper removal capability, which could help to obtain more stable diesel oil [Peiselt, 2004]. They also have been successfully tested for the adsorption of organic pollutants [Aksu, 2005]. Chitin has also been utilized by *Clostridium paraputrificum* M-21 to produce hydrogen gas. This gas is considered to be a potential source of alternative energy [Evvyernie, 2000; Morimoto, 2005].

10.2. Chitooligosaccharides

Oligomers of chitin and chtitosan can be obtained both chemically and enzymatically [Aam, 2010]. Their degree of polymerization (DP) is usually < 20. The CHOS accelerates the wound healing effects of poly vinyl alcohol if it is used in the early stages of the healing process [You, 2004]. It could inhibit the growth of *Actinobacillus*, indicating that it has antimicrobial activity [Choi, 2001]. The CHOS causes an increase in biophotons emission from suspension-cultured rice cells. Photon emissions from rice cells elicited by *N*-acetylchitohexaose are closely associated with the ROS-generating system, and are regulated by Ca^{2+} signaling and protein phosphorylation *via* phosphatidic acid, an intermediate of phospholipid signaling [Kageyama, 2006]. CHOS play an important role in defense mechanisms of plants against microbial invasion [Day, 2001]. They could also promote carrot somatic embryos survival [De Jong, 2003]. Chitin fragments can desensitize the perception system of tomato, which can lead to improvement of the defense mechanism in tomato cells [Felix, 1998]. Chitin in the form of lipo-chitin can induce the formation of nodule in soybean root [Minami, 1996]. Rice is another important plant in which chitin fragments have been shown to boost the defense system [Ito, 1997; Okada, 2001].

CHOS have shown inhibitory effects on tumor growth and metastasis of lung cancer in mice [Shen, 2009]. *N*-acetylchitohexaose (chitin derivertive) and chitohexaose (chitosan derivertive) also induce production of interleukins 1 and 2 and, consequently, help to improve the function of macrophages, natural killers, cytotoxic T cells, and polymorphonuclear leukocytes, in defense mechanisms. There are many reports indicating antitumor activity of CHOS [Wang, 2008]. The oligosaccharides that contain *N*-acetylglucosamine play an important role in the interaction between HIV and T-helper during pathogenesis of this virus [Mizuochi, 1999]. CHOS was used as analogue to study lysozyme [Chipman, 1967]. Thus, these materials can be used to study protein-carbohydrate interaction and associated enzymatic activities.

Production of specific CHOS by chemical methods is challenging due to the need to selectively protect and manipulate chemically quite similar saccharide donors and acceptors. For this reason, development of methods for enzymatic synthesis of oligosaccharides is desirable. The transglycosylation activity of family 18 chitinases [Fukamizo, 2001; Aguilera, 2003; Boer, 2004; Chou, 2006; Taira, 2010] is of special interest because there are potential applications for CHOS [i.e., homo- or hetero-oligomers of glucosamine (GlcN) and *N*-acetylated glucosamine (GlcNAc)], especially in the food, medical, and agriculture fields [Tharanathan, 2003; Rinaudo, 2006]. The bioactivities of CHOS are thought to depend on a combination of oligomer length like DA and P_A (sequence) [Aam, 2010]. Transglycosylating chitinases have the potential to play a central role in the development of new well-defined mixtures of CHOS with new or improved biological activity, by coupling smaller CHOS building blocks to each other or to other functional groups [Zakariassen, 2011].

11. Conclusion

Several aspects of chitinases have stimulated research in the study of biochemical, regulatory, and molecular aspects of chitinolytic enzyme systems. Researchers are now directing their studies toward the discovery and engineering of novel enzymes that are more robust with respect to their kinetics and the identification of residues or domain which involved in hydrolysis/TG. In the future, protein engineering will offer the possibility of generating chitinases with entirely new functions such as improved TG. Much attention has being paid to chitinase research, but its substrate seems to have been neglected or, at least, underestimated. If we are going to use the huge amount of chitin produced annually in nature, chitin research needs an increase in funding to develop equipment to harvest chitin from its diverse sources, improve the purity of obtained chitin, and produce novel materials with new applications from this environmentally friendly biopolymer.

Acknowledgments

The research work in ARP's lab on chitin, chitosan and chitooligosaccharides is currently supported by Department of Biotechnology (DBT), Government of India in the form of major research project. The authors thank DST-FIST, UGC-CAS and DBT-CREBB for the support to the Department of Plant Sciences, University of Hyderabad. PP and PVSRNS thank DST and DBT for the fellowship, SND and JMP thank CSIR for senior research fellowship.

References

Aam BB, Heggset EB, Norberg AL, Sørlie M, Vårum KM, Eijsink VGH (2010) Mar Drugs 8:1482-1517

Adachi W, Sakihama Y, Shimizu S, Sunami T, Fukazawa T, Suzuki M, Yatsunami R, Nakamura S, Takenaka A (2004) J Mol Biol 343:785-795

Aguilera B, Ghauharali-van der Vlugt K, Helmond MTJ, Out JMM, Donker-Koopman WE, Groener JEM, Boot RG, Renkema GH, van derMarel GA, van Boom JH, Overkleeft HS, Aerts JMFG (2003) J Biol Chem 278:40911-40916

Aksu ZM (2005) Process Biochem 40:997-1026

Arai N, Shiomi K, Yamaguchi Y, Masuma R, Iwai Y, Turberg A, Kolbl H, Omura S (2000) Chem Pharm Bull (Tokyo) 48:1442-1446

Aronson NNJr, Halloran BA, Alexeyev MF, Zhou XE, Wang Y, Meehan EJ, Chen L (2006) Biosci Biotechnol Biochem 70:243-251

Aronson NNJr, Halloran BA, Alexyev MF, Amable L, Madura JD, Pasupulati L, Wort C, Patrick van Roey P (2003) Biochem J 376:87-95

Atkins EDT (1985) J Biosci 8:375-387

Austin PR, Brine CJ, Castle JE, Zikakis JP (1981) Science 212:749-753

Bhattacharya D, Nagpure A, Gupta RK (2007) Crit Rev Biotechnol 27:21-28

Bhushan B, Hoondal, GS (1998) Biotechnol Lett 20:157-159

Boer H, Munck N, Natunen J, Wohlfahrt G, Söderlund H, Renkonen O, Koivula A (2004) Glycobiol 14:1303-1313

Bokma E, Rozeboom HJ, Sibbald M, Dijkstra BW, Beintema JJ (2002) Eur J Biochem 269:893-901

Brameld KA, Goddard WA (1998) J Am Chem Soc 120:3571-3580

Cantarel BL, Coutinho PM, Rancurel C, Bernard T, Lombard V, Henrissat B (2009) Nucleic Acids Res 37:233-238

Cederkvist HF, Parmer MP, Vårum KM, Eijsink GHV, Sørlie M (2008). Carbohydr Polym 74:41-49

Cheng CY, Chang CH, Wu YJ, Li, YK (2006) Exploration of glycosyl hydrolase family 75, a chitosanase from *Aspergillus fumigatus*. J Biol Chem 281:3137-3144

Chipman DM, Grisaro V, Sharon N (1967) J Biol Chem 242:4388–4394

Choi BK, Kim KY, Yoo YJ, Oh SJ, Choi JH, Kim CY (2001) Int J Antimicrob Agents 18: 553-557

Chou YT, Yao S, Czerwinski R, Fleming M, Krykbaev R, Xuan D, Zhou H, Brooks J, Fitz, L, Strand J, Presman E, Lin L, Aulabaugh A, Huang X (2006) Biochemistry 45: 4444-4454

Dahiya N, Tewari R, Hoondal GS (2006) Appl Microbiol Biotechnol 71:773-782

Das SN, Neeraja Ch, Sarma PVSRN, Prakash JM, Purushotham P, Kaur M, Dutta S, Podile AR (2011) Exploitation of chitinolytic bacteria to chitin waste management. In: Satyanarayana, T., Johri, B.N., Prakash, A. (Eds.), Microorganisms in Environmental Management. Spinger Inc., Netherlands.

Davies G, Henrissat B (1995) Structure 3:853-859

Day RB, Okada M, Ito Y, Tsukada K, Zaghouani H, Shibuya N, Stacey G (2001) Plant Physiol 126:1162-1173

De Jong AJ, Heidstra R, Spaink HP, Hartog MV, Meijer EA, Hendriks T, Schiavo FL, Terzi M, Bisseling T, Van Kammen A, De Vries SC (1993) Plant Cell 5:615-620

Dev A, Mohan JC, Sreeja V, Tamura H, Patzke GR, Hussain F, Weyeneth S, Nair, Jayakumar R (2010) Carbohydr Polym 79:1073-1079

Eijsink VGH, Vaaje-Kolstad G, Vårum KM, Horn SJ (2008) Trends Biotechnol 26:228-235

Eneyskaya EV, Golubev AM, Kachurin AM, Savel'ev AN, Neustroev KN (1997) Carbohydr Res 305:83-91

Evvyernie D, Yamazaki S, Morimoto K, Karita S, Kimura T, Sakka K, Ohmiya K (2000) J Biosci Bioeng 89:596-601

Felix G, Baureithel K, Boller T (1998) Plant Physiol. 117:643-650

Fujita K, Sato R, Toma K, Kitahara K, Suganuma T, Yamamoto K, Takegawa K (2007) J Biochem 142:301-306

Fukamizo T, Honda Y, Goto S, Boucher I, Brzezinski R (1995) Biochem J 311: 377-383

Fukamizo T, Sasaki C, Schelp E, Bortone K, Robertus JD (2001) Biochemistry 40:2448-2454

Hamel LP, Beaudoin N (2010) Planta 232:787-806

Henrissat B, Davies G (1997) Curr Opin Struct Biol 7:637-644

Hoell IA, Dalhus B, Heggset, EB, Aspmo SI, Eijsink VGH (2006) FEBS J 273:4889-4900

Horn SJ, Sikorski P, Cederkvist JB, Vaaje-Kolstad G, Sørlie M, Synstad B, Vriend G, Vårum KM, Eijsink VGH (2006) Proc Natl Acad Sci USA 103:18089-18094

Horn SJ, Sørbotten A, Synstad B, Sikorski P, Sørlie M, Vårum KM, Eijsink VG (2006) FEBS J 273:491-503

Huskisson EC (2008) J Int Med Res 36: 1161-1179

Inaba T, Ohgusi T, Iga Y, Hasegawa E (1984) Chem Pharmac Bull 32:1597-1603

Iseli B, Armand S, Boller T, Neuhaus JM, Henrissat B (1996) FEBS Lett 382:186-188

Ishihara M, Obara K, Nakamura S, Fujita M, Masuoka K, Kanatani Y, Takase B, Hattori H, Morimoto Y, Ishihara M, Maehara T, Kikuchi M (2006) J Artif Organs 9:8-16

Ito Y, Kaku H, Shibuya N (1997) Plant J 12:347-356

Izumida H, Imamura N, Sano H (1996) J Antibiot (Tokyo) 49:76-80

Je JY, Kim SK (2006) Bioorg Med Chem Lett 16:1884-1887

Kageyama C, Kato K, Iyozumi H, Inagaki H, Yamaguchi A, Furuse K, Baba K (2006) Plant Physiol Biochem 44:901-909

Kamiyama K, Onishi H, Machida Y (1999) Biol Pharm Bull 22:179-186.

Kim J, Yun S, Ounaies Z (2006) Macromolecules 39:4204-4206

Kobayashi S, Kiyosada T, Shoda Si (1997) Tetrahedron Lett 38: 2111-2112

Kurita K, Ishii S, Tomita K, Nishimura S, Shimoda K (1994) J Polym Sci Part A: Polym Chem 32:1027-1032

Larsson AM, Anderson L, Xu B, Munoz IG, Usón I, Janson JC, St albrand H, St ahlberg J (2006) J Mol Biol 357:1500-1510

Li DC (2006) Mycopathol 161:345-360

Lü Y, Yang H, Hu H, Wang Y, Rao Z, Jin C (2009) Glycoconj J 26:525-534

Ly HD, Withers S G (1999) Annu Rev Biochem 68:487-522

Minami E, Kouchi H, Carlson RW, Cohn JR, Kolli VK, Day RB, Ogawa T, Stacey G (1996) Mol Plant Microbe Interact 9:574-583

Minami S, Suzuki H, Okamoto Y, Fujinaga T, Shigemasa Y (1998) Carbohydr Polym 36:151-155

Mizuochi T, Nakata M (1999) J Infect Chemother 5:190-195

Morimoto K, Kimura T, Sakka K, Ohmiya K (2005) FEMS Microbiol Lett 246: 229-234

Muzzarelli RAA (2010) Mar Drugs 8:292-312

Neeraja Ch, Anil K, Purushotham P, Suma K, Sarma P, Moerschbacher BM, Podile AR (2010) Crit Rev Biotechnol 30:231-241

Ngo DN, Lee SH, Kim MM, Kim SK (2009) J Funct Foods 1:188-198

Ohnuma T, Numata T, Osawa T, Mizuhara M, Lampela O, Juffer AH, Skriver K, Fukamizo T (2011) Planta 234:123-137

Ohnuma T, Numata T, Osawa T, Mizuhara M, Vårum K, Fukamizo T (2011b) Plant Mol Biol 75:291-304

Okada M, Matsumura M, Shibuya N (2001) J Pl Physiol 158:121-124

Okamoto Y, Yano R, Miyatake K, Tomohiro I, Shigemasa Y, Minami S (2003) Carbohydr Polym. 53:337-342

Omura S, Arai N, Yamaguchi Y, Masuma R, Iwai Y, Namikashi M, Turberg A, Kolbl H, Shiomi K (2000) J Antibiot 53:603-608

Peiselt da Silva KM, Pais da silva MI (2004) Colloids Surf A 237:15-21

Perugino G, Trincone A, Rossi M, Moracci M (2004) Trends Biotechnol 22:31-37

Razdan A, Pettersson D (1994) Br J Nutr 72:277-288

Rinaudo M (2006) Prog Polym Sci 31:603-632

Sakaguchi T, Horikoshi T, Nakajima A (1981) Agric Biol Chem 45:2191-2195

Salvatore S, Heuschkel R, Tomlin S, Davies SE, Edwards S, Walker-Smith JA, French I, Murch SH (2000) Aliment Pharmacol Ther 14:1567-1579

Samain E. *In*: J.P. Kamerling (Ed.) (2007) Production of Oligosaccharides in Microbes, Comprehensive Glycoscience, 1.2: Synthesis of Carbohydrates, Elsevier 923.

Sasaki C, Yokoyama A, Itoh Y, Hashimoto M, Watanabe T, Fukamizo T (2002) J Biochem 131:557-564

Seidl V, Huemer B, Seiboth B, Kubicek CP (2005) FEBS J 272:5923-5939

Shaikh FA, Withers SG (2008) Biochem Cell Biol 86:169-177

Shen KT, Chen MH, Chan HY, Jeng JH, Wang YJ (2009) Food Chem Toxicol 47:1864-1871

Sørbotten A, Horn SJ, Eijsink VG, Vårum KM (2005) FEBS J 272:538-549

Suginta W, Vongsuwan A, Songsiririthigul C, Svasti J, Prinz H (2005) FEBS J 272:3376-3386

Synstad B, Gåseidnes S, Van Aalten DMF, Vriend G, Nielsen JE, Eijsink, VGH (2004) FEBS J 271:253-262

Tabudravu JN, Eijsink VA, Gooday GW, Jaspar M, Komander D, Legg M, Synstad B, van Aalten DM (2002) Bioorg Med Chem. 10: 1123-1128

Taira T, Fujiwara M, Dennhart N, Hayashi H, Onaga S, Ohnuma T, Letzel T, Sakuda S, Fukamizo T (2010) Biochim Biophys Acta 1804: 668-675

Taira T, Hayashi H, Tajiri Y, Onaga S, Uechi G, Taira T, Hayashi H, Tajiri Y, Onaga S, Uechi G, Iwasaki H, Ohnuma T, Fukamizo T (2009) Glycobiology 19:1452-1461

Terwisscha van Scheltinga AC, Armand S, Kalk KH, Isogai A, Henrissat B, Dijkstra BW (1995) Biochemistry 34:15619-15623

Terwisscha van Scheltinga AC, Hennig M, Dijkstra BW (1996) J Mol Biol 262:243-257

Tews I, Scheltinga AC, Terwisscha V, Perrakis A, Wilson KS, Dijkstra BW (1997) J Am Chem Soc 119:7954-7959

Tharanathan RN, Kittur FS (2003) Crit Rev Food Sci Nutr 43:61-87

Umekawa M, Huang W, Li B, Fujita K, Ashida H, Wang LX, Yamamoto K (2008) J Biol Chem 283:4469-4479

Usui T, Matsui H, Isobe K (1990) Carbohydr Res 203:65-77

Vaaje-Kolstad G, Horn SJ, van Aalten DM, Synstad B, Eijsink VG (2005) J Biol Chem 280:28492-28497

Vaaje-Kolstad G, Houston DR, Riemen AH, Eijsink VGH van Aalten DM (2005) J Biol Chem 280:11313–11319

Vaaje-Kolstad G, Vasella A, Peter MG, Netter C, Houston DR, Westereng B, Synstad B, Eijsink VG, van Aalten DM (2004) J Biol Chem 279:3612-3619

Vaaje-Kolstad G, Westereng B, Horn SJ, Liu Z, Zhai H, Sørlie M, Eijsink GVH (2010) Science 330:219-222

Van Aalten DMF, Komander D, Synstad B, Gåseidnes S, Peter MG, Eijsink VGH (2001) Proc Natl Acad Sci USA 98:8979-8984

Vocadlo DJ, Davies GJ, Laine R, Withers SG (2001) Nature 412:835-838

Wang SL, Lin HT, Liang TW, Chen YJ, Yen YH, Guo SP (2008) Bioresour. Technol 99:4386-4393

Williams SJ, Withers SG (2000) Carbohydr Res 327:27-46

Yamanoi T, Tsutsumida M, Oda Y, Akaike E, Osumi K, Yamamamoto K, Fujita K (2004) Carbohydr Res 339:1403-1406

Yin H, Zhao X, Du Y (2010) Carbohydr Polym 82:1-8

You Y, Park WH, Ko BM, Min BM (2004) J Mater Sci Mater Med 15:297-301

Zakariassen H, Aam BB, Horn SJ, Vårum KM, Sørlie M, Eijsink VG (2009) J Biol Chem 284:10610-10617

Zakariassen H, Hansen MC, Jøranli M, Eijsink GVH, Sørlie M (2011) Biochemistry 50:5693-5703

Zhao Y, Park RD, Muzzarelli RAA (2010) Marine Drugs 8:24-46.

Chapter - 20

Advances in Solid State Fermentation for the production of Ligno-cellulolytic Enzymes

O.V.S. Reddy*, P.R. Mohan and B. Ramesh

Department of Biochemistry, Sri Venkateswara University,
Tirupati 517 502, India
E-mail: ovsreddy@yahoo.com

ABSTRACT

Enzymes are biocatalysts and the most important products obtained through microbial sources. There is a tremendous application of enzymes in large number of industrial processes in the areas of industrial, environmental and food biotechnology. Current developments in biotechnology are yielding new applications for enzymes. Solid state fermentation (SSF) has become very attractive and alternative to submerged fermentation (SmF) for production of ligno-cellulolyitc enzyme production. The type of bioreactor design also increases the enzyme production. The selection of an adequate support for performing solid state cultivation is essential, since the success of the process depends on it. Low cost, renewable, agro–industrial residues offers a potential substrates for enzyme production through the action of different microor-ganisms. There is an increasing demand for cellulases and xylanases in the market for various applications and the industrial uses of cellulases and xylanases has been developed rapidly. Microbial enzymes are biocatalysts that have interesting characteristics such as action under mild conditions, stability in organic solvents, high substrate specificity and reign- and enantio-selectivity. Although SmF has been commonly used in the biotechnological industry in the last decades, SSF has been reassessed in the last years, especially because of the possibility of using low-cost agro-industrial residues as raw materials.

Key words: Solid state fermentation, bioreactor, lignolytic enzymes, Cellulases, Xylanases

1. Introduction

Biotechnological conversion of biomass is potentially sustainable approach to develop novel bioprocesses and products. Enzymes are among the most important bio-products and are being utilized in a large number of processes in the areas of industrial, environmental and food biotechnology. Moreover, current developments in biotechnology are yielding new applications for enzymes (Pandey *et al.*, 2004). Microbial enzymes have become the focal biocatalysts due to their complex nature and wide spread industrial applications. There are a wide variety of microorganisms involved in enzyme production including aerobic and anaerobic bacteria, whiterot, softrot fungi and anaerobic fungi and yeast. In filamentous fungi, actinomycetes and in aerobic bacteria, enzymes are mostly secreted as free molecules. Most of the enzymes exploited for industrial applications are from filamentous fungi such as *Trichoderma, Penicillium, Fusarium, Humicola, Phanerochaete*, etc., where a large number of enzymes are encountered. Though the filamentous growth form causes difficulties in mass transfer compared to yeast or bacterial growth, efficient technologies have been developed for antibiotic, organic acid and native enzyme production from filamentous fungi (Wiebe, 2003).

In general, enzymes are produced either by submerged (SmF) and solid state fermentation (SSF) process. Both of these, solid-state fermentation (SSF) reproduces the natural microbiological processes like composting and ensiling. In industrial applications this natural process can be utilized in a controlled way to produce a desired product. In addition, it presents several advantages over the traditionally employed SmF (Table 1). Solid state fermentation (SSF) can be of special interest in processes where crude fermented product could be used directly as enzyme source. This is partly because processes involving SSF have lower energy requirements, produce less wastewater and are environmentally friendly as they resolve the problem of solid wastes disposal (Pandey, 2003).

Table 1. Advantages and disadvantages of SSF over SmF

Advantages	Disadvantages
Higher yields in a shorter time period	Problems in scale-up
Better oxygen circulation	Difficult control of process parameters (pH, heat, nutrient, conditions, etc.)
It resembles the natural habitat for filamentous Fungi	Less knowledge of the SSF process by west scientists.
Less effort in downstream Processing	Higher impurity product, increasing recovery product costs
Wild-type strains of microorganisms perform better in SSF than do genetically modified ones, reducing energy and cost requirements	

(Adopted from Robinson *et al.*, 2001)

By utilizing the low cost agricultural residues SSF adds on to economic feasibility of the process it solves the problem of its disposal which otherwise cause pollution. There has been a continuous extension of SSF arena, for the development of bioprocesses, such as bioremediation and biodegradation of hazardous compounds, biological detoxification of agro-industrial residues, biotransformation of crops and crop-residues for nutritional enrichment, biopulping, and production of value-added products, such as biologically active secondary metabolites, including antibiotics, alkaloids, plant growth factors, enzymes, organic acids, biopesticides, including mycopesticides and bioherbicides, biosurfactants, biofuel, aroma compounds, etc. (Pandey, 2003).

The cost of production in SmF systems is however high and it is uneconomical to use them in many of the processes such as enzyme production which necessitates reduction in production cost by employing alternative methods such as SSF. One of the advantages often cited for SSF processes is that enzyme titers are higher than in SmF, when comparing the same strain and fermentation broth (Viniegra-Gonz´alez, 1998), however, there are no papers explaining the reason for this. In SSF the productivity per reactor volume is much higher compared with that of submerged culture (Grajek, 1987). Also, the operation cost lower, because simpler plant, machinery and energy are required (Roche *et al.*, 1994). Recently, Viniegra-Gonz´alez *et al.* (2003) compared the productivity of three fungal enzymes, invertase, pectinase and tannase, using SSF and SmF techniques. Thus, they reported that the higher titres found in SSF than in SmF were due to SSF cultivation works as a fed batch culture with fast oxygenation but slow sugar supply and, in addition, SSF has the added advantage of being a static process without mechanical energy expenditures. Moreover, Castilho *et al.* (2000) performed a comparative economic analysis of solid-state and submerged processes for the production of lipases by *Penicillium restrictum*. They found that for a plant producing 100m^3 lipase concentrate per year, the process based on SmF needed a total capital investment 78% higher than the one based on SSF and its product had a unitary cost 68% higher than the product market price. These results showed the great advantage of the SSF due to its low cost. Several reviews on SSF have already been reported (Robinson *et al.*, 2001; Raimbault, 1998; Pandey *et al.*, 2000). They considered the general aspects of SSF and its application to the development of several bioprocesses. Although an economical and high yielding process, SSF suffers from certain drawbacks such as control of pH during growth and metabolism of the organism and operational problems during scale up. However, the SSF has been used by many researchers because of its advantages than SmF.

Tolan and Foody (1999) have described about the different facets of commercial production of cellulases using SmF whereas Cen and Xia (1999) and Holker *et al.* (2004) have evaluated the potential of SSF for production of cellulases and other commercially important products using fungi. The selection of suitable strain for the required purpose depends upon a number of factors, in particular upon the nature of substrate and environmental conditions. Generally, hydrolytic enzymes, e.g. cellulases, xylanases, pectinases, etc. are produced by fungal cultures, since such enzymes are used in nature by fungi for their growth. Filamentous fungi are preferred for commercially important enzymes production, because the level of the enzymes produced by these cultures is higher than those obtained from yeast and bacteria (Bakri *et al.*, 2003). *Aspergillus niger* and *Trichoderma viride* are the most important and safe microorganisms for industrial use (Camacho and Aguilar, 2003; Seyis and Aksoz, 2003). These had been in use already for many decades to produce extracellular enzymes (Schuster *et al.*, 2002) with homologous and heterologous proteins due to high capacity of their protein secretion machinery (Iwashita, 2002; Pandey *et al.*, 2004).

SSF generally employee the cultivation of microorganisms on moist solid supports, either on inert carriers or on insoluble substrates that can, in addition, be used as carbon and energy source. The fermentation takes place in the absence or near absence of free water, thus being close to the natural environment to which microorganisms are adapted (Pandey *et al.* 2000). More generally, SSF can be understood as any process in which substrates in a solid particulate state are utilized (Mitchell *et al.* 2000). The aim of SSF is to bring the cultivated fungi or bacteria into tight contact with the insoluble substrate and thus to achieve the highest substrate concentrations for fermentation.

This technology results, although so far only on a small scale, in several processing advantages of significant potential economic and ecological importance as compared with SmF. However, there are also several disadvantages of SSF, which have discouraged use of this technique for industrial production. The main obstructions are due mainly to the build-up of gradients of temperature, pH, moisture, substrate concentration or pO2 - during cultivation, which are difficult to control under limited water availability.

Enzyme production is an increasing field of Biotechnology. Most enzyme manufacturers produce enzymes by sub merged fermentation (SmF) techniques. However, in the last decades there has been an increasing trend towards the utilization of the solid-state fermentation (SSF) technique to produce several enzymes. SSF is known from ancient times (approximately 2600 b.c.). Table 2 gives a brief summary of the historical evolution of SSF for production of various products (Robinson *et al.*, 2001).

Though historically known since centuries, SSF gained a fresh attention from researchers and industries all over the world since recent few years, mainly due to few advantages it offers over liquid (submerged) fermentation, particularly in areas of solid waste management, biomass energy conservation and its application to produce high value–low volume products such as biologically active secondary metabolites, etc., apart from the production of food, feed, fuel and traditional bulk chemicals (Pandey *et al.*, 2007a; Pandey *et al.*, 2007b). Capability of genetic manipulation of fungal strains has broadened horizon for SSF enabling the technology for the production of recombinant proteins and value added chemicals. Literature survey reveals that much work has been carried out on the production of enzymes of industrial importance, like proteases, cellulases, ligninases, xylanases, pectinases, amylases, glucoamylases, etc.; and attempts are also being made to study SSF processes for the production of inulinases, phytases, tannases, phenolic acid esterases, microbial rennets, aryl-alcohol oxidases, oligosaccharide oxidases, tannin acyl hydrolase, a-L-arabinofuranosidase, etc. using SSF process. The present paper shows an overview of the use of SSF technique for the production of ligno-cellulolytic enzymes.

Table 2. Historical evaluation of SSF for various products

Period	Development
2,600 BC* BC in Asia (recorded history 1,000 BP**) 2,500 BP	Bread making by Egyptians.
	Cheese making by *Penicillium roqueforti*
	Fish fermentation/preservation with sugar, starch, salts, etc
	Koji process.
7th Century	Koji process from China to Japan by Buddhist priests.
18th Century	Vinegar from pomace, Gallic acid used in tanning, printing, etc
1860-1900	Sewage treatment.
1900-1920	Fungal enzymes (mainly amylases), kojic acid.
1920-1940	Fungal enzymes, gluconic acid, rotary drum fermenter, citric acid.
1940-1950	Fantastic development in fermentation industry. Penicillin production by SSF and SmF.
1950-1960	Steroid transformation by fungal cultures.
1960-1980.	Production of mycotoxins, protein enriched feed.
1980-present	Various other products like alcohol, gibberellic acid.

*BC, before Christ, **BP, before present (Adopted from Robinson *et al.* 2001).

2. General features of SSF

Solid-state fermentation (SSF) has been defined as the fermentation process which involves solid matrix and is carried out in absence or near absence of free water (Pandey *et al.*, 2000) and the substrate must possess enough moisture to support growth and metabolism of the microorganism. The solid matrix could be either the source of nutrients or simply a support impregnated by the proper nutrients that allows the development of the microorganisms. However, all the fermentation processes used in ancient times were based on the principles of solid-state fermentation technology; history indicates that it was lost in oblivion in western countries after 1940 due to emergence of submerged fermentation technology.

Perhaps SSF was neglected because development of wonder drug, penicillin took place in submerged fermentation (SmF), which was having enormous importance at that time. Research related to SSF always continued, though in isolated pockets. During 1950–1960, steroid transformation was reported using fungal culture and reports on mycotoxin production employing SSF appeared during 1960–1970, which enabled SSF to attain another milestone and it continued with the reports on production of protein enriched cattle feed by SSF utilizing agro-industrial residues, thus offering a unique process development for value addition of these low cost residues which are also considered as pollutant to some extent (Pandey, 2003). Typical examples of SSF are traditional fermentation such as Japanese "koji", Indonesian "tempeh" and French "blue cheese". SSF processes have shown to be particularly suitable for the production of enzymes by filamentous fungi (Pandey *et al.*, 1999a; Moo-Young *et al.*, 1983), since they reproduce the natural living conditions of such Fungi (Pandey *et al.*, 1999b). SSF processes may involve the pure culture of organisms, or the culture of several pure strains inoculated simultaneously or sequentially, while in some processes a "self-selected" microflora arises from the original microflora (e.g., in composting) or from a specially prepared traditional inoculum (Mitchell *et al.*, 2006). Success of SSF depends on the selection of adequate support to perform. Pastrana *et al.* (1995) and Murado *et al.* (1996) studied the production of gibberellic acid and amylases on solid-state culture using systems composed of maize cob and particles of polyurethane foam soaked, in both cases, in an amylaceous effluent. In both processes, they determined that the most important factors to take into account to select a support were particle size, porosity and chemical composition.

A number of agro-industrial residue substrates have been employed for the cultivation of microorganisms to produce host of enzymes through SSF. Some of the substrates that have been used included sugar cane bagasse, wheat bran, rice bran, maize bran, gram bran, wheat straw, rice straw, rice husk, soyhull, sago hampas, grapevine trimmings dust, saw dust, corncobs, coconut coir pith, banana waste, tea waste, cassava waste, palm oil mill waste, aspen pulp, sugar beet pulp, sweet sorghum pulp, apple pomace, peanut meal, rapeseed cake, coconut oil cake, mustard oil cake, cassava flour, wheat flour, corn flour, steamed rice, steam-pretreated willow, starch, etc. (Mitra *et al.*, 1996; Selvakumar *et al.*, 1996; Babu and Satyanarayana, 1996; Nigam and Singh, 1996; Pandey and

Radhakrishna, 1993; Pandey *et al.*, 1994; Pandey *et al.*, 1995; Selvakumar *et al.*, 1998; Benjamin and Pandey, 1997; Tengerdy, 1996).

In addition to this, availability and cost are also criteria of great importance. The wide range of solid materials used in SSF can be classified into two great categories: inert materials, which only act as an attachment place for the microorganism and non-inert materials, which not only function as an attachment place but also supply some nutrients to the microorganism (Durand *et al.*, 1993; Ralph, 1976). The latter, due to their double role, are also named support–substrate. These materials are typically starch or lingo-cellulose-based agricultural products or agro-industrial sources such as grains and grain by products (Rodriguez Couto *et al.*, 2003). In addition, the utilization of this kind of support helps in solving both the economic and the environmental problems caused by their disposal. The composition of this type of support should be selected attending to the enzyme is going to be produced, since it may contain substances acting as inducers for such enzyme.

Thus, for example, the production of lignin peroxidase was favored by the utilization of organic wastes rich in lignin (Lorenzo *et al.*, 2002), whereas employing organic wastes rich in cellulose stimulated laccase production and the use of rice, which has a high contain in starch, led to the production of amylases (Rodriguez Couto, 1999; Cruz *et al.*, 2000). Table 3 shows the composition of the typical organic wastes employed in the production of ligninolytic enzymes by SSF techniques.

Table 3. Chemical composition of several organic wastes commonly used as supports for ligno-cellulolytic enzyme production

Waste	Lignin (%)	Cellulose (%)	Hemi-cellulose (%)	Reference
Barley bran	21.4	23	32.7	Cruz *et al.*, 2000
Cane bagasse	14	33	22	Martin *et al.*, 1998
Corn cobs	20.3	31.7	31.4	Cruz *et al.*, 2000
Corn leaves	12.6	37.6	34.7	Cruz *et al.*, 2000
Hardwood	20.3	45.8	30.7	Martin *et al.*, 1998
Grape seeds	43.54	7.10	31.13	Rodriguez, 1999
Grape stalks	22.94	29.95	35.33	Rodriguez, 1999
Oat straw	18	49.3	25	Martin *et al.*, 1998
Rice straw	17	35	25	Martin *et al.*, 1998
Softwood	29.5	43.8	34.3	Martin *et al.*, 1998
Wheat straw	15	30	21	Martin *et al.*, 1998

Selection of process parameters and their optimization are other key aspects of SSF. This part includes physiochemical and biochemical parameters such as particle size, initial moisture content, initial pH, relative humidity, incubation, temperature, aeration, age and size of inoculum, supplementation of nutrients and trace elements, supplementation of additional carbon source and inducers, extraction of product and its purification

SSF has found increased application in the production of antibiotics, surfactants, biocides and enzymes. The latter include ligninolytic enzymes, especially using agro-wastes as a support (Table 4). Lignin peroxidase (LiP, EC-1.11.1.7), manganese peroxidase (MnP, EC-1.11.1.13) and laccase (EC-1.10.3.2) are the most important lignin-modifying enzymes. LiP and MnP are heme-containing glycoproteins requiring hydrogrn peroxide as an oxidant. LiP oxidizes nonphenolic lignin structures by abstracting one electron and generating cation radicals, which are then decomposed chemically. MnP oxidizes Mn (II) to Mn (III), which then oxidizes phenolic compounds to phenoxy radicals. This leads to the decomposition of the lignin substructures. Laccase, a copper containing oxidase, utilizes molecular oxygen as the oxidant and oxidizes phenolic components to phenoxy radicals. (Wong, 2009). However, there are few reports dealing with the production of ligninolytic enzymes by SSF on inert supports despite product recovery is less complicated than employing natural supports because the extracellular product can be easily extracted from the inert support and products are obtained with fewer impurities (Ooijkaas *et al*., 2000), but medium costs are higher. Therefore, although product recovery and purification processes are more expensive employing natural supports their utilization supposes a reduc-tion in production costs and usually much higher activities are obtained (Rodrıguez Couto *et al*., 2004). Hence, an economical evaluation of the overall process should be done in order to determine its feasibility to a specific purpose. This system is especially suitable for the production of high-value products like enzymes and other fermentable products.

Table 4. Ligninolytic enzymes produced by SSF technique

Support	Microorganism	Ligninolytic enzymes	References
Bagasse	*Polyporus BH1, Polyporus BW1*	Ligninase	Nigam *et al*.,1987
	T. versicolor	MnP, laccase	Pal *et al* ., 1985
	Pleurotus ostreatus, Phanerochaete chrysosporium	Laccase, MnP, LiP	Pradeep and Datta, 2002
Ballico seed	*Botryosphaeria* sp.	Laccase	Barbosa *et al*., 1996
Banana waste	*P. ostreatus, Pleurotus sajor-caju*	Laccase, LiP	Reddy *et al*., 2003
Canola roots	*Cyathus olla*	Laccase, MnP	Shinners-Carnelley *et al*., 2003
Corn	*Lentinus edodes* strain CS-495	Laccase	Annibale *et al*., 1996

	Sporotrichum sp. *Pleurotus pulmonaris*	Lignocellulo- lytic enzymes Laccase	Li *et al.*, 2000 Tychanowicz *et al.*, 2004
Cotton	*P. ostreatus, Pleurotus cystidiosus,* *P. pulmonarius, Pholiota nameko,* *P. chrysosporium, Funalia trogii*	Laccase, HRP, MnP, SOD.	Jaszek *et al.*, 1998
		Laccase, pero- xidase, Cellulase	Sik and Nyayar, 1998
Grape	*Coriolus versicolor, Panus tigrinus* *P. chrysosporium,*	Ligninase	Golovleva *et al.*, 1987
	Daedaleopsis confragosa, *Marasmius allaceus,* *P. chrysosporium.*	Laccase, MnP	Elisashvili *et al.*, 2001
	Aspergillus niger	Glycosidase	Huerta-Ochoa *et al.*, 2003
Olive mill wa- ste water	*P. tigrinus*	Laccase, MnP	Fenice *et al.*, 2003
Perlite	*P. ostreatus*	MnP, VP	Cohen *et al.*, 2002
Sawdust	*Rigidoporus lignosus*	MnP	Galliano *et al.*, 1991
	Coriolus hirsutus	Laccase, MnP	Elisashvili *et al.*, 2001
Wheat bean	*Pleurotus eryngii,* *P. pulmonarius*	MnP	Martiinez *et al.*, 1996
Wheat bran	*P. pulmonarius* *P ostreatus, P. chrysosporium* *Fomes sclerodermeus*	Laccase Laccase, MnP, LiP Laccase, MnP	Marques de Souza *et al.*, 2002 Shinners-Carnelley *et al.*, 2003 Papinutti, *et al.*, 2003
Wheat straw	*C. versicolor, P. tigrinus,* *P. chrysosporium* *P. radiata* *Pleurotus* sp. *P. chrysosporium* *Pleurotus* sp.2, *Trametes gallica* *P. chrysosporium* *P. ostreatus* *P. pulmonarius*	Ligninase LiP, MnP, laccase Laccase, MnP LiP, MnP MnP, laccase LiP, MnP Laccase Laccase	Golovleva *et al.*, 1987 Vares *et al.*, 1995 Lang *et al.*, 1996 Del Pilar-Castillo *et al.*, 1997 Xie, *et al.*, 2001 Fujian *et al.*, 2001 Baldrian *et al.*, 2002 Marques de Souza *et al.*, 2002
Wood	*Bjerkandera* sp. strain BOS55	LiP, MnP	Mester *et al.*, 1998
	Ceriporiopsis subvermispora	MnP, laccase, cellulase	Ferraz *et al.*, 2003
	P. ostreatus, P. chrysosporium	Hemicellulase, Laccase, MnP, LiP	Pradeep and Datta, 2002

3. Important features of SSF

There are various important factors that produce immense impact on success of a particular technology hence, needed to be considered for the development of any bioprocesses and so is the SSF. It includes selection of microorganism and substrate, optimum process parameters and also purification of the end product, which has been a challenge for this technology.

Fungi and yeast were termed as suitable microorganisms for SSF according to the theoretical concept of water activity, where as bacteria have been considered unsuitable. Still, availability of several research articles (Gupta *et al.*, 2008; Mukherjee *et al.*, 2008; Chinn *et al.*, 2007) proves that bacterial cultures can also be well manipulated and managed for SSF process even for scarcely produced tannase enzyme (Sabu *et al.*, 2006). *Bacillus thuringenesis* production was standardized by SSF on wheat bran to obtain maximum toxin and was found to be cost effective (Vimala Devi *et al.*, 2005).

The important aspects which should be considered before starting a solid-state fermentation include the selection of suitable microorganism and substrate, optimization of process parameters and isolation and purification of the product. SSF follows the same steps as in the SLF, namely substrate preparation, inoculum preparation, cultivation within a bioreactor and finally the downstream processing and recovery of the product. Fungi and yeast seem to be the favorable one than bacteria because of their ability to grow on the substrates, which are low in water content. Due to high water activity requirement, bacterial cultures might not be suitable for SSF. However, Pandey (1992) reported that bacterial culture can be well managed and manipulated for SSF process. It seems that the high yield in SSF as compared to SLF is due to the growth of the microorganisms similar to their natural habitat, resulting in higher metabolic activities.

The establishment of the relationship between the physiology of the microorganisms and the physico-chemical factors is the aim for the development of proper models. These factors include temperature, pH, aeration, water activity and moisture, bed properties, nature of solid substrate employed, etc. Among several critical factors moisture and nature of solid substrate employed are the most important factor affecting SSF processes. Selection of moisture depends on microorganism employed and also the nature of substrate. Fungi needs lower moisture, 40–60% moisture could be sufficient but selection of substrate depends upon several factors mainly related with cost and availability and thus may involve the screening of several agro-industrial residues (Singhania *et al.*, 2009).

4. Challenges and advantages of SSF

Today's environment is rapidly changing, where we can experience constant technological advancement backed by innovation playing major catalyst in this race. SSF appears to possess several biotechnological advantages, though at present mostly on a laboratory scale, such as higher fermentation productivity, higher end-concentration of products, higher product stability, lower catabolic repression, cultivation of micro-organisms specialized for the water-insoluble substrates or mixed cultivation of various fungi, and last but not least, lower demand on sterility due to the low water activity used in SSF (Holker *et al.*, 2004). Viniegra-Gonzalez *et al.* (2003) have attempted to develop a general approach for the comparison of productivity of enzymes employing SSF and SmF and have tried to explain the reason for higher production in SSF. Higher biomass, high enzyme production and lower protein breakdown, contributes to the better production in SSF. Scale up, purification of end products and biomass estimation are the major challenges that led the researchers to thrive hard to find the solutions.

Scale up in SSF has been a limiting factor since long but recently with advantage of biochemical engineering a number of bioreactors have been designed which could overcome the problems of scale up and to an extent also the on-line monitoring of several parameters, as well as heat and mass transfer. Separation of biomass is a big challenge in SSF, which is essential for the kinetic studies. However, certain indirect methods are there such as, glucosamine estimation, ergosterol estimation, protein (kjeldahl) estimation, DNA estimation, dry weight changes and CO_2 evolution, but all of them have their own weaknesses. Recently digital image processing has been developed as a tool for measuring biomass in SSF. The images were acquired by stereomicroscope and a digital camera and processed using KS400 software (Couri *et al.*, 2006). In recent times estimation of oxygen intake and carbon dioxide evolution rate are considered to be most accurate for the determination of growth of the microorganism (Pandey *et al.*, 2007a; Pandey *et al.*, 2007b).

Although, product recovery and purification processes are more expensive employing natural supports, their utilization supposes a reduction in production costs and usually much higher activities are obtained (Rodrıguez Couto and Sanroman, 2005). Hence, an economical evaluation of the overall process should be done in order to determine its feasibility to a specific purpose. This system is especially suitable for the production of high-value products like enzymes. There are particular applications where concentrated end products with high titers are required rather than degree of purity, for example, bioconversion of biomass, requires concentrated crude cellulase, in leather industries crude protease are enough to remove the hairs from the leather.

5. Advances in biochemical engineering aspects of solid-state fermentation

There are major challenges that need to be addressed for the successful implementation of SSF technology, such as scale up in SSF, which is a major bottle neck; providing adequate heat and mass transfer within the substrate bed, to monitor on-line several key processes parameters and also to mix the bed adequately without damaging the microorganisms as well as the substrate particles of bed (Khanahmadia *et al.*, 2006). Heat generation due to metabolic activities of the microorganisms, which is desirable in composting, is often fatal for biotechnological processes because a large part of the enzymes produced during the fermentation can be heat denatured at the end of the process (Muller dos Santos *et al.*, 2004). High moisture percentage results in low substrate porosity which in turn prevents oxygen penetration, whereas, low moisture content may lead to poor accessibility of nutrients resulting in hampered microbial growth (Pandey, 2003). The consequences of these difficulties for bioreactor performance have been well explored (Mitchell *et al.*, 2000; Mitchell *et al.*, 2003; Rodnguez Couto and Sanroman, 2006). Water relations in SSF system is one among the crucial factor, which needs to be critically evaluated. Water activity (αw) of the substrate has determinant influence on microbial activity.

Importance of αw has been widely studied by researchers as, the type of microorganism that can grow in SSF are determined by αw. The αw of the medium has been attributed as a fundamental parameter for mass transfer of the water and solutes across the microbial cells. The control of this parameter could be used to modify the metabolic production or excretion of a microorganism (Rodríguez León *et al.*, 2001).

Bioreactor design

For the majority of fermentation products, SSF gives better yields and is easier to apply. It is notoriously difficult to control the fermentation conditions in SSF; these difficulties are already apparent at small scale in the laboratory and are exacerbated with increase in scale. However, there are particular circumstances and products for which SSF technology is appropriate. For example, a desire to reuse solid organic wastes from agriculture and food processing rather than simply discarding them leads naturally to the use of SSF. Further, some microbial products, such as fungal enzymes and spores, amongst others, are produced in higher yields or with better properties in the environment provided by SSF systems. With recognition of this potential of SSF, a revival of interest began in the mid- 1970s. However, the theoretical base for SSF bioreactor technology only began to be established around 1990. However, the efficient performance of SSF bioreactors will only be achieved through: (1) the use of mathematical models in making design and operating decisions for bioreactors and (2) The application of control theory (Mitchell *et al.*, 2006). Two categories of bioreactor exist for the SSF processes: (i) at laboratory-scale, using quantities of dry solid medium from a few grams up to few kilograms, (ii) at pilot and industrial-scale, where several kilograms up to several tons are used. The first category comprises many designs, more or less sophisticated, while the second category, which is used mainly at industrial level, is markedly less varied. With in each category, some of the bioreactors can operate in aseptic conditions In spite of strong resurgent of SSF in last few years, design of bioreactor aspect have not been given enough attention from the researchers, yet; there are certain path-breaking developments.

Commonly used SSF bioreactors can be divided into four types based on type of aeration or the mixed system employed. These are tray, packed-bed, horizontal drum and fluidized bead having their own advantages and disadvantages, which promoted the necessity to develop novel bioreactors with better design. Although scale-up methods for SmF are well developed (Pandey *et al.*, 2001), ranging from successful rules-of-thumb to semi-fundamental methods, these methods could not be applied directly to SSF bioreactors: due to the differences in the physical structures of the systems, heat removal is a major consideration in the design of SSF bioreactors and has been the limiting factor for commercial viability of SSF since long. Several bioreactor designs have been developed in an attempt to combat this problem, but only a few have been used at large-scale, and information available about their performance is increasing in the form of original research articles as well as review. Durand (2003) has given relevant information on various designs of bioreactor for SSF. Review on modeling in SSF has thrown light on various designs which could solve the major problem of heat and mass transfer (Mitchell *et al.*, 2003).

Evaporative cooling has been incorporated for large scale operation of rotating drum bioreactor (RDB). The mass transfer coefficient for evaporation from the bran bed to the head space has been determined by measuring outlet water vapour concentrations for a RDB containing wet wheat bran (Hardin *et al.*, 2002). Sangsurasak and Mitchell (1998) suggested that evaporation can remove as much as 78% of the heat from the bed during the time of peak heat generation, even when bed is aerated with saturated air. This is due to the increase of water vapor pressure at increasing temperature. It results in large moisture losses and drying of the solid. A new generation of small reactors was developed by an INRA-team in France. Such reactors have a working volume of about 1 l, a relative humidity probe, a cooling coil on the air circuit and a heating cover for the vessel. Each reactor is automatically controlled by a computer (Giovannozi Sermanni and Tiso, 2007). A novel bioreactor with two dynamic changes of air (including air pressure pulsation and internal circulation) can increase mass and heat transfers, as well as improve the porosity within the substrate. The reactor can be significant in exploiting numerous socioeconomic advantages of solid-state fermentation (Hongzhang *et al.*, 2002).

Continuous processes are emerging in SSF like SmF, making it economically viable. There have been reports on development of continuous SSF process for the production of fungal tannase. A laboratory scale prototype reactor, with the specific aim in operating continuously with solid substrates and without inoculation of the feed was built and monitored successfully (Lagemaat and Pyle, 2001). Various bioreactor designs and their use for protein production under solid state fermentation have been discussed along with their advantages and disadvantages (Robinson and Nigam, 2003). Various products by different microorganisms employing different agro-industrial residues, using different reactors has been listed in Table 5.

Table 5. Different reactors for microorganisms cultivation for production of various fermentative products

Microorganism/ biocatalyst	Substrate	Product	Type of bioreactor	References
Aspergillus niger	Buckweed seed	Spores	Packed bed column	Kashyap *et al.*, 2002
A. niger	Cassava bagasse	Citric acid	Column fermenter	Di Luccio *et al.*, 2004
A. oryzae	Wheat bran + coconut oil cake (3:1)	Neutral metallo protease	Static flask	Suryanarayan *et al.*, 2003
Bacillus amyloliquefacians ATCC 23842	Ground nut oil cake + wheat bran (1:1, w/w)	Alpha amylase	Static flask	Fujian *et al.*, 2002
Bacillus subtilis	Rice bran	Iturin A	Tray	John *et al.*, 2006
B. thuringiensis	Wheat bran and bean cake powder	Bt wet powder	Cylindrical steel container	Vimala Devi *et al.*, 2005
Lactobacillus delbrueckii	Cassava bagasse	Lactic acid	Static flask	Rodríguez Couto *et al.*, 2004
Lactobacillus sp.	Tamarind seed powder	Tannase	Static flask	Sabu *et al.*, 2006
Monascus purpureus	Jack fruit seed	Pigment	Static flask	Shih *et al.*, 2008

Mortierella alpina	Rice bran	Polyunsaturated fatty acid	Column reactor	Luciana Vandenberghe *et al.*, 2000
Penicillium simplicissimum	Soybean cake	Lipase	Static flask	Ramachandran *et al.*, 2005
P. simplicissimum	Olive oil cake + sugarcane bagasse (1:1,w/w)	Lipase	Fixed bed reactor	Anisha *et al.*, 2008
Rhizopus oryzae NRRL 1891	Coconut oil cake + sesame oil cake (1:1, w/w)	Phytase	Static flask	Singhania *et al.*, 2007 b
Streptomyces griseoloalbus	Soy-bean floor	Alpha galactosidases	Static flask	Ramachandran *et al.*, 2007
Streptomyces clavuligerus	Wheat rawa with cotton seed, cake and sunflower cake	Cephamycin C	Static flask	Hung-Der and Shang-Shyng, 2008
Thermoascus auranticus	Wheat straw	Cellulase	Perforated drum bioreactor	Xiong *et al.*, 2007
Trichoderma reesei	Wheat bran	Cellulase	Static flask	Gangadharan *et al.*, 2005
T. harzianum TUBF 927	Colloidal chitin	Chitobiase	Static flask	Pandey and Soccol, 2000
Zygosaccharomyces rouxii 2547 NRRL-Y	Sesame oil cake	L-Glutaminase	Static flask	Sumantha *et al.*, 2005
Lipase from *Rhizomucor miehei*	–	Hexyl laurate	Packed-bed bioreactor	Babitha *et al* ., 2007
Erwinia sp.	Sucrose + free and calcium alginate immobilized cells	Palatinose	Packed bed reactor	Kota *et al.*, 1999
Microbial biofilm developed without prior inoculation	Clay beads + glucose-based synthetic wastewater	Organic acids and hydrogen	Horizontal packed-bed bioreactor	Hama *et al.*, 2007
Lipase producing *Rhizopus oryzae*	Cuboidal polyurethane foam biomass support particles	Bio-diesel fuel	Packed-bed reactor	Sukumaran *et al.*, 2005

Hydrogen producing sludge	Carbohydrate substrates enriched poly-ethy-lene–oc-tane elastomer immobilized anaerobic sludge	Hydrogen and ethanol	Fluidized-bed and packed-bed reactor	Singhania *et al.*, 2007
Clostridium tyrobutyricum JM1	Polyurethane foam enriched with glucose medium	Biological hydrogen	Fixed-bed bioreactor	Wu *et al.*, 2007

Modeling in SSF

The concept of modeling in SSF is the search for mathematical expressions that represent the system under consideration. These expressions have the objective to establish the relationships or functions between two different variables that characterize the system. This is to as certain the validity of the system, to establish different parameters that characterize a particular process, and to find appropriate mechanisms for development and control of the process (Pandey *et al.*, 2001).

Modeling of bioreactors employed in SSF processes can play a crucial role in the analysis, design and development of bioprocesses based on the phenomena; which encompass a variety of applications ranging from the production of enzymes to the treatment of agro-industrial residues. It can significantly reduce the number of wet experiments, which in turn saves time. Modeling in SSF has been playing major role to solve the problem of mass and heat transfer (Hardin *et al.*, 2002). A mathematical model has been developed for a packed bed SSF bioreactor employing the N-tanks in series approach, used to analyze production of protease by *A. niger* (Sahir *et al.*, 2007).

Another simple mathematical model has been used by Khanahmadiaet al. (2006), to quantify performance of continuous solid-state bioreactors having two different solid substrate flow patterns, plug flow and completely mixed flow, where plug flow shown to have superior performance when high product concentration is needed. Hamidi-Esfahani *et al.* (2004), has investigated modeling and optimization of simultaneous influence of temperature and moisture on microbial growth through a quantitative description.

A model has been analyzed, for the effect of microbial biomass on the isotherm of the fermenting solids in SSF (Marques *et al.*, 2006). Existing models of SSF processes describe coupled substrate conversion and diffusion and the consequent microbial growth neglecting many of the significant phenomena that are known to influence SSF. As a result, available models fail to explain the generation of numerous products that form during any SSF process and the outcome of the process in terms of the characteristics of the final product. Important issues that need to be resolved for improved modeling of SSF have been discussed in a review by Rahardjo *et al.* (2006). Information regarding modeling in SSF has been limited due to unavailability of suitable method for direct measurement of growth of microorganism due to difficulty to separate microorganism from the substrate and determination of rate of substrate utilization. Among the several approaches to tackle this problem, an important one has been to use a synthetic model substrate. It is well known that the fermentation kinetics is extremely sensitive to the variation in ambient and internal gas compositions. So, the cellular growth of the micro-organisms can be determined by measuring the change in gaseous compositions inside the bioreactor.

The importance of models in SSF, as in any chemical or biochemical process lies in the facts that they allow the establishment of the parameter values that explain a particular process, the basis to evaluate the process, including economically, the way in which a process could be scaled-up, and, the design and control criteria. The developments of models in SSF focused the following main problems: (i) the representation of the microbial activity (kinetic patterns and thermodynamic concerns), (ii) studies of the problems of heat and mass transfer in the solid systems, (iii) the connection between the two above systems and (iv) the selection of the best type of the fermenter.

6. Applications of SSF

In industrial applications solid-state fermentation can be utilized in a controlled way to produce the desired product. Development of bioprocesses, such as bioremediation and biodegradation of hazardous compounds, biological detoxification of agro-industrial residues, bioconversion of biomass, biotransformation of crop-residues for nutritional enrichment, biopulping, and production of value-added products, such as biologically active secondary metabolites, including antibiotics, alkaloids, plant growth factors, enzymes, organic acids, biopesticides, including mycopesticides and bioherbicides, biosurfactants, biofuel, aroma compounds,etc. have witnessed unprecedented growth of SSF. Misinterpreted' low-technology' system; appear to be a promising one for the production of value-added 'low volume-high cost' products such as biopharmaceuticals. SSF processes offer potential advantages in bioremediation and biological detoxification of hazardous and toxic compounds (Pandey and Soccol, 2000). Enzymes such as, phytase, amylase, inulinase, cellulase, protease, alpha-galactosidase, lipase, tannase, laccase, chitinase, L-glutaminase, lipase (Samantha *et al.*, 2005; Binod *et al.*, 2007; Kashyap *et al.*, 2002;

Xiong *et al.*, 2007; Ramachandran *et al.*, 2007; Fujian *et al.*, 2002; Anisha *et al.*, 2008; Gangadharan *et al.*, 2005; Cavalcanti *et al.*, 2005; Singhania *et al.*, 2007a; Ramachandran *et al.*, 2004; Ramachandran *et al.*, 2005; Rodríguez Couto *et al.*, 2004; Di Luccio *et al.*, 2004) organic acid, such as, lacticacid, citric acid and bio-ethanol, antibiotic such as cephamycin C and other metabolites as polyunsaturated acid, iturin A, pigment, hexyl laureate, palatinose, and also spores, have been successfully produced employing SSF (John *et al.*, 2006; Luciana Vandenberghe *et al.*, 2000; Shih *et al.*, 2008; Hung-Der and Shang-Shyng, 2008; Babitha *et al.*, 2007; Kota and Sridhar 1999; Chang *et al.*, 2007).

Spores are designated as small reservoirs of metabolites and can be used as biocatalyst for bioconversion reactions, also as biocontrol agents. Spores production is the only process where SSF dominates over SmF in all aspects such as, better yield, morphology, high stability and various other properties (Ramachandran *et al.*, 2007).

Biorefineries has added more value to SSF, as biomass is the only foreseeable source of energy to meet needs of the future generation, which adds to the importance of agro-residual waste (sukumaran *et al.*, 2005; Singhania *et al.*, 2007b). Cellulase production by SSF using agro-industrial residues for bio-fuel applications is in great demand. Employing packed bed, anaerobic packed bed and fluidized bioreactor, production of hydrogen, organic acids, ethanol and bio-diesel has been successfully materialized using either solid substrate or solid support (Hama *et al.*, 2007; Wu *et al.*, 2007; Jose Leite *et al.*, 2008; Jo *et al.*, 2008). Bioremediation is again a very important aspect where SSF has proved its credibility. Various bioreactors have been employed to detoxify hazardous chemicals as well as to decolorize synthetic dyes which are major chemical pollutants of the present environment. Research articles are available which shows potential of SSF for example; to decolorize a synthetic dye (congo red), a continuous decolorization system rice hull-*Schizophyllum* sp. in a packed-bed bioreactor has been successfully employed (Li *et al.*, 2008) also the anerobic decolourisation of azo dye Acid Orange (AO7) was studied in a continuous up flow stirred packed-bed reactor (USPBR) filled with biological activated carbon (BAC) (Mezohegyi *et al.*, 2008). A bench-scale packed-bedbioreactor equipped with a net draft tube riser for liquid circulation and oxygenation (PB-ALR) has been constructed for the aerobic biodegradation of the fungicide and defoliant 2, 4, 6-trichlorophenol (2, 4, 6-TCP) (Jesus *et al.*, 2009). Cells of Methylibium petroleiphilum PM1 were immobilized in gel beads to degrade methyl tert-butyl ether (MTBE) (Chen *et al.*, 2008). Thus there is hardly any bioprocess untouched by SSF.

Valuable secondary metabolites may be produced by SSF on agro-industrial residues and by-products. Biocon India manufactures three pharmaceuticals of fungal origin (lovastatin, cyclosporin and mycophenolic acid) by SSF on optimized wheat bran media in a new kind of bioreactor called the Plafractor TM (Suryanarayan *et al.*, 2001).

6.1 Enzyme production

Today, the enzymes are commonly used in many industrial applications, and the demand for more stable, highly active and specific enzymes is growing rapidly. It was estimated that in 1995, the world sale of industrial enzymes would be 1.0 billion US dollars, while the world market for industrial enzymes is expected to be in the range between 1.7 and 2.0 billion US dollars by the year 2005 (Godfrey and West, 1996a). Active research on cellulases and related polysaccharides began in the early 1950s, owing to their enormous potential to convert lignocellulose, the most abundant and renewable source of energy on Earth, to glucose and soluble sugars (Coughlan, 1985a, b; Mandels, 1985; Reese, 1976; Reese and Mandels, 1984.). About 2.9×10^3 million tons of lignocellulosic residues are available from cereal crops and 3×10^3 million tons from pulse and oilseed crops. Also 5.4×10^2 million tons is produced annually from plantation crops worldwide (FAOSTAT, 2006). Extensive basic and applied research during the 1970s and 1980s demonstrated that the enzyme-induced bio-conversion of lignocellulose to soluble sugars was rather difficult and uneconomical (Coughlan, 1985a; Ladisch *et al.*, 1983; Mandels, 1985; Ryu and Mandels, 1980). Enzymatic degradation of cellulose by fungi is of great biologic as well as economic importance. It constitutes one of the necessary steps in the balance of opposing synthetic and degradative forces in the carbon cycle, and is a major limitation to the usefulness of wood, paper, pulp, cotton, rayon, cellophane, and a host of other cellulosic materials of great and diverse utility. Biotechnology of cellulases and hemicellulases began in early 1980s, first in animal feed followed by food applications (Chesson, 1987; Thomke *et al.*, 1980; Voragen, 1992; Voragen *et al.*, 1980 & 1986). Subsequently, these enzymes were used in the textile, laundry as well as in the pulp and paper industries (Godfrey, 1996; Wong and Saddler, 1992 & 1993). During the last two decades, the use of cellulases, hemicellulases and pectinases has increased considerably, especially in textile, food, brewery and wine as well as in pulp and paper industries (Godfrey and West, 1996b; Harman and Kubicek, 1998; Uhlig, 1998).

6.1.1 Production of Cellulases

Cellulases are a complex enzyme system, comprising endo-1, 4-ß-D-glucanase (EC-3.2.1.4), exo-1, 4-ß-D-glucanase (exocellubiohydrolase, EC-3.2.1.91) and ß-D-glucosidase (ß-D-glucosidase glucanhydrolase, EC-3.2.1.21). These enzymes, together with other related enzymes, viz. hemicellulases and pectinases, are among the most important group of enzymes that are employed in the processing of lignocellulosic materials for the production of feed, fuel and chemical feedstocks. The demand for cellulases is consistently on the rise due to its diverse applications. Currently, industrial demand for cellulases is being met by production methods using submerged fermentation (SmF) processes, employing generally genetically modified strains of *Tichoderma.* The cost of production in SmF systems is however high and it is uneconomical to use them in many of its aforesaid processes. This therefore, necessitate reduction in production cost by deploying alternative methods, for example the SSF process. There are a wide variety of microorganisms involved in cellulase production including aerobic and anaerobic bacteria (Gilkes *et al.*, 1991; Kumar *et al.*, 2004; Thirumale *et al.*, 2001; Chung-Yi *et al.*, 2009) white rot and soft rot fungi (Tanaka *et al.*, 2009; Shreshta *et al.*, 2009; Lo *et al.*, 2010) and anaerobic fungi (Dashtban *et al.*, 2009; Ljungdahl, 2008). In filamentous fungi, actinomycetes and in aerobic bacteria, cellulases are mostly secreted as free molecules. Most of the cellulases exploited for industrial applications are from filamentous fungi such as *Trichoderma, Penicillium, Fusarium, Humicola, Phanerochaete*, etc., where a large number of cellulases are encountered (Lo *et al.*, 2010, De Siqueira *et al.*, 2010; Mathew *et al.*, 2008; Javed *et al.*, 2007; Bak *et al.*, 2009).

Though the filamentous growth form causes difficulties in mass transfer compared to yeast or bacterial growth, efficient technologies have been developed for antibiotic, organic acid and native enzyme production from filamentous fungi (Wiebe, 2003). These filamentous fungi produce cellulase complexes with all the three classes of enzymes at different proportion needed for the complete hydrolysis of cellulose. *Trichoderma reesei* is one among the most potent cellulase producers studied in detail. It produces two cellobiohydrolases (CBH I and CBH II) and the two endoglucanases (EG1 and EG2), in a rough proportion of 60:20:10:10, which together can make up to 90% of the enzyme cocktail; while seven ß- glucosidases-BGLI-BGLVII secreted by this fungus typically makes up less than 1% (Lynd *et al.*, 2002; Herpoë l-Gimbert *et al.*, 2008; Aro *et al.*, 2005)

Tengerdy (1996) compared cellulase production in SmF and SSF systems. While the production cost in the crude fermentation by SmF was about \$ 20/Kg, by SSF it was only \$ 0.2/Kg if *in situ* fermentation was used. The enzyme in SSF crude product was concentrated; thus it could be used directly in such agro biotechnological applications as silage or feed additive, ligno–cellulosic hydrolysis, and natural fibre (e.g. jute) processing. A number of reports have appeared on microbial cellulase production in recent years (Table 6). Nigam and Singh (1996) have reviewed processing of agricultural waste in SSF systems for cellulase production. They stated that with the appropriate technology, improved bioreactor design, and operation controls; SSF may become a competitive method for the production or cellulases. They also enumerated advantages of cellulase production together with the factors affecting the cellulase production in SSF systems. In a recent study on the lignolytic system of *Cerrena unicolor* 062–a higher basidiomycete – upon supplementation of the medium with carbon sources and phenolic compounds in SSF system, it was observed that the growth of *Cerrena unicolor* 062 could be regulated by the exogenous addition of these compounds.

Table 6. Microbial cultures employed for the production of cellulase with other enzymes in solid state fermentation systems

Substrate	Microorganisms	Enzymes	References
Bagasse	*Trichoderma reesei* + *Aspergillus niger*	Cellulase	Duenas *et al.*, 1995
Coconut coir pith	*Aspergillus niger*	Cellulase, β –glucosidase	Muniswaran *et al.*, 2004
Grapevine trimming dust	*Correna unicolor*	Cellulase, xylanase, ligninase	Zakariasvilli *et al.*, 1993
Rice husk	*Penicillium citrunum*	Cellulases	Kuhad *et al.*, 1993
Cellulose, Starch	*T. viride, A.niger*	Cellulase	Desgranges *et al.*, 1990
Ligno-cellulosic materials	*Lentinus edodus*	Various enzymes	Mistra *et al.*, 1990
Bagasse	Strains of Basidiomycetes	Cellulase, ligninase	Nigam *et al.*, 1986
Cellulosics	*T. reesei*	Cellulase	Chahal *et al.*, 1991
Agro waste	*A. niger*	Cellulase, β–glucosidase	Madamwar *et al.*, 1989
Wheat bran + rice straw, spent wheat bran	*Trichoderma* sp., *Botrius* sp., *Sporotrichum* sp.,	Cellulase, β–glucosidase, xylanase	Shamala *et al.*, 1987a
Wheat straw	*Neurospora crassa*	CMCase, β –glucosidase,	Macris *et al.*, 1987
Rice straw, spent wheat bran	*Botrius* sp., *Sporotrichum* sp.,	β –glucosidase, xylanase	Shamala *et al.*, 1987b
Wheat bran	*T.reesei*, *Sporotrichum* sp.,	Cellulase	Kim *et al.*, 1985
Wheat straw + wheat bran	*T. harzianum*	Cellulase	Deschamps *et al.*, 1985
Bagasse, wheat bran, rice bran	*Aspergillus* sp.	Cellulase, β–glucosidase	Gupte *et al.*, 1994
Rice straw	*Aspergillus niger* isolate	Cellulase, CM Case, β–glucosidase, xylanase	Mohan *et al.*, 2011

The efficiencies of the degradation of cellulose and lignin were dependent on the nature and concentration of the compounds added

(Elisashavili *et al.*, 1998). Sun *et al.* (1997) developed a novel fed-batch SSF process for cellulase production which could overcome the problems associated with high initial nutrients concentration while retaining advantages from the high total effective salt concentration.

A pan bioreactor, requiring a small capital investment, was developed for SSF of wheat straw (Chahal *et al.*, 1996; Awafo *et al.*, 1996). High yields of complete cellulase system were obtained in comparison to those in the SmF. A complete cellulase system is defined as one in which the ratio of the ß-glucosidase activity to filter paper activity in the enzyme solution is close 1.0. The prototype pan bioreactor however required further improvements so that optimum quantity of the substrate could be fermented to obtained high yields of complete cellulase system per unit space. Table 6 presents the production of cellulolyitc enzymes along with other enzymes by different fungal cultures on different substrates.

Cellulase-Market scenario: The demand for cellulases is consistently on the rise due to its diverse applications. There are several companies involved in cellulase production for textile detergent, paper industries and other industries. Globally, there are two major players known for cellulase production for biomass conversion—"Genencor" and "Novozyme". Both the companies has played a significant role in bringing down the cost of cellulase several folds by their active research and are continuing to bring down the cost by adopting novel technologies. Recently, Genencor has launchedAccelerase®1500, a cellulase complex intended specifically for lignocellulosic biomass processing industries (Genencor, 2010). It is claimed to be more cost-effective and efficient for bioethanol industries than its predecessor—Accelerase®1000. Accelerase®1500 is produced with a genetically modified strain of T. reesei. This enzyme preparation is claimed to contain higher levels of ß-glucosidase activity than all other commercial cellulases available today, so as to ensure almost complete conversion of cellobiose to glucose (Genencor, 2010). Genencor has also launched Accelerase® XY accessory xylanase enzyme complex that enhances both xylan (C5) and glucan (C6) conversion when blended with other Accelerase® enzyme products (Genencor, 2010). Similarly, Accelerase® XC is an accessory xylanase/ cellulase enzyme complex that contains a broad profile of hemicellulose and cellulase activities and enhances both xylan (C5) and glucan (C6) conversion when blended with other Accelerase® enzyme products (Genencor, 2010). Also, Accelerase®BG is an accessory β-glucosidase enzyme that enhances glucan (C6) conversion when blended with cellulase products (Genencor, 2010). There are several potential cellulases which may prove to be effective for biomass hydrolysis when supplemented with ß -glucosidase, indicating the importance of Accelerase® BG. Novozymes also have a diverse range of cellulase preparations available based on application as Cellusoft®AP and Cellusoft®CR for bioblasting in textile mills, Carezyme® and Celluclean for laundry indetergent, Denimax® 601l for stonewash industry at low temperature as well as many others specific for particular application (Bioenergy, 2010). Novozyme also announced the availability of cellulase preparation specifically for biomass hydrolysis last year, though no information is available on the source of production as well as availability in the market. Amano Enzyme Inc. in Japan and MAP's India in India are another enzyme industry actively involved in cellulase production. Though, most of the enzyme producing companies worldwide is involved in production and marketing of cellulases for diverse applications, there are very few of them who develop cellulases for biomass conversion, the most successful of them probably being Genencor and Novozyme.

Recently, the bacterial strain B39, previously isolated from poultry manure compost in Taichung, Taiwan, was identified through 16S rRNA gene sequencing and phylogentic analysis to be a novel cellulose-degrading *Paenibacillus* sp. strain. A high-molecular weight (148 kDa) cellulase, possessing both CMCase and Avicelase activities, was found to be secreted by this isolate into the media. The CMCase activity of the newly isolated cellulase was much higher than the activity on Avicel or filter paper and this cellulase was found to have maximum CMCase activity at 60°C, pH 6.5. Due to the promising thermo stability and slight acidic tolerance of this enzyme, it has good potential for industrial use in the hydrolysis of soluble cellulose as well as activity on microcrystalline sources of cellulose (Wang *et al.*, 2008).

Desirable characteristics of the cellulase preparations for bioconversion: Biological conversion of cellulosic biomass to fuels and chemicals offers the high yields to products vital to economic success and the potential for very low costs. Lot of specific features, such as higher catalytic efficiency on insoluble cellulosic substrates, increased stability at elevated temperature and at a certain pH, and higher tolerances to end-product inhibition are desirable in cellulases for biomass conversion (Maki *et al.*, 2009).The organism *Bacillus* strain M-9 has the potential to produce highly thermostable and acid/alkalistable endoglucanase which could have potential applications for wide range of industries. novel enzyme compositions using newly identified and isolated *C.lucknowense* enzymes, including CBH Ib CBH IIb, EG II, EG VI, β-glucosidase, and xylanase II in conjunction with previously identified enzymes CBH Ia, CBH IIa (previously described as Endo 43), and EG V.

These enzyme compositions demonstrate an extremely high ability to convert lignocellulosic biomass (e.g., Avicel, cotton, Douglas fir wood pretreated by organosolvents) to glucose. CBH Ia and IIb, which both have a cellulose-binding module (CBM) displayed a pronounced synergism with three major endoglucanase (EG II, EG V, EG VI) from the same fungus in hydrolysis of cotton as well as a strong synergy with each other.

The enzyme compositions are effective in hydrolysis of the lignocellulosic biomass (Maki *et al.*, 2009).

Acidic cellulases are usually more desirable for bioconversion since in situations where acidic pretreatment of biomass is employed or when the enzyme has to work in cocktail with *T. reesei* enzymes whose optimal activity is in the close to 5, a pH range of 4–6 is preferred. Most of the cellulases tried or recommended or for bioconversion perform better at pH range 4–6. Accelerase®1500 has an optimal pH of 4.6–5.0 but it gets inactivated below 4.0 or above pH 7.0. Usually 50°C temperature is employed for hydrolysis by cellulases (Maki *et al.*, 2009), even Accelerase®1500 works well at temperature ranging from 50 to 65°C though at lower ends of the temperature range, the effective life time of enzyme is higher. Most of the filamentous fungi such as *T. reesei* and *A. niger* produce acidic cellulases, which are suitable for biomass hydrolysis.

Advancement in research in bioprocess of cellulases: The combustion of petroleum-based fossil fuels has become a concern with respect to global climate change due to accelerated carbon emissions. Due to the Burning of fossil fuels, the rising cost of fuels, production of low titers of cellulase has always been a major concern and thus several workers are trying to improve the production titers by adopting multifaceted approaches, which include the use of better bioprocess technologies, using cheaper or crude raw materials as substrates for enzyme production, bioengineering the microorganisms etc. (Sukumaran *et al.*, 2005; Lynd *et al.*, 2002). The ability to secrete large amounts of extracellular protein is characteristic of certain fungi and such strains are most suited for production of higher levels of extracellular cellulases. One of the most extensively studied fungi is *Tricoderma reesei,* which converts native as well as derived cellulose to glucose. Most commonly studied cellulolytic organism include: Fungal species- *Tricoderma, Humicola, Penicillium, Aspergillus;* Bacteria: *Bacillus, pseudomonds, cellulomonas* and *Actrinomycetes- streptomyces, Actinomucor*. Present knowledge regarding the mechanism of enzymatic digestion of cellulose by fungal degradation of cellulosic materials now generally accepted provides that the actively growing hyphae secrete extracellular enzymes that diffuse through a film of water coating the surfaces of the material. These enzymes hydrolyze the long cellulose molecules to cellobiose. This product diffuses into the fungous cells where it is further hydrolyzed to glucose and metabolized by intracellular enzymes to provide the energy and substance needed by the organism for continued growth.

6.1.2 Production of xylanases

Xylanase enzyme catalyzes the hydrolysis of 1,4-β-D-xylosidic linkages in xylan, a hemicellulose in pulp. Xylanases have been reported to be produced by a vast variety of prokaryotes and eukaryotes including bacteria, fungi, protozoa, insects, crustaceans and germinating seeds (Dekker and Richards, 1976). Cellulases and xylanases (endo-1-4-β-D-xylanase, EC-3.2.1.8) however, find applications in several other areas, like in textile industry for fibre treatment and in retting process. Xylanases find specific application in jute fibre up gradation also. Although xylanases produced by fungi, yeast and bacteria, filamentous fungi are preferred for commercial production as the levels of the enzyme produced by fungal cultures are higher than those obtained from yeast or bacteria.

In many microorganisms, xylanase activity has generally been found in association with cellulases, β-glucosidase or other enzymes, although there are many reports that have described in SSF systems, production of cellulase free-xylanses (Table 7). Haltrich *et al.* (1996) reviewed the different factors that influence xylanase production by fungi. In view of the considerable commercial importance of enzymes, it was emphasized that efforts should be directed towards enhanced enzyme production with reduced associated costs.

Table 7. Microbial cultures employed for the production of xylanase with other enzymes in solid state fermentation systems

Substrates	Microrganisms	Enzyme	References
Bagasse	*T. reesei, A. niger, A. pheonics*	Xylanase	Gutieera-Correa *et al.*, 1998
Wheat straw	*Phlebia radiate, P. eryngii*	Xylanase, laccase, aryl-alcohol oxidase	Guadalix *et al.*, 1997
Wheat bran	*Bacillus licheniformis*	Xylanase	Archana *et al.*, 1997
Wheat straw	*P. sajor caju*	Xylanase, cellulase, β-glucosidase	Bisaria *et al.*, 1997
Rice straw + cotton stalks + Kernal	*L. edodus*	Xylanase, laccase β-xylosidase	Giovannozziserm-anni *et al.*, 1996
Rice straw, soy-bean hull, whe-at bran, birch-wood, oat spent, kraft pulp, beet pulp	*A. sojae*	Xylanase	Kimura *et al.*, 1995
Wheat/rice straw, bagasse, rice husk	*Melanocorpus albomyces*	Xylanase, "acetoesterase	Jain *et al.*, 1995
Wheat bran, apple pomace, sugar beet pulp, Wheat straw	*Chaetomium globosum, A. niger*	Xylanase	Wiacek-Zychlinska *et al.*, 1994

Rice straw, corn hull, corncobs, bagasse, wheat bran	*A. fumigatus*	Xylanase, xylosidase	Kitpreechavanich *et al.*, 1992
Wheat bran	*Thermoascus aurantiacus*	Xylanase	Alam *et al.*, 1994
Apple pomace	*A. niger, A. fumigatus, T. viride*	Xylanase, CMCase	Bhalla *et al.*, 1993
Coffee processing plant waste	*Thermomonospora* sp.	Xylanase, cellulase, α-arabinofuranosidase, β-xylosidase	Srivastava *et al.*, 1993
Solka floc, wheat bran, barley straw, oat straw, wheat straw	*Taleromyces emersonii*	Xylanase	Tuohy *et al.*, 1991
Wheat bran	*A .awamori*	β-xylosidase	Kurakake *et al.*, 1997
Wheat bran	*T. harzianum*	β-xylosidase	Ximenes *et al.*, 1996
Wheat bran	*A. fumigatus*	β-xylosidase, β-glucosidase	Kitpreechavanich *et al.*, 1984

Archana and Satyanarayana (1997) described a SSF process for the production of thermostable xylanase by thermophilic *Bacillus licheniformis*. Enzyme production was 22-fold higher in SSF system than in SmF system. Cai *et al.* (1997) also reported production of a thermostable xylanase in SSF system. Enzyme produced in SSF system was more thermostable than in SmF system. Jain (1995) too described a SSF process for the production of xylanase by thermostable *Melanocorpus albomyces*.

Alam *et al.* (1994) using SSF process, isolated a thermostable cellulase free xylanase produced by *T. lanuginosa*. Addition of 0.7 % xylan induced enzyme production to an extent of 28%. The enzyme was stable at 70°C. A thermostable xylanase preparation from *Humicola* sp. showed the temperature optima at 75°C (Da-Silva *et al.*, 1994). Srivastava (1993) reported a xylanase from *Thermomonospora* sp., which was stable at 80°C. Touhy and Coughlan (1991) compared thermostable xylanase production on various substrates by a strain of *Telaromyces emersonii* in liquid culture and SSF systems. The latter showed higher enzyme activity compared to former, but liquid culture resulted in greater yields (U/g substrate).

Several authors have compared the performance of various microbial strains, grown on different substrates (individual or in combination and reported varying results. Wiacek-Zychlinska *et al.* (1994) compared xylanase production by *C. globosum* and *A. niger* on four different substrates. Although activities obtained by *A. niger* were higher than those from the other microbial cultures, but high-spore production by the *A.niger* strain could result in problems for a pilot plant or large-scale process.

Agro-industrial residues are used as substrates for the cultivation of microorganisms to produce xylanolytic enzymes. Table 7 represents the production of xylanase enzyme on different substrates.

Xylanase market scenario: The increasing of lignocellulosic material in the environmental is leads to pollution that's why these renewable carbon source can be utilized for the production of xylanases. several companies can utilize the these chiefly available in the earth, for the production of variety of xylanases for the use of many industries like pulp, poultry,textile, paper krafting and bleaching, feed etc.

The DSM Nutritional product enzyme Ronozyme® wx is an outstanding xylanase designed to meet the requirements of modern poultry and pig feeding. Characterized by exceptional heat stability, a broad pH range and a high, consistent level of xylanase activity, Ronozyme® WX can be used as the sole NSP enzyme in diets based on wheat, rye and triticale or these cereals in combination with barley and cereal by-products. Ronozyme® WX is derived from *Thermomyces-lanuginosus* spp., an organism that thrives in hot, dry conditions. This benefit is passed on, ensuring that the enzyme can remain active even at high temperatures. Ronozyme® WX is an endoxylanase which hydrolyses arabinoxylans and xylans in feeds. Plant polysaccharides (long-chain carbohydrates) play an important role in feedstuffs. They can function as a major nutritional component in the diet (e.g. starch) but a number of non-starch polysaccharides (NSP's) are detrimental to non-ruminants such as poultry and pigs. Poultry and pigs are not capable of digesting these NSP's due to a lack of the relevant enzymes in their digestive tracts (DSM nutritional products, 2011).

ALKX (L/P) wide range Xylanase enzyme was a Xylan hydrolyzing Xylanase enzyme with heat and pH stability derived from a genetically modified strain of Aspergillus. Xylans are polysaccharides composed of ß-1, 4- linked Xylopyranose units. They are one of the major constituents of plant cell walls and account for more than 30% of the dry weight of terrestrial plants. The enzyme falls into the category of endo- ß-1, 4-xylanase which is of prime importance for hydrolysis of Xylan polymers in Cereal Grains and Lignocellulosic Biomass. The enzyme breaks Xylan to short chain xylo-oligosaccharides of varying length (Enzyme-India, 2011).

Desirable characteristics of the xylanase preparations for bioconversion: Xylan, the second most abundant polysaccharide and a major component in plant cell wall consists of ß-1, 4-linked xylopyranosyl residues. The plant cell wall is a composite material in which cellulose, hemicellulose (mainly xylan) and lignin are closely associated. 1-2 Three major constituents of wood are cellulose (35-50%), hemicellulose (20-30%)

- a group of carbohydrates in which xylan forms the major class- and lignin (20-30%).

The importance of xylanases is not bound to the paper and pulp industry and there are other industries with equal importance of applicability. Potential applications of xylanases also include bioconversion of lignocellulosic material and agro-wastes to fermentative products, clarification of juices, improvement in consistency of beer and the digestibility of animal feed stock etc.

Advancement in research in bioprocess of xylanases: Xylanolytic enzymes from microorganism have attracted a great deal of attention in the last decade, particularly because of their biotechnological potential in various industrial processes (Lynd *et al*., 2002), such as food, feed, and pulp and paper industries. Appreciable quantities of xylan are present in materials released from wood during pulping and pulp processing. It is presently regarded as waste and often is deposited in streams and rivers where it is ecologically harmful. Considerable amounts are also present in agricultural residues.

The conversion of xylan to useful products represents part of our efforts to strengthen the overall economics of the processing of the lignocellulose biomass, and also to develop new ways of energy production from renewable resources (Biely, 1985; Kuhad and Singh, 1993). xylanase can be produced by the fungal strains- *Aspergillus awamori*, *Fusarium oxysporum*, *Thermomyces lanuginosus*; Bacteria- *Bacillus SSP*-34, *Clostridium absonum* (Subramanyam *et al*., 1997).

Currently used bioprocesses and the status of xylanase production based on literature are provided in A direct comparison of the xyalanase yields and activities in these reports are impossible due to the differences in the methods of assays like Fpase, Cmcase and the way the activities are being expressed despite the fact that there is an IUPAC approved method of assay for xylanase activity determination. Moreover, there is no way of comparing xylanase produced by SSF and SmF directly.

Engineered/artificial cellulases and xylanases: When enzymes attack polymeric substrates at different sites and create new sites for each other, they act synergistically, i.e., their combined activity is greater than the sum of the individual activities. Enhancing enzyme synergy to improve industrial processes is an important biotechnological challenge. Enhancing the synergy among different glycoside hydrolases (i.e., cellulases, xylanases, and other plant cell wall polysaccharide-degrading enzymes) could have a major impact on reducing environmental pollution and improving bioenergy production. Dilute acid pretreatment of hard woods tends to yield residual biomass that is high in native cellulose and very low in hemicellulosic sugars. In the early 1990s, NREL researchers determined that a 90:9:1 mixture of a cellobiohydrolase from *T. reesei* (CBH I), a thermal-tolerant endoglucanase from *A. cellulolyticus* (EI), and a β-D-glucosidase was capable of nearing the performance observed for a leading commercial *T. reesei* preparation at comparable protein loading (i.e., end-point saccharification of cellulose in pretreated yellow poplar after 120 h).

This discovery was an important breakthrough because the option to engineer cellulase systems for specific pretreated biomass feedstocks became apparent for the first time. *T. reesei* and other filamentous fungi produce non-complexed cellulases. Cellulase engineering for non-complexed cellulase systems contains three major research directions : (1) rational design for each cellulase, based on knowledge of the cellulase structure and the catalytic mechanism (Schulein, 2000; Wilson, 2004) (2) Expression cassette and directed evolution for each cellulase, in which the improved enzymes or ones with new properties were selected after random mutagenesis and/or molecular recombination (Cherry and Fidantesf, 2001; Arnold, 2001; Hibbert *et al.*, 2005; Shoemaker *et al.*, 2003) and (3) the reconstitution of cellulase cocktails active on insoluble cellulosic substrates, yielding an improved hydrolysis rate or higher cellulose digestibility (Sukumaran *et al* 2009; Boisset *et al.*, 2001; Himmel *et al.*, 1999; Kim *et al.*, 1998; Zhang and Lynd, 2004). Improvements in specific cellulase activities for non-complexed cellulase mixtures can be implemented through cellulase engineering based on rational design or directed evolution for each component of cellulase, as well as its reconstitution (Zhang *et al.*, 2006). Artificial cellulase designing appears to open a new avenue to this field and seems to be more promising for creating cellulases with desired features. Highly thermostable class II cellobiohydrolase (CBH II) chimeras were developed by structure guided recombination of three fungal class II cellobiohydrolase (CBH II) (Heinelman *et al.*, 2009). Increasing the thermostability and improvement in activity of CBH II were two major goals. Optimized mixtures of cellulase for biomass conversion could be formulated once inventories of stable cellulases are available.

Synergism among enzymes of the free-cellulase-producing, thermophilic, aerobic bacterium *Thermobifida fusca* has received much attention, owing to the potential industrial advantages of these enzymes, including high thermostability properties, broad pH range, and high activity. In this context, synergism among *T. fusca* xylan-degrading enzymes (Zhang *et al.*, 2006; Heinelman *et al.*, 2009). Designer cellulosomes are precision-engineered multienzyme complexes in which the molecular architecture and enzyme content are exquisitely controlled. This system was used to examine enzyme cooperation for improved synergy among *Thermobifida fusca* glycoside hydrolases. Two *T. fusca* cellulases, Cel48A exoglucanase and Cel5A endoglucanase and two *T. fusca* xylanases, endoxylanases Xyn10B and Xyn11A were selected as enzymatic components of a mixed cellulase/xylanase containing designer cellulosome.

The resultant mixed multienzyme complex was fabricated on a single scaffoldin subunit bearing all four enzymes. Conversion of *T. fusca* enzymes to the cellulosomal mode followed by their subsequent incorporation into a tetravalent cellulosome led to assemblies with enhanced activity (~2.4-fold) on wheat straw as a complex cellulosic substrate. The enhanced synergy was caused by the proximity of the enzymes on the complex compared to the free-enzyme systems. The hydrolytic properties of the tetravalent designer cellulosome were compared with the combined action of two separate divalent cellulase- and xylanase-containing cellulosomes. Significantly, the tetravalent designer cellulosome system exhibited an ~2-foldenhancement in enzymatic activity compared to the activity of the mixture of two distinct divalent scaffoldin-borne enzymes. These results provide additional evidence that close proximity between cellulases and xylanases is key to the observed concerted degradation of the complex cellulosic substrate in which the integrated enzymes complement each other by promoting access to the relevant polysaccharide components of the substrate. The data demonstrate that cooperation among xylanases and cellulases can be augmented by their integration into a single designer cellulosome (Bachuman and Mc Carhy, 1991).

6.2 Production of other high-value byproducts by SSF

Currently a number of products such as organic acids, amino acids, vitamins and a number of bacterial and fungal polysaccharides such as xanthan are produced by fermentation using glucose as the base substrate but theoretically these same products could be manufactured from "lignocellulose waste". Based upon the predicted catabolic pathway and the known metabolism of *Phanerochaete chrysosporium* of lignin, (Ribbons *et al.*, 1987) presented a detailed discussion of the potential value added value added products which could be derived from lignin. Vanillin and gallic acid are the two most frequently discussed monomeric potential products which have attracted interest. Vanillin extraction from *Vanilla* pods costs between $ 1200 to $ 4000 per kilogram whereas synthetic vanillin costs less than $15 per kilogram (Walton *et al.*, 2003).

Vanillin is used for various purposes including being an intermediate in the chemical and pharmaceutical industries for the production of herbicides, anti-foaming agents or drugs such as papaverine, L-dopa and the anti microbial agent, trimethoprim. It is also used in household products such as air-fresheners and floor polishes (Walton *et al.*, 2003). The high price and limited supply of natural vanillin have necessitated a shift towards its production from other sources (Priefert *et al.*, 2001). Hemicelluloses are of particular industrial interest since these are a readily available bulk source of xylose from which xylitol and furfural can be derived. Xylitol used instead of sucrose in food as a sweetner, has odontological applications such as teeth hardening, remineralisation, and as an antimicrobial agent, it is used in chewing gum and toothpaste formulations (Roberto *et al.*, 2003). The yield of xylans as xylitol by chemical means is only about 50-60% making xylitol production expensive. Various bioconversion methods, therefore, have been explored for the production of xylitol from hemicellulose using microorganisms or their enzymes (Nigam *et al.*, 1995). Furfural is used in the manufacture of furfural phenol plastics, varnishes and pesticides (Montané *et al.*, 2002). Over 200 000 tones of furfural with a market price of about $1700 per ton (Montané *et al.*, 2002) is annually produced (Zeitch, 2000).

7. Conclusions

Accumulation of abundant lignocellulosic substrates may cause environmental pollution. However the advantage of SSF offers the using of these renewable sources for the production of value added products in which enzymes plays critical role in the conversion of many intermdietes into valuble products such ethanol, lactic acid, citric acid etc. Using of low cost agricultural byproducts will also reduce the production cost of enzymes a main drawback in enzyme production industries. Ligno-cellulolytic enzymes offers numerous applications in many industries paper, pulp, detergent, food, bioethanol industries, etc. It has been well established that enzyme titres produced in SSF systems are much higher than the achieved in SmF ones. Artificial/engineered cellulases and enzyme cocktails rich in glucose tolerant β-glucosidase has been proved successful for increasing the rate or efficiency of hydrolysis of biomass so as to prove the technology economically feasible. Inspite of advanatages SSF over SmF it also need for improving the bioreactor design to increase the enzyme production.

Acknowledgements

We are thankful to UGC, New Delhi for providing funds. We also wish to thank Dr. S.C. Basappa, former Deputy Director and Scientist, Central Food Technological Research Institute (CFTRI), Mysore, for his encouragement and critical comments on the manuscript.

References

Alam M, Gomes I, Mohiuddin G and Hoq MM (1994) Enzyme Microbiol. Technol. 16: 298-302.

Anisha GS, Rojan PJ, Nicemol J, Niladevi KN and Prema N (2008) Food Chem. 111: 631–635.

Annibale AD, Celleti D, Felici M, Di Mattia E and Giovannozzi- Sermani G (1996). Acta Biotechnol. 16: 257–270.

Archana A and Satyanarayana T (1997). Enzyme Microbiol.Technol. 21: 12-17.

Arnold FH (2001). Nature. 409:253–7.

Aro N, Pakula T and Penttilä M (2005) FEMS Microbiol. Rev. 29: 719-739.

Awafo VA, Chahal DS, Simpson BK and Le GBB (1996) Appl. Biochem. Biotechnol. 57/58: 461– 470.

Babitha S, Soccol CR and Pandey A (2007) Bioresour. Technol. 98: 1554–1560.

Babu KR and Satyanarayana T (1996) J.Sci.Ind.Res. 55: 464-467.

Bachuman SL and Mc Carhy AJ (1991) Appl. Environ. Microbial. 57:2121–2130.

Bak JS, Ko JK, Choi I-G, Park Y-C, Seo J-H and Kim KH (2009) Biotechnol. Bioeng. 104:471–82.

Baldrian P and Gabriel J (2002) Folia Microbiol. 47:385–390.

Barbosa TCP, Luz AP, Bedin ML and Gabilan NH (1996). Int. Biodeterior. Biodegrad. 37: 122.

Benjamin S and Pandey A (1997) Acta Biotechnol. 17: 241-251.

Bhalla TC and Joshi M (1993) Indian J. Microbiol. 34 253–255.

Biely P (1985) Microbial xylanolytic systems. Trends Biotechnol. 3:286–290

Binod P, Sandhya C, Suma P, Szakacs G and Pandey A (2007) Bioresour. Technol. 98: 2742–2748.

Bioenergy Website: 01/14/2010 http://www.bioenergy.novozymes.com/cellulosic-biofuel/ 01/14/2010.

Bisaria R, Madan M and Vasudevan P (1997) Bioresour. Technol. 59: 5-8.

Boisset C, Petrequin C, Chanzy H, Henrissat B and Schulein M (2001) Biotechnol. Bioeng. 72:339–345.

Cai J, Wu K, Zhang J, He X and Pang R (1997). Ind. Microbiol. 27: 1-4.

Camacho NA and Aguilar OG (2003). Appl. Biochem. Biotechnol.104: 159-172.

Castilho LR, Polato CMS, Baruque EA, Sant'Anna Jr. GL and Freire DMG (2000) Biochem. Eng. J. 4: 239–247.

Cavalcanti EAC, Gutarra MLE and Freire DMG (2005) Braz. Arch. Biol. Technol. 48: 79–84.

Cen P and Xia L (1999) Adv. Biochem. Eng./Biotechnol. 65: 69–92.

Chahal DS (1991) Production of *Trichoderma reesei* cellulase system with high hydrolytic potential by solid-state fermentation. In: *Comstock J,* ed. ACS symposium series, vol. 460. American Chemical Society, Washington, DC, pp 111-122.

Chahal PS, Chahal DS and Le GBB (1996) Appl. Biochem. Biotechnol. 57/58: 432–442.

Chang SW, Shaw JF, Yang CK and Shieh C (2007). Process Biochem. 42: 1362–1366.

Chen D, Chen J, Zhong W and Cheng Z (2008) Bioresour.Technol. 99:4702–4708.

Cherry JR and Fidantsef AL (2003) Curr. Opin. Biotechnol.14:438–43.

Chesson A (1987) Supplementary enzymes to improve the utilization of pigs and poultry diets. In: Haresign W, Cole DJA, editors. Recent advances in animal nutrition. London: Butterworths. pp. 71–89.

Chinn MS, Nokes SE and Strobel HJ (2007) Bioresour. Technol. 98: 2184–2193.

Chung-Yi W, Yi-Ru H, Ng C-C, Chan H, Lin H-T, Wen-Sheng T and Shyu YT (2009) Enzyme Microbiol.Technol. 44:373–9.

Cohen R, Persky L, Hazan-Etian Z, Yarden O and Yitzhak H (2002) Appl. Biochem. Biotechnol. 102–103: 415–429.

Coughlan MP (1985) Biochem Soc Trans. 13:405–406.

Couri S, Merces EP, Neves BCV and Senna LF (2006) J. Microsc. 224: 290–297.

Cruz JM, Dominguez JM, Dominguez H and Parajo JC (2000) Food Biotechnol. 14: 79–97.

Dashtban M, Schraft H and Qin W (2009). Int. J. Biol. Sci. 5:578–95.

Da-Silva R, Yim DK and Park YK (1994) J. Ferm. Bioeng. 77:109-111.

De Siqueira FG, De Siqueira EG, Jaramillo PMD, Silveira MHL, Andreaus J, Couto FA, Batista LR and Filho EXF (2010) The potential of agro-industrial residues for production of holocellulase from filamentous fungi. Int. Biodeter. Biodegr. 64: 20–6.

Dekker RFH and Richards GN (1976) Adv. Cabohydr. Chem. Biochem. 32: 277-352.

Del Pilar-Castillo M, Ander P and Stenstrom J (1997) Biotechnol. Technol. 11: 701–706.

Deschamps F, Giuliano C, Aster MC and Roussos S (1985) Biotechnol.Bioeng.27: 1385-1389.

Desgranges C and Durand A (1990) Enzyme Microbiol.Technol.12: 546-548

Di Luccio M, Capra F and Ribeiro NP (2004) Appl. Biochem. Biotechnol. 113: 173–180.

Duenas R, Tengerdy RP and Gutierrez-correa M (1995). World J. Microbiol. Biotechnol.11: 333-337.

Durand A (2003) Bioreactor designs for solid state fermentation. Biochem. Eng. J. 13: 113–125.

Durand A, Renaud R, Almanza S, Maratray J, Diez M and Desgranges C (1993) Biotechnol. Adv. 11: 591–597.

Elisashavili VI, Daushvili LP, Zakariashvili NG, Kachlishvili ET, Kiknadeze MO and Tsuishvili KA (1998) Microbiology. 67: 33–37.

Elisashavili V, Parlar H, Kachlishvili E, Chihua D, Bakradze M, Kokhreidze N and Kvesitadze G (2001) Adv. Food Sci. 23: 117–123.

FAOSTAT (2006) FAO statistical databases. http://faostat.fao.org/.

Fenice M, Giovannozzi SG, Federici F and Annibale AD (2003) J. Biotechnol. 100: 77–85.

Ferraz A, Cordova AM and Machucan A (2003) Enzyme Microbiol. Technol. 32: 59–65.

Fujian X, Hongzhang C and Zuohu L (2001) Bioresour. Technol. 80: 149–151.

Fujian X, Hongzhang C and Zuohu L (2002) Enzyme Microbiol. Technol. 30: 45–48.

Galliano H, Gas G, Seris JL and Boudet AM (1991) Enzyme Microbiol. Technol.13: 478–482.

Gangadharan D, Sivaramakrishnan S and Nampoothiri KM (2005Food Technol. Biotechnol.44: 269–274.

Genencor website: dated 01/14/2010 http://www.genencor.com/wps/wcm/ connect/ genencor/genencor/products and services/business development/ biorefineries/ products/accellerase product line en.htm.

Gilkes NR, Henrissat B, Kilburn DG, Miller Jr RC and Warren RA (1991) Microbiol. Rev.55:303–315.

Giovannozi Sermanni G and Tiso N (2007) Aspects of design of bioreactors in SSF, in: A. Pandey, C.R. Soccol, C. Larroche (Eds.), Current Developments in Solid-state Fermentation, Springer Science/Asiatech Publishers, Inc., New York, USA/New Delhi, India, pp. 117–144.

Giovannozzi-sermanni G, D-Annibale A, Cretini C and Cappelletto PL (1996) High-yield pulps from enzymically-treated non-woody plants. Mededelingen–Faculteit Landbouwkundige en Toegepaste Biolosche Wetenschappen (Universiteit gent). 61: 2129-2132

Godfrey T (1996) Textiles In: Godfrey T, West S, editors. Industrial enzymology, 2nd ed. London: Macmillan Press. pp. 360–71.

Godfrey T and West S (1996a) Introduction to industrial enzymology. In: Godfrey T, West S, eds. Industrial enzymology, 2nd ed. London: Macmillan Press. pp. 1–8.

Godfrey T and West S (1996b) Industrial Enzymology, 2nd ed. London: Macmillan Press.

Golovleva LA, Malt'seva OV, Leont'evskii AA, Myasoedova NM and Skryabin GK (1987) Dokl. Akad. Nauk SSSR 294: 992–995.

Grajek W (1987) Appl. Microbiol. Biotechnol. 26: 126–129.

Guadalix ME, Almendros G, Matrinez AT, Gonzalez-vila FJ and Lankes U (1997) Bioresour. Technol.60: 245-249.

Gupta S, Kapoor M, Sharma KK, Nair LM and Kuhad RC (2008) Bioresour. Technol. 99: 937–945.

Gupte A and Madamwar D (1994) High strength cellulase and beta-glucosidase formation from *Aspergillus* sp. under solid state fermentation. In: Solid State Fermentation. Ed. A. Pandey, Wiley Eastern Publishers, New Delhi, pp. 130-133.

Gutieera-CorreaM and Tengerdy RP (1998) Biotechnol. Lett. 20: 45-47.

Haltrich D, Nidetzky B, Kulbe KD, Steiner W and Zupancic S (1996). Bioresour. Technol. 58:137–161.

Hama S, Yamaji H, Fukumizu T, Numata T, Tamalampudi S, Kondo A, Noda H and Fukuda H (2007). Biochem. Eng. J. 34: 273–278.

Hamidi-Esfahani Z, Shojaosadati SA and Rinzema A (2004). Biochem. Eng. J. 21: 265–272.

Hardin MT and Howes T Mitchell DA (2002) J. Biotechnol. 97: 89–101.

Harman GE, Kubicek CP (1998) *Trichoderma* and *Gliocladium*: Enzymes, biological control and commercial applications, Vol. 2. London: Taylor & Francis Ltd, p. 393.

Heinelman PC, Snow D, Wu I, Nguyen C, Villalobos A, Govindarajan S, Minshull J and Arnold FH (2009) PNAS.106 :5451–5452.

Herpoë l-Gimbert I, Margeot A, Dolla A, Jan G, Mollé D, Lignon S, Mathis H, Sigoillot JC, Monot F and Asther M (2008). Biotechnol. Biofuels. 1:18.

Hibbert EG, Baganz F, Hailes HC, Ward JM, Lye GJ, Woodley JM and Dalby PA (2005) Biomol. Eng. 22:11–9.

Himmel ME, Ruth MF and Wyman CE (1999) Curr. Opin. Biotechnol. 10:358–64.

Holker U, Hofer M and Lenz J (2004) Appl. Microbiol. Biotechnol. 64: 175–189.

Hongzhang C, Fujian X, Zhonghou T and Zuohu L (2002). J. Biosci. Bioeng. 93: 211–214.

Huerta-Ochoa S, Nicolas-Santiago S, Acosta-Hernandez M, Dayanara W, Prado-Barragan LA, Gutierrez-Lopez GF, Garcia- Almendarez BC and Regalado-Gonzalez C (2003) Production and partial purification of glycosidases obtained by solid state fermentation of grape pomace using *Aspergillus niger* 10, in: G.F. Gutierrez-Lopez, G.V. Barbosa-Canovas (Eds.), Food Science and Food Biotechnology, CRC Press LLC, Washington, 2003, pp. 119–138.

Hung-Der J and Shang-Shyng Y (2008) Bioresour. Technol. 99: 6181–6189.

Iwashita K (2002) J. Biosci. Bioeng. 94: 530-535.

Jain A (1995) Process Biochem. 30: 705-709.

Jaszek M, Malarczyk E and Leonowicz A (1998) Investigation of ligninolytic enzymes during solid state fermentation of cotton wastes by selected strains of basidiomycetes, in: Proceedings of the 7th International Conference on Biotechnology in the Pulp and Paper Industry, Vancouver, 16–18 June, 1998.

Javed MM, Khan TS and Haq I-U (2007) Electron. J. Environ. Agric. Food Chem. 6:2290–2296.

Jesus GD, Romano-Baez FJ, Leyva-Amezcua L, Juarez-Ramirez C, Ruiz-Ordaz N and Galindez-Mayer J (2009) J. Hazard. Mater. 161: 1140–1149.

Jo JH, Lee DS, Park D and Park JM (2008) Bioresour. Technol. 99:6666–6672.

John RP, Nampoothiri KM and Pandey A (2006) Process Biochem. 41: 759–763.

Jose Leite AC, Bruna Fernandes S, Pozzi E, Barboza M and Zaiat M (2008) Int. J. Hydrogen Energy. 33: 579–586.

Kashyap P, Sabu A, Pandey A, Szakacs G and Soccol CR (2002) Process Biochem. 38: 307–312.

Khanahmadia M, Mitchell DA, Beheshtic M, Roostaazadc R and Sánchezd RL (2006). Chem. Eng. Sci. 61: 2675–2687.

Kim E, Irwin DC, Walker LP and Wilson DB (1998) Biotechnol. Bioeng. 58:494–501.

Kim JH, Hosobuchi M, Kishimoto M, Seki T, Yoshida T, Taguchi H, Ryu D and Doo Y (1985) Biotechnol. Bioeng. 27: 1445- 1450.

Kimura I, Sasahara H and Tajima S (1995) J.Ferment. Bioeng. 80: 334–339.

Kitpreechavanich V, Hayashi M and Nagai S (1984) J. Ferment. Technol. 62: 63-69.

Kitpreechavanich V, Srisuk W and Lotong N (1992) Prog. Biotechnol. 7: 519-522.

Kota KP and Sridhar P (1999) Process Biochem. 34: 325–328.

Kuhad RC and Singh A (1993) World J. Microbiol. Biotechnol. 9: 100-101.

Kumar B, Trivedi P, Mishra AK, Pandey A and Palni LMS (2004) Microbiol. Res. 159:141–6.

Kurakake M, Osada S and Komaki T (1997) Biosci. Biotechnol. Biochem.61:2010–2014.

Ladisch MR, Lin KW, Voloch M and Tsao GT (1983) Enzyme Microbiol. Technol. 5:82–100.

Lagemaat V and Pyle DL (2001). Chem. Eng. J. 84: 115–123.

Lang E, Nerud F, Novotna E and Martens R (1996) Folia Microbiol. 41: 489–493.

Li X and Jia R (2008) Bioresour. Technol. 99: 6885–6892.

Li X, Liun Y and Ji L (2000). Henong Xuebao. 14: 99–103.

Ljungdahl LG (2008) Ann. NY Acad. Sci. 1125: 308–21.

Lo C-M, Zhang Q, Callow NV and Ju L-K (2010) Bioresour. Technol. 101:717–23.

Lorenzo M, Moldes D, Rodr´ıguez Couto S and Sanrom´an A (2002) Bioresour. Technol. 82:109–113.

Luciana Vandenberghe PS, Soccol CR, Pandey A and Lebeault JM (2000). Bioresour. Technol. 74: 175–178.

Lynd LR, Weimer PJ, van Zyl WH, Pretorius IS (2002) Microbiol. Mol. Biol. Rev. 66:506.

Macris BJ, Kekos D, Evangelidou X, Galiotou-Paanayotou MG and Rodis P (1987) Biotechnol. Lett.9: 661-664

Madamwar D, Patel S and Parikh H (1989).. J. Ferment. Bioeng. 67: 424-426.

Maki M, Leung KT and Qin W (2009) Int. J. Biol. Sci. 5: 500-516.

Mandels M (1985) Applications of cellulases. Biochem. Soc. Trans. 13:414–5.

Marques BC, Barga MC, Balmant W, De Lima Luz Jr. LF, Krieger N and Mitchell DA (2006) Food Technol. Biotechnol. 44: 457–463.

Marques de Souza CG, Zilly A and Peralta RM (2002) J. Basic Microb. 42: 83–90.

Mart´ınez MJ, Ru´ız-Due˜nas FJ, Guill´en F and Mart´ınez AT (1996) Eur. J. Biochem. 237: 424–432.

Mathew GM, Sukumaran RK, Singhania RR and Ashok P (2008) J. Sci. Ind. Res. 67:898–907.

Mester T, Sierra-Alvarez R and Field J (1998) Holzforschung. 52: 351–358.

Mezohegyi G, Bengoa C, Stuber F, Font J, Fabregat A and Fortuny A (2008) Chem. Eng. J.143: 293–298.

Mishra C and Leatham G (1990) J.Ferment .Bioeng. 69: 8-15.

Mitchell DA, Berovic M and Krieger N (2000). Adv. Biochem. Eng. Biotechnol. 68:61–138.

Mitchell DA, Krieger N and Berovic M (2006) Solid-State Fermentation Bioreactors: Fundamentals of design and operation. Springer-Verlag Berlin Heidelberg.

Mitchell DA, Von Meien OF and Krieger N (2003). Biochem. Eng. J. 13: 137–147.

Mitra P, Chakraborty R and Chandra AL (1996) J. Sci. Indus. Res.55: 439-442.

Mohan PR, Ramesh B and Reddy OVS (2011) Res. J. Microbial.1-12.

Montané D, Salvadó J, Torras C and Farriol X (2002) High-temperature dilute-acid hydrolysis of olive stones for furfural production. Biomass Bioenerg. 22:295-304.

Moo-Young M, Moreira AR and Tengerdy RP (1983) Principles of solid state fermentation, In: J.E. Smith, D.R. Berry, B. Kristiansen (Eds.), The Filamentous Fungi, Edward Arnold Publishers, London, pp. 117–144.

Mukherjee K, Adhikari H and Rai SK (2008) Biochem. Eng. J. 39:353–361.

Muller dos Santos M, Souza das Rosa A, Dal'Boit S, Mitchell DA and Krieger N (2004) Bioresour. Technol. 93: 261–268.

Muniswaran PKA, Selvakumar P and Charyulu NCL (1994) J. Chem. Technol. Biotechnol. 60:147-151.

Murado MA, Gonz´alez MP, Torrado A and Pastrana LM (1996) Process Biochem. 32: 35–42.

Nigam P and Singh D (1996) *J. Sci. Ind. Res.* 55: 457–463

Nigam P, Pandey A and Prabhu KA (1987) Biol. Wastes. 21: l-10.

Nigam P and Singh D (1995) Process Biochem. 30:117- 124.

Ooijkaas LP, Weber FJ, Buitelaar R, Tamper J and Rinzema A (2000) Trends Biotechnol. 18: 356–360.

Pal M, Calvo A, Terron MC and Gonzalez AE (1995) World J. Microbiol. Biotechnol. 11: 541–545.

Pandey A (1992) Recent developments in solid state fermentation. Process Biochem. 27: 109-117.

Pandey A (2003) Solid-state fermentation. Biochem. Eng. J. 13: 81–84.

Pandey A and Radhakrishna S (1993). Process Biochem. 28: 305-309.

Pandey A and Soccol CR (2000) J. Sci. Ind. Res. 59: 12–22.

Pandey A, Ashakumary L and Selvakumar P (1995) Bioresour. Technol. 51: 217-220.

Pandey A, Ashakumary L, Selvakumar P and Vijayalakshmi KS (1994) World J. Microbiol. Biotechnol. 10: 485-486.

Pandey A, Azmi W, Singh J and Banerjee UC (1999a) Types of fermentation and factors affecting it, in: V.K. Joshi, A. Pandey (Eds.), Biotechnology: Food Fermentation, Educational Publishers, New Delhi, pp. 383–426.

Pandey A, Selvakumar P, Soccol CR and Nigam P (1999b) Curr. Sci. 77: 149–162.

Pandey A, Selvakumar P, Soccol CR and Nigam P (2004) Solid-state fermentation for the production of industrial enzymes. 1-22.

Pandey A, Soccol CR and Larroche C (2007a) Introduction, in: A. Pandey, C.R. Soccol, C. Larroche (Eds.), Current Developments in Solid-state Fermentation, Springer Science/Asiatech Publishers, Inc., New York, USA/New Delhi, India, pp. 3–12.

Pandey A, Soccol CR and Larroche C (2007b) Current Developments in Solid-state Fermentation, Springer Science/Asiatech Publishers, Inc., New York, USA/New Delhi, India, pp. 13–25.

Pandey A, Soccol CR, Leo JAR and Nigam P (2001) Solid-state Fermentation in Biotechnology. Asiatech Publishers, Inc., New Delhi, pp. 221.

Pandey A, Soccol CR, Nigam P, Soccol VT, Vandenberghe LPS and Mohan R (2000a) Bioresour. Technol. 74 81–7487.

Papinutti VL, Diorio LA and Forchiassin F (2003) J. Ind. Microb. Biotechnol. 30:157–160

Pastrana LM, Gonzalez MP, Pintado J and Murado MA (1995) Enzyme Microbiol. Technol. 17:784–790.

Pradeep V and Datta M (2002) Appl. Biochem. Biotechnol. 102–103: 109–118

Priefert H, Rabenhorst J and Steinbüchel A (2001) Biotechnological production of vanillin. Appl. Microbiol. Biotechnol. 56:296-314.

Rahardjo SPY, Tramper J and Rinzema A (2006) Biotechnol. Adv. 24: 161–179.

Raimbault M (1998), Electron. J. Biotechnol. 1: 1-15.

Ralph BJ (1976) Food Technol. Aust. 28: 247–251.

Ramachandran S, Patel AK, Nampoothiri KM, Francis F, Nagy V, Szakacs G and Pandey A (2004). Bioresour. Technol. 93: 169–174.

Ramachandran S, Roopesh K, Nampoothiri KM, Szakacs G and Pandey A (2005). Process Biochem. 40: 1749–1754.

Ramachandran S, Fontanille P, Pandey A and Larroche C (2007) Food Technol. Biotechnol. 46: 190–194.

Reddy GV, Ravindra BP, Komaraiah P, Roy KRRM and Kothari IL (2003) Process Biochem. 38: 1457–1462.

Reese ET (1976) History of the cellulase program at the US Army Natick development centre. Biotechnol. Bioeng. Symp. 6:9–20.

Reese ET and Mandels M (1984) Rolling with the time: production and applications of *Trichoderma reesei* cellulase. Annual Report of Fermentation Processes. 7:1–20.

Ribbons RW (1987) Chemicals from lignin. Phil. Trans. R. Soc. Lond. Ser. A. 321:485-494.

Roberto IC, Mussatto SI and Rodrigues RCLB (2003) Indust. Crops Prod. 17:171-176.

Robinson T and Nigam P (2003) Bioreactor design for protein enrichment of agricultural residues by solid state fermentation. Biochem. Eng. J. 13:197–203.

Robinson T, McMullan G, Marchant R and Nigam P (2001) Bioresour. Technol. 77: 247–255.

Roche N, Desgragnes C and Durand A (1994) J.Biotechnol. 38: 43–50.

Rodriguez Couto S (1999) Producci´on de enzimas ligninol´ıticos en cultivos en estado semi-s´olido por *Phanerochaete chrysosporium* y *Phlebia radiata*, Ph.D. Thesis, University of Vigo, Spain.

Rodriguez Couto S, Rosales E, Gund´ın M and Sanrom´an MA (2004) J. Food Eng. 64: 423–428.

Rodriguez Couto S and Sanroman MA (2005) Biochem. Eng. J. 22: 211–219.

Rodriguez Couto S and Sanroman MA (2006) J. Food Eng. 76: 291–302.

Rodriguez Couto S, Rodrıguez S, Gallego PP and Sanroman A (2003) Acta Biotechnol. 23: 62–64.

Rodriguez León JA, Soccol CR, Singhania RR, Pandey A, Sturm W, Alberto Jr Letti L, Vandenberghe LS and Rodriguez Fernández DE (2001) Solid-state fermentation, in: A. Pandey, C. Larroche, C.R. Soccol, C.G. Dussap (Eds.), Current Developments in Solid-state Fermentation, Asiatech Publishers, Inc., New Delhi, India, pp. 554–569.

Ryu DD and Mandels M (1980) Enzyme Microbiol. Technol. 2:91–101.

Sabu A, Augur C, Swati C and Pandey A (2006). Process Biochem. 41: 575–580.

Sahir AH, Kumar S and Kumar S (2007). Biochem. Eng. J. 35: 20–28.

Sangsurasak P and Mitchell DA (1998) Biotechnol. Bioeng. 60: 739–749.

Schulein M (2000) Biochim Biophys Acta-Protein Struct. Mol. Enzymol. 1543:239–52.

Schuster E, Dunn-Coleman N, Frisvad JC and Van Dijck PWM (2002) Appl. Microbiol. Biotechnol. 59: 426-435.

Selvakumar P, Ashakumary L and Pandey A (1996) J. Sci. Indus. Res., 55: 443-449.

Selvakumar P, Ashakumary L and Pandey A (1998) Bioresour. Technol. 65: 83-85.

Seyis I and Aksoz N (2003) Microbiologica. 26: 89–95.

Shamala TR and Sreekantiah KR (1987a). Enzyme Microbiol. Technol.9: 97-101.

Shamala TR and Sreekantiah KR (1987b) Enzyme Microbiol.Technol. 8:178-182.

Shih IL, Kuo CY, Hsieh FC, Kao SS and Hsieh C (2008) J. Chin. Inst. Chem. Eng. 39: 635– 643.

Shinners-Carnelley TC, Szpacenko A, Tewari JP and Palcic MM (2002). Phytoprotection. 83: 31–40.

Shoemaker HE, Mink D and Wubbolts MG (2003) Science. 299:1694–7.

Shrestha P, Khanal SK, Pomettoiii AL and Van Leeuwen J (2009) J. Agric. Food Chem. 57:4156–61.

Sik S and Nyayar AU (1998) Turkish J. Biol. 22: 287–298.

Singhania RR, Patel AK, Soccol CR and Pandey A (2009). Biochem. Eng. J. 44:13–18.

Singhania RR, Soccol CR and Pandey A (2007b) Application of tropical agro-industrial residues as substrates for solid-state fermentation processes, in: A. Pandey, C.R. Soccol, C. Larroche (Eds.), Current Developments in Solid-state Fermentation, Springer Science/Asiatech Publishers, Inc., New York, USA/ New Delhi, India, pp. 412–442.

Singhania RR, Sukumaran RK and Pandey A (2007a). Appl. Biochem. Biotechnol. 142: 60–70.

Srivastava KC (1993) Biotechnol. Adv. 11: 441-465.

Subramaniyan S, Prema P and Ramakrishna, SV (1997). J. Basic Microbiol. 37: 431.

Sukumaran RK, Singhania RR and Pandey A (2005). J. Sci. Ind. Res. 64:832–844.

Sukumaran RK, Singhania RR, Mathew GM and Pandey A (2009) Renew. Energ. 34 :421–424.

Sumantha A, Sandhya C, Szakacs G, Soccol CR and Pandey A (2005) Food Technol. Biotechnol. 43: 313–319.

Sun T, Liu BH and Li ZH (1997) J.Chem.Technol. Biotechnol. 69: 429–432.

Suryanarayan S (2001) New frontiers in solid-state fermentation, in: Proceedings of the International Conference on New Horizons in Biotechnology, Trivandrum, India, .April 18–21,

Suryanarayan S (2003). Biochem. Eng. J. 13: 189–195.

Tanaka H, Koike K, Itakura S and Enoki A (2009) Enzyme Microbiol. Technol. 45: 384–90.

Tengerdy RP (1996) J. Sci. Ind. Res, 55:313-316.

Thirumale S, Swaroopa Rani D and Nand K (2001) Process Biochem. 37:241–5.

Thomke S, Rundgreen M and Hesselman K (1980) The effect of feeding high-viscosity barley to pigs. In: Proceedings of the 31st meeting of the European Association of Animal Production, Commission on Animal Production. Munich, Germany, p. 5.

Tolan JS and Foody B (1999). Adv. Biochem. Eng. Biotechnol. 65: 41–67.

Touhy MG and Coughlan MP (1991). Bioresour. Technol. 39:131–137.

Tychanowicz GK, Zilly A, Marques de Souza CG and Peralta RM (2004) Process Biochem. 39: 855–859.

Uhlig H (1998) Industrial enzymes and their applications, New York: John Wiley & Sons, Inc., pp. 435.

Vares T, Kalsi M and Hatakka A (1995) Appl. Environ. Microb. 61:3515–3520.

Vimala Devi PS, Ravinder T and Jaidev C (2005) J. Invertebr. Pathol. 88:163–168.

Viniegra-Gonz´alez G (1998) Strategies for the selection of mold strains geared to produce enzymes on solid substrates, in: E. Glindo, O.T. Ram´ırez (Eds.), Advances in Biprocess Engineering II, Kluwer Academic Publishers, Dordrecht, pp. 123–136.

Viniegra-Gonz´alez G, Favela-Torres E, Noe Aguilar C, De Jes´us Romero-G´omez J, D´ıaz-God´ınez G and Augur C (2003) Biochem. Eng. J. 13: 157–167.

Voragen AGJ (1982). Fruit Processing. 7:98–102.

Voragen AGJ, Heutink R and Pilnik W (1980) J. Appl. Biochem. 2:452–68.

Voragen AGJ, Wolters H, Verdonschot-Kroef T, Rombouts FM and Pilnik W (1986) Effect of juice-releasing enzymes on juice quality. In: International Fruit Juice Symposium, The Hague (NL), May 1986. Zurich: Juris Druck Verlag, pp. 453–62.

Walton NJ, Mayer MJ and Narbad A (2003) Molecules of interest: Vanillin. Phytochem. 63. 505-515.

Wang CM, Shyu CL, Ho SP and Chiou SH (2008). Lett. Appl. Microbiol. 47:46-53.

Wiacek-Zychlinska A, Czakaj J and Sawicka-Zukowska R (1994). Bioresour. Technol. 49: 13-16.

Wiebe MG (2003). Mycologist. 17:140–4.

Wilson DB (2004) Chem. Rec. 4:72–82.

Wong DWS (2009). Appl. Biochem. Biotechnol. 157:174–209.

Wong KKY and Saddler JN (1993) Applications of hemicellulases in the food, feed, and pulp and paper industries. In: Coughlan MP, Hazlewood GP, eds. Hemicellulose and hemicellulases. London: Portland Press. pp.127–43.

Wong KKY and Saddler JN (1992) *Trichoderma* xylanases, their properties and applications. In: Visser J, Beldman G, Kusters-van Someren MA, Voragen AGJ, eds. Xylans and xylanases. Progress in Biotechnology, Vol. 7. Amsterdam: Elsevier, pp. 171–86.

Wu KJ, Chang CF and Cheng JS (2007) Process Biochem. 42: 1165–1171.

WWW. DSM nutritionalproducts.com, 2011.

WWW. Enzyme-India.com, 2011.

Xie J, Sun X, Ren L and Zhang YZ (2001). Sheng Wu Kung Ch. Eng. Hsueh Pao. 17:575–578.

Ximenes, FQ Silveira and Filho EX (1996). Curr. Microbiol. **33:** 71–77.

Xiong C, Jinhua W and Dongsheng L (2007) Biochem. Eng. J. 34: 179–184.

Zakariashvili NG and Elisashvili VI (1993). Mikrobiologiya. 62:876-881.

Zeitch KJ (2000) The chemistry and technology of Furfural and its many by-products. Elsevier, pp. 358.

Zhang Y-HP and Lynd LR (2004) Biotechnol. Bioeng. 88:797–824.

Zhang YHP, Himmel ME and Miclonz JR (2006) Biotechnol. Adv. 24:452–481.

Chapter - 21

Marine Actinobacteria- A rich source of bioactive compounds

T. Ramana, R. Haritha and K. Siva Kumar*

Department of Biotechnology, Andhra University,
Visakhapatnam – 530 003, India.
E-mail: ramanaterli@yahoo.com

ABSTRACT

The marine-derived actinobacteria have become recognized as a major source of novel secondary metabolites with unusual structures and properties. This interest in marine actinobacteria is related to the ecological features of habitat in which marine actinobacteria thrive, which impacts their metabolic functions enabling their chemical biodiversity. The spectacular technological instrumentation and mechanization that is now available will lead to a fuller understanding of marine actinobacterial biodiversity and will inevitably translate into an increase in the discovery and development of novel bioactive compounds from marine actinobacterial sources in the future. This review deals with the research and development work investigating the marine actinobacterial diversity and their secondary metabolite production and applications in biotechnology.

Key words: Marine actinobacteria, diversity, bioactive compounds

1. Introduction

Actinobacteria are Gram positive bacteria containing high G + C content in the genetic material, were ascertained as the potential source of secondary metabolites of pharmaceutical importance and recycling of metabolites from the degradation of biopolymers such as keratin, lignocellulose, and chitin (Ellaiah *et al.*, 2004, McCarthy, 1987). After the discovery of actinomyin much of the research was shifted towards actinobacteria producing bioactive compounds and antitumor agents in addition to enzymes of industrial interest (Tanaka & Omura, 1990). Among the potential sources of natural products, actinobacteria have proven to be a particularly prolific resource with a surprisingly small group of taxa accounting for the vast majority of bioactive compounds discovered (Jensen, 2005a). The representative genera of actinobacteria include *Streptomyces, Actinomyces, Arthrobacter, Corynebacterium, Frankia, Micrococcus, Micromonospora* and several others. The screening of microbial natural products continues to represent an important route to the discovery of novel chemicals for the development of new therapeutic agents, and evaluation of the potential of lesser-known and/or new bacterial taxa (Bull *et al.*, 2000).The recent study by Berdy revealed that among the presently known bioactive microbial metabolites, over 10000 compounds were isolated from actinomycetales species, 34% from *Streptomyces* and 11% from the rare actinobacteria (*Micromonospora, Actinomadura, Salinospora, Saccharopolyspora* and *Streptosporangium* (Berdy, 2005). The most frequent producers, the *Streptomyces* species produces 7600 compounds (74% of all actinomycetales), While the rare actinobacteria represent 26%, altogether 2500 compounds.

A significant amount of effort has been focused on the successful isolation of novel actinomycetes from terrestrial sources for drug screening programs from the past fifty years. Recently the rate of discovery of new compounds from terrestrial actinobacteria has decreased, whereas, the rate of reisolation of known compounds has increased (Fenical and Jensen, 2006). The demand for new antibiotics and novel enzymes continues to grow due to the rapid spread of antibiotic-resistant pathogens causing life-threatening infections and the ongoing need for enzyme based industrial bioprocess in the pharmaceutical, processed foods and allied fermentation industries (Subramani Ramesh and Narayanasamy Mathivanan, 2009). Although considerable progress is being made within the fields of chemical synthesis and engineered biosynthesis of antimicrobial compounds, nature still remains the richest and the most versatile source for new bioactive compounds (Baltz, 2006; Koehn & Carter, 2005; Pelaez, 2006). Though the first report of mycelium-forming actinomycetes from marine sediments appeared several decades ago it is only more recently that marine-derived actinobacteria have become recognized as a source for collecting an inventory of diverse enzymes, novel antibiotics and anti-cancer agents with unusual structures and properties (Weyland, 1969; Antonio Trincone, 2010; Jensen *et al.*, 2005b).

Thus the unexplored or underexploited marine habitats were selected as rich sources of new bacterial taxa producing novel bioactive metabolites (Lam, 2006). There has been considerable progress on the detection and identification of marine actinobacteria since they were reported to produce novel bioactive compounds (Fenical, 1993; Copp, 2003; Shin *et al.*, 2003). It is necessary to continue to screen for new metabolites and to evaluate the potential of the new and improved compounds for their future use (Kurtboke, 2005).

In the present review, emphasis on different species of marine actinobacteria and their metabolite production and application in biotechnology are given in detail.

2. Distribution and diversity of marine actinobacteria

So far the terrestrial soil was the most predominant and widely exploited source for actinobacteria. A very little is known about the microbial diversity of marine sediments, which is an inexhaustible resource that has not been properly exploited. However, the full potential of this domain as the basis for biotechnology, particularly in India, remains largely unexplored. Morethan 70% of our planet's surface is covered by oceans, which has greatest biodiversity than the terrestrial environment. As marine environmental conditions are extremely different from terrestrial ones, it is assumed that marine actinobacteria have different characteristics from those of terrestrial counterparts and, therefore, might produce different types of bioactive secondary metabo-lites. The living conditions to which marine actinobacteria had to adapt during evolution range from extremely high pressure and anaerobic conditions to high acidic conditions. These extreme conditions have reflected in genetic and metabolic diversity of marine actinobacteria, which remain largely unknown.

An uncertainty exists regarding the indigenous status of the actinobacteria present in marine environment. This scepticism arised due to the fact that the terrestrial bacteria produce resistant spores that are known to be transported from land into sea, where they can remain available but dormant for many years (Bull *et al.*, 2000). Thus, it has been frequently assumed that actinobacteria isolated from marine samples are merely of terrestrial origin. Early evidence supporting the existence of marine actinobacteria came from the description of *Rhodococcus marinonascene*, the first marine actinomycete species to be characterized (Helmke and Weyland, 1984). However, these early findings did not generate enough excitement to stimulate the search for novel actinobacteria in the marine environment.

In spite of improvements being made in the cultural methods for the isolation of marine actinobacteria, many of these organisms remain unculturable and have to be detected by using molecular techniques (Stach, 2003). In addition to molecular techniques several statistical tools should be used in order to establish an extent of actinobacterial richness and to define genetic diversity within and between actinobacterial populations in the marine environment (Martin, 2002).

The diversity of actinobacteria in the marine environment usually is investigated with culture-dependent and/or culture-independent techniques. The approach to culture dependent diversity surveys is based upon the use of sediment pre-treatments, taxon-selective isolation media, preliminary circumscription and dereplication of isolates (Maldonado, 2004). Meanwhile, a cultivation-independent methodology, metagenomics involves the modern molecular PCR-based studies, in which DNA is directly isolated from environmental samples and cloned into suitable vectors to construct complex genomic libraries. These libraries can be analysed for novel genes or pathways using sequence-based techniques or through screening of the enzymatic activities produced in surrogate hosts (Natsumi, 2006). Metagenomic based high-throughput approaches now opened up the road to analyze and screen the genetically and metabolically rich marine actinobacterial communities in their entirety (Yanjuan, 2008).

Marine actinobacteria were widely distributed in biological sources such as fishes, molluscs, sponges, seaweeds, mangroves, besides seawater and sediments. Actinobacteria genera identified by cultural and molecular techniques from different marine ecological niches include *Actinomadura, Actinosynnema, Amycolatopsis, Arthrobacter, Blastococcus, Brachybacterium, Corynebacterium, Dietzia, Frankia, Frigoribacterium, Geodermatophilus, Gordonia, Kitasatospora, Micromonospora, Micrococcus, Microbacterium, Mycobacterium, Nocardioides, Nocardiopsis, Nonomurea, Psuedonocardia, Rhodococcus, Saccharopolyspora, Salinispora, Serinicoccus, Solwaraspora, Streptomyces, Streptosp-orangium, Tsukamurella, Turicella, Verrucosispora* and *Williamsia* (Jensen, 2005b; Stach *et al.*, 2004; Ward and Bora, 2006; Das, 2006).

Yoshida *et al.* (2008) studied the actinobacterial community structures in marine and freshwater environments (the Pacific Ocean, East China Sea, Tokyo Bay, and Arakawa River) by a culture-independent molecular method to clarify spatial and seasonal distributions. The denaturing gradient gel electrophoresis (DGGE) and non metric multidimensional scaling analysis, followed by sequencing analysis revealed that most of the predominant actinobacterial strains were uncultured and were quite different from well-known culturable strains, such as the *Streptomyces, Micromonospora, Microbispora, Salinispora,* and *Actinoplanes* groups.

3. Secondary metabolites from marine actinobacteria

Since the antibiotic research started more than half a century ago by the discovery of penicillin by Fleming; the research in the field of bioactive compound has been a fascinating, exciting, continuously changing and developing adventure. The interest towards the antibiotic research was generally increasing and declining with success and failures to scope the clinical needs and new enabling technology. The forties and early fifties were called as the "golden era" of the antibiotics, when almost all groups of antibiotics have been discovered. The bioactive compounds reported in this period were mainly isolated from *Streptomyces* species.

After all these years there is an exponential increase in the number of new metabolites (mainly non – antibiotic compounds, analogous and minor compounds) which were produced by the rare actinobacteria isolated from marine source. Above 500 species of *Streptomyces* account for 70-80% of relevant secondary metabolites, with small contributions from rare actinobacteria such as *Saccharopolyspora, Amycolatopsis, Micromonospora* and *Actinoplanes* (Dharmaraj, 2010). These fastidious organisms produce the most diverse and most unique, unprecedented, complicated compounds exhibiting excellent antibacterial, antiviral and antitumor potency with low toxicity. Presently more than 50 rare actinobacteria are known as producers of the 2500 bioactive compounds (Berdy, 2005). These bacteria have proven to be efficient producers of bioactive metabolites which have a wide range of activities such as anti-bacterial, antifungal, antitumour, anticancer, cytotoxic, etc. depending on the chemical nature; these bioactive substances are classified as follows:

Quinones

Quinones are compounds having a fully conjugated cyclic dione structures. Quinones represent an important group of isoprenoid lipids which occur in the cytoplasmic membrane of most prokaryotes. Marine actinobacteria is especially rich in biologically active quinone compounds

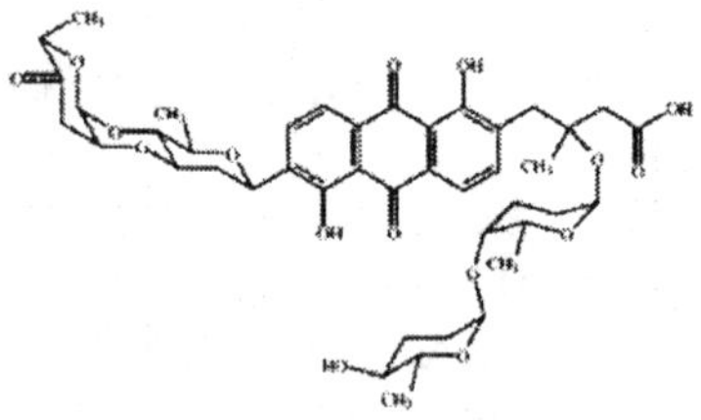

Himalomycin A 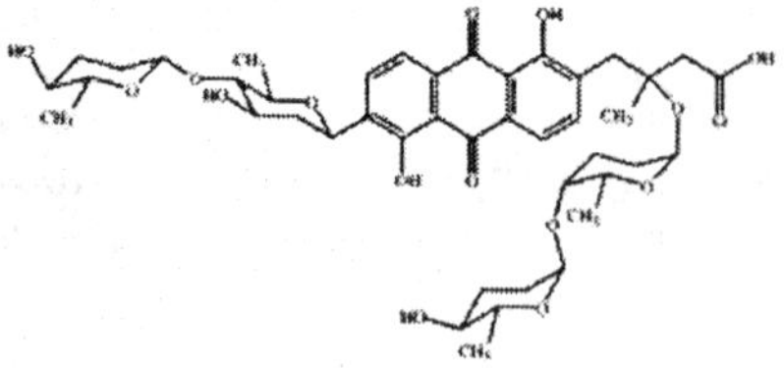 **Himalomycin B**

Himalomycin A and himalomycin B are the complex C-glycosides obtained from *Streptomyces* sp. 6921 isolated from marine sediments of Mauritius and exhibited strong antibacterial activity (Maskey *et al.*, 2003). They are the precursors of the anthracycline antibiotics.

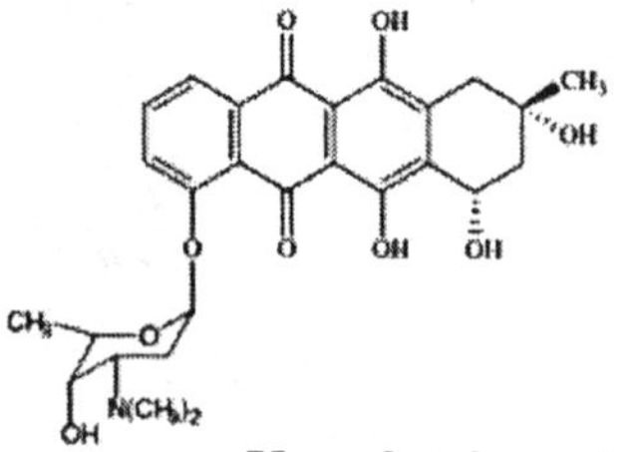

Komodoquinone A 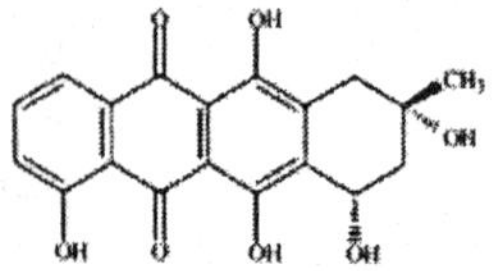 **Komodoquinone B**

Komodoquinone A and its chromophore, komodoquinone B were anthracycline antibiotics isolated from the fermentation broth of *Streptomyces* sp. KS3, they showed neuritogenic activity. They induce cell differentiation in the neuroblastoma cell line, Neuro2A and arrests cell cycle at the G1 phase (Itoh *et al.*, 2003)

Macrolides

The macrolide antibiotics contain a large macrocyclic lactone ring to which one or more deoxy sugars may be attacned. It binds to the 50S subunit of bacterial ribosomes, inhibiting the translocation stage of protein synthesis. This type of action makes it a bacteriostatic rather than a bacteriocidal antibiotic. The *Streptomyces* sp. M491, a marine actinobacterium produces a macrolide antibiotic named chalcomycin (Wu *et al.*, 2007).chalcomycin B, another macrolide antibiotic was isolated from the culture broth of a marine *Streptomycete* isolate B7064 (Asolkar *et al.*, 2002).

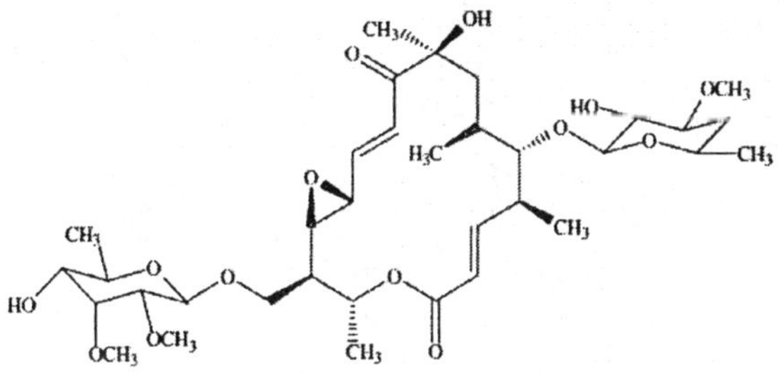

Chalcomycin

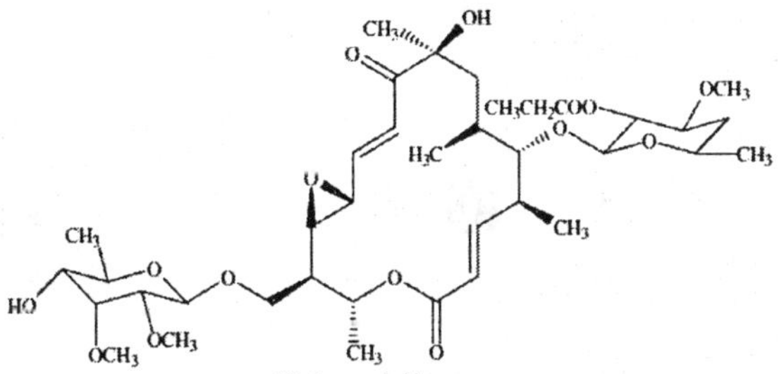

Chalcomycin B

Terpenes

Terpenes are derived biosynthetically from units of isoprene (2-methyl- 1, 3- butadiene), which has the molecular formula C_5H_8. Based on the number of isoprene units joined together, they are classified as monoterpenes, sesquiterpenes, diterpenes, sesterterpenes, triterpenes, and tetraterpenes.

The *Streptomyces* sp. NPS008187 isolated from marine sediment collected in Alaska, synthesized three new pyrrolosesquiterpenes, glyciapyrroles A, B and C They exhibited antibacterial activity(Macherla *et al.*, 2005). A Marine derived *Streptomyces* sp. QD518 produced a bioactive terpene called as selinane sesquiterpene, showed anticancer properties (Wu *et al.*, 2006). T-Muurolol sesquiterpenes were isolated from *Streptomyces* strain M491 derived from marine sediment Qingdao coast, China. These sesquiterpenes of the T-muurolol family were tested for their cytotoxycity against 37 human tumour cell lines, but only 15-hydroxy-T-muurolol showed moderate cytotoxic activity(Ding *et al.*, 2009).

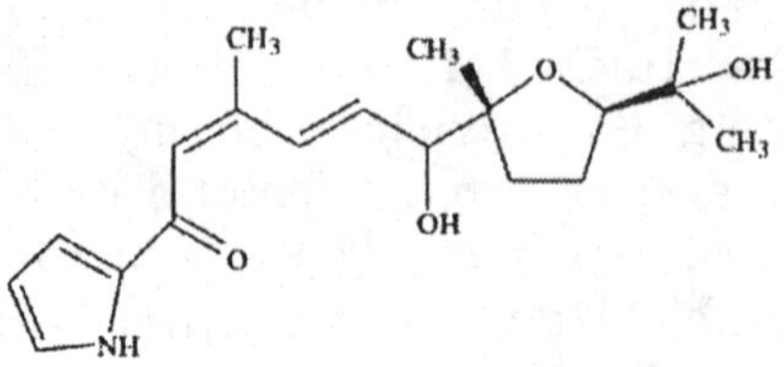

Glyciapyrrole A

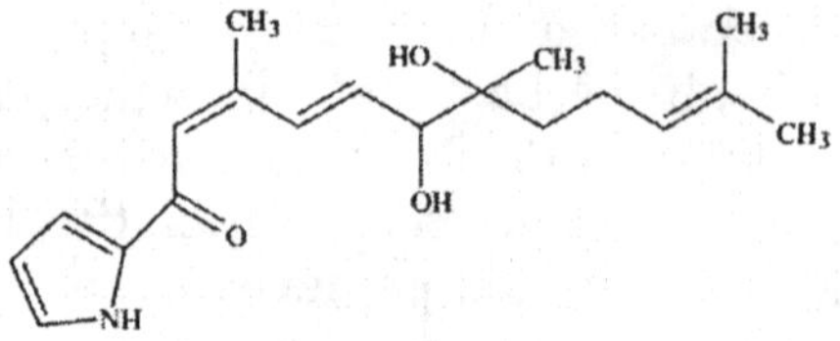

Glyciapyrrole B

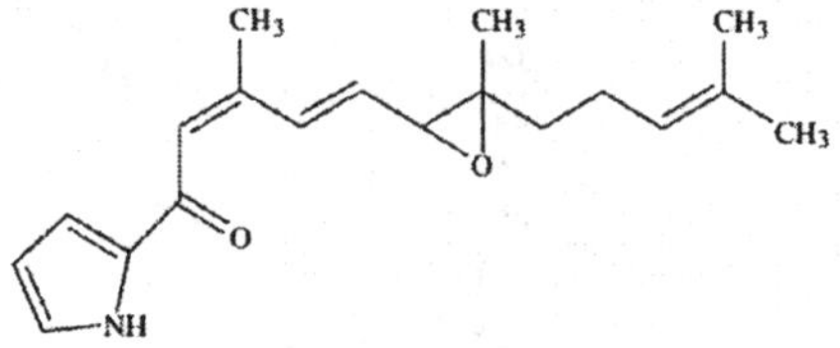

Glyciapyrrole C

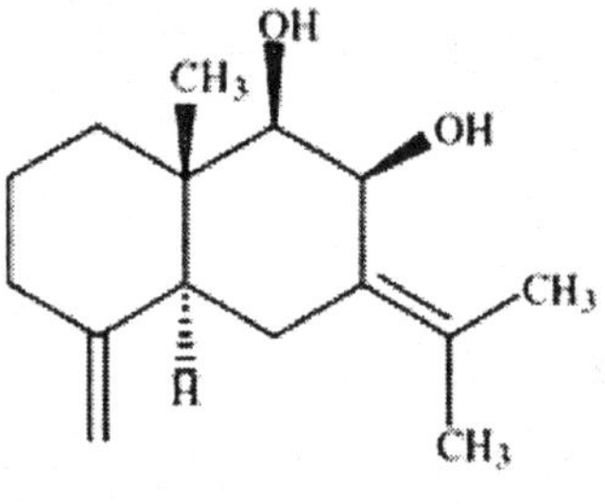

Selinane sesquiterpene

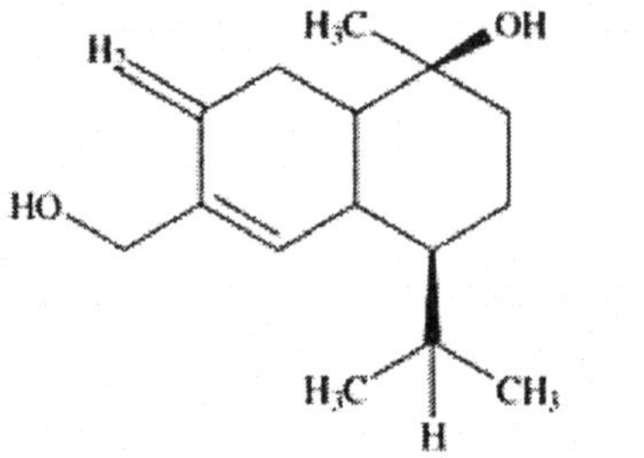

15-hydroxy T-Muurolol

Polyketides

Polyketides are a large and structurally diverse group of natural products synthesised by multifunctional or mono or bi-functional enzymes called polyketide synthases (PKSs). Many actinobacterial isolates from deep oceans contain non-ribosomal polyketide synthetase (NRPS) and polyketide synthetase (PKS) pathways, the hallmarks of secondary metabolite production (Salmon *et al.*, 2003).

The *Streptomyces* sp. JP95 isolated from the marine ascidian *Aplidium lenticulum* at Heron Island, Australia was found to biosynthesise the polyketide, griseorhodin A. Another polyketide Daryamides A, B and C are cytotoxic polyketides isolated from culture broth of a *Streptomyces* strain, CNQ-085 (Asolkar *et al.*, 2006). These bioactive compounds showed weak to moderate cytotoxicity against the human colon carcinoma cell line HCT-116 and very weak antifungal activities against *Candida albicans*. Recently three new polyketides, phaeochromycins F, G and H were obtained from the culture broth of marine *Streptomyces* sp. DSS-18. These compounds exhibited cytotoxic effect (Li *et al.*, 2008).

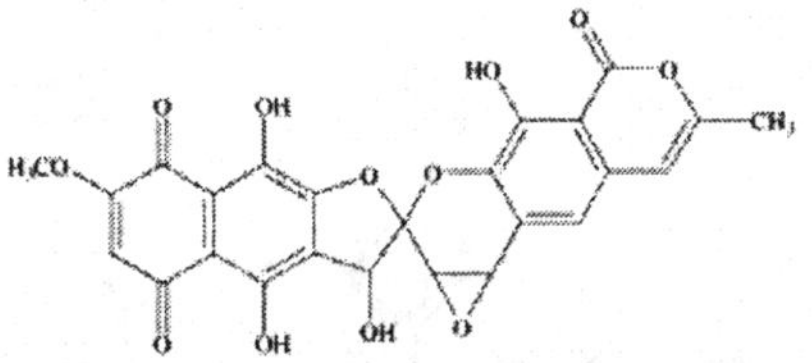

Griseorhodin A

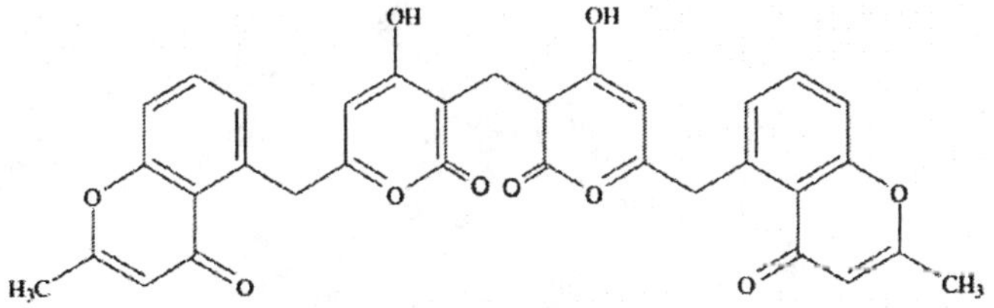

Phaeochromycins F

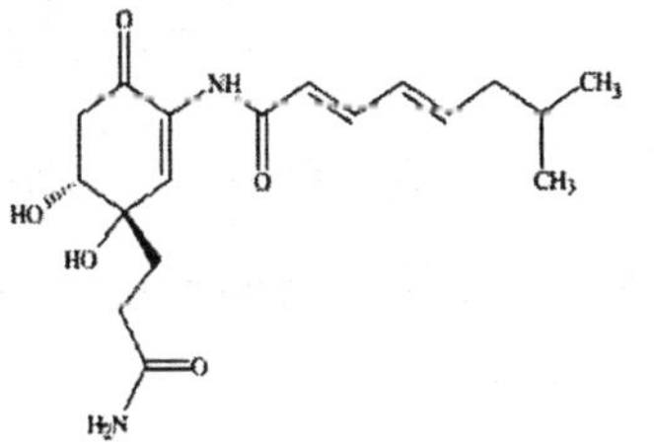

Daryamides A 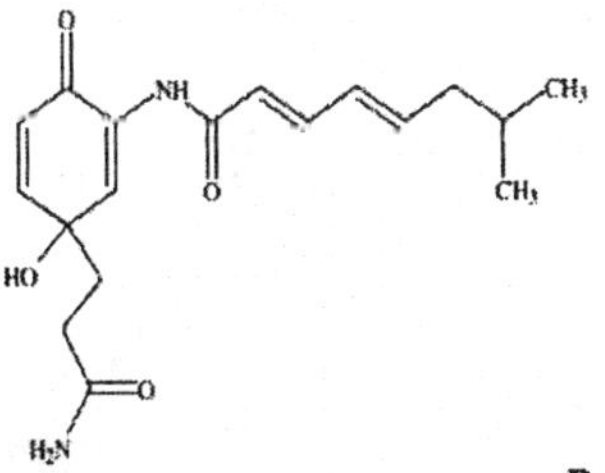 **Daryamides B** 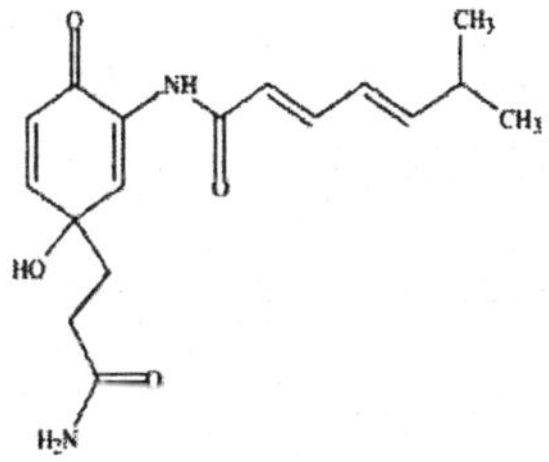 **Daryamides C**

Peptide antibiotics

Salinamides A and B are bicyclic depsipeptides produced by a *Streptomyces* sp. CNB-091 isolated from the jelly fish *Cassiopeia xamachana*. These metabolites were used as antibiotic and anti-inflammatory agents (Moore *et al.*, 1999). Two new dipeptide derivatives, 8-amino-[1, 4] diazonane-2, 5-dione and leucyl-4-hydroxyproline were isolated from marine *Streptomycesacrimycini* (Hernandez *et al.*, 2004).

8-Amino-[1,4] diazonane-2,5-dione

Salinamide B

Leucyl-4-hydroxyproline

Piericidins

Piericidins C_7 and C_8 are the two new members of the piericidin family, isolated from a marine *Streptomyces* sp. Piericidins C_7 and C_8 cytotoxicity against rat glia cells transformed with the adenovirus E1A gene and Neuro2A mouse neuroblastoma cells (Hayakawa *et al.*, 2007).

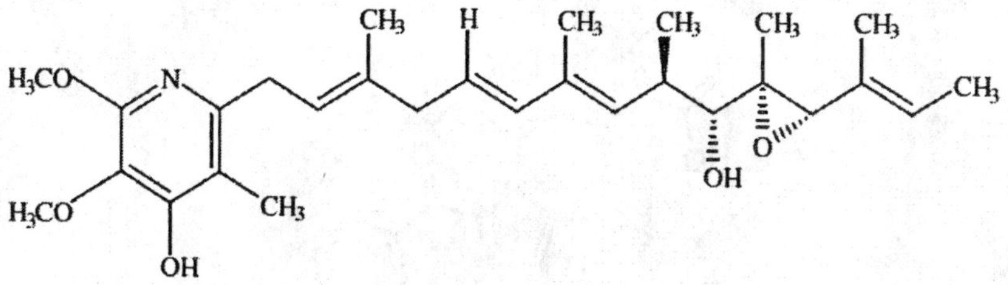

Piericidin C7

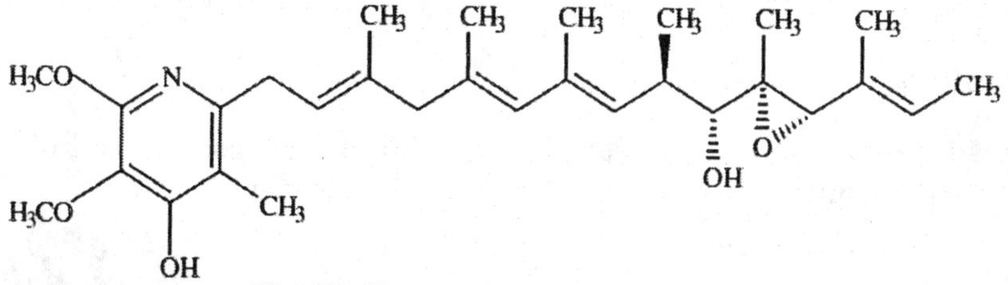

Piericidin C8

Methylpyridine

Streptokordin is a new cytotoxic compound of the methylpyridine, produced by *Streptomyces* sp. KORDI-3238 (Jeong *et al.*, 2006). Streptokordin exhibited significant cytotoxicity against seven human cancer cell lines.

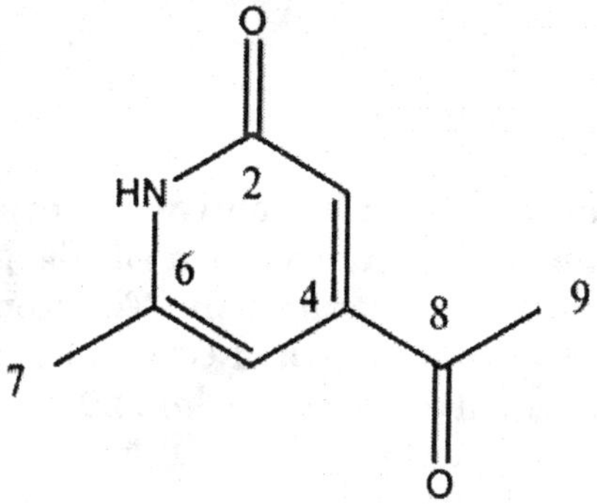

Streptokordin

Marinopyrroles

Marine *Streptomyces* strain CNQ-418 obtained from a marine sediment sample collected near La Jolla, California, produced marinopyrroles A and B. They possess potent antibiotic activities against methicillin-resistant *Staphylococcus aureus*(Hughes *et al.*, 2008).

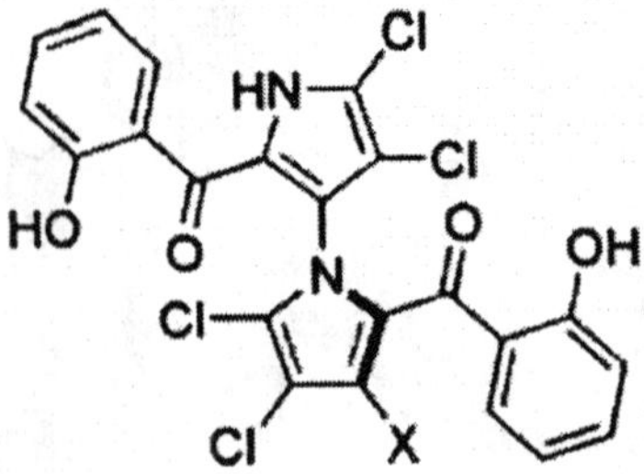

Marinopyrrole A (X = H) and Marinopyrrole B (X = Br)

Sisomicin

Marine *Streptomyces* strain GB-2 isolated from coastal soil of China produced the antibacterial substance sisomicin (Lu *et al.*, 2009).

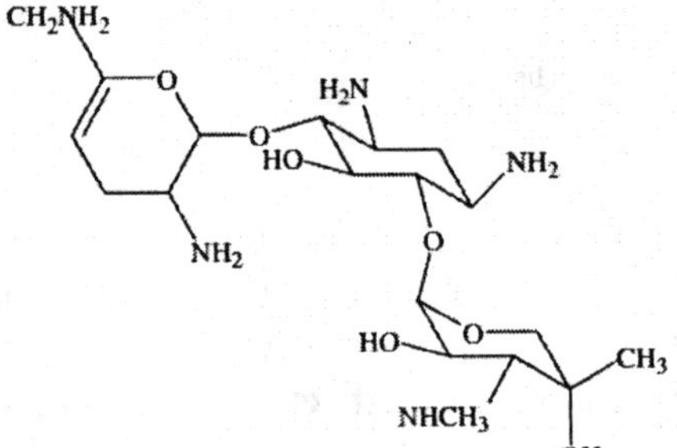

Sisomicin

Triazolopyrimidine

The *Streptomyces* sp. Merv8102 derived from sediment samples of Paltium coast on the Mediterranean Sea of Egypt produced a novel triazolopyrimidine antibiotic, essramycin. The compound is antibacterially active against Gram-positive and Gram-negative bacteria, while it showed no antifungal activity(El-Gendy *et al.*, 2008).

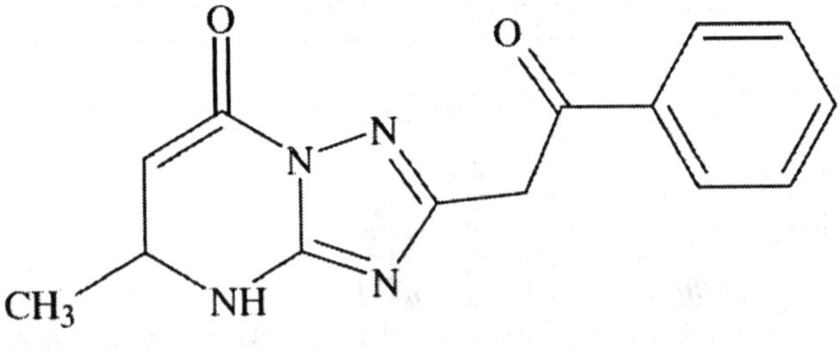

Essramycin

Streptochlorin

Streptochlorin was isolated from the fermentation broth of a marine *Streptomyces* strain 04DH110. Streptochlorin exhibited significant antiproliferative activity against human cultured cell lines (Jae *et al.*, 2007).

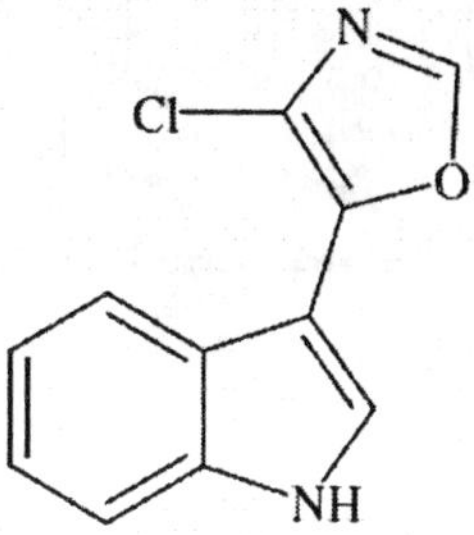

Streptochlorin

Mechercharmycin

New cytotoxic substances named mechercharmycin A and mechercharmycin B were isolated from marine-derived *Thermoactinomyces* sp. YM3-251(Kanoh *et al.*, 2005). Mechercharmycin A contain a cyclic peptide-like compound bearing four oxazoles and a thiazol rings, and its linear congener, mechercharmycin B. Mechercharmycin A exhibited relatively strong antitumor activity, whereas mechercharmycin B exhibited almost no such activity.

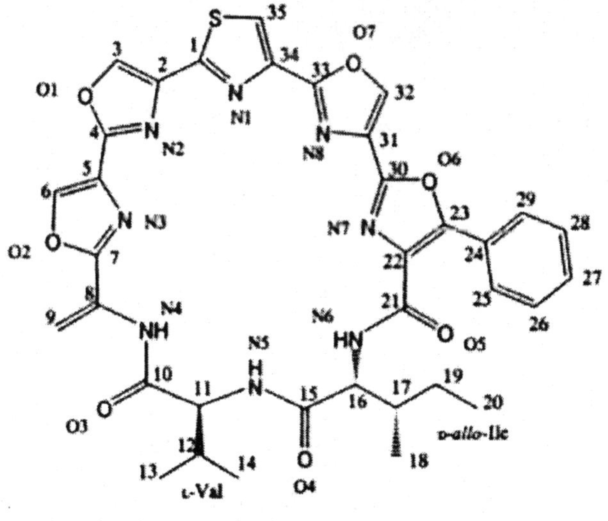

Mechercharmycin A

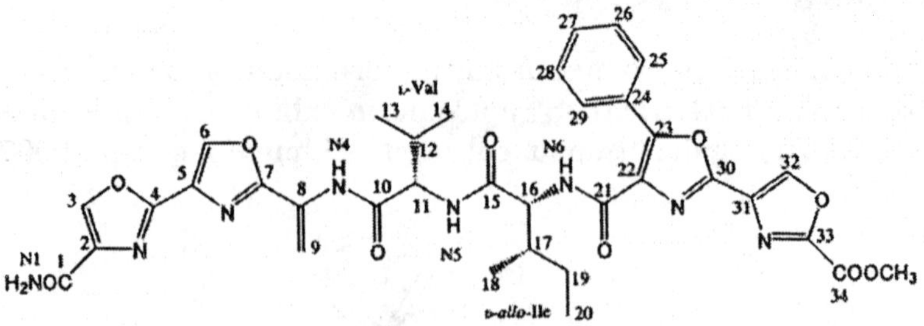

Mechercharmycin B

Aloesaponarin

The *streptomycete* isolate M097 isolated from Jiaozhou Bay in China was found to produce aloesaponarin II a typical plant anthraquinone structure (Cui *et al.*, 2008).

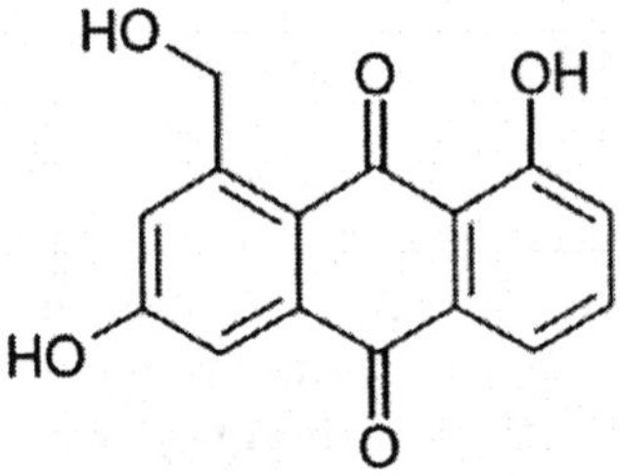

1, 6- dihydroxy-8-hydroxymethyl-anthraquinone

Marinomycin

Marinomycins A and B are antitumor-antibiotics reported from marine actinobacteria *Marinophilus* (Jensen *et al.*, 2005b)

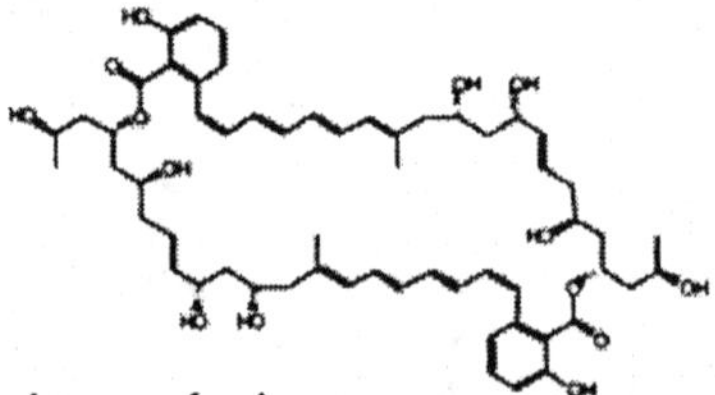

Marinomycin A

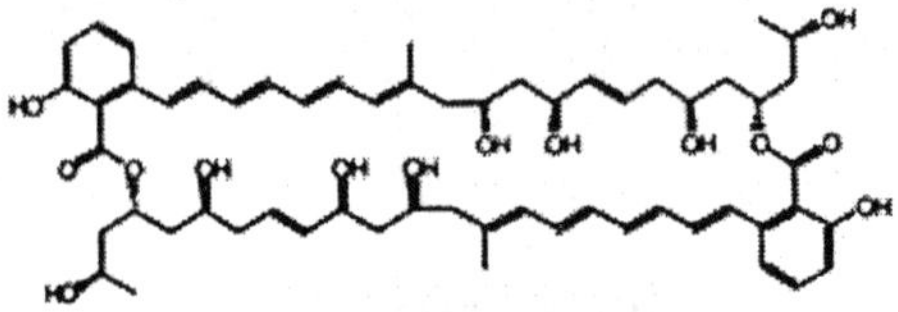

Marinomycin B

Abyssomicins

A rare actinobacteria *Verrucosispora* strain AB- 18-032 which was isolated from a sediment sample collected from the Sea of Japan at a depth of 289 m. Produced bioactive compound abyssomicins - a potent inhibitor of the folic acid biosynthesis. Abyssomicin C strongly inhibits Gram-positive bacteria, in particular multiresistant and vancomycin resistant *Staphylococcus aureus* strains (Riedlinger *et al.*, 2004).

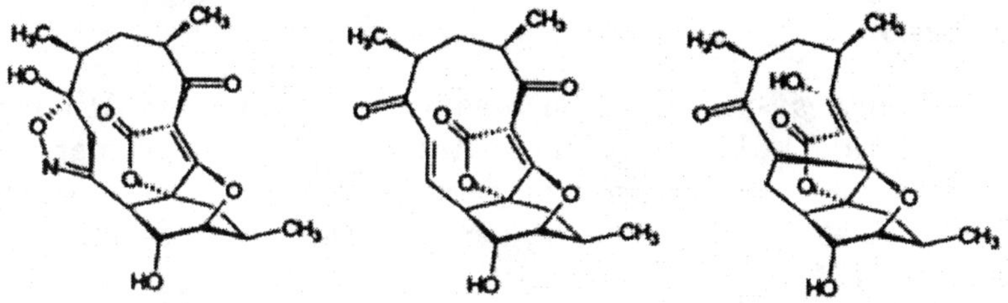

Abyssomicin B **Abyssomicin C** **Abyssomicin D**

Ikarugamycin

The macrolactam antibiotic ikarugamycin was obtained from the fermentation broth of *Micromonospora* strain originated from marine sediment of a Norwegian fjord. The compound has antibacterial, antiprotozoal and antiulcer properties (Riedlinger *et al.*, 2004).

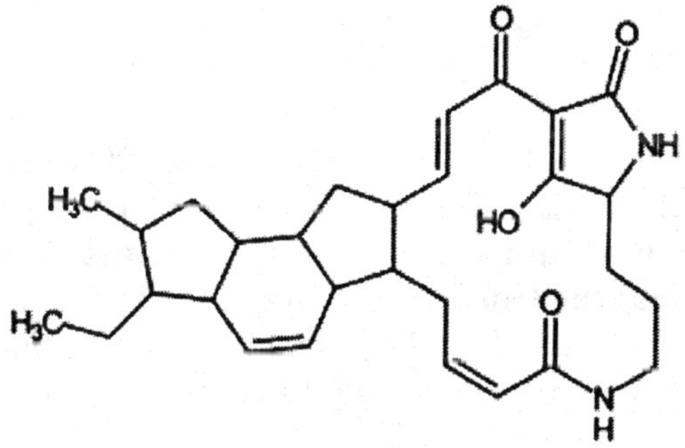

Ikarugamycin

Albidopyrone

A α-pyrone containing secondary metabolite produced by *Streptomyces* sp. NTK 227, a strain isolated from Atlantic Ocean sediment and found to be a member of the *Streptomyces albidoflavus* 16S rRNA gene clade. It showed inhibitory activity against protein-tyrosine phosphatase B (Hohmann *et al.*, 2009a)

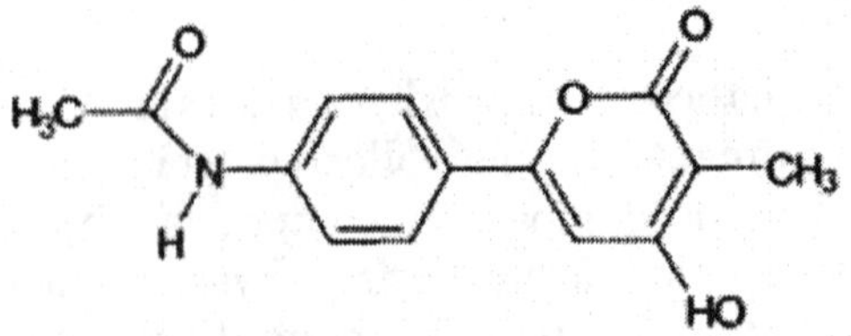

Albidopyrone

Benzoxazine

Streptomyces sp. NTK 935 was isolated from an Atlantic Ocean sediment core (3, 814 m) at the southern edge of the Canary Basin. It produced a new benzoxazine compound which had a strong inhibitory activity against the enzyme glycogen synthase kinase 3-beta (Goodfellow and Fiedler, 2010).

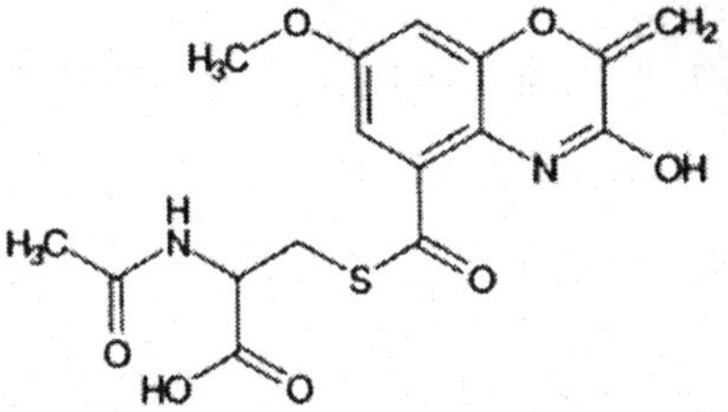

Benzoxazine

Caboxamycin

This new benzoxazole antibiotic was obtained from the *Streptomyces sp.* NTK 937, isolated from sediment collected from the Canary Basin. Caboxamycin showed inhibitory activity against both Gram-positive bacteria and against the tumour cell lines gastric adenocarcinoma (AGS), hepatocellular carcinoma (Hep G2) and breast carcinoma cells (MCF7). The antibiotic also showed an inhibitory activity against the enzyme phosphodiesterase.(Hohmann *et al.*, 2009b)

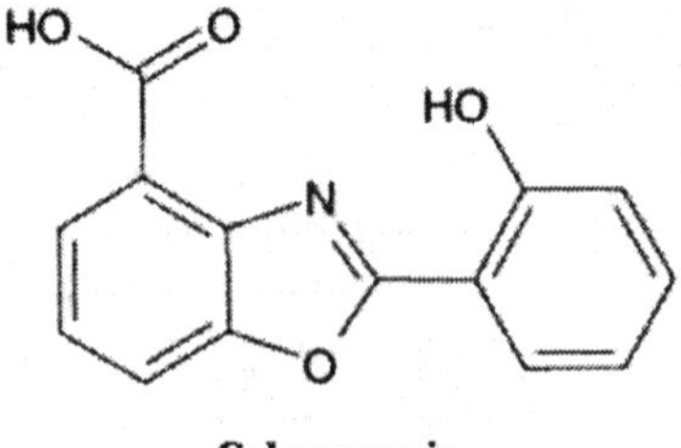

Caboxamycin

Dermacozines

A group of phenazine compounds called dermacozines, isolated from the culture filtrates of *Dermacoccus*, an actinobacteria isolated from sediment sample collected from the Challenger Deep (10898 m) of the Mariana Trench. They show antitumour, antiprotozoal and free radical scavenging activities (Abdel-Mageed, 2010).

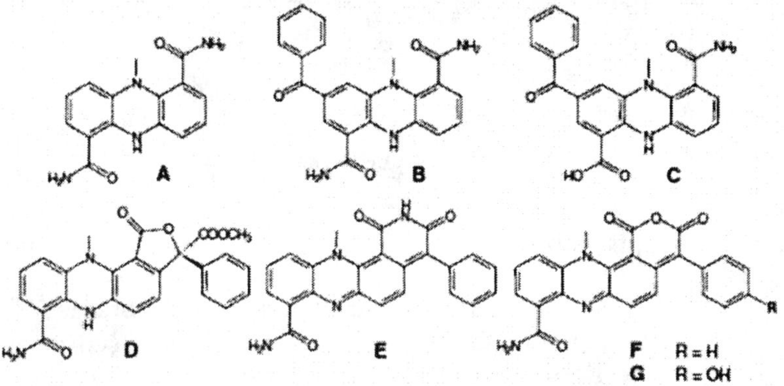

Structures of dermacozines extracted from Dermacoccus sp. MT1.1 and MT1.2

Lipocarbazoles A1–A4

New secondary metabolites with a carbazole moiety and an alkyl side chain were isolated from the cell extracts of *Tsukamurella pseudospumae* strain Acta 1857, an organism isolated from activated sludge foam collected at Stoke Bardolph Water Reclamation Works, near Nottingham, UK. These compounds exhibited strong free radical scavenging activity (Schneider *et al.*, 2008). The same secondary metabolite pattern was detected in *Tsukamurella* strains isolated from the sediment collected from the Challenger Deep of the Mariana Trench (Goodfellow and Fiedler, 2010)

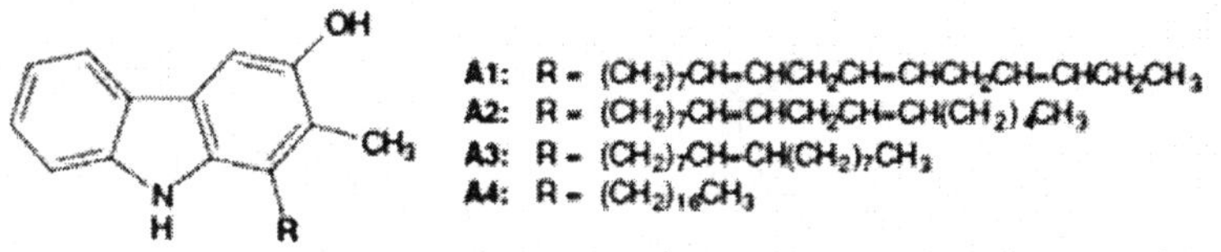

Lipocarbazoles

Lysolipin

Streptomyces sp. NTK 963 was isolated from the deep-sea sediment sample collected from the Canary Basin. It produced a secondary metabolite lysolipin, an interesting antibiotic with strong antitumor and antibacterial activity. (Goodfellow and Fiedler, 2010))

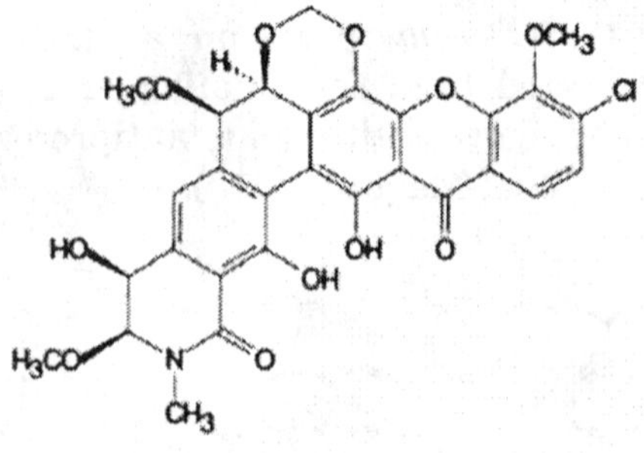

Lysolipin I

Proximicins A, B and C

These novel aminofuran-type antibiotics were obtained from the extracts of *Verrucosispora* strain MG-37, isolated from a sediment sample collected from the Raune Fjord, Norway at a depth of 250 m. proximicin A, was also detected in the abyssomicin producer *Verrucosisporamaris* AB-18- 032.(Fiedler *et al.*, 2008)

1. Proximicins exhibit weak antibacterial activity but have a strong cytostatic effect against various human tumour cell lines. All of the proximicins showed significant growth inhibitory activity towards gastric adenocarcinoma (AGS) and hepatocellular carcinoma (Hep G2) though breast carcinoma cells (MCF 7) were less sensitive. The proximicins were found to arrest AGS cells in the G0/G1 phase of the cell-cycle and increase the level of cell-cycle regulatory proteins p53 and p21.(Schneider *et al.*, 2008)

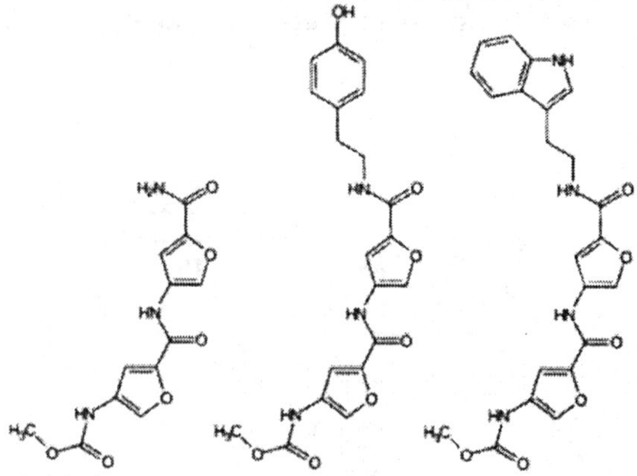

Proximicin A **Proximicin B** **Proximicin C**

4. Enzymes from marine actinobacteria

Proteases

Proteases account for about 60% of the total amount of the enzymes sale in the world with 2/3 of this amount, being of microbial origin (Johnvesly *et al.*, 2002; Mehta *et al.*, 2006). Proteases are used for various industrial applications, such as laundry detergents, leather preparation, food industry and protein recovery or solubilization, organic synthesis and with the advent of new frontiers in biotechnology, the spectrum of protease application has expanded into many new fields, such as clinical, medicinal and analytical chemistry. (Xiao-Ge Chen *et al.*, 2004; Ramesh *et al.*, 2007; Cudic and Fields, 2009; Southan, 2001; Hooper, 2002). Sahu *et al.* studied the extra-cellular enzyme (amylase, lipase, protease, cellulase and chitinase) activities of actinobacteria isolated from the sediment samples of the Vellar estuary. The *S. clavuligerus* strain isolated from the western coast of India, exhibited the production of alka-line protease (Thumar and Singh, 2007a). Sahu *et al.* (2007a) screened actinobacterial strains from gut contents of the tiger shrimp, *Penaeus monodon*. Out of the 17 strains tested, the strain *Streptomyces galbus* showed protease activity. *Streptomyces* sp. isolated from coastal mangrove sediment exhibited alkaline protease activity (Kathiresan and Maniva-nnan (2007). Thermostable alkaline protease was characterized from marine *S. fungicidicus* MML1614 (Ramesh *et al.*, 2009). Debananda *et al.* (2009) isolated alkali thermotolerant actinomycetes from marine samples which produced extracellular protease. *Streptomyces albidoflavus* isolated from marine samples exhibited high proteolytic activity (Hanaa *et al.*, 2010). *Streptomyces alboniger* isolated from the Kodiyakarai coastal sediments of Bay of Bengal exhibited multiple enzyme activity such as amylase, cellulase and protease activities (Manivasagan *et al.*, 2010).

L-Asparaginase

L-asparaginase is an anti-neoplastic agent used in the lymphoblastic leukaemia chemotherapy (Fisher *et al.*, 2002). L-asparaginase has received increased attention in recent years for its anticarcinogenic potential (Amena *et al.*, 2010). The clinical action of this enzyme is attributed to reduction of L-asparagine, so that tumor cells unable to synthesize this amino acid are selectively killed by L-asparagine deprivation. Mathew *et al.* (1994) isolated the antagonistic *Streptomycetes* associated with clam *Villorita cyprinoides*, which exhibited potent L-asparaginase activity. Dhevendaran & Annie (1999) studied the L-asparaginase activity of *Streptomycetes* isolated from sediments of Veli estuarine lake along Kerala coast. *Streptomyces plicatus* isolated from the alimentary canal of the marine fish, *Gerres filmentosus* exhibited high L-asparaginase activity (Koshey *et al.*, 1997). *Streptomyces* sp. associated with *Therapon jarbua* and *Villorita cyprinoids* of Veli Lake exhibited L-asparaginase activity (Dhevendaran and Anithakumari (2002)). Marine Streptomyces strains (PDK2, PDK7) isolated from South Indian coast, produced L-asparaginase and exhibited cytotoxic effect on JURKAT cells (Acute T cell leukemia) and K562 cells (chronic myelogenous leukemia) (Dhevagi and Poorani (2006). *S. aurantiacus* isolated from Bhitarkanika mangrove ecosystem of Orissa, India exhibited different extracellular activity like phosphate solubilisation, lipase and L-asparaginase production (Gupta *et al.*, 2007). Sahu *et al.* (2007b) isolated 40 species of actinobacteria from the different parts of three species of marine fishes from the Vellar estuary. The strains were tested for their L-asparaginase activity and among them, six strains showed significant L-asparaginase activity. Sahu *et al.* (2007c) also partially purified L-asparaginase enzyme from *Streptomyces canus* and studied the anti-leukemic activity in mice. *Streptomyces* marine isolates S3, S4 and K8 showed the production of extracellular anti-leukemic enzyme L-asparaginase obtained by solid-state and submerged fermentation (Basha *et al.*, 2009). Selvakumar Dharmaraj (2011) studied the production of L- asparaginase production by *Streptomyces noursei* MTCC 10469, isolated from marine sponge *Callyspongia diffusa*. Marine *Streptomyces* sp. strain EPD27 exhibited high levels of extracellular L-asparaginase enzyme production (Poorani *et al.*, 2009).

L-Glutaminase

L-glutaminase has attracted much attention in both pharmaceutical and food industrial applications. In food industry, L-glutaminase is used as a flavor enhancer by increasing glutamic acid content in food through hydrolysis of L-glutamine to L-glutamic acid and ammonia. It also used in enzyme therapy for cancer especially for acute lymphocytic leukemia (Sabu *et al.*, 2000). Another important application of L-glutaminase is in biosensors for monitoring the glutamine levels in mammalian and hybridoma cells. Umamaheswary *et al.* (2005) isolated 40 strains of actinobacteria from the estuarine fish, *Mugil cephalus* using Kuster's agar medium. Out of 40 strains tested, the strain *Streptomyces galbus* showed good L-glutaminase activity. Sivakumar *et al.* (2006) isolated actinobacterial strains from skin, gills and gut contents of the estuarine fish, *Chanos chanos*. Out of 20 strains tested, *Streptomyces rimosus* showed the highest L glutaminase activity. Balagurunathan *et al.* (2010) isolated L-glutaminase producing actinomycetes from marine sediments. Marine alkalophilic *Streptomyces* sp-SBU1 isolated from Cape Comorin coast exhibited extracellular L-glutaminase activity.

Lipase

Lipases are efficient in various reactions such as esterification, transesterification and aminolysis (Babu *et al.*, 2008). Lipases are valuable biocatalysts, because they act under mild conditions and are highly stable in organic solvents, show broad substrate specificity (Pierre *et al.*, 2000; Amare *et al.*, 2003). Commercially lipases are used in detergents, paper production, cosmetic production, food flavoring, organic synthesis and some other industrial applications (Kobayashi *et al.*, 2008; Chi *et al.*, 2009). Kokare *et al.* (2007) isolated lipase producing marine *Streptomyces* sp. S1 from Alibag, Janjira and Goa coastal regions of India. *Streptomyces griseochromogenes* isolated from the sediment exhibited highest lipase activity (Gunalakshmi *et al.*, 2008). Mo *et al.* (2009)purified a novel extracellular phospholipase C from a marine *Streptomycete*, which hydrolyzed only phosphatidylcholine. *Streptomyces* sp. VITDDK3-a novel halotolerant actinobacteria isolated from Saltpan produced extracellular lipase (Deepika Lakshmipathy *et al.*, 2010). Seghal *et al.* (2010) isolated marine *Brevibacterium aureum* MSA13 capable of producing extracellular lipase. Marine *Streptomyces* species B3 exhibited potent lipase activity (Abhijit *et al.*, 2011)

Cellulase

Cellulases are used in textile industry, detergent, paper and pulp industry, improving digestibility of animal feeds and food industry (Gusakov *et al.*, 2000; Kottwitz and Schambil, 2005). The growing concern about the shortage of fossil fuels, the emission of green house gases has resulted in an increased focus on bioethanol from lignocellulosics and the possibility to use celluloses to perform the enzymatic hydrolysis of lignocellulosic material (Zaldivar *et al.*, 2001). *Streptomyces* sp. collected from the Bay of Bengal exhibited extra cellular cellulase activity (Chandramohan *et al.*, 1972). Balasubramanian *et al.* (1979) isolated *Streptomycetes* from the digestive tract of marine borer. Cellulolytic actinomycetes from marine sediments were isolated by Veiga *et al.* (1983). Balagurunathan & Subramanian (2001) isolated *Streptomyces* strains from the littoral sediments of Parangipettai coastal waters. These strains exhibited chitinase, protease and cellulase activities. Antagonistically active extra-cellular cellulase enzyme producing actinomycetes in the alimentary canal of estuarine fishes was isolated by Kundu *et al.* (2006a). Murugan *et al.* (2007) isolated actinobacteria from the gut contents of the estuarine finfish *Mugil cephalus* and examined for their cellulase activity. The strain *Streptomyces actuosus* showed maximum cellulase activity. *Streptomyces ruber* isolated from marine sediments produced cellulase using agro-industrial residues (Nermeen *et al.*, 2010).

Chitinase

Chitinolytic enzymes have been widely used in various processes including the agricultural, biological and environmental fields. The antifungal activity and highly biocompatible quality make chitinase and its derivatives particularly useful for biomedical applications, such as wound healings, cartilage tissue engineering, drug delivery, and nerve generation (Shi *et al.*, 2006; Yan *et al.*, 2006). The biodegradable and antifungal properties of chitinase are also useful for environmental and agricultural uses, food technology and cosmetics (Lin and Lin, 2005) Pisano *et al.* (1992) determined the bioactivity of chitinolytic actinomycetes of marine origin. Marine actinomycetes isolated from the alimentary canal of estuarine fishes exhibited chitinolytic activity (Kundu *et al.*, 2006b). Marine *Streptomyces* sp. DA11 isolated from South China, found to be associated with sponge *Craniella australiensis*, produced the enzyme chitinase and showed antifungal activities against *Aspergillus niger* and *Candida albicans* (Han *et al.*, 2009). A potent chitinolytic *Streptomyces champavatii* AZ-1 isolated from water and sediments of the Red sea coast produced chitooligosaccharides (Alaa *et al.*, 2010). Kim-Chi Hoang *et al.* (2011) studied the chitinolytic activities of *Streptomyces* sp. TH-11.

Enzymes possessing novel chemical and stereochemical properties

Polycyclic aromatic hydrocarbons (PAHs) are a class of organic pollutants present in the marine environment. They are composed of two or more fused aromatic rings. Persistence, ability to bioaccumulate and carcinogenicity of PAHs are all important aspects of interest concerning the need for biotransformation of these compounds (Jorgensen *et al.*, 2008). Results of experiments using various di-substituted naphthalenes such as dimethylnaphthalenes using the cells of *Escherichia coli* expressing aromatic dihydroxylating dioxygenase genes of marine bacteria, *Nocardioides* sp. KP7 were reported. The 1,4-dimethylnaphthalene was predominantly bioconverted into 1,4-dihydroxymethylnaphthalene, which possess applications as starting materials for the synthesis of industrially useful chemicals (Shindo *et al.*, 2007).

Reaction of transformation of 1, 4-dimethylnapthalene using the cells of *Escherichia coli* expressing aromatic dihydroxylating dioxygenase.

aromatic dihydroxylating dioxygenase

1,4-dimethylnaphthalene → yield: 3.3% + ca. 60%

Other hydrolases

Several other enzymes have also been identified from marine actinobacteria. Poornima *et al.* (2007) investigated the α- amylase activity of actinomycete strain AE-19 isolated from shrimp pond. Chakraborty *et al.* (2009) reported a novel α-amylase from marine *Streptomyces* sp. D1 by using media containing 2% sucrose, 0.35% peptone and 0.15% of malt extract. Sahu *et al.* (2007d) studied total actinobacteria and phosphate solubilizing actinobacterial population density from the different sediment samples of the Vellar estuary. *Streptomyces galbus,* exhibited good phosphate solubilizing activity. *S. aurantiacus* isolated from Bhitarkanika mangrove ecosystem of Orissa, India exhibited different extracellular activity like phosphate solubilisation, lipase and L-asparaginase production. An alginate-degrading bacterium classified as *Streptomyces* sp. ALG-5 was isolated from marine seaweeds of South Korea (Kim *et al.*, 2009)

5. Enzyme inhibitors

Enzyme inhibitors are molecules that bind to enzymes and decrease their activity. The binding of an inhibitor can stop a substrate from entering the enzyme's active site and/or hinder the enzyme from catalysing its reaction. A few enzymes inhibitors have been reported from marine *Streptomyces*.

α -Amylase inhibitors

An α-amylase inhibitor was isolated from *S. corchorusii* subsp. *rhodomarinus*. subsp. nov (Imada *et al.*, 1988). The α-amylase inhibitor can be used for determination of amylase isozyme activities, the purification of amylases, and the control of such carbohydrate-dependent diseases as diabetes, obesity or hyperlipemia.

Pyroglutamyl peptidase inhibitors

Pyroglutamyl peptidase acts by catalyzing the release of pyroglutamyl residues that block terminal amino acids in many proteins and peptides. This enzyme is widely distributed in animals, plants and microorganisms. Pyrizinostatin is an inhibitor of pyroglutamyl peptidase, isolated from culture of *Streptomyces* sp. SA-2289 (Aoyagi *et al.*, 1992). Pyrizinostatin, therefore, may be used for elucidating the mechanisms of various diseases caused by the enzyme.

N-acetyl - β glucosaminidase inhibitors

N-Acetyl-glucosaminidase is an enzyme that catalyzes the hydrolysis of glycosidic linkages as an exoglycosidase and releases N-acetyl-glucosamine from glycoprotein. The activity of this enzyme is increased markedly in patients with diabetes, leukemia or cancer. Therefore, identification of inhibitors of this enzyme might lead to the elucidation of the causative agent of those diseases. Pyrostatin A and B are inhibitors of N-acetyl- ß glucosaminidase, produced by *Streptomyces* sp. SA-3501 (Aoyama *et al.*, 1995; Imada, 2004; Imada, 2005).

β-Glucosidase inhibitors

ß -Glucosidases are involved in tumor metastasis and the human immunodeficiency virus and therefore, have potential for use as drug targets. ß-Glucosidase inhibitors have been produced by actinobacteria from the marine environment (Imada and Okami, 1995).

6. Conclusion

The diversity of actinobacteria varies with the ecological niches of marine ecosystem and due to the existence of the extreme conditions in the marine environment, actinobacteria with different morphological features, have ability to produce potent, unique, modified, unprecedented bioactive secondary metabolites. The complex natures of these bioactive compounds enable them to be used as potent therapeutic agents to treat the emerging challenging diseases and to have significant applications in the biotechnology industries for the benefit of mankind to lead a better life.

REFERENCES

Abdel-Mageed *et al* (2010). *Org Biomol Chem* 8:2352–2362.

Abhijit Khopade *et al* (2011) *Journal* of *Colloid and Interface Science* (In press). DOI: 10.1016/j.jcis.2011.11.009.

Alaa. El-Dein, Hosny MS, El-Shayeb NA, Amira Abood, Abdel-Fattah AMA (2010). *Australian Journal* of *Basic and Applied Sciences* 4(4) 615-63.

Amare G, Thomas D, Steffen BP, Per HN (2003) *Water Res* 37:3652–3657.

Amena S, Vishalakshi N, Prabhakar M, Dayanand A, Lingappa K (2010). *Brazil J Microbiol* 41: 173-178.

Antonio Trincone (2010). Journal of Molecular Catalysis B: Enzymatic 66 :241–256.

Aoyagi *et al* (1992) *J Antibiot* (Tokyo), 45: 1795–1796.

Aoyama *et al* (1995). *J Enzyme Inhib* 8:223–232.

Asolkar RN, Jensen PR, Kauffman CA, Fenical W (2006) *J Nat Prod* 69:1756–1759

Asolkar RN, Maskey RP, Helmke E, Laatsch H (2002) *J Antibiot* 55:893–898

Babu J, Pramod WR, George T (2008) *Biotechnol Adv*, 26 457–470.

Balagurunathan R, Radhakrishnan M, Somasundaram ST (2010) *Australian Journal* of *Basic and Applied Sciences* 4(5): 698-705.

Balagurunathan R, Subramanian A (2001) *Ad Bios* 20(II):71–76.

Balasubramanian T, Laksmanaperumalsamy P, Chandramohan D, Natarajan R (1979) *Indian J mar Sci* 8:11–13.

Baltz RH (2006) *J. Ind. Microbiol. Biotechnol.*33: 507-513.

Basha SN, Rekha R, Komalal M, Ruby S (2009). *Tropical J Pharmaceutical Res* 8: 353-360.

Berdy J (2005) *J. Antibiot.* 58(1): 1-26.

Bull AT, Ward AC, Goodfellow M (2000) *Microbiol Mol Biol Rev* 64:573–606

Chakraborty *et al* (2009) *J Mol Catal B Enzym.* 58 17–23.

Chandramohan D, Ramu S, Natarajan RC (1972) *Curr Sci* 41:245–246.

Chi Z, Zhang T, Liu G, Li J, Wang X (2009) *Biotechnol Adv* 27:236–255.

Copp BR (2003). *Nat. Prod. Rep.* 20: 535–557.

Cudic M, Fields GB (2009). *Curr Protein Pept Sci* 10(4):297-307.

Cui HX, Shaaban KA, Schiebel M, Qin S, Laatsch H (2008) *World J Microbiol Biotechnol* 24:419–421.

Das S, Lyla PS, Ajmal Khan S (2006) *Curr Sci* 25:1325–1335

Debananda S. Ningthoujam, Pintubala Kshetri, Suchitra Sanasam, Salam Nimaichand (2009) *World Applied Sciences Journal* 7 (7): 907-916.

Deepika Lakshmipathy T, Arun Prasad AS, Krishnan Kannabiran (2010) *Advances in Biological Research* 4(2):108-115.

Dharmaraj S (2010). *World J. Microbiol. Biotechnol.*, 26: 2123-2139.

Dhevagi P, Poorani E (2006). *Ind J Biotech* 5: 514-520.

Dhevendaran K, Anithakumari YK (2002) *Indian J mar Sci* 39:155–159.

Dhevendaran K, Annie K (1999) *Indian J mar Sci* 28:335–337.

Ding L, Pfoh R, Ruhl S, Qin S, Laatsch H (2009) *J Nat Prod* 72:99–101

El-Gendy MMA, Shaaban M, Shaaban KA, El-Bondkly AM, Laatsch H (2008) *J Antibiot* 61:149–157

Ellaiah P, Ramana T, Bapi Raju KVVSN, Sujatha P & Uma Sankar A (2004) *Asian J Microbiol Biotech Env Sci* 6:53–56.

Fenical W (1993). *Chem. Rev.* 93: 1673–1683.

Fenical W, Jensen PR (2006) *Nat Chem Biol* 2: 666-673.

Fiedler *et al* (2008) *J Antibiot* 61:158–163.

Fiedler *et al* (2005) *Antonie van Leeuwenhoek* 87:37–42.

Fisher SH, Wray Jr LV (2002) *J Bacteriol* 184: 2148-2154.

Goodfellow M, Fiedler HP (2010) *Antonie van Leeuwenhoek* 98:119–142.

Gunalakshmi *et al* (2008) *Research J* of *Microbiology* 3:273-81.

Gupta N, Mishra S, Basak UC (2007) *Malaysian J Microbiol* 3:7–14.

Gusakov *et al* (2000) *Appl Biochem Biotechnol* 88:119-126.

Han Y, Yang B, Zhang F, Miao X, Li Z (2009) *Mar Biotechnol* 11:132–140.

Hanaa A. El-Shafei, Mohamed S. Abdel-Aziz, Mohamed F. Ghaly, Ahmed A. H. Abdalla (2010). Proceeding of fifth scientific environmental conference, Zagazig Uni., 125 – 142.

Hayakawa Y, Shirasaki S, Shiba S, Kawasaki T, Matsuo Y, Adachi K, Shizuri Y (2007) *J Antibiot* (Tokyo) 60:196–200

Helmke E, Weyland H (1984) *Int J Syst Bacteriol*, 34:127-138.

Hernandez ILC, Macedo ML, Berlinck RGS, Ferreira AG, Godinho MJL (2004) *J Braz Chem Soc* 15:441–444

Hohmann *et al* (2009a). *J Antibiot* 62:75–79.

Hohmann *et al* (2009b) *J Antibiot* 62:99–104.

Hooper NM (2002) Proteases: A Primer. In: Proteases in Biology and Medicine Essays in Biochemistry, Hooper, N.M. (Ed.). Portland Press, London, pp: 1-8.

Hughes CC, Davo AP, Jensen PR, and Finical W (2008) *Org Lett.* 10(4): 629-631

Imada C (2004) *Mar Biotechnol* 6:193–198.

Imada C (2005) *Antonie van Leeuwenhoek* 8759–63.

Imada C, Okami Y (1995) *J Mar Biotechnol* 2:109–113.

Imada C, Simidu U (1988). *Nippon Suisan Gakkaishi* 54:1839–1845.

Itoh T, Kinoshita M, Aoki S, Kobayashi M (2003) J Nat Prod 66:1373–1377

Jae SH, Jeong HS, Lee HS, Park SK, Kim HM, Kwon HJ (2007) *J Microbiol Biotechnol* 17:1403–1406

Jensen PR, Mincer TJ, Williams PG and Fenical W(2005b) *Antonie van Leeuwenhoek* 87:43–48

Jensen, PR, Gontang E, Mafnas C, Mincer TJ, Fenical W (2005a) *Environ. Microbiol*, 7: 1039-1048.

Jeong SY, Shin HJ, Kim TS, Lee HS, Park SK, Kim HM (2006) *J. Antibiot.* 59(4): 234–240

Johnvesly B, Manjunath BR, Naik GR (2002) *Bioressour Technol* 82 (1): 61-64.

Jorgensen *et al* (2008) *Mar Environ Res* 65:171–186.

Kanoh K, Matsuo Y, Adachi K, Imagawa H, Nishizawa M, Shizuri Y (2005) *J. Antibiot.* 58(4): 289–292.

Kathiresan K, Manivannan S (2007) *Res J Environ Sci* 1(4): 173-178.

Kim DE, Lee EY, Kim HS (2009) *Mar Biotechnol* 11:10–16.

Kim-Chi Hoang *et al* (2011). *Int J Mol Sci* 12: 56-65.

Kobayashi *et al* (2008) *Extremophiles* 12:519–527.

Koehn FE & Carter GT (2005) *Nat. Rev. Drug. Discov.,* 4: 206-220.

Kokare CR, Kadam SS, Mahadik KR, Chopad BA (2007) *Indian Journal* of *Biotechnology* 6:78-84.

Koshey A, Dhevendaran K, Georgekutty MI, Natarajan P (1997) *J Mar Biotechnol* 5: 181-185.

Kottwitz B, Schambil F (27 January, 2005) US Pat, 2005002047.

Kundu S, Sahu MK, Sivakumar K, Kannan L (2006a) *Asian J Microbiol Biotech Envi Sc* 8: 707–710.

Kundu S, Sahu MK, Sivakumar K, Kannan L (2006b) *Asian J Microbiol Biotech Envi Sc* 8: 811–815.

Kurtboke DI (2005). *Anton. Leeuw.* 1:19-28.

Lam KS (2006) Current Opinion in Microbiology 9:245–251

Li J, Lu CH, Zhao BB, Zheng ZH, Shen YM (2008) *J Org Chem* 4:1–5

Lin CC, Lin HL (2005) *J Hazard Mater* 122:7–15.

Lu Y, Dong X, Liu S, Bie X (2009) *Mar Biotechnol* 11:717–724

Macherla VR, Liu J, Bellows C, Teisan S, Nicholson B, Lam KS, Potts BCM (2005) *J Nat Prod* 68:780–783

Maldonado *et al* (2004) *Antonie van Leeuwenhoek* 87: 11–18

Manivasagan *et al* (2010) *African Journal* of *Microbiology Research* 4(14): 1550-1559.

Martin AP (2002) *Appl. Environ. Microbiol.* 68: 3673–3682.

Maskey RP, Helmke E, Laatsch H (2003) J Antibiot 56:942–949

Mathew A, Dhevendaran K, Georgekutty M, Natarajan P (1994). *Indian J mar Sci* 23:204 208.

McCarthy AJ (1987) *FEMS Microbiol Rev* 46:145 163.

Mehta VJ, Thumar JT, Singh SP (2006) *Bioresource Technology* 97: 1650–1654.

Mo SJ, Kim JH, Cho KW (2009) *Biosci Biotechnol Biochem* 73: 2136-2137.

Moore BS, Trischman JA, Seng D, Kho D, Jensen PR, Fenical W (1999) *J Org Chem* 64:1145–1150

Murugan *et al* (2007) *J Sci Ind Res India* 6:388–393.

Natsumi *et al* . *J. Proteome Res.* 2006, *5*, 1979–1987.

Nermeen *et al* (2010) *African Journal* of *Biotechnology* 9(38):6355-6364.

Pelaez F (2006) *Biochem. Pharmacol.*,71: 981-990.

Pierre V, Jean MM, Jean G, Michael JH (2000). *J Mol Catal B-Enzym* 9:113-148.

Pisano MA, Sommer MJ, Taras L (1992) *Appl Microbiol Biotechnol* 36:553–555.

Poorani E, Saseetharan MK, Dhevagi P, *International Journal* of *Integrative Biology*, 7:3 (2009) 150-155.

Poornima R, Sahu MK, Pushpavalli V, Sivakumar K (2007) National Seminar on Bioprospecting of Marine Resources with special reference to Marine Natural Products and Drug Discovery", Department of Oceanography and Coastal Area Studies, Alagappa University, Thondi,148p.

Ramesh C. Kasana, Sudesh K. Yadav (2007) *Current Microbiology* 54:224–229.

Ramesh S, Rajesh M, Mathivanan N (2009) *Bioprocess Biosyst Eng* 32:791–800.

Riedlinger *et al* (2004) *J. Antibiot.* 57: 271–279.

Sabu A, Keerthi TR, Rajeevkumar S, Chandrasekaran M (2000) *Process Biochem* 35(7): 705-710.

Sahu MK, Poorani E, Sivakumar K, Thangaradjou T & Kannan L (2007c) J Environ Biol 28 (3):645–650.

Sahu MK, Sivakumar K, Kannan L (2005) *J Aqua Biol* 20(2):142–144.

Sahu MK, Sivakumar K, Kannan L (2007a) *Natl Acad Sci Lett* 30:61–65.

Sahu MK, Sivakumar K, Poorani E, Thangaradjou T, Kannan L (2007b) *J Environ Biol* 28:465–474.

Sahu MK, Sivakumar K, Thangaradjou T, Kannan L (2007d) *J Environ Biol* 29:4

Salmon CE, Magarvey NA and Sherman DH (2003) *Nat Prod Rep* 21:105–121

Schneider *et al* (2008). *Angew Chem Int Ed* 47:3258– 3261.

Seghal Kiran *et al* (2010) *Bioresource Technology* 101:2389–2396.

Selvakumar Dharmaraj (2011). *Iranian Journal* of *Biotechnology* Vol. 9, No. 2.

Shi C, Zhu Y, Ran X, Wang M, Su Y, Cheng T (2006) *J Surg Res* 133:185–192.

Shin J, Seo Y, Lee HS, Rho JR, Mo SJ (2003). *J. Natl. Prodt.* 66: 883–884.

Shindo *et al* (2007). *J Mol Cata. B Enzym.* 4877-83.

Sivakumar K, Sahu MK, Manivel PR, Kannan L (2006) *Ind J Exp Biol* 44: 256-258.

Southan C (2001). *FEBS Letters* 498:214-218.

Stach *et al* (2003). *Environ Microbiol* 5:828–841

Stach JEM, Maldonado LA, Ward AC, Bull AT, Goodfellow M (2004) *Int J Syst Evol Microbiol* 54:191–194.

Subramani Ramesh and Narayanasamy Mathivanan (2009). World J Microbiol Biotechnol 25:2103–2111

Tanaka Y & Omura O (1990) *Actinomycetologica* 4: 13–14

Thumar JT, Singh SP (2007a). *Brazil J Microbiol* 38:766–772.

Umamaheswary *et al* (2005). *Environment & Ecology* 23:942–947.

Veiga M, Esparis A, Fabregas J (1983) *Appl Environ Microbiol* 46:286–287.

Ward AC and Bora N (2006) Curr Opin Microbiol 9:279–286

Weyland H (1969). *Nature* (Lond) 223: 858.

Wu SH, Fotso S, Li F, Qin S, Kelter G, Fiebig HH, Laatsch H (2006) *J. Antibiot.* 59(6): 331–337.

Wu SJ, Fotso S, Li F, Qin S, Laatsch H (2007) *J Nat Prod* 70:304–306

Xiao-Ge Chen *et al* (2004) *Extremophiles* 8:489–498.

Yan J, Li X, Liu L, Wang F, Zhu TW, Zhang Q (2006) *Artif Cells Blood Substit Immobil Biotechnol* 3427–39.

Yanjuan Xin, Jianyu Huang, Maicun Deng and Wei Zhang (2008). Antonie van Leeuwenhoek. 94:533–542.

Yoshida A, Seo Y, Suzuki S, Nishino T, Kobayashi T, Sato NH, Kogure K, Imada C (2008) *Mar Biotechnol* 10:554–563

Zaldivar J, Neilson J, Olsson L (2001). *Appl Microbiol Biotechnol* 56:7-34.

Chapter - 22

Recent Trends in Exploration of Microbial Diversity

S. Ram Reddy, P. Pavan Kumar and N. Kishore**

Department of Microbiology, Kakatiya University, Warangal -506009, A P
*Department of Microbiology, Palamuru University,
Mahabubnagar-50 001, A P
E-mail: sanditi_ramreddy@yahoo.com

ABSTRACT

The knowledge of microbial diversity is not only important for basic and applied scientific research, but also to understand the link between diversity and community structure and function. Given the understanding that many microorganisms cannot be coaxed readily into culture, recently vigorous attempts are being made to understand the microbial diversity employing new tools and techniques. The strategies for microbial biodiversity analysis may be conveniently divided into two general categories: (i) those relying on laboratory cultivation or incubation and (culture dependant), (ii) those based on the direct extraction and analysis of indicator molecules such as nucleic and fatty acids (culture independent). The culture based methods reveal only a minor fraction (0.1% to 10%) of the microorganisms. Moreover, it offers a very limited insight into the spatial distribution of the microorganisms. Culture independent methods include: community level physiological profiling (CLPP), fatty acid methyl ester analysis (FAME), phospholipids fatty acid (PFLA) analysis, molecular based techniques that in turn include: G+C content, nucleic acid reassociation and hybridization, DNA microarrays and PCR based methods. PCR targeting the 16S rDNA has been extensively used to study prokaryotic diversity and allows identification of prokaryotes as well as the prediction of phylogenetic relationship. PCR amplified products are separated and analyzed through DGGE/TGGE, RFLP, ARDRA, T-RFLP, ARISA etc. All these methods have widened our horizons of understanding of microbial diversity of different ecosystems. However, no techniques are without limitations or biases. A holistic view of a microbial community in an ecosystem can be achieved by integrated approaches. Not withstanding the drawbacks, molecular methods still generate valuable information on microbial diversity and community structure taking into account both culturable and nonculturable fractions of microorganisms.

Metagenomics, the genomic analysis of a population of microorganisms that provides an access to the physiology and genetics of uncultured organisms, has emerged as a powerful technique in recent times. Metagenomic analysis involves isolating DNA from an environmental sample, cloning the DNA into a suitable vector, transforming the clones into a host bacterium and screening the resultant transformants. Metagenomic analysis has several advantages over culture or PCR methods. Recent developments in metagenomics have drastically changed the perception of microbiologists for biodiversity.

Key words: Microbial diversity, Molecular techniques, PCR, Metagenomics

1. Introduction

Biodiversity has been defined as the range of significantly different types of organisms and their relative abundance in an assemblage or community. The diversity has also been defined according to information theory, as the amount of and distribution of information in an assemblage or community (Atlas, 1984). Traditionally, biodiversity is based upon the species as a unit. Species diversity consists of two components, species richness and species evenness (distribution) (Ovreas, 2000). Our knowledge of microbial diversity is is very poor in part of our inability to study microorganisms (Table 1). Torsvik *et al.* (1990a, b) estimated that in one gram of soil there are 4000 different bacterial 'genomic units' based on DNA-DNA reassociation. It has been estimated about 5000 bacterial species have been described (Pace, 1997, 1999). Approximately 1% of the soil bacterial population can be cultured by standard laboratory practices. In a study Kellenberger (2001) approximately 10^7 cells were counted in 1g of soil, but only 0.1 percent of the cells were culturable. It is not known it this 0.1 to 1% is representative of the bacterial population (Tosvik *et al.* 1998). An estimated 1,500,000 fungal species exist in the world (Giller *et al*, 1997) and many of these cannot be cultured by current standard laboratory methods (van Elsas *et al.*, 2000). It is difficult to study the diversity of microorganisms when it is not understood how to categorize or identify the species present. Nevertheless, our perception of microbial diversity has improved enormously over the last few decades (Pavan Kumar *et al.*, 2009). Analysis of rRNA gene sequences has resulted in more than tripling of the number of identifiable bacterial division (Fig. 1). In this review, we discuss the different methodologies used to understand the microbial diversity.

2.. Culture dependant methods

2.1 Plate count method

The classic approach to enumerate microorganisms in environment has been culture dependant techniques combined with a simultaneous or subsequent differentiation of the isolates based on batteries of physiological and biochemical tests. However, culture-dependant methods do not accurately reflect the actual bacterial community structure, but rather the selectivity of growth media for certain bacteria. Furthermore, (i) all the techniques that rely on cultivation are time consuming and expensive because of physiological and biochemical tests involved; (ii) after many generations necessary to form plate colonies, the organism may deviate from its physiology and possibly even from genotype mix, of the population nature, (iii) only a minor fraction of the microorganisms can be cultivated using standard techniques and (iv) it offers a very limited insight into the spatial distribution of the microorganisms (Theron and Cloete, 2000). In addition, plate culture favors those micro-organisms with fast growth rates and those fungi that produce large numbers of spores. All of these limitations can influence the apparent diversity of microbial community.

Table 1. Known and estimated number of microbial species and of some "higher" organisms

Group	Known species	Estimated total	Percentage of known species
Viruses	5,000	130,000	4
Bacteria	4,760	40,000	12
Fungi	69,000	1,500,000	5
Algae	40,000	60,000	67
Protozoa	30,800	100,000	31
Insects	800,000	6-10,000,000	8-13
Fish	19,000	21,000	90
Birds	9,198		~ 100
Mammals	4,170		~ 100

2.2 Community level physiological profiling (CLPP)

Gardand and Mills (1991) developed a technique using a commercially available 96-well microtitre plate to assess the potential functional diversity of diversity of the bacterial populations through sole source carbon utilization (SSCU) patterns, Gram-negative (GN) and Gram-positive (GP) plates are available from Biolog (Hayward, CA, USA, www.biology.com) and each contains 95 different carbon sources and one control well without a substrate. Alternatively, researchers can use plates containing the growth medium and a tetrazolium salt from Biolog, and add site specific carbon sources to analyze their samples (Becker and Stottmeister, 1998). Biolog method can also be extended to fungi, however, without tetrazolium salt (Classen *et al.* 2003). This method has been used successfully to assess potential metabolic diversity of microbial communities in a variety of habitats (Konopaka *et al.*, 1998; Graysson *et al.*, 1998) Derey *et al.* 1999). Biolog method can also be used in conjunction with denaturing gradient gel electrophoresis (DGGE) (Rolling *et al.* 2000).

Advantages of Biolog method include (i) fast, (ii) highly reproducible (iii) relatively inexpensive (iv) differentiate microbial communities (v) generate large amount of data and (vi) options of using bacteria (GP and GN) and fungal plates. Disadvantages are (i) it represents culturable fraction of community, (ii) favours fast growing organisms, (iii) favours organisms capable of utilizing available carbon sources, (iv) potential metabolic diversity but not in situ diversity and (v) sensitive to inoculum density. Despite disadvantages, CLPP is useful when studying functional diversity of soils and is a valuable tool especially when used in conjunction with other methods.

3. *IN SITU* METHODS

3.1 Fatty acid methyl ester (FAME) analysis

This method provides information on the microbial community composition based on groupings of fatty acids (Ibekwe and Kennedy, 1998). Fatty acids make up of relatively constant proportion of the cell biomass and signature fatty acids exist that can differentiate major taxonomic groups within a community. Therefore, a change in the fatty acid profile would represent a change in the microbial population. In this method, fatty acids are extracted directly from soil, methylated and analyzed by gas chromatography (Ibekwe and Kennedy, 1999). FAME profiles of different soils can be compared using multivariate analysis. This method will detect changes in the composition of the bacterial and/or fungal community. This method has been used to study microbial community composition and population changes due to chemical contaminants (Kelly *et al.* 1999) and agricultural practices (Bossio *et al.* 1998).

The advantages of FAME analysis are: (i) there is no need to culture the microorganism (ii) one can follow changes in the population of specific organisms or communities. The disadvantages include: (i) if fungal spores are used, a lot of material is required, (ii) can be influenced by external factors and (iii) results can be confounded by other organisms (Zeller, 1999).

3.2 Phospholipids fatty acid (PLFA)

Phospholipids fatty acid (PLFA) technique is based on the extraction, fractionation, methylation and chromatography of the phospholipids component of soil lipids (Zelles, 1999; Pankhurst *et al.*, 2001). Phospholipids are thought to be related to the viable component of soil microflora because they are present as important components of membranes of living cells and break down rapidly when the cells die, and they cannot survive long enough to interact with soil colloids (Zelles, 1999). In addition, they have another prerequisite of a biomarker: they make up a fairly constant proportion of the biomass of organisms. Changes in the phospholipid profiles are generally related to variation in the abundance of microbial groups and can be interpreted by reference to a database of pure cultures and known biosynthetic pathways (Zelles, 1999). Direct extraction of PLFAs or whole fatty acids (WSFAMEs) from soil does not permit detection at the species level and can be used to estimate only gross changes in community structure. However, identific-ation of species by fatty acid analysis is possible with standard cultural-based media and databases (MIDI, Newark, Delaware). Zelles (1999) listed many fatty acids isolated from specific microbial groups (Table 2). Bacterial biomass can be calculated by summing up several fatty acids (15:0, a15:0, i15:0, i16:0, i17:0, cy17:0, cy19:0 and 16:1o7c), whereas fungal biomass can be obtained from the 18:2o6c fatty acid (Frostegård & Bååth, 1996).

Table 2. Biomarker fatty acids (modified from Zelles, 1999)

Group	Biomarker
Bacteria	
Gram-positive	Branched chain fatty acids (br 17 : 0*, br 18 : 0, i17 : 0, a17 : 0, i16 : 0, i16 : 1, 10Me16 : 0, 10Me17 : 0), iso and anteiso isomers of 15 : 0, MUFA < 20% (16 : 1ω9, 16 : 1ω7c, 16 : 1ω5, 18 : 1ω7, 19 : 1)[13]
Gram-negative	Cyclopropane fatty acids (cy17 : 0, cy19 : 0)[46], β-hydroxy fatty acids present in cell wall[64,65], MUFA > 20% (16 : 1ω9, 16 : 1ω7c, 16 : 1ω5, 18 : 1ω7, 19 : 1).
Fungi	PUFA 18 : 2ω6 (linoleic acid)[14]
Actinomycetes	10Me16 : 0, 10Me17 : 0 and 10Me18 : 0 (ref. 66)
Anaerobic bacteria	NEL-PLFA, e.g. Plasmalogens (*Clostridia*), sphingolipids (*Bacteroides*, *Flavobacterium*)
Sulphate reducing bacteria (*Desulfobacter* sp.)	cy17 : 0 and 10Me16 : 0 without high level of 10Me 18 : 0
Methanogens	
Type I	16 : 1ω8
Type II	18 : 1ω8 (refs 67–69)

PLFA provides a broad diversity measurement of microbial community at the phenotypic level. As PLFAs- the quantitative measurable biomolecules in soil microorganisms are rapidly turned over and represent the current living community, both qualitatively and quantitatively, they seem to have high indicative value to assess and monitor the microbial community structure, physiological and stress state. However, there are few limitations and pitfalls associated with PLFA biomarker analysis, which may limit its use at the regional and global scale. It does not reveal any information at the species-level, archae bacteria cannot be determined using this method and databases for interpretation of biomarkers are centered on fatty acids from microorganisms from pure cultures. Nevertheless, the full potential of PFLA as a bioindicator of environment monitoring and assessment at higher scales of resolution is certainly growing as databases and novel methods focusing on functions are being developed.

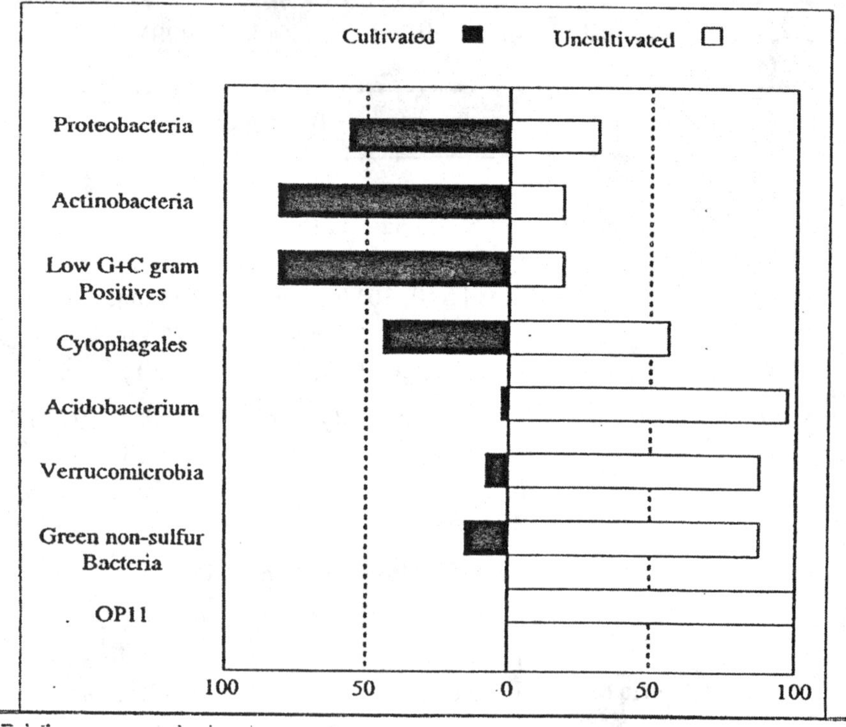

Fig. 1. Relative representation in selected cosmopolitan bacterial divisions of 16SrRNA sequences from cultivated and uncultivated organisms.

3.3 Fluorescent antibody technique

Fluorescent antibody techniques offered major advances in out-ecological studies of microorganisms (Wetzel, 1990). This method allows direct identification and enumeration of individual bacteria in environmental samples without requirement of prior growth in culture media. The technique involves the use of specific antibodies raised against surface markers of defined pure cultures that are either labeled directly with fluorescent dye molecules or via a fluorescent secondary antibody. Although this approach has yielded important insights into the spatial distribution of microorganisms, it has the following limitations (Amann *et al*, 1992); (i) expression of the antigen may be influenced by environmental factors; (ii) false positive and false negative results may be obtained due to cross-reactivity or lack of reaction; (iii) non-specific binding of antibodies may result in high levels of background fluorescence; and (iv) production of specific antibodies requires a pure culture of the organism of interest.

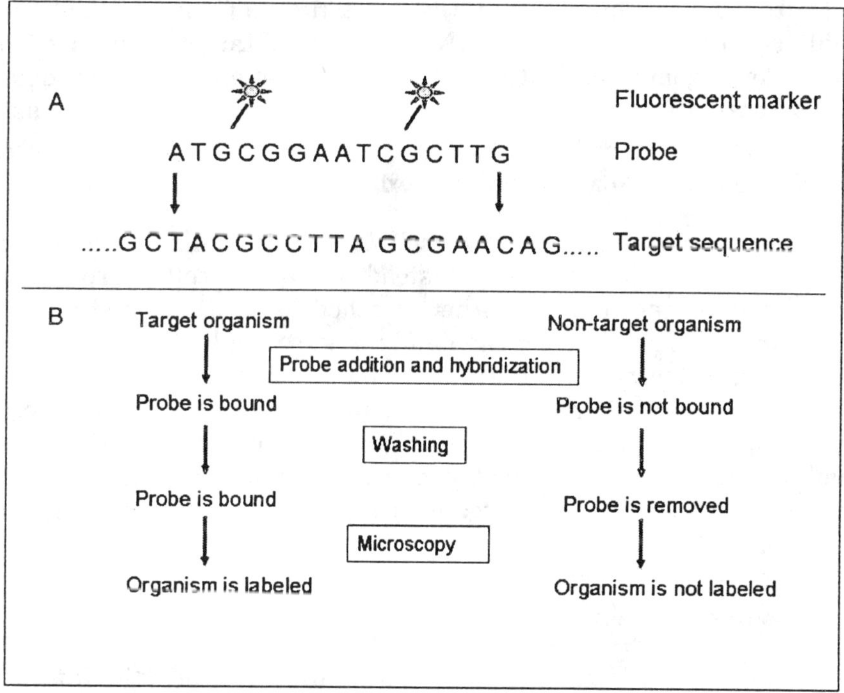

Fig. 2. Schematic diagram of FISH (Fluorescent in-situ hybridization). Principle of probe hybridization with target sequence (A) and labeling and detection of target organisms in an environmental sample.

3.4 Fluorescent in-situ hybridization (FISH)

FISH also known as oligonucleotide probes and can be used to determine the location of target microbial species or strains in an environment sample (Assmus, 1997). In this technique environmental samples such as root pieces are fixed on a slide and the DNA is denatured to separate the two strands. The cell membrane is then perforated chemically to allow the entrance of the probe into the cell and thus binding to the DNA. Oligonucleotide probes, 10-20 bp long complimentary to target regions of genome that are specific for a kingdom, family or genus are labeled with a fluorescent marker. During hybridization, the probe will bind to the single stranded DNA of the target organism (Fig. 2). The fluorescent signal of the probes or antibodies, which reflects the distribution of the target organism in the sample, can be detected with confocal laser microscope which gives three dimensional image of the sample. The images can be analyzed with image analysis software.

The advantages of FISH are: (i) in situ detection up to strain level giving the exact location in the soil or on roots; (ii) several species can be detected simultaneously; (iii) quantification is possible with appropriate software. On the other hand the limitations of this method are: (i) relatively time consuming, (ii) DNA sequence of target group needs to be known for oligonucleotide design; (iii) only target organisms are detected; (iv) confocal laser microscopes required for this technique are expensive.

3.5 Whole cell in situ hybridization

The basic principle of whole cell *in situ* hybridization is similar to FISH, however intact cells, instead of lysed cells, are used for hybridization. The cells are either attached to gelatin-coated microscope slides or hybridized in supension and immersed in hybridization solution containing a fluoresecently labeled oligonucleotide. Following incubation at the hybridization temperature, the sample is viewed by epifluorescence microscopy. Fluorescently labeled probes have also been used in combination with flow cytometry for rapid, automated identification and counting or collection of microorganisms in hybridized cell suspensions (Wallner *et al*. 1997).

3.6 Reporter genes

A range of reporter genes have been used to study the interactions between certain soil and rhizosphere microorganisms. Three reporter genes commonly used are: lux (bioluminescence), *inaZ* (ice nucleation) and *gfp* (green fluorescent protein). The principle of reporter genes is that they are inserted downstream from the gene of interest so that they will be expressed at the same time as the native gene. They can be inserted in two different ways: (i) associated with a housekeeping gene, which is always expressed in active cells, (ii) associated with functional genes. The expression of the reporter gene can be quantified and used as an indicator of presence of an organism or functional gene expression. The resulting organisms are genetically modified reported genes that are to be used in natural soil should not be normally found in native organisms because the background expression by the native microflora would make the detection of the reporter organism impossible. Due to strict regulations for the use of genetically modified organisms the use of reporter organism is limited to experiments under controlled conditions.

3.7 Thymidine incorporation

The use of ^{3}H-labelled thymidine to study growth rates of soil bacteria was developed by Baath & Coworkers (Baath and Johensson, 1990). The principle of the method is that bacteria are extracted from the soil and ^{3}H-labelled thymidineis added to the cell suspension. Actively growing bacteria will incorporate^{3}H-labelled thymidine into the newly formed DNA. After incubation and washing, the radioactivity of the cell suspension is determined and the incorporation rate can be used as indicator for growth rate. The assumption being that the growth rate during the test reflects that in the environment from which they were extracted. However, thymidine incorporation is specific for bacteria (Diaz-Ravina *et al.*, 1996) and bacterial species differ in thymidine incorporation rate therefore only an average value can be obtained.

4. Nucleic acid extraction based methods

4.1 Guanosine plus cytosine (G+C) content

Differences in the guanine plus cytosine (G+C) content of DNA can be used to study the bacterial diversity (Nusslein and Tiedje, 1999). It is based on the principle that microorganisms differ in their G+C content and that taxonomically related groups only differ between 3% and 5% (Tiedje *et al.*, 1999). This method provides a coarse level of resolution as different taxonomic groups may share the same G+C range. Adavatages of G+C analysis are that: (i) it is not influenced by PCR biases, (ii) it includes all DNA extracted, (iii) it is quantitative and (iv) it can cover rare members in the microbial populations. However, the major limitation of this method is it requires large quantities of DNA.

4.2 DNA reassociation

DNA reassociation is a measure of genetic complexity of the microbial community and has been used to estimate diversity (Torsvik *et al.*, 1990a,b, 1996). In this method, community DNA isolated from environment is denatured and then allowed to reanneal. The principle of DNA renaturation kinetics is that the rate of hybridization is proportional to the concentration of complimentary DNA sequences and inversely proportional to the total length of different sequences in the sample (Chelm, 1982). Experimentally, DNA reassociation is measured over time and the fraction of reassociated DNA (C/Co) is expressed as a function of Cot, where Co is the initial molar concentration of nucleotides in single-stranded DNA and t is the time in seconds. The plot of this relationship is referred to as a Cot curve. The reaction rate constant K, can be expressed as $1/Cot_{1/2}$, $t_{1/2}$ denotes the time in seconds required for 50% reassociation. Under defined conditions, $Cot_{1/2}$ is proportional to the complexity of the DNA. The conclusion from DNA reassociation studies is generally that natural ecosystems are very complex.

The advantages of these methods are: (i) total DNA extracted, (ii) not influenced by PCR biases and, (iii) DNA or RNA can be studied in situ. However, one concern in the interpretation of DNA reassociation estimates is a reduction in the rate of reassociation resulting from impurities in the DNA sample and thus it is dependent on lysing and extraction efficiency.

4.3 Nucleic acid hybridization

Quantitative and quantitative estimates of community structure can be made using tailor made oligonucleotide probes that bind to targets with a wide range of specificities from domain to strain. The most important hybridization technique is dot-blot hybridizations of extracted nucleic acids (Stahl *et al.*, 1988). For quantitative dot-blot hybridization the samples are lysed to release nucleic acids. Quantification of 16S rRNA can then be obtained by dot-blot hybridization of a directly isolated nucleic acid mixture with universal and specific nucleotide probes. The relative abundance is calculated by dividing the amount of specific probe bound to a given sample by the amount of hybridized universal probe. The relative abundance of a given rRNA might reflect changes in the abundance of the cell population of interest or in the cellular rRNA content. The relative abundance of a rRNA thus can be interpreted as the relative importance of a defined species in terms of actual metabolic activity or potential metabolic activity, but cannot be directly translated into cell numbers (Amann *et al.*, 1997).

4.4 DNA microarrays

In recent times, DNA-DNA hybridization has been used together with DNA microarrays to detect and identify bacterial species (Cho and Tiedje, 2001) or to assess microbial diversity (Greene and Voordouw, 2003). This could be valuable in bacterial diversity since a single assay can contain thousands of DNA sequences with high specificity. The microarrays can either contain specific target genes or can contain a sample of environmental "standards" representing different species found in the environmental sample.

Reverse sample genome probing (RSGP) is a method used to analyze microbial community composition of the most dominant culturable species and uses genome microarrays. RSGP has four steps: (i) isolation of genomic DNA from pure cultures; (ii) Cross-hybridization testing to obtain DNA fragments with less than 70% cross hybridization (DNA fragments with greater than 70% cross hybridization are considered to be the same species, (iii) preparation of genome assays into a solid support, and (iv) random labeling of a defined mixture of total community DNA and internal standard. This method has been successfully used to analyze microbial communities in oil fields, contaminated soils. In general, the species need to be cultured, but in principle cloned DNA fragements of uncuturables could be used. RSGP is a useful technique when diversity is low. If diversity is high, then cross hybridization can be a problem or interpretation of the results is difficult. Further, it only detects the most abundant species.

In microarrays, using genes or DNA fragments instead of genomes on the microarrays offers the advantages of eliminating the need to keep cultures of organisms growing as genes can be cloned into plasmids or PCR used to continually amplify the DNA fragments. In addition, fragments would increase the specificity of hybridization ones the use of genomes and functional genes in the community could be assessed.

4.5 PCR based approaches

Analysis of genomic DNA (or RNA) reveals the clear distinction between species, genera or kingdom. Methods involving DNA are usually easier to perform than those for RNA. However, fingerprinting of DNA is much more difficult than RNA. A first approach to fingerprinting was based on the generation of a profile of low molecular weight RNAs (5s rRNA, tRNAs) extracted from the environment. The profile is characteristic for individual organisms or even for natural microbial populations and the pattern obtained from a complex community serves as a direct measure of the community structure.

A second technique is based on the generation of 16S rDNA profiles that indicate the structural diversity of the bacterial communities. The 16S RNA approaches defined by Woese and Colleagues in 1980's has already been applied to a fair number of cultured bacteria and one can find over 30,000 sequences of small ribosomal subunits in the nucleic acids data banks. The under lying assumption of the 16S rRNA approach is that the sequential structure of the molecule is the authentic record of the evolutionary events of the molecule (i.e gene) and the organism at the same time (Woese, 1987). The required conditions for such a molecule to be truly an authentic representative of its bearing organism, like the ancient nature, ubiquity, lack of the selection pressure, and of the lateral spread, are all met by the small ribosomal subunit. Additionally, the molecule has a manageable length and a detailed analysis for a wide range of relationships (Fig. 3).

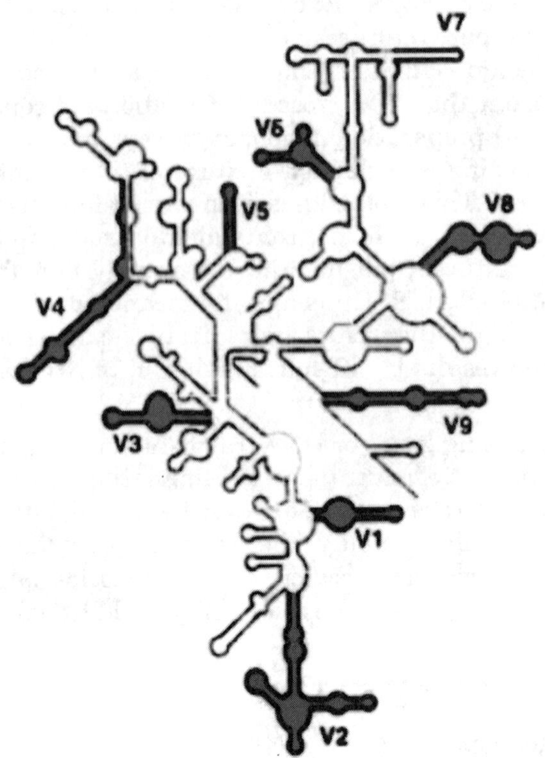

Fig. 3. A 16srRNA molecule (*E.coli*) showing nine variable regions

PCR targeting the 16S rDNA has been used extensively to study prokaryote diversity and allows identification of prokaryotes as well as the prediction of phylogenetic relationships. 18S rDNA and internal transcribed spacer (ITS) regions are increasingly used to study fungal communities (Reddy *et al.*, 2005). Target DNA (16S, 18S or ITS) is amplified using universal or specific primers and the resulting products are separated in different ways (Tables 3 & 4).

Table 3. Bacteria specific 16SrRNA gene sequence /PCR primers

Code	Sequence 5'-3'	N	Ta	B	A	E	Matches
E8F	AGAGTTTGATCCTGGCTCAG	20	55	+++	+	-	2424
E9F	GAGTTTGATCCTGGCTCAG	19	53	+++	+	++	2741
E334F	CCAGACTCCTACGGGAGGCAGC	21	65	+++	-	-	13,172
E341F	CCTACGGGIGGCIGCA	16	51	+++	+	-	16,685
E786F	GATTAGATACCCTGGTAG	18	47	+++	+	-	12,616
E533R	TIACCGIIICTICTGGCAC	19	56	+++	+	++	18,724
E926R	CCGICIATTIITTTIAGTTT	20	50	+++	+++	++	19,950

E939R	CTTGTGCGGGCCCCCGTCAATTC	23	71	+++	-	-	8620
E1115R	AGGGTTGCGCTCGTTG	16	47	+++	-	-	9052
E1541R	AAGGAGGTGATCCANCCRCA	20	57	+++	+	-	1355

Primers described as bacteria-specific were submitted to the 'check probe' facility of the Ribosomal Database Project (http://www.rdp.cme.msu.edu/) to check for bacterial specificity. The total number of matches are given, and matches for Bacteria (B), Archaea (A), Eucarya (E) are represented as follows: (-, no matches; +, <25 matches; ++, 25-100 matches; +++,<100 matches). Primer numbering relates to *E. coli* position complementary to the 5' end of the primer. Last letter denotes direction: Forward and Reverse. Ta=(4x(G+C)+2x(A+T))-5M=A or C; R=A or G; W=A or T; S=C or G; Y=C or T; K=G or T; N=A or G or C or T.

Table 4. Summary of PCR primers that have been used for assessing fungal diversity in soil extracts

PCR primer	Primer sequence (5'-3')	Genomic target
NS1	GTAGTCATATGCTTGTCTC	18S rDNA
NS2	GGCTGCTGGCACCAGACTTGC	18S rDNA
NS3	GCAAGTCTGGTGCCAGCAGCC	18S rDNA
NS8	TCCGCAGGTTCACCTACGGA	18S rDNA
ITS1	TCCGTAGGTGAACCTGCGG	ITS
ITS2	GCTGCGTTCTTCATCGATGC	ITS
ITS4	TCCTCCGCTTATTGATATGC	ITS
ITS1F	CTTGGTCATTTAGAGGAAGTAA	ITS
ITS4B	CAGGAGACTTGTACACGGTCCAG	ITS
2234C	GTTTCCGTAGGTGAACCTGC	ITS
3126T	ATATGCTTAAGTTCAGCGGGT	ITS
ITS4A	CGCCGTTACTGGGGCAATCCCTG	ITS
EF3	TCCTCTAAATGACCAAGTTTG	18S rDNA
EF4	GGAAGGGRTGTATTTATTAG	18S rDNA
Fung5	GTAAAAGTCCTGGTTCCCC	18S rDNA
nu-SSU-0817	TTAGCATGGAATAATRRAATAGGA	18S rDNA
nu-SSU-1196	TCTGGACCTGGTGAGTTTCC	18S rDNA
nu-SSU-1536	ATTGCAATGCYCTATCCCCA	18S rDNA
FR1	AICCATTCAATCGGTAIT	18S rDNA
FF390	CGATAACGAACGAGACCT	18S rDNA
PN3	CCGTTGGTGAACCAGCGGAGGGATC	ITS
PN34	TTGCCGCTTCACTCGCCGTT	ITS
Fun18S1	CCATGCATGTCTAAGTWTAA	18S rDNA
Fun18S2	GCTGGCACCAGACTTGCCCTCC	18S rDNA

4.6 Denaturing gradient gel electrophoresis (DGDE)/Temperature gradient gel electrophoresis (TGGE)

DGGE and TGGE are two similar methods for studying microbial diversity. These techniques were originally developed to detect point mutations in DNA sequences. Muyzer *et al* (1993) expanded the use of DGGE to study microbial genetic diversity. DNA is extracted from soil samples and amplified using PCR with universal primers targeting part of the 16S or 18S rRNA sequences. The 5'end of the forward primer contains a 35-40 base pairs GC clamp to ensure that at least part of the DNA remains double stranded. This is necessary so that separation on a polyacrylamide gel with a gradient of increasing concentration of denaturants (Formamide or urea) will occur based on melting behaviour of the double strandard DNA. On denaturation, DNA melts in domains, which are sequence specific and will migrate differentially through the polyacrylamide gel. Theoretically, DGGE can separate DNA with one base pair difference. TGGE uses the same principle as DGGE except the gradient is the temperature rather than chemical denaturants. DGGE/TGGE has been used to assess the diversity of bacteria and fungi in rhizosphere and other polluted environments (Kirk *et al.*, 2004).

DGGE/TGGE has the advantages of being reliable, reproducible, rapid and somewhat inexpensive. Multiple samples can also be analyzed concurrently (Muyzer, 1999). Limitations of DGGE/TGGE include PCR bases, laborious sample handling. It is estimated that DGGE can only detect 1-2% of the microbial population represent dominant species present in an environment sample. Further one band can represent more than one species (Co-migration).

4.7 Single strand conformation polymorphism

Like DGGE/TGGE, this technique was originally developed to detect known or novel polymorphisms or point mutations in DNA. Single stranded DNA is separated on a polyacrylamide gel based on differences in mobility caused by their folded secondary structure (Lee *et al.*, 1996). When DNA fragments are equal size and no denaturant is present, folding and hence mobility will be dependant on the DNA sequences. SSCP has all the same limitations of DGGE. Further, some ssDNA can form more than one stable conformation. Therefore, one sequence may be represented by more than one band on the gel. SSCP has been used to measure succession of bacterial communities (Peter *et al.*, 2000) and AMF species in roots (Kjoller and Rosendahl, 2000).

4.8 Restriction Fragment Length Polymorphism (RFLP)

RFLP was developed to differentiate isolates based on sequence differences. However, it can also be used to assess microbial community composition without prior isolation of individual species (Liu *et al.*, 1997). After DNA extraction, a DNA fragement is amplified by PCR. This fragment is then incubated with restriction enzymes in the so called digest. The resultant fragments are then separated according to their size in an agarose gel. A modification of RFLP is terminal (T-RFLP). The principle is that PCR is performed with one or both primers labeled with a fluorescent label. After restriction enzyme digest, detection of the labeled fragements can be automated (Clement *et al.*, 1998). RFLP patterns of a given isolate can contain from one to more than 10 bands and they will vary with the restriction enzyme used. RFLP patterns from soil communities may be very complex and can be digitized and analyzed by multivariate analysis. RFLP has been used to the genetic variation on agricultural soils and effect of soil emendments on soil bacterial communties (Pesaro *et al.*, 2003; Harvey *et al.*, 2000).

RFLP is eay to perform, however the limitations are: (i) labor intensive if several enzymes are used, (ii) different enzymes will give different results making the interpretation difficult and (iii) banding patterns from environmental samples can be very complex.

4.9 Amplified ribosomal DNA restriction analysis (ARDRA)

ARDRA is a DNA fingerprinting technique based on restriction enzyme digestions and agarose gel electrophoresis of PCR – amplified 16S rRNA genes using primers for conserved regions (Tiedje *et al.*, 1999). The amplified portions of 16S rRNA genes from a mixed microbial population might be of similar sizes with a particular set of oligonucleotide primers, but have small differences at the nucleotide level. One way of detecting these differences is to restrict the PCR-amplified product with restriction endonucleases and examine the pattern of restriction fragements (Smith *et al.*, 1997). The analytical power of this procedure derives from the fractionation step. The separation of DNA fragements derived from different populations requires that they differ in sequence at the restriction endonuclease sites, or differ in length of DNA flanked by common restriction sites. The separated DNA fragments may also be transferred to filters for hybridization with probes specific for an organism of interest i.e. are subjected to Southern transfer. By design, the probe for each organism will hybridize to a fragment of a different size and thus can be used to resolve different environmental populations.

4.10 Ribosomal intergenic spacer analysis (RISA)/Automated ribosomal intergenic spacer analysis (ARISA)

Similar in principle to RFLP and T-RFLP, RISA and ARISA provide ribosomal based fingerprinting of the microbial community. In RISA and ARISA, the intergenic spacer (IGS) region between the 16S and 23S ribosomal subunits is amplified by PCR, denatured and separated through PAGE under denaturing conditions. This region may encode tRNAs and is useful for differentiating between bacterial strains and closely related species because of heterogeneity of the IGS length and sequence (Fisher and Triplett, 1999). In RISA, the sequence polymorphisms are detected using silver stain whereas in ARISA the forwarded primer is fluoresecently labeled and is automatically detected. Both these methods provide highly reproducible bacterial community profiles but RISA requires large quantities of DNA is more time consuming, silver staining is insensitive and resolution tends to be low. ARISA increases the sensitivity of the method and reduces the time but is still subject to the traditional limitations of PCR. RISA has been used to compare microbial diversity in soil (Ranjard *et al*. 2000).

4.11 Real time PCR

Normal PCR is not quantitative because the DNA concentration is measured only once after completion of the 20-35 cycles. Even then, after completion of the amplification program the concentration of DNA is not or weakly related to the initial DNA concentration, especially in samples which contain a large number of different species. Real time-PCR allows quantification of the initial target DNA concentration because the concentration of amplified DNA is measured after each cycle. The measurement of increase in ds DNA can be based on the dye that fluoresces when bound to ds DNA. Quantification is based on the fact that the time until the copy number of given gene sequence increases experimentally is negatively correlated with the original gene copy number. Thus, the higher the original concentration of the target sequence in the sample, the earlier the experimental increase will occur. Real time-PCR has been used to determine the population density of fungi (Filion *et al.*, 2003) or methane oxidizers (Seghers *et al.*, 2003) in soil.

The advantages of real time PCR are: (i) quantification of target DNA in environmental samples and, (ii) once established and calibrated relatively easy to perform. However, limitations are: (i) it is expensive because it needs special real-time thermalcycler (ii) calibrations can be difficult especially in environmental samples.

4.12 Other types of PCR

PCR can also be used in several different ways to fingerprint bacteria. One method uses a single arbitrary primer and is known as AP-PCR. A random or arbitrary primer (~10bp) can insert into a bacterial genome at multiple sites and thus give rise to multiple amplification products of variable size. When separated by electrophoresis, the products result in a banding pattern, or fingerprint, unique to the organism under investigation. However, when AP-PCR is used, it is critical that all PCR components and conditions result in variable banding patterns.

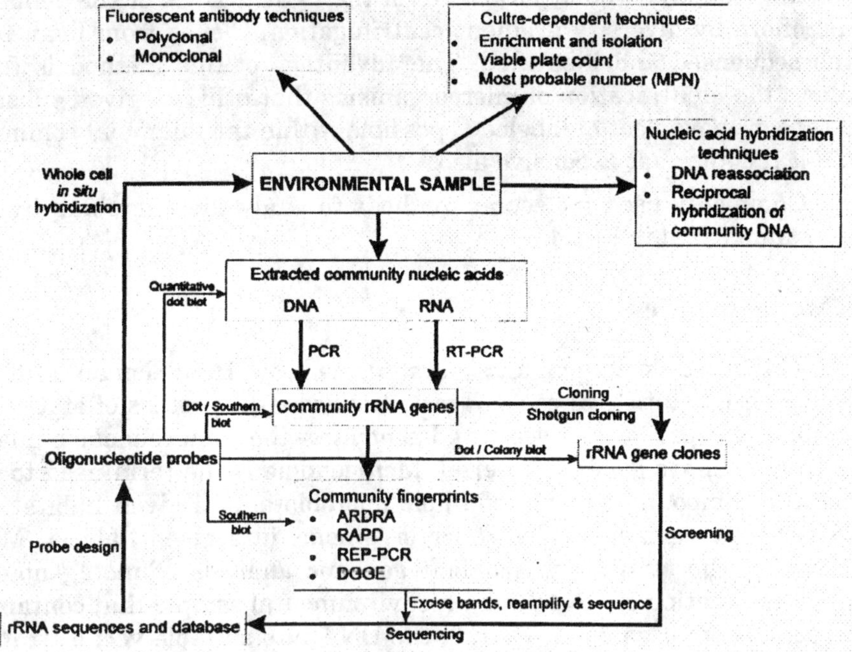

Fig. 4. Commonly used molecular approaches in microbial ecology (Theron and Cloete, 2000).

A useful method for a fine-scale composition of differences between bacterial populations is the use of rep-PCR (repetitive extragenic palindromic-PCR). Differences in genomic fingerprints of chromosome structure among bacterial strains can be obtained with this method, using specific primers (i.e. REP, BOX or ERIC) that bind to randomly interspersed repetitive DNA sequences around the chromosome (Tiedje *et al*, 1999). Because the rep-PCR is most effectively used when it is targeted to populations selected to be certain phenotypic or taxonomic group, mixed natural communities of bacteria in contaminated soils have not been studied by this method. Recently, Roane and Pepper (2000) used enterobacterial repetitive intragenic consensus polymerase chain reaction (ERIC-PCR) to genetically distinguish the bacterial isolates from soils contaminated with different levels of Cd.

4.13 Stable isotope probing

Radajewski *et al*. (2000) were the first to report the use of stable isotopes in soil microbial ecology. They added $^{13}CH_3H$ to the soil in order to assess methanol utilizing bacteria in soil. The labeled ^{13}C is incorporated into the DNA of those species that utilize CH_3OH. After DNA extraction from the soil, the heavy ^{13}C-DNA from the methanol utilizing bacteria was separated from the light ^{12}-DNA of the remaining microflora by diversity gradient centrifugation. Clones from heavy DNA was sequenced and identified. The advantage of this method is that it allows the identification of microorganisms that utilize a given substrate. However, movement of labeled C possible within the microbial community and it is somewhat expensive also.

Commonly used molecular methods to study the microbial diversity are summarized in Fig. 4.

5. Metagenomics

The word 'metagenomics' first coined by Handelsman (1998) is derived from statistical concept *meta*-analysis (the process of statistically combining separate analyses) and *genomics*- the comprehensive analysis of an organism's genetic material. Metagenome is the term used to describe the compound genome of whole microbiota, of a given habitat. It is the collective genomes of all microorganisms in a given habitat. Metgenomics is the culture-independent genomic analysis of metagenome. It involves isolation of DNA from an environmental sample that contains an assemblage of organisms, cloning the DNA into suitable vector, transforming the clones into bacterium and screening the resultant transformants (Fig. 5). The clones can be screened for phylogenetic markers or "anchors" such as 16S rRNA and *rec* A or for other conserved genes by hybridization or multiplex PCR or for expression of certain traits or they can be sequenced randomly. Notwithstanding the individual targeted approaches, the ultimate goal of metagenomics is to understand the uncultured world and to gain insight into groups of prokaryotes that are otherwise entirely unknown. In simplest terms it provides a window on the richness of the uncultured world (Rappe and Giovannoni (2003).

Metagenomic analysis has several advantages over culture or PCR-based methods. For example, it (i) provides access to uncultured microorganisms, (ii) does not require prior knowledge of gene sequences, and (iii) recovers complete genes. Recent developments in metagenomics have drastically changed the perception of microbiologists' biodiversity. It redefined the concept of genome and accelerated the arte of gene discovery. Applications of metagenomics are immense. The heart of this approach is that it has provided us an access to the genome heterogeneity of both culture-dependent and culture independent microorganisms. Metagenomics has already opened new avenues of research by enabling unprecedented analysis of genome and evolution in environmental contexts and providing accesses to far more microbial diversity. For more details the reader is suggested to refer review article by Pavan Kumar *et al*. (2009b) and another article in this volume by V V Lakshmi.

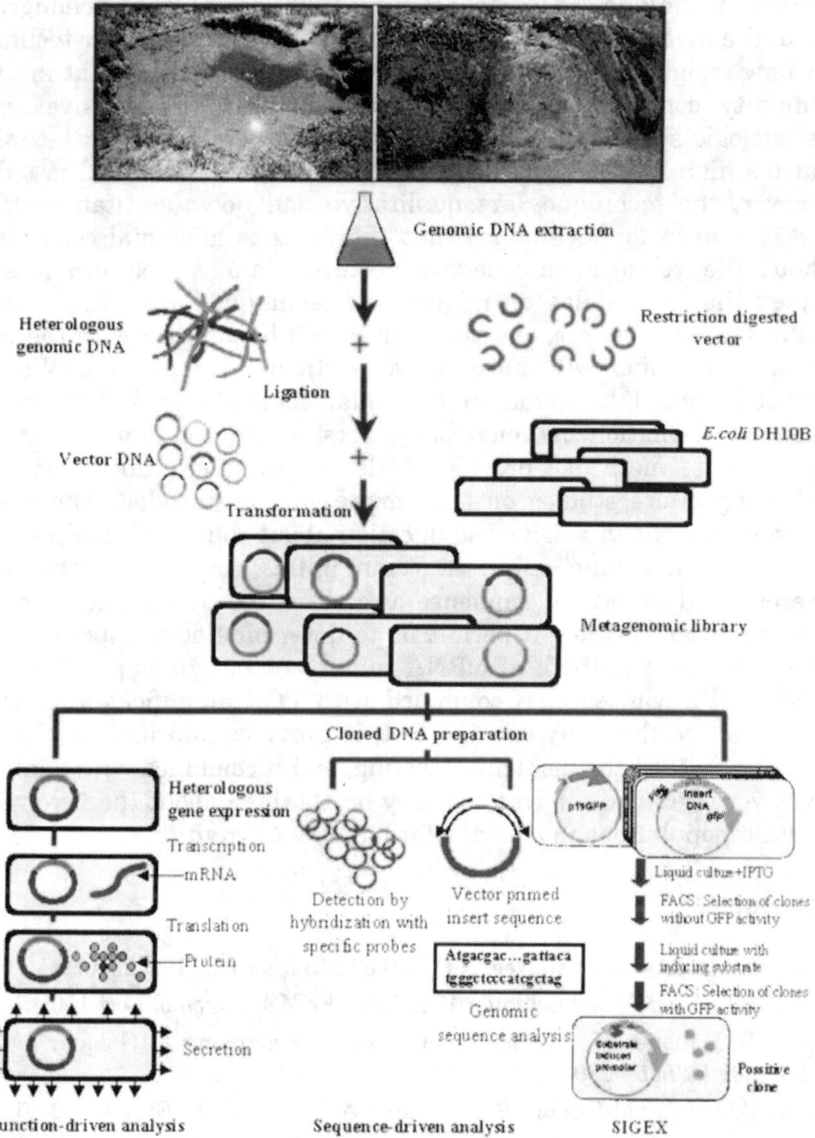

Fig. 5. Metagenomic approach for the analysis of environmental DNA (Pavan Kumar *et al.*, 2009)

6. Conclusions

The study of biodiversity is motivated by interest in the heterogeneity and variability of organisms. Because of the immense member of microorganisms on Earth and the wide range of habitats they occupy, the potential heterogeneity and variability of microorganisms are probably larger than for any other group of organisms. The application of molecular techniques to unravel the microbial diversity has changed our view of the diversity and evolution of microbial life. Molecular techniques offer new insight into the 'black box' of an environment so that microbial community composition and microbial activities can be investigated. Microbiologists are well aware of this potential. Molecular techniques require a high level of expertise and sophistication and are often tedious. Moreover, the techniques are qualitative and not quantitative. Hence, there is a need for techniques that characterize microbial communities without the reliance on selective culturing and which are less com complex than molecular techniques. No techniques are without limitations or biases, however, and the user should be familiar with the effects that a given bias will have on a particular application. Whatever drawbacks would be found in molecular methods, still they generate valuable information on microbial diversity and community structure. DNA based methods provide little information about microbial functioning. More studies on gene expression are needed. These would also be important to resolve the question about the relationship between biodiversity and functioning in communities. In this regard mRNA centered studies are of immense use. The recent advances in RNA extraction from soil might permit us to determine active species in soil. Due to the short half-life of mRNA this is currently not possible, but it has several advantages. If combined with PCR amplification it will be more sensitive than any enzyme test, it provides information about the metabolic state at the moment of testing, and it combined with analysis of rDNA, very detailed information may be obtained about the involvement of certain populations in a particular metabolic activity.

References

Amann R, Glockner FO and Neef A (1997) FEMS Microbiol Rev 20:191

Amann R, Ludwig W and Schleifer KH (1992) FEMS Microbiol Lett 100:45

Assmus B, Hausner M, Grenier D, Schloter M, Hartmann A, Goldner M (1997) Microbiol 20:155-159

Assmus B,Schloter M,Hutzler P, Hartmann A (1997) Microb Ecol 33: 32-40

Atlas, RM (1984). Diversity of microbial communities. In Marshall, KC (Ed.) Advances in Microbial Ecology, Vol.7, Plenum New York, pp.1-47.

Becker PM, Stottmeister U (1998) Can J Microbiol 44:913-919

Biith E, Johansson T (1990) Plant Soil 126:133-136

Bossio DA, Scow KM, Gunapala N, Graham KJ (1998) Micro Ecol 36:1-12

Chelm BK (1982) DNA-DNA renaturation Kinetics, in Methds in Chloroplast molecular Biology, Edelman, M, Hallick, RB and Chua HH, Eds, Elseivier Biomedical, New York, pp.301

Cho JC, Tiedje JM (2001) Appl Environ Microbiol 67:3677-3682

Classen AT, Boyle SI, Haskins KE, Overby ST, Hart SC (2003) FEMS Microbiol Ecol 44:319-328

Clement BG, Kew LE, De Bord KL, Kitts CL (1998) J Microbiol Meth 31:135-142

Derry AM, Staddon WJ, Kevan PG and Trevors JT (1999) Biodivers Conserv 8:205-221

Diaz-Ravina M, Prieto A, Biithe (1996) Soil Biol Biochem 28:419-426

Filion M, Arunand M and J Jabaji-Hae H (2003) J Microbiol MEth 1735:1-10

Fisher MM, Triplett EW (1999) Appl Environ Microbiol 65:4630-4636

Frostegard A and Baath E (1996) Biol Fertili Soils 22:59-65

Garland JL, Mills AL (1991) Appl Environm MIcrobiol 57:2351-2359

Giller KE, Beare MH, Lavelle P, Izac AMN, Swift NJ (1997) Appl Soil Ecol 6:3-16.

Grayston SJ, Wang S, Campbell CD, Edwards AC (1998) Soil Biol Biochem 30:369-378

Greene EA, Voordouw G (2003) J Microbiol Methods 53 :211-219

Griffiths BS, Ritz K, Ebblewhite N, Dobson G (199) Soil Biol Biochem 31:145-153

Handelsman *et al* (1998) Chem Biol 5:245-249

Harvey PR, Butterworth PJ, Hawke BG, Pankhurst CE (2000) Plant Pathol 49 :619-627

Ibekwe AM, Kennedy AC (1998) FEMS Microbiol Ecol 26:151-163

Ibekwe AM, Kennedy AC (1999) Plant Soil 206:151-161

Jennifer L Kirk, Lee A Beaudette, Miranda Hart, Peter Moutoglis, John N Klironomos, Hung Lee, Jack T Trevors 2004 J microbial Methods 58:169-188

Kellenberger E (2001) EMBO Rep 2:5-7

Kelly JJ, Haggblom M, Tate III RL (1999) Soil Biol Biochem 31:1455-1465

Kjoller R, Rosendahl S (2000) Plant Soil 226:189-196

Konopka A, Oliver Jr L, Turco RF (1998) Microbiol Ecol 35:103-115

Lee DH, Zo YG, Kim SJ (1996) Appl Environ Microbiol 62:3112-3120

Liu WT, Marsh TL, Cheng H, Forney LJ (1997) Appl Environ Microbiol 63:4516-4522

Muyzer G (1999) Curr Opin Microbiol 2:317-322

Muyzer G, Wall EC and Uitterlinden AG (1993) Appl Environ Microbiol 59:695-700

Nusslein K, Tiedje JM (1999) Appl Environ Microbiol 65:3622-3626

Overeas L (2000) Ecol Lett 3:236-257.

Pace NR (1997) Science 276:734-740

Pace NR (1999) ASN News 65:328-333

Pavan Kumar *et al* (2009) In Bio-Nano-Geo Sciences (Ed.) Alok Srivastava & Ispita Roy Ane Books Pvt. Ltd.

Pavan Kumar P, Kishore N and Ram Reddy S (2009) In a Text book of Biotechnology Ed.Chanhan AK and A. Varma. IK International Publishing House Pvt. Ltd.

Pesaro M, Widmer F, Nicollier G, Zeyer J (2003) Soil Biol Biochem 35:1049-1061

Peters S, Koschinsky S, Schwieger F, Tebbe CC (2000) Appl Environ Microbiol 66:930-936

Radajewski S, Ineson P, Parkeh NR and Murell JC (2000) Nature 403:646-649

Ranjard L, Brotheir E, Nagaret S (2000) Appl Environ Microbiol 66:5334-5339

Rappe MS and Giovannoni SJ (2003) Ann Rev Microbiol 57:369-394

Reddy SR, Pavan Kumar P and Reddy SM (2005) Curr Sci 89(10):1699-1709

Roane TM and Pepper IL (2000) Microbiol Ecol 38:358-364

Rolling WFM, Van-Breukelen BM, raster M, Goeltom MT, Groen J (2000) Micro Ecol 40:177-188

Seghers *et al* (2003) FEMS Microbiol Ecol 46:139-146

Smith C, Leeglang P and Wemars K (1998) FEMS Microbiol Ecol 23:249

Stahl DA, Flesher B, Mansfield H and Montagomery L (1997) Appl Environ Microbiol 54:1079

Theron J and Cloete (2000) Critical Reviews Microbiol 26(1):37-57

Tiedje JM, Asuming-Brempong S, Nusslein K, Marsh TL, Flynn SJ (1999) Appl Soil Ecol 13:109-122

Torsvik V, Goksoyr J, Daae FL (1990). Appl Environ Microbiol 56:782-787.

Torsvik V, Salte K, Soerheim R, Goksoeyr J (1990) Appl Environ Microbiol 56:776-781.

Trosvik V, Daae FL Sandaa RA and Ovreas L (1998) J Biotechnol 64:53-62

Van Elsas JD, Frois-Durate G, Keijzer-Wolters A, Smit E (2000) J Microbiol Methodol 43:133-151

Wallner G, STeinmentz I, Bitter-Suermann D and Amann R (1997) Appl Environ Microbiol 63:423

Waltner G, Fuchs B, Springs S, Beisker W and Amann R (1997) Appl Environ Microbiol 63:4223

Witzel KP (1990) Approaches to bacterial population dynamics, in Aquatic Microbial Ecology:Biochemical and Molecular approaches, Overbeek J and Chrost, RJ, Eds. Springer-Verlag, New York, pp.96

Woes CR (1987) Microbiol Rev 51:221-271

Zelles L (1999) Biol Fertil Soils 29:111-129.

Chapter - 23

Biotransformations by thermophilic fungi- A green chemistry technology for drug synthesis: A mini review

G. Shyam Prasad, S. Girisham and S.M. Reddy*

Department of Microbiology, Kakatiya University,
Warangal-506 009 Andhra Pradesh, India,
*E-mail: Shyamprasad1919@yahoo.com

ABSTRACT

Though biotransformations are mediated by microbial cells, plant cells, animal cells and by cell free enzymes, microbial process is far more efficient and economical due to their rapid growth and high metabolic rate. This process besides having many advantages over chemical process makes possible to produce compounds otherwise not possible by chemo-synthetic methods. Exploiting of variety of enzymes from thermophilic bacteria are being steadily increasing in biotransformations mainly as replacements for thermolabile enzymes despite thermophilic fungi are more versatile and fermentatively more efficient in biotransformation process. The importance and the advantages of biotransformations by thermophilic fungi over mesophiles and thermophilic bacteria are discussed.

Key words: Biotransformation, thermophilic fungi, cell free enzymes, fermentation process.

1. Introduction

The increasing demand of drugs has raised the number of pharma and chemical industries, releasing hundreds of tonnes of hazardous waste into the environment every hour of the day. Thus global environment is threatened with large number of synthetic chemicals which are hazardous to health of a man. It is expected that even the fetus in the womb is exposed directly to these pollutants. Many of these chemicals enter the food chain and get circulated. Biotransformation is an alternate key safe technology in the provision of chiral intermediates and active pharmaceutical ingredients. It is the process by which an organism or its isolated enzyme brings out minor chemical change on compounds that are not part of their metabolism and result in the formation of novel or more useful product than its parent compound that is difficult or impossible to obtain by conventional chemical methods (Prasad *et al.*, 2010). Almost every type of chemical reaction is possible with microbial transformation. The interest in bioconversion is mainly because the product of the process is more useful or valuable than the precursor used. Further this process is simple, hazard free, minimizes the problems of isomerization, racemization, epimerization and rearrangement that generally occur during chemical process, renewable or recycled many times, economical and ecofriendly. Biotransformation is not a new phenomenon, it has been practiced since immemorial times (Table 1) and interest in this area is increasing steadily due to various reasons. The mechanism involved in microbial transformation may be "xenobiotic transformations" in which the substrate is completely alien to the microorganism. This type of transformations utilizes substrate induced enzyme systems that may be part of the bio-degradation or even defense mechanism of the organism or the transformation may be biogenetically directed, sometimes known as analogue biosynthesis. In this transformation, the substrate is an analogue of a biosynthetic intermediate and the transformation relies on the flexibility of the existing enzymatic pathway. This process is now being exploited in pharmaceutical, agriculture, flavour and fragrance, food and chemical industries. Biotransformation reactions are the subject of increasing interest in the pharmaceutical industry because of the demand for enantiomerically pure compounds (Schulze and Wubbolts, 1999). This process facilitates for introducing functional groups into various positions of organic compounds and to generate structural diversity in a chemical library (Prasad *et al.*, 2008). This process has become a part and parcel of green chemistry. Classification of chemical reactions catalyzed by microbial enzymes is précised in Table 2. Though biotransformation reactions employing mesophilic microbial cultures are well documented (Prasad *et al.*, 2009), such reactions employing thermophilic organisms are limited (Table 3).

Table 1. Milestones in industrially relevant biotransformation and biocatalytic process (Laresche and Meyer, 2006)

Year	Process
500 BC-	Vinegar production
800 BC-	Casein hydrolysis with chymosin for cheese production
1670 -	"Orlran" process for the industrial bio-oxidation of ethanol to acetic acid
1680 -	Antony van leevenhoek first observed microorganisms with his microscope

1897 -	Buchner discovered yeast enzymes converting sugar into alcohol
1934-	Regioselective bio-oxidation of sorbite to sorbose for Reichstein Vitamin C synthesis
1940 -	Sucrose inversion using an invertase
1950-	Bioconversion of steroids
1970-	Hydrolysis of penicillin to 6-aminopenicillanic acid
1973-	First successful genetic engineering experiments
1974-	Glucose to fructose isomerisation with immobilized glucose isomerase
1985-	Enzymatic process for production of acrylamide
1990-	Hydrolysis of protease (trypsin) of porcine insulin to human insulin
1995-	Biotransformation plant 3000 tons per year for bioconversion of nicotinonitrile to nicotinamide

Since then products such as lotrafiban, paclitaxel, xemilofiban, rennin inhibitor, ACE inhibitor, Omapatrilat, 2-Quinoxaline carboxylic acid (Shraathof *et al.*, 2002), acrylamide, aspartame, L-carnitine, L-dopa, lysine,7-ACA,6-APA,(S) naproxen, glucose-fructose syrup, vitamin C, citric acid etc. (Flickinger *et al.*, 2010) are being produced in bulk quantity in industries through process of biotransformation.

Table 2. Chemical reactions mediated by microbial enzymes

Reactions Process	Chemical reactions
Oxidations	Hydroxylation, epoxidation, dehydrogenation of C-C bonds, Oxidation of alcohols and aldehydes, oxidative degradation of alkyl, carboxy alkyl or oxoalky chains. Oxidative removal of substituents, oxidative deamination, oxidation of hetero-functions, oxidative ring fission.
Reductions	Reduction of organic acids, aldehydes, ketones and hydrogenation of C=C bonds, reduction of hetero-functions, dehydroxylation, reductive elimination of substituents.
Hydrolysis	Hydrolysis of esters, amines, amides, lactones, ethers, lactams etc.
	Hydration of C=C bonds and epoxides.
Condensation	Dehydration, O and N-acylation, glucosidation, esterification, lactonization, amination.
Isomerization	Migration of double bonds or oxygen functions. Racemization, Rearrangements.
Formation of	C=C bonds or hetero-atom bonds

Table 3. Biotransformation reactions mediated by thermophilic microorganisms

Name of the Microorganism	Substrate	transformation product	Reference
I. Fungi			
Acremonium alabamensis	cholesterol	cholestenone	Satyanarayana & Chavant (1987)
	Stigmastadienone	stigmasterol & sitosterol	
Talaromyces emersonii	cholesterol	cholestenone	
Humicola fitscoatra	Progesterone	androst-4-ene-3,17-dione	Zohri & Abdel (1999)
		testosterone	
		testololactone	
H. grisea	Progesterone	androst-4-ene-3,17-dione	
		testosterone	
		testololactone	
H. hyalothermophila	Progesterone	11α-hyd roxyprogesterone	
		11ß-hydroxyprogesterone	
		17 α -hydroxyprogesterone	
		21-hydroxyprogesterone	
		11 α, 17 α-dihydroxypr-ogesterone	
		11 α, 17 α, 21-t rihydroxy-progesterone (epicortisol)	
		11ß, 17 α, 21-trihydrox3' progest erone (cortisol)	
Sporotrichum thermophile	Ferulic acid	vanillic acid	Topakas *et al.* (2003)
Rhizomucor tauricus	progesterone,	dihydroxylation	Christy *et al.* (2007)
	androst-4-en-3,17-dione,	dihydroxylation	
	testosterone,	Monohydroxylation	
	pregnenolone	dehydroxylation	
	dehydroepiandrost erone	Monohydroxylation	

II. Bacteria			
Bacillus sp	Progesterone	6α-hydroxyprogesterone 6β-hydroxyprogesterone 20α- and 20β-dihydroprogesterone.	Kelvin *et al.* (1992)
B.thermoglucosidasius	Progesterone	6α- and 6β-hydroxyprogesterone, androstenedione testosterone	Sideso *et al.* (1998)
B stearothermophilus	Progesterone	20α-Hydroxyprogesterone 6β-hydroxyproge-sterone 6α-hydroxyproge-sterone 9,10-seco-4-pregnene-3, 9, 20-trione	Awadi *et al.* (2001)
B. stearothermophilus	Progesterone	6α, 20 α-dihydroxyprogesterone 6 β, 20 α-dihydroxyprogesterone 5α-pregnane-3,6,20-trione 6-dehydroprogesterone monohydroxylated metabolites B-Seco compound	Awadi *et al.* (2002)
B. stearothermophilus	Testosterone	androst-4-en-3,17-dione 6a-hydroxyandrost-4-en-3,17-dione 6b-hydroxyandrost-4-en-3,17-dione 6a-hydroxytestosterone 6b-hydroxytestosterone	Awadi *et al.* (2003)
Geobacillus stearothermophilus	Chenodeoxycholic acid	cholic acid, methylcholate methylchenodeo-xycholate, 3α-hydroxy-7-oxo-5β-cholanic acid, 7α-hydroxy-3-oxo-5β-cholanic acid	Awadi *et al.* (2001)

2. Thermophilic fungi

Thermophilic fungi, extremophilic organisms, are often proposed as source of industrially relevant enzymes, because of their thermostable and innate properties (Huber and Stetter, 1998). Thermophilic fungi, a small assemblage in mycota have unique mechanism of growing at an elevated temperature. These are distinguished into thermophilic and thermotolerant forms on the basis of their minimum and maximum temperature limits for growth (Cooney and Emmerson, 1964). The thermophilic fungi grows at or above 20°C and attain maximum growth at or above 50°C, and the thermotolerant forms opt 20 to 55°C for growth. Thermophily in fungi is not as extreme as in eubacteria or archaea, some species of archea are able to grow near or above 100°C in thermal springs, solfatara fields, or hydrothermal vents (Brock, 1995; Blo hl *et al.*, 1997). Thermophilic fungi provide scientists with valuable experimental material for investigations of the mechanisms which, although allowing their growth at moderately high temperatures, limit it beyond 60 to 62°C (Tansey and Brock, 1972). Thermophilic fungi are the chief components of the microflora that develops in heaped masses of plant material, piles of agricultural and forestry products and other places of accumulations of organic matter where in the warm, humid, and aerobic environment provides the ideal conditions for their growth (Miehe, 1907; Allen and Emmerson, 1949). They constitute a heterogeneous physiological group of various genera in the mastigomycotina, ascomycotina, deuteromycotina and aganomycetes (Mouchacca, 1997) (Table 4).

Table 4. Taxonomic status and cardinal temperature of some thermophilic fungi

Fungus (Present nomenclature)	T_{Opt} (°C)	T_{max} (°C)
Ascomycotina		
Chaetomium mesopotamicum	45	52
C. thermophile	45-55	58-61
Coonemeria aegyptiaca	40	55
C. crustaceam	40	<60
C. verrucosa	30-40	55
Corynascus thermophilus	50	60
Melanocarpus albomyces	45	57
M. thermophilus	35	50
Talaromyces byssochlamydioides	40-45	>50
T. emersonii	40-45	55
T. thermophilus	45-50	60
T. aurantiacus	49-52	61
Thielavia pingtungia	40	>50
T. terrestris	40-45	52

T. australiensis	35-40	50
Deuteromycotina		
Canariomyces thermophila	45	
Dactylomyces thermophiles	40-45	
Malbranchia cinnamomea	45	57
Myceliophthora hinnulea	40-45	>50
M. thermophila	45-50	55
Myriococcum thermophilum	45	53
Paecilomyces varioti	50	55
Scytalidium thermophilum	40	58
Stilbella thermophila	35-50	55
Thermomyces ibadanensis	42-47	61
T. lanuginosus	45-50	60
T. stellatus	40	50
Mastigomycotina		
Rhizomucor miehei	35-45	57
R. pusillus	35-45	55

The first of the known thermophilic fungi, *Mucor pusillus*, was isolated from bread over a century ago by Lindt (1886). A little later, Tsiklinskaya discovered another thermophilic fungus, *Thermomyces lanuginosus*, growing on potato which had been inoculated with garden soil (Tsiklinskaya, 1899). Initially, because of the exothermic reactions of the saprophytic, mesophilic microflora present therein, the temperature of the substratum rises to 40°C, resulting in warm environment favoring the germination of spores of thermophilic microflora, and eventually the latter outgrows the mesophilic microorganisms. In the process, the temperature of the substratum is raised further to 60°C or even higher.

Thermophilic fungi can be grown in minimal medium with high metabolic rate and growth yields comparable to those of mesophilic fungi. Thermophilic fungi have a powerful ability to degrade polysaccharide constituents of biomass. The properties of their enzymes differ not only among species of same genus but also among strains of the same species. Their extracellular enzymes display temperature optima for activity that are close to or above the optimum temperature for the growth of organism. In general, their enzymes are more heat stable than those of the mesophilic fungi. Some extracellular enzymes from thermophilic fungi are being produced commercially, and still a few others have commercial prospects.

Thermostable enzymes offer robust catalyst alternatives, able to withstand the often relatively harsh conditions of industrial processing.

Thermostable polymer-degrading enzymes such as amylases, pullulanases, xylanases, proteases and cellulases are expected to play an important role in food, chemical, pharmaceutical, paper, pulp, and waste-treatment industries. Cellular components of thermophilic organisms (enzymes, proteins and nucleic acids) are also thermostable. Apart from tolerant to high temperature, they are also known to withstand denaturants of extremely acidic and alkaline conditions. Thermostable enzymes are highly specific and thus have considerable potential for many industrial applications. The use of such enzymes in maximizing reactions accomplished in the food and paper industry, detergents, drugs, toxic wastes removal and drilling for oil are being studied extensively (Haki and Rakshit, 2003). Though there are many reports on large application of bacterial thermophilic enzymes in industries (Table 5) only limited reports are available on the utilization of thermophilic fungal enzymes in industries, especially in biotransformation process despite the fact that these enzymes are versatile in their activity.

Table 5. Some of the industrially important enzymes from thermophilic bacteria and actinomycetes

Source	Enzyme	Temp optimum	pH optimum
I. Bacteria			
Thermophilic bacterium TSSB-6	α-amylase	80°C (50-100°C)	8.0(6.0-9.0)
Bacillus subtilis JS-2004	α-amylase	70°C	8.0
B.thermoleovorans CCR11	Lipase	60°C	9.0-10.0
Thermophilic bacterium TSSB-4	Alkaline, Cellulase	70°C (50-100°C)	8.0 (6.0-9.0)
Thermophilic bacterium HS-08	Serine	65°C (60-70°C)	7.5 (7.0-9.0)
Anoxybacillus flavithermus EHP 1	Cellulase	75°C (65-75°C)	7.5 (7.0-9.0)
Pyrococcus furiosus	Xylanase	100°C	6.0
Actinomycetes			
Thermobifida.fusca NTU22	α-amylase	60°C	7.0
Streptomyces transformant T3-1	Cellulase	50°C	6.5
Thermoactinomycetes sp.HS682	Protease 70°C	1.0	
Streptomyces rimosus R6-554W Lipase	50°C	9.0-10.0 (4.0-10.0)	
Thermomonospora fusca	Xylanase	60°C (60-80°C)	7.0(6.0-8.0)

3. Mechanism of thermal adaptations

Adaptations of thermophilic microorganisms to high environmental temperature have been under extensive study for years (Kumar and Nussinov, 2001; Hicky and Singer, 2004). The various components of the cell adopted to their extreme living conditions includes not only DNA, RNA and proteins but also other small molecules such as membrane lipids. The best studied cell components of thermophiles are proteins (Kumar and Nussinov, 2001). Proteins from thermophiles have characteristic amino acid compositions compared with mesophiles (Fukuchi and Nishikawa, 2001; Singer and Hickey, 2003) which is mainly due to the differences in the composition on protein surface (Fukuchi and Nishikawa, 2001). The surface regions of thermophilic proteins have fewer (non-charged) polar amino acids and more charged amino acids, and these charged residues result in an increased number of intramolecular salt bridges. Decreased sequence length is another feature of thermophilic proteins (Thompson and Eisenberg, 1999). They tend to be shorter than their mesophilic homologs, and the main cause of this length reduction is the deletions in the loop regions. Thermophiles are also reported to contain proteins which are thermostable and resist denaturation and proteolysis (Kumar and Nussinov, 2001). Specialized proteins known as chaperonins are produced by these organisms, which help, after their denaturation to refold the proteins to their native form and restore their functions (Everly and Alberto, 2000). The DNA of thermophiles contains a reverse DNA gyrase which produces positive super coils in the DNA (Lopez, 1999). This raises the melting point of the DNA (the temperature at which the strands of the double helix separate) to atleast as high as optimum temperature for organisms growth. Thermophiles also tolerate high temperature by using increased inter-actions that non-thermotolerant organisms use, namely, electrostatic, Disulphide Bridge and hydrophobic interactions (Kumar and Nussinov, 2001). Thermal adaptations of RNA molecules are also known. rRNA and tRNA molecules of thermophilic bacteria have higher G+C contents than mesophiles (Galtier and Lobry,1997). Because the GC base pair forms more hydrogen bonds than the AT base pair, higher G+C content in the double-stranded stem region improves thermostability of the RNA molecules. Additionally, mRNA of thermophiles contain more purine nucleotides (Lao and Forsdyke, 2000; Paz *et al.*, 2004). A likely explan-ation of this phenomenon is that it is to prevent aggregation of mRNA molecules (Lao and Forsdyke, 2000). Furthermore, the cell membrane of thermophiles is made up of saturated fatty acids. The fatty acid provides a hydrophobic environment for the cell and keeps the cell rigid enough to live at elevated temperatures (Herbert and Sharp, 1992). The archae, which compose most of the hyperthermophiles, have lipids linked with ether on the cell wall. This layer is much more heat resistant than a membrane formed of fatty acids (De Rosa *et al.*, 1994).

4. Different methods of biotransformation process

***(i) Biotransformations with growing cultures*:** In this process, the substrate is added to the growth medium at the time of inoculation or during later growth of microbial growth. Here, growth and biotransformation takes place simultaneously. The major advantage of this technique is its ease of handling which makes it suitable for large series of screening experiments. Incubation times are relatively short and there is the potential for induction of the desired enzyme activity during growth in presence of the substrate. On the other hand if substrate exerts an inhibitory effect on microbial growth addition of the substratum has to be postponed until the organism has reached exponential phase of growth. Some of the advantages of this method are précised below:

a. Microbial biocatalysts provide access to a range of biotransformations for which no cell free enzymes are available.

b. Co factor recycling is not needed

c. Growing cells are often cheaper than the isolated enzyme systems.

On the other hand disadvantages of this process include:

Side reactions due to uncontrolled metabolism may give rise to more than onr product.

Expensive equipment and tedious work up.

***(ii) Biotransformations with previously grown cultures*:** In this method, the process of growth and biotransformation are separated. Biomass of microorganism cultivated under optimized growth conditions is used after separation from culture broth through centrifugation or filtration and can be stored in the refrigerator till it is used. When it is needed an appropriate amount of biomass is resuspended in a simple transformation medium consisting usually of an aqueous buffer solution with optimized pH and optimum amount of substrate. The main advantages of this process are growth and bioconversion can be independently optimized. Growth inhibitory effect of the substrate or product is eliminated. The optimum cell concentration for bioconversion can be adjusted, the risk of microbial contamination is reduced and product isolation is facilitated due to the simple composition of the transformation medium.

***(iii) Biotransformations with purified enzymes*:** This process is used when the permeability of the substrate through the cell membrane is insufficient or if undesired side reactions take place due to the presence of other enzyme systems. Some of the advantages of this process are:

1. Enzymes are very efficient catalysts and are environmentally acceptable.

2. Enzymes act under mild conditions of temperature (20-40°C) and in the pH range of 5-8
3. It helps to minimize undesired side reactions.
4. As enzymes are compatible, several biocatalytic reactions can be performed in one flask. It also facilitates sequential reactions in multienzyme systems.
5. Enzymes have high chemo-selectivity, regio-selectivity and enantio-selectivity. However, some disadvantages of this method include
   a. Enzymes are provided by nature in one enantiomeric form.
   b. Enzymes require narrow operation parameters. If a reaction proceeds only slowly, under given parameters, there is only a narrow scope of alteration as it leads to deactivation of the protein.
   c. Enzymes display their highest catalytic activity in water, which is the least desired solvent for most of organic compounds. But shifting of enzymatic reaction from an aqueous to an organic medium reduces its activity.
   d. Many enzymatic reactions are prone to substrate or product inhibition, which causes enzyme to cease to work at higher substrate or product concentration.
   e. Enzyme may cause allergic reactions.
   f. Recycling of cofactor is necessary for some enzymes.

(iv) Biotransformation with immobilized cells or enzymes: In this process, the biocatalyst can be removed easily from reaction broth and it is repeatedly used or to retain it within the reaction vessel and to run the process continuously. Furthermore, immobilization procedure enhances operational stabilities of catalysts in many instances.

(v) Biotransformation with liquid two-phase systems: Cells and enzymes display their activities in aqueous media. Since enzymes act on dissolved substrates, the biotransformation of lipophilic substrates is limited by their solubility in the reaction medium. The solubility of lipophilic compound in water can be improved to a certain extent by the use of detergents or water-miscible solvents. However, higher concentration of the solubilized substrate together with the solubility mediators might inhibit or even poison the biocatalyst. Such problems can be circumvented by adding a second water-immiscible phase, eg. a water-immiscible organic solvent with high solubility for the lipophilic substrate.

Thus, the biocatalysts can remain in the aqueous phase, while the organic phase serves as a reservoir for the lipophilic substrate and product. The transformation takes place either in the aqueous phase where the substrate concentration is determined by its solubility in water or at the interface between the two phases.

An ideal organic solvent used for this purpose should be immiscible with water, exhibit high solubility for the substrate and product in question and not inhibit the activity of the biocatalyst. Examples of organic solvents successfully used for liquid phase system are: n-alkanes, cyclohexane, toluene, benzene, chloroform, carbon tetrachloride, methylene dichloride, ethyl or butyl acetate, and diethyl or dibutyl ether.

High concentration of lipophilic substrate/product in the organic phase continuous slow release of the substrate into the aqueous phase and with drawl of the product by organic phase maintains the biotransformation and minimizes inhibitory effects on the biocatalyst. Easy phase separation facilitates the recovery of the product from the organic phase and the repeated use of the biocatalyst in the aqueous phase is the advantages of this process.

5. Isolation of biotransformation products

Biotransformation products contained in the reaction medium are either recovered from the whole broth or from the supernatant after removal of the biomass by filtration or centrifugation. First steps of product isolation aim at the concentration and/or fractionation of the product. Depending on the physical and chemical properties of the product to be recovered and on the chemical composition of the whole broth, the following methods are taken in to consideration: extraction, ion exchange, adsorption to polymeric resins, precipitation and distillation. The resulting raw product can be further purified by chromatographic techniques, fractionated distillation, crystallization, decolorization and drying. Small product quantities can in many cases be readily obtained by applying preparative gas chromat-ography or HPLC.

6. Advantages of thermophiles over mesophiles

Enzymes of thermophilic organisms show thermostability in contrast to their counterparts, mesophilic enzymes. Thermophilic enzymes offer many major advantages over mesophiles. Once expressed in mesophilic hosts, thermophilic and hyperthermophilic enzymes are easier to purify by heat treatment. Their thermostability is associated with a higher resistance to chemical solvents.

The main advantages of performing biocatalytic processes at higher temperatures are reduced risk of microbial contamination, lower viscosity of medium, improved transfer rates, and improved solubility of substrates (Bruce *et al.,* 1991; Lasa and Berenguer, 1993). Apart from high temperature, they are also known to withstand denaturants of extremely acidic and alkaline conditions. These enzymes are highly specific and thus have considerable potential for many industrial applications. The use of such enzymes in maximizing reactions in the food and paper industry, detergents, drugs, toxic wastes removal, and drilling for oil is being explored extensively (Haki and Rakshit, 2003). Biotransformation of albendazole, an anti-helmintic drug with a mesophilic fungus *Cunninghamella blakesleeana* NCIM 687 gave three metabolites (Prasad *et al.*, 2008) (Fig. 1), while with thermophilic fungus *Rhizomucor pusillus* NRRL 28626 gave four metabolites (Prasad *et al.*, 2011) (Fig. 2) which clearly reveals superiority of thermophilic fungi in the production of novel compounds which are more needed in novel drug discovery.

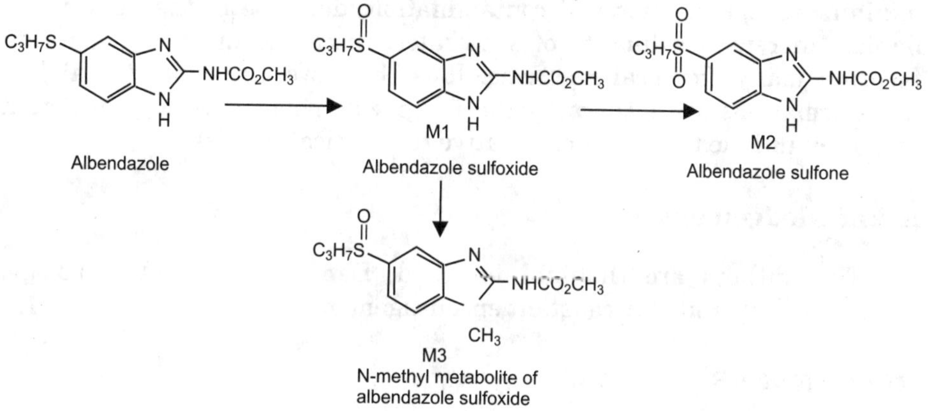

Albendazole
M1 Albendazole sulfoxide
M2 Albendazole sulfone
M3 N-methyl metabolite of albendazole sulfoxide

Fig 1 Biotransformation of albendazole by mesophilic fungus *C.blakesleeana*

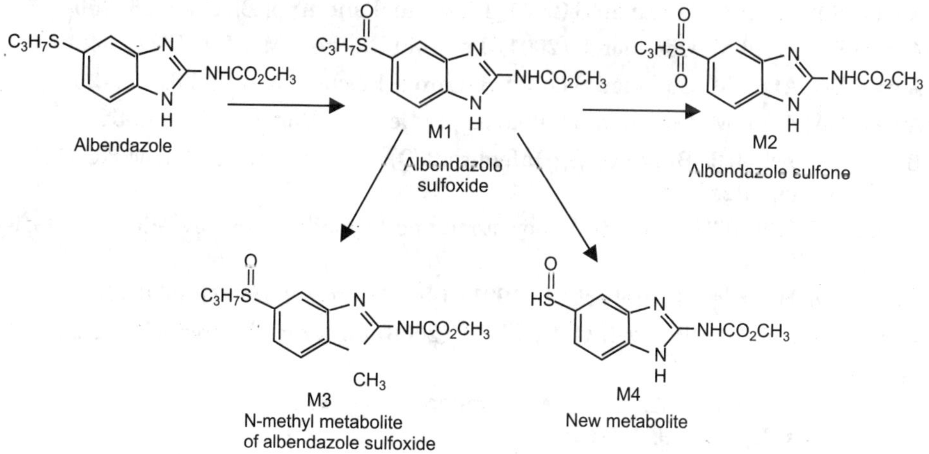

Albendazole
M1 Albendazole sulfoxide
M2 Albendazole sulfone
M3 N-methyl metabolite of albendazole sulfoxide
M4 New metabolite

Fig. 2. Biotransformation of albendazole by thermophilic fungus *R. pusillus*

7. Conclusions

The production of enantiomers of drug intermediates is increasingly receiving attention in the pharmaceutical industry. Biocatalysis provides an added dimension and an enormous scope to prepare pharmaceutically useful chiral compounds than through organic chemistry. The use of different classes of enzymes for the catalysis of wide variety of chemical reactions to generate a wide variety of chiral compounds. In the recent past decade, progress in biochemistry, protein chemistry, molecular cloning, random and site-directed mutagenesis, directed evolution of biocatalysts and fermentation technology have opened up unlimited access to a variety of cell free enzymes and microbial cells in the production of novel compounds through biotransformation. Due to various reasons isolation and maintenance of pure culture of thermophilic fungi have been less explored and needs to be strengthened in Biotransformation process through thermophiles has many advantages over mesophilic fungi like minimizing risk of microbial contamination, decrease in the viscosity of liquids, increased solubility of substrates, increased mass transfer rate favoring many chemical reactions. It is therefore, anticipated that biotransformations with thermophilic fungi will be increasingly exploited and often unique tool as an alternative to chemical processes.

Acknowledgement

The authors are thankful to Head, Department of Microbiology, Kakatiya University for constant encouragement.

REFERENCES

Allen P J, Emerson R (1949) Ind Eng Chem 41:346-365.

Awadi SA, Afzal M, Oommen S (2001) Biotechnol and Appl Biochem 58: 250-255.

Awadi SA, Afzal M, Oommen S (2001) J steroid Biochem Mol Biol 78:493-498.

Awadi SA, Afzal M, Oommen S (2002) J steroid Biochem Mol Biol 82:251-256.

Awadi SA, Afzal M, Oommen S (2003) Appl Microbiol Biotechnol 62:48-52.

Blo hl E, Rachel R, Burggraf S, Hafenbradl D, Jannasch HW, Stetter KO (1997) Extremophiles 1:14-21.

Brock T D (1995) The road to Yellowstone and beyond. Annu Rev Microbiol 49:1-28.

Bruce L Z, Henrik K N, Robert LS (1991) J Ind Microbiol Biotechnol 8:71-81.

Christy AH, Mills PW, Dedi C, Dodd HT (2008) J Steroid Biochem Mol Biol 108: 155-163.

De Rosa M, Morana A, Riccio A, Gambacorta A, Trincone A, Incani O (1994) Biosens Bioelectr 9: 669-675.

Everly C, Alberto J (2000) Front Bioscien. 5: 780-786.

Fukuchi S, Nishikawa K (2001) J Mol Bio 309:835-843.

Galtier N, Lobry JR (1997) J Mol Evol 44: 632-636.

Haki GD, Rakshit SK (2003) Bioresour Technol 89: 17-34.

Hickey DA, Singer G A (2004) Genome Biol 5:117.

Huber H, Stetter KO (1998) J Biotechnol 64: 39-52.

Kelvin ES, Williams RAD, Sideso O (1992) FEMS Microbiology Letters 92:29-34.

Kumar S, Nussinov R (2001) Cell Mol Life Sci 58:1216-1233.

Lao PJ, Forsdyke DR (2000) Genome Res 10: 228-236.

Lasa I, Berenguer J (1993) Microbiologia SEM 9:77-89.

Leresche J, Meyer H P (2006) Org Proc Res Dev10: 572-580.

Leuenberger HGW (1990) Pure Appl Chem 62: 753-768.

Lindt W (1886) Arch Exp Pathol Pharmakol 21: 269-298.

Lopez G (1999)DNA. J Mol Evol 46: 439-452.

Mouchacca J (1997) Cryptogamie Mycol. 18:19-69.

Paz A, Mester D, Baca I, Nevo E, Korol A (2004) Proc Natl. Acad Sci USA 101: 2951-2956.

Prasad GS, Girisham S, Reddy SM, Srisailam K (2008) World J Microbiol Biotechnol 24:1565-1571.

Satyanarayana, Chavant (1987) Folia Microbiologica 32:354-359.

Schulze B, Wubbolts M G (1999) Curr Opin Biochemical 10: 609-615.

Shyam Prasad G, Girisham S, Reddy SM (2010) Indian J Expt Biol 48:415-420.

Shyam Prasad G, Girisham S, Reddy SM (2011) Appl Biochem Biotechnol 165: 1120-1128.

Sideso O, Williams RAD, Welch, SG, Smith KE (1998) J Steroid Biochem Mol Biol 67:163-169.

Singer GA, Hickey DA (2003) Gene 317: 39-47.

Straathof A JJ, Panke S, Schmid A (2002) Current Opinion in Biotech. 13: 548-556.

Tansey M R, Brock T D (1972) Proc Natl Acad Sci USA 69:2426-2428.

Thompson M J, Eisenberg D (1999) J Mol Biol 290:595-604.

Topakas E, Kalogeris D, Kekos B J, Macris P, Christakopoulos (2003) Food Sci technol 36:561-565.

Tsiklinskaya P (1899) Ann Inst Pasteur Paris 13: 500-515.

Zohri A A, Abdel-galil MSM (1999) Folia Microbiol 44: 277-282.

Chapter - 24

Algae and the Soil Ecosystem

B. Digamber Rao

Department of Botany, Kakatiya University, Warangal-506 009 (A.P)
E-mail: bdr_kvc@yahoo.co.in

ABSTRACT

Algae constitute a dynamic component of the soil, which add and increase the availability of the crop nutrients. They help in improving the soil structure and amend the chemical nature of the soil. Algae form a rapidly multiplying cover crop of microscopic plants which are capable of colonizing almost all types of soils with ease. Algal crusts on the soil surface add organic matter, reduce evaporation from the soil surface, and remove compaction. Proliferation of algae increase the productivity of the soil through encouraging the growth of other soil microorganisms, increasing the availability and ensuring better crop nutrient management, amending the physical and chemical nature of the soil and increasing the nutrient utilization efficiency of the plants. Augmentation of the algal component of the soil through introduction of selected forms leads to bioenrichment of the soil which enhances this influence and ensures sustained long range agroecological security.

Key words: Soil algae, Microbial interaction, Soil characters.

1. Introduction

The group Algae is heterogenous assemblage of organisms ranging from a motile unicell to complicated heterotrichous thallus construction with both prokaryotic and eukaryotic celluar organization. Although, they are predominantly true hydrophytes, aerial and subaerial forms are not rare. The soil inhabiting algae are known as edaphic algae. Those colonizing the moist soil surface are referred as Saphophytes and play an important role in soil conditioning and vitality of the soil.

The algae, especially the blue-greens are able to survive prolonged periods of drought. Successful revival of algae has been possible even after 80 years of preservation in dried form. Many of the xeric forms are able to survive temperatures exceeding 100°C when dormant in the soil.

Forms which do not form resting propagules, are able to resist the dry and warm conditions because of the presence of mucilage, viscous cytopkesus, absence of vacuoles, aggregation of several trichomes pigment adaptation.

Ability of many forms to withstand adverse ecological conditions, capacity to thrive well in hostile environments and respond to the onset of dry conditions by entering into a dormant resistant state, have distinguished this group as pioneers of plant succession. Both in xerosere and hydrosere, algae appear as the first plant community leading to the establishment of higher plant commu nities. Algae play an important role in primary and secondary plant community succession by acting as an integral part of the colonial synusium.

Algae constitute an important component of the ecosystem. They account for about 40 percent of the total carbon fixed annually on this planet. The spontaneous fertility of the soil of the tropical rice fields is largely because of the activity of the soil of the authochthonous algae. But it is difficult to believe in the effectiveness of a "Cover Crop" which one can hardly see. One has to have perseverance to realise the effect of this important self regenerating system which has sustained cumulative effect. The present day strategy is to increase the number of these good" Microbugs" in the soil to make the soil productive through adding and increasing the availability of plant nutrients.

2. Rock crust and other lithophytes

Algae as the oligotrophic primary producers are the pioneers inhabiting bare rock surfaces and then allowing further plant succession by way of accumulation of organic and inorganic nutrients. These forms are also desiccation and thermotolerant. Although reports were made on lithophytic algae and cyanobacteria at Rajasthan, but these were concerned primarily with damp rocks only showing cyanobacteria *Aphanothece, Apanocapsa, Chroococcus,* (coccoid). *Phormidium, Lyngbya,* (non heterocystous) *Nostoc, Anabaena, Tolypothrix, Scytonema* (hetero-cystous filamentous) and chlorophytes, namely, *Cladophora, Pleurococcus* and *Oedogonium* (Srivastava, 1993). No attempt could be traced where algae was cultivated from dried rock crusts and identified. Friedmann and associates reported *Chroococcidiopsis* as the most desiccation resistant cyanobacteria known. It is found in the inner spaces of porous rock (endolithic), areas dominated by desert pavements, lives on lower surface of translucent stones and pebbles half embedded in the soil (hypolithic). The short periods of rain and the metabolic activity suffice its resource maintenance including DNA repair for the rest of the year (Friemann and Galun, 1974). Chasmoendolithic forms are these inhabiting fissures and cracks of rocks, whereas those found in the internal structural cavities of porous rocks are called cryptoendolithic such as the one mentioned above.

Gloeocapsa is the dominant endolithic form of deserts. It is reported from Nagev Desert, Death Valley and Sonoran Desert. *Trebouxia* is the lichen growing in these endolithic habitats (Friendmann and Ocampo, 1976).

3. Common soil algae

Interpretation of edaphic algae as soil microorganisms is more widespread today, because of the realization that, as a component of the soil microflora, algae act as reservoirs for the plant nutrients, as organisms influencing the soil structure and the activity of other microorganisums and as agents for the incorporation of organic carbon and nitrogen through photosynthesis and nitrogen fixation. However, still our understanding of the ecophysiological nature of the soil algae is based upon the laboratory experiments and empirical observations.

While typical aquatic and edaphic algal forms do exist, no clear demarcation can be made between the two types, as many forms can generally occupy the border line. Forms like *Chroococcus, Microcystis, Anabaena, Wollea* and *Oscillatoria* amongst blue-greens and *Chlamydomonas, Chlorella, Cladophora and Rhizoclonium* amongst green algae are strict hydrophytes, while *Fritschiella, Vaucheria, Scytonema* and many species of *Calothrix, Nostoc, Westiellopsis* and *Hapalosiphon* grow only on moist and shaded soil surface. Algae like *Chlorococcum, Plectonema, Phormidium, Nostoc, Calothrix, Cylindrospermum* and Diatoms occupy the intermediate position as they are capable of growing both in water as well as on the soil.

4. Distribution of soil algae

Microalgae are ubiquitous components of the soil microflora. They are cosmopolitan in distribution as they require only little moisture and diffused light for their growth. However, being photoautotrophic in nature, they have restricted distribution along the soil profile. In fallow fields, algae are generally found upto about 2-3 cm depth. In light and sandy soils, some unicellular forms like *Chlorella, Chlamydomonas, Chlorococcum* and Diatoms may be carried to the deeper layers by the percolating water. Grasslands, during wet seasons, generally show very good growth of *Scytonema* and *Nostoc*.

This microstratification of algal populations in the soil profiles has earlier been reported by Cameron and Blank, 1996: Nordin and Blinn, 1972: Friedmann and Galun, 1974). Over a period of time, these forms have been seen resulting in the formation of soil crust sufficiently thick and rich in organic matter to support the growth of moss plants and small grasses during the cooler and moist months of the year.

In cultivated fields, the algae are found even at 20cm depth because of the turning of the soil during plougning. However, one does not expect them to multiply inside the soil because of the limitation of light. Thus, the algal activity influences only the top few centimeter layer of the soil, showing pronounced effect on the surface layer.

The texture and chemical nature of the soil greatly influences the algal distribution. Sandy soils show restricted algal incidence. High moisture content and water holding capacity encourage algal growth. Presence of clay has been shown to enhance the algal incidence in the soil. Temperate climate generally favours green algae and tropical conditions favour blue-greens. Likewise, low pH accelerates the growth of green algae and at high pH blue-greens proliferate. At alkaline pH, the predominance of diatoms and cyanobacteria is due to increased availability of organic and inorganic nutrients (Lund, 1945, 1947, Mitra, 1961). Predominance of green algae at lower pH may be due to the inability of other algae to live at lower levels of nutrient supply (Metting, 1981).

The redox potential of the soil indirectly regulates the incidence of the algae by controlling the availability of nutrients. Although the algae are autotrophic, presence of organic matter increases the algal incidence through increasing the availability of carbon dioxide as a consequence of decomposition especially at low pH. Presence of humic substances stimulates the algal population in the soil without acting as a direct source of nutrients (Lee and Bartlett, 1976). High organic matter content increases algal incidence through increasing the moisture content of the soil (Cameron and Blank, 1966, Friedman and Galun, 1974).

Vegetation cover in the fields also regulates the occurrence of algal communities. In a rice field, crop canopy encourages the occurrence of forms like *Calothrix, Scytonema, Tolypothrix* and *Aulosira,* which flourish well on moist and shaded soil surface. The same fields just after transplantation, had been seen harbouring green algae like *Hydrodictyon, Spirogyra, Cladophora* and *Rhizoclonium* because of the absence of shading by the crop plants and availability of comparatively more of combined nitrogen.

Cultivation practices also affect the distribution of prevalence of the soil algae. Rice fields soils, because of the flooded conditions and shade provided by the crop canopy, provide simplified ecosystem which encourages climax communities, favouring the growth of only the adapted algal forms. Generally these soils harbor a very rich cyanobacterial flora. Conversely, soils from fields where only dry crops are grown, the soil algae show tremendous species diversity and very rarely a particular alga or a group of algae are found to dominate the niche.

Quantification of algal incidence in the soil has been difficult in the absence of an appropriate method. Direct count and MPN method can be used for algal enumeration but these methods are cumbersome and give only approximation. Dilution, plating and counting of colonies scores over these methods but is invariably made difficult by indiscrete and spreading algal colonies. Pour plating can solve the problem to some extent but this again does not yield information on community composition and algal predominance. Kolte and Goyal (1985) suggested a method for numerical quantification of the distributional pattern of algae. In this method, the proportionate occurrence of different algae in enrichment culture as observed under a light microscope is converted to numerical values as dominant (5), very common (4), common,(3), rare (2) and very rare (1). The total number of soil samples from an area harbouring a particular alga is taken as "Numerical total". The numerical total is multiplied with the numerical values (1-5), depending upon the preponderance of the algal to obtain "Rated total". The rated total is then divided with the total number of soil samples collected form the area and converted to percentage to arrive at "Rated Score" for the alga in a given area. This rated score is used for comparing the proportionate distributional pattern of different algae in the area.

5. Influence on soil properties

Soil is a dynamic system comprising of physical, chemical and biological components. While the physical and chemical status of the soil has a bearing on the fertility level of the soil, The productivity of the soil largely depends upon the availability of the nutrients which in turn is regulated by the microbial population. The microbial denizens of the soil actually feed the plants.

Algal growth greatly influences the physical and chemical nature of the soil. Nutritional independence for carbon in general and for both carbon and nitrogen in the case of nitrogen fixing blue-green algae, make algae an important component of the soil microflora. Algae add substantial quantities of organic matter through primary production. Out of the global net primary production of 2×10^{11} tonnes of organic matter, 3.2×10^{10} tonnes, amounting to 40 per cent, is accounted by the algae of aquatic and terrestrial habitats. Algal growth not only narrows the C : N ratio making the soil more fertile, but also increases the humus content of the soil. The enhanced adsorption by the soil organic matter has been shown to reduce the effect of toxicants.

There are reports on the occurrence of other obligate photoautotrophic and chemoautotrophic soil microorganisms, but the sporadic reference to these forms indicate that they are probably not important in the soil ecology. However, a number of facultative photoheterotrophic and chemoheterotrophic algae capable of utilizing simple carbohydrates and organic acids present in the soil substantially affect the process of nutrient transformation in the soil (Metting, 1981).

Physical nature

Tillage leads to mechanical destruction of the soil structure and promotes microbial degradation of the organic matter by increasing the oxygen availability. While the latter increases the nutrient availability to the crop plants, the former reduces soil tilth.

Most of the soil algae possess mucilaginous sheath and have profusely branced thalli which help in binding the soil particles together. This cementing effect increases the size of the soil aggregates. Algal crust on the soil surface minimizes soil erosion, increases the soil aggregate size and optimizes aeration, water movement, root development and fertilizer utilization. Use of algae like *Chlamydomonas mexicana* (Metting and Rayburn,1983) palmelloid green algae (Lewin, 1956, 1977), and blue-green algae (Roychoudhury *et al.,* 1979; Subhashini and Kaushik, 1981) has been found to be very successful in increasing the size of the soil aggregates.

Soil compaction is the ultimate process observed in fields under intensive cultivation and repeated and heavy irrigation. This is caused by the break down of the soil structure creating High Bulk Density and robbing the soil of the pore space. Puddling is also aimed at compacting the soil to reduce water percolation in a rice field. The Turf grass in the Golf courses is periodically sprinkled with water to keep it cool. This creates microaerophilic conditions and formation of " Blank layer" in the sub –soil zone leading to compaction of the soil.

Fred Williams of lowa State University, Ames, USA (Personal communication) found that addition of 0.02g dried algal polysaccharide to 100g soil increased the water stable aggregates of 0.1 mm diameter from 44 to 600g. Contrary to the general belief, standing water on the top of the soil is not necessary for optimum algal activity. Studies have shown that the inoculated algae multiply and bloom in the water film around the soil particles which carry upto 40 per cent or more moisture in the top 1 inch of the soil. The polysaccharides synthesized by the algae make the soil more crumbly by increasing aggregate size which allows (i) faster rainfall absorption, (ii) increased aeration (iii) accelerated microbial performance and (iv) easier root penetration.

Schaefer (1986) using a Bcaring Ring Cone Penetrometer showed that algal inoculation of Corn field considerably reduced the soil compaction. However, as expected the effect was more in the top soil layer. The reduction in the top 5cm layer ranged between 31-43 per cent and at 10cm depth it was reduced to 24 percent. The reduction in the soil compaction was reflected in the crop yield also.

5.1 Chemical nature

A flooded rice field represents macro-as well as microenvironments. The macroenvironment is represented by the photic zone or the oxidized layer constituted by the water column and the soil-water interphase. The aphotic zone or the reduced layer lies below the oxidized layer and comprises of reduced soil, plough layer and the sub soil. The photic zone has positive redox potential while in the reduced zone the redox potential is negative with Eh > 300 mVm^{-1}. Abundance of algae in the photic zone is helped by the availability of light and nitrate nitrogen, the latter happens to be the most preferred source of nitrogen for algae. The heterocystous cyanobacteria contribute substantial amount of biologically fixed nitrogen in this zone. In the reducing conditions of the aphotic zone, the non- heterocystous cyanobacteria form the most active nitrogen contributing biological system.

5.2 Soil reaction and salinity

Flooding and puddling have a buffering effect on the soil pH. Irrespective of the original soil pH, these operations bring it to equilibrium pH of 6.5-7.0 in about 2-3 weeks (Sanchez 1972). In acid soils, the pH increases due to reduction of iron and manganese hydroxides releasing OH^- ions. In alkaline soils, the pH decreases because of the increase in the partial pressure of carbon dioxide. Although, creation of reduced conditions increases the stability of ammonium ions and availability of elements like iron, manganese and phosphorus, continued availability of the latter results in the release of non- reducible exchangeable cations like K^+, Ca^{++} and Na^+, leading to increase in the conductivity and partial pressure of carbon dioxide in the soil.

One of the reasons of reduction in the soil salinity may be the control of redox potential by the inoculated algae through the liberation of oxygen which indirectly checks the release of Na^+ and K^+ ions into the soil medium. The algae also reduce salinity through adsorption of these ions onto their cell surface and mucilaginous envelope. Algalisation of saline-sodic soils brings down the pH, reduces electrical conductivity and exchangeable sodium and increases hydraulic conductivity (Subhashini and Kaushik, 1981).

Although algae medicated amelioration of salinity in the soil is an easy method, the treatment effect is not stable as the adsorbed ions are released back into the soil medium unless the algal biomass is not removed from the ecosystem. Nevertheless, during the crop growth cycle, algal growth helps in the creation of microenvironments devoid of salinity stress, in and around the root zone of the rice plants.

5.3 Phosphorus

Phosphorus is an indispensible macroelement for plant growth as it forms an important constitutent of a veriety of vital enzymes. The availability of this element in the soil is greatly limited by the soil pH. In the soil it tends to become bound as calcium phosphate especially at high pH. Although, the exact mechanism is not known, many algae have been found to be capable of solubilizing the insoluble phosphate in the soil (Bose *et al.,* 1971).

Amongst the component algae of the mixed inoculums, *Nostoc sp.* and *Tolypothrix tenuis* were found to be better solubilisers of the insoluble phosphate. Screening of a larger number of algae may lead to the identification of still efficient phosphate solubilising algal forms which may become an added attribute of the inoculum.

5.4 Nitrogen

Algae add to the nitrogen content of the soil through the process of biological nitrogen fixation. These diazotrophic algae possess the enzyme nitrogenase, which brings about the reduction of molecular nitrogen to ammonia. Part of this fixed nitrogen is added to the soil through excreation and the rest at the time of autolysis of the algal cells. It is estimated that algal inoculation of rice fields adds 15-25 kg N ha^{-1} $season^{-1}$ (Goyal 1987,1993,1996).

The enzyme nitrogenase gets repressed in presence of the added nitrogen. Ammonium form of nitrogen competitively suppresses the process of nitrogen fixation. The inhibition is both at synthesis and functional level of the enzyme. However, the algal nitrogenase, unlike the bacterial nitrogenase, gets depressed once the combined nitrogen, the growth of the cyanobacteria continues metabolizing the available combined nitrogen. One the level of combined nitrogen falls below a threshold value (>40 ppm), the nitrogenase activity is regained. Thus, algae increase the availability of nitrogen not only through the addition but also by presenting the loss of part of the added nitrogen through metabolisation and form a source of linear availability of nitrogen for the crop plants.

6. Microbial interactions

The microbial component of the soil plays a very important role in nutrient transformation and mobilization in the soil. The soil microorganisms actually feed the plants through mobilization of various nutrients. Variety of organisms like bacteria, actinomycetes, mycorrhiza, fungi, algae, protozoa, crustaceae, nematodes, annelids and viruses constitute the microbial world of the soil. These organisms interact at various levels and tend to attain equilibrium under a given set of conditions.

The algae-microbes interaction can be symbiotic, antagonistic synergistic, competition, ammensalism, comensalism, parasitism, neutralism and protocooperation. Algae liberate variety of compounds like amino acids, vitamins, enzymes, carbohydrates, organic acides, polypeptides, antibiotics and phenolic which generate the above inter-action responses

(Foog, 1971; Whitton, 1965; Chapman and Rae,1969; Venkatar-aman *et al.*, 1974; Jones, 1988).

Auto and allo-inhibition amongst algae is not a rare phenomenon and has been reported in *Chlorella vulgaris, Nostoc punctiforme, Chlamydomonas reinhardii* and *Pandorina morum* (Patterson and Harris, 1983). Autostimulation of the reproductive process amongst algae has been reported by Pandey and Talpasayi (1980) and Sharma (1984).

Antimicrobial factors were indentified in the cultures of *Chlorella* sp. (Pratt and Fong 1940). *Isochrysis galbana* (Bruce and Duff, 1967), marine algae (Hornsey and Hide 1976) and *Skeletonema* sp. And *Phaeodactylon tricarnatum* (Cooper *et al.*,1983). Many bactria find an ideal lodging in the mucilaginous sheath of cyanobacteria. *Azotobatcer* was isolated from the sheath of cyanobacteria like *Nostoc, Gloeocapsa* and *Rivularia* (Jones, 1930). *Azotobacter* was found to fix more nitrogen in the cultures containing actively growing *Chlorella* cells (Lipman and Teakle, 1925). Goyal and Venkataraman (1965) reported that some cellular components of the fungus *Alternaria* sp. stimulated the growth of the blue- green alga *Nostoc* sp.

7. Conclusion

The soil ecosystem is our heritage and has a limited regeneration capacity. It is a dynamic system. The more we caress and cooperate with the living things in it, the more support we will get from this. The ever deteriorating soil health is our creation, as we did not do natural resource accounting and failed to recognize the importance of environment management. Increased use of chemical inputs in agriculture has lead to considerable biodepletion of the soil. All these factors have contributed to stagnated crop yield not to speak of contamination of ground water and resource depletion. The present day negative nutrient balance in the soil can be corrected only through integrated nutrient management involving judicious use of chemical fertilizers organic manures and bioinoculants. In addition, soil vitality also has to be preserved, which seems possible only through the use of economical and eco friendly bioinoculants.

REFERENCES

Aiyer, R.S., Aboobekar, V.O., Venkatraman, G.S. and Goyal, S.K., 1971, *Phykos* 10:34-39.

Bose, P. Nagpa, U.S., Venkataraman, G.S. and Goyal, S.K. 1971. *Curr.Sci.* 40:165-166.

Bruce, D.L. and Duff, D.C.B.1967. *J.Gen. Microbiol*, 48: 293-298.

Cameron. R.E. and Blank, G.B. 1966. *J.P.C. Tech. Report.* Pp.32-971.

Chapman, G, and Rae, A.C. 1969. *Mar. Biol.* 3: 341-351

Fogg, G.E. 1971. *Arch, Hydrobiol.5*:1-25.

Friedmann, E.L. and Galun, M, 1974. Desert Biology II (Ed. Brown, G.W. Jr. Academic Press, New York.

Fried Mann, E.I. and Ocampu, R. Science 193, 1247-1249.

Goyal, S.K. 1987 J. *Maharastra Agri. Univ.* 12:11-19.

Goyal, S.K. 1993, Proc. Indian natl, Sci. Acad,B 59: 295-302.

Goyal S.K, 1996. In agrochemicals: Food security and environmental protection. Ed. Roy, N.K. Pub. AP . Publications, New Delhi. Pp.163-174.

Goyal S.K. and Venkataraman, G.S. 1965, *Curr.Sci. 34(5)*:152-153.

Homsey, I.S. and Hide, D. 1976 *Br. Phycol. J. 11*:63-67.

Jones, A.K. 1988. (Eds. Rogers. L.J. and Gallon, J.R.) 257-281, Oxford Science Pub.

Jorgensen, E.G. 1962, *Physiol. Pl. 15:* 530-545.

Kolte, S.O. and Goyal S.K. 1985. *Phykos* 24:156-161.

Lee, Y.S. and Bartlett, R.J.1976. *Soil, Sci. Soc. Amer. Proc 40*: 876-879.

Lewin, R.A. 1977. *Can. J. Microbiol.2*:665-672.

Lewin R.A. 1977. Centros Invest, Baja Calif Scripps. Inst Oceanogr 3:33-35.

Lipman C.B. and Teakle, L.J.H. 1925. *J.Gen. physiol. 7:* 509-511.

Lund, J.W.G. 1945. *New Phytol 44:* 196-219.

Lund, J.W.G.1947. *New Phytol.46:* 35-60.

Metting, B. 1981. *Bot. Rev. 47*:195-312.

Metting B. and Raybum, W.R. 1983. *Soil Sci. Soc. Am. J. 47*:682-685.

Mitra, A.K. 1961. *Indian J.Agri. Sci. 21:* 357-373.

Nordin, R.N. and Blinn, D.W. 1972. *Proc. North Dakota Acad. Sci. 25*:8-17.

Pandey, R.K. and Talpasayi, E.R.S. 1980. *Ind. J. Bot. 3:* 128-133.

Pratt, R. and Fong.J. 1940. *Amer. J. Bot. 27*:431-438.

Rao.D.L.N. and Burns, R.G. 1991. *Biol, Fert. Soils 11*:306-312.

Roychoudhury, P. Kaushik, B.D. Krishnamurthy, G.S.R. and Venkataraman, G.S. 1979. *Curr. Sci. 48*:454-455.

Sanchez, P.A. 1972. Bull No. 213, North caroling Agriculture Experiment station, Raleigh, USA, pp. 1-31

Shaefer, J. 1986. Soil Technologies Inc. Fairfield, Lowa, USA.

Sharma, P.1984. Arch. Microbiol. 139:196-201.

Srivastava. P.J. Phytol. Res.6 (1): 13-17.

Subhashini, D. and Kaushik, B.D. 1981. *Aust. J. Soil. Res. 19:361-366.*

Venkartraman, G.S. Goyal, S.K. Kaushik, B.D. and Roychoudhury, P. 1974 Algae: Form and function, Today and Tomorrows, New Delhi, pp. 343-352.

Whitton, B.A. 1965. *J. Gen Microbial* 40:1-11.

William R.B. and Lewin, R.A. 1985. *Plant Soil 88:* 159-169.

Chapter - 25

Management of post harvest contamination of moulds and mycotoxins in food grains

A.K. Roy, N. L. Mandal and S. Roy

PG Department of Botany, T.M. Bhagalpur University, Bhagalpur-812 007.
E-mail: botanyakr@yahoo.co.in

ABSTRACT

The warm and humid agro-climatic conditions of India make food grains susceptible to contamination of moulds and mycotoxins both in pre- and post harvest stages i.e., from field to storage. Several post harvest management strategies viz., proper drying, hygienic storage and application of physical, chemical and biological-agents are immensely helpful for prevention of moulds and mycotoxins contamination in different food grains.

Key words: Mycotoxins, Food grains, Moulds, Post harvest

1. Introduction

Mycotoxins comprising structurally diverse and chemically unrelated group of naturally occurring fungal metabolites have been strongly implicated as chemical precursors of toxicity in human and animals (Chauhan, 2004). The toxicity syndrome (mycotoxicoses) resulting from the intake of mycotoxin by man and animals is known due to its toxigenic, carcinogenic, mutagenic, teratogenic and estrogenic properties (Bennet, 1987). Though mycotoxicoses have been known for a long time, however, it got first recognition probably as ergotism in Europe under the name "Holy fire" which was resulted by the intake of contaminated grain with sclerotia of *Claviceps purpurea*. Despite the fact, mycotoxicoses were remained as "neglected diseases" until the out break of Turkey -X disease in Great Britain in the year 1960s which led to a multidisciplinary approach to investigate the causes of the problem in aflatoxin contaminated ground nut. At present more than 300 different mycotoxins produced by about 150 fungi are known, showing a large variety of chemical structure and their effect. Although many toxic metabolites have been isolated from laboratory cultures of moulds that occur on agricultural products so far only eight viz., **aflatoxins, ochratoxin, citrinin, patulin, trichothecene, zearalenone, Fumonisins** and **Cyclopiazonic acid** are found to have possible significant occurrence in naturally contaminated foods and feeds (Table 1).

Table 1. Natural contamination of mycotoxins and their effects

Mycotoxins	Mycotoxin producing fungi	Food commodities	Biological effects
Aflatoxins (B_1B_2,G_1,G_2)	*Aspergillus flavus* & *A.parasiticus*	Peanut, Corn, Wheat, Rice, Herbal druga and milk	Hepatotoxic, carcinogenic
Ochratoxin	*A.ochraceus* & *Penicillium viridicatum*	Cereal grains, Bean, peanuts, Wild seeds of medicinal use	Nephrotoxic
Citrinin	*P.citrinum* *P.viridicatum*	Wheat, barley, rice, Corn and herbals	Nephrotoxic
Patulin	*P.patulum*	Mouldy feed, Apple, Wheat straw residue	Edema, death of cattle, paralysis of motor nerve
Trichothecenes	*Fusarium tricinctum* & *F. solani*	Corn, Wheat, Cattle feed, mixed feed	Dermal necrosis, haemorrhages, digestive disorder
Zearalenone	*F.graminiarum*	Corn, mouldy hay, Commercial feed	Volvo vaginitis, estrogenic effect, atrophy of testis, ovaries, abortion
Fumonisins	*F. moniliforme*	Groundnut, Poultry feed	Leukoenuphalomalasa, abortion, carcinogenic
Cyclopiazonic acid	*P. cyclopium*	Corn, peanut, kodo millet	Human intoxication, muscle necrosis, haemorrhage and edema

2. Natural contamination of mycotoxin

Natural contamination of mycotoxins in agricultural commodities is a worldwide problem but greater in tropics and temperate zones. An overview of reports on mycotoxin contamination clearly indicates that practically every edible material including fruits, vegetables, cereals, pulses, spices and plant drugs too is not spared by moulds contaminated and mycotoxin elaboration (Abramson 1998; Mayer *et al.,* 2003; Reddy and Hayes 2001; Roy *et al.,* 1988; Roy 1989; Sinha *et al.,* 2003; Sweta *et al.,* 2010; Windham, and Williams 1998). India being a tropical country has great diversity in agro climatic condition due to unseasonable rains, high temperature and humidity, which adversely affect the both standing crops and stored food materials by favouring the growth of moulds. Mycotoxigenic moulds contaminate the agricultural commodities starting from pollination of flower to germination of seeds. Generally food grains have been reported to be good host for a number of fungi including the species of *Aspergillus, Penicillium and Fusarium* etc. which are well known producers of various mycotoxins (Nepote *et al.*, 1997; Ngoko *et al.*, 2001; Sinha *et al.*, 2003). *Fusarium* species viz., *F. chlamydosporum, F. semitectum, F. graminearum, F. poe,* etc. have also been reported to produce deoxynivalnol, zearaleonone mycotoxins in the in several food substrate. Hesseltine *et al.* (1976) recorded predominant *Aspergillus* and *Penicillium* isolates from surface sterilized kernels placed on the malt agar plate.

3. Management of mycotoxin contamination

The mycotoxigenic fungi survive in plant residues and produce abundant spores. The spores are carried by the wind and infect food crops, usually through insect wounds. Prevention of natural occurrence of mycotoxin in various food grains by mycotoxigenic fungi and their detoxification are cumbersome process, therefore, 100% safe food and feed supply is not effective. However, it may be possible either by preventing moulding of substrates or detoxifying the mycotoxin contaminated foods. In order to manage the mycotoxin problems in various foods and feeds substrate, different methods viz., i) Cultural ii) Physical iii) Chemical and iv) Biological control have been adopted in the recent past however, more investigations are still required for its sustenance.

3.1 Cultural practices

Pre-harvest measures in which following Cultural practices are adopted:

3.1.1 Reduce plant stress: Use recommended production practices to minimize plant stress and maximize yields. These include insect, weed, and disease control practices, and the use of recommended plant populations and fertility practices. When possible, irrigation during pollination can decrease predisposition of the crop to aflatoxin. Plant corn as early as possible, and plant several different hybrids of different pollination periods to reduce the chance of environmental stress at pollination in at least part of the crop. Care should be taken to store uncontaminated hybrids away from those that are contaminated.

3.1.2 Early harvesting and drying: Harvesting food grains when it is above the 20 percent moisture content and drying it within 24 to 48 hours to moisture content no greater than 14 percent greatly reduces the infection, growth, and toxin production by *Aspergillus* and other mycotoxigenic fungi.

3.1.3 Avoidance of damaged food grains: Damaged food grains are more likely to become infected with molds both in the field and in storage. Hybrid selection and insect control can play a role in reducing food grains.

3.2 Physical practices

Physical practices are mainly Post-harvest practices. The important post-production/post-harvest operations for maize and other grains are drying, storage, shelling, and milling. Despite technological advances, solar drying, not mechanical drying, is commonly practiced

3.2.1 Field drying: When the growing crop has reached maturity it begins to dry. If left in the fields it will eventually become dry enough to store. However, the longer it stays in the field the more likely it will become infested by storage insects, damaged by rodents and other animals, or be stolen. Therefore, it is best to transport the crop to the homestead as quickly as possible where it can be dried safely. Drying crops should never be placed directly on the ground as they may get wet or come into contact with moulds in the soil.

3.2.2 Sun drying: Grain can be dried by spreading it in thin layers in the sun. It should be stirred occasionally to ensure that all the grain is exposed to the sun. The grain should never be spread out directly on the ground because of the risk of drawing up moisture from the soil. Spreading grain across a concrete plinth will speed drying and will also reduce contamination from dirt. Better still, spreading grain on a polythene or cloth sheet, or empty sacks, enables it to be gathered quickly and protected from rain. A black polythene sheet will also help by absorbing heat from the sun and so speed up drying.

3.2.3 Seed storage: Many farmers retain their own seed each year. Soon after harvest, seed grain is selected from the crop that has just been brought home. Usually, the seed is of the best visible quality and farmers try to take good care of it while in store. Generally, relatively small quantities are kept back and these cannot be kept in the stores normally reserved for food grain. Any reduction in the quality of seed, such as insect and mould damage and shrivelling through water loss or heat exposure, will reduce the germination potential and therefore reduce the size of the next crop. Seed storage therefore demands extra careful management. Containers for storage of seed should be sealable and just large enough for the quantity to be stored. The grain must be dry and free of insect pests. If seed is stored in an airtight (hermetic) container, such as in a bottle with a tight fitting lid, insects in the grain will die from the combined effects of reduced levels of oxygen and poisoning by the accumulated carbon dioxide produced as they respire. Grain may be stored in sealed pots or gourds, or in tins or bottles. The grain must be dry; otherwise it will go mouldy or ferment, as these containers do not allow water to be lost to the outside. These methods are fine for keeping grain free from pests as long as the container is well sealed. Plastic containers with screw-on lids are ideal for seed storage. Larger quantities of maize seed are frequently stored by suspending entire cobs from the roof leaves or overhang of the house; they may also be suspended.

Storage problems are not simply restricted to what actually happens to the grain once it has been stored. Deterioration is caused by many factors that occur throughout the production cycle. The analysis of problems must include an assessment of the following:

(a) Crop husbandry: poor cultivation practices (e.g. inadequate weeding) may allow pests to survive and provide a source of infestation.

(b) Type of seed cultivated: some grain varieties, especially high-yielding ones, are more susceptible to insect damage. Open, loose varieties of sorghum and millet are more easily attacked by birds than tight-head varieties.

(c) Time of harvest: the longer the crop is allowed to stand in the field the more likely it is to be attacked by pests and diseases. The crop should be transported to the homestead as quickly as possible after harvest.

(d) Climatic conditions: rains at harvest time make drying difficult and grain will be prone to mould invasion. Drought conditions may also favour attack by some moulds.

(e) Post-harvest techniques (threshing, drying, winnowing): poor handling after harvest may damage grains making them vulnerable to pest attack; delayed or inadequate drying will also encourage mould and insect damage.

(f) Storage methods: types of storage structures (traditional and improved), pest control and storage management all determine the quality and shelf-life of the stored crop.

(g) Processing methods: flour has a shorter shelf-life than whole grains. When cereals and pulses are milled, the flour must be stored properly to prevent moisture uptake and insect infestation.

(h) Marketing and sale: where crops are intended for sale, the location of the farm, transport requirements and access problems must be considered.

(i) Causes and extent of losses: at which points in the post-production chain are significant losses occurring? What is causing these losses? What actions can the farmer take to reduce them to acceptable levels?

3.3 Chemical practices

The most effective way of preventing mould growth is to dry the grain as quickly as possible to a moisture content that is low enough for safe storage. Safe storage is achieved when moulds cannot develop. For cereals and pulses in tropical and subtropical countries this point is achieved when the moisture content is below 13-14 percent, for groundnuts and other oilseeds below 7 percent. Once the grain is dry it must be kept dry while in store. Several Chemical and botanical practices have been adopted in recent past for avoiding fungal infestation (Allameh *et al.*, 2002). Some of the important chemical measure are widely adopted in recent past are as follows.

3.3.1 Box fumigation: Fumigation kills insects inside grains with phosphine gas. Phosphine is released from aluminium phosphide tablets when these are exposed to air. *The gas is lethal even in low concentrations to humans and livestock.* Gas is released from the tablets within 20 minutes in some climates. It is therefore, essential that only trained personnel undertake fumigations. Untrained personnel must not handle the tablets. Fumigations are only effective if carried out in gas-tight enclosures. Traditional farmers' stores, even mud plastered or mud brick buildings, are not gas-tight and must not be used for fumigation. Leakage of gas is extremely dangerous for the farmer's family and animals and does not kill insects in the grain. Small-scale fumigation can be carried out under gas-tight covers such as thick polythene sheets, or in containers that can be easily sealed, such as polypropylene water tanks or purpose-built metal tanks. Farmers producing many tonnes of grain can hire trained extension personnel or private pest control operators to fumigate their crops at the homestead as long as the area to be fumigated can be made gas-tight. Alternatively, farmers may be able to take their produce to a fumigation centre where it can be fumigated under controlled conditions by trained personnel who can guarantee that the grain will be free of insects after treatment. A good fumigation, which lasts from seven to ten days, will kill all insects present but it will not provide long-lasting protection against infestation. It may be necessary to use the grain immediately or to treat it with insecticide to ensure that reinfestation does not occur.

3.3.2 Chemical treatment (insecticides): Insecticides, when properly applied to stored cereals and pulses as dusts or sprays can offer long-term protection against insect attack. Insecticides are usually applied to dry threshed grain. This minimizes the amount of insecticide required and provides good protection. It can be difficult and expensive to apply insecticides to maize stored on the cob or to other unthreshed cereals and pulses, and treatments may not be very effective in preventing insect damage.

3.3.3 Smoking: unthreshed cereals and pulses can be stored on platforms or in the loft of the house above a fire. The smoke and heat from the fire may kill insects or drive them out of the grain. The method is not always effective; in particular the larger grain borer will not be killed.

3.4 Biological practices

Recognition of the need to control mycotoxin contamination of foods and feeds different researchers have applied their ideas to eliminate or to detoxify different mycotoxin from susceptible crops via physical and chemical means (Basappa,1983; Bilgrami, 1990 Goldflat, 1971) however, the most promising and safest approach is through biological means where living micro organism such as bacteria, fungi, actinomycetes, yeasts and some botanicals are used to control these toxigenic strains either by the phenomenon of antibiosis, competition or parasitism. Prevention of mycotoxins contamination is thus prerequisite and their detoxification is secondary factor of its management. Decontamination and detoxification are two main strategies to control mycotoxin elaboration in foods, feeds and plant drugs. The contribution of earlier workers (Jung *et al.*, 2000; Kessel *et al.*, 2002; Mandal *et al.*, 2007; Romina and Miriam. 2006) on this aspect cannot be ruled out. These methods undoubtedly show good result but are costlier affairs, time consuming and also alter the quality of the product. Pollution hazards by the use of chemicals/ pesticides also cannot be ruled out. Biological control is ultimate mechanism against mycotoxin management which offers both potential strategies. The concept of Biological control based on the microbial antagonism to the plant pathogen/ toxigenic strains has opened a new vista under Control Program. Its application in the suppression or elimination of harmful microbes has been found significantly useful. In biological control strategy, living microbes (antagonist) are used to control harmful micro organism either by producing some toxic metabolites or causing disease in them without any ecological imbalance. The phenomenon of biological control is based on antibiosis (inhibition or destruction of one organism by the metabolic product of other), competition (injurious effect of one organism on other because of the utilization or removal of the same resource) or parasitism (where antagonist parasites on pathogen).

Therefore, attempts have been made in recent past to control mycotoxin problem in foods and feeds and herbal drugs by the application of bio agents (Chourasia & Sinha 1994; Cotty, 1990; Kumar & Roy 2001; Roy & Chourasia 1990; Rosic *et al.*, 1991) have also reported earlier that microorganisms can inactivate or remove the mycotoxin. *Flavobacterium auratiacum* and selected acid producing moulds have been utilized for successful removal of aflatoxin form liquid culture media. Transformation of aflatoxin B_1 to aflatoxicol (less toxic then aflatoxin B_1) was observed in case of *Corynobacterium rubrum* (Mann & Rehm, 1976) and several sps of moulds belonging to the genera of *Aspergillus, mucor, Trichoderma* etc. (Detroy & Hesseltine, 1969; Mitsuo *et al.,* 1990; Robertson *et al.,* 1970).

3.4.1 Antagonistic activity of microbes against mytoxigenic fungi: The entire biological control program is based on the isolation of various microbes from different sources and their screening *in vitro* condition against various pathogenic microbes. During the screening of microbes as antagonists it is also imperative to study their mechanism because they interact in different manners such as parasitism, competition and antibiosis. It is also essential to notice that the degree of antagonistic potential of microbes is varied from species to species and also to strain. It is also important that some of the antagonists strain possesses more than one mode of action and also effective against many pathogens while other antagonist strains are specific to some pathogens. Sharma and Paul (1998) outlined some characteristics of an ideal biocontrol microbes are: i) Survival for prolonged periods in soil either in active or in passive forms. ii) Multiplication in laboratory conditions should be simple and in expensive. iii) be specific to the target organism if possible. iv) be active under the required environmental conditions. v) be efficient, economical and not be health hazards during handling and applications. In the last few years emphasis on the research and practice of biological control is steadily increasing as an important component of an integrated pest management in certain areas, which helps in reducing the dependency of fungicides and other chemicals for sustainable ecological damage. Thus, the careful screening is most successful steps for the success of Biocontrol program.

3.4.2 Through fungi: Fungi are extremely diverse group of organisms with about 75, 00,000 species were widely distributed in every ecosystem. Among all the sps of fungi, only limited number of sps is used as antagonists in biological control program. Biological control has played a major role in disease management since the beginning of agriculture. Farmers learn disease management practice through careful screening of antagonists by trial and error that takes advantage of natural biological control, even though they may not recognize its contributions to the well being of their crops. A critical analysis of the literature on biological control reveals a skewed research towards soil borne pathogens, as the response has been more positive in this area as compared to foliar pathogens. Among all microorganism, fungi are one of the potential candidates for biological control programme and several antagonist fungi viz. *Trichoderma, Gliocladium, Penicillium, Aspergillus, Talaromyces, Trichothecium* sps inhabited in rhizosphere and other ecosystem have been exploited for the control of plant diseases. Several excellent examples of biological control of plant pathogens are available in India through fungi and are successfully implemented. Among the fungi, *Trichoderma* has been extensively used to manage a variety of plant pathogens. Several strains of *Trichoderma* have been commercially exploited and are available for use by the farmers. *T. harzianum* and *T. viridae* applied as seed dressing, effectively reduces mortality of ground nut seedlings due to stem rot and collar rot (Nagaraju and Urs 1998). Several sps of *Fusarium* viz. *F. semitectum*, *F. chlamydosporum* and non pathogenic *F. oxysporum* have been reported as bio control agents against various phyto pathogenic fungi (Gill and Chahal 1988; Rao and Thakur 1988). Hyperparasitic fungi that attack other fungi are known for several plant pathogens including some rusts, powdery and downy mildews. The antagonistic potential of fungi is achieved either by the production of chitinase or other cell wall degrading enzymes, production of antifungal chemical or competition for nutrients between pathogens and antagonist. Parasitism and hyper parasitism are very common in fungal isolates. Similarly mycotoxin producing fungi is greatly influenced by other co existing fungi competing for nutrients and space. Antagonistic effect of different fungi on aflatoxin production was also reported in different food commodities by earlier workers (Dubey and Patel, 2001; Mehan, 1992; Singh *et al.*, 1989; Singh and Faull, 1982; Stack and Pettit, 1985).

3.4.3 Through Rhizobacteria: Over the past one century research has repeatedly demonstrated that out of diverse microorganism rhizobacteria can act as natural antagonists of various pathogenic microbes (Cook, 2000; Saxena *et al.*, 2000). Isolates of *Pseudomonas*, *Bacillus*, *Agrobacterium*, *Erwinia*, *Enterobacter* and *Serratia* have been isolated from different rhizosphere and were reported to have an important role in biological control of plant pathogens (Cook and Baker 1983). The biocontrol bacteria provide plant protection through induction of host plant defence mechanism, elimination of plant signals that trigger pathogen development or competition for nutrients or by production of antagonists compounds such as antibiotics, siderophore, cyanide and hydrolytic enzymes. The major success in the field of biological control through bacteria was the control of crown gall caused by *Agrobacterium tumefaciens* using *Azatobactor* strains 84 (Kerr, 1972). Since then several bacterial antagonists were proved for successful controlling of many post harvested diseases (Pathak, 1997). Among the bacterial antagonists sps of *Pseudomonas* are known to be highly potential. Fluorescent *Pseudomonas* has been reported to be antagonistic against many soil born and foliar pathogens. Podile and Prakash (1996) have studied in detail the mode of action of *Bacillus subtilis* AF strain, effective against *A. niger*. The bacterial cell adhered to the fungal mycelium, multiplied *in situ*, colonized and lysed mycelial surface. Extracellular chitinilytic enzymes of microorganism have a potential to suppress the activities of the pathogens by degrading the chitin in their cell walls and thus protected the plants from disease (Krishna *et al.,* 2003) Bacterial isolates viz. *Pseudomonas fluorescens*-2 and *Bacillus* isolates are found effective in reducing the population of *A. flavus*. These isolates also showed plant growth promoting efficacy (Weller 1988). Seed treatment with biocontrol agents is the easiest way of introducing sufficient quantity of these antagonists on the seed and protects against seed borne and soil borne diseases, via a variety of mechanisms.

4. Conclusion

Good quality of whole grain is less likely to suffer mould attack than poor quality damaged grain. Grain that shows signs of infestation or damage should be separated out before harvest and either discarded, reserved for animal feed or, depending on the type and extent of damage, kept for immediate consumption. The grain should be cleaned well and any damaged grain removed. The grain should be well dried. Good drying is essential. Damp grain will become mouldy. If the crop is to be dried in the field, it should not be heaped directly on the ground in order to prevent moisture being taken up. Similarly, when grain is dried at the homestead it must be kept off the soil. Throshed grain can be dried on polythene sheeting, grass matting or sacks placed on the ground in the sun. Unthreshed crops, (cobs, heads and pods) can be dried in the same way although it is better to dry them on platforms or frames or in cribs. Solar dryers can be used for grain if these are available in the village. The crop should be carefully handled once it has dried to ensure that grain remains intact. This will restrict problems from secondary insect pests and moulds. Crop quality can be improved through good handling, storage and pest management. However, the costs of these interventions must be assessed against their potential benefits. Once the crop is in store its future quality will depend on how well it is managed. If the store is in poor repair and allows easy access to rodent insect pests and moulds to the grain which might deteriorate quickly. Out of several management practices biological control is ultimate mechanism against management of mould and mycotoxin from food grains and offers potential strategies in future.

5. Future prospects

The survival of mycotoxigenic fungi in nature and their natural contamination in food commodities are very common. Out of several pre and post harvest management practices, discovery of antagonist microbes such as bacteria, fungi and other has major role to manage the aflatoxin contamination but some limitation are also here like:-

1. Identification and selection of effective strain of natural antagonists.

2. To understand the biology, ecology, physiology, genetic behavior of bio control agents.
3. To identify most efficient host genotype-symbiont strains of antagonists.
4. To develop mass culture technique for field application of antagonists.
5. To demonstrate and assess the full benefits of bio control agents under field conditions.
6. To educate the public for effective utilization of bio control agent.
7. Seed bacterization and seed coating with antagonists micro organism to reduce the incidence of *A. flavus* and *A. parasiticus* in soil for check the aflatoxin production.
8. Identification and evaluation of aflatoxin inhibitor gene for reducing and checking the aflatoxin production
9. Making transgenic plants which have potential to resistance against *A. flavus and A. parasiticus.*
10. Identification of detoxifying gene for aflatoxin control and making gene library.

Fortunately, the expansion of integrated pest management strategy has encouraged management approach with integrated multiple control practice, one of which may be the biological control. Research workers should pay due attention on the aflatoxin management using potential biocontrol agent in a concerned ecosystem.

References

Abramson, D. (1998) In mycotoxins in Agriculture and food safety Marcel Dekker Inc., New York, pp-255-278.

Allameh, A., Abyanch M.R., Shams M.R.,.Rezaee M.B and.Kaimand KJ (2002).. *Mycopathologia*, 154:79-84.

Basappa, S.C. (1983) Physical methods of detoxification of aflatoxin contaminated materials. In Mycotoxins in food and feed (Eds.K.S.Bilgrami, T.Prasad and K.K.Sinha) Allied Press, Bhagalpur pp. 251-273.

Bennet, J.W. (1987) *Mycopathologia*. 100: 3-5.

Bilgrami, K.S.and Chaudhary (1990) *Indian Phytopath*.43: 38-42.

Chauhan, R. K.S (2004).J.Bot. Soc. 83 (4): 11-21

Chourasia, H.K. and Sinha, R.K. (1994) *J.Food Sci, Technol*. 31: 362.

Cook, R. J. (2000) *Annu.Rev.Phytopathol*. 38: 95-116.

Cook, R. T. and H. F. Baker (1983) The nature and practice of Biological control of plant pathogens. A.P.S. St. Paul, M.N. 539 pp.

Cotty P.J. (1990) Plant disease 74:233-235.

Detroy, R.W. and Hesseltine, C.W. (1969) *Can. J. Biochem*. 48: 830-832

Dubey, S.C. and B.Patel (2001) *Indian Phytopath.* 54(2): 206-209.

Gill, K.S. and Chahal, S.S. (1988) *Plant.Dis.Res.* 3: 64-65.

Goldflat L. A. (1971) *J American oil Chem Soc* 48: 605-610.

Hesseltine, C.W., Shotwell O.L., Kwolek W.F., Lillehoj E.B.,. Jackson W.L and R.J. Bothast (1976). Mycologia 68 (2) : 341 – 353.

Jung, I.C., Hong K.W. and Kang K.J. (2000). *J.Microbiol. Biotechnol.*, 10(2):154-160.

Kerr, A. (1972) *J. Appl. Bacteriol.* 35: 493-497.

Kessel, G. J. T., Haas, B. H. De., Werf, Woke Van Der and Jurgen, Kohl. (2002).. *Mycological Research.* 106(6): 716-728.

Krishna Kishore,G.,Pandey,S.,Naryana,R.J. and Podile,A.R. (2003) ISOR National Seminar: Stress management in oilseeds. Jan 28-30, 2003, pp.19-20.

Kumar S and Roy AK. 2001. *J Indian Bot Soc.* 80: 149-151.

Mandal, N. L., Singh, A. N., Roy S. and Roy A. K. (2007). *J. Indian Bot Soc* 86 (3&4): 60-63

Mann, R. and H. Rehm (1976). *European. J. Appl. Microbiol.* 2: 297-306.

Mayer, P., Färber and Geisen R.. (2003). *Appl. and Environ. Microbiol.,* 69: 1154-1158.

Mitsuo, N., M. Satoshi, S.Kazuo, F.Kengi, N.Taichiro and K.Nabuhika (1990) *Appl. Environ. Microbiol.* 56: 1465-1470.

Nagaraju, P. and Urs, S.D. (1998) Current Research University of Agricultural Science, Bangalore. 27: 137-139.

Nepote M. C., Epintelli and Saubois A. (1997). Archivos, Latinoawericanos de Nutricion 47 : 262 – 264.

Ngoko, Z., Marasas W.F.O., Rheeder J.P., Shephard G.S., Wingfield M.J. and cardwell K.F. (2001). Phytoparasitica 29 : 352 – 360.

Pathak, V.N. (1997) *Indian Phytopath.* 50: 161-185.

Podile, A.R. and Prakash, A.P. (1996) *Canadian Journal of Microbiology*, 42: 533-538.

Rao, V. P. and Thakur, R.P. (1988) *Indian Phytopath.* 41: 567-574.

Reddy CS, Hayes WA (2001). Food borne toxicants. In Hayes, W.A. Principles and methods of toxicology, Taylor and Francis Philadelphia, U.S.A. pp 508 – 514.

Robertson, J.A.,.Teunisson D.J and.Bourdeaux G.J (1970). *J.Agric.Food Chem.* 18: 1090-1091.

Romina VB and Miriam GE. 2006. *Plant management Science.* 62(3): 242-251.

Rosic, J. L., M. Skrinjar and S. Markor (1991) *Mycopathologia.* 113: 117-119.

Roy AK and Chourasia HK. 1990. *J Gen Appl Microbial.* 36: 59-62.

Roy, A. K. (1989) *J.Indian Bot.Soc.* 68: 149-153.

Roy. A.K., Sinha K. K. and Chourasia H. K. (1988) *Applied and Environmental microbiology*, 54(3): 842-843.

Saxena, A. K. Pal, K.K. and Tilak, K.V.B.R.(2000) In: *Biocontrol Potential and its Exploitation in Sustainable Agriculture*, Kluwer Academic/Plenum Publisher, New York,pp.25-37.

Sharma, P.D. and Paul, P.K. (1998). In: *New Trends in Microbial Ecology*, (Eds. Bharat Rai and M.S.Dkhar), Dept. of Botany, NEHU, Shillong and ISCNR, Dept.of Botany,BHU,Varanasi,pp.272-286.

Singh, J. and Faull, J.L. (1982) Antagonism and biological control. In: *Biocontrol of Plant disease*, Vol.II (eds. K.G.Mukerji and K.L.Garg). CRC Press, Inc. 40: 167-176.

Singh, P, Ahmad S.K.and Sita Bhagat (1989). *J.Indian bot.Soc.* 68(1989): 169-171.

Sinha, K.K., Roy M.K.and Batsa R. (2003). *JNKVV Res. J.*, 37(1):72-77.

Stack, J.P. and Pettit, R.E, (1985) *Proc. Am. Peanut Res. Educ. Soc.*, 16, 45.

Sweta Roy, Mandal N. L., Sinha K. K. and. Roy A. K (2010) *J. Mycol Pl Pathol.* 40 (2): 224-226.

Weller, D.M. (1988) *Annual Rev. Phytopathol.*, 26: 379-407.

Windham, G.L. and W.P. Williams (1998). Plant Disease, 82:281-284

PART-B:

RESEARCH PAPERS

Chapter - 1

Molecular characterization and Biological Control of *Sclerotium rolfsii* Causing Stem rot of Groundnut (*Arachis hypogaea* L.)

N.P. Eswara Reddy and T.M. Hemalatha

Department of Plant Pathology, S.V. Agricutural College,
Tirupati-517502, Andhra Pradesh

ABSTRACT

Twenty isolates of *Sclerotium rolfsii*, the incitant of stem rot of groundnut collected from Chittoor and Kadapa districts of Rayalaseema region were assessed for morphological, Pathological and Genetic variability. Majority of the isolates were fast growing and varied in sclerotial production and size. The pathogenicity tests of these isolates showed that isolates Csr9, Ksr19 and Ksr20 were highly virulent. Genetic variability was analyzed using RAPD and ITS-PCR. All the isolates showed amplification of 650-700 bp by ITS-PCR. A total of 221 reproducible polymorphic RAPD bands ranging from 0.1-2.5 kb were produced. Dendrogram generated using UPGMA results showed the formation of two main clusters based on the RAPD profiles. The *Trichoderma* isolates GRSF10 and GRSF15 and the fungicidal compatible bacterial antagonists viz. GSE1, GSE5, GRE29, GRS15, GRS21, GRS23, GGS6 and GGS8 were found to be effective in completely inhibiting the growth of *S. roflsii*.

Key words: *Sclerotium rolfsii,* Biological Control, Groundnut, Molecular characterization

Introduction

Groundnut (*Arachis hypogaea* L.) is a major legume and important oil seed crop in India, covering nearly half of the area under oil seed production. In Groundnut, many diseases occur at different growth stages, among these stem rot caused by *Sclerotium rolfsii* Sacc. is one of the important disease with typical symptoms which include yellowing and wilting of branches, presence of mustard seed like sclerotia at the site of infection (Asghari and Mayee, 1991).

To understand the present plant disease situations and for effective management of diseases, it is essential to study the variability in plant pathogens. The information on variability among the groundnut isolates of *S. rolfsii* is limited. The cultural and morphological variability among the pathogen population in different regions is highly influenced by environment. Hence, the significant variations need to be defined which is possible by utilizing the nucleic acid based marker studies.

Management of soil borne disease by chemical means is difficult and has already proved to be harmful to the environment. Use of native potential biocontrol agents is hence found to be an alternate approach, being effective, eco-friendly and biodegradable. Hence, this paper mainly focused on molecular characterization and biological control of *Sclerotium rolfsii* causing stem rot of Groundnut (*Arachis hypogaea* L.) and their application in present scenario.

Materials and Methods

Morphological and Cultural variability

Roving survey was conducted in Chittoor and Kadapa districts of Rayalaseema region of Andhra Pradesh for collection of groundnut plants infected with *S. rolfsii*. The pathogen was isolated from the groundnut plants showing typical symptoms of stem rot disease by tissue segment method (Rangaswami and Mahadevan, 1999) on Potato Dextrose Agar (PDA). Axenic cultures of the pathogen were obtained by single hyphal tip method and maintained on PDA throughout the present Investigation. The pathogen was identified as *S. rolfsii* based on its mycelial and sclerotial characteristics through standard mycological keys (Barnett and Hunter, 1972)

A study was conducted to document the variation among the isolates of *S. rolfsii* in Morphological and cultural characteristics like colony growth, growth rate, number, size, colour and arrangement of Sclerotia on PDA medium.

Pathogenic variability

Plastic pots of 15 cm diameter each containing 3 kgs of soil were used for raising the host plants. Groundnut seedlings (Var. TCGS-888) of 10 days old from directly sown seeds were used for inoculation. Five plants were maintained for each pot. The pathogen was artificially inoculated with dried Sclerotia germinated for 36-48 hr on PDA at 28°C and 1 cm disc of germinated sclerotia along with the actively growing mycelium was used as the inoculum. Observations were recorded in terms of seedling mortality 15 days after inoculation (DA1) and the experiment was repeated thrice.

Percent Disease Incidence (PDI) was computed by using the following formula:

$$PDI = \frac{\text{Number of diseased plants}}{\text{Total number of plants}} x100$$

Isolation and Identification of fungicidal tolerant potential Biocontrol Agents

Seed and Root endophytes were isolated by the method followed by Kishore *et al.* (2005). The potential biocontrol agents were identified by dual culture technique. The fungicide tolerant potential biocontrol agents were recovered/identified by poisoned food technique for fungi and spectrophotometric method for bacteria.

Molecular characterization of *S. rolfsii* isolates

The molecular variability among the isolates of *S. rolfsii* was studied by using Random Amplified Polymorphic DNA (RAPD), ITS region of rDNA and internal transcribed spacer - Restriction fragment length polymorphism (ITS-RFLP).

DNA extraction

The total genomic DNA of *S. rolfsii* isolates was extracted from vegetative mycelium using the procedure given by Murray *et al.*, (1980). The quality and quantity of DNA was observed on 1%agarose gel.

RAPD profiles through Polymerized Chain Reaction (PCR)

Five different random primers belong to Operon "A" series *viz.*, OPA-01, 12, 17, 18 and 20 (Operon technologies Inc.) were used for generating polymorphism among the isolates under the study. The experiment was repeated thrice and results were reproducible. The primer sequences used in RAPD technique are given below.

Sequence of Oligonucleotide primers used in RAPD

Sl. No.	Operon	Sequence
1.	OPA-01	5' – CAG GCC CTT C-3'
2.	OPA-12	5' – TCG GCG ATA G-3'
3.	OPA-17	5' – GAC CGC TTG T-3'
4.	OPA-18	5' – AGG TGA CCG T-3'
5.	OPA-20	5' – GTT GCG ATC C-3'

PCR amplifications were carried out in 0.2 ml eppendorf tubes with 25 μl reaction mixture which consists of 2.5 μl of 10x Taq buffer, 2.5 μl of 25 mM $MgCl_2$, 2.5 μl of primer (1 picomolar / μl), 0.5 μl of 100 mM dNTP mix, 0.2 μl of *Taq* polymerase enzyme (conc. 1 U μl^{-1}) and 16.8 μl of sterile PCR water (Genei, Bangalore) and 2 μl (4-50 ng) of DNA sample. Amplification was carried out by 5 minutes of initial denaturation at 94°C followed by 40 cycles of denaturation of 94°C for 1 minute; annealing at 37°C for 1 minute; extension at 72°C for 2 minutes with final extension at 72°C for 5 minutes. Amplified PCR products were subjected to 1.0 per cent agarose gel electrophoresis with 1.0 x TBE as running buffer.

Scoring and Data Analysis

The RAPD data was used to calculate Jaccard's similarity coefficient for each pair wise comparison. The coefficients were calculated In Silico using the following formula:

Similarity coefficient = a/n

Where,

a = Number of matching band for each pair of comparisons

n = Total number of bands observed in two samples.

The similarity coefficients were subjected to Unweighted Pair-Group Method on Arithmetic average (UPGMA) cluster analysis to group the isolate based on their overall similarities. Statistical Package for Social Sciences (SPSS) was used for the cluster analysis and sub-sequent dendrogram preparation.

PCR amplification of ITS region

PCR amplification of Internal Transcribed Spacers (ITS) region of rDNA was performed using universal primers ITS-1 (5' – TCC GTA GGT GGA CCT GCG G-3') as forward primer and ITS-4 (5' – TCC TCC GCT TAT TGA TAT GC-3') as reverse primer (White *et al.*, 1990) in eppendorf PCR master cycler. Amplification was carried out in 0.2 ml eppendorf tubes with 25 μl reaction mixture containing 2.5 μl of 10x Taq buffer, 2.5 μl of 25 mm $MgCl_2$, 2.0 μl of ITS primer (0.6 picomolar/μl), 2.0 μl of ITS-4 primer (0.6 picomolar/μl), 0.5 μl of 100 mM dNTP mix, 0.125 μl of Taq polymerase (0.5 u/μl) and 14.37 μl of sterile PCR water (Genei, Bangalore) and 3 μl (40-50 ng) of DNA sample.

The PCR amplification was carried out by 35 cycles, of which denaturation at 94°C for 1 minute, annealing at 56°C for 1 minute and extension at 72°C for 1.5 minutes with initial denaturation at 94°C for 4 minutes before cycling and final extension at 72°C for 6 minutes after cycling.

ITS – RFLP of *S. rolfsii* isolates

Amplified PCR products of ITS region for the isolates under the study were digested with three different restriction enzymes *viz., Alu*I, *Hinf*I and *Mse*I. Digestion was carried out with 20 μl reaction mixture which contains 7 μl of ITS-PCR product, 1 μl of enzyme (10 U/μl), 2 μl of 10X enzyme buffer and 10 μl of sterile PCR water. The digestion was carried out overnight at 37°C for *Alu*I, *Hinf*I and 65°C for *Mse*I in a water bath. Restricted products were analyzed on 3.0 per cent agarose gel.

Results and Discussion

Morphological and cultural variability

The pathogen was isolated from the collar region of infected plants using tissue segment method and maintained on PDA for further studies. Based on Morphological characters the pathogen was identified as *S. rolfsii* sacc. Similar reports regarding to Morphcological characters were observed by Harinath Naidu (2000) and Rajani (2004).

There was a considerable variation among the isolates in their total growth and growth rate (Table 1). The maximum radial growth was recorded in the isolates CSR2, CSR4, KSR12, KSR14, KSR15, KSR16, KSR17 and KSR18 (90.0mm) followed by CSr9, CSr10 (83.00mm). The least growth was observed in KSr 19 (62.67).

Regarding growth rate, Maximum growth rate was observed in the isolate KSr15 (34.50 mm/day) followed by KSr12 (33.85 mm / day) while least growth rate was observed in KSr19 (15.55 mm/day). Sharma *et al.* (2002) reported that the growth rate of different isolates of *S. rolfsii* varies from 25.31 mm/day.

The maximum radial growth of 90.00 mm of *S. rolfsii* was recorded in the isolates CSr2, CSr3, CSr4, KSr12, KSr14, KSr15, KSr16, KSr17 and KSr18 (Akram *et al.*, 2007 in Chickpea; Kulkarni *et al.*, 2008 in potato) while minimum was observed in KSr19 (62.67 mm).

Variation with regard to sclerotial size and number was observed among the different isolates. The maximum size of the sclerotia was observed in the isolate KSr 16 (2.2 mm), followed by CSr 8, KSr 14, KSr 15, KSr 16 and KSr 18 (2.0 mm). The minimum size of the sclerotia was observed in the isolate CSr 4 (0.90 mm).

Number of sclerotia per plate was highest recorded in KSr 18 (571/plate) followed by CSr 4 (385/plate). The least number was observed in CSr 6 (30/plate).

Variation in sclerotial colour was also observed. Most of the isolates produced reddish brown sclerotia, some produced light brown and dark brown sclerotia. These results also in agreement with that of Akram *et al.*

(2007), Sharma *et al.* (2002), Kulkarni *et al.* (2008), Rajalakshmi (2002) and Arunasri (2001).

Pathogenic variability

The seedling mortality was recorded 15 days after pathogen inoculation. The pathogenicity reactions ranged from 20-100 percent among the isolates (Table 1).

Table 1. Percent disease incidence (PDI) of stem rot in field and the cultural, morphological and *in vitro* pathogenic variability among the isolates of *Sclerotium rolfsii*.

Isolate	Village	PDI in field(%)	Time taken for disease expression (DAI)	PDI in glasshouse (%)	Total growth (mm)	Growth rate (mm/day)	Colony type	Time taken for sclerotia production (days)	Colour of sclerotia	No. of sclerotia (per plate)	Size of sclerotia (mm)
CSr 1	Kottapalli	11.15	15	40.00 (39.23)	73.00	22.63	Fluffy	16	RB	81	1.42
CSr 2	Kalikiri	6.70	8	53.33 (46.92)	90.00	30.35	Compact	13	RB	138	1.80
CSr 3	Teegalakona	15.00	10	40.00 (39.23)	90.00	31.25	Compact	15	RB	347	0.92
CSr 4	Taragonda	20.00	11	53.33 (46.92)	90.00	31.20	Compact	9	DB	385	0.90
CSr 5	Yellampalli	2.00	10	46.67 (43.08)	68.50	17.62	Fluffy	18	LB	53	1.50
CSr 6	Kothapalli	5.00	7	60.00 (50.77)	66.00	25.96	Fluffy	15	LB	30	1.50
CSr 7	K. Kuruvapalli	30.00	16	40.00 (39.23)	78.00	23.83	Fluffy	15	DB	130	1.31
CSr 8	Mudivedu	10.00	12	53.33 (46.92)	69.50	23.63	Fluffy	18	RB	58	2.00
CSr 9	Nallacheruvupalli	1.00	8	100.00 (90.00)	83.00	20.80	Fluffy	13	RB	185	1.63
CSr 10	Peddapalem	3.00	8	26.66 (31.00)	83.00	22.88	Compact	11	DB	281	1.13
KSr 11	Battalapalli	5.60	12	93.33 (81.14)	72.17	23.33	Fluffy	18	RB	64	0.95
KSr 12	Gollapalli	5.80	10	86.67 (72.29)	90.00	33.85	Compact	16	RB	242	1.12
KSr 13	Nagalaguttavaripalli	8.30	16	20.00 (26.57)	82.00	24.13	Fluffy	17	LB	257	1.48
KSr 14	Kattuluru	85.00	8	60.00 (50.77)	90.00	32.60	Compact	15	LB	168	2.00
KSr 15	Gangammapeta	5.70	12	20.00 (26.57)	90.00	34.50	Fluffy	13	RB	56	2.00
KSr 16	Animela	30.00	15	26.67 (31.00)	90.00	29.45	Fluffy	14	RB	81	2.20
KSr 17	V.N. Palli	13.30	15	53.33 (46.92)	90.00	30.35	Compact	14	DB	151	1.50
KSr 18	Vuruturu	10.00	7	80.00 (63.43)	90.00	30.25	Compact	16	LB	571	2.00
KSr 19	Sunkesula	10.00	10	100.00 (90.00)	62.67	15.54	Compact	12	LB	190	1.00
KSr 20	Telagandlapalli	5.00	6	100.00 (90.00)	67.17	24.16	Fluffy	17	LB	122	1.62
CD (0.05)				9.912	0.71						
S.Em ±				3.4683	0.25						

The isolates CSr4, KSr19 and KSr20 exhibited maximum disease incidence (100%) followed by KSr11 (93.33%) the lowest disease incidence (20%) was recorded in case of KSr13 and KSr15 isolates. The isolate KSr20 has taken only 6 days for seedling mortality whereas the isolate CSr7 has taken 16 days.

Shukla and Pandey (2008) showed the pathogenic variability of 10 *S. rolfsii* isolates recovered from diseased parts of *Parthenium*. The isolate Par#02 showed maximum disease incidence (80%) whereas the isolate Par#10 showed lowest (30%).

Isolation and identification of fungicide tolerant potential biocontrol agents

Biological control using *Trichoderma* and plant growth promoting rhizobacteria (PGPR) especially antagonistic *Pseudomonas* sp. appears to be a potential management tool for reducing the severity of several soil borne plant pathogens.

In the present investigation, a total of 122 antagonistic microflora were isolated from different habitats viz. Rhizosphere (GRS), Geocarposphere (GGS), Root (GRE) and Seed endophytes (GSE) and evaluated against highly virulent isolate of *S. rolfsii* (KSr20) by dual culture technique. *Trichoderma* isolates, GRSF10 and GRSF15 were found to be effective in inhibiting the growth of *S. rolfsii* to an extent of 75.5% and 74.4% respectively (Table 2).

Table 2. *In vitro* evaluation of the efficacy of potential microflora against *S. rolfsii*

S.No	Potential isolate	Percent inhibition over control (%)
1	GRSF10	75.55
2	GRSF15	74.40
3	GSE-1	100.0
4	GSE-5	100.0
5	GRE-2	100.0
6	GRS-15	100.0
7	GRS-21	100.0
8	GRS-23	100.0
9	GGS-6	100.0
10	GGS-8	100.0
11	GSE-37	99.60
12	GSE-36	98.43
13	GSE-33	96.97

Among the antagonistic bacteria isolated against *S. rolfsii* 8 isolates GSE-1, GSE-5, GRE-2, GRS 15, GRS21, GRS23, GGS 6 and GGS8 were effective in completely inhibiting the growth of the pathogen followed by GSE37, GSE36 and GSE33. Similarly, Kishore *et al.* (2005) evaluated antagonistic bacteria from six habitats of groundnut for their broad-spectrum antifungal activity against 8 fungal pathogens including *S. rolfsii*. *Pseudomonas* sp. GRS 175, *P. aeruginosa* GPS 21, GSE 18, GSE 19 and GSE 30 were highly antagonistic.

The potential bacterial isolates were tested for their compatibility with four commonly used fungicides viz., Carbendazim, Copper oxychloride, Thiophenate methyl and mancozeb.

The fungicide compatibility results indicated that most of the potential biocontrol agents were compatible with mancozeb followed by Copper oxychloride. (Durai, 2004). (Data not shown). Field evaluation of the efficacy of fungicidal tolerant potential bio control agent as apart of Integrated Disease Management (IDM) is being in progress.

Molecular characterization of *S. rolfsii* isolates

Characterization of *S. rolfsii* isolates by RAPD

The genetic variability among the eight isolates of *S. rolfsii* in the study was analyzed by RAPD (Almeida *et al.*, 2001). Five random primers *viz.,* OPA-01, OPA-12, OPA-17, OPA-18 and OPA-20 generated reproducible polymorphism among the isolates of *S. rolfsii* (Fig. 1). Amplified products with all the primers have shown polymorphic and distinguishable banding pattern indicating the genetic diversity among all isolates of *S. rolfsii.* A total of 221 reproducible and scorable polymorphic bands ranging approximately as low as 100 bp to as high as 2500 bp were generated with 5 primers among the eight isolates.

The primer OPA-01 amplified two unique fragments of approximately 100 bp and 250 bp in case of KSr 19 isolate. Primer OPA-12 amplified specific bands of 1400 bp and 1500 bp in case of KSr 15 isolate. Moreover, 600 bp fragment was absent in case of CSr 1 isolate compared to other isolates. OPA-17 primer yielded a specific fragment approximately 200 bp in case of isolates CSr 9 and KSr 17 and absent in all other isolates. The 700 and 650 bp bands were specifically amplified in case of isolate KSr 20 showing the polymorphism among the isolates. The primer OPA-18 amplified a unique band of approximately 500 bp in case of CSr 1, 575 bp in case of CSr 9 and 750 bp in case of KSr 20. Primer OPA-20 amplified a fragment of 1000 bp and 1050 bp both in case of KSr 15 and KSr 17 isolates. Relationships among the isolates were evaluated by cluster analysis of data based on similarity matrix. The dendrogram (Fig. 2) was generated using UPGMA package based on Ward's Squared Equalidean Distance method. Based on the results obtained, all the 8 isolates were grouped into two main cultures.

Cluster I contains four isolates *viz.,* CSr 9, KSr 17, CSr 1 and KSr 19 of which the first two and last two form two separate sub-clusters *viz.,* cluster Ia (CSr 9and KSr 17); Ib (CSr 1 and KSr 19). Cluster II contains four isolates *viz.,* CSr 2, CSr 10, KSr 15 and KSr 20.

Jaccard's similarity co-efficient among the 8 isolates were calculated to establish their genetic relationships. The similarity index values ranged from 0.00 to 100 per cent indicating the presence of a high range of variability at nucleic acid level among the eight *S. rolfsii* isolates.

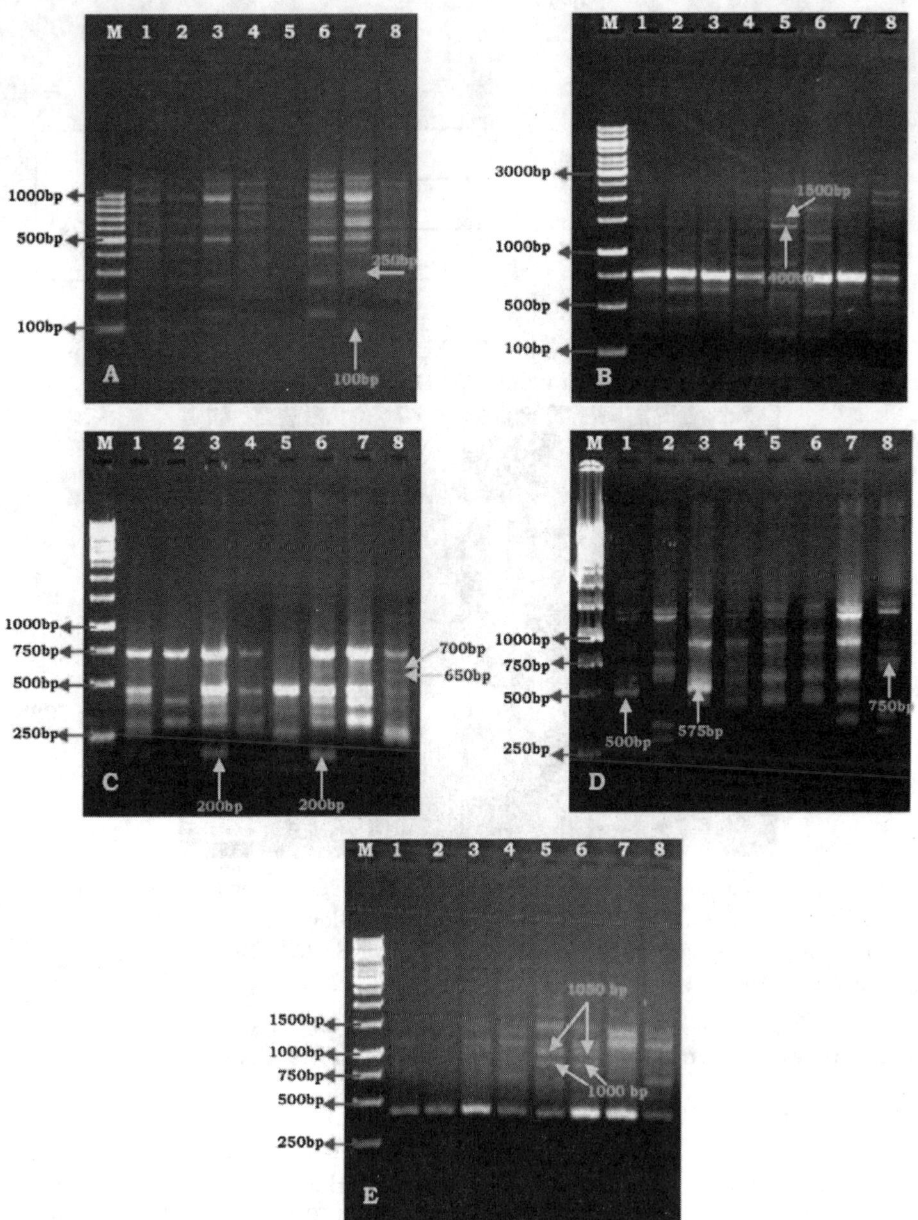

Fig. 1. Random Amplified Polymorphic DNA (RAPD) patterns of *S. rolfsii* isoaltes with random primers.
M: 1kb DNA ladder; A: OPA-01; B: OPA-12; C: OPA-17; D: OPA-18; E: OPA-20; Lane 1-8 are *S. rolfsii* isolates, CSr 1, CSr 2, CSr 9, CSr 10, KSr 15, KSr 17, KSr 19 and KSr 20 respectively

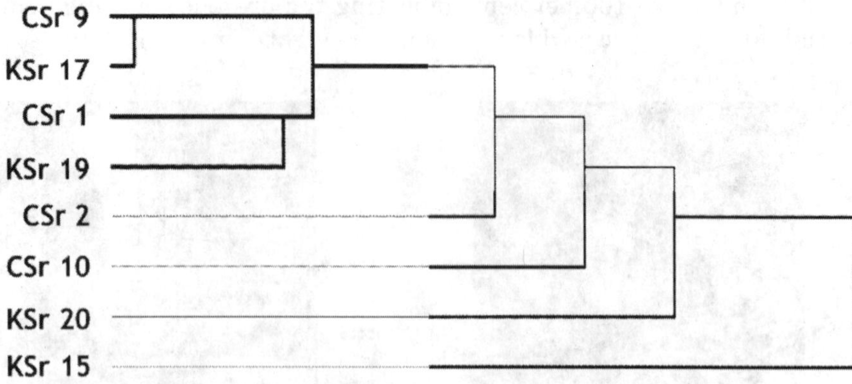

Fig. 2. Dendrogram generated using UPGMA analysis showing polymorphism between isolates of *S. rolfsii* using RAPD markers

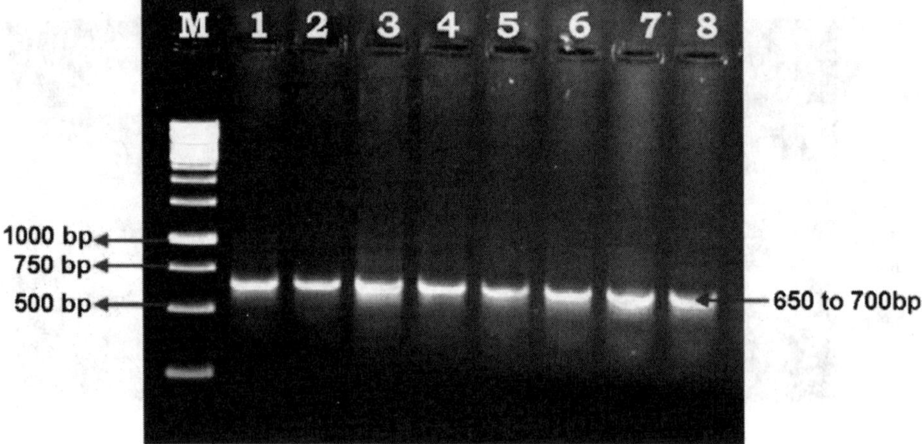

Fig. 3. Amplification product of Internal Transcribed Spacer (ITS) with ITS-1 an ITS-4 ribosomal DNA primers M=1 kb DNA ladder Lanes 1to 8 represents *S. rolfsii* isolates, CSr 1, CSr 2, CSr 9, CSr 10, KSr 15, KSr 17, KSr 19 and KSr 20 respectively.

Characterization of *S. rolfsii* isolates by ITS-PCR

Genus Specific primers *viz.,* ITS-1 and ITS-4 were used for PCR amplification of ITS region of rDNA cluster which includes ITS-1, 5.8S and ITS-2 regions of all the eight isolates. Both the primers produced amplified product size of approximately 650-700 bp in all the isolates under the study (Fig. 3). These results confirmed that all the isolates belong to genus *Sclerotium*. Adandonon *et al.* (2005) also studied genetic variation among *S. rolfsii* isolates of cowpea from Benin and South Africa by using mycelial compatibility and ITS rDNA sequence data and obtained approximately 700 bp which is specific for *S. rolfsii.*

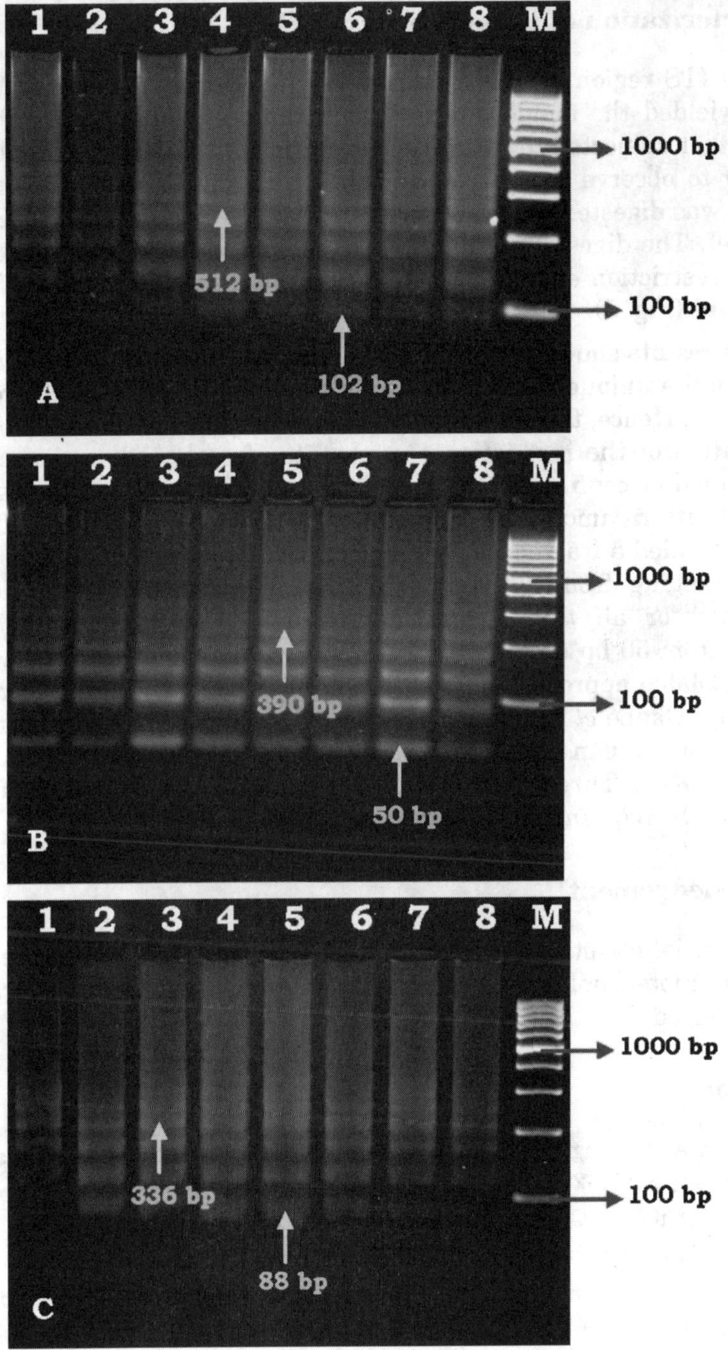

Fig. 4. Restriction enzyme digestion of polymerase chain reaction-amplified ribosomal DNA of internal transcribed spacer (ITS) region from *S. rolfsii*; A: A*lu* I; B: H*inf* I; C: M*se* I; M: 100 bp DNA ladder; Lane 1-8 are *S. rolfsii* isolates, CSr 1, CSr 2, CSr 9, CSr 10, KSr 15, KSr 17, KSr 19 and KSr 20 respectively.

Characterization of isolates by ITS-RFLP

The ITS region of rDNA amplified with specific primers (ITS-1 and ITS-4) yielded the single band of approximately 650-700 bp. The band was further subjected to restriction analysis with different endonucleases in order to observe the polymorphism among the *S. rolfsii* isolates. The product was digested with 3 different restriction enzymes *viz., Alu*I, *Hinf*I and *Mse*I. The digested products were separated in 3.0 per cent agarose gel and restriction endonuclease digestion pattern was compared among all isolates (Fig. 4).

The results showed that the restriction enzymes *viz., Alu*I, *Hinf*I and *Mse*I has the unique restriction site in the amplified ITS-rDNA region of all isolates. Hence, the fragments of different size were obtained based on the position of the restriction enzyme site of each endonuclease. These results further confirmed that there was no polymorphism in restriction banding pattern among the isolates for the above enzymes used. The *Alu*I enzyme yielded 5 fragments for all isolates approximately with molecular weight ranging from 102 bp to 512 bp. The *Hinf*I enzyme yielded 5 fragments for all the isolates approximately with molecular weight ranging from 50 bp to 390 bp. The *Mse*I enzyme yielded 4 fragments for all the isolates approximately with molecular weight ranging from 88 bp to 336 bp. Okabe *et al.* (1998) divided 67 isolates of the southern blight fungus from Japan into five groups based on ITS-RFLP analysis of nuclear rDNA. Three groups were reidentified as *S. rolfsii* and two resembled *S. delphinii* in RFLP patterns

Acknowledgement

Financial assistance received by N.P. Eswara Reddy from the Department of Biotechnology, Government of India, New Delhi is gratefully acknowledged.

References

Adandonon A, Aveling T A S, Merwe N A and Sanders G 2005 Australian Plant Pathology 34: 19-25.

Akram A, Iqbal S M, Qureshi R A and Rauf C A 2007 Mycopath 5: 23-28

Ansari M M and Agnihotri S K 2000 Indian Phytopathology 53: 65-67.

Arunasri P 2003 Management of collar rot disease of crossandra (*Crossandra infundibuliformis* (L.) Nees.) incited by *Sclerotium rolfsii* Sacc. M.Sc.(Ag.), Thesis submitted to Acharya N.G. Ranga Agricultural University, Hyderabad, Andhra Pradesh.

Asghari M A and Mayee C D 1991 Indian Phytopathology 44: 328-332.

Barnett H L and Barry B Hunter 1972 Illustrated genera of imperfect fungi. Burgess Publishing Company, Minnesota.

Durai M 2004 Management of charcoal rot of sesame (*Sesamum indium* L.); incited by *Macrophomina phaseolina* (Tassi) Goid, M.Sc.(Ag.), Thesis submitted to Acharya N.G. Ranga Agricultural University, Hyderabad, Andhra Pradesh

Harinath Naidu 2000 Indian Phytopathology 53: 496-497.

Kishore G K, Pande S and Padile A R 2005 Letters in Applied Microbiology 40: 260-268.

Kishore GK, Pande S, and Podile AR 2005 Canadian Journal of Microbiology 51:123-132.

Kulkarni S, Kulkarni VR, Gorawar MM and Hedge YR 2008 Technological Bulletin-6.

Okabe I, Arakawa M and Matsumoto N 2001 Mycoscience 42:107-113.

Rajalakshmi 2002 Studies on variability among the isolates of *Sclerotium rolfsii* M.Sc.(Ag.), Thesis submitted to Acharya N.G. Ranga Agricultural University, Hyderabad, Andhra Pradesh.

Rajani T 2004 Integrated management of *Scleortium rolfsii* Sacc. the incitant of stem rot of groundnut. M.Sc.(Ag.), Thesis submitted to Acharya N.G.Ranga Agricultural University, Hyderabad, Andhra Pradesh.

Rangaswami G and Mahadevan A 1999 Diseases of crop plants in India. Prentice Hall of India Pvt. Ltd., NewDelhi pp: 6079.

Sarma BK, S ingh UP and Singh KP 2002 Mycologia 94(6): 1051-1058.

Shokes FM, Rhogalski K, Gorbet DW, Brenneman TB and Berger DA 1996 Peanut Science 23: 124-128.

Shukla R and Pandey AK 2007 Journal of Mycology and Plant Pathology 37(3): 514-518.

Chapter - 2

Growth and biomass studies in ectomycorrhizal fungus *Pisolithus albus* (Cook & Massee) Priest

Radhika, K.P., Mohan, V. and Rodrigues, B.F.*

Department of Botany, Goa University, Goa 403 206, India.
Corresponding author, E-mail: *felinov@gmail.com

ABSTRACT

The pure mycelial culture of ectomycorrhizal basidiomycetes *Pisolithus albus* (Cook & Massee) Priest was obtained from context tissue culture of freshly harvested fruit bodies associated with the host *Acacia auriculiformis* A. Cunn. The culture was grown using different compositions of modified MNM medium aimed to select the best composition favouring rapid growth and higher mycelial biomass production. Pelletized colonies were successfully obtained on the selected modified medium under liquid shaken conditions. The pellets thus obtained were tested for viability and vigour on solid MMN medium successfully. Somatic compatibility tests by pairing the wild type strain and the pellet-derived strains established positive identity. Experiments carried out to obtain higher biomass yield on same medium indicated maximum yield after 21 days.

Keywords: Basidiomycetes, Ectomycorrhizal fungus, Pisolithusalbus, *Acacia auriculiformis*.

Introduction

Ectomycorrizal (ECM) symbiosis occurs in about 3000 plant species by around 5000 species of fungi, mainly Basidiomycetes and Ascomycetes (Smith and Read, 1997; Estradrodorres and Santiago Martinez, 2003). Forest species depend on ECM symbiosis and diverse species of tree including pines, would not grow without it. Inoculation in nurseries with selected ectomycorrhizal fungi is important in improving the establishment of plantations in the field (Marx *et al.*, 1994).

Pisolithus, a gastromycete genus, has a worldwide distribution and forms an ectomycorrhizal association with a wide range of woody plants (Marx, 1977) including members of Pinaceae, Myrtaceae, Fagaceae,

Mimosaceae, Dipterocarpaceae and Cisteaceae. Isolates of *Pisolithus* are some of the most commonly used in forestry, with growth stimulation reported for several tree species including eucalyptus, pines and acacias (Garbaya *et al.*, 1988, Marx *et al.*, 1977, Duponnois and Ba, 1999). *Pisolithus* have proved particularly effective in improving plant growth on drier soils with high soil temperature (Momoh and Gbadegesign, 1980, Marx *et al.*, 1985) and isolates of this fungus also alleviated Aluminium sensitivity in *Pinus* (Cumming and Weinstein, 1990). In India, *Pisolithus* population are wide spread, although they are found mostly in Eucalyptus plantations and were identified as *Pisolithus albus* (Cooke and Mass) M.J. Priest, nom.prov. based on the ITS sequence analysis (Singla *et al.*, 2004). *Pisolithus albus* (Pis-pea, lith-stone, alb-white) or white dye ball fungus, belongs to a group of fungi commonly referred to as Earth Balls. It fruits late in summer and the fruit body is initially firm, and the inner tissue has a tightly packed cellular structure of varying shades from yellow to orange.

Major interest in this species has been due to its success in inoculating forest plants. Traditional sources of ectomycorrhizal inoculum used in nurseries have been spores and mycelial structures. In mycelial propagation, growth of the etomycorrhizal fungus strain is affected by culture conditions, principally pH, temperature and composition of the media. Within a single species, isolates from different origin exhibit morphological variations and cause differential effects in plant growth when used as inoculants (Castellano and Molina, 1989). In this paper, we have attempted to study the variations in mycelial growth before mass production of inoculum, to characterize the strains in vitro on Solid MNM media at different pH and to evaluate the growth of the strains in liquid media under constant agitation for higher biomass yield.

Materials and Methods

Biological material

The fruit bodies of *Pisolithus albus* (Cooke and Mass) M.J. Priest, nom.prov. were collected from *Acacia auriculiformis* from Sanquelim in North Goa.

Culture under different condition

Solid medium

Pisolithus albus isolated from the fruit bodies were cultured on solid Melin Norkrans modified following Marx and Kenney (1984), (MNM) in different media compositions at different pH. The wild strains grown in different media composition were checked for growth colony diameter (GD) after 7, 14, 21 and 28 days of growth in growth incubator at 28°C in dark. Growth in diameter was calculated as the average (cm) of three measurements that have taken through the centre of inoculated fragment. Three replicates were maintained for each composition.

Liquid medium

14-day-old mycelia grown in solid MNM medium, which showed rapid growth, were used for experiment in liquid MNM medium for pelletization under constant agitation in rotary shaker at 62 rpm at 28°C. Mycelial strains were taken from the periphery of the medium. The pellets obtained were then tested for viability on solid MNM medium at pH 5.7.

Somatic compatibility

Somatic compatibility test was carried out in solid MNM medium using the 7-day-old mycelial strain (solid medium) and pellet (liquid medium) culture in growth incubator at 24°C in dark. The pH was maintained at 5.7 in both solid medium and liquid medium. Mycelia strain (6 mm) grown on the periphery of the nutrient medium and pellets (0.9 cm diameter) formed in liquid medium under constant agitation in rotary shaker at 62 rpm at 28°C were transferred to solid MNM medium and were placed at 5 cm apart. Three replicates were maintained.

Biomass

Biomass studies were carried out in MNM liquid medium using the 7-day-old wild strain of *Pisolithus albus* mycelia grown in solid medium at pH 5.7. Biomass experiments were carried out in rotary shaker at 62 rpm at 28^{0}C Growth diameter (GD), colour, texture, fresh weight and dry weight of the pellet were recorded after every 7 days. Three replicates were maintained for each culture.

Results and Discussion

Culture in solid media

The present study showed the MNM medium composition with pH 5.7-favoured rapid growth and higher mycelial biomass compared to other medium with pH 5.4 and 5.5. During the initial stage of growth, medium with pH 5.4 recorded higher growth diameter between 7-14 days of growth but a significant increase in growth diameter was recorded in medium with pH 5.7 between 21-28 days of growth **(Fig 1, Plate 1a).** The isolated strain was identified using the mycelial morphology, which showed the presence of yellowish brown velvety mycelia with concentric zones. Numerous clamp connections occurred throughout the mycelium of 21-day-old culture. Similar observations were made earlier in *P.*

tinctorius (Hile and Hennen, 1969). In the culture, the strain produced a yellowish brown pigment, which diffused through the medium causing a change from their original colour. A similar morphology was described by Marx *et al.* (1984) and Garcia Rodriguez *et al.* (2006) in *Pisolithus tinctorius.*

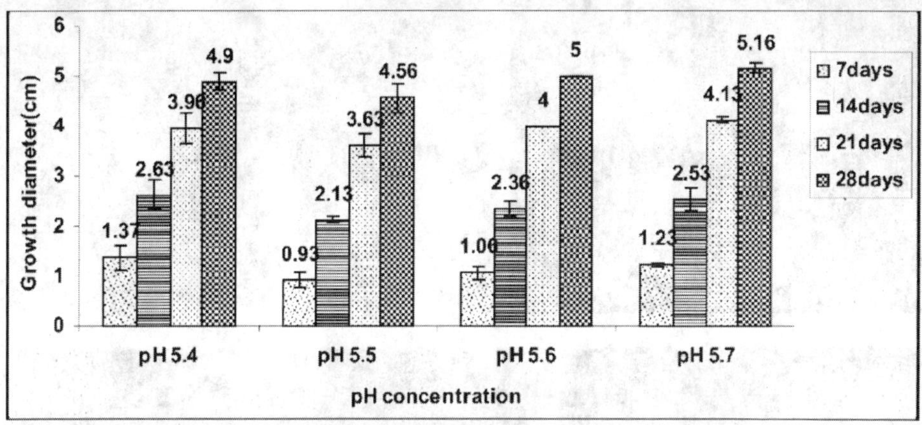

Fig. 1. Mycelial growth diameter at different pH values after 7, 14, 21 and 28 days of growth in MNM medium.

Culture in Liquid media

Pellatization of the mycelial strains were obtained after 21 days of growth in liquid medium at pH under constant agitation in rotary shaker at 62 rpm at 28^{0}C. Six to eight pellets were obtained in each flask **(Plate 1b).** Similar observations were reported in *P.tinctorius* ((Hile and Hennen, 1969). Abundant dark brown mycelium to golden yellowish brown with velvety appearance, abundant hyphae tangled in to ball or concentric spherules growing in all directions. As in the solid medium, in liquid medium the strains consistently produced a yellowish brown pigment, which diffused in the medium causing a change from its original colour.

Somatic Compatibility

Somatic compatibility between the mycelia and pellet were obtained after 14 days of growth in solid MNM medium **(Plate 1d).** Free hyphal intermingling at the interface of the colonies occurred after 14 days of growth in all the replicate plates. In *P. arhizus* somatic compatibility was shown to occur between heterokaryon- homokaryon pairings (Kope 1992). Pellets in the medium showed overnight growth compared to the mycelial strain. Somatic recombination has been shown to be a supplement to the sexual cycle with the generation of novel genotypes and the maintenance of genetic variability in the population (Raper, 1966), but mycelial interactions expressed in pure culture may be differently expressed when influenced by the symbiotic partners. Pellets of *P. tinctorius* showed faster growth and than mycelial strains in solid MNM media after 7, 14, 21 & 28 days of growth when inoculated separately (**Fig. 2, Plate 1c**).

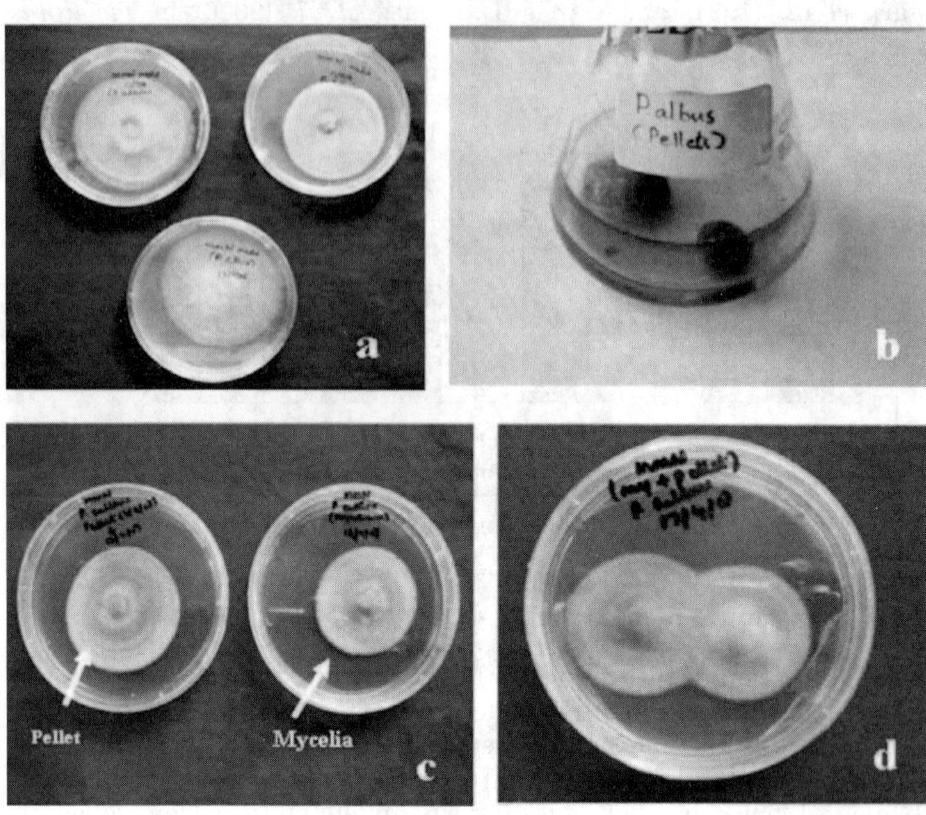

Plate 1. (a) *P. albus* raised from mycelium after 28 days of growth (b) Pellet formation in liquid medium after 21 days of growth (c) Growth of pellet and mycelia after 28 days (d) Somatic compatibility of mycelium and pellet in *P. albus* after 28 days of growth.

Biomass Studies

The biomass study indicated that higher biomass yield was obtained in 21-day-old pellets formed in liquid MNM medium. Dry weight and fresh weight of the pellets were also recorded. There was a gradual increase in the number of pellets formed from 7-21 days of growth period, but after this period there was a decrease in the number of pellets formed. The fresh weight of the total pellets increased from 7 (95.77 ± 3.55) to 21 (502 ± 89.2) days of growth and then gradually decreased. Colour of the pellets was yellowish throughout the growth study and the texture was smooth. Total number of pellets formed was found to be maximum during the 21st day (19.6 ± 9.6) in the liquid MNM medium. This shows that there is a gradual increase in the growth the formation of pellets in liquid MNM medium under constant agitation but after 21st days of growth there is a lag phase in the growth of the pellet and after 28 days the growth ceases. There was also significant change in the colour of the medium from light yellowish from 7th day to dark yellowish 28th day after culturing in liquid medium. The pH of the media has been observed from 5.7 to 3.9. This reduction in pH has been attributed to the excessive uptake of cations such as ammonium (Giltrap and Lewis, 1981) or to the release of organic acids by the ectomycorrhizic mycelia (Tsantrios *et al.*, 1991). Decrease in pH of solid and liquid media have been reported in association with *P. tinctorius* culture (Hung and Trappe, 1983; Vazquez-Garcia *et al.*, 2002). This decrease has been recorded in the culture of *Laccaria laccata* (Kuez, 1996), *Coenoccocum geophilum*, *Suillus bovines*, *Rhizopogon reoselus* (Glitrap and Lewis, 1981), *Lactarius deliciosus* and *Trichoderma striatum.*

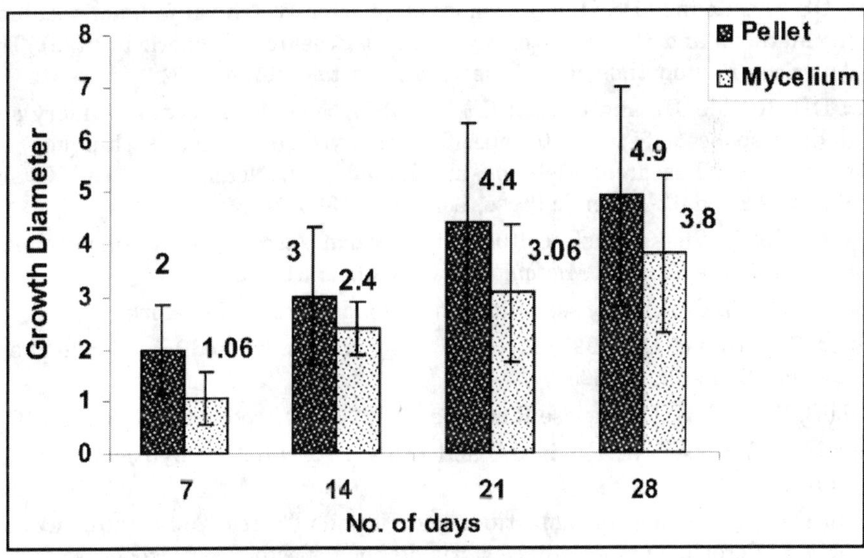

Fig. 2. Diamctor of pellet and mycelia (in cm) after 7, 14, 21 and 28 days of growth on solid MNM media.

References

Cumming JR and Weinstein LH (1990) Plant Soil 125:7-18.

Castellano MA and Molina R (1989) Mycorrhizae. In: The Container Tree Nursery Manual. Vol. 8. Landis TD, Tinus RW, McDonald SE, Barnett BP (eds.) United States Department of Agriculture, Washington. pp.101-167.

Duponnois R and Ba AA (1999 For Ecol Manage 24: 151-157.

Estrada-Torres A and Santiago-Martinez YG (eds) (2003) Avances en el Estudio de al Ectomicorriza en el Estado de Tlaxcala, Mexico. Universidad Autonoma de Tlaxcala. Ixtacuixtla, Tlaxcala, Mexico, pp. 76.

Garcia Rodriguez JL, Perez-Moreno J, Aldrete A, Cetino-Alcala VM and Vaquera-Huerta H (2006) Agrociencia 665-676.

Garbaye J, Delwaulle JC and Diagana D (1988) For Ecol Manage 24: 151-157.

Glitrap NJ and Trappe JM1983. New Phytol 87:669-675.

Hile N and Hennen JF 1969. Mycologia 61:195-198.

Hung L and Trappe JM (1983) Mycologia 75: 234-241.

Marx DH, Cordell CE, Kenney DS, Mexal JG, Artman JD, Riffle JW and Molina RJ (1984) Forest Sci. 30 (Monograph 25):1-101.

Marx DH (1997) Can J Microbiol 23:217-223.

Marx DH, Bryan WC and Cordell CE (1977) For Sci 16:363-373.

Momoh ZO and Gbadegesign RA (1980) Field performance of *Pisolithus tinctorius* as a mycorrhizal fungus of pines in Nigeria. In: Mikola P (ed.) Tropical Mycorrhizal Research. Oxford: Clarendon. 72-79.

Marx DH, Hedin A and Toe SFP (1985) For Ecol and Manage 13: 1-25.

Marx DH and Kenney DS (1984) Production of ectomycorrhizal fungus inoculum. In: Methods and Principles of Mycorrhizal Research. Schenck NC (ed). The American Phytopathological Society. Minnesota, USA. pp. 131-146.

Marx DH, Ruehle JL, and Cordell CE (1994) Methods for studying nursery and field response of trees to specific ectomycorrhiza. In: Techniques for Mycorrhizal Research: Methods in Microbiology. Norris JR, Read D and Varma AK (eds). Academic Press, London. pp.383-411.

Kuez C (1996) Shake flask culture of *Laccaria laccata,* an ectomycorrhizal basidiomycete. Applied Microbiol.Biotechnol. 45:319-326.

Raper JR (1966) Genetics of sexuality fungi. Ronald Press. New York.

Smith SE and Read DJ (1997) Mycorrhizal symbiosis. (2nd. ed.) Academic Press. London, England. pp. 605.

Singla S, Reddy MS, Marmeisse R and Gay G (2004) Microbiol Res 159:203-210.

Tsantrizos Y, Kope HH, Fortin JA and Ogilvie KK (1991) Phytochemistry 30: 1113-1118.

Vazquez- Garcia A, Santiago-Martinez and Estrada Torres (2002) Influencia del pH en el crecimiento de quince cepas de hongos ectomicorrhizogenos. Anales del Instituto de Biologia, Universidad Nacional Autonoma de Mexico, Serie Botanica 73: 1-15.

Chapter - 3

Optimization studies of Manganese peroxidase by *Stereum ostrea*

***K. Praveen* [1], *B. Viswanath* [2], *H. Pallavi* [1], *G. Venkata Subba Reddy* [1], *and B. Rajasekhar Reddy*[1*]**

Department of Microbiology, Sri Krishnadevaraya University,
Anantapur -515 055,
2. Department of Environmental Science,
Global College of Engineering and Technology, Kadapa -516 165

ABSTRACT

The extracellular enzyme production of manganese peroxidase (MnP) was studied in *Stereum ostrea* (*S. ostrea*) isolated from wood-logs under shaking and inducing conditions. The organism was cultivated with different amino acids at 0.2% concentration, organic solvents at 5% and inducers at 0.2% concentration. Among the compounds tested in this study, leucine, methanol and lignosulfonic acid favored the maximum production of the lignolytic enzyme - manganese peroxidase.

Key words: white-rot fungi, manganese peroxidase, amino acids, inducers, organic solvents

Introduction

Lignin is an abundant natural aromatic polymer occurring in the woody tissue of higher plants. Due to its hydrophobicity and complex random structure lacking regular hydrolysable bonds, lignin is poorly biodegraded by microorganisms. The best degraders of lignin are white-rot fungi that produce laccase (Lac) (EC. 1.10.3.2), and peroxidases- lignin peroxidase (LiP) (1.11.1.4) and manganese peroxidase (MnP) (EC. 1.11.1.13) [Kirk, 1987]. These enzymes are involved in the initial attack of the lignin polymer [Hammel, 1993]. Certain organisms of white-rot fungi – *Phanerochaete chrysosporium, Trametes versicolor, Pleurotus ostreatus, Bjerkandera adusta, Irpex lacteus, Lentinus tigrinus, Phanerochaete sordida, Phlebia radiata, Pleurotus eryngii,* and *Poliporus cialatus* have drawn more attention for lignolytic enzymes [Praveen, 2011]. *Stereum ostrea* known as false turkey-tail fungus is a white-rot fungus belonging to family Stereaceae and most common in hard wood forests and is relatively characterized for lignolytic enzymes. *S. ostrea* has been recently shown to secrete laccase into extracellular medium [Viswanath, 2008]. There was confusion prevailing about ability of *S. ostrea* to secrete MnP in extracellular medium in view of contradictory results of two studies [Buswell, 1987; Vander Woude, 1993]. According to our recent study [Praveen, 2011], *S. ostrea* has been recognized to produce all three main lignolytic enzymes - MnP, LiP and Lac. MnP is one of the most frequently encountered peroxidase among white-rot fungi [Praveen, 2011], and a glycosylated heme-protein [Kirk, 1987]. The catalytic cycle is similar to that of other peroxidases. The principal role of Mn^{2+} dependent peroxidase is related to oxidation of Mn^{2+} to Mn^{3+}, which then binds to an appropriate ligand, diffuses from the enzyme, and in turn oxidizes phenolic substrates [de Jong, 1992]. In the present study, attempts have been made to understand culture conditions for growth of *S. ostrea* in the liquid medium in the presence of additives – different amino acids, inducers and organic solvents with an intention of improving MnP titres.

Materials and Methods

Microorganism and growth medium

S. ostrea was kindly supplied by Prof. M.A. Singaracharya, Department of Microbiology, Kakatiya University, Andhra Pradesh, India and was isolated from wood logs. The reference culture, *Phanerochaete chrysosporium* was obtained from IMTECH, India. The fungal cultures were maintained at 30°C on 2% Koroljova-Skorobogat'ko medium [Koroljova-Skorobogat'Ko, 1998] because of good growth [Praveen, 2011]. The maintenance medium was prepared according to Koroljova *et al.* [Koroljova-Skorobogat'Ko, 1998] containing the following composition (g/l): 3.0 peptone, 10.0 glucose, 0.6 KH_2PO_4, 0.001 $ZnSO_4$, 0.4 K_2HPO_4, 0.0005 $FeSO_4$, 0.05 $MnSO_4$, 0.5 $MgSO_4$, 20.0 agar (pH 6.0).

Culture conditions of fungal cultures

The composition of production medium was the same as used in the maintenance medium except agar. Production of MnP enzyme by *S. ostrea and P. chrysosporium* was carried out on this medium containing selected substances/additives. Additives – amino acids (L - glutamic acid, L-phenyl alanine and L- leucine) at the final concentration of 0.2% (w/v), organic solvents (methanol, ethanol and n-butanol) at the final concentration of 5% (v/v), inducers (ferulic acid or lignosulfonic acid) at the final concentration of 0.2% (w/v) and Tween-80 at the final concentration of 0.2% (v/v) were selected for their influence on production of MnP. Sterile production medium was dispensed in sterile 250-ml Erlenmeyer flasks at the rate of 50 ml per flask. Flasks were divided into different groups. All flasks in a group received only one additive/ substance at the specified concentration. Flasks in one group without receipt of any additive served as control. Fungal mycelial suspension was prepared by adding 2ml of sterile distilled water to the freshly (7 day) grown slants of two fungal cultures. The homogenized mycelial suspe-nsion was used to inoculate 50 ml of medium in 250 ml Erlenmeyer flasks at the rate of one milliliter of homogenous mycelial suspension per each flask in all groups. The flasks were incubated at 30ºC in a temperature controlled gyratory shaker (ORBITEK- Chennai, India) at 150 rpm for a period of 10 days. All the flasks with growing cultures of *S. ostrea* and *P. chrysosporium* were withdrawn on at regular intervals for processing with analytical methods.

Analytical methods

Fungal biomass estimation: The cultures - *S. ostrea* and *P. chrysosporium* in the flasks at regular intervals were aseptically filtered through pre-weighed Whatman No-1 filter paper to separate mycelial mat and culture filtrate. The filter paper along with mycelial mat was dried at 70°C in an oven until constant weight and this weight was recorded. Difference between the weight of the filter paper bearing mycelial mat and weight of only filter paper represented biomass of fungal mat. Fungal growth was expressed in terms of g/flask. Filtrate collected was used as a source of MnP.

Protein estimation: An aliquot of culture filtrates of *S. ostrea* and *P. chrysosporium* with appropriate dilution was used for estimation of soluble protein content according to Lowry *et al.* [1951]. Bovine serum albumin was used as protein standard.

Assay of manganese peroxidase enzyme: One gram/liter phenol red, 2mM H_2O_2, 2mM $MnSO_4$, 0.5% BSA, 250mM sodium lactate (pH 4.5), were included in 20mM sodium succinate buffer (pH 4.5). Enzyme activity was assayed by incubating the above assay mixture with aliquot of the culture filtrate at room temperature of 30°C. Appropriate controls were simultaneously set up. The changes in absorbance of the reaction mixture containing phenol red was monitored at 610 nm (ε = 22,000 M^{-1} cm^{-1}) for five minutes of incubation in Spectra scan UV-2600 Double beam UV-Visible spectrophotometer (Chemito 2600). The enzyme activities were calculated using an extinction coeffic-ient of 22,000 M^{-1} cm^{-1} and expre-ssed in International Units (IU), where one unit of enzyme activity is defined as the amount of enzyme that oxidizes one μ mole of substrate in one minute [Kuwahar, 1984].

Results and Discussion

Since media composition and growth conditions are important factors affecting the production of extracellular lignolytic enzymes and consequently lignin degradation [Niku-Paavola, 1990], the influence of substances/additives/inducers on manganese-dependent peroxidase production by fungal cultures in the culture medium was examined in the present study. Though cultures were monitored for growth, secretion of extracellular protein and MnP at 2-day interval for period of 10 days in all experiments, only results of 8th-day incubation are presented in the present study because of peak production at this sampling time.

Influence of amino acids on MnP production by fungal cultures

Cultivation of *S. ostrea* and *P. chrysosporium* on Koroljova medium supplemented with different amino acids (L-glutamic acid, L-phenyl alanine and L-leucine) did not improve the growth of both cultures as reflected by biomass in comparison to biomass of the same cultures on unsupplemented medium (control) (Fig. 1).

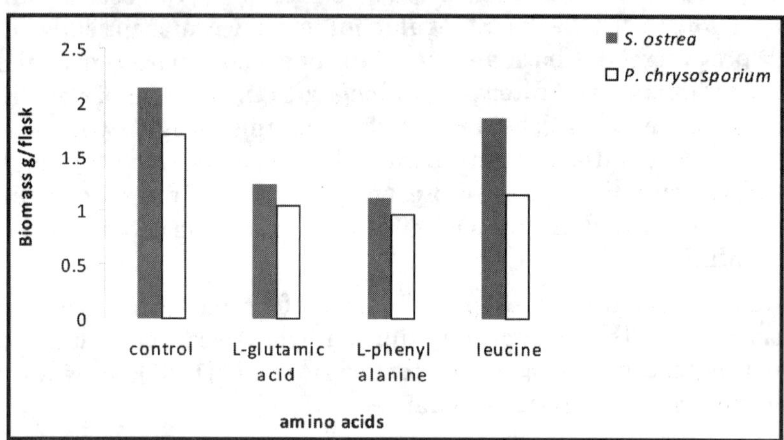

Fig. 1 Influence of amino acids on biomass of fungal cultures on 8th day of incubation

Growth of *S. ostrea* was relatively higher than that of *P. chrysosporium* on medium with/without supplementation. The medium used for growth of organisms in the present study already contained peptone as in the case of control and yields of biomass of the cultures from control flasks were higher. It appears that supplementation of amino acids to the medium did not provide any benefit to growth of cultures as peptone in Koroljova medium was sufficient enough to meet nitrogen demand of the cultures for growth requirements of the cultures. Similarly, growth of other lignolytic cultures in on glucose and yeast extract medium with peptone was fast [Kamitsuji, 2004]. Inclusion of aromatic amino acids in Koroljova medium did not have any positive influence on the growth of the same cultures used in the present study [Viswanath, 2007].

Growth of both organisms on the medium without supplementation resulted in secretion of protein into the medium in considerable proportions (Fig. 2). Among amino acids supplemented in this study, only L-leucine caused more release of extracellular protein of 715 and 600 µg/ml into the medium by *S. ostrea* and *P. chrysosporium* on 8th day of incubation, respectively (Fig. 2).

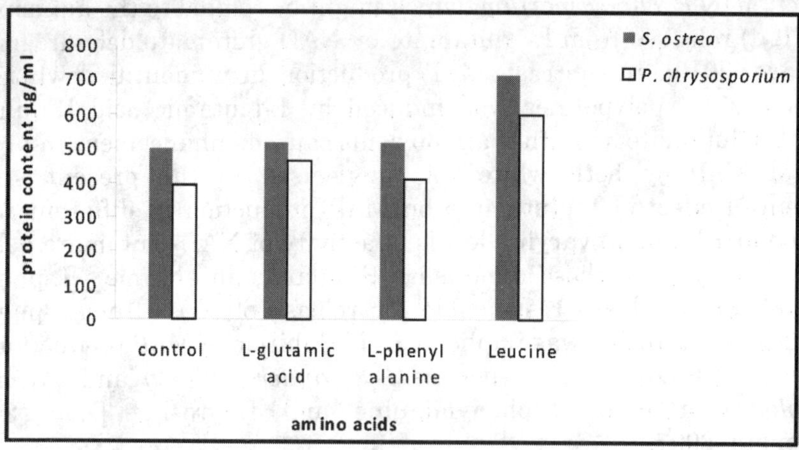

Fig. 2. Influence of amino acids on extracellular protein of fungal cultures on 8th day of incubation

Unlike on parameters – growth and secretion of extracellular protein, supplementation of three amino acids to the medium had a positive effect on MnP production by both fungal cultures (Fig. 3). Yields of MnP obtained upon growth of the fungal cultures – *S. ostrea and P. chrysosporium* on the medium supplemented with amino acids were higher than yields of MnP by the same cultures on unsupplemented medium. The fungal cultures – *S. ostrea and P. chrysosporium* gave yields of MnP to the extent of 2.8 and 1.8 U/ml on the supplemented medium as against 2.0 and 0.94 U/ml on unsupplemented medium on 8th day of incubation, respectively (Fig. 3).

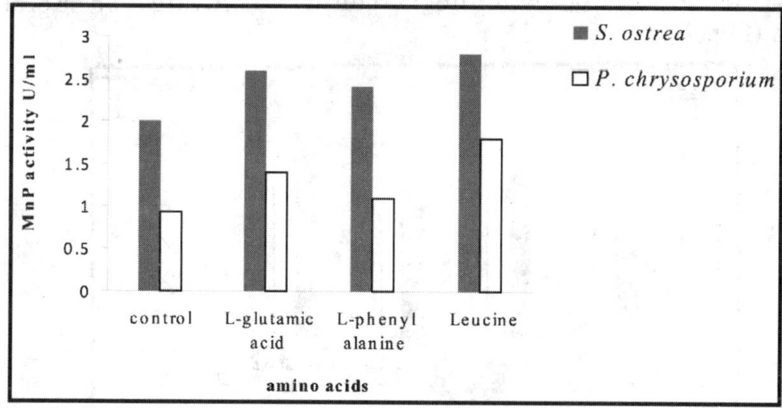

Fig. 3. Influence of amino acids on manganese peroxidase activity of fungal cultures on 8th day of incubation

Inclusion of amino acids, in particular, L-glumate had varied effect on the ligninolytic enzyme production by white rot fungi belonging to polyporales [Akamatsu, 1996; Fenn, 1981]. L-glutamic acid was the most effective amino acid regarding the suppression of the ligninolytic enzyme production by *P. chrysosporium*, which might be related to the ammonium ion (NH_4^+) released from L- glutamate by NAD-glutamate dehydrogenase [Fujihara, 2010]. In contrast, MnP production in unidentified white rot fungus L-25 of polyporales was induced by L-glutamic acid [Fujihara, 2010]. L-Glutamate and other amino acids caused enhancement in production of MnP by both white rot fungi used in the present study. Differential effect of L-glutamate on MnP production in different fungi could be attributed to varying levels of activity of NAD-glutamate dehydrogenase enzyme in those organisms. Similarly, an enzyme - L-phenylalanine ammonia-lyase responsible for release of NH_4^+ from L-phenylalanine and L-tyrosine was implicated in inhibition of MnP production in different white rot fungi - such as *Bjerkandera adusta* and *Trametes versicolor* exerted by L-phenylalanine and L-tyrosine [Xue, 2007; MacDonald, 2007].

Influence of inducers on MnP production by fungal cultures

In view of known beneficial effects of inducers and detergents on production of enzymes by certain organisms, inducers and detergents were tested in Koroljova medium to find out any improvement in MnP production by less characterized white rot fungus – *Stereum ostrea*. An enhancement in MnP production due to addition of inducers to the culture medium may facilitate purification and utilization of enzyme for application. Different inducers viz., Tween-80, ferulic acid and lignosulfonic acid at 0.2% in Koroljova medium were screened to find out their influence on biomass, extracellular protein content and MnP activity by *S. ostrea* and *P. chrysosporium*. Lignosulfonic acid exerted positive influence on growth of both fungal cultures cultivated on Koroljova medium (Fig. 4).

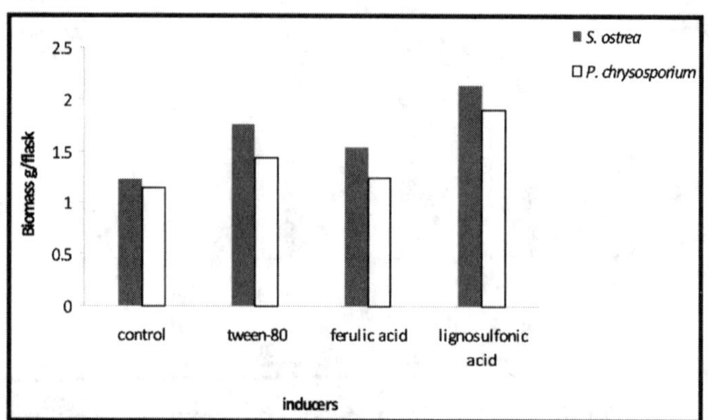

Fig. 4 Influence of inducers on biomass of fungal cultures on 8th day of incubation

Addition of lignosulfonic acid to the medium also caused maximum release of extracellular protein by *S. ostrea* and *P. chrysosporium* on 8th day of incubation (Fig. 5).

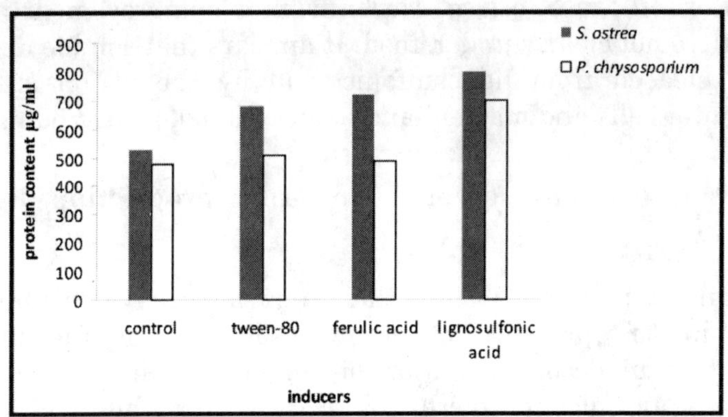

Fig. 5 Influence of inducers on extracellular protein of fungal cultures on 8th day of incubation

An induction effect on MnP production by lignosulfonic acid was more pronounced in *S. ostrea* with yields of 6.21 U/ml on 8th day of incubation in comparison to yields of 2.5 U/ml from medium without amendment (Fig. 6).

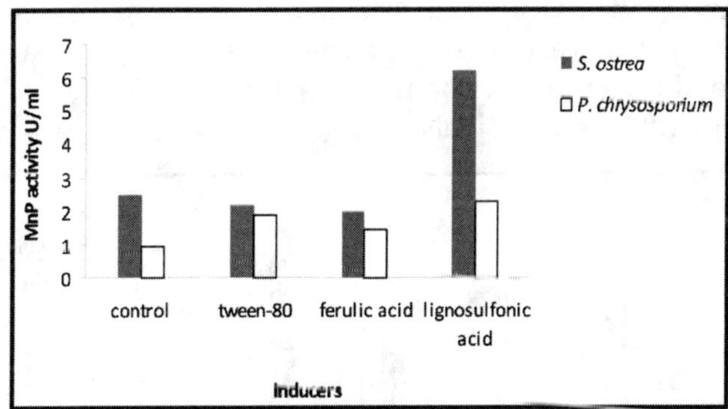

Fig. 6 Influence of inducers on manganese peroxidase activity of fungal cultures on 8th day of incubation

Lignolytic enzymes may either be inducible or constitutively expressed in lignolytic organisms [Yaver, 1996; Gianfreda, 1999]. These organisms have been classified based on enzyme production pattern. In the constitutive or non-inducible group, levels of production of enzymes was not influenced by inducers whereas in the inducible group, levels of enzyme activity were altered in response to inducers. Many compounds, that have been reported to improve lignolytic enzyme production in white-rot fungi included cellobiose [Lu, 1996], copper sulphate [Palmieri , 2000], gallic acid, cycloheximide, xylidine [Eggert, 1996], ferulic acid [Bollag,

1984] Mn^{2+} and Tween – 80 [Urek , 2007]. White-rot fungi were very diverse in their response to inducers on production of lignolytic enzymes. In the present study, lignosulfonic acid caused 2-fold increase in the production of MnP in *S. ostrea*. Mechanisms for improvement in enzyme production are not clearly understood. It appears that soluble degradation products, released from lignosulfonic acid by the action of lignolytic enzymes, enter cells and may enhance expression of genes coding MnP.

Influence of organic solvents on MnP production by fungal cultures

As lignolytic cultures and their enzymes have to function in situations in the presence of organic solvents for the purpose of bioremediation in decontamination of xenobiotics in polluted samples, impact of organic solvents (methanol, ethanol and n-butanol) on growth, capacity of secretion of extracellular protein and MnP production by *S. ostrea* and *P. chrysosporium* was assessed.Growth of *S. ostrea* and *P. chrysosporium* in Koroljova medium in the presence of organic solvents except methanol was comparable to biomass of the respective culture on the same medium without organic solvents on 8th day of incubation (Fig. 7). 2.30 g of biomass of *S. ostrea* from 50 ml of medium with methanol was obtained as against 1.75g of biomass of the same culture on the medium without organic solvents on the 8th day of incubation. Corresponding figures of biomass in respect of *P. chrysosporium* on the medium with methanol and the medium without any solvent on 8th day of incubation were 1.84g and 1.24g, respectively.

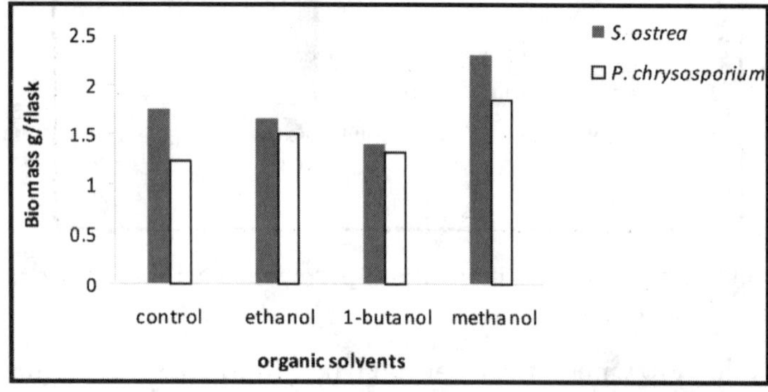

Fig. 7 Influence of organic solvents on biomass of fungal cultures on 8th day of incubation

Secretion of extracellular protein by both fungal cultures was not affected by the presence of organic solvents in the medium (Fig. 8). Release of protein into medium with and without organic solvents by both cultures occurred within a range of 500-740 μg/ml and 425-600 μg/ml (Fig. 8).

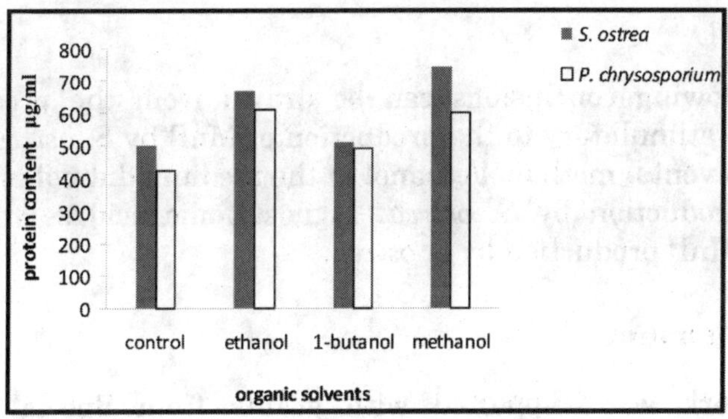

Fig. 8 Influence of organic solvents on extracellular protein of fungal cultures on 8th day of incubation

The presence of organic solvents in the medium was beneficial to MnP production by both fungal cultures. Titres of MnP by *S. ostrea* and *P. chrysosporium* on medium in the presence of organic solvents were higher than yields of MnP by the same cultures on the same medium without organic solvents on 8th day of incubation (Fig. 9).

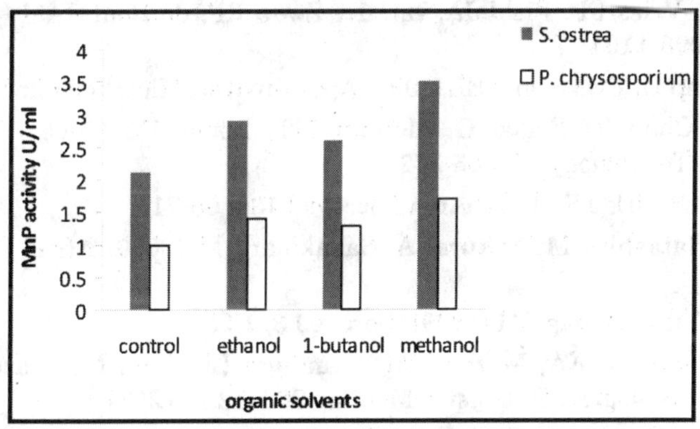

Fig. 9 Influence of organic solvents on manganese peroxidase activity of fungal cultures on 8th day of incubation

A system based on MnP with use of different water miscible organic solvents (acetone, methyl-ethyl-ketone, methanol and ethanol) for the degradation of polycyclic aromatic hydrocarbon (PAH) - anthracene was reported by [Eibes, 2005]. Comparing the maximum solubilisation of anthracene and the minimum loss of MnP activity, acetone was the best co-solvent for an acetone concentration of 36% (v/v). Other studies show the use of peroxidases for enzyme-catalyzed polymerization reactions due to the advantages over traditional polymerization, namely to improve the control of the polymer structure [Singh, 2002]. The results of the present study confirm safe use of organic solvents at concentration of 5% v/v without any toxicity for production of MnP by the fungal cultures.

Conclusions

The following conclusions can be drawn from the present study. Leucine was stimulatory to the production of MnP by *S. ostrea*. Inclusion of organic solvents- methanol/ethanol in the medium did not affect growth and MnP production by *S. ostrea*. Lignosulfonic acid served the best inducer for MnP production by *S. ostrea*.

Acknowledgement

This work was supported with grants from Special Assistance Program of UGC, New Delhi and fellowship to Mr. K. Praveen from BSR Scheme of UGC, New Delhi.

References

Akamatsu, Y, Shimada M (1996) FEMS Microbiol Lett 145: 83-86

Bollag JM, Leonowicz A (1984) Appl Environ Microbiol 48: 849-854

Buswell JA, Odier E (1987) Crit Rev Biotechnol. 6 1-60

de Jong E, de Vries FP, Field JA, van der Zwan RP, de Bont JAM (1992) Mycol Res 12: 1098-1104

Eggert C, Temp U, Eriksson KEL (1996) Appl Environ Microbiol 62: 1151-1158

Eibes G, Lú-Chau T, Feijoo G, Moreira MT, Lema JM (2005) Enzyme and Microbial Technology 37: 365-372

Fenn P, Choi S, Kirk TK (1981) Arch Microbiol 130: 66-71

Fujihara S, Hatashita M, Sakurai A, Sakakibara M (2010) Afr J Biotechnol 9: 725-731

Gianfreda L, Xu F, Bollag JM (1999) Biorem J 3: 1-25

Hammel KE, Jensen KA, Mozuch MD, Landucci LL, Tien M, Pease EA (1993) Journal of Biological Chemistry Medline 268: 1227-12281

Kamitsuji, H, Honda Y, Watanabe T, Kuwahara M (2004) Appl Microbiol Biotechnol 65:287-294

Kirk TK, Farrell RL (1987) Annu Rev Microbiol 41: 465-505

Koroljova-Skorobogat'Ko OV, Stepanova EV, Gavrilova VP, Morozovaa OV, Lubimova NV, Dzchafarova AN, Jaropolov AI, Makower A (1998) Biotechnol Appl Biochem.28: 47-54

Kuwahar M, Glenn JK, Morgan MA, Gold MH (1984) FEBS Lett 169: 247-250

Lowry O, Rosebrough NJ, Farr AL, Randall RJ (1951) *J Biol Chem* 193: 265-275

Lu SXF, Jones CL, Lonergan GT (1996) Correlation between fungal morphology and laccase expression under the influence of cellobiose induction. Proceedings of the 10th International Biotechnology symposium and 9th International Symposium on yeasts; Sydney, Australia, Poster session 1

MacDonald MJ, D'Cunha GB (2007) Biochem Cell Biol 85: 273-282

Niku-Paavola M.L, Karhunen E, Kantelinene A, Viikari L, Lundell T, Hatakka A (1990) J Biotechnol 13: 211-221

Palmieri G, Giradina P, Bianco C, Fontallella B, Sannina G (2000) Appl Environ Microbiol 66:920-924

Praveen K, Viswanath B, Usha KY, Pallavi H, Reddy GVS, Naveen M, Reddy BR (2011) lignolytic enzymes of a mushroom – *Stereum ostrea* isolated from wood logs. In Press

Singh A, Kaplan DL (2002) Journal of Polymers and the Environment 10: 85-91

Urek RO, Pazalioglu NK (2007) Braz Arch Biol Technol 50:913-920.

Vander Woude MW, Boominathan K, Reddy CA (1993) Arch Microbiol. 160: 1-4

Viswanath B, Chandra MS, Kumar KP, Reddy BR (2008) DBPBMB. 2: 19-25

Viswanath, B (2007) Production, purification and characterization of laccase from *Stereum ostrea*. Thesis submitted to Sri Krishnadevaraya University.

Xue Z, McCluskey M, Cantera K, Ben-Bassat A, Sariaslani FS, Huang L (2007) Enzyme Microb Technol 42: 58-64

Yaver, DS, Xu F, Golightly EJ, Brown KM, Brown SH, Rey MW, Schneider P, Halkier T, Mondorf K, Dalboge H (1996) Appl Envir Microbiol 62: 834-841.

Chapter - 4

Synergistic effects of arbuscular mycorrhizal fungi and plant growth promoting rhizobacteria on sorghum

B. Bhadraiah and K.V.B.R. Tilak

Department of Botany, Osmania University, Hyderabad – 500 007

ABSTRACT

Initial pot experiments on seed inoculation of sorghum (*Sorghum bicolor*) with *Pseudomonas fluorescens* demonstrated increased dry matter production, with the cultivar CSH 5 responding better than CSH 6. The response of sorghum to nitrogen was linear upto 80 kg N ha^{-1} under uninoculated treatments. No difference between uninoculated and inoculated treatment at 80 kg Nha^{-1} was noticed. Inoculated plants showed 17 fold higher pseudomonad population than the roots of uninoculated control plants.

Growth and P uptake of sorghum on a non-sterile, phosphorus-deficient soil were improved by soil inoculation with the arbuscular-mycorrhizal (AM) fungi (*Acaulospora*, *Gigaspora margarita* and *Glomus fasciculatum*) in combination with seed inoculation with *P. fluorescens*. The responses were significant with *P. fluorescens* + *G. margarita* and *P. fluorescens* + *G. fasciculatum*.

Keywords: Arbuscular mycorrhizal fungi, PGPR, synergestic effect, Sorghum

Introduction

Plant growth promoting rhizobacteria (PGPR) include a diverse group of free-living soil bacteria that can stimulate the growth of plants by one or more of different mechanisms (Kloepper *et al.* 1988, Glick and Pasternak, 2003, Dey *et al.* 2005, Tilak *et al.* 2010, Angaw *et al.* 2011). They are extensively studied for their potential applications in agriculture. Garbaye and Bowen (1989) demonstrated that some bacteria closely associated with the fungus directly played a positive role in ectomycorrhiza formation. They were called MHB (mycorrhization helper bacteria). Most MHB and PGPR belong to the same taxonomic groups; fluorescent pseudomonadas and sporulating bacilli. AM fungi and rhizosphere microorganisms interactions have been mainly observed on selected non-pathogenic microorganisms involved in nutrient transformations (Azcon and Barea 1975, Subbarao *et al.*, 1985, Tilak *et al.*, 2005, 2006). The objective of studying these AM fungi-rhizobacterial interactions have focused upon obtaining an additive effect on plant growth by microbial inoculants either singly or in a combined form.

The present study was conducted to (1) evaluate the response of sorghum to *Pseudomnas fluorescens* inoculation, alone and in combination with graded levels of fertilizer nitrogen in pots, and (2) to find out the combined effects of seed inoculation with *P. fluorescens* and soil inoculation with AM fungi on the growth and P uptake of sorghum.

Materials and Methods

Fluorescent pseudomonads were isolated from the roots of 18 cultivars of sorghum grown in semi-arid soils by using King 's medium (King *et al*. 2010).

Seeds of sorghum cultivars CSH 5 and CSH 6 were inoculated with a carrier-based (soil+FYM 1:1) culture of *P. fluorescens* containing efficient strains following the method outlined by Tilak *et al*. (2010). Uninoculated seeds were coated with sterilized (heat-killed) carrier-based culture.

Two pot culture experiments were conducted using 12 kg unsterilized soil of pH 7.3. In the first experiment, an interaction of two sorghum cultivars (CSH 5 and CSH 6) with 3 strains of *P. fluorescens*(P1, P12 and P18) singly and in combination was studied. A basal dose of 40 kg N ha^{-1} was applied to all the treatments. Phosphorus (P_2O_5) and potassium (K_2O) were applied at the rates of 40 kg ha^{-1}. In the second experiment , the combined effects of seed inoculation with *P. fluorescens* and soil inoculation with three AM fungi (Acaulospora sp., *Gigaspora margarita* and *Glomus fasciculatum*) were studied on sorghum cultivar CSH 5. Nitrogen was applied at the rate of 40 kg N ha^{-1} and potassium (K_2O) at 60 kg ha^{-1}. In both pot experiments, dry matter produced by 75-day-old plants (preflowering stage) was determined. In all experiments ,N was applied as urea, P_2O_5 as single super phosphate (SSP) and K_2O as potassium chloride.

Enumeration of the *Pseudomonas* was done 20 days after sowing in the rhizosphere -soil samples and roots separately. (King *et al*, 1954). The total P content in plants (root, shoot and grain) was determined (Jackson, 1967) and P uptake per plant was calculated. The results were subjected to statistical analyses following analysis of variance and critical difference was calculated (Fisher, 1958).

Results and Discussion

In general, seed inoculation with culture(s)of *P. fluorescens* brought in appreciable increase in dry matter production of sorghum cultivars,

CSH 5 and CSH 6.The maximum mean yield was recorded by the culture P 12 even though there was no significant difference between various cultures. The response of cultivar CSH 5 to *P. fluorescens* (strain P 12) inoculation was better than that of CSH 6.

The response of sorghum to N was linear upto 80 kg N ha^{-1} under no –inoculation treatments while with inoculation, yield increases were noticed only upto the 60 kg N ha^{-1} level (Table 1). No differences could be noticed between the inoculated and uninoculated treatments under the highest level of N (80 kg N ha^{-1}). The data in Table 1 indicate that the increased dry matter yield obtained by the application of low levels of N could be achieved by inexpensive inoculation.

Table 1. Dry matter (g pot^{-1}) of sorghum as affected by inoculation with *Pseudomonas fluorescens* in combination with fertilizer nitrogen [1]

Treatment	Nitrogen (kg ha^{-1})					
	0	20	40	60	80	Mean
Without inoculation	135	147	158	166	173	156
With inoculation	146	152	172	178	170	164
Mean levels of nitrogen	140.5	149.5	165.0	172.0	171.5	
CD (P=0.05)[2]		12.60				5.68
CV (%)		7.3				

1: Average of 4 replications

2. Significantly (P=.05) different from corresponding uninoculated controls

Inoculation with *Pseudomonas* brought an increase of 8-fold in its rhizosphere population (Table 2). Similarly plant roots of inoculated plants contained 17-fold higher bacterial population than the roots of uninoculated plants. Sterilization of the roots with chloramine-T for 30 minutes sharply decreased the number of pseudomonads indicating that many of the bacteria were associated with the root tissue That this decrease was of the order of a 1000-fold in treated plants as against 100-fold in uninoculated ones showed that the root of inoculated plants harboured a substantial number of bacterial cells on the outer root tissue.The application of PGPR's performed better without N application indicating thereby that the organism could establish in the rhizosphere and promote the growth of plants by producing growth promoting sustances (Kloepper *et al.*1988, Angaw *et al.* 2011). The beneficial effects of PGPR on sorghum may be attributed to the increased population of rhizosphere bacteria. The evidence from this study indicated that *Pseudomonas* may be favoured in the rhizosphere of an inoculated sorghum crop. But the overall analysis of data points the possibility of increasing the populations of pseudomonads in the rhizosphere as well as in the roots of a sorghum crop by population of selected strain(s).

Table 2. Numbers of *Pseudomonas* associated with the rhizosphere and roots of sorghum as affected by inoculation with *Pseudomonas* strain P12.

Treatment	Population of *Pseudomonas* / g soil
Rhizosphere	
No I_o [1]	2.8×10^3
NO I_1 [2]	2.2×10^4
Root	
No I_o (unsterilized)	2.1×10^4
No I_o (sterilized)	1.8×10^2
No I_1 (unsterilized)	3.5×10^5
No I_1 (sterilized)	3.5×10^2

1. No I_o Uninoculated control without fertilizer nitrogen
2. NO I_1 Without fertilizer N but inoculated with *Pseudomonas*

Table 3. Response of sorghum CSH 5 to inoculation with *Pseudomonas fluorescens* and arbuscular mycorrhizal (AM) fungi

Treatments	Shoot biomass (g plant^{-1})	Root biomass (g plant^{-1})	Total phosphorus content (%)	Phosphorus uptake (mg plant^{-1})
1.Uninoculated control (no AM and no *P.fluorescens*)	14.5	0.18	0.10	14.5
2. *P. fluorescens* (P12)	18.5	0.32[2]	0.12	21.9
3. *Acaulospora* sp.	17.2	0.22	0.15	25.8
4. *Gigaspora margarita*	17.5	0.25	0.15[2]	25.5
5. *Glomus fasciculatum*	17.8	0.22	0.16[2]	26.7
6. 3 + 2	17.8	0.25	0.15[2]	26.5
7. 4 + 2	20.5[2]	0.34[2]	0.15[2]	28.5[2]
8. 5 + 2	20.8[2]	0.37[2]	0.17[2]	32.5[2]
	CD (P= 0.05)			
P. fluorescens (*P.f*)	NS[3]	0.12	NS	NS
AM fungi (AM)	NS	NS	0.04	NS
P.*f* x AM	2.12	0.11	0.03	5.3

1. Average of 6 replicated pots, each pot had 4 plants
2. Significant increase over corresponding control
3. NS = Not significant

Soil inoculation with mycorrhizal fungi increased the dry matter of roots, shoots, total P content in plant and phosphorus uptake by sorghum. Seed inoculation with *P. fluorescens* in conjunction with *Gigaspora margarita* or *Glomus fasciculatum* produced significantly more dry matter content of roots, shoots, P content of shoots and P uptake of sorghum than the non-mycorrhizal treatments. The significant increase in dry matter production and root biomass by a combinationof seed inoculation with *P. fluorescens* and soil inoculation with Gigaspora *margarita* and/or *Glomus fasciculatum* could be attributed to an increased P transport by the mycorrhizal fungus aided by an increased root biomass (Table 3).

Acknowledgements

One of the authors (KVBRT) is grateful to The National Academy of Sciences, India for providing him financial assistance as Senior Scientist, Platinum Jubilee Fellowship.

References

Angaw Tsigie, Saxena AK, Tilak KVBR. 2011. *Biol Fertil Soils* **47**:971-974.

Azcon R, Barea JM. 1975. *Pl Soil* **43**: 609-619.

Dey R, Pal KK, Tilak KVBR. 2005. Interaction of plant growth promoting rhizobacteria and AM fungi : Present status and future prospects. In: *Basic Research and Applications* of *Mycorrhizae* (eds. GK Podila and AKVarma), J.K.Inernational Pvt. Ltd., New Delhi, India, pp. 83-108.

Fisher RA. 1958. Statistical methods for research workers. 13th ed. Oliver and Boyd, London

Garbaye J, Bowen JD. (1989) *New Phytol* **112**:383-388.

Glick BR, Pasternak JJ.(2003)Plant growth promoting bacteria.In: *Molecular Biotechnology. principles and applications* of *recombinant DNA.* (eds Glick BR, Pasternak JJ) ASM Press Washington DC,USA, pp.436-454.

Jackson ML. 1967. *Soil chemical analysis*. Prentice Hall, New Delhi, India.

King EO, Ward MK, Raney DE 1954. *J Lab Clin Medicine* **44**: 301-307.

Kloepper JW, Hume DT, Scher FM, Singleton C, Tipping B, Laliberte M, Franley K, Kutchaw T, Simonson C, Lifshitz R, Zaleska I, Lee L . 1988. *Plant Dis The Amer Phytopath Soc* **72** : 42-46

Okon Y, Bloemberg GV , Lugtenberg BJJ. 1998. Biotechnology of biofertilization and phytostimulation.. In: *Agricultural Biotechnology* (ed. Altma) Marcel Deker Inc., New York, pp 327-349.

Subba Rao NS, Tilak KVBR, Singh CS. 1985. *Soil Biol Biochem* **17(1)** : 119-121.

Tilak KVBR, Ranganayaki N, Manoharachari C .2006. *Eur J Soil Sci* **57**: 67-71.

Tilak KVBR, Pal KK, Dey R. 2010 . Microbes for Sustainable Agriculture. IK International Pub Ltd ,New Delhi,India, pp 200.

Tilak KVBR, Ranganayaki N, Pal KK, De R, Saxena AK, Nautiyal CS, Mittal S, Tripathi AK, Johri BN. 2005. *Current Sci* **89(1)**:1-15.

Chapter - 5

Effect of carbon and nitrogen sources on the production of lignolytic enzymes by *Trametes versicolor* under submerged culture condition

Samatha Burra, Ram Prasad Metuku, Shiva Krishna Pabba and M.A. Singara Charya

Department of Microbiology, Kakatiya University, Warangal-506 009

ABSTRACT

The effect of various carbon and nitrogen sources on production of laccase and lignin peroxidase (LiP) by *Trametes versicolor* was investigated. Among all sources of carbon and nitrogen, mannitol and peptone were proved the best sources for lignolytic enzymes production respectively. The maximal laccase yield of *T.versicolor* was 686 U/ml in mannitol and LiP yield was 564 U/ml in peptone. Laccase and LiP activities of the organism decreased with supplementation of defined medium by inorganic nitrogen sources. Peptone followed by yeast extract appeared to be the best nitrogen sources for laccase as well as LiP accumulation.

Key words: *Trametes versicolor*, laccase, lignin peroxidase

Introduction

White rot fungi are efficient degraders of the major components of lignocellulosic materials including lignin, cellulose and hemicellulose by producing ligninase, cellulase and hemicellulase respectively (Ruhl *et al.*, 2011, Papinutti and Forchiassin 2007). They are among the important enzymes that have attracted tremendous attention in recent years due to their important applications in different industries (Patel *et al.*, 2009). The genus *Trametes* includes medicinal species that belong to group of white rot fungi due to their ability to produce extracellular lignolytic enzyme.

The lignolytic enzymes are produced during the secondary metabolism under conditions of limited nitrogen. Several hypotheses on the role of solvents in laccase production have been formulated: (i)

increase in membrane permeability and promotion of protein secretion, (ii) inhibition of melanin formation and consequent increase in phenolic monomers, and (iii) activation of an oxidative stress which could be indirectly responsible for laccase induction (Hammel., 1997, Stajic *et al.*, 2004). In basidiomycetes, extra cellular laccases are constitutively produced in small amounts and they are affected by many typical fermentation factors such as medium composition, carbon and nitrogen ratio, pH, temperature, aeration rate, etc. (Mikiashvili *et al.*, 2005, Galhaup *et al.*, 2002, Ravankar and Lele, 2006). Culture condition and medium composition also play a major role on enzyme expression. Addition of nitrogen to certain fungal cultures increased the degradation of lignin related compounds.

The aim of this study was to evaluate for the significance of various carbon and nitrogen sources for oxidative lignolytic enzyme production by *T. versicolor*.

Materials and Methods

Collection and identification: Fruiting body of the *T. versicolor* was collected from Kothagudem timber depot, Andhra Pradesh, India. *Trametes versicolor* formerly known as *Coriolus versicolor* and *Polyporus versicolor* is an extremely common polypore mushroom which can be found throughout the world. *T. versicolor* is commonly called Turkey Tail because of its resemblance to the tail of the wild turkey. *T. versicolor* is recognized as a medicinal mushroom in Chinese medicine under the name yunzhi. In China and Japan *T. versicolor* is used as in immuno adjuvant therapy for cancer. *T. versicolor* belongs to Polyporaceae family and Polyporales order.

The morphological characteristics of the fruiting bodies of *T. versicolor* have been described (Fungi Research Institute 1993). The pileus is fan shapped with a wavy margin and colored concentric zones.

Organism maintenance and inoculum preparation: The culture was maintained on malt-extract agar medium (Malt extract 15g; Dipotassium hydrogen phosphate 1g; Ammonium chloride 1g; 1N citric acid 15ml; Agar agar 20g; Distilled water 1000ml) which is used as control medium. Inoculum of *T. versicolor* was prepared from mycelia grown on the same media incubated at 25°C for 4–6 days. From the plate 7 mm diameter mycelial discs were used as the inoculum.

Culture conditions: Incubation was carried out statically at 25 ± 1°C in 125 ml Erlenmeyer flask containing 50 ml of the medium inoculated with 7mm agar plug from 6 day old mycelia grown on malt-extract agar. Periodic harvesting of the mycelia was performed using the filter paper. Culture filtrates were used as enzyme source.

Carbon sources used: Glucose, fructose, sucrose, lactose, galactose, mannitol, starch, cellulose, sorbitol and maltose used in 0.02% and 0.05% concentration in Malt Extract medium.

Nitrogen sources used: Ammonium sulphate, beef extract, gelatin, tryptone, glycine, potassium nitrate, peptone, yeast extract, sodium nitrate and casein used in 0.02% and 0.05% concentration in malt extract medium.

Enzyme assays: White rot fungus was grown in malt extract broth and incubated for 7 and 14 days. Activities of enzymes laccase (E.C.1.10.3.2) (Coll *et al.*, 1993) and lignin peroxidase (E.C.1.11.1.14) (Sarkanen and Ludwig, 1977] were measured using guaiacol as substrate. For these enzymes, one activity unit was defined as the amount of enzyme necessary to oxidize one µ mol of substrate per minute.

Results and Discussion

The role of different carbon sources on lignolytic enzyme production of *T. versicolor* was evaluated. All tested compounds ensured good growth of fungi providing biomass accumulation. However, enzyme activity significantly depended on the carbon source in the medium. The polysaccharides appeared to be poor substrates for laccase production while easily metabolizable compounds ensured comparatively high laccase activity. Total ten carbon sources are used in the study and all sources showed positive effect on Laccase accumulation in the culture liquid of *T.versicolor*. The laccase activity of *T. versicolor* reached 588 U/ml i.e., it was 5-fold higher than that in fungus cultivation in the control medium (Fig. 1) and Lip activity reached 564 U/ml (Fig. 2). However, their positive effects on enzyme accumulation were due to a higher biomass production. Vahidi *et al.* (2004) reported when simple sugars were used as carbon sources, increased production of laccase and lignin peroxidase in *M. leptocephoda* at pH 5.5 and 25°C. Many previous studies have proved that both the nature and concentration of nitrogen sources are powerful nutrition factors regulating lignolytic enzyme production by wood-rotting basidiomycetes [Buswell *et al.*, 1995; Elisashvili *et al.*, 2001]. In our study, all nitrogen sources were added to the medium in two concentrations i.e., 0.02% and 0.05%. None of them significantly changed the final pH of medium. Among them, peptone and yeast extract exhibited significant influence on Laccase secretion giving a 5-fold increase compared with the control medium (Fig. 3). Utilization of peptone sharply (4-fold) accelerated the secretion of LiP, while medium supplementation with ammonium sulphate, glycine and potassium nitrate inhibited this enzyme accumulation with the concentration of 0.02% and 0.05%. In addition, all organic nitrogen sources tested favored comparatively high LiP and laccase accumulation by *T. versicolor*. Buswell *et al.*, (1995) reported that the production of lignin modifying enzymes is strongly affected by the nature and amount of nutrient especially nitrogen and trace element in the growth medium.

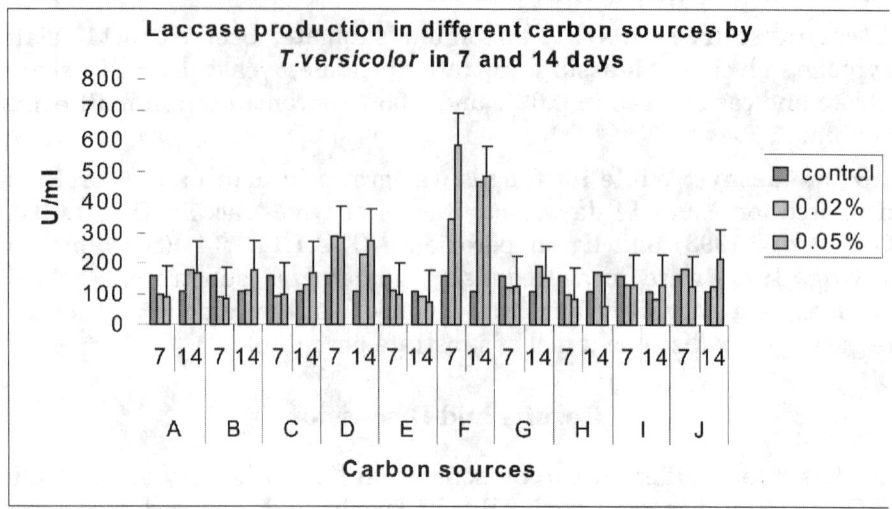

Fig 1:

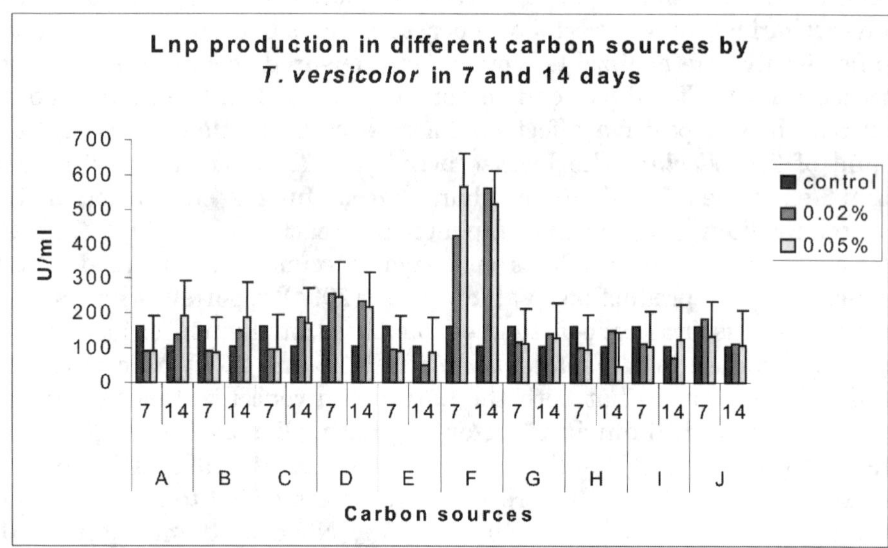

Fig 2:

Carbon sources: A- Glucose, B- Fructose, C- Sucrose, D- Lactose, E- Galactose, F- Mannitol, G- Starch, H-Cellulose, I- Sorbitol and J-Maltose.

Supplementation of peptone and yeast extract enhanced the LiP and laccase production 4-fold, while fungus cultivation in the presence of ammonium sulfate sharply inhibited enzyme secretion by *T. versicolor* (Fig 3). High nitrogen source containing media gave the highest laccase activity in *Lentinus edodes*, *Rigidoporus lignosus*, and *Trametes pubesc-*

ens while the nitrogen limited conditions enhanced enzyme production in *Pycnoporus cinnabarinus*, *P. sanguineus*, and *Phlebia radiate* [Mester TA and Field AJ 1997, Gianfreda *et al.*, 1999]. Johnsy and Kaviyarasan (2011) showed that among the different carbon sources tested fructose supported maximum (63.51 U/ml) laccase activity, while beef extract supported the maximum activity (64.31 U/ml) among the different nitrogen sources tested.

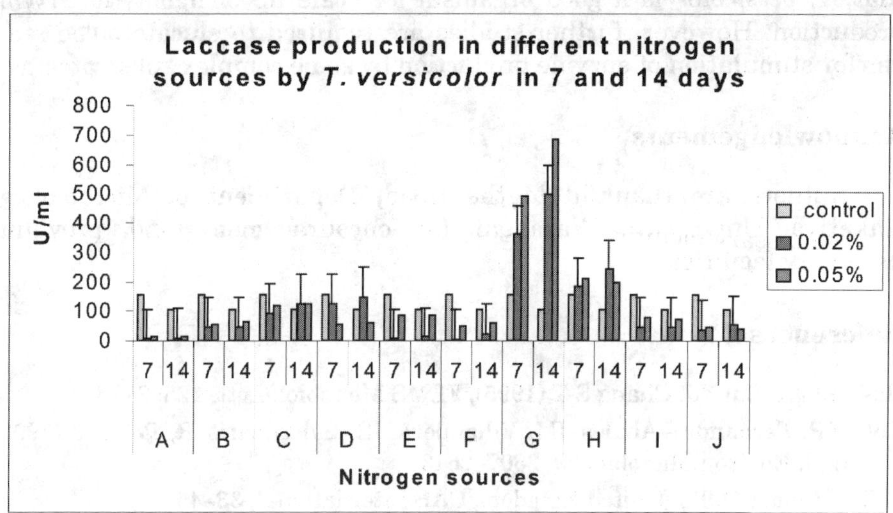

Fig 3:

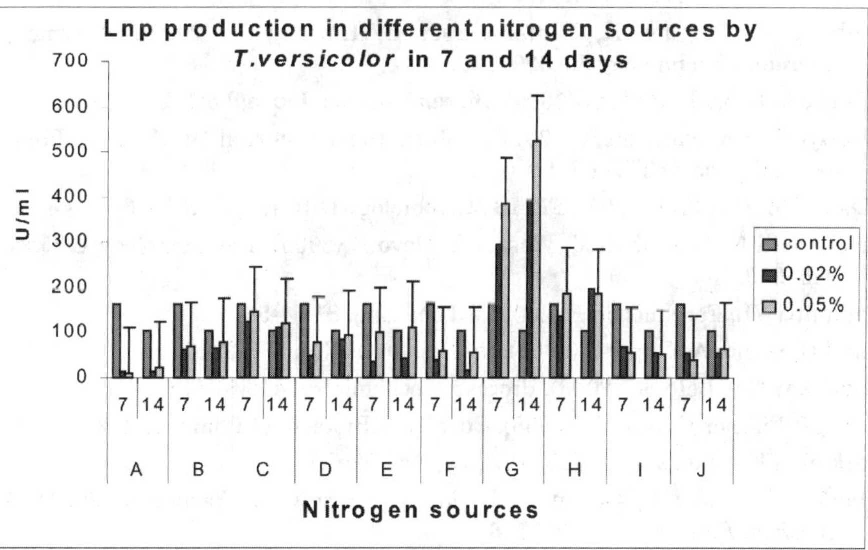

Fig 4:

Nitrogen sources: A-Ammonium Sulphate, B-Beef Extract, C-Gelatin, D-Tryptone, E-Glycine, F-Potassium nitrate, G-Peptone, H-Yeast Extract, I-Sodium nitrate, J- Casein.

Conclusions

This study emphasised the need to explore many more fungi to evaluate the real potentials of the production of lignolytic enzymes. By proper selection of the different carbon and nitrogen sources for fungus growth and target enzyme synthesis may play an important role in the development of an efficient technology. The results obtained here reveal that, *T. versicolor* is a good organism for scale up of lignolytic enzyme production. However, further studies are required to elucidate the reasons for stimulation of enzyme production by some complex substrates.

Acknowledgements

Authors are thankful to the Head, Department of Microbiology, Kakatiya University, Warangal for encouragement and providing necessary facilities.

References

Buswell JA, Cai YJ, Chang S-T (1995). FEMS Microbiol. Lett. 128: 81–88.

Coll MP, Fernandez–Abalos JM, Villanueva JR, Santamaria R, Perez P (1993). Appl. Environ.Microbiol 59: 2607–2613.

E.K. Hammel(1997). United Kingdom, CAB International, 33–45.

Elisashvili V, Parlar H, Kachlishvili E, Chichua D, Kvesitadze G (2001). Adv. Food Sci. 23: 117–123.

Galhaup C, Wagner H, Hinterstoisser B, Haltrich D (2002). Enzyme and Microbial Technology 30: 529–536.

Gianfreda L, Xu F, Bollag J(1999). Bioremediation Journal 3:1–25.

Johnsy G, Kaviyarasan V (2011). International Journal of Current Pharmaceutical Research, Vol 3, Issue 4.

Mester TA, Field AJ (1997). FEMS Microbiology Letters 155:161–168.

Mikiashvili N, Elisashvil V, Wasser S, Nevo E (2005). Biotechnology Letters 27: 955–959.

Papinutti VL, Forchiassin F (2007). J.Food Eng. 81:54-59.

Patel H, Gupte A, Gupte S (2009). *BioResources* 4(1): 268-284.

Ravankar MS, Lele SS (2006). Process Biochemistry. 41:581-588

Ruhl M, Fischer C, Kuer U (2008). Curr.Tren.Biotechnol.Pharmacol. 2:478-492.

Sarkanen KV, Ludwig CH (1970).*Wiley*, New York.

Staji´c M, Persky L, Cohen E, Hadar Y, Brˇceski I, Wasser SP(2004). *Appl Biochem Biotechnol*, 117:155–64.

Vahidi H, Kobarfard F, Nawjoyan F (2004). African Journal of Biotechnology 3(11): 606-609.

Chapter - 6

Optimization of growth parameters for cellulase production by an alkaline *Fusarium* sp. under submerged fermentation conditions

B Sreedevi, S Jeevan Chandra and S Ram Reddy

Department of Microbiology, Kakatiya University,
Warangal-506 009, A.P., India
E-mail: sanditi_ramreddy@yahoo.com

ABSTRACT

Cellulose degrading organisms are being employed for the conversion of native cellulosic materials into soluble sugars. In the present investigations, a cellulolytic fungus *Fusarium* sp. (FC-11) was isolated from soil sample. An effort was made to optimize the growth parameters for the maximum production of cellulase. This strain was observed to exhibit significant amounts of cellulase activity at alkaline pH 8.5, incubation period of 5 days and temperature of 50°C. Further, results also indicated that maximum cellulase was produced with carboxymethyl cellulose as the cellulosic substrate, yeast extract as the nitrogen source and Tween-80 as surfactant.

Key words: Cellulases, Fusarium sp (FC-11), Carboxymethyl cellulose, Yeast extract, Tween-80.

Introduction

The potential use of cellulosic materials as renewable sources of carbon was recognized and accelerated only after cellulose degrading enzymes or cellulases were identified (Bhat and Bhat, 1997). Cellulose bioconversion through enzymatic hydrolysis to obtain sugars, which can be fermented to produce liquid fuels, has received global attention during the last few years (Mehta *et al.*, 2004). Cellulases, which have received attention from the point of cellulose degradation are primarily of microbial origin. The importance of cellulolytic microbes for the production of glucose from cellulose is well recognized (Sharma *et al.*, 1995). Cellulases are produced by wide variety of bacteria, fungi, actinomycetes, aerobes, mesophiles and thermophiles (Vyas *et al.*, 2004). Recently, alkaline cellulases have gained commercial importance because of their potential applications in textile, paper/pulp and detergent industries. These enzymes are found to be useful in denim washing (Kierulff, 1997), biopolishing of textiles (Peterson *et al.*, 1992) as well as in deinking operations during recycling of mixed office waste paper (MOW) (Jeffries *et al.*, 1994) Alkaline cellulases are also used as a component of detergents (Susmo, 1997). In all these applications, the enzyme should be active in alkaline conditions and at elevated tempera-tures. Very few fungal sources are identified for the industrial production of cellulases. In the present investigations, an attempt was made to optimize the growth parameters for the maximum production of cellulases by *Fusarium* sp (FC-11) under submerged fermentation conditions.

Materials and Methods

Microorganism: The alkaline fungal strain was isolated from soil sample and identified as *Fusarium* sp on the basis of microscopic observations. It is maintained by periodic transfer on potato dextrose agar slants at 28 ± 2°C.

Media and culture conditions : The fungus was grown in M-I medium (Reese and Mendels, 1963). The basal M-I medium contained (g/*l*): KH_2PO_4, 2.0; $(NH_4)_2SO_4$, 1.4; $CaCl_2.2H_2O$, 0.3; $MgSO_4$. $7H_2O$, 0.3; urea, 0.3; proteose peptone, 0.25; yeast extract, 0.2; cellulose or CMC, 10.0; trace element solution, 1ml [$FeSO_4.7H_2O$, 5mg /*l*; $MnSO_4.7H_2O$, 5.6 mg/*l*; $ZnSO_4.7H_2O$, 3.34 mg/*l*; $CoCl_2.2H_2O$, 2 mg/*l*] and Tween-80, 1 ml. The pH of the medium was adjusted after autoclaving by using separately sterilized 1M Na_2CO_3. 45 ml of Reese and Mandels medium was taken in Erlenmeyer flask and incubated with 5 ml of culture filtrate and incubated for 1 week at 28±2°C under submerged fermentation conditions. After incubation, the samples were withdrawn aseptically and centrifuged at 8000rpm for 20 min. The filtrate obtained was used as a source of crude enzyme.

Effect of cultural conditions on cellulose production: Effect of initial pH (5.5, 6.5, 7.5, 8.5, 9.5 and 10.5), temperature (30°C, 40°C, 50°C and 60°C), incubation period (3, 5, 7 and 9 days), cellulosic substrates (cellulose, filter paper, CMC and jute powder), nitrogen sources (KNO_3, NH_4Cl, $(NH_4)_2SO_4$, $(NH_4)NO_3$, $(NH_4)HPO_4$, yeast extract, urea and peptone) and surfactants (Tween-20, Tween-80, Triton-X and SDS) were studied on cellulose production by *Fusarium* sp (FC-11).

Assay: The activities of CMCase and FPase were assayed in terms of release of reducing sugars by 3, 5-dinitrisalicylic acid (DNS) method (Miller, 1959; Miller *et al.*, 1960). An aliquot (0.5 ml) of appropriately diluted culture filtrate was mixed with 0.5ml of 1% CMC in 0.05M sodium phosphate buffer, pH 7.0. The reaction mixture was incubated for 30 min at 60^{0}C. The released reducing sugars were estimated by DNS method. For filter paper activity (FPase), 50 mg of Whattman No.1 paper was added to 1 ml of phosphate buffer of pH 7.0 and 1 ml diluted enzyme sample. The reaction mixture was incubated for 1 hour and the released reducing sugars were estimated. One unit of CMCase, FPase activity was defined as amount of enzyme which liberates 1μ mole glucose/min under standard assay conditions.

Statistical analysis: Values are mean of three replicates significant at $p < 0.001$ according to Fisher's one way analysis. Later the means were compared with each other following Fisher's protected least significance different test L.S.D (at $p \leq 0.01$ and $p \leq 0.05$).

Results and Discussion

In the present investigations, on the basis of morphology and microscopic observations the isolated strain was tentatively identified as *Fusarium* sps (FC-11). From the data presented in the table it is evident that cellulase yields is depend on a complex relationship involving a variety of factors like nutrient resources (carbon source and cellulose quality), pH, temperature, presence of inducers, medium additives, aeration, growth time etc(Immanueul *et al* ., 2006).

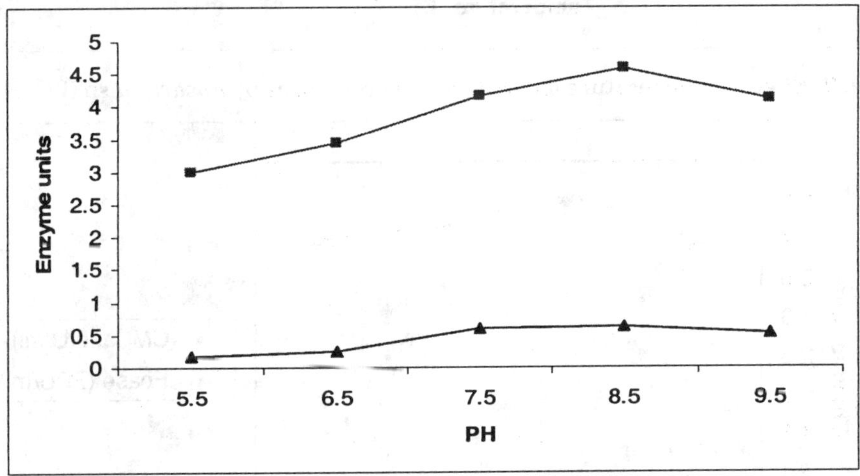

Fig. 1: Effect of initial pH on production of cellulases by *Fusarium* sp (FC-11)

A critical perusal of Fig. 1 reveals that initial pH has a direct effect on the cellulase production. FC-11 exhibited significant amounts of cellulase (CMCase: 4.5 U/ml; FPase: 0.62 FU/ml) ($p < 0.01$) at alkaline pH 8.5. Vyas *et al.* (2004) reported an alkalotolerant *Fusarium* strain which secreted high amount of cellulolytic enzymes when grown under submerged fermentation conditions. Kang and Rhee (1995) reported alkaline active carboxymethyl cellulases from *Cephalosporium* strain.

Wagarchuk *et al.*, 1994) found that paper industry needs those cellulases, which are active above pH 6.5 – 7.0.

The present isolate FC-11 exhibited significant amounts of CMCase (4.58 U/ml) and FPase (0.62 FPU/ml) at temperature 50^{0}C ($p < 0.01$) (Fig. 2). Vyas *et al.* (2004) reported a *Fusarium* strain which showed a maximum activity at temperature 60^{0}C. The optimum temperature for *Humicolafuscoatra* was found to be 45-50^{0}C. (Rajendran *et al.*, 1994)

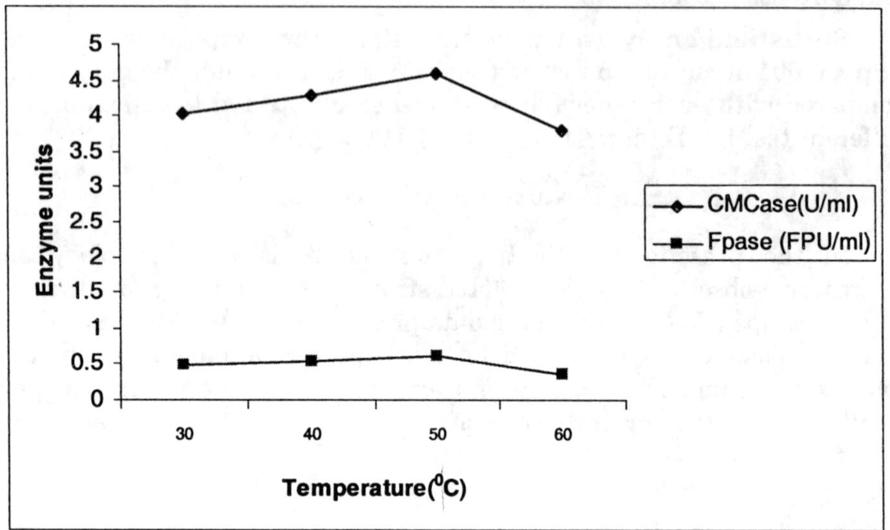

Fig. 2: Effect of temperature on production of cellulases by *Fusarium* sp (FC-11)

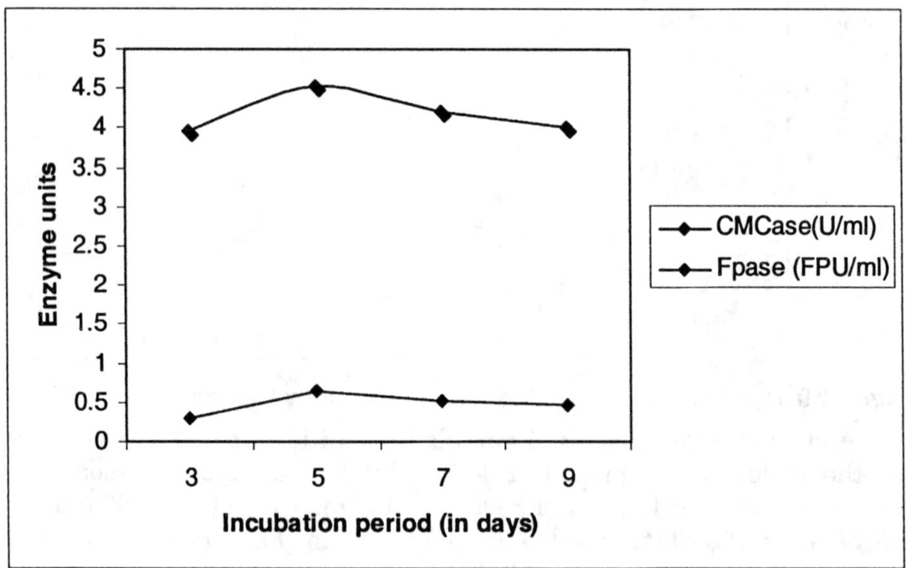

Fig. 3: Effect of incubation period on production of cellulases by *Fusarium* sp (FC- 11)

In general, maximum enzyme production was recorded after one week of incubation. In our observations, the isolate FC-11 exhibited maximum CMCase activity (4.52U/ml) ($p < 0.05$) and FPase activity (0.65 FPU/ml) ($p < 0.01$) at an incubation period of only 5 days (Fig. 3) incubation. Upon further incubation, there was a sharp decline in the cellulase activity. This may be due to depletion of nutrients in the medium. The decrease in activity may be either due to lower level of cellulase production or recalcitrant nature of the residual cellulose. (Paul, 1992)

A critical perusal of Table 1 reveals that out of different cellulosic substrates, CMC was found to be superior over other substrates in inducing the cellulase production. Maximum CMCase activity of 4.62 U/ml and FPase activity of 0.55 FPU/ml was detected in presence of CMC cellulosic substrate ($p < 0.001$). Similar to the present investigations, Narasimha *et al.* (2006) reported CMC to be the best substrate for cellulase production. Similar to the present observations, Ekperigin (2007) reported that a higher production of cellulase when CMC served as a substrate may be as a result of induction of enzyme. Among carbon sources, cellulose, cellobiose and a mixture of carboxymethyl cellulose and cellobiose induced maximal endogluconase production by *Trametes* sp. Levin and Forchiassin (1995). Similarly, induction of cellulase production by *Humicola fuscoatra* in presence of cellulosic substrates has been reported by Rajendran *et al.* (1994).

Table 1. Influence of cellulosic substrates on production of cellulases by *Fusarium* sp (FC-11)

Cellulosic Substrates	Growth	CMCase# (U/ml)	FPase* (FPU/ml)
Cellulose	++	3.76	0.45
Filter paper	++	3.61	0.21
CMC	+++	4.62	0.55
Jute powder	+	3.14	0.16
L.S.D. 0.05		0.22	0.11
0.01		0.31 0.15	

+ indicates little growth; ++ indicates moderate growth;
+++ indicates abundant growth

Carboxy methyl cellulase (CMCase) is expressed in terms of units. One unit is the amount of enzyme releasing 1 μ mole of reducing sugar from Carboxy methyl cellulose per min.

*Filter paperase (FPase) is expressed in terms of filter paper units. One unit is the amount of enzyme in the culture filtrate releasing 1 μ mole of reducing sugar from filter paper per min.

Values are mean of three replicates significant at $p < 0.001$ according to Fisher's one way ANOVA analysis

Various researchers have shown that different organic and inorganic nitrogen sources such as yeast extract (Ganguly and Mukherjee, 1995), soyameal (Gomes *et al*, 2000) influence the cellulase production. Results pertaining to the effect of various nitrogen sources on cellulase production by the isolate FC-11 are presented in Table 2. It is evident from table 2 that FC-11 exhibited significant amounts of CMCase (4.68 U/ml) ($p < 0.001$) and FPase (0.58 FPU/ml) ($p < 0.001$) when the medium was supplemented with yeast extract as nitrogen source. Similar to the present observations, Vyas *et al.* (2004) reported a *Fusarium* sp which exhibited maximum cellulase activity in presence of yeast extract.

Table 2. Influence of nitrogen sources on production of cellulases by *Fusarium* sp (FC-11)

Nitrogen	Growth	CMCase (U/ml)	FPase FPU/ml)
KNO_3	++	3.12	0.26
NH_4Cl	+	1.48	0.18
$(NH_4)_2NO_3$	+	1.85	0.18
$(NH_4)_2SO_4$	+++	4.21	0.39
$(NH_4)HPO_4$	++	4.11	0.37
Yeast extract	+++	4.68	0.58
Urea	++	3.11	0.42
Proteose peptone	++	2.58	0.38
L.S.D. 0.05		0.213	0.09
0.01		0.31	0.12

+ indicates little growth; ++ indicates moderate growth;
+++ indicates abundant growth

Carboxy methyl cellulase (CMCase) is expressed in terms of units. One unit is the amount of enzyme releasing 1 μ mole of reducing sugar from Carboxy methyl cellulose per min.

*Filter paperase (FPase) is expressed in terms of filter paper units. One unit is the amount of enzyme in the culture filtrate releasing 1 μ mole of reducing sugar from filter paper per min.

Values are mean of three replicates significant at $p < 0.001$ according to Fisher's one way ANOVA analysis

The effect of different surfactants i.e. Tween-20, Tween-800, Triton-X and SDS on cellulase production by the isolate FC-11 was investigated (Table 3) ($p < 0.001$). A critical analysis of table 3 reveals that among different surfactants employed in the medium, Tween-80 was found to be the best surfactant. In presence of Tween -80, maximum CMCase activity was 4.46 U/ml and FPase activity was 0.51 FPU/ml and found to be significant of $p < 0.01$. Domingues *et al.*, (2000) reported that Tween-80 influences the morphology of *Trichoderma reesei* Rut C-30 as well as the enzyme production. Helle *et al.* (2004) reported that Tween-80 increases the rate of cellulase activities.

Table 3: Influence of surfactants on production of cellulases by *Fusarium* sp (FC-11).

Surfactants	CMCase (U/ml)	FPase (FPU/ml)
Tween-20	4.02	0.40
Tween-80	4.46	0.51
Triton-X	3.81	0.41
SDS	3.08	0.36
L.S.D. 0.05	0.09	0.04
0.01	0.12	0.05

Carboxy methyl cellulase (CMCase) is expressed in terms of units. One unit is the amount of enzyme releasing 1 μ mole of reducing sugar from Carboxy methyl cellulose per min.

*Filter paperase (FPase) is expressed in terms of filter paper units. One unit is the amount of enzyme in the culture filtrate releasing 1 μ mole of reducing sugar from filter paper per min.

Values are mean of three replicates significant at $p < 0.001$ according to Fisher's one way ANOVA analysis

Out of these investigations, it is concluded that the isolate, *Fusarium* sp. (FC-11) exhibited significant amounts of cellulase production at alkaline pH (8.5) and at high temperature (50^{0}C) indicating thermotolerant nature which can be exploited for various industrial applications.

Acknowledgements

Thanks are due to Head, Department of Microbiology for encouragement and providing necessary laboratory facilities.

References

Bhat, M.K. and S. Bhat (1997). *Biotechnol. Adv.***15**: 583-620.

Domingues, F.C., J.A. Queiroz, J.M. Cabral and L.P. Fonseca (2000). *Enzyme Microb. Technol.* **26**: 394

Ekperigin, M.M. (2007). *Afr. J. Biotechnol* **6(1)**: 28-33.

Ganguly, R. and S. Mukherjee (1995). *J. Microbiol. Biotechnol.* **10**: 47-58.

Gomes, I., J. Gomes, D.J. Gomes and W. Steiner (2000). *Appl. Microbiol. Biotechnol.* **53**: 461-468.

Helle, S.S., S.J.B Duff and D.G. Cooper (2004). *Biotechnol. Bioeng.* **42**: 611- 617.

Immanuel, G., R. Dhanusha, P. Prema and A. Palavesam (2006). *Int. J. Env.Sci. and Technol.* **3(1)**: 25-34.

Jeffries, T. W., J.H. Klungness, S. Marguerite and K.R. Cropsey (1994). *J.Tappi.* **77(4)**: 173-179.

Kang, M.K. and Y.H. Rhee (1995). *Biotech. Lett.* **17**: 507-512.

Kierulff, J.V. (1997). *Textile Horizons.* **8**: 33-36.

Levin, L. and F. Forchiassin (1955). *Rev. Argent. Microbiol.* **27**: 11-20.

Mehta, A., A. Singh, S. Ferosh and P. Mehta (2004). *Ind. J. Microbiol.* **44(1)**: 47-52.

Miller, G.L. (1959). *Anal. Chem.* **31**: 726-428.

Miller, G.L., R. Blum, W.E., Gelnnon and A. Burton (1960). *Anal. Biochem.***2**: 127-132.

Narasimha, G., A. Sridevi, B., Vishwanath, M. Subash Chandra and B. Rajashekar Reddy (2006). *Afr. J. Biotechnol.* **5**: 472-476.

Paul, J. (1992). *Ind. J. Microbiol.* **32**: 127-147

Peterson, G.C., G.A. Screws and D.M. Quadroon, (1992). *Canadian Textile Journal*, 31-35.

Rajendran, A., P. Gunasekaran and M. Lakshmanan (1994). *Ind. J. Microbiol.* **34(3)**: 289-295.

Reese, E.T. and M. Mandels (1963). Methods in carbohydrate chemistry, Academic press, New York, London, **3**: 139-143.

Sharma, A., C.L.S.N. Rao, B. Ball and S.K. Hasija (1995). *J. Ind. Bot. Soc.* **75**: 122

Susmo, I. (1997). *Extremophiles.* **1**: 61- 66.

Vyas, A., D. Vyas and K.M. Vyas (2004a). *J. Bot. Soc.* **39**: 66-71.

Vyas, S., A. Ahmas and A. lachke (2004b). Microbiology and Biotechnology for Sustainable Development (Ed. P.C. Jain) New Delhi. 283-292

Wakarchuk, W.W., W.L. Sung, R.L., Campbell, A. Cunningham, D.C. Watson and M. Yaguchi (1994). *Protein Eng.* **7**: 1379-1386.

Chapter 7

In vitro activity of four antifungal drugs against clinical isolates of two *Trichophyton* species

K.R. Reddy and S. Ram Reddy*

Department of Microbiology, Prathima Institute of Medical Sciences, Karimnagar- 505417 A.P.

*Department of Microbiology, Kakatiya University, Warangal - 506 009. AP

ABSTRACT

Investigations on antifungal drug activity were carried out on 90 clinical isolates of *Trichophyton rubrum* and *Trichophyton mentagrophytes* with four antifungal drugs, viz, griseofulvin, fluconazole, itraconazole and terbinafine as suggested by NCCLS M27-A (1997) document by broth macrodilution method to standardize *in vitro* antifungal susceptibility testing and to find out the MIC of the drugs. In this study, terbinafine was found to be the most efficient drug for all isolates. Terbinafine had the lowest MIC range of 0.001 μg/ml to 0.09 μg/ml and MIC_{50} was low at 0.005 μg/ml and MIC90 was also low at 0.04 μg/ml against *T. rubrum* and MIC range of 0.001 μg/ml to 0.19 μg/ml with a MIC50 of 0.01 μg/ml and MIC90 at 0.09 μg/ml against *T. mentagrophytes*. Itraconazole showed antifungal activity superior to that of fluconazole, with a MIC range of 0.04 μg/ml to 1.56 μg/ml, with MIC_{50} 0.19 μg/ml and MIC90 at 1.56 μg/ml against *T. rubrum* and MIC range of 0.04 μg/ml to 1.56 μg/ml, with MIC50 at 0.19 μg/ml and MIC90 at 0.78 μg/ml against *T. mentagrophytes*. Griseofulvin had a MIC range of 0.15 μg/ml to 5.07 μg/ml with MIC50 at 1.26 μg/ml and MIC90 at 2.53 μg/ml against *T. rubrum*; and MIC range of 0.31 μg/ml to 5.07 μg/ml with MIC50 at 1.26 μg/ml and MIC90 at 2.53 μg/ml against *T. mentagrophytes*. Fluconazole showed a high MIC range of 0.19 μg/ml to 50 μg/ml and MIC50 was high at 1.56 μg/ml and MIC_{90} was also high at 12.5 μg/ml against *T. rubrum*; and a high MIC range of 0.09 μg/ml to 25.0 μg/ml, with MIC50 at 1.56 μg/ml and MIC90 at 12.5 μg/ml towards *T. mentagrophytes*.

Key words: NCCLS, Antifungal drugs, Minimum Inhibitory Concentration (MIC), *Trichophyton* species

Introduction

Cases of dermatophytoses have increased over the past few decades. At the same time, a number of newer less toxic antifungal drugs have become available for clinical use. The increased use of antifungals, often for prolonged periods, has led to the recognition of the phenomenon of acquired antifungal resistance among previously susceptible strains or species and consequently to the increased incidence of infections with less common species.

The rapid increase in fungal infections and the growing number of new antifungal agents[2] indicate an pressing need for rapid and. accurate methods for antifungal susceptibility testing (Andriole, 2000; Cormican and Pfaller, 1993). The present. study describes and compares the in vitro susceptibility of clinical isolates of *T. rubrum*, and *T. mentagro-phytes* against four antifungal drugs, namely to terbinafine (Gupta and Shear, 1997), itraconazole (Pfaller *et al.*, 2000) fluconazole (Agarwal *et al.*, 1996) and griseofulvin (Artis *et al.*, 1981).

Material and Methods

i. Isolation and identification of the isolates

A total of 90 isolates of *T. rubrum* and 37 isolates of *T. mentagrophytes* were obtained from 500 clinically diagnosed patients of tinea (ringworm) infection attending as out patients at Department of Skin & Venereology at Government General Hospital, Kolar, Karnataka during the period 2000-2004. They were identified by conventional morphological, cultural and biochemical methods including urease test (Bahuguna and Kushawaha, 1989), *in vitro* hair perforation test (Chander, 2002), pigment test, rice grain test (Jessup *et al.*, 2000) The species isolated and tested (and numbers of isolates of each species) are as follows: Trichophyton rubrum (n = 90), *T. mentagrophytes* (n = 37).

ii. Antifungal drugs

Four antifungal drugs namely griseofulvin, fluconazole, itraconazole and terbinafine were selected and tested for their activity. Griseofulvin stock solution was prepared in 70% ethanol, fluconazole in distilled water and itraconazole and terbinafine in dimethyl sulfoxide that were stored at -20°C to -70°C. The concentrations in the stock solutions were 100 times the final concentration of each compound. Further dilutions of each antifungal agent were prepared using yeast nitrogen broth as diluents (NCCLS, 1997). The final concentrations ranged from 0.03 μg/ml to 81.25 μg/ml for griseofulvin; 0.047 μg/ml to 50 μg/ml for fluconazole; 0.02 μg/ml to 25 μg/ml for itraconazole and 0.0005 μg/ml to 6.25 μg/ml for terbinafine

Testing was performed by a broth macrodilution method (Norris *et al.*, 1999) following the recommendation of the National Committee for Clinical Laboratory Standards (NCCLS) M27-A (1997). In brief stock inocula of the *T. rubrum* and *T. mentagrophytes* strains were prepared from 7 to 14 day cultures grown on Sabouraud's dextrose agar (SDA) with chloramphenicol (Espingel-Ingroff and Kerkering, 1991). After the appearance of the sufficient growth 2 to 3 ml sterile normal saline (0.95%) was added and the suspensions were made by gently scraping the colony with the tip of a sterile Pasteur pipette. The resulting suspended mixture was withdrawn and transferred to a sterile tube. Heavy particles of the suspension, when present, were allowed to settle for 15 minutes at room temperature and the upper homogenous suspension was used for further testing. The suspensions were mixed with a vortex mixer for 15 seconds and adjusted with sterile normal saline to match an opacity of 0.5 McFarland's standard. The inoculum size was adjusted to between 1.0 x 10^6 and 5.0 x 10^6 spores/ml by microscopic enumeration with a cell counting haemocytometer (Neubauer chamber). In some instances where fungi do not readily produce conidia, small portion of the mycelial growth was harvested and gently homogenized in 2 ml of sterile saline using tenbroeck tissue grinder and resulting suspensions were adjusted to an opacity of 0.5 Mc Farland standard[14] by adding sterile saline (Fernandez-Torres *et al.,* 2001). All standardized inocula were plated on SDA before the test to check the viability of the fungus. 0.3 ml of fungal inocula were added to the different drug dilutions. A control tube (broth without any drug but inoculated with the fungus) was included with each test. Tubes were incubated at 35°C in a BOD incubator until growth appeared in the drug-free control tube. Incubation ranged 6 to 20 days. The highest dilution of the drug, which inhibited the fungal growth, was taken as the MIC (Minimum Inhibitory Concentration). MIC50 was calculated by taking the drug concentration, where fifty percent of isolates are inhibited. Similarly MIC90 was calculated with drug concentration where ninety percent of the isolates were inhibited.

Table 1. *In vitro* activities of four antifungal agents against *T. rubrum* and *T. mentagrophytes*

Species (No. of isolates)	MIC	Drug concentration (μg/ml)			
		Griseofulvin	Fluconazole	Itraconazole	Terbinafine
T. rubrum (90)	Range	0.15-5.07	0.19-50.0	0.04-6.25	0.001-0.09
	MIC_{50}	1.26	1.56	0.19	0.005
	MIC_{90}	2.53	12.5	1.56	0.04
T. mentagrophytes (90)	Range	0.31-5.07	0.09-25.0	0.04-1.56	0.001-0.19
	MIC_{50}	1.26	1.56	0.19	0.01
	MIC_{90}	2.53	12.5	0.78	0.09
Average	MIC_{50}	1.26	1.56	0.19	0.075
	MIC_{90}	2.53	12.5	1.17	0.065

The minimum inhibitory concentration (MIC50 and MIC90S) (Perea *et al.*, 2001) (Table 1) of griesofulvin, fluconazole, itraconzole and terbinafine are compared to determine the efficacy and dosage of the drug for treatment of dermatophytoses.

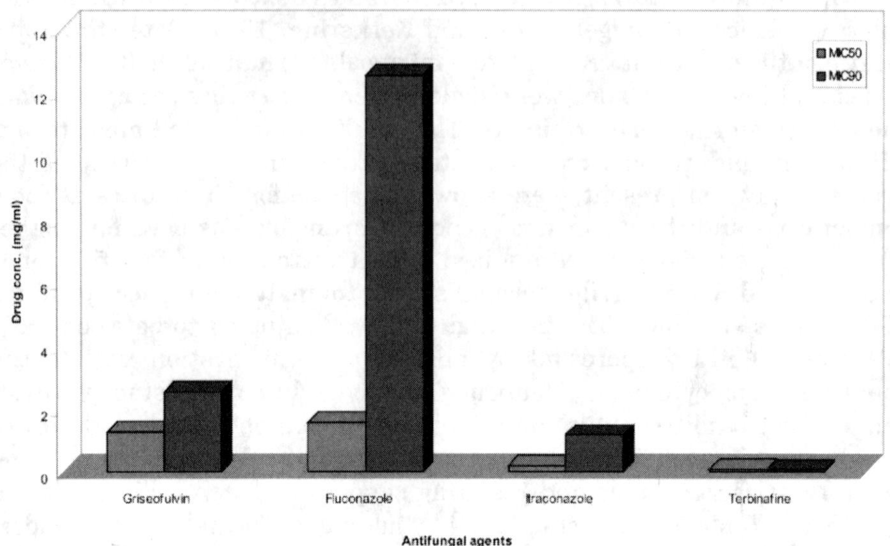

Fig. *In vitro* activities of four antifungal agents (MIC50 and MIC90)

Results and Discussion

A critical perusal of the table reveals that the MIC ranges of all the 90 isolates of *T. rubrum* and 37 isolates of *T. mentagrophytes* tested, terbinafine had the lowest MIC range of 0.001 (µg/ml to 0.09 µg/ml towards *T. rubrum* and 0.001 µg/ml to 0.19 µg/ml for *T. mentagrophytes*, followed by itraconazole with a MIC range of 0.04 µg/ml to 6.25 µg/ml for *T. rubrum* and 0.04 µg/ml to 1.56 µg/ml towards *T. mentagrophytes*. Griseofulvin showed a MIC range of 0.15 µg/ml to 5.07 µg/ml against *T. rubrum* and 0.31 µg/ml to 5.07 µg/ml towards *T. mentagrophytes*. Fluconazole showed a high MIC range of 0.19 µg/ml to 50.0 µg/ml towards *T. rubrum*: and 0.09 µg/ml to 25.0 µg/ml against *T. mentagrophytes*.

The MIC_{50} of terbinafine was low at 0.005 µg/ml towards *T. rubrum* and 0.01 µg/ml for *T. mentagrophytes*; of itraconazole was 0.19 µg/ml for both *T. rubrum* as well as *T. mentagrophytes*. In case of griseofulvin, it was at 1.26 µg/ml for both *T. rubrum* and *T. mentagrophytes*. For fluconazole the MIC was high at 1.56 µg/ml for both *T. rubrum* and *T. mentagrophytes*. The MIC90 of terbinafine was low at 0.04 µg/ml for *T. rubrum* and 0.09 µg/ml for *T. mentagrophytes*. For itraconazole the values are at 1.56 µg/ml against *T. rubrum* and 0.78 µg/ml against *T. mentagrophytes*. Griseofulvin has exhibited same MIC90 values (2.53 µg/ml) for both *T. rubrum and T. mentagrophytes.* Whereas for fluconazole, MIC90 was high at 12.5 μg/ml for *T. rubrum* as well as *T. mentagrophytes.*

Thus it is evident from the present study that terbinafine appears to be the most efficient drug for all the isolates of *T. rubrum* and *T. mentagrophytes.* Among the azoles, itraconazole was the second best drug against all *T. rubrum* and *T. mentagrophytes* isolates and found to be the better drug compared to fluconazole. Griseofulvin appears to be still a potent drug for the management of dermatophytoses caused by *T. rubrum* and *T. mentagrophytes.*

The present study also indicates that the broth macrodilution method can be adopted for in vitro antifungal sensitivity testing, as it is simple, reproducible, cost effective and easy to perform in a routine clinical microbiology laboratory.

Acknowledgements

We thank Dr. B. Prabhakar Rao, Dean; Dr. P. Satyavathi Devi, Principal; Dr. Y.K. Mastan Rao, Professor & Head, Microbiology Department, Prathima Institute of Medical Sciences for their support and encouragement and to the Professor & Head, Department of Microbiology, Kakatiya University, Warangal, A.P. for encouragement.

References

Agarwal, P.B., A. Narang and P. Kumar (1996) Indian J. Paediatr. 63: 775-780

Alexander, B.D. and J.R. Perfect (2000) Drugs 54: 657-678

Andriole, V.T. (2000) Int. J. Antimicrob. Agents 19: 317-322

Artis, W.M., B.M. Odle and H.E. Jones (1981) Arch. Dermatol. 117: 16-19

Bahuguna, S. and R.K. Kushawaha (1989) Mycoses 32: 340-343

Chander, J. (2002) Text book of Medical Mycology, 2nd edition, Mehta publishers, New Delhi.

Cormican, M.G. and M.A. Pfaller (1993) J. Antichemother. 38: 561-578

Espingel-Ingroff, A. and T.M. Kerkering (1991) J. Clin. Microbiol. 29: 393-394

Fernandez-Torres, B., A.J. Carillo, E. Martin, A. Del Palacio, M.K. Moore, A. Valverde, M.Serrano and J. Guarrow (2001) Antimicrob. Agents Chemother. 45: 2524-2528

Gupta, A. K. and N. H. Shear (1997) J. Am. Acad. Dermatol. 37: 979-998

Jessup, C.J. Warner, N. Isham, I. Hasan and M.A. Ghannoum (2000) J. Clin. Microbiol. 38: 341-344

National Committee for Clinical Laboratory Standards (1997). M27-A, NCCLS, Villanova,Pa

Norris, H.A., B.E. Elewski and M.A. Ghannoum (1999) J. Amer. Acad. Dermatol. 40(6): S9-S13

Perea, S., A.W. Fothergill, D.A. Sutton and M.G. Rinaldi (2001) J. Clin. Microbiol. 39: 385-388

Pfaller, M.A., S.A. Messer, K. Millis and A. Bolmstoum (2000) J. Clin. Microbiol. 38: 3359-3361.

Chapter - 8

Silver Ions induced alterations in the Energy Transfer Process in Photosystem II of the Cyanobacterium, *Spirulina platensis*

Ayya Raju, M., Sudheer, S.D. and Murthy, S.D.S.

Department of Biochemistry, S.V.University, Tirupati-517 502, A.P. INDIA
E-mail: sdsmurthy@rediffmail.com.

Abstract

Exposure of intact cells of *Spirulina* to silver ions (5-15 μM) caused decrease in the phycocyanin absorption and loss in the phycocyanin fluorescence emission indicating the altered energy transfer from phycocyanin to chlorophyll *a* under *in vivo* conditions. The *in vitro* measurements of spectral properties of phycobilisomes isolated from silver treated cells, clearly demonstrated that phycobilisome energy transfer is the major target for silver action. Thus phycocyanin is the major susceptible pigment protein for silver ion action which is evidenced from the decrease of fluorescence emission of phycocyanin.

Key words: Absorption; Allophycocyanin; Cyanobacteria; Fluorescence; Phycocyanin.

Introduction

Excess use of fungicides and pesticides and release of industrial effluents is responsible for the accumulation of heavy metals in the aquatic environment. The increase in of heavy metal toxicity in water bodies results in impairment of plant growth and loss of biomass production (Prasad, 2004). Therefore there is a need for the indicator protein to identify the presence of heavy metal toxicity. Cyanobacteria are photosynthetic prokaryotes which respond to the variety environmental stresses (Foy *et al.*, 1978; Clijsters and Van Assche, 1985; Murthy, 1991). These organisms contain specialized pigment protein called phycobiliproteins, as light harvesting complexes (Bryant *et al.,* 1979). They are phycoerythrin (PE), phycocyanin (PC) and allophycocyanin (APC).

Murthy *et al* (1989) showed that the above pigment proteins are sensitive and exhibit spectral alterations in *Spirulina* under mercury stress. Therefore in this investigation an attempt has been made to analyze the effect of silver ions (5-15 μM) on spectral properties of phycobiliproteins in *Spirulina platensis* both and under *in vivo* and *in vitro* conditions.

Materials and Methods

Spirulina platensis trichomes were grown in Zarrouk's medium (Zarrouk, 1966) at 25 ± 2^0 C under continuous illumination (15 Wm^{-2}). *In vivo* experiments were conducted by incubating the cells with silver (Ag) (5-15 μM) for 10 min under continuous stirring. For *in vitro* studies the cells were treated with Ag ions (3 μM) for 12 h and the phycobilisomes (PBsomes) have been isolated. The PBsomes isolated were according to the method of Gantt *et al.* (1979) with slight modifications. The PBsomes were recovered from the 1.0 M region as an intense blue band. Sucrose was removed from the isolated PBsomes by using dialysis, with against 0.75 K. PO_4 (pH 7.0) PBsomes and intact cells both were used for spectral measurements. The absorption spectra of intact cell suspension and PBsomes were taken by using a Hitachi - 557 double beam, spectroph-otometer as described by Murthy *et al.* (1989). The emission spectral of cell samples and PBsomes were measured by using Perkin-Elmer spectrofluoremeter. Cells equivalent to 15 μg of Chl were used for spectral measurements, where as PBsomes equivalent to 30 μg of protein was used for both absorption as well as fluorescence emission measurements.

Results and Discussions

Initially after giving the silver ions (5-15 μg) treatment for 10 min, the absorption characteristics of different pigment proteins present in the control cells of *Spirulina platensis* was measured. The peak at 679 nm is due to the absorption of Chl *a*; at 621 nm is due to the absorption of PC and 480 nm and carotenoids is due to a peak at 438 nm is due to the soret band of Chl *a* (Fork and Mohanty, 1986). The treatment of intact cells of *Spirulina* with different concentrations silver (5-15 μg) caused drastic decrease in phycocyanin absorption by marginally affecting the Chl *a* and carotenoid absorption. In addition there is a 7 nm blue shift of PC indicating the structural alterations in PBsomes (Fig. 1). Similar observations were made by Murthy *et al.* (1989) in *Spirulina* under mercury stress. Since silver affected the phycocyanin absorption quite extensively further studies were made by measuring room temperature phycocyanin fluorescence. The control cells excited with 545 nm light beam exhibited an emission peak at 658 nm which indicates that the energy is transferred from PC to Chl *a* (Singhal *et al.,* 1981; Fork and Mohanty, 1986). Figure 2 shows the phycocyanin fluorescence emission spectra of silver treated *Spirulina* intact cells. With the treatment of Ag ions drastic decrease was noticed in the fluorescence emission intensities and with 10 μM of silver ions almost 50% of loss in the fluorescence intensity of phycocyanin was observed. It clearly demonstrates that selected heavy metal (Ag) induced alterations in the energy transfer from PC to Chl *a* by inducing the structural changes in the phycobiliproteins.

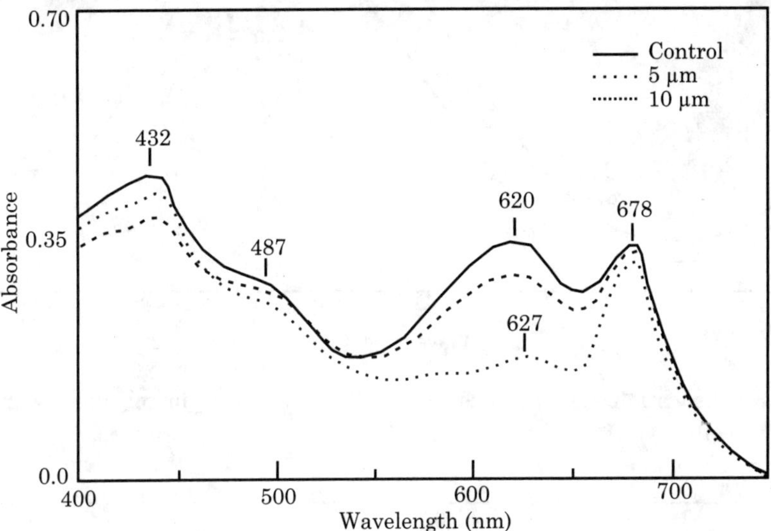

Figure 1. Effect of Ag ions on the absorption spectra of intact cells of *Spirulina platensis*

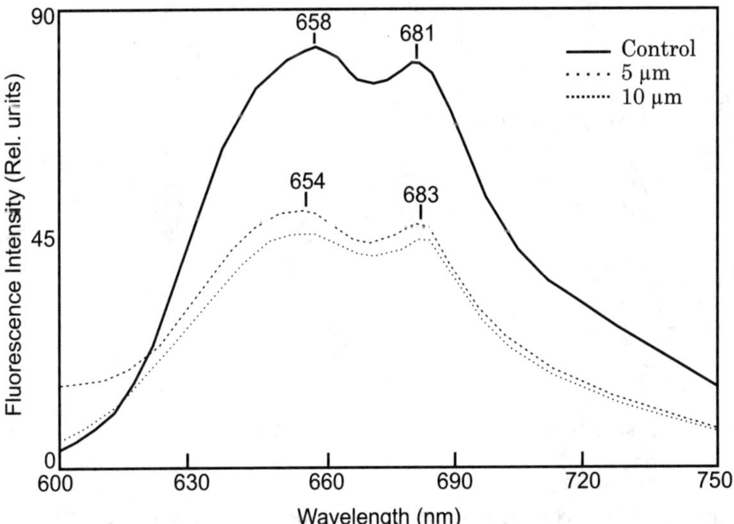

Figure 2. Effect of Ag on phycocyanin fluorescence emission spectra of the intact cells

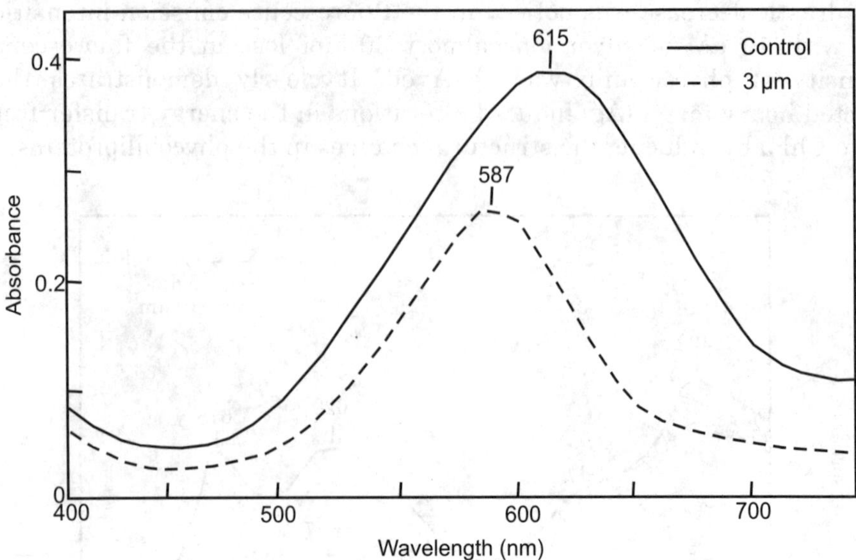

Figure 3: Absorption spectra of isolated PBsomes from control and Ag treated cells.

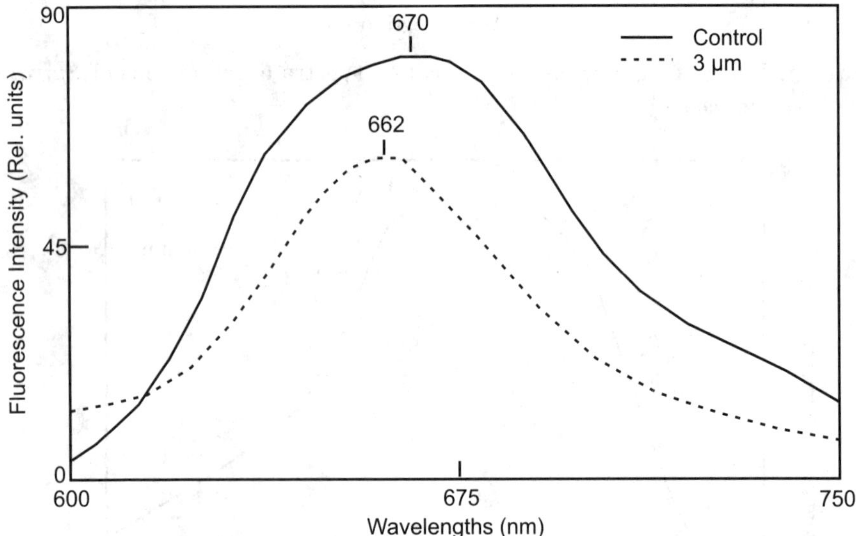

Figure 4: Fluorescence emission spectra of isolated PBsomes from control and Ag treated cells

To correlate the results of *in vivo* experiments with *in vitro* experiments, phycobilisomes have been isolated from control as well as Ag treated cells by using sucrose density gradient. After removal sucrose the spectral properties have been measured (Fig. 3 and 4). The absorption spectra of PBsomes exhibit a main peak at 615 nm. The PBsomes isolated from 3 μM of Ag treated cells, caused a decrease in the absorption of PC by 40% and shifted the peak position from 615 nm to 600 nm. Since the absorption properties are related to the fluorescence emission of PC, PBsome samples which were isolated from silver exposed (3 μM) *Spirulina* cells were used for the measurement of phycocyanin fluorescence emission. Silver is able to cause 42% decrease in the fluorescence intensity and blue shift in the emission peak from 670 nm to 662 nm. The decrease in the fluorescence intensity indicates the change in the energy transfer and blue shift gives information about structural changes in the PBsomes (Fig. 4).

Similar reports were made by Murthy *et al.* (1989) during the toxic effect of mercuric chloride ($HgCl_2$) on the spectral properties of phycobiliproteins in the same organism. Thus silver is able to cause alterations in the energy transfer from PC to Chl *a* in the *Spirulina* both under *in vivo* as well as *in vitro* conditions by inducing changes in PBsomes.

Acknowledgement

S.D.S. Murthy is thankful to University Grants Commission for providing Major Research Project (2011-2014).

References

Bryant DA and Stirewalt VL (1990) FEBS Lett 259: 273-280.

Clijsters H, Van Assche, F (1985). Photosynth Res 7: 31-40.

Fork DC and Mohanty P (1986) Fluorescence and other characteristics of blue-green algae (Cyanobacteria), red algae and cryptomonads. In: Light Emission by Plants and Bacteria (Govindjee, Amsez, 1. and Fork, D.C., eds.) pp. 451-496, Academic Press, New York.

Foy CD, Chaney RL and White MC (1978) Ann Rev Plant Physiol **29**: 511- 566. doi: 10.1146/annurev.pp.29.060178.002455

Gantt E, Lipschutz CA, Grabowski J and Zimmerman BK (1979) Plant Physiol 63: 615-620.

Murthy SDS (1991) Ph.D thesis, Jawaharlal Nehru University, New Delhi.

Murthy SDS, Sabat SC and Mohanty P (1989) Plant Cell Physiol 30: 1153–1157.

Prasad MNV (2004) Heavy metal stress in plants: from Biomolecules to ecosystems. Vol-1: pp. 462, Springer-Verlag, Berlin.

Singhal GS, Mohanty P and Govindjee (1981) Z. Pflanzen-physiol. 103: 217-228.

Zarrouk C (1966) Ph.D. Thesis, University of Paris, Paris.

Chapter - 9

Succession of mycotoxigenic fusaria in relation to storage structures

Narasimha Rao, K.V. Koteswer Rao, S. Girisham and S.M. Reddy

Department of Microbiology, Kakatiya University,
Warangal – 506 009, A.P., India.

ABSTRACT

Influence of storage structures on the incidence of species of *Fusarium* and biochemical composition of maize, paddy and sorghum during one year storage in basket, gunny bag and pot was analysed. Species of *Fusarium* decreased both qualitatively and quantitatively while incidence of species of *Aspergillus* and *Pencillium* increased with the progress of one year storage. Proteins, free fatty acids and phenols increased, while free amino acids and sugars decreased with the progress of storage period.

Key words : *Fusarium, Aspergillus, Penicillium*, Storage structures.

Introduction

The storage of food grains either for human consumption during scarcity or for sowing the future crops, can be traced back to the bablical times. Though there is a tremendous advancement in the storage of grains in advanced countries like USA, the most traditional way of storage is being still practiced in our country. Chohan *et al.* (1972) reported that the condition of stored wheat in aerated bins was comparatively better than the wheat stored in non-aerated and unventilated bin. Girisham and Reddy (1985) and Komuraiah and Reddy (1986) during their study on the influence of storage structures on pearl-millet and methi seeds respectively concluded that the closed tins supported more mycoflora than the aerated basket. Similarly Rajakumari and Reddy (1992) reported that mould infestation and seed deterioration of fennel was more in seeds stored in tin. According to Chary and Reddy (1988) the seeds in pot were comparatively more mould infested which they attributed to the porous nature of the pot. Such variation in the seed mycoflora with the storage structure was also recorded by Schroedar and Hein (1986), Prasad *et al.* (1988) and Prasad and Pathak (1987). Though there are reports on the nutritional changes in relation to storage structure (Prasad *et al.*, 1988; Pushipinder Paul Singh *et al.*, 1986; Prasad and Kumar 1987; Hareesh *et al.*,1996) and mycotoxin production (1986), a wide gap still exists between what is observed and what is the fact. In the present investigations, effect of storage structures used by tribals of Godavari belt on the paddy, maize and sorghum was studied.

Materials and Methods

Three kilograms of freshly harvested seeds of paddy, maize and sorghum were stored in Gunny bag, basket and pot under laboratory conditions for one year (October 2001 to September 2002) and seed mycoflora of each sample was analysed at 3 months interval by blotter technique. Three layers of blotters (size equivalent to the petri plate) soaked in water were placed in petri plate and excess water was drained. Fixed number (25-30) of seeds were placed at equal distance in each petri plate. Petri plates thus prepared were incubated at 27-29°C for 7 days. Nearly 300-400 seeds were examined for qualitative and quantitative estimation of fungi associated with the seeds. The fungi appearing on the seed surface were isolated and identified with the help of standard monographs after recording morphological features. Fungi which failed to grow on culture medium were identified with the help of several slides prepared from the seed surface itself. Percentage of incidence of individual fungus was calculated by the following formulae. Special attention was paid for isolation, purification and identification of different species of *Fusarium*.

$$\text{Percentage of incidence} = \frac{\text{No. of colonies of species in all the species}}{\text{Total no. of colonies of all the species in all the plates}} \times 100$$

Simultaniously, changes in biochemical composition of paddy, maize and sorghum were also analysed by standard methods such as proteins (Lowry *et al.*,1951), free amino acids (Moore and Stein, 1948), free fatty acids (AOAC,1960) and reducing sugars (Dubois *et al.*,1951).

Results and Discussion

Four species of *Fusarium,* besides large number of other fungi including species of *Aspergillus* and *Penicillium* could be recorded on the seed surface of paddy (Table 1). However, the incidence of species of *Fusarium* was considerably low with no definite trend quantitatively. *F. moniliforme* followed by *F. oxysporum* were most frequent. Seeds stored in bag and gunny bag supported more population of *F. moniliforme. F. graminarum* disappeared after 3 months of storage in pot, while seeds stored in basket could support its growth up to 9 months of storage. However, its population decreased gradually. *F. solani* could be traced up to 9 months of storage on seeds stored in basket, while on seeds stored in pot and gunny bag its incidence decreased from 3rd month onwards. The population of *A. flavus, A. niger* followed by *A. fumigatus* was high, during initial stages only.

Table 1. Effect of storage structure on the incidence of fungi on Paddy

		Nov			Jan			Mar			May			July			Sep		
	Intial day	1	2	3	1	2	3	1	2	3	1	2	3	1	2	3	1	2	3
Fusarium moniliforme	5.6	2.7	1.6	0.8	3.7	4.2	--	0.8	--	--	2.2	4.2	--	1.8	2.2	0.8	1.2	0.1	--
F. oxysporum	4.2	1.2	0.8	--	0.9	0.4	--	0.6	--	--	3.2	1.6	--	1.2	1.6	--	0.9	--	--
F. solani	2.8	0.8	--	--	0.6	--	--	0.2	--	--	1.2	1.4	0.3	--	--	--	--	--	--
F. graminarum	1.2	1.1	0.1	--	0.2	--	--	0.1	--	--	--	--	--	--	--	--	--	--	--
Aspergillus. flavus	11.6	14.2	13.7	12.6	17.9	14.3	11.2	8.7	6.8	4.1	2.1	4.3	2.4	1.9	3.6	1.8	0.8	2.1	1.2
A. niger	6.7	4.8	6.7	4.3	0.8	2.9	0.8	1.2	4.8	2.0	1.2	4.2	1.4	1.6	3.1	1.2	0.6	1.5	0.9
A. fumigatus	9.7	6.8	4.2	4.8	6.4	11.3	6.8	7.9	4.3	3.1	4.2	6.7	2.4	3.8	5.2	1.8	2.1	4.6	1.2
A. japanicus	--	--	--	0.2	--	--	0.6	--	--	0.3	-	--	0.1	--	--	0.4	--	--	0.7
Chaetomium globosum	1.4	--	--	--	--	--	--	--	--	--	--	--	--	--	--	--	--	--	0.2
Cladosporium cladosporoides	--	1.4	--	1.6	2.7	--	--	2.8	4.7	--	2.7	4.8	--	2.4	3.6	--	1.6	2.2	--
Curvularia lunata	2.8	1.2	--	--	--	--	1.4	--	--	--	--	--	--	--	--	--	--	--	--
Drechslera halodes	--	--	--	--	--	--	1.2	--	--	0.6	--	--	--	--	--	--	--	--	--
Mucor spp	2.2	1.2	--	0.3	1.2	4.2	1.4	--	0.3	0.5	--	1.2	1.7	--	0.9	1.4	--	0.6	1.1
Myrothecium roridum	--	--	--	--	--	1.2	1.4	--	--	0.8	--	1.4	0.8	--	1.2	0.6	--	0.9	0.4
Penicillium spp	11.2	4.7	6.2	4.2	2.9	4.2	2.9	4.8	2.6	8.7	4.8	4.2	2.7	3.6	3.9	2.4	2.8	36.2	1.8
R. stolonifer	1.4	--	--	--	--	4.2	--	1.4	--	--	1.5	--	0.7	1.2	--	0.6	1.8	--	--
T. viride	6.8	4.7	6.2	9.2	4.8	--	--	1.4	--	--	1.9	--	--	1.3	--	--	0.9	--	--
T. roseum	--	--	1.8	--	--	--	--	--	1.4	--	--	--	2.4	--	--	--	--	--	--
Sterile mycelium	--	--	--	--	--	1.2	--	--	1.0	--	--	--	--	--	--	--	2.4	--	--

1 = Basket, 2 = Gunny bag; 3 = Pot

Significant change in biochemical composition due to mould activity was recorded which varied with the storage sturctures and the biochemical constituent (Table 2). Phenols recorded a progressively increase due to storage. Phenol content was high in seeds stored in pot and low in seeds stored in gunny bag. Proteins also recorded increase with the progress of storage period tried. The increase was very significant in seeds stored in basket. Free fatty acids increased with the storage period, while reducing sugars and FAA recorded decrease with progress of storage period.

Table 2. Effect of storage structures on biochemical composition of paddy during one year storage (Oct. 2005 to Sept. 2006)

Storage structure	Storage Period	Phenols (mg/gr)	Proteins (mg/gr)	FFA (mg/100 mg)	Reducing sugars (mg/gr)	FAA (mg/g)
	Initial day	0.172	0.492	0.08	0.592	0.123
Pot	November	0.185	0.583	0.11	0.566	0.108
	January	0.190	0.610	0.19	0.543	0.083
	March	0.205	0.693	0.28	0.440	0.043
	May	0.234	0.751	0.36	0.454	0.038
	July	0.280	0.889	0.44	0.289	0.023
	September	0.298	0.901	0.49	0.263	0.014
Basket	November	0.187	0.614	0.14	0.533	0.990
	January	0.198	0.680	0.23	0.504	0.830
	March	0.216	0.703	0.29	0.483	0.518
	May	0.238	0.781	0.34	0.421	0.324
	July	0.254	0.853	0.48	0.310	0.128
	September	0.271	0.956	0.53	0.289	0.025
Gunny bag	November	0.165	0.493	0.10	0.561	0.310
	January	0.172	0.518	0.17	0.481	0.140
	March	0.178	0.534	0.25	0.423	0.114
	May	0.183	0.581	0.32	0.403	0.112
	July	0.205	0.634	0.39	0.323	0.102
	September	0.219	0.845	0.44	0.210	0.098

Four species of *Fusarium* besides large number of fungi including species of *Aspergillus* and *Penicilluim* could be recorded on the seed surface of maize (Table 3). Interestingly, the incidence fusaria was very low and they could not be traced in many observations. The incidence of

A. niger followed by *A. terreus* and *A. japanicus* was common and could be recorded on grains stored in all the three storage structures. The population of species of *Aspergillus* increased with the progress of the incubation periond. Among the species of *Fusarium, F. moniliforme* was dominant and recorded on seeds stored in all the three storage structures. *F. moniliforme* was more during initial stage and decreased with the progress of storage. It was completely eliminated from the seed surface of maize grains stored in bag. *F. oxysporum,* which was recorded during initial stages of storage on seeds stored in different storage structures, decreased and disappeared during subsequent observations periods. However, it could be recorded on the seeds stored in basket during November and May. It was also traced on seeds stored in bag. *F. solani* and *F. equisetti* could be recorded only during August. However, *F. equisetii* was recorded on maize grains stored in pot, even upto 9 months of storage. It was also recorded on seeds stored in bag after 6 months storage. *R. stolanifer* was recorded through out the observation period with varying percentage of incidence. Species of *Penicillium* and *Curvularia* were also recorded during observation period. *T. viride,* one of the biocontrol agents, could be recorded only sporadically.

Table 3. Effect of storage structures on the incidence of fungi on maize grains

Name of the fugas	August – 2005			November-2005			February–2006			May - 2006		
	Pot	Bas	Bag	Pot	Bas	Bag	Pot	Bas	Bag	Pot	Bas	Bag
Fusarium moniliforme	33.6	28.4	23.6	18.3	21.4	17.8	3.4	2.8	6.4	2.8	3.2	--
F. oxysporum	6.9	1.4	3.4	--	2.8	--	--	--	--	--	1.4	2.1
F.solani	--	1.8	1.4	--	--	--	--	--	--	--	--	--
F. equiseti	4.4	5.8	1.2	1.4	--	1.8	1.1	--	--	--	--	--
Aspergillus flavus	--	1.4	0.8	--	--	--	--	--	--	--	--	--
A.fumigatus	--	0.3	--	--	0.8	1.3	--	--	--	--	--	--
A. niger	10.8	16.0	17.4	16.4	14.3	15.6	23.4	28.4	20.3	30.4	36.8	23.9
A. japanicus	--	1.4	2.3	4.8	2.9	1.3	8.6	3.9	9.4	9.8	14.3	12.8
A. terreus	1.8	3.6	4.8	12.2	16.3	9.8	18.4	18.8	13.4	17.9	28.3	19.3
Alternaria alternata	--	0.8	--	1.3	--	--	--	--	--	--	--	--
Curvularia lunata	1.6	--	--	6.4	8.6	9.1	6.3	4.8	3.4	8.6	9.1	1.8
Drechslera halodes	--	1.3	2.8	--	--	--	--	--	--	--	--	--

Chaetomium globosum	0.0	1.4	--	1.4	2.6	--	--	--	--	--	--	--
Cladosporium cladosporoides	3.8	1.9	--	4.8	9.3	1.8	1.3	--	--	--	--	--
Pencillin spp.	--	1.3	4.8	--	1.8	1.4	--	--	--	--	3.1	--
(P.citrinum,	5.8	4.3	8.3	1.4	3.8	3.9	1.6	3.4	--	6.1	3.8	1.3
P.chrysogenum, P. purpuro-genum)	2.4	1.3	--	--	1.4	--	1.6	3.8	3.9	1.8	2.4	4.9
Rhizopus stolonifer	8.4	9.2	6.4	9.3	8.1	8.4	14.3	18.4	9.6	8.9	16.4	13.8
Trichoderma viride	--	1.8	--	--	1.4	1.8	--	--	--	1.3	--	--
Sterile mycelia	1.8	2.3	4.6	14.3	16.3	14.8	19.4	16.8	28.3	14.8	17.9	11.8

Pot – Pot; Bas – Basket; Bag - Gunny Bag

Phenols and proteins increased with the storage (Table 4). There was not much difference in total phenols which increased with the advancement of observation period. The increase in proteins was comparatively more in seeds stored in gunny bag. Free fatty acids, which are an index for seed deterioration, also increased. The increase was maximum in seeds stored in pot followed by basket. Reducing sugars decreased with storage time. Comparatively more significant change in seeds stored in gunny bag was recorded. Free amino acids also decreased progressively with progress of the incubation period.

Table 4. Biochemical changes in maize grains stored in different storage condition structures Employed by the tribals of Godavari belt (August, 2005 – May, 2006)

Storage structure	Time of storage	Phenols (mg/g)	Proteins (mg/g)	Free fatty acids (mg/g)	Reducing sugars (mg/g)	Free amino acids (mg/g)
	Initial day (0 day)	0.78	47.30	0.58	284	1.63
Pot	August	0.84	48.30	0.86	271	1.16
	November	1.10	64.10	1.51	254	1.05
	February	1.48	82.40	2.13	239	0.88
	May	1.96	92.10	2.74	228	0.67
Basket	August	0.89	51.30	0.71	278	1.08
	November	0.97	66.40	0.94	268	0.82

	February	1.17	88.30	1.21	257	0.59
	May	1.61	96.40	2.05	246	0.41
Gunny Bag	August	0.80	54.30	0.86	240	1.20
	November	1.12	73.20	1.41	231	0.93
	February	1.51	96.40	1.53	215	0.81
	May	1.82	108.90	1.90	201	0.53

Altogether, 4 species of *Fusarium* were recorded on seeds of sorghum (Table 5). The population of *F. moniliforme* was more, which decreased with advancement of storage period. *F. oxysporum* was also recorded up to 6 months of storage in different storage sturctures. *F. solani* recorded on seeds stored in bag after 6 months of storage. *F. equiseti* could be recorded after 9 months of storage on the seeds stored in pots, while the seeds stored in other storage structures recorded low percentage of incidence. *A. flavus* followed by *A. fumigatus* and *A. niger* were recorded with comparatively high percentage of incidence. The population of these fungi increased, while population of *Fusarium* decreased with the progress of storage in different storage structures.

Table 5. Effect of storage structures on the incidence of fungi on sorghum grains

Name of the fungas	August – 2005			November – 2005			February – 2006			May – 2006		
	Pot	Bas	Bag	Pot	Bas	Bag	Pot	Bas	Bag	Pot	Bas	Bag
Fusarium moniliforme	30.4	20.8	26.8	18.6	14.3	20.4	--	12.4	6.9	1.3	2.8	0.8
F. oxysporum	3.6	4.3	8.1	1.4	6.3	3.8	--	--	1.6	--	--	--
F. solani	--	--	1.4	--	--	2.1	--	--	--	--	--	--
F. equseti	1.3	--	--	1.7	2.8	3.4	2.3	--	--	1.8/	2.1	--
Aspergillus flavus	34.4	23.1	14.6	38.6	30.3	28.4	28.4	17.6	20.3	29.3	20.6	28.6
A. fumigatus	3.4	19.8	18.4	--	8.3	20.4	9.8	13.9	12.6	--	11.3	10.8
A. niger	12.4	10.4	11.6	9.8	1.3	0.8	16.8	18.9	10.4	23.4	24.6	28.9
A. japanicus	--	--	16.4	--	--	--	--	--	--	--	--	--
A. terreus	--	--	1.3	--	--	--	--	--	--	--	--	--
Alternaria alternata	--	--	--	1.4	--	2.1	3.1	1.8	2.3	--	--	--
Curvularia lunata	--	1.8	0.4	2.8	6.4	1.8	6.4	1.3	--	--	--	--
Drechslera halodes	--	--	--	--	1.4	--	--	--	--	--	--	--
Chaetomium globosum	1.3	6.4	3.2	2.1	1.8	--	1.3	--	--	4.8	3.9	--

Cladosporium cladosporoides	--	--	--	1.8	2.8	1.3	--	3.9	4.2	6.4	8.3	1.4
Penicillium spp. (*P. citrinum, P.chrysogenum, P.purpurogenum)*	2.8	6.8	4.3	--	--	2.8	--	1.3	2.1	10.4	6.8	3.4
Rhizopus stolonifer	13.4	14.8	20.6	4.9	3.1	1.4	8.4	9.3	8.9	7.8	13.4	11.3
Trichoderma viride	--	--	--	--	--	--	1.4	--	--	--	--	1.3
Sterile mycelia	2.8	2.3	4.7	1.8	4.8	4.9	16.8	24.3	21.4	26.8	19.3	16.8

The phenols increased with progress of storage, till the end of observation, while seeds stored in gunny bag, showed a decline during later 9 months of storage (Table 6). The protein content increased significantly with progress of incubation period in the seeds stored in all 3 storage structures employed. Seeds stored in basket showed minimum changes, when compared with seeds stored in gunny bag. Free fatty acids increased with the progress of incubation period. It was comparatively more in seeds stored in pot and gunny bag. Reducing sugars progressively decreased during the storage. Decline was maximum when seeds were stored in gunny bag, followed by basket. Free amino acids decreased progressively in seeds stored in basket, followed by gunny bag. The seeds stored in pot recorded minimum change in amount of free amino acids.

Table 6. Biochemical changes in sorghum vulgare seeds stored in different storage structures (August 2004 to May 2005)

Storage structure	Time of storage	Phenols (mg/g)	Proteins (mg/g)	Free fatty a cids (mg/g)	Reducing sugars (mg/g)	Free amino acids (mg/g)
	Initial day (0 day)	0.24	8.40	0.20	199.40	2.18
Pot	August	0.38	9.80	0.29	176.40	1.81
	November	0.71	11.80	0.31	150.10	1.43
	February	0.99	14.40	0.39	136.30	0.98
	May	1.31	17.80	0.58	109.80	0.48
Basket	August	0.49	10.80	0.20	184.30	1.43
	November	0.70	9.90	0.23	143.40	0.91
	February	0.89	12.80	0.28	116.80	0.52
	May	0.99	16.80	0.33	99.70	0.23
Gunny Bag	August	0.26	9.40	0.24	118.40	1.71
	November	0.41	14.90	0.30	140.50	1.20
	February	1.64	15.80	0.46	108.30	0.99
	May	1.01	19.80	0.56	79.40	0.34

Acknowledgements

Thanks are due to Head, Department of Microbiology and UGC, New Delhi for providing facilities and financial assistance.

REFERENCES

AOAC (1960) Official methods of analysis of the association of agricultural chemists, Benzamin, Franken Station, Washington, 832.

Chary MP, Reddy SM (1988) Geobios New Rept. **7**; 190-192.

Chohan JS, Dhanraj, KS, Sunand MS, Warachi KS (1972) Bull Gr. Technol. **10** : 256-262.

Dubois M, Gilles K, Hamitton JK, Rebers PA, Smith F (1951) Nature **168:** 167.

Girisham S, Reddy SM (1985) Giobios New Reptr., 4 : 126-129.

Hareesh Vardhan Rao P, Giridhar P, Reddy SM (1996) J. Ind. Bot. Soc. **75**: 313-314.

Komaraiah M, Reddy SM (1986) Proc. Natl. Acad. Sci. **56** : 298-299.

Lowry OH, Rose Brough NJ, Farr L, Randall RJ (1951) J. Biochemical Chem. **193**: 265-275.

Moore S, Stein WH (1948) J. Biol. Chem. **176** : 367-388

Prasad T, Kumar S (1987) J. Indian Bot. Soc. **66**: 388-391.

Prasad T, Pathak SS (1987) Indian Phytopath. **40**: 39-46.

Prasad T, Sinha RK, Shivendra Kumar (1988) J. Indian Bot. Soc. **67**: 194-196.

Punam Jeswal, 1986. J. Indian Bot. Soc. 65: 101-106.

Pushipinder Paul Singh, Bedi PS, Dhoeja Singh (1986) Indian Phytopath. **39**: 593-596.

Rajakumari D, Reddy SM (1992) *Proc.* Nat. Acad. Sci. Indian **62**: 89-94.

Schroeder HW, Hein H (1986) Appl. Microbiol. **15** : 441-449.

Chapter - 10

Production of protease on different synthetic media by three fruit-rot fungi of banana

Supriya Sarkar, S. Girisham and S.M. Reddy

Department of Microbiology, Kakatiya University, Warangal, (A.P.)-506 009

ABSTRACT

Production of protease by three fruit-rot fungi-*Macrophomina phaseolina* R-4242, *Fusarium oxysporum* QJC-1403 and *Nigrospora oryzae* NRRL 54030 isolated from the decaying fruits of banana (*Musa paradisiaca* L.) on different synthetic media was investigated.

Present investigation reveals that all the three pathogens were able to produce protease which, however, varied with the pathogen and the medium tried. *M. phaseolina* could produce maximum protease in medium C, followed by medium B. *F.oxysporum* also preferred medium C for the production of protease when compared to the other media. While *N. oryzae* recorded maximum protease activity in medium E and F. Rest of the media supported intermediate degree of protease activity for all the three pathogens.

Key words: *Musa paradisiaca*, protease, *Macrophomina phaseolina, Fusarium oxysporum, Nigrospora oryzae*, synthetic media.

Introduction

Banana fruits are consumed worldwide. It is an important commercial fruit crop of India, accounting for about 37% of the total fruit production. Fruits at the post-harvest stage are exposed to the unnatural conditions which predispose the fruits for the attack of fungi. Pathogenic fungi establishes themselves in the host by secreting extracellular enzymes which breakdown the host tissues (Ajayi *et al.*, 2007).

Plant membrane contains nearly 30-35% protein and they are in the globular form with high portion of molecules in the configuration of α-helix. Plant cell walls are also reported to contain hydroxy proline rich glycoproteins (Bateman and Bashman, 1976). Plant pathogens are known to produce proteinases to degrade proteins of the membrane. Proteinases which hydrolyse peptide bonds in proteins also catalyse the formation of peptide bonds and amino transfer reaction and hydrolyse ester linkages involving amino acids (Barras *et al.*,1994). Proteinases have been distinguished into two types: (1) Endo-peptidases (EC 3.4.21-99) act on peptide bond randomly in the interior of protein chain. (2) Exo-peptidases (EC 3.4.11-19) act only on peptide bonds which are adjacent to a free α-amino or a free carbonyl group and release small peptides. Johnson and Brown (1974) classified proteases into four categories.

(a) Acid proteases

(b) Alkaline proteases

(c) Neutral proteases

(d) Thiol proteases

The role of proteases in altering the functional properties of proteins has been reviewed by Chen *et al.* (1999). Kuc (1962) reported the synergistic action of proteinase and pectinases in the tissue maceration of disease caused by *Cladosporium cucumerinum*. Prasad *et al.* (1988), Babu and Reddy (1988), Sakririsat *et al.* (1991), Singh and Shukla (1999), Robert *et al.* (2001), Adejuwon and Olutiola (2005), Pekkarinen *et al.* (2000), Shumi *et al.* (2004) and Fabiana *et al.* (2005) have also reported the production of protease by several pathogenic fungi studied by them. Dhawan (1980) felt that proteases play a major role in the dissolution of middle lamella and other structural wall proteins, so as to facilitate the establishment of pathogen. Mandakini and Shastry (1983) have tried to correlate protease secretion and pathogenicity of *Alternaria alternata*. Fungi are reported to differ in the production of proteases both at inter specific (1983) and intra specific level (Raghavan *et al.*, 1979; Bhagavan Reddy, 1986; Madhukar and Reddy,1988). In view of the above facts, an attempt has been made to study the protease production by three fruit-rot fungi of banana in relation to their pathogenicity.

Materials and Methods

Organism and culture conditions

The three fruit-rot fungi employed for this research work were isolated from banana fruits in the Department of Microbiology, Kakatiya University, Warangal. The fungi were routinely grown and maintained on Asthana and Hawker's agar slants. They were subcultured from the old culture onto fresh agar slants whenever it was to be used. Five day old culture of the organism served as inoculum.

Protease activity was assayed in the following media:

(1) Asthana and Hawker's medium A (Glucose 5.0 g, potassium nitrate 3.5 g, potassium dihydrogen orthophosphate 1.75 g, magnesium sulphate 0.75 g & distilled water -1000 ml) **(A)**

(2) Medium A + 1% banana fruit extract **(B)**

(3) Medium A + 1% banana unripe fruit extract **(C)**

(4) Medium A + 1% banana leaf extract **(D)**

(5) Mineral protein medium (starch 10g, peptone 0.5 g, potassium dihydrogen orthophosphate 0.5 g, magnesium sulphate 0.5 g, egg albumin 2.0 g & distilled water 1000 ml) **(E)**

(6) Glucose-peptone medium (Glucose 20.0 g, peptone 4.5 g, asparagine 4.5 g, potassium dihydrogen orthophosphate 3.4 g, magnesium sulphate1.9 g, sodium chloride 0.01g and distilled water 1000 ml) **(F)**

The fungi (*M. phaseolina*, *F. oxysporum* and *N. oryzae*) were grown in 25 ml of sterilized medium taken in 100 ml Ehrlenmeyer conical flasks and incubated at room temperature (27± 2°C). The pH of the medium was adjusted to 5.6 with the help of 0.1 M HCl. At the end of 5, 10 and 15 days of incubation period, a set of flasks were harvested on previously dried and weighed Whatman filter paper No. 42. The filter papers with the mycelium were dried at 65-75°C for 48 hrs. They were weighed to a constant weight in an analytical balance after cooling to room temperature in a desiccator.

Enzyme assay

Protease activity was assayed as suggested by Keay and Wrildi (1970). Two grams of casein was dissolved in 20 ml of 1N NaOH solution at room temperature in a 250 ml beaker by working the paste against the side of the beaker with a spatula. To the above mixture 50 ml of distilled water was added and the solution was stirred. The pH was adjusted to desired value by drop wise addition of dilute phosphoric acid (1 part of 85% phosphoric acid diluted with 3 parts of distilled water). After equilibrium the solution was transferred to a volumetric flask and the volume was made upto 100 ml with distilled water and this solution was used as substrate.

One ml of dialyzed enzyme sample was incubated in a test tube at 37°C for 5 min. Then 1 ml of 2% casein solution was added and the reaction mixture was incubated at 37°C. After 10 min the reaction was terminated by adding 2 ml of 0.4 M TCA (Trichloroacetic acid). The reaction mixture was again incubated for 20 min. At the end of incubation period, it was filtered through Whatman filter paper No. 1 into a clean test tube. One ml of this filtrate was added to 5 ml of sodium carbonate (0.4 M) and 1 ml of dilute FC reagent and incubated at 37°C for 30 min. At the end of the incubation period the absorbency was measured at 660 nm. The blank was prepared by the same procedure but TCA was added at 'zero' time and casein after 10 min. One unit of protease activity was defined as that quantity of enzyme which produced TCA soluble fragments giving blue colour equivalent to 0.5 mg/ml of tyrosine under conditions of assay.

Protein content of the culture filtrates was determined by Lowry *et al.* (1951) using Bovine Serum Albumin (BSA) as standard. Five ml of culture filtrate was taken into a clean test tube and 30% TCA was added for protein precipitation and centrifuged for 20 min at 1800 xg for sample preparation.

Table 1a. ANOVA of protease production on different synthetic media by three fruit-rot fungi of banana

Sources of variation	Sum of Squares	df	Mean Square	P - value	F	Result
Between Groups	1423274.003	107	13301.626	< 0.001	8.491	S
Within Groups	338385.348	216	1566.599			
Total	1761659.351	323				

S-significant

The statistical analysis performed on the results obtained were calculated and ANOVA (SPSS 12.0 version) revealed the significance of the experiment (Table 1 a).

Results and Discussion

From Table 1 and 1a it is clear that all the three fungi under investigation produced proteases in one or other medium. The degree of enzyme production varied both with the fungus as well as with the medium.

Mineral protein medium (medium E) followed by Glucose-peptone medium (medium F) were good substrates for protease secretion by all the three pathogenic fungi. Secretion of protease increased with the progress of incubation period in medium C. *M. phaseolina* could produce maximum protease in medium C followed by medium B. Medium A was poor substratum for supporting the protease activity of the above fungi. Rest of the media supported intermediate amount of protease production.

F. oxysporum and *N. oryzae* were comparatively poor in protease production. *F. oxysporum* also preferred medium C for production of protease. However, *N. oryzae* showed maximum protease activity in medium E and F in the mid way of the incubation period. Chauhan and Nagar (1979) reported the specific influence of substratum on protease secretion by *Fusarium moniliforme* var. *subglutinous* and *Mucor hiemalis* isolated from maize kernels.

Table -1 Production of Protease*, mycelial growth (in mg) and pH changes during the growth of three fruit-rot fungi of banana (*M. paradisiaca* L.) on different synthetic media

Medium	Days of incu-bation	*M. phaseolina* R-4242				Specific activity	*F. oxysporum* QJC-1403				Specific activity	*N. oryzae* NRRL:54030				Specific activity
		Dry wt (in mg)	pH	Total protein (mg)	Prote-ase (in units)	U/mg Protein	Dry wt (in mg)	pH	Total protein (mg)	Prote-ase (in units)	U/mg Protein	Dry wt (in mg)	pH	Total prot-ein (mg)	Protease (in units)	U/mg Protein
Asthana & Hawker's medium A [A]	5	47.9	5.6	70.5	2.0	0.03	65.6	6.1	80.6	2.5	0.03	63.9	5.6	92.0	0.9	0.01
	10	82.8	6.4	98.3	2.3	0.02	104.5	6.3	110.0	4.0	0.04	108.3	5.8	210.6	2.2	0.02
	15	93.0	7.0	45.1	1.0	0.02	131.2	7.0	40.0	1.0	0.02	134.3	6.7	100.0	4.2	0.04
Medium A + 1% banana fruit extract [B]	5	66.9	5.7	100.5	10.2	0.10	130.6	6.0	140.0	10.5	0.08	103.5	5.6	120.6	2.4	0.02
	10	115.2	6.3	310.5	15.8	0.05	176.0	6.8	100.7	7.0	0.07	134.2	6.2	285.0	12.0	0.04
	15	130.9	6.7	55.0	11.0	0.20	209.9	7.1	95.0	3.0	0.03	169.4	7.0	110.9	10.0	0.10
Medium A + 1% unripe banana fruit extract [C]	5	90.6	5.8	115.0	12.0	0.10	82.9	5.7	210.8	12.0	0.06	94.3	6.2	120.4	4.0	0.03
	10	106.5	6.4	322.4	16.0	0.05	109.3	6.5	200.0	10.5	0.05	173.7	6.7	298.5	19.0	0.06
	15	145.9	6.8	60.5	18.0	0.29	194.7	7.2	150.6	8.0	0.05	214.3	7.0	102.4	15.0	0.15
Medium A + 1% banana leaf extract [D]	5	68.9	5.8	75.2	0.9	0.01	92.7	6.2	200.0	10.8	0.05	71.4	5.8	110.0	5.0	0.03
	10	84.2	6.6	290.0	13.0	0.04	137.2	6.7	115.9	7.5	0.06	102.9	6.0	240.0	8.1	0.04
	15	140.6	7.0	55.0	10.0	0.18	150.0	7.0	80.0	2.9	0.04	135.2	6.9	55.0	10.0	0.02
Mineral protein medium [E]	5	76.9	5.6	95.0	11.0	0.20	50.2	5.7	310.0	15.0	0.05	61.2	5.7	108.9	5.0	0.05
	10	94.5	6.0	190.0	12.0	0.06	74.4	5.9	280.4	12.5	0.04	82.9	6.0	240.6	22.0	0.09
	15	110.3	6.8	150.6	14.0	0.09	150.8	6.0	107.0	10.0	0.09	113.1	6.4	84.0	14.0	0.17
Glucose-peptone medium [F]	5	156.3	5.6	80.0	9.0	0.21	157.6	5.9	110.0	4.0	0.04	149.0	5.6	160.0	2.5	0.02
	10	174.0	6.3	148.0	12.5	0.08	238.4	6.3	140.0	7.6	0.05	177.5	5.9	310.0	20.0	0.16
	15	276.0	6.9	115.5	10.0	0.09	269.3	6.9	80.1	3.0	0.03	318.4	6.5	110.0	12.0	0.11

*Expressed in units, one unit of protease is that quantity of enzyme which produced TCA soluble fragments giving blue colour equivalent to 0.05 mg/ml of tyrosinase

Mycelial growth was maximum in medium E and F followed by medium C, B, D and A for all the three pathogens under investigation. The pH drift was towards alkaline side by 10th day of incubation and final pH recorded was near neutral in all the media tried. Satyanarayana and Jain (1979) and Sastry (2002) observed maximum protease activity between pH 5.0 and 7.0 by fungi studied by them. No positive correlation could be observed between mycelial growth and protease secretions.

Low protease production in protein containing medium and maximum protease activity in medium containing fruit extract suggests constitutive nature of protease produced by *M. phaseolina* and *F. oxysporum*. Further increased activity in raw fruit extract in comparison to ripe fruit suggests its limited role in fruit-rot of banana. On the other hand, *N. oryzae* secreted more enzyme in protein containing medium which suggests the adaptive nature of its protease. Similarly Robert *et al.* (2001) have reported that variation in *invitro* protease activity is related to the virulence of *Aspergillus flavus* isolates.

Medium C followed by medium E was responsible for maximum specific activity of protease. However, specific activity of protease varied both with the pathogen as well as with the medium. The more specific activity of protease of *M.phaseolina* and *F. oxysporum* was recorded in medium containing extract of unripened banana fruit. On the other hand, specific activity of protease in protein containing medium was low in these fungi. The behaviour of *N. oryzae* was different as its protease activity was more in protein containing medium. In general a positive correlation could be observed between protease production and its specific activity. Sakririsat *et al.* (1991), Singh and Shukla (1999) have reported the constitutive nature of protease of the fungi studied by them.

Conclusions

In the present study, it was observed that all the three fruit-rot fungi were able to produce proteases. However, they differed significantly in producing protease on the different medium tried. Studies on enzyme production by phytopathogenic fungi are complicated by the presence of plant, particularly by the presence of plant enzymes and microbial enzyme inhibitors that occur in the plants. The most practical way to study the production of these enzymes is by culturing them on artificial growth media that contain no plant or enzyme inhibitors produced by the plant.

This study may serve to increase the understanding of different substrates that effect the production and role of fungal proteases in phytopathogenicity. Studies on production of hydrolytic enzymes using several plant materials are in progress in our laboratory.

Acknowledgement

The authors thank Head, Department of Microbiology, Kakatiya University, Warangal (A.P.) for providing necessary facilities.

References

Adejuwon AO, Olutiola PO (2005) Plant Pathol.J.4:43-45.

Ajayi AA, Adejuwon AO, Olutiola PO (2007) *J. Plant Sci.* **2**:216-221.

Babu KJ, Reddy SM (1988) Indian J Mycol. and Plant Pathol.**17**:220-223.

Barras F, Gijseman F. Van, Chatterjee AK (1994) Ann. Rev. Phytopathol. **32**:201-234.

Bateman DF, Bashman HG (1976) Plant Pathology.**11**:316-345.

Bhagavan Reddy P (1986) Studies on pre- and post-harvest diseases of cucurbitaceous vegetables. Ph.D. thesis, Kakatiya University, Warangal.

Chauhan SK, Nagar ML (1979) Study of pectolytic, cellulolytic and proteolytic enzymes secreted by *Fusarium moniliforme* var. *subglutinous* and *Mucor hiemalis* pathogenic to maize kernels. 'Physiology of Parasitism'symp.p.171-187.

Chen ZY, Brown RL,.Damann KE, Cleveland Thomas E (1999) Phytopathology **89**:S15.

Dhawan S (1980) *Curr Sci*. **49**:361-362.

Fabiana G, M, Simene R, Maria A F C,Cristina G M, Rosane M P (2005) Braz. J. Microbiol. 36:7-11.

Johnson GL and Brown JL (1974) Biophysica.Acta. **370**:530-540.

Keay L,Wrildi BS (1970) Biotechnol and Bioengineering **72**:179-212.

Kuc J (1962) Phytopathology **52**: 961-963.

Lowry OH, Rosebrough NJ, Farr AL, Randall RJ (1951) J Biological Chem. **193**: 265-275.

Madhukar J, Reddy SM (1988) . Plant and Environment.**4**: 69-75.

Mandakini Patil, Shastry NV (1983) . J.Indian Inst. Sci. Sec. Bio.Sci. **64**:111-120.

Pekkarinen A, Mannonen L, Jones BL, Niku-Paavola ML (2000) J. Cereal. Sci. **31**:253-261.

Prasad BK, Shanker V, Narayan N, Kishore A, Dayal (1988) Indian Phytopath. **41**:578-580.

Raghavan U, Saksena SB, Vyas KM (1979) *In vivo* production of protease by fourteen isolates of *"Botrydiplodia theobromae"*. In *Physiology* of *Parasitism*, (G.P.Agrawal and K.S.Bilgrami, eds.), pp. 229-234, Today and Tomorrow's, New Delhi.

Reddy SR., Reddy SM (1983) Physiol. Eco.**8**:69-72.

Robert L.Brown, Zhi-Yuan Chen, Thomas E.Cleveland, Peter J. Cotty and Jeffery W Cary (2001) J Food Protec. **64**:401-404.

Sakririsat W, Hoppe H (1991) *J. Phytopath.* **131**:161-173.

Sastry VRSSS (2002) Studies on post-harvest diseases of fruits and vegetables Ph.D. thesis, Dept.of Botany, Kakatiya University, Warangal.

Satyanarayana,T, Jain S (1979) 'Physiology of Parasitism'symp.p.189-195.

Shumi W, Hossain Md.Towhid, Anwar MN (2004) Pak. J. Biolog. Sci .7:312-317.

Singh SN, Shukla P (1999) Indian Phytopath. 52:64-65.

Chapter - 11

Enhanced ethanol production incorporating process improvement strategies in molasses fermentations

N. Kishore Babu, K. Arunakar[#], K. Balakrishnan[$] and G. Seshagiri Rao[@]

Department of Biochemistry, Andhra University, Visakhapatnam 530 003
[$]Biotechnology Group, BIT, Sathyamangalam 638 401
[#]Biotechnology Dept., ANITS Engg. College, Visakhapatnam 531 162
[@]Corresponding : drgsrao@yahoo.com

ABSTRACT

Many cost effective strategies were proved to be leading to enhanced ethanol production during molasses fermentations. Strains of yeast exhibiting better tolerance to higher levels of alcohol in the broth and higher initial concentrations sugar was the most used option. Sequential batch processing resulting in nutrient balancing and incorporation of yield boosting nutrients were the choice of the methods used for enhanced ethanol yield in this study.

Our experiments indicated as high as 2 ± 1% (v/v) yield enhancements for ethanol in sequential batch process. This was run continuously for 10 cycles with appropriate nutrient supplementations. The viability and activity of the yeast cells were found to be positively influenced by the incorporation of nutrient stimulants. These yield increasing strategies when worked in tandem, it results in promising, easy to implement and economically viable strategies.

Key words: Ethanol, Molasses, Fermentations

Introduction

As the demand for ethanol is showing a progressive rise over the years, there is an imminent requirement of high yielding production strains and economically viable process realization for the production of ethanol. The quest for cheaper rate ethanol has been pursued actively by research groups (Suresh *et, al.*, 1999, 1999a, Kiran Sree *et, al.*, 2000, Marcos Antonia *et, al.*, 2006, Wouter *et, al.*, 2007). There are many proven process corrections for attaining enhanced ethanol production in molasses based fermentations. Use of specially selected strains of yeast exhibiting better tolerance to ethanol and initial sucrose levels is one such reliable approach for additional yield in ethanol fermentations.

Faster fermentation rates for ethanol production have been reported by following various strategies such as (1) higher cell densities per unit fermentation volume (Choudary and Chincholkar, 1996) (2) enhancing ethanol tolerance (Krish and Sezajani, 1997) (3) recycling of biocatalyst population (Nolan *et, al.,* 1994) etc. Yield enhancements can be thought of by employing sequential batch processes with well designed nutrient balancing. Certain micronutrients/ micronutrient preparations have been successful in contributing to better ethanol outputs (Nolan *et al.,* 1994).

In our investigation, regular bakers yeast (NCIM 3570) and specially isolated yeast were used for knowing the comparative significance in terms of yeast growth and activity. Yeasts were grown in both batch and sequential batch (10 cycles) formats. Appropriate nutrient supplementation was envisaged in the latter. The yield boosting was maximum when the aforesaid strategies were used in combination with indigenously made organic micronutrient stimulants.

Material and Methods

Microorganisms: An indigenously isolated strain of *Saccharomyces cerevisiae* Tejas (NCIM 3640) was used in this study along with a standard strain of bakers yeast (NCIM 3570). The principal investigator's team isolated this yeast culture from toddy sap employing the regular procedure of serial dilution and plating (in yeast development agar (YDA)) followed by incubation at 30°C. The toddy sap isolate was subsequently deposited with NCIM (NCL, CSIR), Pune, India.

Preservation and sub-culture: Standard stock cultures where prepared in 50% aqueous solution of glycerol and the same was stored at -20°C. Working stocks were prepared as agar slants (YDA) and stored at 4°C after incubation at room temperature for 2-3 days.

Inoculum Development: Yeast cells were transferred from agar slants as two loopful of culture per 50 mL of sterilized YD broth held in 250 mL E.M. flask. The seed culture flasks were incubated in environmental shaker set to 150 rpm and 30±1°C for 24 hours. After 24 hours, the cells from seed flasks were subjected to centrifugation (4,000 rpm, 10 min at 4°C). Such centrifuged cell pellets were used as inoculum. Hence, inoculum mentioned in this study comprise of the total cell pellets obtained at the end of 24 hr of cultivation from 50 mL of seed culture developed aerobically as explained above. Cells from 50 mL were used for inoculating every 1L production medium.

Fermentation: Fermentation experiments were carried out with commercially available cane molasses collected locally from the sugar factory (GMR Distilleries, Srikakulum). Molasses was characterized by subjecting molasses samples to analyses such as total reducing sugars, pH, volatile acids etc. by following the standard analytical procedures.

Initial studies were carried out using both the yeast cultures separately in 150 mL plastic screw cap bottles containing 100 mL each of 1.080 molasses medium with 1% each of urea & DAP (pH was adjusted to 4.5). Static cultures were maintained at room temperature for 36 h. In a parallel set of experimentation, the same was repeated with the addition of a nutrient mix (indigenously prepared organic extract) at 0.05% (v/v) level. The samples were collected for distillation followed by estimation of ethanol using gas chromatography.

Three two liter plastic reactors with provision for sending in or drawing out medium at every 200 mL levels (200 mL, 400 mL, 600 mL......) were used for this experimentation. All such extended provisions from the reactor (6mm dia) were provided with manually controlled ball valves. Two peristaltic pumps were used per reactor, one for sending in medium and other to draw out fermenting broth at regular intervals. The three tanks were entitled as A, B and C. In reactor A & B diluted molasses of specific gravity 1.080 was used at 1 liter level to initiate the fermentation on day 0. The p^H of molasses medium was adjusted to 4.5 using 1 N NaOH. The fermentation was carried out under anaerobic conditions as static process with no agitation and aeration provided during the course of the experimentation. At every 24 hrs interval, 200 mL of fermented broth was drawn out of the reactors and the same quantity of specified concentration (1.020 to A and 1.060 to B) of fresh medium was sent into reactor. The table 01 indicates the feeding and medium withdrawal schedules followed for various reactors. All the molasses media used for various reactors were supplemented with 1% each of DAP and urea.

If the various samples ports available in the reactors can be named as S1, S2, S3, ...to S9 (S1 here refers to sample port at 200 mL level, S2 that at 400 mL level, S3 refers to that at 600 mL and so on), at 24 hrs interval, 200mL of fermented broth was drawn out from S1; at 48 hrs the sample (200mL) was derived from S2; at 72 hrs from S3 and so on (cyclically). The addition of fresh medium (200 mL each time) was always from S9. The reactor A was a typical batch fermentor with constant volume of medium inside the reactor at any given point of time between day 1 and day 10. Reactor B, on the other hand was showing varying levels of medium from day 1 to day 10 (the pattern of flow in to the reactor and withdrawals for reactor B is indicated in Table 1.

Reactor C was essentially working as a cascade reactor for reactor B and was operational only between day 5 and day 10 i.e. reactor A, B and C were all sequential batch reactors. Reactors A & B were operated for a period of 10 days wereas reactor C was operational only for last six days of the 10 day duration of the experiment. Besides this, reactor C was supplemented with the indigenously made organic extract at 0.05% (v/v) level once in every 24h. The other general nutrient substitution to molasses medium (common to all reactors) consist of commercial grade Urea and DAP (1% wt/v each) at every 40 hr interval during the entire duration of experimentation (240 hrs).

Analytical procedures: Initial molasses medium characterization for Total Reducing Sugars (TRS) and pH were carried out using standard analytical procedures (Breckler, 1915). The samples of fermented broth derived at various intervals during the period of experimentation were analyzed for alcohol levels using Gas Chromatography (column: Porapak P; temperatures: injector (120°C), oven, (100°C), detector (140°C), carrier gas: nitrogen) and expressed as percentage of alcohol (% v/v alcohol). Periodic samples were subjected to alcohol estimation after distilling the samples. Samples at every 48 hr interval were centrifuged (4000 rpm, 10 minutes, at 10°C) and 2 mg of pellets were dispersed in 1 mL of distilled water and observed under microscope for cell viability using standard procedure of methylene blue staining. Average number of viable cells present in five microscopic fields was expressed as percentage of viability.

Results and Discussion

Initial fermentative studies indicated the superiority of the toddy sap isolate of yeast for higher ethanol production in 36 h. In the usual static fermentation of molasses medium with urea and DAP, NCIM 3640 yielded 5.18% and 6.26% for 1.080 and 1.120 molasses media respectively (Table 2). This was confirming superiority of the local isolate.

Table 1. The pattern of medium flow into the reactors and the withdrawals indicated along with the strength of molasses media used for various reactors.

Reactor name	A			B			C		
Days	Additions, mL		With-drawal vol, mL	Additions, mL		With-drawal vol, mL	Additions, mL		With-dawal vol, mL
	Stre-ngth	Vol		stren-gth	Vol		stren-gth	Vol	
D0	1.080	1000	-	1.080	1000	-	-	-	-
D1	1.200	200	200	1.160	400	200	-	-	-
D2	1.200	200	200	1.160	400	200	-	-	-
D3	1.200	200	200	1.160	400	200	-	-	-
D4	1.200	200	200	1.120	400	200	-	-	-
D5	1.200	200	200	1.160	400	300@ +200	1.160 From B	200 300	-
D6	1.200	200	200	1.160	400	300@ +200	1.160 From B	200 300	-

D7	1.200	200	200	1.160	400	300@ +200	1.160 From B	200 300	200
D8	1.200	200	200	1.160	400	300@ +200	1.160 From B	200 300	200
D9	1.080	200	200	1.060	400	300@ +200	1.060 From B	200 300	200
D10	-	-	200	-	-	200	-	-	200

Reactor C has nutrient boosters added at 500 microlitre per litre of broth for every 24 h.
Strength here refers to specific gravity of the molasses medium
@300 mL for addition to reactor C & 200 mL for sample processing

The indigenously prepared organic booster mix when supplemented in the production medium (1.120), both strains of yeast gave better production with NCIM 3640 producing superior yield in this trial too (Table 2).

Table 2. Ethanol production by yeast cultures in molasses media and the influence of an indigenously made nutrient mix on the biosynthesis of ethanol

Isolate type	Regular yeast (NCIM 3570)	Toddy sap isolate of yeast (NCIM 3640)
Molasses mix	% Ethanol (V/V)	
1.080	4.66	5.18
1.120	5.92	6.26
1.120 with nutrient mix (500 μ L / L)	6.28	6.96

Table 3. Ethanol concentrations obtained from various samples from reactors (A,B&C) operated as sequential batch reactors

Day	Ethanol Conc. % (v/v)		
	A	B	C
1	4.40	4.37	-
2	8.20	7.30	-
3	7.94	8.32	-
4	7.89	8.30	-
5	7.94	7.76	-
6	7.44	7.60	
7	7.20	7.33	7.59
8	7.12	7.25	7.33
9	6.10	6.92	7.26
10	5.50	6.30	6.62

All the three reactors produced better ethanol levels than the static flasks used in the preliminary studies with the same local isolate of yeast (NCIM 3640). The typical batch fermentations yielded just 5.18% for

1.020 medium and 6.25% for 1.120 medium. Even in the presence of yield booster the ethanol concentration was only 6.96 (Table 2). While in sequential batch process, the yield was as high as 8.2% for reactor A and 8.32 for reactor B. Beyond a level if the medium sugar levels are increased, the rate of production of ethanol is not proportionately increasing for this isolate.

The ethanol concentrations depicted in table 03 clearly depicts the superiority of the sequential batch process for ethanol production. The findings from the reactor C when read in comparison with the corresponding findings for reactor B, it is clear the inductive role of the yield booster used in this study. 1.080 medium is better suited for the sequential batch process with NCIM 3640 than 1.120 medium. The viability of cells observed under microscope also indicated more viable cells in the slides made with 1.080 medium than 1.120 medium especially after 5th day of process. Ernandes *et al* (1990) reported the diversified response of various isolates of yeasts for medium sugar levels, rate of ethanol production and cell viability etc.

References

Breckler AM 1915 *J. Ind. Engg. Chem.* 7(1), 37-38.

Choudary AB and SB Chincholkar. 1996. *Ind J Microbiol* 36: 75-83.

Ernandes JR, Matulionis M, Cruz SH, Bertolini MC & Laluce C 1990 *Biotechnol Lett.,* 12(6), 463-68.

Kiran Sree N, Sridhar M, Suresh K, Banat IM, and Venkateswar Rao L. 2000. High alcohol production by repeated batch fermentation using an immobilized osmotolerant *Saccharomyces cerevisiae*

Krisch J and B Szajani. 1997. Biotechnol Lett 19: 525-528.

Marcos Antonio das Neves, Toshinori Kimura, Naoto Shimizu and Kiwamu Shiiba. 2006. *Brazilian Archives* of *Biol and Technol*, 49 (3): 481-490.

Nolan AM, N Brady Barron, T McAree, D Smith, L McHale and AP McHale. 1994. *Biotechnol Lett* 16: 849-852.

Suresh K, N Kiran Sree and L Venkateswar Rao. 1999. Biores Technol 68: 301-304.

Wouter Wisselink H *et al* . 2007. *Appl. Environ. Microbiol*, 73 (15): 4881-4891.

Chapter - 12

Effect of surfactants, organic acids and solvents on mycelial growth and exopolysaccharide production by *Fomitopsis feei*

S.V.S.S.S.L. Hima Bindu N, Krishna Bolla, Sridevi Boddireddy and M.A. Singara Charya

Department of Microbiology, Kakatiya University, Warangal – 506 009, Andhra Pradesh, India

ABSTRACT

Addition of different surfactants, organic acids and solvents was investigated to study their effect on mycelial biomass and biopolymers production by *Fomitopsis feei*. On the other hand, all organic acids had bad effect on exopolysaccharides production, while tartaric acid and oxalic acid had shown to increase cell growth and the maximal cell concentration. EPS production was found to decrease with all the tested solvents but mycelial growth was increased with hexane (10.11 ± 0.01 g/lit.), chloroform (7.22 ± 0.02 g/lit.) and toluene (6.83 ± 0.17 g/lit.). The addition of surfactants shown to have reduced exopolysaccharide production except Tween 80 and Polyethyleneglycol 600 (5.67 ± 0.33 g/lit.). But the mycelial growth was increased two fold compared to control. The results obtained were useful in better understanding the regulation and optimization of *Fomitopsis feei* culture for efficient production of cell mass and polysaccharides in the submerged culture.

Key words: *Fomitopsis feei*, surfactants, organic acids, solvents, exopolysaccharides, submerged culture

Introduction

For many years, interest has concentrated on polysaccharides produced by numerous microorganisms, especially mushrooms, because of their various biological and pharmacological activities [Kim, 2002; Lee, 1996]. These include immunostimulating, antitumor, hypoglycaemic activities, as dietary supplements for the enhancement of stamina as a therapeutic for curing coughs and for blood circulatory problems or as a tonic promoting longevity and improving the quality of life [Yang, 2000; Kuo, 1996]. Submerged culture method offers a feasible alternative for the production of high quality mushroom products with immunostimulating polysaccharides. Although mycelial fermentations of medicinal mushrooms have often been reported, this mushroom has not been reported in submerged culture. In order to achieve a high specific mycelial growth and high EPS productivity, it is necessary to optimize the nutritional conditions and to accelerate biomass and EPS productions. One of the problems with producing EPS via submerged cultures of mushroom is low productivity, which makes it difficult to utilize mushroom EPS via mass production. Many investigators have endeavored to increase the production of microbial metabolites using some stimulating agents, including fatty acids, surfactants, vegetable oils, and organic solvents [Reese, 1969; Stasinopoulos, 1990; Haq, 2003; Flores, 1994; Choi, 2004]. These stimulation agents are known to mediate cell permeabilization by disorganizing the cell membrane and/or directly affecting the level of enzyme synthesis involved in product formation, thereby contributing to enhanced production of the target products [De Leon, 2003]. However, little information is currently available on the effects of physico-chemical properties of organic solvents on cell permea-bilization processes [Liu, 2000; Smet, 1978]. Moreover, in comparison with bacterial and yeast cells, there are limited reports concerning enhancement of permeability of fungi to improve the productivity of specific metabolites. In order to search for a proper additive to accelerate mycelia biomass and EPS productions in *F.feei* culture, the additives with stimulatory effects, including organic acids, solvents and surfactants were employed to test their stimulatory effects on both mycelial growth and EPS production in *F.feei* culture.

Material and Methods

Organism

The macrofungus *F. feei* was collected from the wood logs at Pakhal forest (latitude 18°1'18.82"N and longitude 80°9'22.16"E) Warangal district, Andhra Pradesh, India. The tentatively identified fungus was phylogenetically confirmed by molecular analysis of D2 region of 28S rDNA as *Fomitopsis feei* (AY 515327.1).

Culture media

All along the duration of the experiment, fungus was maintained on Malt extract agar medium (MEA (g/L): Malt extract 15, Dipotassium hydrogen phosphate 1, Ammonium chloride 1, 1N Citric acid 15 ml; Agar 20) and stored at 4°C and sub cultured every fortnight. Common production medium is used for broth culture which contains (g/lit.):

Peptone – 1, Yeast extract – 2, Dipotassium hydrogen phosphate – 1, Magnesium sulphate hepta hydrate – 0.2, Ammonium sulphate – 5, Glucose – 20 and pH – 6.0.

Chemicals used

Four types of surfactants such as Triton X 100 (Polyethylene glycol octylphenol ether), Tween 20 (Polyoxyethylene sorbitan monolaurate), Tween 80 (polyoxyethylene sorbitan monooleate) and PEG 600 (Polyethyleneglycol), ten types of organic acids such as acetic acid, ascorbic acid, citric acid, gallic acid, α-keto glutaricacid, maleic acid, oxalic acid, propionic acid, tartaric acid and valeric acid and seven types of organic solvents such as toulene, acetone, chloroform, hexane, dicholoromethane, ethyl acetate, dimethyl sulphoxide were procured from Sigma chemicals (Mumbai).

Inoculation and incubation

The surfactants, solvents and organic acids were added separately to production media in order to test their effect on biopolymers as well as biomass production. These experiments were constructed by adding solvents (0.05%) and surfactants at 0.5% concentration before sterilization of media, while organic acids were added at 0.05% and were incubated at 28°C for 7 and 14 days.

Mycelial dry weight and Exopolysaccharide (EPS) determination

The mycelial dry weight was measured after repeated washing (with distilled water) of the mycelial pellet, obtained after filtration and then drying at 70°C for 15 min. To measure EPS, filtrates collected were mixed with isopropanol (1:4 ratio), stirred vigorously, then left overnight at 4°C. The precipitated EPS were centrifuged at 10,000 rpm for 15 min. and the supernatants were discarded and pellet were dried and weighed for the exopolysaccharide determination by modified method of Dubois [1956].

Statistical analysis: Analyses were performed in triplicate and the data were recorded as Mean ± standard deviations using Ms Excel, 2003.

Results

The results of mycelial biomass and exopolysaccharide production, pH and total sugars were shown in Table 1. Highest mycelial biomass (13.57 ± 0.04 g/lit.) was recorded from Tween 80, followed by Tween 20 (9.65 ± 0.01 g/lit.) on the day 14. Poly ethyleneglycol 600 given good mycelial biomass (8.53 ± 0.01 g/lit.) and exopolysaccharide production (5.67±0.33 g/lit) in 7 days but they were decreased after 14 days. Triton X 100 was not favourable for both mycelial biomass and exopolysaccharide production. Overall, surfactants did not stimulate the highest production of exopolysaccharides compared to control. Tween 80 enhanced the production of the exopolysaccharides (5.67±0.33 g/lit.) by means of either providing an additional energy source to the fungus thus saving glucose for polymerization into exopolysaccharides, or directly interacting with cell membranes where the glucan synthase complexes are located, thereby enhancing the release of the exopolysaccharides into the extracellular fluid. The high concentration of surfactant might damage the cell membrane or interact with other bio-compounds in cell and then resulted in low cell growth. However, Tween 80 with longer side chain of surfactant might also be able to diffuse the cell wall with less concentration of surfactants. Total sugar content was also high with Tween 80 (338.33 ± 4.41 μg/ml) followed by PEG 600 (338.33±4.41 μg/ml) but were low compared to control. Tween 80 would increase cell growth instead of polysaccharides production. The less carbon chain of surfact-ants exhibited higher diffusion rate in cell wall of *F. feei.* The high concentration of surfactant might damage the cell membrane or interact with other bio-compounds in cell and then resulted in low cell growth. However, Tween 80 and Span 80, with longer side chain of surfactants, might also be able to diffuse through the cell wall with less concentration of surfactants.

Table 1. Effect of surfactants on pH, mycelial biomass, exopolysaccharides production and total sugar content of *Fomitopsis feei*

Type of Surfactant (0.5%)	Final pH		Mycelial biomass (g/lit.)		Exopoly-saccharides (g/lit.)		Total sugars (μg/ml)	
	7	14	7	14	7	14	7	14
Triton X 100	6.32 ± 0.01	6.32 ± 0.01	0.25 ± 0.01	1.35 ± 0.01	0.83 ± 0.33	0.67 ± 0.17	11.67 ± 1.67	130.00 ± 2.89
Tween 20	5.99 ± 0.01	2.17 ± 0.01	5.36 ± 0.00	9.65 ± 0.01	4.00 ± 0.58	1.67 ± 0.17	168.33 ± 1.67	80.00 ± 2.89
Tween 80	4.97 ± 0.01	2.25 ± 0.00	5.44 ± 0.04	13.57 ± 0.04	5.67 ± 0.33	2.67 ± 0.17	338.33 ± 4.41	235.00 ± 2.89
PEG 600	2.46 ± 0.01	2.12 ± 0.01	8.53 ± 0.01	6.65 ± 0.01	5.67 ± 0.33	1.83 ± 0.17	328.33 ± 4.41	315.00 ± 2.89
Control	1.64 ± 0.01	1.74 ± 0.01	6.88 ± 0.00	6.05 ± 0.00	5.50 ± 0.29	4.00 ± 0.88	638.33 ± 4.41	828.33 ± 6.01

The effect of organic acids on the production of mycelial biomass and EPS was resulted in Table 2. Ten different types of organic acids were tested for their effect, among those tartaric acid (7.08±0.01 g/lit.) increased the growth of *F.feei* but there was no EPS production compared to control followed by oxalic acid (6.97±0.03 g/lit.). Valeric acid did not support both EPS and mycelial biomass compared to control. Although all organic acids did not support the EPS production, they were beneficial to mycelia growth in the present study. In contrast, all the organic acids expressed negative results on EPS production.

Table 2: Effect of organic acids on pH, mycelial biomass, exopolysaccharides production and total sugar content of *Fomitopsis feei*

Type of Organic acids (0.05%)	Final pH		Mycelia biomass (g/lit.)		Exopoly-saccharides (g/lit.)		Total sugars (μg/ml)	
	7	14	7	14	7	14	7	14
Acetic acid	2.32 ± 0.01	1.95 ± 0.00	5.14 ± 0.01	5.16 ± 0.02	0.50 ± 0.01	1.00 ± 0.00	40.00 ± .2.89	89.67 ± 0.33
Ascorbic acid	2.16 ± 0.01	1.74 ± 0.00	5.65 ± 0.01	6.12 ± 0.01	0.90 ± 0.06	0.93 ± 0.07	117.67 ± 1.45	98.33 ± 1.20
Citric acid	2.02 ± 0.01	1.77 ± 0.01	5.24 ± 0.00	5.64 ± 0.01	0.50 ± 0.00	0.50 ± 0.00	61.67 ± 1.67	80.67 ± 0.67
Gallic acid	2.47 ± 0.01	1.93 ± 0.00	4.88 ± 0.00	6.45 ± 0.01	0.52 ± 0.02	1.00 ± 0.00	60.00 ± 2.89	98.33 ± 1.67
α-Keto glutari-cacid	2.26 ± 0.00	1.96 ± 0.01	4.89 ± 0.01	6.60 ± 0.03	0.48 ± 0.02	0.52 ± 0.02	78.33 ± 1.67	71.67 ± 1.67
Maleic acid	2.18 ± 0.01	2.11 ± 0.01	6.76 ± 0.01	6.10 ± 0.01	0.48 ± 0.02	1.00 ± 0.00	71.67 ± 1.67	21.67 ± 1.67
Oxalic acid	2.13 ± 0.03	1.97 ± 0.01	4.85 ± 0.01	6.97 ± 0.03	0.50 ± 0.00	0.60 ± 0.08	71.67 ± 1.67	70.67 ± 0.67
Propionic acid	2.32 ± 0.01	1.85 ± 0.00	5.64 ± 0.00	6.85 ± 0.01	0.48 ± 0.02	0.52 ± 0.02	21.67 ± 1.67	80.00 ± 2.89
Tartaric acid	2.17 ± 0.01	1.97 ± 0.01	5.67 ± 0.01	7.08 ± 0.01	0.48 ± 0.02	0.53 ± 0.03	65.00 ± 2.89	88.33 ± 4.41
Valeric acid	3.04 ± 0.00	2.78 ± 0.00	3.29 ± 0.01	3.12 + 0.00	0.48 ± 0.02	0.53 ± 0.03	43.33 ± 1.67	85.00 ± 2.89
Control	1.64 ± 0.01	1.74 ± 0.01	6.84 ± 0.03	6.05 ± 0.00	5.47 ± 0.03	4.33 ± 0.33	636.67 ± 6.67	823.33 ± 3.33

Seven types of solvents were tested for biomass and EPS production by *F.feei* (Table 3). Among these hexane (10.11±0.01 g/lit.) supported the growth of *F.feei* in 14 days of incubation compared to control followed by chloroform (7.22 ± 0.02 g/lit.). Exopolysaccharide production was also less compared to control. Ethyl acetate containing medium showed (3.60±0.06 g/lit.) EPS production followed by acetone (2.83±0.17 g/lit.). Total sugars content was less compared to control. Hence, from this study it was clear that solvents did not support the mycelial and EPS production compared to control.

Table 3. Effect of solvents on pH, mycelial biomass, exopolysaccharides production and total sugar content of *Fomitopsis feei*

Type of Solvent (0.05%)	Final pH		Mycelial Biomass (g/lit.)		Exopoly-saccharides (g/lit.)		Total sugars (μg/ml)	
	7	14	7	14	7	14	7	14
Toulene	4.22 ± 0.01	1.98 ± 0.00	3.17 ± 0.03	6.83 ± 0.17	2.17 ± 0.17	1.33 ± 0.17	100.00 ± 2.89	98.67 ± 1.33
Acetone	4.68 ± 0.00	2.09 ± 0.00	3.95 ± 0.00	6.11 ± 0.01	2.83 ± 0.17	1.17 ± 0.17	30.00 ± 2.89	113.33 ± 3.33
Chloroform	2.88 ± 0.01	2.57 ± 0.01	5.67 ± 0.03	7.22 ± 0.02	2.55 ± 0.03	1.50 ± 0.00	88.33 ± 1.67	118.33 ± 1.67
Hexane	2.74 ± 0.00	2.04 ± 0.01	5.88 ± 0.01	10.11 ± 0.01	1.67 ± 0.17	1.17 ± 0.17	205.00 ± 2.89	55.00 ± 2.89
Dichloro-Methane	2.98 ± 0.01	2.43 ± 0.04	4.85 ± 0.01	6.11 ± 0.01	1.33 ± 0.17	0.67 ± 0.17	463.33 ± 1.67	235.00 ± 2.89
Ethyl Acetate	2.98 ± 0.00	2.56 ± 0.00	5.85 ± 0.01	6.79 ± 0.01	3.60 ± 0.06	0.50 ± 0.01	95.00 ± 2.89	108.33 ± 1.67
Dimethyl Sulphoxide	3.05 ± 0.01	2.64 ± 0.00	5.79 ± 0.03	6.74 ± 0.03	3.00 ± 0.00	1.17 ± 0.17	41.67 ± 1.67	131.67 ± 1.67
Control	1.62 ± 0.01	1.73 ± 0.01	6.88 ± 0.01	6.05 ± 0.00	5.53 ± 0.03	4.17 ± 0.17	635.00 ± 2.89	825.00 ± 2.89

Discussion

Galindo and Salcedo [1994] stated that the use of small amounts of some surfactants (such as Tween 80, Tween 40 and Triton X-100) or oils is a promising way for improving xanthan production. Two surfactants namely, Triton X-100 (0.1 g/L) and Tween 80 (0.3 g/L) improved xanthan production by 1.22 and 1.15 fold after 4 days incubation period as compared to control (without surfactants) [Abdelhady, 2000]. Miller and Robyt [1994] realized that inactivation occurred in the presence of high concentrations (greater than 0.1%) of Triton X-100 in *Leuconostoc mesenteroides* NRRL B-512F. Despite the finding that exopolysaccharide synthesis by the fungus *Acremonium persicinum* and the bacterium *Xanthomonas campestris* were stimulated by the addition of vegetable oils, organic acids and surfactants to the growth medium [West, 1995]. The stimulation of EPS production, or its release into the extracellular medium, could be related to an interaction between the fatty acyl chains and the fungal cell membrane phospholipids [Angelova, 1999]. The addition of Tween 80 was evaluated with the objective of increasing the production of botryosphaeran, an exopolysaccharide (EPS) of the (1,3;1,6)-b-D-glucan type produced by the fungus *Botryosphaeria rhodina* MAMB-05 [Cassiano, 2007]. Tween 20 and Span 20 addition were shown to have serious inhibition on cell growth of *G. frondosa* and also on polysaccharides production [Chienyan Hsieh, 2007]. PEG with a molecular weight of 8,000 Da was found to be an effective stimulator for both mycelial biomass and EPS productions in *G. umbellata* submerged culture [Hua-Bing Chen, 2010].

Yang *et al.* [2000] reported that some fatty acids have promotional effects on biomass and exopolysaccharide productions. They observed that oleic acid at a level of 0.15g/100ml led to a significant increase in biomass while palmitic acid showed a great improvement on polysaccharide production. The addition of certain organic acids such as succinate, pyruvate and α-ketoglutarate, stimulated xanthan production, but excess concentrations of these organic acids inhibited xanthan formation [Souw, 1980]. Growth of *Aspergillus oryzae* was entirely inhibited by capric acid, caproic acid, propionic acid, and acetic acid, all with carbon chain lengths shorter than 10 [Fan-Chiang Yang, 2000]. Propionic acid and lactic acid increased pullulan production to only 16.3 and 28.6%, respectively. In case of pullulan production, increasing organic acids concentration more than 0.2 g/L led to a gradual decrease in pullulan production and growth of *A. pullulans* [El-Tayeb 2006].

Butler and Cundliffe [2001] established the enhanced tylosin production using dimethylsulfoxide (DMSO)-supplemented culture in *Streptomyces fradiae*. They demonstrated that DMSO had a significant effect on deoxyhexose metabolism, while enhancing tylosin production. However, the nature of the DMSO effect was unclear. Higher EPS production and less inhibition of mycelial growth compared to control culture were achieved in the media supplemented with toluene and chloroform after 4 days of fermentation. But in our study both were not stimulated the production of exopolysaccharides. This difference might be because of the time of addition of the solvents. Cells should be permeabilized without lysis or destruction of the whole inner organization. Nevertheless, the cells are obviously no longer viable after permeabilization by treatment with organic acids. This is an important reason why solvents have to be supplemented in the late stage of growth phases [Jong-Min Lim, 2006]. The use of solvents for the enhanced EPS production may not be advantageous when permeabilized cells are to be used in the food or pharmaceutical industry. The use of organic solvents is known to be a relatively effective method for cell permeabilization, because organic solvents are less expensive than other stimulating agents and may be eliminated by simple evaporation [Choi, 2004; Smet, 1978]. The representative organic solvents that enhance cell membrane permeability are toluene, chloroform, ethanol, diethyl ether, and dimethyl sulfoxide [De Leon, 2003). A few research reports have appeared in the literature to explain the effect of toluene on the permeability of the cell membrane [De Leon, 2003; Smet, 1978; Kondo, 2000].

Conclusions

The feasibility of using surfactants, solvents and organic acids for the mycelial growth and polysaccharides production of *F. feei* was investigated in this study. It can be concluded that surfactants such as Tween 80 and Tween 20, solvents such as hexane and toluene and organic acids such as oxalic acid and tartaric acid could be favorable used as an additive for the mycelial growth and Tween 80 for EPS production by *F. feei*.

Acknowledgements

Authors are thankful to the Head, Department of Microbiology, Kakatiya University, Warangal for providing necessary facilities. One of the authors, S.V.S.S.S.L. Hima Bindu N is greatful to University Grants Commission, New Delhi for providing financial assistance.

References

Abdelhady Hemmat M (2000) Annals Agric Sc Moshtohor 38(3): 1433 – 1451

Angelova B, Schmauder HP (1999) J Biotechnol 67:13–32

Butler AR, Cundliffe E (2001) J Indi Microbiol Biotechnol 27: 46-51

Cassiano C. Silva, Robert F.H. Dekker, Rui Se´rgio S.F. Silva, Maria de Lourdes Corradi da Silva, Aneli M. Barbosa (2007) Process Biochem 42: 1254–1258

Chienyan Hsieh, Hui-Liang Wang, Chien-Cheng Chen, Tai-Hao Hsu, Mei-Hua Tseng (2007) Biochem Eng J.

Choi KO, Song SH, Yoo YJ (2004) Biotechnol Bioprocess Eng 9:147–50

De Leon A, Garcia B, Barba de la rosa AP, Villasenor F, Estrada A, Lopezrevilla R (2003) Process Biochem 39:301–5

Dubois M, Giles KA, Hamilton JK, Rebers PA, Smith F (1956) Anal Chem 28:350–356

El-Tayeb TS, Khodair TA (2006) Res J Agric Biol Sci 2(6):483-492

Fan-Chiang Yang, Yn-Fuu Ke, Shanq-Shin Kuo (2000) Enz Microb Technol 27:295–301

Flores MV, Voget CE, Ertola RJJ (1994) Enz Microb Technol 16:340–6

Galindo E and Salcedo, Ramirez ME (1994) Appl Microbiol Biotechnol 40:634-637

Haq IU, Ali S, Qadeer MA, Iqbal J (2003) J Microbiol Biotechnol 86:227–33

Hua-Bing Chen, Hung-Chang Huang, Chih-I Chen, Yet-Pole I, Yung-Chuan Liu (2010) Bioprocess Biosyst Eng 33:401–406

Jong-Min Lim, Jong-Won Yun (2006) Process Biochem 41:1620–1626

Kim DH, Yang BK, Jeong SC, Park JB, Cho SP, Das S, Yun JW, Song CH (2002) Biotechnol Lett 2(23):513-7

Kondo A, Liu Y, Furuta M, Fujita Y, Matsumoto T, Fukuda H (2000) Enzyme Microb Technol 27:806–11

Kuo YC, Tasi WJ, Shiao MS, Chen CF, Lin CY (1996) Am J Chin Med 24:111-25

Lee JH, Cho SM, Kim HM, Hong ND, Yoo ID (1996) J Microbiol Biotechnol 6:52-5

Liu Y, Fujita Y, Kondo A, Fukuda H (2000) J Biosci Bioeng 89:554–8

Miller AW, Robyt JF (1994) Appl Environ Microbiol 60(8): 2723–2731

Reese ET, Maguire A (1969) Appl Microbiol 17:242–5

Smet MJ, Kingma J, Witholt B (1978) Biochim Biophys Acta 506:64–80

Souw P, Demain A (1980) J Ferment Technol 58(5):411 – 416

Stasinopoulos SJ, Seviour RJ (1990) Biotechnol Bioeng 36:778–82

West PT, Reed Hamer B (1995) Microbios 83: 249 – 259

Yang BK, Ha JY, Jeong SC, Das S, Yun JW, Lee YS, Choi JW, Song CH (2000) J Microbiol Biotechnol 10:784-8

Yang F, Ke Y, Kuo S (2000) Enz Mic Technol 27:295-301.

Chapter - 13

Effect of phenols on growth and penitrem B production by *Penicillium aurantiogriseum*

M. Surekha, V. Krishna Reddy, S. Kiran, A. Rajendar Reddy and S.M. Reddy

Department of Botany, Kakatiya University, Warangal-506 009, A.P., India

ABSTRACT

Phenols were screened for their effect on growth and penitrem B production by *Penicillium aurantiogriseum*. P-chloronitrophenol and pentachlorophenol have shown total inhibitory effect on Penitrem B production. On the other hand catechol, phenol and resorcinol were ineffective in checking the penitrem Production by *P. aurantiogriseum*. Rest of the phenolic compounds were responsible for inhibition of production of penitrem B by *P. aurantiogriseum* to a varying degree.

Key words: Phenols, Penitrem B, *Penicillium aurantiogriseum*.

Introduction

Phenols which are secondary metabolic products of plants are reported to be fungicidal. Pan *et al.* (1985) Mahadevan *et al.* (1988) Vinceno Lattanzio *et al.* (2006) have reviewed the role of phenolic compounds in disease resistance. Premalatha Singh (1983) Beekrum *et al.* (2003) and Rathnavathi *et al.* (2007) have attributed the inhibitory effect of natural plant products in the control of aflatoxin production by *Aspergillus parasiticus* to variety of phenols. Krishna Reddy (1988) and Neri *et al.* (2005) have also reported the efficacy of certain phenols in preventing the production of cyclopiazonic acid by *Penicillium griseofulvum*. Hence, in the present investigations, effect of some phenols on the growth and penitrem B production by *P. aurantiogriseum* was studied and the results are presented in this paper.

Materials and Methods

Effect of some phenols (as listed in Table 1) on growth and penitrem B production by *P. aurantiogriseum* was studied by adding these substances to the medium at different concentration aseptically before inoculation of the fungus. Fifty ml of Richards medium (KNO_3 10g; $KHPO_4$ 1g; $MgSO_4$ 2.5g; Sucrose 35g; $FeCl_2$ traces and distilled water 1000ml pH 5.5) apportioned in 250 ml Erlenmeyer conical flasks was sterilized at 15 lbs pressure, for 30 min. Different phenols were added aseptically just before inoculation of the fungus. Flasks thus inoculated were incubated at 27±2°C for 15 days. At the end of incubation period, cultures were harvested and the biomass produced was determined. The pH of the culture filtrate was also determined.

Penitrem B produced was extracted from dried fungal mycelium by employing Soxhlet apparatus for 24 hrs. Penitrem B present in the extract was estimated colorimetrically as suggested by Hou *et al.* (1970).

Results and Discussion

From Table 1, it is evident that *O*-nitrophenol, *P*-nitrophenol, *P*-chloronitrophenol and pentachloraphenol were responsible for total inhibition of penitrem B production.

Table 1. Effect of phenols on growth and penitrem B production by *Penicillium aurantiogriseum*

Name of the compound	Conc. (μg/ml)	Final PH	Dry wt. (mg/ml)	Penitrem B (mg/ml)
Phenol	50.00	6.20	7.63	3.87
	75.00	6.00	7.54	3.62
	100.00	6.00	7.12	3.01
O-cresol	50.00	7.20	7.33	3.53
	75.00	7.50	7.01	3.00
	100.00	7.80	6.40	2.00
M-cresol	50.00	7.50	7.36	3.68
	75.00	7.50	6.84	3.15
	100.00	7.80	5.80	2.30
O-chlorophenol	50.00	7.20	6.83	3.14
	75.00	7.50	6.12	3.00
	100.00	7.50	5.24	2.64
Penta chlorophenol	5.00	7.50	4.38	0.90

	10.00	7.00	4.01	0.58
	15.00	7.20	3.83	0.24
	25.00	6.50	--	Nil
O-nitrophenol	50.00	6.80	3.10	0.20
	75.00	5.50	-	Nil
	100.00	5.50	-	Nil
P-nitrophenol	15.00	6.80	3.02	0.25
	25.00	6.50	3.03	0.14
	50.00	5.50	-	Nil
	75.00	5.50	-	Nil
P-chloro nitrophenol	5.00	7.50	4.38	0.90
	10.00	7.00	3.81	0.90
	15.00	7.00	3.10	0.24
	25.00	6.00	--	Nil
Catechol	50.00	7.50	8.88	4.00
	75.00	7.50	8.80	3.89
	100.00	7.50	8.83	3.80
Resoreinol	50.00	7.00	8.13	3.54
	75.00	7.20	7.94	3.20
	100.00	7.50	7.90	3.20
Guaiacol	50.00	7.50	8.30	4.10
	75.00	7.20	8.00	4.00
	100.00	7.00	8.00	4.00
Control	-	6.00	8.80	4.06

M-cresol and *O*-cresol have inhibited the fungal growth at comparatively higher concentrations. Of all the phenolic compounds investigated pentachlorophenol and *P*-chloronitrophenol were effective even at the lower concentration (5µg/ml). Catechol, phenol and resorcinol were ineffective in checking the penitrem B production. Guaiacol was also ineffective in preventing either the growth of the fungus or penitrem B production.

Pentachlorophenol, *O*-nitrophenol, *P*-nitrophenol and *P*-chloronitrophenol inhibited the growth of the fungus, which was proportional to the increase in concentration of compound. The pH changes were minimal and the final pH of the media was near neutral. A positive correlation was observed between the vegetative growth and penitrem B production.

Acknowledgement

The authors are grateful to the Head Department of Botany for providing facilities and to UGC, New Delhi for financial assistance.

References

Beeekrum S., R. Govinden., T. Padayachee and B Odhav (2003). *Food Addit Contain*. 20: 490-493.

How C.T., A. Ciegler and Hesseltine C.W. (1970). *Anal Bio Chem*. 37: 422-428.

Krishna Reddy V. (1988). Studies on mycotoxin producing fungi associated with maize (*Zea mays* Linn). Ph.D *Thesis*. Kakatiya University.

Mahadevan A and P.T Kalaichevan (1978). *Curr.Sci*. 57: 193-194.

Neri F., M. Mari and S. Brigati (2005). Plant pathology. Vol. 55(1): 100-105.

Pan S.B., A. Mukherjee., S.B. Ganguly, A.Mitra and A. Bhattacharya (1985). Z Pflanzenzucht. 90(4): 392-395.

Premalatha Singh (1983). Proc. Symp. Mycotoxin in food and feed. Bhagalpur. 307-313.

Ratnavathi C.V., R.B. Sashidhar (2007). Inhibitory effect of phenolics extracted from sorghum genotypes on *Aspergillus parasiticus* (NRRL 2999) growth and Aflatoxin production. Journal of the science of food and agriculture. Vol: 87. 1140-1148.

Vinceno Lattanzia., M. Veronica., T. Latfanzio and Angela Cardinali (2006). Phytochemistry: Advances in Research. 23-67.

Chapter - 14

Biochemical and molecular based identification of plant parasitic nematodes of Telangana region

*Anilkumar Vemunoori, Rajesh Yarra, Mahendar Porika and Sadanandam Abbagani**

Department of Biotechnology, Kakatiya University, Warangal – 506 009, India.

ABSTRACT

Root knot nematodes belonging to the genus Meloidogyne are the most economically important group of plant parasitic nematodes. Their worldwide distribution, extensive host range and interaction with other plant pathogens. In Telangana region, some important commercial crops are mostly grown. It is affected by a multitude of pests including, plant parasitic nematodes, which are a major growth constraint, particularly on sandier soils. In contrast, other nematodes are beneficial in that they are pathogenic to certain insect pests. Knowing which of the plant parasitic nematodes are causing the most damage and which of the entomopathogenic nematodes might be exploited to control insect pests is obviously important when developing control strategies. Traditional methods of identification of these nematodes relied on the use of morphological characteristics. However, continued advances in Biochemical and molecular biology have resulted in the development of newer, potentially more powerful, means of identification. The objectives of this work were to develop biochemical and molecular biological techniques to identify plant parasitic nematodes using Isoenzymes, protein profiling and species specific primers. The obtained results of single nematodes allowing discrimination among the three root-knot species of Telangana region. Biochemical and molecular techniques found to be a useful tool for discriminating between the important Meloidogyne species in this study.

Key words: *Meloidogyne* spp, Esterase, Protein profiling, RFLP, species-specific primer, ISSR, diagnostics.

Introduction

Plant nematodes account for an estimated 14% of worldwide plant losses, which translates into almost $100 billion dollars annually. By far, root-knot nematodes are the most common and destructive nematode pathogens. They produce some of the most dramatic symptoms and can substantially reduce crop yields. Root-knot nematodes are found in all agricultural regions worldwide. They can survive in temperate climates and can devastate crops grown in the tropics. Most root-knot nematodes also have extremely wide host range. Although it is difficult to ascertain the number of hosts for any one root knot nematode species, it is likely that some root-knot nematodes can survive on hundreds of different plant species. It is extremely difficult to control root knot nematode problem, particularly if they can survive on weeds. In addition, root knot nematodes have repeatedly been shown to predispose their host plants to infection by other crop pathogens, increasing the potential for crop loss.

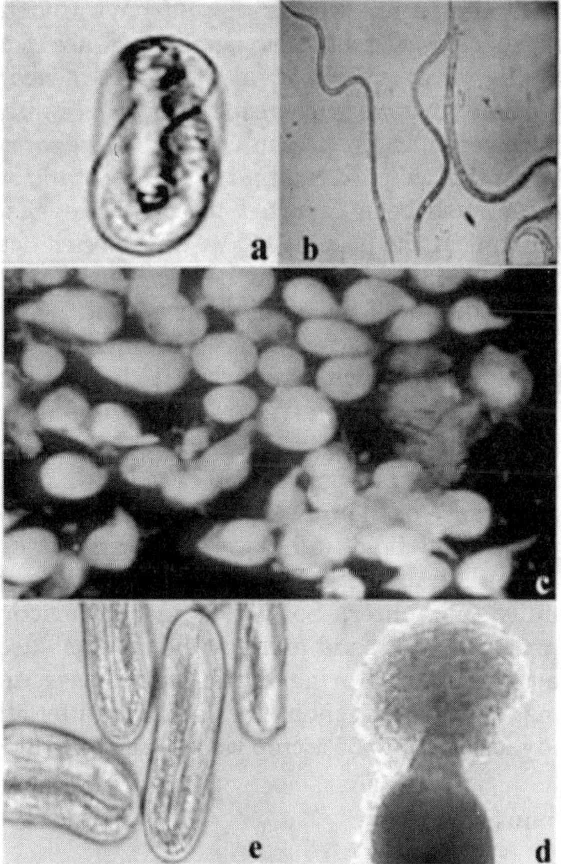

Fig. 1. Microscopic view of life cycle of a root-knot nematode (*Meloidogyne* sp.) a & e - Infective first stage molt (J_1); b- Infective second stage juvenile (J_2); c- Adult female nematodes; d- Egg mass extruded from adult female.

Therefore, the objectives of this experiment were to find out intensity of soil infestation by root knot nematodes, to determine their effect on vegetative growth and yield of some important commercial crops, and to identify *Meloidogyne* species and their management present in these prevalent areas of Telangana region.

Materials and Methods

Sampling nematode communities and extraction for bio assays

The present study results indicate that root-knot nematodes are fairly widespread in the study area. All the four most common species viz. *Meloidogyne javanica, M. incognita, M. arenaria and M. hapla* were found associated with several crop plants. In addition to them there were many populations which could not be assigned to any of these four species and could possibly belong to one or more already described or entirely new species which the future workers would have Codetermine. Above mentioned four species of root-knot nematodes are considered most common species (Taylor *et al* 1982). Among them *M. incognita* has been ranked first by these authors with respect to geographical distribution and host range. However, in the present survey *M. javanica* was found to be more widely spread than *M. incognita* in the study area On global basis *M. incognita* has been reported to constitute about 64% of the total(Sasser,1979) but in the present study, this species -was recovered in 32% of the samples. This variance may be ascribed to the particular soil characteristics and climate of the study area in other parts of Telangana such as Warangal, Karimnagar and Khammam. These areas are characterized by 120 samples with sandy soil with sand comprising about 70% whereas in most part of the NWFP soil contains about 50% of sand (Taylor *et al*, 1982). Out of 12 regions made arbitrarily for the purpose of survey, Warangal and karimnagar regions occupied the first rank with respect to the occurrence of the root knot nematodes. In some localities of the region, majority of the fields (59%) were found infested. As this region is surrounded by hills and the winter wind prevents frost, tomato were grown successfully as winter. Soil of Nalgonda district, which also contains large proportion of sand and gravels, is also susceptible to root knot nematode infestation. During the survey, losses up to 80% were recorded in some fields in nalgonda. It was followed by other regions, in the present study losses up to 55% crop were found in some regions.

Collection of root samples

A total number of 12 sites were surveyed (Table 1). From which a total of 150 samples from tomato (*Solanum Lycopersicon*) plants infected with root knot nematodes (15 samples from each site) were collected.

Table 1. Infested root materials of different localities of Telangana region

S. No	Locality	Source of material	Species identified
1	Warangal-University feild	Tomato	Mj
2	Muchharla Nagaram	Tomato, Brinjal	Mj
3	Hasanparthy	Tomato, Tobacco	Mj
4	Taralapally	Tomato, Tobacco	Mj
5	Old Bus depot road	Tomato	Mj
6	Karimnagar-Ramnagar/chinthapally	Tomato	Mj,Mh
7	Khammam-kinnerasani Road	Tomato	Mi
8	Nalgonda-Noothankal	Tomato	Mi,
9	Adilabad-Boath	Tomato	Mj
10	Nizamabad-Baddipally	Tomato	Mj
11	Sangareddy-lingampally	Tomato	Mi
12	Icrisat (agricultural field)	Tomato	Mi

M.j: *Meloidogyne javanica:* M.i: *Meloidogyne inconita;* M.h: *Meloidogyne hapla.*

Sampling from tomato plants

Sampling from tomato plants was done by uprooting the whole plant from soil by using a spade. Effort was made to remove the entire root system by digging carefully around the roots before uprooting. After excising the aerial portion of entire rooted plant and the adhering soil was removed from root system using running tap water .The roots were cut and reduced to desirable size and were stored in separate polythene bags. All the bags were tied, labeled and stored in -20°C until use.

One hundred grams of soil and roots of each sample were processed for the isolation of nematodes. The Nematodes were isolated by using sieving and decanting method.

Problems associated in the identification based on morphological characters of *Meloidogyne* species, such as intra-species merging of morphological characters, rarely-seen characters, and lack of apparent differences between species, has encouraged much interest in the utilization of biochemical techniques as a complementary, routine method for the identification of root-knot nematodes by Enzyme phenotypes, through staining of polyacrylamide gel electrophoresis have become a reliable, less subjective approach to identification of root knot nematodes (Hussey, 1985). The first study on root-knot nematode isozyme and protein profile stability and its utilization in the identification of root-knot nematodes was conducted by Dickson *et al.* (1970), who established the usefulness of disc-electrophoresis in identification of root-knot nematodes.

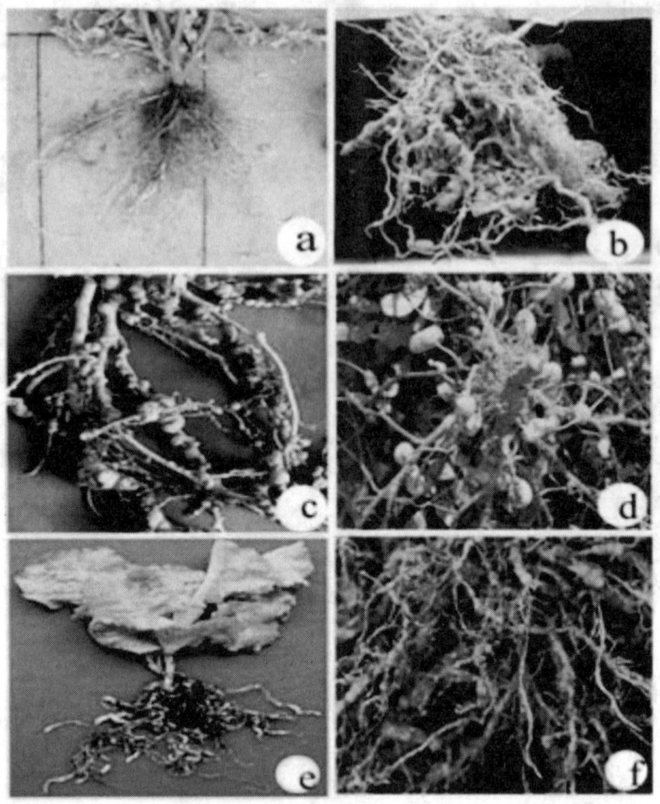

Fig. 2. Typical gall symptoms on infested roots of various plants. (a) Non infested roots of tomato (control); (b) Infested tomato roots with gallu; (c) Infested roots of brinjal; (d) Infested roots of peanuts; (e) Infested roots of tobacco; (f) Infested root of cotton.

Isozyme analysis

Root-knot nematode females, dissected from different galls of the root systems, were used for isozyme analyses by electrophoresis, using PAGE (Poly acrylamide gel electrophoresis) apparatus. *Meloidogyne* spp. were identified by comparing specimen Mdh (Malate dehydrogenase) and Est (Esterase) phenotypes to those of previously published root-knot nematode species (Dalmasso and Berge, 1978; Esbenshade and Triantaphyllou, 1985c, 1990; Fargette, 1987a, 1987b; Hussey *et al.*, 1972; Yongfang *et al.*, 1998). At least 26 females from each root sample were examined for species identification with the poly acrylamide gel electrophoresis. The samples were processed on the same day of female extraction, while females extracted for isozyme were stored at -4 °C for up to six months.

Samples were loaded on native poly acryl amide gel electrophoresis for Isoenzyme analysis.

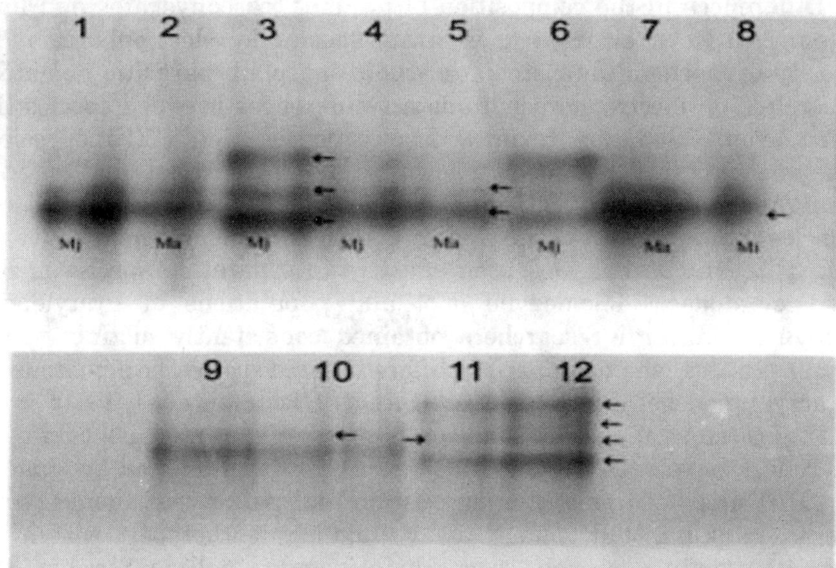

Mj - *Meloidogyne javanica*

Ma - *Meloidogyne arenaria*

Mi - *Meloidogyne incognita*

Fig. 3. Esterase Isozyme patters of Root Knot Nematodes.

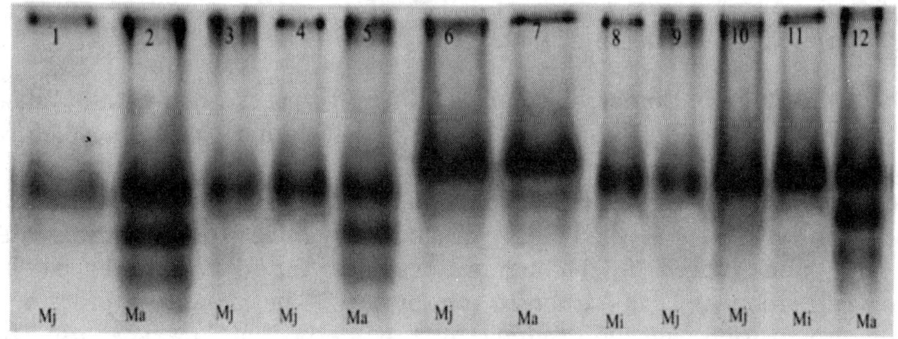

M.j - *Meloidogyne javanica*
M.a - *Meloidogyne arenaria*
M.i - *Meloidogyne incognita*

Fig. 4. Malate dehydrogenase isozyme patterns Root Knot Nematodes.

Protein Profiling

Differences in the composition of proteins, carbohydrates and lipids of nematodes can be used to separate species by electrophoresis. The majority of biochemical systematic studies on plant-parasitic nematodes have relied on electrophoresis to detect differences between species (Fox & Atkinson, 1986, for review). Isoelectric focusing (IEF) of soluble nematode proteins has been shown to be a practical and effective technique for the identification of closely related species and members of a species complex. Most of the work has been done on cyst-nematodes, but the IEF technique has also been effective in separating species of root-knot nematodes, Meloidogyne spp. and populations of Pratylenchus brachyurus. Several researchers obtained consistently distinct soluble protein profiles when comparing different races/species of nematodes by polyacrylamide gel electrophoresis (PAGE) (Dickson et.al, 1970; evans, 1971). Proteins of species and pathotypes of several globodera and Heterodera have been compared using PAGE. Results found by (Jones et ah, 1970) supported an earlier suggestion that pathotypes should receive specific ranking, but failed to distinguish pathotypes within the subsequent defined species, G rostochiensis and G. pallida. Stanton, J.M., Mc.Nicol (1974) verified that two pathotypes of *H. avenae*, which were distinguished by their ability to reproduce on barley (*Hordeum vulgare* L.), were electrophoretically identical and were considered subspecifically related. Franco (1979) observed that protein patterns from different G. rostochiensis populations were identical, but differences were found in patterns from British and Peruvian populations of G. pallida. Also, species of Aphelenchoides were differed in protein banding patterns resolved by SDS-PAGE (Ibrahim et ah, 1994). Therefore, in this study we were interested to compare protein patterns of *Meloidogyne spp.* by PAGE.

Preparation of protein extract

Adult female Nematode samples (100 mg) were washed with distilled water and suspended in freshly prepared cold buffer containing 0.05 M tris-HCL, 10% sucrose, 0.9% NaCl, 2% triton X-100, 5% β-mercaptoethanol and 2% SDS. Nematode samples were macerated in this solution with a pre-cooled Remi tissue homogenizer (Remi equipments, India) and placed in boiling water bath for 5 min followed by centrifugation at 12,000 g for 10 min at 4 °C. The amount of protein in the preparation was determined by the method of Lowry *et al.* The resultant soluble protein preparation was immediately introduced in the electrophoretic cell. The SDS-PAGE system resolved some slight variations in overall electrophoresis mobility of protein samples of nematodes of the 12 different areas of telangana districts.

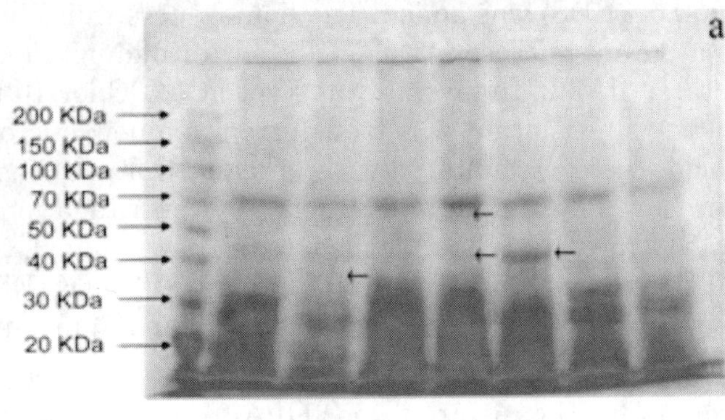

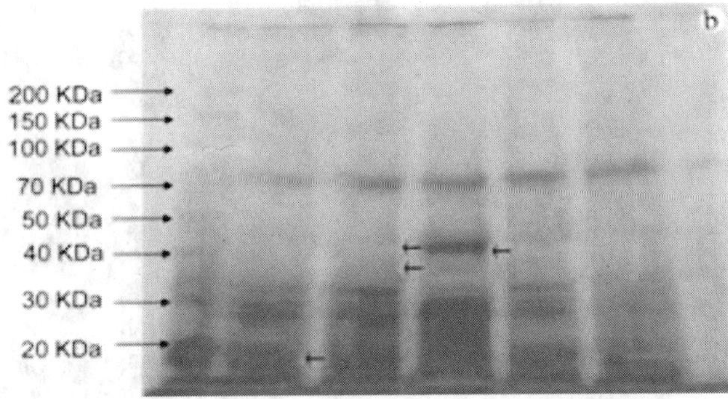

Fig. 5. Total soluble protein profiles of Root Knot Nematodes.

Molecular diagnosis of single nematode DNA using RAPD profiling

The objectives of this study was to establish a rapid and effective screening procedure to detect root knot nematodes (*Meloidogyne spp*), by selecting a good random sequence 10-mer primer (OPA) for identification of the ten Meloidogyne spp. in telangana region, and to establish the better extraction method of template DNA from an individual adult female in order to obtain reproducible RAPD patterns

Extraction of single nematode DNA

This is the slightly modified method of Cenis (Cenis, J. L.1993 and Adam, M.A.M., Phillips) for extraction of DNA from single adult females (Orui, Y. 1996). Single adult females were hand-picked from roots of host plant, and washed in sterile plastic dish (9- cm diameter) containing sterile water. They were ruptured with a pestle on a slide glass, and 50 µl of DNA extraction buffer (200 mM Tris-HCl: pH 8.5, 250 mM NaCl, 25 mM EDTA and 0.5% SDS) was added. The homogenate thus obtained was transferred to a 0.5 ml microcentrifuge tube. After adding 75 μl of 3 mM potassium acetate: pH 5.2, the lysate was kept at -20°C for 10 min. After centrifugation, the supernatant was transferred to a new microcentrifuge tube, nucleic acids were precipitated with an equal volume of isopropanol at room temperature for 30 min, and pelleted by centrifugation at 13,000 rpm for 15 min. The pellet was washed with 70% ethanol, dried, and dissolved in TE buffer (10 mM Tris-HCl: pH 8.0,1 mM EDTA). The concentration of template DNA was adjusted to 100 ng/μl for further PCR based molecular diagnosis.

The PCR amplification using primer OPA-01 showed species-specific RAPD patterns which allowed identification of four Meloidogyne spp: *M. incognita, M.arenaria, M. javanica,* and *M. hapla.*

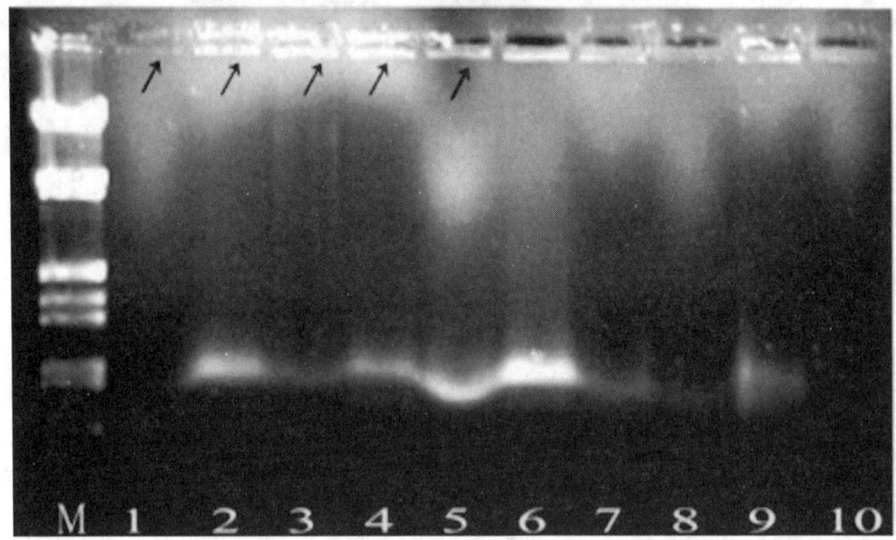

Fig. 6. Single adult female nematode DNA on 0.8% Agarose gel

Identification of nematode using species specific primers

The species-specific sequence tagged sites were selected with primer pairs listed in Table 2. Species-specific PCR products were produced by the designed primer pairs and resulted in unique band(s) for each of the four species. The species-specific primers produced a single band (1 500 bp) for the *M. hapla* isolates. Among the seven *M. hapla* isolates, a single 950 bp band was amplified. In *M. incognita* reactions, there were doublet bands (both 1 350 bp, differing by 20 bp in size) produced from each of the eleven *M. incognita* isolates. For *M. javanica*, a single band (1 650 bp) was produced in four out of the five *M. javanica* isolates.

Table 2 . The species-specific PCR primer pairs for the four *Meloidogyne* spp.

Species	Sequence of Primer Pairs (5'- 3')
M. arenaria	TCGAGGGCATCTAATAAAGG
	GGGCTGAATATTCAAAGGAA
M. hapla	GGCTGAGCATAGTAGATGATGTT
	ACCCATTAAAGAGGAGTTTTGC
M. incognita	TAGGCAGTAGGTTGTCGGG
	CAGATATCTCTGCATTGGTGC
M. javanica	CCTTAATGTCAACACTAGAGCC
	GGCCTTAACCGACAATTAGA

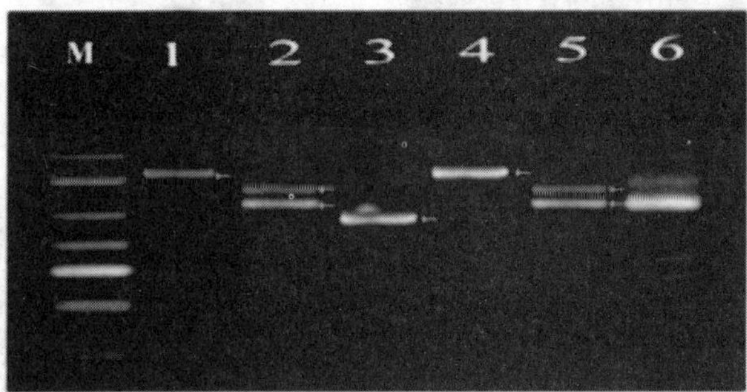

Fig. 7. Species specific PCR electrophoretic pattern of single adult root knot nematodes

Evaluation of Genetic Polymorphism using ISSR markers

Further more genetic polymorphism among the six different species (identified by Isozyme patterns and by using species specific RAPD markers) was carried out using ISSR-PCR markers.

ISSR-PCR was conducted in a reaction volume of 15 µl containing 30 ng template DNA, 0.2 µmol/L primer, 200 µmol/L each dNTP, 10 mmol/L Tris-Cl (pH 8.3), 50 mmol/L KCl, 2.0 mmol/L MgCl2, and 1 U of Taq polymerase. The PCR amplification conditions were set as initial denaturation at 94°C for 5 min, 40 cycles of denaturation at 94°C for 1 min, annealing at 50°C for 1 min, extension at 72°C for 2 min, and a final extension at 72°C for 7 min. The PCR was performed in 96-well plate thermal cycler (Eppendorf, Germany). The amplified products were mixed with loading dye (0.4 g/ml sucrose and 2.5 mg/ml bromophenol blue), resolved on 18 mg/ml agarose gel in 0.5× Tris borate EDTA (TBE) buffer under room temperature at constant voltage of 100 V, and detected by ethidium bromide staining (0.5 mg/ml). The gels were documented under ultraviolet light using gel documentation unit from Bio-Rad. Each amplified product was scored as 1 or 0 depending on its presence or absence. The frequency of microsatellite polymorphism was calculated based on the presence or absence of common bands. The genetic associations among the nematodes from different regions were evaluated manually by counting the total number of bands/alleles (similar and dissimilar) observed with all the primers tested.

The amplified products of template DNA extracted from single female nematode using ISSR primers were found to useful for the assessment of genomic polymorphism among the species.

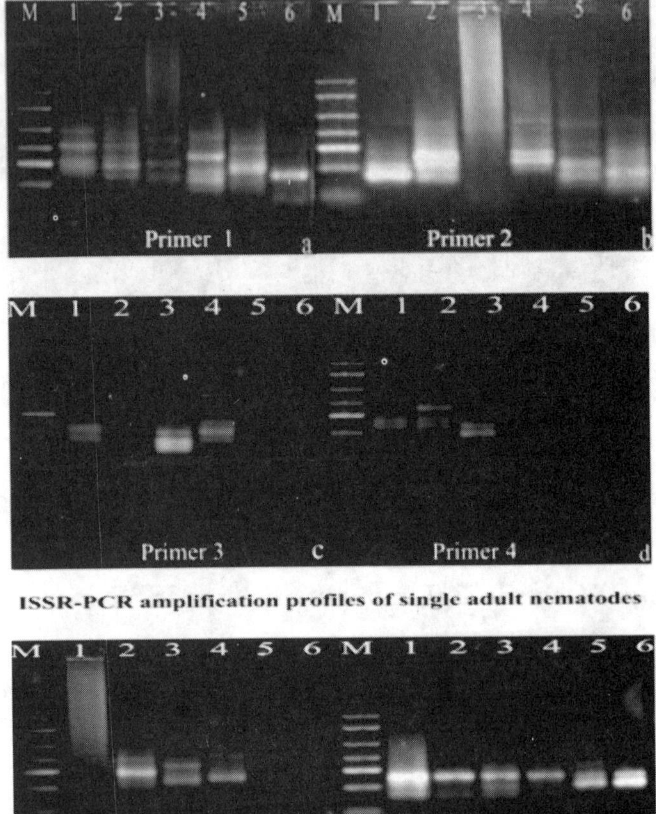

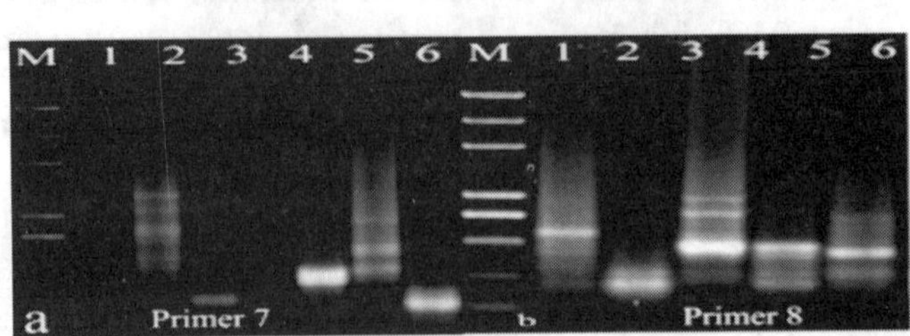

Fig. 8. ISSR-PCR amplification profiels of single adult nematodes.

Results and Discussion

Biochemical and molecular methods for identification of Meloidogyne species are now widely used and, in some cases, essential for species diagnosis. They cannot, however, be used with confidence to identify all Meloidogyne species. A clear understanding of species boundaries and adequate sampling of known species across their geographic range are lacking. A problem associated in the identification based on morphological characters of Meloidogyne species, such as intra-species merging of morphological characters, rarely-seen characters, and lack of apparent differences between species, has encouraged much interest in the utilization of biochemical techniques as a complementary, routine method for the identification of root-knot nematodes by Hansen and Buecher, 1970. Enzyme phenotypes, unveiled through staining of polyacrylamide gel slabs following electrophoresis have become a reliable, less subjective approach to identification of root knot nematodes (Hussey, 1985). The first study on root-knot nematode isozyme and protein profile stability and its utilization in the identification of root-knot nematodes was conducted by Dickson *et al*. (1970), who established the usefulness of disc-clcctrophoresis in identification of root-knot nematodes.

Several solanaceae members infested with *Meloidogyne* spp.were identified as hosts via the speciation of root-knot nematode females that were dissected directly from their hosts roots. These females were identified primarily by evaluation of esterase (Est) and malate dehydrogenase (Mdh) isozyme phenotypes, unveiled following polyacrylamide gel electrophoresis. Resolved isozyme phenotypes indicate that Tomato (*Solanum lycopersicum*), Brinjal (*Solanum melongena*), Tobaco (*Nicotiana tobacum)* are hosts to three *Meloidoigyne* spp. In the present studies, the major root-knot nematode species identified were *M. incognita*, *M. javanica*, *M. arenaria* and *M. hapla* and they were not determined earlier. Mixed populations were found in 40% of the samples. *Meloidogyne arenaria* was found with (Mdh Rm 0.18, 0.20, 0.22, Est Rm 0.38, 0.40), *Meloidogyne javanica* was found (Mdh Rm 0.21, Est Rm 0.42, 0.50, 0.54), *M. incognita* distinguished from other species by one band each of Esterase and Malate dehydrogenase had an average R_f of Mdh Rm 0.19, Est Rm 0.38.The fastest migrating esterase of *M. incognita* had an average of R_f OF 0.41, where as R_f of 0.46 was obtained for the fastest migrating esterase of *M.arenaria.* In addition, *Meloidogyne* spp. that could not be identified on the basis of their Est and Mdh isozyme phenotypes.

Differences in the composition of proteins, carbohydrates and lipids of nematodes can be used to separate species by electrophoresis. The majority of biochemical systematic studies on plant-parasitic nematodes have relied on electrophoresis to detect differences between species.The SDS-PAGE system resolved some slight variations in overall electrophoresis mobilities among the 12 different areas of telangana districts (Figure 5), although the lower portion of the gel was not discernible photographically, because several bands were very dark, but better resolution was obtained for slowly migrating proteins. A single extra protein band was visible in the lane 6 and lane 11 was absent from other protein profiles. Total 9 to15 proteins ranging from 25 to 125 kDa in molecular mass in Root knot nematode adult females proteins but very little further analysis of these proteins was carried out. It is possible that further work would find other races of Root knot nematodes with distinct protein profiles as analyzed by PAGE and also, the present molecular techniques might give rise to a new method for distinguishing the population within *Meloidogyne* spp.

The PCR amplification using primer OPA-01 showed species-specific RAPD patterns which allowed identification of four Meloidogyne spp: *M. incognita, M. arenaria, M. javanica,* and *M. hapla*. The PCR amplification patterns were obtained by using single primer for ten samples from telangana region. In these, three Meloidogyne species examined were successfully identified using the primer OPA 01. Intraspecific variations in RAPD patterns were observed among the 10 isolates of *Meloidogyne* spp. Using the template DNA prepared with the miniprep, *M. javanica* was distinguished from *M. arenaria* by the presence of 660 bp band, and the presence and absence of a weak band of 480 bp or 620 bp, respectively. Sub races A-1 and A-2 of *M. arenaria* can be distinguished from each other based on the presence of 1,400 bp band. In this study, primer OPA-01 alone produced species-specific RAPD patterns which discriminated three *Meloidogyne* species in Telangana region.

The species-specific sequence tagged sites were selected with primer pairs listed in Table 1. Species-specific PCR products were produced by the designed primer pairs and resulted in unique band(s) for each of the four species (Figure 7). The species-specific primers produced a single band (1 500 bp) for the *M. hapla* isolates. Among the seven *M. arenaria* isolates, a single 950 bp band was amplified (Figure 7). In *M. incognita* reactions, there were doublet bands (both 1 350 bp, differing by 20 bp in size) produced from each of the eleven *M. incognita* isolates. For *M. javanica*, a single band (1 650 bp) was produced in four out of the five *M. javanica* isolates.

Further more genetic polymorphism among the six different species (identified by Isozyme patterns and by using species specific RAPD markers) was carried out using ISSR-PCR markers. The amplified products of template DNA extracted from single female nematode using ISSR primers were found to useful for the assessment of genomic polymorphism among the species.

The unprecedented movement of plant derived commodities due to globalization, impact of climate change on geographical distribution, behavior of known population with regard to consequent changes in land use and agricultural practices pose a new threat to agricultural production either from new or recently distributed species of plant parasitic nematodes. Molecular techniques may be the only practical approach to recognize these modern identification methods must provide accuracy, speed, reliability and affordability. Further research in this area will lead to develop suitable strategies for Plant parasitic nematode management.

Acknowledgement

Anil Kumar Vemunoori, is thankful to Andhra Pradesh Council for Science and Technology (APCOST) for providing partial financial assistance in the born of A.P.Y.S. Award. The authors are also thankful to Prof. Michael Jones, Murdoch University, Perth, Australia, for the scientific guidance.

References

Abad P, Favery B, Rosso MN, Castagnone-Screno P (2003) *Molecular Plant Pathology* 4, 217-224

Adam, M.A.M., Phillips, M.S. and Blok, V.C. (2005) *Nematology* 7, 599–609.

Brito, J., Powers, T.O., Mullin, P.G., Inserra, R.N. and Dickson, D.W. (2004) *Journal* of *Nematology* 36, 232–240.

Brito, J.A., Kaur, R., Cetintas, R., Stanley, J.D., Mendes, M.L., McAvoy, E.J., Powers, T.O. and Dickson, D.W. (2008) *Nematology* 10, 757–766.

Dalmasso, A. and J.B. Berge. 1978. Journal of Nematology 4:323-332.

Dickson, D.W., D. Huisingh, and J.N. Sasser. 1971 Journal of Nematology 3:1-16.

Dickson, D.W., J.N. Sasser, and D. Huisingh. 1970. Journal of Nematology 2:286-293.

Eisenback, J.D. 1985. Diagnostic characters useful in the identification of the four most common species of root-knot nematodes (*Meloidogyne* spp.). Pp. 95-112 *in* J.N. Sasser and C.C. Carter, eds. An advanced treatise on *Meloidogyne*, vol. 1.Biology and control. North Carolina State University Graphics, Raleigh.

Evans, A.A.F. 1971.. .International Journal of Biochemistry 2:72-79.

Fargette, M. 1987a.. Revue de Nématologie 10:39-43.

Fargette, M. 1987b. Revue de Nématologie 10:45-56.

Goodey, J.B., M.T. Franklin, and D.J. Hooper. 1965. The nematode parasites of plants catalogued under their hosts. Third ed. Commonwealth Agricultural Bureaux, Farnham, UK. :491-496.

Haseeb, A., R. Pandey, and A. Husain. 1984.Current Science 53:1151-1152.

Hirschmann, H. 1985. The genus *Meloidogyne* and morphological characters differentiating its species. Pp. 79-92 *in* J.N. Sasser and C.C. Carter, eds. An advanced treatise on *Meloidogyne*, vol. 1. Biology and control. North Carolina State University Graphics, Raleigh

Ibrahim, S.K., Davies, K.G. and Perry, R.N. (1996) *Physiological and Molecular Plant Pathology* 49, 7988

Ibrahim, A.A.M., and F.A. Al-Yahya.2002. Journal of Agricultural Research 47:157-167.

Orui, Y., 1998.. Appl. Entomol. Zool., 33: 43-51.

Rammah, A., and H. Hirschmann. 1988. *Meloidogyne mayaguensis* n. sp. (Meloidogynidae), a root-knot nematode from Puerto Rico. Journal of Nematology 20:58-69.

Sasser, J. N., and C. C. Carter. 1995. Overview of the international *Meloidogyne* project 1975-1984. Pp. 18-24 *in* J.N. Sasser and C.C. Carter, eds. An advanced treatise on *Meloidogyne*, vol. 1. Biology and Control. North Carolina State University Graphics, Raleigh.

Stanton, J.M., McNicol, C.D. and Steele, V. (1974) *Australian Plant Pathology* 27, 112–115.

Thorne, G. E. 1961. Principles of nematology. McGraw-Hill Book Co., Inc., New York.

Taylor AL, Sasser JN, 1982. Bilogy, identification and control of root –knot nematodes (*Meloidogne* spp.) IMP Publication, Raleigh, North Carolina.

Williamson, V.M. (1991). Molecular techniques for nematode species identification. Pp. 107-123 in W.R., Nickle, ed. Manual of agricultural nematology. New York: Marcel Dekker.

Williams, J.G., Hanafey, M.K., Rafalski, J.A. and Tingey, S.V. (1993). Genetic analysis using random amplified polymorphic DNA markers. Academic Press, Inc. Pp. 704-741.

Yongfang, C., W. Jianyu, X. Xianqi, and Y. Shengfu. 1998.. Acta Phytopathologica Sinica 28:73-77.

Zarina, B., and M. Abid. 1995.. Pakistan Journal of Nematology 13:49-50.

Zijlstra, C. (1997). *Fundamental and Applied Nematology* 20, 505–511.

Chapter - 15

Influence of AM fungi on soil borne pathogens and growth of two agroforestry tree species

P. Srinivas, G. Laxma Reddy and S.M. Reddy

Department of Biotechnology, Kakatiya University,
Warangal- 506 009, A.P., India

ABSTRACT

Efficacy of *Glomus mosseae, G. fasciculatum, G. etunicatum, G. aggregatum, Giga margarita* and *Giga gigantea* in the control of *Rhizoctonia solani* and *F. moniliforme* soil-borne pathogens of *Acacia nilotica* and *Albizzia lebbeck* was investigated. In general inoculation of AM fungi not only decreased the root colonization of both the pathogens under investigation but also promoted the growth of both the plants under study.

Key words: AM fungi, soil- borne pathogens, *Albizzia lebbeck, Acacia nilotica*.

Introduction

The report of antagonistic activity of AM endophytes against phytopathogenic fungi, stimulated the researches on possible exploitation of these fungi in the control of soil- borne pathogens. Large number of workers have reported the successful control of soil-borne diseases by employing AM fungi (Caron *et al.,* 1986; Ramiraj *et al.,* 1988 and Dar *et al.,* 1997; Vyas *et al.,* 1998; Thygesen *et al.,* 2004; Demir and Akkopru, 2007; Chandanie *et al.*; 2009) About and Robbson (1982) thus aptly used the term 'Biocide for AM fungi'. Although AM fungi do not protect the plant roots from soil-borne pathogens directly as ectomycorrhizae does but these play an important role indirectly that *Glomus fasciculatum* inoculated plants showed high incidence of wilt disease caused by *Verticillium* spp. than non-mycorrhizal plants in the cotton crop. The resistance against root-pathogens or soil-borne pathogens of mycorrhizal plants may be attributed to high accumulation of total phenol and orthodihydric phenols (Chandel, 1991).

Perusal of the literature reveals that most of the work on AM fungus and pathogens interaction have been investigated on agricultural and horticultural Crops (Gianinazzi *et al.,* 1990; McAllister *et al.,* 1995; Joseph *et al.,* 1997) Only limited information is available on tree species (Hedge and Rai 1987; Ramaiah and Bagyaraj; 1989; Farquhar and Peterson 1991; Aysan and Demir 2009; Al-Askar and Rashad 2010). Thus a more detailed and systematic studies involving more number of tree specieses are required before reaching any decissive conclusions. Therefore, the present investigations were aimed on the interaction of two soil borne pathogens and AM fungi in relation to two forest trees study.

Materials and Methods

Interaction of six AM fungal species (*Glomus mosseae, G. fasciculatum, G. agreegatum, G. etunicatum, Gig. margarita* and *Gig. gigantea* and two plant pathogens (*Rhizoctomia solani* and *Fusarium moniliforme*) in relation to two agroforestry trees (*Albizzia lebbeck* and *Acacia nilotica)* was studied. AM fungal spores were maintained on Bajra seedlings in earthen pots were employed as AM inoculum. Sand soil mixture (1:1) was sterilized at 1516 pressure for two consecutive days for 30 minutes each day. The autoclaved soil mixture was filled in small sterilized earthen pots. Seven days old culture of *Fusarium monitiforme* and *Rhizoctonia solani* were employed as inoculum.The fungal mycelium containing spore suspension (10 ml/per pot) was added to the soil along with different AM fungal sps. Surface sterilized seeds of test plants were sown in such earthen pots and watered at regular intervals with measured quality of water. Hoagland nutrient solution without phos-phorus was employed. The pots were placed under the uniform day light. Suitable controls were maintained. At the end of 45 days of growth plants were harvested to measure plant growth and mycorrhizal colonization. The feeder roots of the test plants were collected, washed and immediately preserved in FAA (Formalin: acetic acid; Alcohol 1:1). The clearing and staining of roots by the method suggested by Phillips and Hayman (1970) and the percentage of roof colonization was calculated by the formula suggested by Giovannetti and Mosse (1970). Resting spores of AM fungi consisting of spore carps, chlamydospores, aggospores and soils bone vesicles were extracted by met-seeing and decanting method (Pacioni, 1992). Resting spores were identified with the help of key provided by scherele and perez (1987).

Results and Discussion

Table 1 reveals that the efficacy of different AM fungi in the management of root pathogens of *A. lebbeck*. However, degree of protection varied both with the pathogen and AM fungus. *R. solani* infection resulted in the decrease of AM colonization and spore population. The presence of different fungi in the rhizosphere of *A. lebbeck* resulted in the decreased activity of *R. solani*. However, protective effect of AM fungi varied with the species. *G. fasciculatum, G. etunicatum, G. margarita* and *G. mosseae* suppressed the pathogenic activity of *R. solani* to significant level. On the other hand, *G. aggregatum* failed to suppress *R. solani* to a required level.

Table 1. Interaction of Soil borne pathogens and AM fungi and their effect on the growth of two forest tree species.

Name of the treatment	% of coloniz -ation	No.of Spores/ 100g soil	Height of the Plant (cm)	Biomass (g)		Phosphorus content (mg/plant)	
				Frest wt.	Dry wt.	Shoot	Root
A. lebbeck							
R. Saloni + G. fasiculatum	55.9	123	62.2	10.8	6.8	0.075	0.089
R. Saloni + G. aggregatum	38.9	117	52.3	10.5	7.2	0.042	0.038
R.Saloni + G. mosseae	50.1	97	53.5	9.3	5.4	0.073	0.050
R. Saoloni + G. etrunicatum	52.9	128	60.2	11.8	7.9	0.062	0.058
R. Saloni + Gig. margarita	53.2	118	65.4	14.2	11.3	0.560	0.640
R. Saloni | Gig. gigantea	47.3	113	64.7	16.2	12.3	0.280	0.370
Rhizoctonia saloni	33.3	72	51.0	8.5	4.9	0.014	0.026
G.fasciculatum (control)	61.1	144	58.5	15.2	13.3	0.140	0.280
F.moniliforme + G. fasciculatum	49.2	146	44.7	16.8	12.5	0.168	0.176
F.moniliforme + G. aggregatum	50.6	202	56.5	16.5	13.2	0.172	0.183
F.moniliforme + G. mosseae	40.2	138	36.2	14.2	10.5	0.152	0.174
F.moniliforme + G. etunicatum	57.5	154	46.0	12.6	7.4	0.138	0.159
F.moniliforme + Gig. gigantca	42.8	149	42.4	13.9	9.2	0.132	0.154
F.moniliforme + Gig. margarita	54.1	162	45.7	15.5	11.2	0.123	0.146
Fusarium moniliforme	30.8	79	26.6	10.4	6.2	0.119	0.128
G.fasciculatum (Control)	37.3	232	50.7	18.2	13.5	0.154	0.172
A. nilotica							
R. Saloni + G.fasiculatum	40.5	108	53.3	12.2	7.4	0.032	0.046
R. Saloni + G.aggregatum	50.0	125	61.0	15.8	12.3	0.038	0.053
R.Saloni + G.mosseae	55.6	124	62.0	13.6	6.8	0.026	0.067
R. Saoloni + G. etrunicatum	42.1	102	58.6	10.8	6.3	0.029	0.034
R. Saloni + Gig.margarita	41.6	107	66.0	16.5	11.2	0.320	0.480
R. Saloni | Gig gigantea	56.1	114	72.0	18.3	14.2	0.230	0.430
Rhizoctonia saloni	25.7	83	49.9	11.3	6.8	0.034	0.054

G.fasciculatum (control)	59.0	138	67.9	15.3	12.6	0.690	0.760
F.moniliforme + G. fasciculatum	51.8	177	73.0	18.5	14.6	0.133	0.165
F.moniliforme + G. aggregatum	46.7	106	67.0	25.5	16.2	0.138	0.156
F.moniliforme + G. mosseae	52.7	128	60.0	15.9	12.2	0.136	0.142
F.moniliforme + G. etunicatum	38.3	121	60.1	23.5	14.3	0.139	0.152
F.moniliforme + Gig. gigantea	47.8	136	81.8	26.6	17.8	0.132	0.149
F.moniliforme + Gig. margarita	40.7	132	72.7	19.8	12.5	0.142	0.168
Fusarium moniliforme	27.1	76	42.4	13.2	9.6	0.122	0.134
G.fasciculatum (Control)	63.9	206	79.7	22.4	15.5	0.146	0.178
F-Value	0.986	0.937	1.039	1.034	0.950	1.206	1.162
P-Value	0.485	0.429	0.458	0.462	0.445	0.305	0.341
F-crit	0.543	0.543	1.840	1.840	0.543	1.840	1.840

Addition of different AM fungi resulted in the change in the spore population to a varying level with the species. *G. etunicatum* and *G. fasciculatum* could effectively check the pathogenic activity of *R. solani* on AM spore population, while *G. mosseae* showed the inhibitory effect on spore population and plant biomass production. The inoculation of different AM fungi not only inhibited the pathogenic activity of *R. solani* but also stimulated the growth of *A. lebbeck*. Similarly decreased biomass production by *A. lebbeck,* due to the infection of *R. solani* could be countered by inoculation of different AM fungi. The phosphorus content of *R. solani* infected *A. lebbeck* could be increased by the inoculation of different AM fungi.

F. moniliforme infection also adversely affected the growth and development of *A. lebbeck*. The adverse effect of *F. moniliforme* could be over come by inoculating different AM fungi, *G. etunicatum* followed by *Gigaspora margarita* were most effective in countering the adverse effect of *F. moniliforme, Gig. gigantea* and *G. mosseae* were poor in antagonizing *F. moniliforme*. Rest of the AM fungi was helpful to an intermediate degree. AM spore population decreased in the presence of *F. moniliforme*. Inoculation of *G. aggregatum* was most beneficial, while *G. mosseae* was least. *G. aggregatum* was not only countered the pathogen activity of *F. moniliforme,* but also stimulated the growth of *A. lebbeck*. Similarly adverse effect of *F. moniliforme* on biomass production by *A. lebbeck* could be countered by the inoculation of AM fungi.

R. solani and *F. moniliforme* were responsible for inhibiting the growth and development of *A. nilotica*. Inoculation of different AM fungi was very helpful in countering pathogenic activity of both the fungi under investigation. *G. mosseae* and *Gig. gigantea* not only countered the pathogenic activity of these fungi but also promoted the AM root infection as well as AM spore population. *G. mosseae* and *G. aggregatum* were also responsible for stimulating the growth of *A. nilotica.* Biomass production which was adversely affected under the influence of *R. solani* could be countered by the AM fungal inoculation.

The infection of *A. nilotica* by *F. moniliforme* adversely affected the growth and development of *A. nilotica.* The adverse effect of *F. monilifome* could be overcome by inoculating different AM fungi. *G. fasciculatum* followed by *G. mosseae* were most effective in countering the adverse effect of *F. moniliforme,* while *G. etunicatum* and *Gig. margarita* were poor in antagonizing the pathogenic activity of *F. moniliforme.* Rest of AM fungi was helpful in checking pathogenic activity to an intermediate degree. *Gig. gigantea* not only countered the pathogenic activity of *F. moniliforme,* but also stimulated the growth of *A. nilotica.*

Table 2. Physico – chemical characteristics of soil borne pathogens and AM fungi on the growth of two forest tree species.

Name of the treatment	Soil texture	pH	EC mho/ cm	Organic matter	Available Phosp-horus	Available Potas-sium
A. lebbeck						
R. Saloni + G. fasiculatum	SL	8.3	0.39	0.05	15	216
R. Saloni + G. aggregatum	SL	8.0	0.52	0.01	46	222
R.Saloni + G. mosseae	SL	8.0	0.32	0.02	48	228
R. Saoloni + G. etrunicatum	SL	8.1	0.50	0.06	45	241
R. Saloni + Gig. margarita	SL	7.9	0.42	0.02	39	203
R. Saloni | Gig. gigantea	SL	7.8	0.60	0.01	15	305
Rhizoctonia saloni	SL	8.0	0.41	0.01	44	241
G.fasciculatum (control)	SL	6.0	0.47	0.14	38	152
F.moniliforme + G. fasciculatum	SL	7.5	2.22	0.43	62	267
F.moniliforme + G. aggregatum	SL	8.2	0.55	0.39	56	330
F.moniliforme + G. mosseae	SL	7.6	0.37	0.26	47	178
F.moniliforme + G. etunicatum	SL	8.0	0.45	0.36	54	229
F.moniliforme + Gig. gigantea	SL	7.9	1.35	0.42	64	378
F.moniliforme + Gig. margarita	SL	8.0	1.29	0.38	59	352
Fusarium moniliforme	SL	7.9	1.23	0.23	35	253
G.fasciculatum (Control)	SL	7.5	1.12	0.28	53	292
A. nilotica						
R. Saloni + G. fasiculatum	SL	8.0	0.33	0.08	25	178
R. Saloni + G. aggregatum	SL	7.0	0.38	0.02	35	244

R.Saloni + G. mosseae	SL	8.1	0.04	0.02	18	203
R. Saoloni + G. etrunicatum	SL	7.6	0.40	0.03	49	267
R. Saloni + Gig. margarita	SL	7.9	0.46	0.03	52	248
R. Saloni | Gig. gigantea	SL	7.8	0.40	0.03	43	252
Rhizoctonia saloni	SL	8.0	0.45	0.01	42	229
G.fasciculatum (control)	SL	7.9	0.63	0.02	32	241
F.moniliforme + G. fasciculatum	SL	8.0	0.36	0.03	43	263
F.moniliforme + G. aggregatum	SL	7.8	0.45	0.04	28	267
F.moniliforme + G. mosseae	SL	8.1	0.38	0.05	52	457
F.moniliforme + G. etunicatum	SL	8.0	0.52	0.03	45	242
F.moniliforme + Gig. gigantea	SL	7.9	0.47	0.05	47	241
F.moniliforme + Gig. margarita	SL	7.9	0.62	0.06	46	264
Fusarium moniliforme	SL	8.0	0.32	0.03	38	226
G. fasciculatum (Control)	SL	7.2	1.07	0.03	50	262

Table 2 reveals that the inoculation of soil-borne fungi adversely affected the soil conditions. Not much change in soil pH was recorded due to inoculating of different AM fungi. Inoculation of *Gig. gigantean* to *R. solani* infested plants increased the EC of the rhizosphere soil. Organic matter also increased in the presence of different AM fungi. Available phosphorus varied with the inoculated AM fungi. Similarly the available potassium increased in the presence of different AM fungi. Almost similar trend was observed with other AM fungi in the presence of *F. moniliforme*. Inoculation of *G. mosseae* and with *F. moniliforme* increases the EC of the rhizosphere soil, but decreased organic matter. Maximum available phosphorus was recorded in *Gig. margarita* inoculated soils. *G. fasciculatum* was next in the solubilization of phosphorus. It was least in *Gig. gigantea* inoculated soils. Similarly the available potassium increased in the presence of different AM fungi. Available potassium was more in *Gig. margarita* inoculated soils followed by *Gig. gigatea* and *G. mosseae* in a desending order. It was least in *G. fasiculatum* inoculated soils. Inoculation of *Gig. margarita* increased EC of the rhizosphere soil of *A. nilotica* Organic matter and available potassium increased in the presence of different AM fungi. Maximum available phosphorus was in *G. aggregatum* and *Gig. margarita* inoculated soil, while it was least in *G. mossae* inoculated soils.

Acknowledgements

Thanks are due to the Head, Department of Biotechnology for kind encouragement and providing necessary facilities. Financial assistance received from UGC, New Delhi (F.No. F-39/3022010SR) is also gratefully acknowledged.

References

Abbott, L.K. and Robson, A.D. (1982). *Austr. J. Agric. Res.,* 33: 289-408.

Akthar, M.S. and Z.A. Siddiqui (2008). Arbuscular Mycorrhizal Fungi as Potential Bioprotectants against Plant Pathogens. In: Mycorrhizae: Sustainable Agriculture and Forestry, Siddiqui, Z.A., M.S. Akhtar and K. Futai (Eds.) Springer Netherlands, Dordrecht, The Netherlands, pp: 61-97.

Aysan, E. and S. Demir (2009). Plant Pathol. J., 8: 74-78.

Al- Askar, A.A. and Y.M. Rashad (2010). Plant Pathol. J., 9 : 31-38.

Bhat, Z.A., Bhat, M.A. and Shawl, A.S. (2003). Comparative efficacy of biocontrol gent, botanical extracts and fungicides in the management of chick peak wilt caused by *Fusarium oxysporum* f.sp. *ciceri. Proc. National Seminar on Recent Advances in Plant Science Research,* Dept. of Botany, Kashmir University, Kashmir, Oct.12-14, 2003, pp.45.

Caron, M., Fortin, J.A. and Richard, C. (1986). *Phytopathology,* 76: 942-946.

Chandel Sushma (1991). Studies on vesicular arbuscular mycorrhizae of *sorghum bicolor* (L) Moench. Ph.D. Thesis, Jodhpur University, Jodhpur.

Chandanie, W.A., M. Kubota and M. Hyakumachi (2009). Applied Soil Ecol., 41: 336-341.

Davis, R.M., Menge, J.A. and Erwin, D.C. (1979). Influence of *Glomus fasciculatum* on soil phosphorus on *Verticillium wilt* of cotton. *Phytopathology,* 69: 453-456.

Dar, G.H., Zargar, M.Y. and Beigh, G.M. (1997). *Microbial Ecology,* 34: 74-80.

Demir, S. and A. Akkopru (2007). Using of Arbuscular Mycorrhizal Fungi (AMF) for Biocontrol of Soil- Borne Fungal Plant Pathogens. In: Biological Control of Plant Diseases, Chincholkar, S.B. and K.G. Mukerji (Eds.). Haworth Press, USA., pp: 17-37.

Faraquhar, M.L. and Peterson, R.L. (1991). *Can. J. Bot,* 69: 1372-1382.

Giovannetti, M. and Mosse, B. (1980). *New Phytol.,* 84: 489-500.

Gianianzzi, S., Trouvelot A. and Ginaninazzi-Pearson, V. (1990). Role of mycorrhizas in horticulture crop production. Pleanary lecture at XXII International Horticultural Congress ISHS, Italy, 25-30.

Hegde, S.V. and Rai, P.V. (1987). Application of *Glomus fasciculatum* for control of damping-off caused by *Phythium. Proc.* of *National Workshop* of *Mycorrhizae,* New Delhi, 13-15 March.

McAllister, C.B., Garica-Romera, I., Martin, J., Godeas, A. and Ocampo, J.A. (1995). *New Phytologist,* 129(2): 309-316.

Pacioni, G. (1992). *Methods in Microbial.,* 24: 317-322.

Phillips, J.M. and Hayman, D.S. (1970). *Trans. Br. Mycol. Soc.,* 55: 158-160.

Ramaiah and Bagyaraj, D.J. (1989). *Jr. Phytol. Research,* 2: 1-6.

Ramiraj, B., Dwarkanath Reddy and Shanmugam, N. (1988). *Proc. First Asian Conference on Mycorrhizae,* 26 (Abstr).

Ratti, N., Alam, M., Sharm, S. and Janardhanan, K.K. (1998). *Symbiosis,* 24: 115-126.

Schenck, N.C. and Perez, Y. (1987). Manual for the identification of VA mycorrhizal fungi. INVAM. Florida University, Gainesville, USA., pp. 245.

Thygesen, K., Larsen, J. and Bodker, L. (2004). *European Journal* of *Plant Pathology,* 110(4): 411-419.

Vyas, S.C., Naik, S.L. and Vyas, A. (1998). Interaction between Mycorrhizal fungi and plant Pathogens. Fungi in Biotechnology. Editor, Anil Prakash BHU, Bhopal, pp. 197-200.